$f(t)$	$\mathscr{L}\{f(t)\} = F(s)$
15. $t^n e^{at}, \quad n = 1, 2, 3, \ldots$	$\dfrac{n!}{(s-a)^{n+1}}$
16. $e^{at} \sin kt$	$\dfrac{k}{(s-a)^2 + k^2}$
17. $e^{at} \cos kt$	$\dfrac{s-a}{(s-a)^2 + k^2}$
18. $t \sin kt$	$\dfrac{2ks}{(s^2 + k^2)^2}$
19. $t \cos kt$	$\dfrac{s^2 - k^2}{(s^2 + k^2)^2}$
20. $\sin kt - kt \cos kt$	$\dfrac{2k^3}{(s^2 + k^2)^2}$
21. $\sin kt + kt \cos kt$	$\dfrac{2ks^2}{(s^2 + k^2)^2}$
22. $\sinh kt - \sin kt$	$\dfrac{2k^3}{s^4 - k^4}$
23. $\cosh kt - \cos kt$	$\dfrac{2k^2 s}{s^4 - k^4}$
24. $1 - \cos kt$	$\dfrac{k^2}{s(s^2 + k^2)}$
25. $kt - \sin kt$	$\dfrac{k^3}{s^2(s^2 + k^2)}$
26. $\dfrac{a \sin bt - b \sin at}{ab(a^2 - b^2)}$	$\dfrac{1}{(s^2 + a^2)(s^2 + b^2)}$
27. $\dfrac{\cos bt - \cos at}{a^2 - b^2}$	$\dfrac{s}{(s^2 + a^2)(s^2 + b^2)}$

A First Course in Differential Equations with Applications

The Prindle, Weber & Schmidt Series in Mathematics

Althoen and Bumcrot, *Introduction to Discrete Mathematics*
Brown and Sherbert, *Introductory Linear Algebra with Applications*
Buchthal and Cameron, *Modern Abstract Algebra*
Burden and Faires, *Numerical Analysis*, Fourth Edition
Cullen, *Linear Algebra and Differential Equations*
Cullen, *Mathematics for the Biosciences*
Eves, *In Mathematical Circles*
Eves, *Mathematical Circles Adieu*
Eves, *Mathematical Circles Revisited*
Eves, *Mathematical Circles Squared*
Eves, *Return to Mathematical Circles*
Fletcher and Patty, *Foundations of Higher Mathematics*
Geltner and Peterson, *Geometry for College Students*
Gilbert and Gilbert, *Elements of Modern Algebra*, Second Edition
Gobran, *Beginning Algebra*, Fourth Edition
Gobran, *College Algebra*
Gobran, *Intermediate Algebra*, Fourth Edition
Gordon, *Calculus and the Computer*
Hall, *Algebra for College Students*
Hall and Bennett, *College Algebra with Applications*, Second Edition
Hartfiel and Hobbs, *Elementary Linear Algebra*
Hunkins and Mugridge, *Applied Finite Mathematics*, Second Edition
Kaufmann, *Algebra for College Students*, Third Edition
Kaufmann, *Algebra with Trigonometry for College Students*, Second Edition
Kaufmann, *College Algebra*
Kaufmann, *College Algebra and Trigonometry*
Kaufmann, *Elementary Algebra for College Students*, Third Edition
Kaufmann, *Intermediate Algebra for College Students*, Third Edition
Kaufmann, *Precalculus*
Kaufmann, *Trigonometry*
Keisler, *Elementary Calculus: An Infinitesimal Approach*, Second Edition
Kirkwood, *Introduction to Real Analysis*
Laufer, *Discrete Mathematics and Applied Modern Algebra*
Nicholson, *Linear Algebra with Applications*
Pasahow, *Mathematics for Electronics*

Powers, *Elementary Differential Equations*
Powers, *Elementary Differential Equations with Boundary-Value Problems*
Powers, *Elementary Differential Equations with Linear Algebra*
Proga, *Arithmetic and Algebra*, Second Edition
Proga, *Basic Mathematics*, Second Edition
Radford, Vavra, and Rychlicki, *Introduction to Technical Mathematics*
Radford, Vavra, and Rychlicki, *Technical Mathematics with Calculus*
Rice and Strange, *Calculus and Analytic Geometry for Engineering Technology*
Rice and Strange, *Plane Trigonometry*, Fifth Edition
Rice and Strange, *Technical Mathematics*
Rice and Strange, *Technical Mathematics and Calculus*
Schelin and Bange, *Mathematical Analysis for Business and Economics*, Second Edition
Strnad, *Introductory Algebra*
Swokowski, *Algebra and Trigonometry with Analytic Geometry*, Seventh Edition
Swokowski, *Calculus with Analytic Geometry*, Second Alternate Edition
Swokowski, *Calculus with Analytic Geometry*, Fourth Edition
Swokowski, *Fundamentals of Algebra and Trigonometry*, Seventh Edition
Swokowski, *Fundamentals of College Algebra*, Seventh Edition
Swokowski, *Fundamentals of Trigonometry*, Seventh Edition
Swokowski, *Precalculus: Functions and Graphs*, Fifth Edition
Tan, *Applied Calculus*
Tan, *Applied Finite Mathematics*, Second Edition
Tan, *Calculus for the Managerial, Life, and Social Sciences*
Tan, *College Mathematics*, Second Edition
Venit and Bishop, *Elementary Linear Algebra*, Third Edition
Venit and Bishop, *Elementary Linear Algebra*, Alternate Second Edition
Willard, *Calculus and Its Applications*, Second Edition
Wood and Capell, *Arithmetic*
Wood, Capell, and Hall, *Developmental Mathematics*, Third Edition
Wood, Capell, and Hall, *Intermediate Algebra*
Wood, Capell, and Hall, *Introductory Algebra*
Zill, *A First Course in Differential Equations with Applications*, Fourth Edition
Zill, *Calculus with Analytic Geometry*, Second Edition
Zill, *Differential Equations with Boundary-Value Problems*, Second Edition

FOURTH EDITION

A First Course in Differential Equations with Applications

Dennis G. Zill
Loyola Marymount University

PWS-KENT Publishing Company
Boston

20 Park Plaza
Boston, Massachusetts 02116

PWS-KENT Publishing Company is a division of Wadsworth, Inc.

Portions of this book also appear in *Differential Equations with Boundary-Value Problems*, Second Edition, by Dennis G. Zill, copyright © 1989 by PWS-KENT Publishing Company.

Library of Congress Cataloging-in-Publication Data

Zill, Dennis G.
A first course in differential equations with applications.

Includes index.
1. Differential equations. I. Title.
QA372.Z54 1988 515.3′5 88-22417
ISBN 0-534-91568-X

Sponsoring Editor: David Geggis
Production Coordinator: Robine Andrau
Interior Design: Robine Andrau/Susan Graham
Cover Design: Robine Andrau
Cover Photo: Morton Beebe/The Image Bank
Interior Illustrations: Deborah Schneck
Typesetting: Polyglot Pte Ltd
Cover Printing: New England Book Components, Inc.
Printing and Binding: Arcata Graphics/Halliday

Printed in the United States of America
89 90 91 92 93 — 10 9 8 7 6 5 4 3 2 1

Preface

I originally wrote *A First Course in Differential Equations with Applications* with the idea that, for many colleges and universities, differential equations is basically a continuation of the calculus sequence since it draws immediately on the student's new knowledge of differentiation and integration. As a consequence the text was purposefully written in the direct, no-nonsense style of calculus texts, with an abundance of examples, problems, and applications. The emphasis of the text was on how to solve differential equations and on how to interpret these equations and their solutions in a physical setting. After considering some suggestions for increasing its theoretical content, I decided to adhere to the philosophy of the earlier editions and to keep the level of presentation of this fourth edition the same.

A First Course in Differential Equations with Applications, Fourth Edition, does, however, reflect some of the refinements suggested by instructors, students, and reviewers. The following list gives some of the modifications incorporated into this revision.

- Most of the biographical footnotes have been expanded to include more history.
- Many new problems and examples have been added.
- Chapter openings now include a list of important terms.
- In Section 2.6 Bernoulli's equation has been moved from the exercises into the body of the text.
- Several methods for solving differential equations—for example, the method of undetermined coefficients in Chapter 4—have been outlined in a step-by-step manner.
- A new section on *L-R-C* series circuits has been added to Chapter 5.
- In Section 6.1 the alternative method for solving the Cauchy-Euler equation is now considered in the exposition rather than being relegated to the exercises.
- Section 7.1 has been split into two sections. The inverse Laplace transform is now introduced in Section 7.2.

- More material on the static deflection of beams has been added to Chapter 7.
- Portions of the text have been polished or partially rewritten to increase clarity. For example, in Section 8.6 instructors may find the treatment of solutions of linear systems corresponding to complex eigenvalues more to their liking.
- At the request of some of the users, a table of integrals has been included on the endpapers of the book.

The symbol [O] denotes optional sections—those that can be omitted without loss of continuity. Section 1.2 (Origins of Differential Equations), although not marked as optional, can also be omitted if a faster introduction to method of solution of differential equations is desired. It is recommended, however, that students review the system of units on page 20 before studying Chapter 5.

Answers to odd-numbered problems are given at the end of the text and are indicated by a color tab on the edge of the page. To facilitate access to the answers, the page number on which the answers start is given at the beginning of each exercise set. Colored problem numbers in the exercise sets indicate problems that are worked out in detail in an accompanying Student Supplement authored by Warren S. Wright. A complete solutions manual is available to instructors. Those wishing to receive supplements to the text should contact PWS-KENT Publishing Company.

As was the case with the three previous editions, this text is intended for use in an introductory one-semester or one-quarter course in differential equations. An expanded version of the text, suitable for a two-semester course that contains more material on partial differential equations and boundary-value problems, is available from PWS-KENT Publishing Company under the title *Differential Equations with Boundary-Value Problems*, Second Edition.

In conclusion I would like to take this opportunity to thank those students and professors who, unsolicited, volunteered suggestions, criticisms, corrections, and encouragement. I am also grateful for the help of the production and editorial staffs of PWS-KENT Publishing Company. Finally, I am indebted to the following reviewers for their many helpful remarks: Vincent Connolly, Worcester Polytechnic Institute; Harvey J. Fletcher, Brigham Young University; Philip S. Mulry, Colgate University; Carol O'Dell, Ohio Northern University; Tom Roe, South Dakota State University; and Barbara Shabell, California Polytechnic State University.

Dennis G. Zill
Los Angeles

Contents

CHAPTER 7 Laplace Transform 295

CHAPTER 8 Systems of Linear Differential Equations 353

CHAPTER 9 Numerical Methods 451

CHAPTER 1

Introduction to Differential Equations

IMPORTANT TERMS

ordinary differential equation
partial differential equation
order of the equation
linear equation
nonlinear equation
solution
trivial solution
n-parameter family of solutions
particular solution
singular solution
general solution

The purpose of this chapter is twofold: to introduce the basic terminology of differential equations and to examine, albeit briefly, how differential equations are derived in an attempt to formulate, or describe, physical phenomena in terms of mathematics.

1.1 Basic Definitions and Terminology

In calculus the reader learned that given a function $y = f(x)$, the derivative

$$\frac{dy}{dx} = f'(x)$$

is itself a function of x and is found by some appropriate rule. For example, if $y = e^{x^2}$ then

$$\frac{dy}{dx} = 2xe^{x^2} \quad \text{or} \quad \frac{dy}{dx} = 2xy. \tag{1}$$

The problem that we face in this course is not: given a function $y = f(x)$, find its derivative. Rather, our problem is: if we are given an equation such as $dy/dx = 2xy$, somehow to find a function $y = f(x)$ that satisfies the equation. In a word, we wish to *solve* differential equations.

DEFINITION 1.1 An equation containing the derivatives or differentials of one or more dependent variables, with respect to one or more independent variables, is said to be a **differential equation**.

Differential equations are classified according to the following three properties.

Classification by Type

If an equation contains only ordinary derivatives of one or more dependent variables, with respect to a single independent variable, it is then said to be an **ordinary differential equation**. For example,

$$\frac{dy}{dx} - 5y = 1$$

$$(x + y)\,dx - 4y\,dy = 0$$

$$\frac{du}{dx} - \frac{dv}{dx} = x$$

$$\frac{d^2y}{dx^2} - 2\frac{dy}{dx} + 6y = 0$$

are ordinary differential equations. An equation involving the partial derivatives of one or more dependent variables of two or more independent variables is called a **partial differential equation**. For example,

$$\frac{\partial u}{\partial y} = -\frac{\partial v}{\partial x}$$

$$x\frac{\partial u}{\partial x} + y\frac{\partial u}{\partial y} = u$$

$$\frac{\partial^2 u}{\partial x^2} = \frac{\partial^2 u}{\partial t^2} - 2\frac{\partial u}{\partial t}$$

are partial differential equations.

Classification by Order

The order of the highest derivative in a differential equation is called the **order of the equation**. For example,

$$\frac{d^2y}{dx^2} + 5\left(\frac{dy}{dx}\right)^3 - 4y = x$$

is a second-order ordinary differential equation. Since the differential equation $x^2\,dy + y\,dx = 0$ can be put into the form

$$x^2\frac{dy}{dx} + y = 0$$

by dividing by the differential dx, it is an example of a first-order ordinary differential equation. The equation

$$a^2\frac{\partial^4 u}{\partial x^4} + \frac{\partial^2 u}{\partial t^2} = 0$$

is a fourth-order partial differential equation.

Although partial differential equations are very important, their study demands a good foundation in the theory of ordinary differential equations. Consequently, in the discussion that follows (as well as in the next nine chapters of the text) we shall confine our attention to ordinary differential equations.

A general nth-order, ordinary differential equation is often represented by the symbolism

$$F\left(x, y, \frac{dy}{dx}, \ldots, \frac{d^ny}{dx^n}\right) = 0. \tag{2}$$

The following is a special case of (2).

Classification as Linear or Nonlinear

A differential equation is said to be **linear** if it has the form

$$a_n(x)\frac{d^ny}{dx^n} + a_{n-1}(x)\frac{d^{n-1}y}{dx^{n-1}} + \cdots + a_1(x)\frac{dy}{dx} + a_0(x)y = g(x).$$

It should be observed that linear differential equations are characterized by two properties: (a) the dependent variable y and all its derivatives are of the first degree, that is, the power of each term involving y is 1; and (b) each coefficient depends only on the independent variable x. An equation that is not linear is said to be **nonlinear**.

The equations

$$x\,dy + y\,dx = 0$$

$$y'' - 2y' + y = 0$$

and

$$x^3 \frac{d^3y}{dx^3} - x^2 \frac{d^2y}{dx^2} + 3x \frac{dy}{dx} + 5y = e^x$$

are linear first-, second-, and third-order ordinary differential equations, respectively. On the other hand,

Coefficient depends on y — Power not 1

$$yy'' - 2y' = x \quad \text{and} \quad \frac{d^3y}{dx^3} + y^2 = 0$$

are nonlinear second- and third-order ordinary differential equations, respectively.

Solutions

As mentioned before, our goal in this course is to solve, or find solutions of, differential equations.

DEFINITION 1.2 Any function f defined on some interval I* which when substituted into a differential equation reduces the equation to an identity, is said to be a **solution** of the equation on the interval.

In other words, a solution of a differential equation (2) is a function $y = f(x)$ that possesses at least n derivatives and *satisfies* the equation—that is,

$$F(x, f(x), f'(x), \ldots, f^{(n)}(x)) = 0$$

for every x in I.

* The name of I is purposely left vague. Depending on the context, I could represent (a, b), $[a, b]$, $(0, \infty)$, $(-\infty, \infty)$, and so on.

EXAMPLE 1

The function $y = x^4/16$ is a solution of the nonlinear equation

$$\frac{dy}{dx} - xy^{1/2} = 0$$

on $(-\infty, \infty)$. Since

$$\frac{dy}{dx} = 4 \cdot \frac{x^3}{16} = \frac{x^3}{4},$$

we see $$\frac{dy}{dx} - xy^{1/2} = \frac{x^3}{4} - x\left(\frac{x^4}{16}\right)^{1/2} = \frac{x^3}{4} - \frac{x^3}{4} = 0$$

for every real number.

EXAMPLE 2

The function $y = xe^x$ is a solution of the linear equation

$$y'' - 2y' + y = 0$$

on $(-\infty, \infty)$. To see this, we compute

$$y' = xe^x + e^x \quad \text{and} \quad y'' = xe^x + 2e^x.$$

Observe $$y'' - 2y' + y = (xe^x + 2e^x) - 2(xe^x + e^x) + xe^x = 0$$

for every real number.

Notice that in Examples 1 and 2 the constant function $y = 0$, for $-\infty < x < \infty$, also satisfies the given differential equation. A solution of a differential equation that is identically zero on an interval I is often referred to as a **trivial solution**.

Not every differential equation necessarily has a solution, as we see in the following example.

EXAMPLE 3

(a) The first-order differential equations

$$\left(\frac{dy}{dx}\right)^2 + 1 = 0 \quad \text{and} \quad (y')^2 + y^2 + 4 = 0$$

possess no real solutions. Why?

(b) The second-order equation $(y'')^2 + 10y^4 = 0$ possesses only one real solution. What is it?

Explicit and Implicit Solutions

Solutions of differential equations can be distinguished as **explicit** or **implicit solutions**. We have already seen in our initial discussion that $y = e^{x^2}$ is an explicit solution of $dy/dx = 2xy$. In Examples 1 and 2, $y = x^4/16$ and $y = xe^x$ are explicit solutions of $dy/dx - xy^{1/2} = 0$ and $y'' - 2y' + y = 0$, respectively. A relation $G(x, y) = 0$ is said to define a solution of a differential equation implicitly on an interval I provided it defines one or more explicit solutions on I.

EXAMPLE 4

For $-2 < x < 2$ the relation $x^2 + y^2 - 4 = 0$ is an implicit solution of the differential equation

$$\frac{dy}{dx} = -\frac{x}{y}.$$

By implicit differentiation it follows that

$$\frac{d}{dx}(x^2) + \frac{d}{dx}(y^2) - \frac{d}{dx}(4) = 0$$

$$2x + 2y\frac{dy}{dx} = 0 \qquad \text{or} \qquad \frac{dy}{dx} = -\frac{x}{y}.$$

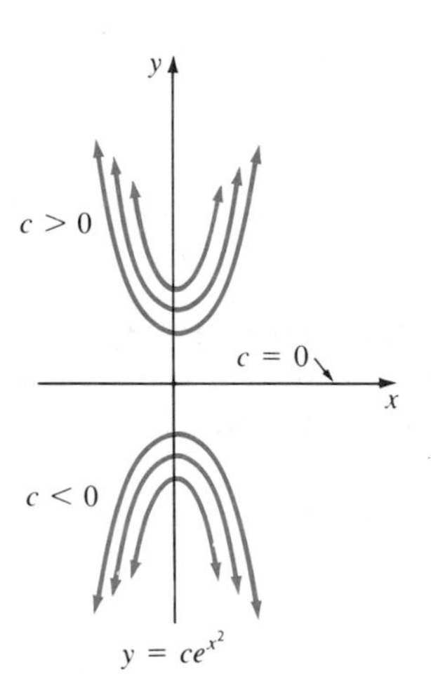

Figure 1.1

The relation $x^2 + y^2 - 4 = 0$ in Example 4 defines two functions: $y = \sqrt{4 - x^2}$ and $y = -\sqrt{4 - x^2}$ on the interval $(-2, 2)$. Also note that any relation of the form $x^2 + y^2 - c = 0$ will *formally* satisfy $dy/dx = -x/y$ for any constant c. However, it is naturally understood that the relation should always make sense in the real number system; thus we cannot say that $x^2 + y^2 + 1 = 0$ determines a solution of the differential equation.

Since the distinction between an explicit and an implicit solution should be intuitively clear, we shall not belabor the issue by always saying "here is an explicit (implicit) solution."

The student should become accustomed to the fact that a given differential equation will usually possess an *infinite* number of solutions. By direct substitution, we can prove that any curve—that is, function—in the one-parameter family $y = ce^{x^2}$, where c is any arbitrary constant, also satisfies (1). As indicated in Figure 1.1, the trivial solution is a member of this family of solutions corresponding to $c = 0$. In Example 2, tracing back through the work reveals that $y = cxe^x$ is a family of solutions of the given differential equation.

EXAMPLE 5

For any value of c the function $y = c/x + 1$ is a solution of the first-order differential equation

$$x\frac{dy}{dx} + y = 1$$

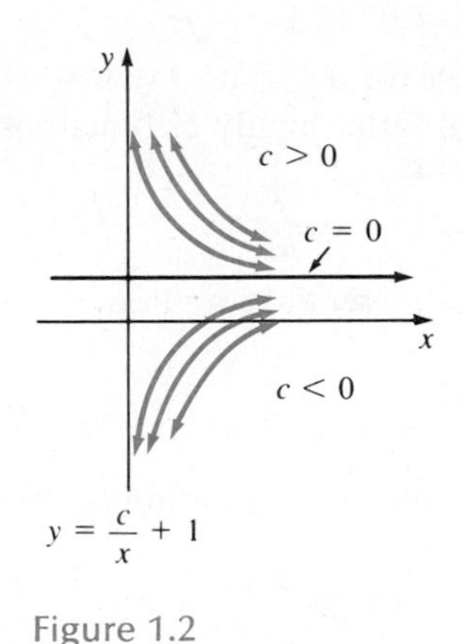

Figure 1.2

on the interval $(0, \infty)$. We have

$$\frac{dy}{dx} = c\frac{d}{dx}(x^{-1}) + \frac{d}{dx}(1) = -cx^{-2} = -\frac{c}{x^2}$$

so that

$$x\frac{dy}{dx} + y = x\left(-\frac{c}{x^2}\right) + \left(\frac{c}{x} + 1\right) = 1.$$

By choosing c to be any real number, we can generate an infinite number of solutions. In particular, for $c = 0$ we obtain a constant solution $y = 1$ (see Figure 1.2).

In Example 5, $y = c/x + 1$ is a solution of the differential equation on any interval not containing the origin. The function is not differentiable at $x = 0$.

EXAMPLE 6

(a) The functions $y = c_1 \cos 4x$ and $y = c_2 \sin 4x$, where c_1 and c_2 are arbitrary constants, are solutions of the differential equation

$$y'' + 16y = 0.$$

For $y = c_1 \cos 4x$, the first and second derivatives are

$$y' = -4c_1 \sin 4x \qquad \text{and} \qquad y'' = -16c_1 \cos 4x,$$

and so

$$y'' + 16y = -16c_1 \cos 4x + 16(c_1 \cos 4x) = 0.$$

Similarly, for $y = c_2 \sin 4x$,

$$y'' + 16y = -16c_2 \sin 4x + 16(c_2 \sin 4x) = 0.$$

(b) The function $y = c_1 \cos 4x + c_2 \sin 4x$ can also be shown to be a solution of the given equation.

EXAMPLE 7

The reader should be able to show that

$$y = e^x, \quad y = e^{-x}, \quad y = c_1e^x, \quad y = c_2e^{-x}, \qquad \text{and} \qquad y = c_1e^x + c_2e^{-x}$$

are all solutions of the linear second-order differential equation

$$y'' - y = 0.$$

Note that $y = c_1 e^x$ is a solution for any choice of c_1, but $y = e^x + c_1$, $c_1 \neq 0$, does *not* satisfy the equation since, for this latter family of functions, we would get $y'' - y = -c_1$.

The next example shows that a solution of a differential equation can be a piecewise-defined function.

EXAMPLE 8

Any function in the one-parameter family $y = cx^4$ is a solution of the differential equation

$$xy' - 4y = 0.$$

We have $xy' - 4y = x(4cx^3) - 4cx^4 = 0$. The piecewise-defined function

$$y = \begin{cases} -x^4, & x < 0 \\ x^4, & x \geq 0 \end{cases}$$

is also a solution. Observe that this function cannot be obtained from $y = cx^4$ by a single selection of the parameter c (see Figure 1.3(b)).

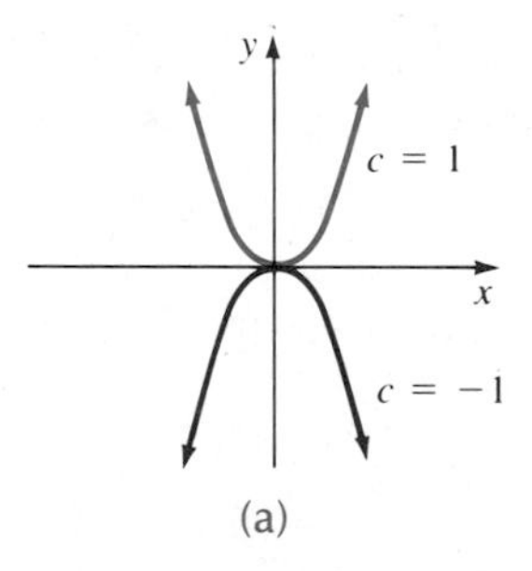

(a)

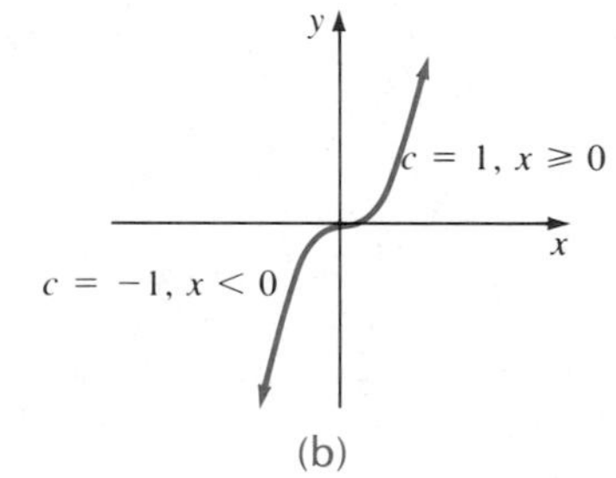

(b)

Figure 1.3

Further Terminology

The study of differential equations is similar to integral calculus.* When evaluating an antiderivative or indefinite integral, we utilize a single constant of integration. In like manner, when solving a first-order differential equation $F(x, y, y') = 0$, we shall usually obtain a family of curves or functions $G(x, y, c) = 0$ containing one arbitrary parameter such that each member of the family is a solution of the differential equation. In fact, when solving an nth-order equation $F(x, y, y', \ldots, y^{(n)}) = 0$, where $y^{(n)}$ means $d^n y/dx^n$, we expect an ***n*-parameter family of solutions** $G(x, y, c_1, \ldots, c_n) = 0$.

* A solution—explicit, implicit, or otherwise (see Problems 43–46)—is sometimes referred to as an **integral curve** or simply as an **integral** of the equation.

A solution of a differential equation that is free of arbitrary parameters is called a **particular solution**. One way of obtaining a particular solution is to choose specific values of the parameter(s) in a family of solutions. For example, it is readily seen that $y = ce^x$ is a one-parameter family of solutions of the simple first-order equation $y' = y$. For $c = 0$, -2, and 5, we get the particular solutions $y = 0$, $y = -2e^x$, and $y = 5e^x$, respectively.

Sometimes a differential equation possesses a solution that cannot be obtained by specializing the parameters in a family of solutions. Such a solution is called a **singular solution**.

EXAMPLE 9

In Section 2.1 we shall prove that a one-parameter family of solutions of $y' - xy^{1/2} = 0$ is given by $y = (x^2/4 + c)^2$. When $c = 0$, the resulting particular solution is $y = x^4/16$. In this case the trivial solution $y = 0$ is a singular solution of the equation since it cannot be obtained from the family for any choice of the parameter c.

If *every* solution of $F(x, y, y', \ldots, y^{(n)}) = 0$ on an interval I can be obtained from $G(x, y, c_1, \ldots, c_n) = 0$ by appropriate choices of the c_i, $i = 1, 2, \ldots, n$, we then say that the n-parameter family is the **general**, or **complete**, solution of the differential equation.

Remark: There are two schools of thought concerning the concept of a "general solution." An alternative viewpoint holds that a general solution of an nth-order differential equation is a family of solutions containing n essential* parameters. Period! In other words, the family is not required to contain all solutions of the differential equation on some interval. The difference in these opinions is really a distinction between the solutions to linear and nonlinear equations. In solving linear differential equations we shall impose relatively simple restrictions on the coefficients; with these restrictions one can always be assured that not only does a solution exist on an interval but that a family of solutions will indeed yield all possible solutions.

Another fact deserves mention at this time. Nonlinear equations, with the exception of some first-order equations, are usually difficult or *impossible* to solve in terms of the standard elementary functions.† Furthermore, if we happen to have a family of solutions for a nonlinear equation, it is not obvious when this family constitutes a general solution. On a practical level then, the designation "general solution" is applied only to linear differential equations.

---

* We won't try to define this concept. But roughly it means: Don't play games with the constants. Certainly $y = x + c_1 + c_2$ represents a family of solutions of $y' = 1$. By renaming $c_1 + c_2$ as c, the family has *essentially* one constant: $y = x + c$. The reader should verify that $y = c_1 + \ln c_2 x$ is a solution of $x^2 y'' + xy' = 0$ on the interval $(0, \infty)$ for any choice of c_1 and $c_2 > 0$. Are c_1 and c_2 essential parameters?

† For example, algebraic functions, exponential and logarithmic functions, trigonometric and inverse trigonometric functions.

Exercises 1.1

Answers to odd-numbered problems begin on page A–1.

Colored problem numbers indicate problems worked out in detail in an accompanying Student Solutions Manual.

In Problems 1–10 state whether the given differential equations are linear or nonlinear. Give the order of each equation.

1. $(1-x)y'' - 4xy' + 5y = \cos x$

2. $x\dfrac{d^3y}{dx^3} - 2\left(\dfrac{dy}{dx}\right)^4 + y = 0$

3. $yy' + 2y = 1 + x^2$

4. $x^2\,dy + (y - xy - xe^x)\,dx = 0$

5. $x^3y^{(4)} - x^2y'' + 4xy' - 3y = 0$

6. $\dfrac{d^2y}{dx^2} + 9y = \sin y$

7. $\dfrac{dy}{dx} = \sqrt{1 + \left(\dfrac{d^2y}{dx^2}\right)^2}$

8. $\dfrac{d^2r}{dt^2} = -\dfrac{k}{r^2}$

9. $(\sin x)y''' - (\cos x)y' = 2$

10. $(1-y^2)\,dx + x\,dy = 0$

In Problems 11–40 verify that the indicated function is a solution of the given differential equation. Where appropriate, c_1 and c_2 denote constants.

11. $2y' + y = 0;\ y = e^{-x/2}$

12. $y' + 4y = 32;\ y = 8$

13. $\dfrac{dy}{dx} - 2y = e^{3x};\ y = e^{3x} + 10e^{2x}$

14. $\dfrac{dy}{dt} + 20y = 24;\ y = \frac{6}{5} - \frac{6}{5}e^{-20t}$

15. $y' = 25 + y^2;\ y = 5\tan 5x$

16. $\dfrac{dy}{dx} = \sqrt{\dfrac{y}{x}};\ y = (\sqrt{x} + c_1)^2,\ x > 0,\ c_1 > 0$

17. $y' + y = \sin x;\ y = \frac{1}{2}\sin x - \frac{1}{2}\cos x + 10e^{-x}$

18. $2xy\,dx + (x^2 + 2y)\,dy = 0;\ x^2y + y^2 = c_1$

19. $x^2\,dy + 2xy\,dx = 0;\ y = -\dfrac{1}{x^2}$

20. $(y')^3 + xy' = y;\ y = x + 1$

21. $y = 2xy' + y(y')^2;\ y^2 = c_1(x + \frac{1}{4}c_1)$

22. $y' = 2\sqrt{|y|};\ y = x|x|$

23. $y' - \dfrac{1}{x}y = 1;\ y = x\ln x,\ x > 0$

24. $\dfrac{dP}{dt} = P(a - bP);\ P = \dfrac{ac_1e^{at}}{1 + bc_1e^{at}}$

25. $\dfrac{dX}{dt} = (2 - X)(1 - X);\ \ln\dfrac{2 - X}{1 - X} = t$

26. $y' + 2xy = 1;\ y = e^{-x^2}\int_0^x e^{t^2}\,dt + c_1 e^{-x^2}$

27. $(x^2 + y^2)\,dx + (x^2 - xy)\,dy = 0;\ c_1(x + y)^2 = xe^{y/x}$

28. $y'' + y' - 12y = 0;\ y = c_1 e^{3x} + c_2 e^{-4x}$

29. $y'' - 6y' + 13y = 0;\ y = e^{3x}\cos 2x$

30. $\dfrac{d^2y}{dx^2} - 4\dfrac{dy}{dx} + 4y = 0;\ y = e^{2x} + xe^{2x}$

31. $y'' = y;\ y = \cosh x + \sinh x$*

32. $y'' + 25y = 0;\ y = c_1 \cos 5x$

33. $y'' + (y')^2 = 0;\ y = \ln|x + c_1| + c_2$

34. $y'' + y = \tan x;\ y = -\cos x \ln(\sec x + \tan x)$

35. $x\dfrac{d^2y}{dx^2} + 2\dfrac{dy}{dx} = 0;\ y = c_1 + c_2 x^{-1}$

36. $x^2y'' - xy' + 2y = 0;\ y = x\cos(\ln x),\ x > 0$

37. $x^2y'' - 3xy' + 4y = 0;\ y = x^2 + x^2\ln x,\ x > 0$

38. $y''' - y'' + 9y' - 9y = 0;\ y = c_1 \sin 3x + c_2 \cos 3x + 4e^x$

39. $y''' - 3y'' + 3y' - y = 0;\ y = x^2 e^x$

40. $x^3\dfrac{d^3y}{dx^3} + 2x^2\dfrac{d^2y}{dx^2} - x\dfrac{dy}{dx} + y = 12x^2;\ y = c_1x + c_2x\ln x + 4x^2,\ x > 0$

In Problems 41 and 42 verify that the indicated piecewise defined function is a solution of the given differential equation.

41. $xy' - 2y = 0;\ y = \begin{cases} -x^2, & x < 0 \\ x^2, & x \ge 0 \end{cases}$

42. $(y')^2 = 9xy;\ y = \begin{cases} 0, & x < 0 \\ x^3, & x \ge 0 \end{cases}$

In Problems 43–46 proceed formally to show that the indicated parametric equations form a solution of the given differential equation.

EXAMPLE 10

$$4\left(\frac{dy}{dx}\right)^2 = y + 2;\quad x = 4t + 1,\quad y = t^2 - 2.$$

SOLUTION Recall from calculus that

$$\frac{dy}{dx} = \frac{dy/dt}{dx/dt} = \frac{2t}{4} = \frac{1}{2}t$$

* Recall that the hyperbolic cosine and hyperbolic sine are defined by

$$\cosh x = (e^x + e^{-x})/2 \quad \text{and} \quad \sinh x = (e^x - e^{-x})/2.$$

and so
$$4\left(\frac{dy}{dx}\right)^2 = 4\left(\frac{1}{2}t\right)^2$$
$$= t^2 = (t^2 - 2) + 2$$
$$= y + 2.$$

43. $\left(\frac{dy}{dx}\right)^3 + 2x\frac{dy}{dx} = 2y + 1;\ x = -\frac{3}{2}t^2,\ y = -t^3 - \frac{1}{2}$

44. $y = xy' + (y')^2;\ x = -2t,\ y = -t^2$

45. $y = xy' + (y')^2 - \ln y';\ x = -2t + \frac{1}{t},\ y = -t^2 - \ln t + 1$

46. $y[1 + (y')^2] = c;\ x = \frac{c}{2}(2\theta - \sin 2\theta),\ y = \frac{c}{2}(1 - \cos 2\theta)$

47. Verify that a one-parameter family of solutions for

$$y = xy' + (y')^2 \qquad \text{is} \qquad y = cx + c^2.$$

Determine a value of k such that $y = kx^2$ is a singular solution of the differential equation.

48. Verify that a one-parameter family of solutions for

$$y = xy' + \sqrt{1 + (y')^2} \qquad \text{is} \qquad y = cx + \sqrt{1 + c^2}.$$

Show that the relation $x^2 + y^2 = 1$ defines a singular solution of the equation on the interval $-1 < x < 1$.

49. A one-parameter family of solutions for

$$y' = y^2 - 1 \qquad \text{is} \qquad y = \frac{1 + ce^{2x}}{1 - ce^{2x}}.$$

By inspection*, determine a singular solution of the differential equation.

50. On page 6 we saw that $y = \sqrt{4 - x^2}$ and $y = -\sqrt{4 - x^2}$ are solutions of $\frac{dy}{dx} = -\frac{x}{y}$ on the interval $(-2, 2)$. Explain why

$$y = \begin{cases} \sqrt{4 - x^2}, & -2 < x < 0 \\ -\sqrt{4 - x^2}, & 0 \le x < 2 \end{cases}$$

is not a solution of the differential equation on the interval.

---

* Translated, this means take a good guess and see if it works.

Miscellaneous Problems

In Problems 51 and 52 find values of m so that $y = e^{mx}$ is a solution of each differential equation.

51. $y'' - 5y' + 6y = 0$

52. $y'' + 10y' + 25y = 0$

In Problems 53 and 54 find values of m so that $y = x^m$ is a solution of each differential equation.

53. $x^2y'' - y = 0$

54. $x^2y'' + 6xy' + 4y = 0$

55. Show that $y_1 = x^2$ and $y_2 = x^3$ are both solutions of

$$x^2y'' - 4xy' + 6y = 0.$$

Are the constant multiples c_1y_1 and c_2y_2, with c_1 and c_2 arbitrary, also solutions? Is the sum $y_1 + y_2$ a solution?

56. Show that $y_1 = 2x + 2$ and $y_2 = -x^2/2$ are both solutions of

$$y = xy' + (y')^2/2.$$

Are the constant multiples c_1y_1 and c_2y_2, with c_1 and c_2 arbitrary, also solutions? Is the sum $y_1 + y_2$ a solution?

57. By inspection determine, if possible, a real solution of the given differential equation.

(a) $\left|\dfrac{dy}{dx}\right| + |y| = 0$ **(b)** $\left|\dfrac{dy}{dx}\right| + |y| + 1 = 0$ **(c)** $\left|\dfrac{dy}{dx}\right| + |y| = 1$

[O] 1.2 Origins of Differential Equations

It would be a shame for a student to pass through a course such as this (as some do) and not have a modicum of appreciation for some of the origins of the subject matter. In the discussion that follows we shall see how specific differential equations arise not only out of consideration of families of geometric curves, but also how differential equations result from an attempt to describe, in mathematical terms, physical problems in the sciences and engineering. It would not be overly presumptive to state that differential equations form the backbone of subjects such as physics and electrical engineering and even provide an important working tool in such diverse areas as biology and economics. Several of the examples and problems in this section will serve as previews of coming attractions for the material in Chapters 3 and 5.

1.2.1 Differential Equation of a Family of Curves

At the end of the preceding section we expressed the expectation that an nth-order ordinary differential equation will yield an n-parameter family of solutions. However, the reader should not get the impression that we can *always* find an n-parameter family of solutions for every conceivable nth-order differential equation. On the other hand, suppose we turn the problem around: starting with an n-parameter family of curves, can we then find an associated nth-order differential equation that is entirely free of arbitrary parameters that represents the given family? In most cases the answer is yes.*

In the discussion of particular solutions we saw that each function in the one-parameter family $y = ce^x$ satisfies the same first-order differential equation $y' = y$. Suppose that we now seek to find the differential equation of the two-parameter family

$$y = c_1e^x + c_2.$$

The first two derivatives are

$$\frac{dy}{dx} = c_1e^x \quad \text{and} \quad \frac{d^2y}{dx^2} = c_1e^x.$$

Thus

$$\frac{d^2y}{dx^2} = \frac{dy}{dx} \quad \text{or} \quad \frac{d^2y}{dx^2} - \frac{dy}{dx} = 0.$$

EXAMPLE 1

By taking two derivatives we find that the differential equation of the two-parameter family of straight lines

$$y = c_1x + c_2$$

is simply

$$\frac{d^2y}{dx^2} = 0.$$

---

* See Problem 30 of this section. You should develop a suspicion that exceptions might exist to "general" discussions unless the points under consideration are summarized by means of a theorem. The hypothesis of the theorem sets the conditions under which the conclusion must always follow.

EXAMPLE 2 Find the differential equation of the family

$$y = cx^3$$

indicated in Figure 1.4.

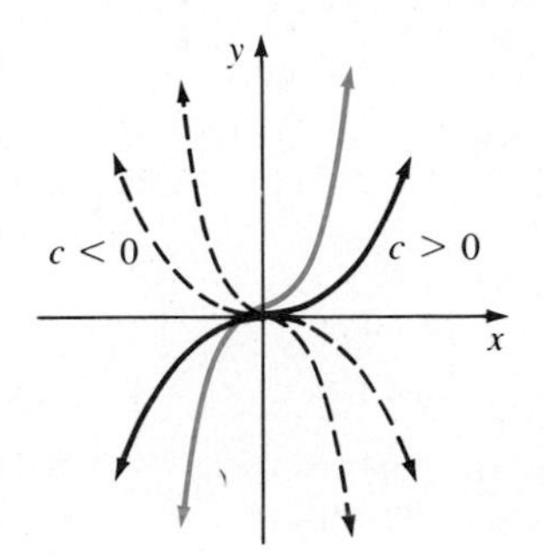

Figure 1.4

SOLUTION We expect a first-order differential equation since the family contains only one parameter. It follows that

$$\frac{dy}{dx} = 3cx^2,$$

but $c = y/x^3$, so that

$$\frac{dy}{dx} = 3\left(\frac{y}{x^3}\right)x^2 = 3\frac{y}{x}.$$

Thus we obtain the lincar first-order equation

$$x\frac{dy}{dx} - 3y = 0.$$

EXAMPLE 3 Find the differential equation of the two-parameter family

$$y = c_1e^{2x} + c_2e^{-2x}. \tag{1}$$

SOLUTION Taking two derivatives we obtain

$$\frac{dy}{dx} = 2c_1e^{2x} - 2c_2e^{-2x}$$

$$\frac{d^2y}{dx^2} = 4c_1e^{2x} + 4c_2e^{-2x}$$

$$= 4[c_1e^{2x} + c_2e^{-2x}].$$

Using (1), we find

$$\frac{d^2y}{dx^2} = 4y \qquad \text{or} \qquad y'' - 4y = 0.$$

EXAMPLE 4

Find the differential equation of the family of circles centered at the origin.

SOLUTION Concentric circles with center at the origin are described by the one-parameter equation

$$x^2 + y^2 = c^2, \qquad c > 0.$$

By implicit differentiation we obtain

$$2x + 2y\frac{dy}{dx} = 0$$

$$\frac{dy}{dx} = -\frac{x}{y} \qquad \text{or} \qquad x\,dx + y\,dy = 0.$$

EXAMPLE 5

Find the differential equation of the family of circles passing through the origin with center on the y-axis.

SOLUTION This family of circles is characterized by the one-parameter equation

$$x^2 + y^2 = cy.$$

Thus

$$2x + 2y\frac{dy}{dx} = c\frac{dy}{dx}.$$

Substituting $c = (x^2 + y^2)/y$ into the last equation and simplifying give

$$(x^2 - y^2)\frac{dy}{dx} = 2xy \qquad \text{or} \qquad (x^2 - y^2)\,dy - 2xy\,dx = 0.$$

EXAMPLE 6

Find the differential equation of the family of parabolas

$$y = (x + c)^2.$$

SOLUTION The first derivative is

$$\frac{dy}{dx} = 2(x + c).$$

From the original equation we have $x + c = \pm y^{1/2}$ so that the differential

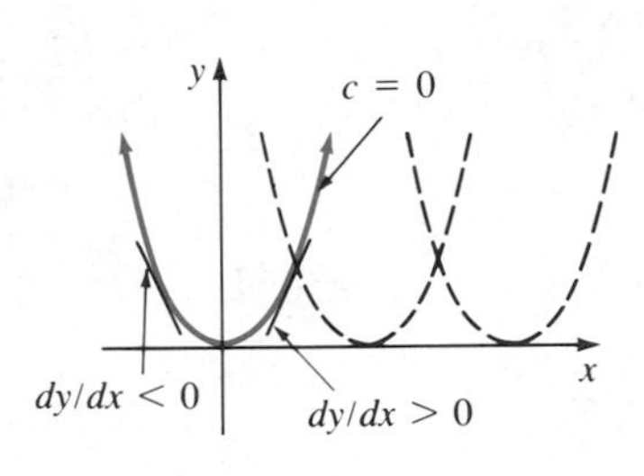

Figure 1.5

equation representing the family is

$$\frac{dy}{dx} = \pm 2y^{1/2} \quad \text{or} \quad \left(\frac{dy}{dx}\right)^2 = 4y.$$

It should be observed that

$$\frac{dy}{dx} = 2y^{1/2} \tag{2}$$

does not describe the complete family since by convention $y^{1/2} \geq 0$. Equation (2) would give the slope only of a right-hand branch ($x > -c$) of any particular parabola. Figure 1.5 illustrates the case when $c = 0$. In this case we could say that $y = x^2$ is a solution of (2) on the interval $(0, \infty)$.

EXAMPLE 7

Find the differential equation of the family

$$y = \frac{2ce^{2x}}{1 + ce^{2x}}.$$

SOLUTION By the quotient rule and algebra, we find

$$\frac{dy}{dx} = \frac{4ce^{2x}}{(1 + ce^{2x})^2} = \frac{y^2 e^{-2x}}{c}.$$

Solving the given equation for c gives $c = e^{-2x}y/(2 - y)$ and thus we obtain

$$\frac{dy}{dx} = y^2 e^{-2x} \frac{1}{\dfrac{e^{-2x}y}{2 - y}} \quad \text{or} \quad \frac{dy}{dx} = y(2 - y). \tag{3}$$

In Examples 6 and 7 the differential equation actually gives a bit more than we bargained for. By inspection we can see that the trivial function $y = 0$ is a solution of (2) and the constant function $y = 2$ is a solution of (3). In neither case is this particular solution a member of the given family.

A two-parameter family of curves can sometimes lead to a rather complicated differential equation.

EXAMPLE 8

Find the differential equation that describes the family of circles passing through the origin.

SOLUTION As Figure 1.6 indicates, the general form of the equation of these circles is

$$(x - h)^2 + (y - k)^2 = (\sqrt{h^2 + k^2})^2$$

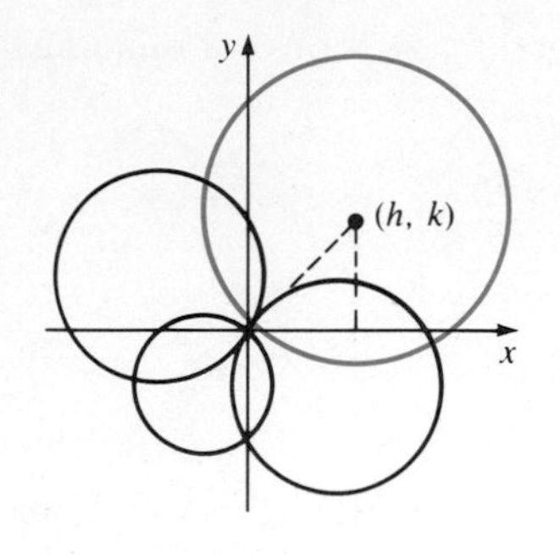

Figure 1.6

or
$$x^2 - 2xh + y^2 - 2ky = 0. \tag{4}$$

Using implicit differentiation twice, we find
$$x - h + yy' - ky' = 0 \tag{5}$$
and
$$1 + yy'' + (y')^2 - ky'' = 0. \tag{6}$$

We then use the original equation of the family (4) to solve for h:
$$h = \frac{x^2 + y^2 - 2ky}{2x}$$

and substitute in (5),
$$x - \frac{x^2 + y^2 - 2ky}{2x} + yy' - ky' = 0. \tag{7}$$

Now solving (7) for k gives
$$k = \frac{x^2 - y^2 + 2xyy'}{2(xy' - y)}. \tag{8}$$

Substituting this latter value in (6) and simplifying yield the nonlinear equation
$$1 + yy'' + (y')^2 - \frac{x^2 - y^2 + 2xyy'}{2(xy' - y)} y'' = 0$$
or
$$(x^2 + y^2)y'' + 2[(y')^2 + 1](y - xy') = 0.$$

Alternatively, we can obtain the last result directly by differentiating (8) by the quotient rule.

1.2.2 Some Physical Origins of Differential Equations

EXAMPLE 9

It is well known that free-falling objects close to the surface of the earth accelerate at a constant rate g. Acceleration is the derivative of velocity, and this, in turn, is the derivative of distance s. Thus if we assume that the upward direction is positive, the statement
$$\frac{d^2s}{dt^2} = -g$$
is the differential equation governing the vertical distance that the falling body travels. The minus sign is used since the weight of the body is a force directed opposite to the positive direction.

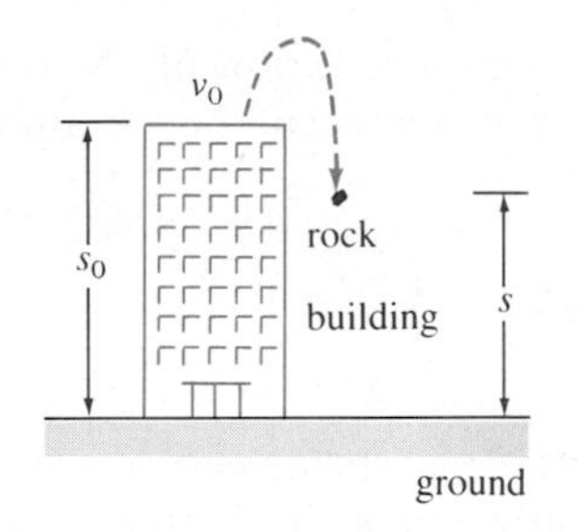

Figure 1.7

If we suppose further that a rock is tossed off the roof of a building of height s_0 (see Figure 1.7) with an initial upward velocity of, say, v_0, then we must solve

$$\frac{d^2s}{dt^2} = -g, \qquad 0 < t < t_1,$$

subject to the side conditions

$$s(0) = s_0, \qquad s'(0) = v_0.$$

Here $t = 0$ is taken to be the initial time when the rock leaves the roof of the building and t_1 is the time required to hit the ground. Since the rock is thrown upward it would naturally be assumed that $v_0 > 0$. This formulation of the problem ignores other forces such as air resistance acting on the body.

EXAMPLE 10

To find the vertical displacement $x(t)$ of a mass attached to a spring we use two different empirical laws: Newton's second law of motion and Hooke's law. The former law states that the net force acting on the system in motion is $F = ma$, where m is the mass and a is acceleration. Hooke's law states that the restoring force of a stretched spring is proportional to the elongation $s + x$, that is, the restoring force is $k(s + x)$, where $k > 0$ is a constant. As shown in Figure 1.8(b), s is the elongation of the spring after the mass has been attached and the system hangs at rest in the *equilibrium position*. When the system is in motion, the variable x represents a directed distance of the mass beyond the equilibrium position. In Chapter 5 we shall prove that when the system is in motion the *net force* acting on the mass is simply $F = -kx$. Thus in the absence of damping and other external forces

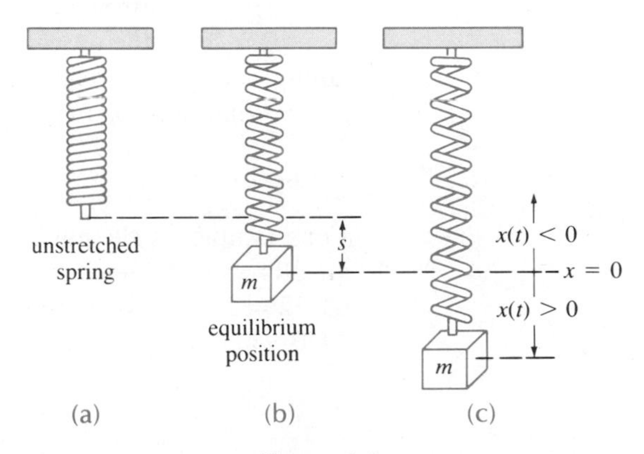

Figure 1.8

that might be impressed on the system, the differential equation of the vertical motion through the center of gravity of the mass can be obtained by equating:

$$m\frac{d^2x}{dt^2} = -kx.$$

Here the minus sign means that the restoring force of the spring acts opposite to the direction of motion, that is, toward the equilibrium position. In practice this second-order differential equation is often written as

$$\frac{d^2x}{dt^2} + \omega^2 x = 0, \tag{9}$$

where $\omega^2 = k/m$.

Units

A word is in order regarding the system of units that are used in describing dynamic problems such as illustrated in the last two examples. Three commonly used systems of units are summarized in the following table. In each system the basic unit of time is the second.

Quantity	Engineering System*	mks	cgs
Force	pound (lb)	newton (nt)	dyne
Mass	slug	kilogram (kg)	gram (g)
Distance	foot (ft)	meter (m)	centimeter (cm)
Acceleration of gravity g (approximate)	32 ft/sec^2	9.8 m/sec^2	980 cm/sec^2

The gravitational *force* exerted by the earth on a body of mass m is called its *weight* W. In the absence of air resistance the only force acting on a freely falling body is its weight. Hence, from Newton's second law of motion, it follows that mass m and weight W are related by

$$W = mg.$$

For example, in the engineering system a mass of 1/4 slug corresponds to an 8-lb weight. Since $m = W/g$, a 64-lb weight corresponds to a mass of $64/32 = 2$ slugs. In the cgs system a weight of 2450 dynes has a mass of $2450/980 = 2.5$ grams. In the mks system a weight of 50 newtons has a

* Also known as the English gravitational system or British engineering system.

mass of $50/9.8 = 5.1$ kilograms. We note that

$$1 \text{ newton} = 10^5 \text{ dynes} = 0.2247 \text{ pounds}.$$

In the next example we derive the differential equation that describes the motion of a *simple pendulum.*

EXAMPLE 11

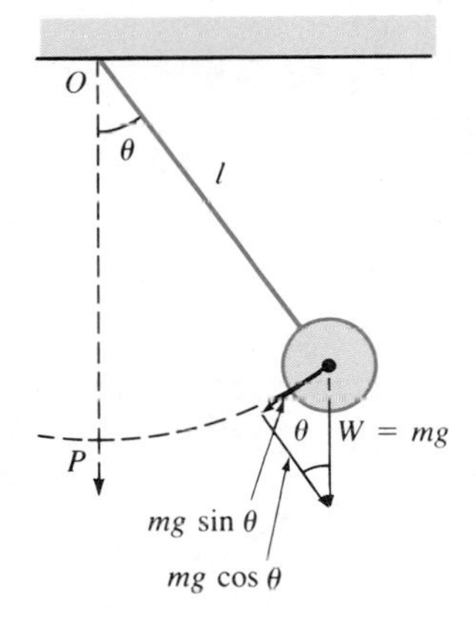

Figure 1.9

A mass m having weight W is suspended from the end of a rod of constant length l. For motion in a vertical plane, we would like to determine the displacement angle θ, measured from the vertical, as a function of time t (we consider $\theta > 0$ to the right of OP and $\theta < 0$ to the left of OP). Recall that an arc s of a circle of radius l is related to the central angle θ through the formula $s = l\theta$. Hence the angular acceleration is

$$a = \frac{d^2s}{dt^2} = l\frac{d^2\theta}{dt^2}.$$

From Newton's second law we then have

$$F = ma = ml\frac{d^2\theta}{dt^2}.$$

From Figure 1.9 we see that the tangential component of the force due to the weight W is $mg \sin \theta$. When the mass of the rod is ignored, we equate the two different formulations of the tangential force to obtain

$$ml\frac{d^2\theta}{dt^2} = -mg \sin \theta \quad \text{or} \quad \frac{d^2\theta}{dt^2} + \frac{g}{l}\sin \theta = 0. \tag{10}$$

Unfortunately the nonlinear equation (10) of the preceding example cannot be solved in terms of the familiar elementary functions (see Exercises 3.3), so usually a further simplifying assumption is made. If the angular displacements θ are not too large, we can use the approximation $\sin \theta \approx \theta$* so that (10) can be replaced with the linear second-order differential equation

$$\frac{d^2\theta}{dt^2} + \frac{g}{l}\theta = 0. \tag{11}$$

* The reader is encouraged to get out his or her calculus text and inspect its table of trigonometric functions and compare the numerical values of $\sin \theta$ with the value of θ in radians. It is also a good idea to look up the Maclaurin expansion for the sine function.

If we set $\omega^2 = g/l$, observe that (11) has the exact same structure as the differential equation governing the free vibrations of a weight on a spring (equation (9)). The fact that one basic differential equation can describe many diverse physical or even economic phenomena is a common occurrence in the study of applicable mathematics.

EXAMPLE 12

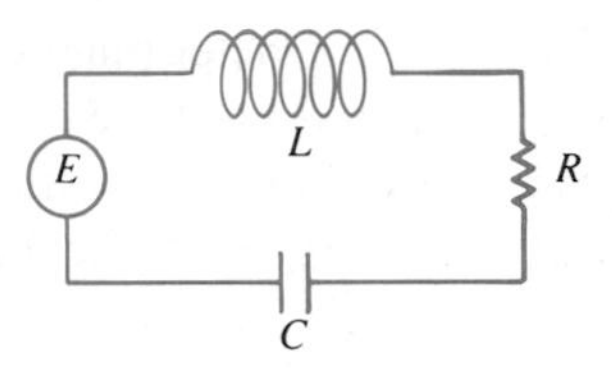

Figure 1.10

Consider the single loop series circuit containing an inductor, resistor, and capacitor, shown in Figure 1.10. Kirchhoff's second law states that the *sum* of the voltage drops across each part of the circuit is the same as the impressed voltage $E(t)$. If $q(t)$ denotes the charge of the capacitor at any time, then the current $i(t)$ is given by $i = dq/dt$. Now it is known that the voltage drops across an

$$\text{inductor} = L\frac{di}{dt} = L\frac{d^2q}{dt^2}$$

$$\text{resistor} = iR = R\frac{dq}{dt}$$

$$\text{capacitor} = \frac{1}{C}q,$$

where L, R, and C are constants called the inductance, resistance, and capacitance, respectively. Therefore, to determine $q(t)$ we must solve the second-order differential equation

$$L\frac{d^2q}{dt^2} + R\frac{dq}{dt} + \frac{1}{C}q = E(t). \tag{12}$$

In the previous example, the side conditions $q(0)$ and $q'(0)$ represent the charge on the capacitor and the current in the circuit, respectively, at $t = 0$. Also, the impressed voltage $E(t)$ is said to be an **electromotive force**, or **emf**. An emf, as well as a charge on a capacitor, causes the current in a circuit to flow. The following table shows the basic units of measurement used in circuit analysis.

Quantity	Unit
Impressed voltage or emf	volt (v)
Inductance L	henry (h)
Capacitance C	farad (f)
Resistance R	ohm (Ω)
Charge q	coulomb
Current i	ampere (amp)

EXAMPLE 13

It seems plausible to expect that the rate at which a population P expands is proportional to the population that is present at any time. Roughly put, the more people there are, the more there are going to be. Thus one model for population growth is given by the differential equation

$$\frac{dP}{dt} = kP, \tag{13}$$

where k is a constant of proportionality. Since we also expect the population to expand we must have $dP/dt > 0$ and thus $k > 0$.

EXAMPLE 14

In the spread of a contagious disease, for example a flu virus, it is reasonable to assume that the rate, dx/dt, at which the disease spreads is proportional not only to the number of people, $x(t)$, who have contracted the disease, but also to the number of people, $y(t)$, who have not yet been exposed—that is,

$$\frac{dx}{dt} = kxy, \tag{14}$$

where k is the usual constant of proportionality. If one infected person is introduced into a fixed population of n people, then x and y are related by

$$x + y = n + 1. \tag{15}$$

Using (15) to eliminate y in (14) then gives

$$\frac{dx}{dt} = kx(n + 1 - x). \tag{16}$$

The obvious side condition accompanying equation (16) is $x(0) = 1$.

Logistic Equation

The nonlinear first-order equation (16) is a special case of a more general equation

$$\frac{dP}{dt} = P(a - bP), \qquad a \text{ and } b \text{ constants}, \tag{17}$$

known as the **logistic equation** (see Section 3.3). The solution of this equation is very important in ecological, sociological, and even managerial sciences.

EXAMPLE 15

Newton's law of cooling states that the time rate at which a body cools is proportional to the difference between the temperature of the body and the temperature of the surrounding medium. If $T(t)$ denotes the temperature of the body at any time t, and T_0 is the constant temperature of the outside medium, it follows that

$$\frac{dT}{dt} = k(T - T_0) \tag{18}$$

where k is the constant of proportionality. Note that when $T_0 = 0$, equation (18) reduces to (13). However, in this case $T(t)$ is decreasing so we want $k < 0$.

Equation (13) also appears in a different context in the next example.

EXAMPLE 16

When interest is compounded **continuously**, the rate at which an amount of money S grows is proportional to the amount of money present at any time—that is,

$$\frac{dS}{dt} = rS, \tag{19}$$

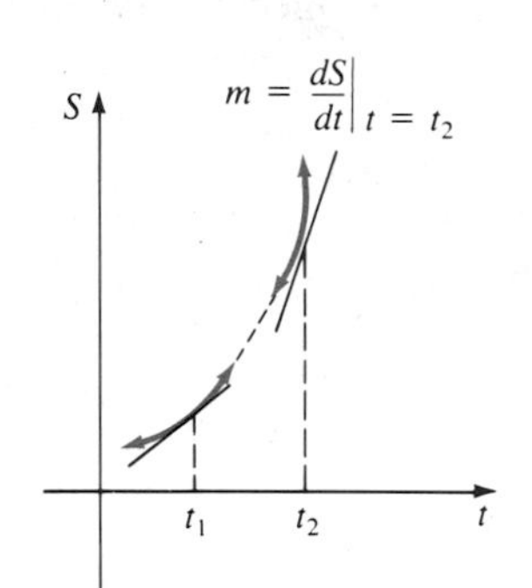

Figure 1.11

where r is the annual rate of interest.* This is analogous to the population growth of an earlier example. The rate of growth is large when the amount of money present in the account is also large. Translated geometrically, this means the tangent line is steep when S is large (see Figure 1.11).

The definition of a derivative provides an interesting derivation of equation (19). Suppose $S(t)$ is the amount accrued in a savings account after t years when the annual rate of interest r is compounded continuously. If h denotes an increment in t, then the interest obtained in the time span $(t + h) - t$ is the difference in amounts accrued:

$$S(t + h) - S(t). \tag{20}$$

Since interest is given by

$$(\text{rate}) \times (\text{time}) \times (\text{principal}), \tag{21}$$

we can approximate the interest earned in this same time period by either

$$rhS(t) \tag{22}$$

* Both dS/dt and r are rates. A ratio such as $(dS/dt)/S$ is often called the *growth rate, specific growth rate, relative growth rate,* or *average growth rate.*

or

$$rhS(t + h). \tag{23}$$

Intuitively (22) and (23) are lower and upper bounds, respectively, for the actual interest (20), that is,

$$rhS(t) \leq S(t + h) - S(t) \leq rhS(t + h)$$

or

$$rS(t) \leq \frac{S(t + h) - S(t)}{h} \leq rS(t + h). \tag{24}$$

Taking the limit of (24) as $h \to 0$ gives

$$rS(t) \leq \lim_{h \to 0} \frac{S(t + h) - S(t)}{h} \leq rS(t),$$

and so it must follow that

$$\lim_{h \to 0} \frac{S(t + h) - S(t)}{h} = rS(t) \qquad \text{or} \qquad \frac{dS}{dt} = rS.$$

EXAMPLE 17

Suppose a suspended wire hangs under its own weight. As Figure 1.12(a) shows, this could be a long telephone wire between two posts. Our goal here is to determine the differential equation governing the shape that the hanging wire assumes.

Let us examine only a portion of the wire between the lowest point P_1 and any arbitrary point P_2 (see Figure 1.12(b)). Three forces are acting on the wire: the weight of the segment P_1P_2, and the tensions $\mathbf{T}_1$ and $\mathbf{T}_2$ in the wire at P_1 and P_2, respectively. If w is the linear density (measured, say, in lb/ft) and s is the length of the segment P_1P_2, its weight is necessarily ws.

Now the tension $\mathbf{T}_2$ resolves into horizontal and vertical components (scalar quantities) $T_2 \cos \theta$ and $T_2 \sin \theta$. Because of equilibrium we can write

$$|\mathbf{T}_1| = T_1 = T_2 \cos \theta, \qquad \text{and} \qquad ws = T_2 \sin \theta.$$

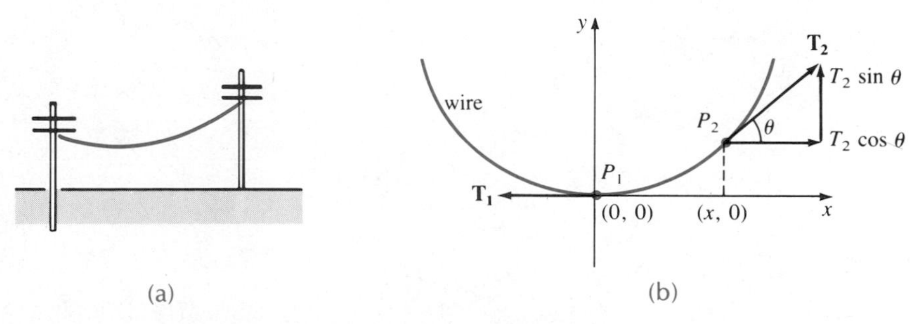

Figure 1.12

Dividing the last two equations, we find

$$\tan\theta = \frac{ws}{T_1}$$

or

$$\frac{dy}{dx} = \frac{ws}{T_1}. \tag{25}$$

Now since the length of the arc between points P_1 and P_2 is

$$s = \int_0^x \sqrt{1 + \left(\frac{dy}{dx}\right)^2}\,dx$$

it follows from one form of the fundamental theorem of calculus that

$$\frac{ds}{dx} = \sqrt{1 + \left(\frac{dy}{dx}\right)^2}. \tag{26}$$

Differentiating (25) with respect to x and using (26) lead to

$$\frac{d^2y}{dx^2} = \frac{w}{T_1}\frac{ds}{dx} \quad \text{or} \quad \frac{d^2y}{dx^2} = \frac{w}{T_1}\sqrt{1 + \left(\frac{dy}{dx}\right)^2}. \tag{27}$$

One might conclude from Figure 1.12 that the shape the hanging wire assumes is parabolic. However, this is not the case; a wire or heavy rope hanging only under its own weight takes on the shape of a hyperbolic cosine (see Problem 12, Exercises 3.3). Recall that the graph of the hyperbolic cosine is called a **catenary**, which stems from the Latin word *catena* meaning "chain." The Romans used the catena as a dog leash. Probably the most graphic example of the shape of a catenary is the 630-ft-high Gateway arch in St. Louis, Missouri (see Figure 1.13).

Figure 1.13 Gateway Arch
Photograph by Frank Siteman/Stock, Boston, Inc.

EXAMPLE 18

In the **theory of learning**, the rate at which a subject is memorized is assumed to be proportional to the amount of material that is left to be memorized. If $A(t)$ represents the amount memorized at time t, and M is the total amount to be memorized, then $M - A$ is the amount of material remaining to be memorized. Hence the rate of memorization is given by

$$\frac{dA}{dt} = k(M - A), \tag{28}$$

where k is positive constant.

Exercises 1.2

Answers to odd-numbered problems begin on page A–1.

[1.2.1] In Problems 1–22 find a differential equation for the given family of curves.

1. $y = c_1 x + 2$
2. $c_1 y + 2x = 3$
3. $y = c_1 e^{-x}$
4. $y = e^x + c_1 e^{-x}$
5. $y^2 - c_1(x + 1)$
6. $c_1(y + 1)^2 = x$
7. $c_1 y^2 + 4y = 2x^2$
8. $c_1 x^2 - y^2 = 1$
9. $y = c_1 + c_2 e^x$
10. $y = c_1 e^{3x} + c_2 e^{-4x}$
11. $y = c_1 \sin \omega t + c_2 \cos \omega t$, where ω is a constant not to be eliminated.
12. $y = c_1 \sin (\omega t + c_2)$, where ω is a constant not to be eliminated.
13. $y = c_1 \sinh kt + c_2 \cosh kt$, where k is a constant not to be eliminated.
14. $y = c_1 e^{kt} + c_2 e^{-kt}$, where k is a constant not to be eliminated.
15. $y = c_1 e^{4x} + c_2 x e^{4x}$
16. $y = c_1 e^x \cos x + c_2 e^x \sin x$
17. $y = c_1 + c_2 \ln x$
18. $y = c_1 x + c_2 x^2$
19. $y = c_1 e^x + c_2 e^{2x} + c_3 e^{3x}$
20. $y = c_1 + c_2 e^x + c_3 x e^x$
21. $r = c_1(1 + \cos \theta)$
22. $r = c_1(\sec \theta + \tan \theta)$
23. Find the differential equation of the family of straight lines passing through the origin.
24. Find the differential equation of the family of circles with centers on the y-axis.
25. Find the differential equation of the family of circles passing through the origin with centers on the x-axis.

26. Find the differential equation of the family of circles passing through (0, − 3), and (0, 3), whose centers are on the x-axis.

27. Find the differential equation of the family of parabolas whose vertex is at the origin but whose focus is on the x-axis.

28. Find the differential equation of the family of tangent lines to the parabola $y^2 = 2x$.

29. Find the differential equation of the family of parabolas whose vertex and focus are on the axis.

30. Show that the two-parameter family $y^2 = 2c_1x^2y + c_2x^4$ yields the first-order differential equation

$$x\frac{dy}{dx} = 2y$$

[1.2.2] In Problems 31–44, derive the approximate differential equation(s) describing the given physical situation.

31. A series circuit contains a resistor and an inductor as shown in Figure 1.14. Determine the differential equation for the current $i(t)$ if the resistance is R, the inductance is L, and the impressed voltage is $E(t)$.

32. A series circuit contains a resistor and a capacitor as shown in Figure 1.15. Determine the differential equation for the charge $q(t)$ on the capacitor if the resistance is R, the capacitance is C, and the impressed voltage is $E(t)$.

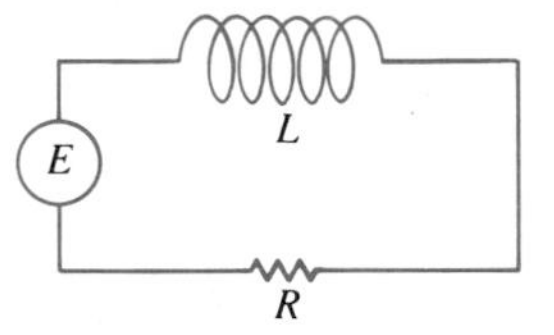

Figure 1.14

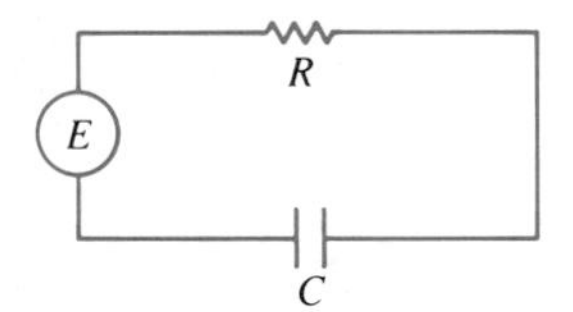

Figure 1.15

EXAMPLE 19

Under some circumstances a falling body B of mass m (such as a man hanging from a parachute) encounters air resistance proportional to its instantaneous velocity, $v(t)$. Use Newton's second law to find the differential equation for the velocity of the body at any time.

SOLUTION Assuming that the downward direction is positive, the sum of the forces acting on the body is

$$mg - kv \tag{29}$$

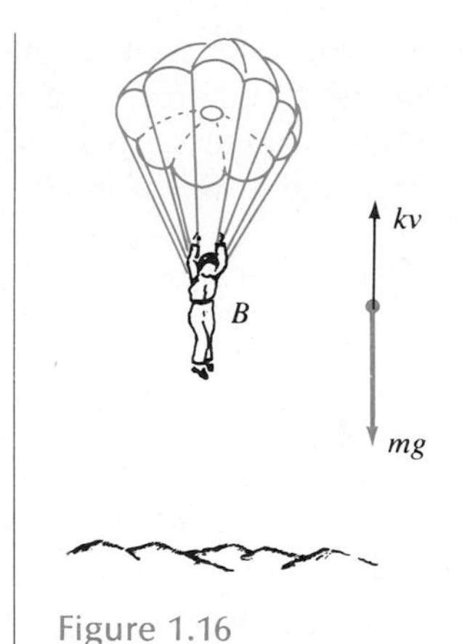

Figure 1.16

where k is a constant of proportionality, and the minus sign indicates that the resistance acts in a direction opposite to the motion (see Figure 1.16). Newton's second law can be written as

$$ma = m\frac{dv}{dt}, \tag{30}$$

where a represents acceleration. Equating (29) and (30) then gives

$$m\frac{dv}{dt} = mg - kv \qquad \text{or} \qquad \frac{dv}{dt} + \frac{k}{m}v = g.$$

33. What is the differential equation for the velocity v of a body of mass m falling vertically downward through a medium offering a resistance proportional to the square of the instantaneous velocity?

34. Determine the differential equation governing the height h, at any time, of water flowing through an orifice at the bottom of a cylindrical tank (see Figure 1.17). Use the fact that the decrease in the volume of the water $-A_1\,\Delta h(\Delta h < 0)$ is the same as the volume of the element of length Δx in a time Δt. Also use the fact that an object falling from rest acquires a velocity $\sqrt{2gh}$ ft/sec, $g = 32$, in h feet. (Where did this come from?)

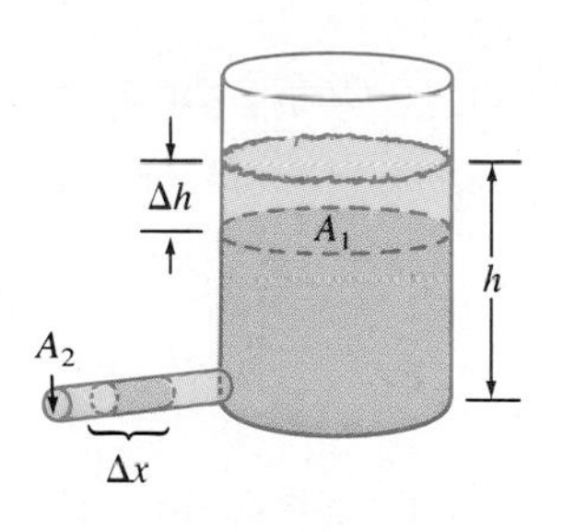

Figure 1.17

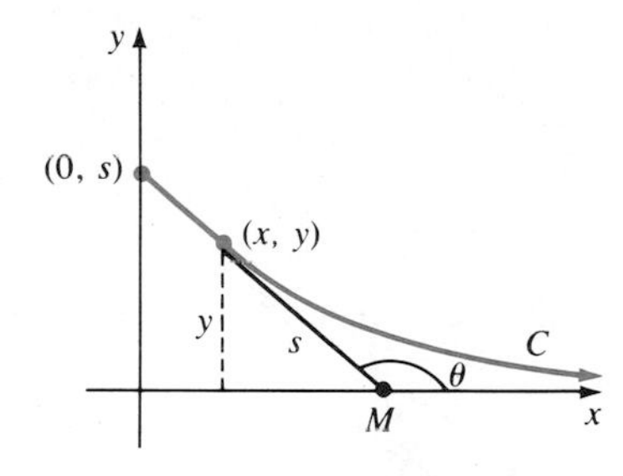

Figure 1.18

35. A man M, starting at the origin, moves in the direction of the positive x-axis, pulling a weight along the curve C (called a **tractrix**) indicated in Figure 1.18. The weight, initially located on the y-axis at $(0, s)$, is pulled by a rope of constant length s, which is kept taut throughout the motion. Find the differential equation of the path of motion. [*Hint:* The rope is always tangent to C; consider the angle of inclination θ as shown in the figure.]

36. A projectile shot from a gun has weight $w = mg$ and velocity $\mathbf{v}$ tangent to its path of motion. Ignoring air resistance and all other forces except its weight, find the system of differential equations that describes the motion. [*Hint:* Use Newton's second law in the x and y direction. See Figure 1.19.]

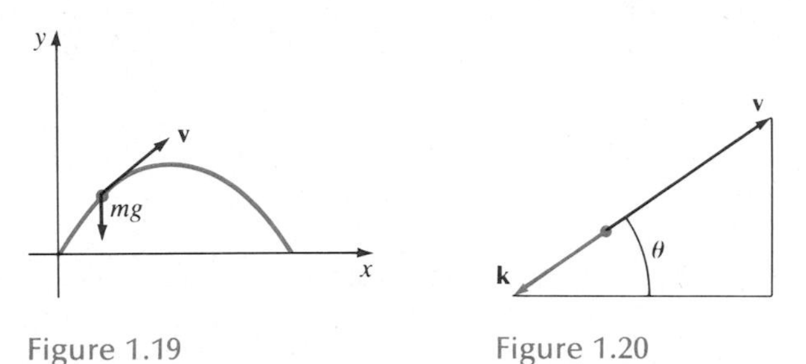

Figure 1.19 Figure 1.20

37. Determine the equations of motion if the projectile in Problem 36 encounters a retarding force $\mathbf{k}$ (of magnitude k) acting tangent to the path but opposite to the motion. [*Hint:* $\mathbf{k}$ is a multiple of the velocity, say $c\mathbf{v}$. See Figure 1.20.]

38. The rate at which a radioactive substance decays is proportional to the amount of the substance remaining at any time. Determine the differential equation for the amount $A(t)$.

39. A drug is infused into a patient's bloodstream at a constant rate r grams/sec. Simultaneously, the drug is removed at a rate proportional to the amount $x(t)$ of the drug present at any time. Determine the differential equation governing the amount $x(t)$.

40. Two chemicals A and B react to form a new chemical C. Assuming that the concentrations of both A and B decrease by the amount of C formed, find the differential equation governing the concentration $x(t)$ of the chemical C if the rate at which the chemical reaction takes place is proportional to the product of the remaining concentrations of A and B.

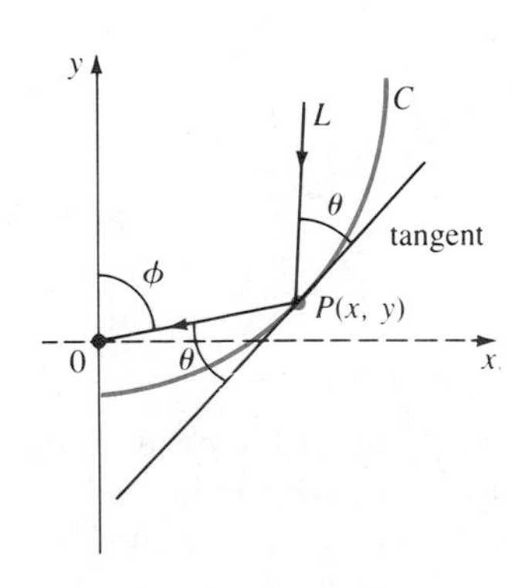

Figure 1.21

41. Light strikes a plane curve C in such a manner that all beams L parallel to the y-axis are reflected to a single point 0. Determine the differential equation for the function $y = f(x)$ describing the shape of the curve. (The fact that the angle of incidence is equal to the angle of reflection is a principle of optics.) [*Hint:* Inspection of Figure 1.21 shows that the inclination of the tangent line from the horizontal at $P(x, y)$ is $\pi/2 - \theta$ and that we can write $\phi = 2\theta$. (Why?) Also, don't be afraid to use a trigonometric identity.]

42. A cylindrical barrel s feet in diameter of weight w lb is floating in water. After an initial depression the barrel exhibits an up and down bobbing

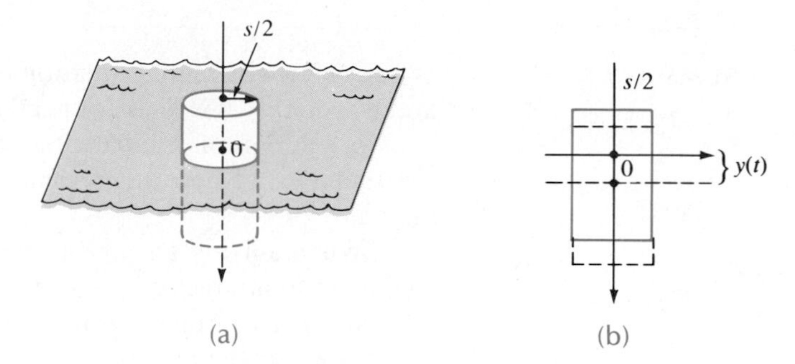

Figure 1.22

motion along a vertical line. Using Figure 1.22(b), determine the differential equation for the vertical displacements $y(t)$, if the origin is taken to be on the vertical axis at the surface of the water when the barrel is at rest. Use Archimedes' principle that the buoyancy, or upward force of the water on the barrel, is equal to the weight of the water displaced, and the fact that the density of water is 62.4 lb/ft^3. Assume that the downward direction is positive.

43. A rocket is shot vertically upward from the surface of the earth. After all its fuel has been expended, the mass of the rocket is a constant m. Use Newton's second law of motion and the fact that the force of gravity varies inversely as the square of the distance to find the differential equation for the distance y from the earth's center to the rocket at any time after burnout. State appropriate conditions at $t = 0$ associated with this differential equation.

44. Newton's second law $F = ma$ can be written $F = d/dt\,(mv)$. When the mass of an object is variable this latter formulation is used. The mass $m(t)$ of a rocket launched upward changes as its fuel is consumed.* If $v(t)$ denotes its velocity at any time, it can be shown that

$$-mg = m\frac{dv}{dt} - V\frac{dm}{dt}, \tag{31}$$

where V is the constant velocity of the exhaust gases relative to the rocket. Use (31) to find the differential equation for v if it is known that $m(t) = m_0 - at$ and $V = -b$ where m_0, a, and b are constants.

---

* It is assumed that the *total* mass:

mass of vehicle + mass of fuel + mass of exhaust gases

is constant. In this case $m(t)$ = mass of vehicle + mass of fuel.

CHAPTER 1 SUMMARY

We classify a differential equation by its type: **ordinary** or **partial**; by its **order**; and by whether it is **linear** or **nonlinear**.

A **solution** of a differential equation is any function, having a sufficient number of derivatives, that satisfies the equation identically on some interval.

When solving an nth-order ordinary differential equation, we expect to find an n-parameter family of solutions. A **particular solution** is any solution free of arbitrary parameters that satisfies the differential equation. A **singular solution** is any solution that cannot be obtained from an n-parameter family of solutions by assigning values to the parameters. When an n-parameter family of solutions gives every solution of a differential equation on some interval, it is then called a **general**, or a **complete**, solution.

Starting with an n-parameter family of curves in the plane, we can find, in *most* cases, an nth-order differential equation representing the family. In the analysis of physical problems, many differential equations can be obtained by equating two different empirical formulations of the same situation. For example, a differential equation of motion can sometimes be obtained by simply equating Newton's second law of motion with the net forces acting on a body.

CHAPTER 1 REVIEW EXERCISES

Answers to odd-numbered problems begin on page A–2.

In Problems 1–4 classify the given differential equation as to type and order. Classify the ordinary differential equations as to linearity.

1. $(2xy - y^2)\,dx + e^x dy = 0$ ______

2. $(\sin xy)\,y''' + 4xy' = 0$ ______

3. $\dfrac{\partial^2 u}{\partial x^2} + \dfrac{\partial^2 u}{\partial y^2} = u$ ______

4. $x^2\dfrac{d^2y}{dx^2} - 3x\dfrac{dy}{dx} + y = x^2$ ______

In Problems 5–8 verify that the indicated function is a solution of the given differential equation.

5. $y' + 2xy = 2 + x^2 + y^2;\ \ y = x + \tan x$

6. $x^2y'' + xy' + y = 0;\ \ y = c_1\cos(\ln x) + c_2\sin(\ln x),\ \ x > 0$

7. $y''' - 2y'' - y' + 2y = 6;\ \ y = c_1e^x + c_2e^{-x} + c_3e^{2x} + 3$

8. $y^{(4)} - 16y = 0;\ \ y = \sin 2x + \cosh 2x$

In Problems 9–14 determine by inspection at least one solution for the given differential equation.

9. $y' = 2x$ ______

10. $\dfrac{dy}{dx} = 5y$ ______

11. $y'' = 1$ ______

12. $y' = y^3 - 8$ ______

13. $y'' = y'$ ______

14. $2y\dfrac{dy}{dx} = 1$ ______

15. Determine an interval on which $y^2 - 2y = x^2 - x - 1$ defines a solution of $2(y - 1)\,dy + (1 - 2x)\,dx = 0$.

16. Explain why the differential equation

$$\left(\frac{dy}{dx}\right)^2 = \frac{4 - y^2}{4 - x^2}$$

possesses no real solutions for $|x| < 2$, $|y| > 2$. Are there other regions in the xy-plane for which the equation has no solutions?

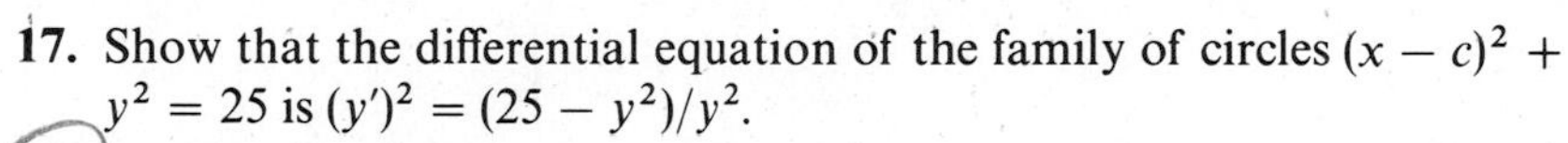

17. Show that the differential equation of the family of circles $(x - c)^2 + y^2 = 25$ is $(y')^2 = (25 - y^2)/y^2$.

18. Find two singular solutions of the differential equation in Problem 17.

19. Find the differential equation that represents the family of straight lines passing through the point (2, 1).

20. Find the differential equation representing the family of circles passing through the origin, with centers on the line $y = x$.

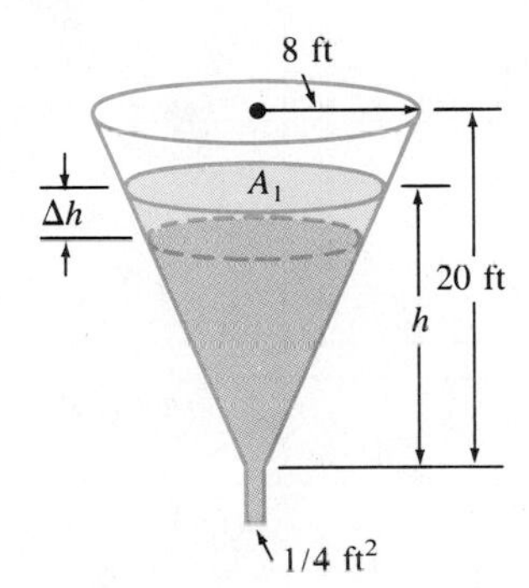

Figure 1.23

21. The conical tank shown in Figure 1.23 loses water out of an orifice at its bottom. If the cross-sectional area of the orifice is $(1/4)\text{ft}^2$, find the differential equation representing the height of the water h at any time (see Problem 34, Exercises 1.2).

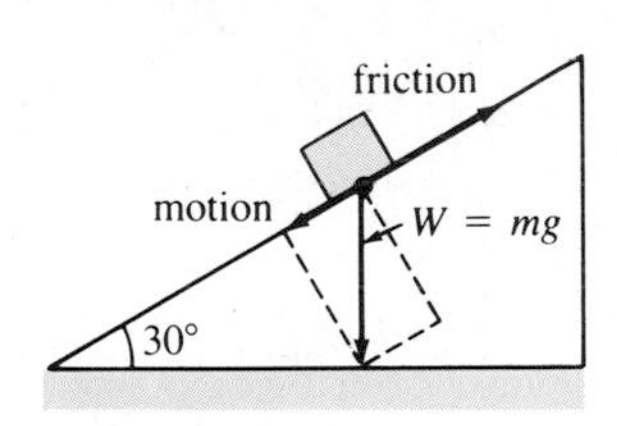

Figure 1.24

22. A weight of 96 lb slides down an incline making a 30° angle with the horizontal. If the coefficient of sliding friction is μ, determine the differential equation for the velocity $v(t)$ of the weight at any time. Use the fact that the force of friction opposing the motion is μN, where N is the normal component of the weight (see Figure 1.24).

CHAPTER 2

First-Order Differential Equations

IMPORTANT TERMS

initial-value problem
initial condition
existence of a solution
uniqueness of a solution
separation of variables
homogeneous function
homogeneous equation
exact differential
exact equation
integrating factor
general solution
linear equation

We are now in a position to solve differential equations. We begin by considering some first-order equations.

2.1 Preliminary Theory

Initial-Value Problem

We are often interested in solving a first-order differential equation

$$\frac{dy}{dx} = f(x, y)^* \tag{1}$$

subject to a side condition $y(x_0) = y_0$, where x_0 is a number in an interval I and y_0 is an arbitrary real number. The problem

$$\begin{aligned} \textit{Solve}: \quad & \frac{dy}{dx} = f(x, y) \\ \textit{Subject to}: \quad & y(x_0) = y_0 \end{aligned} \tag{2}$$

is called an **initial-value problem**. The side condition is known as an **initial condition**.

EXAMPLE 1

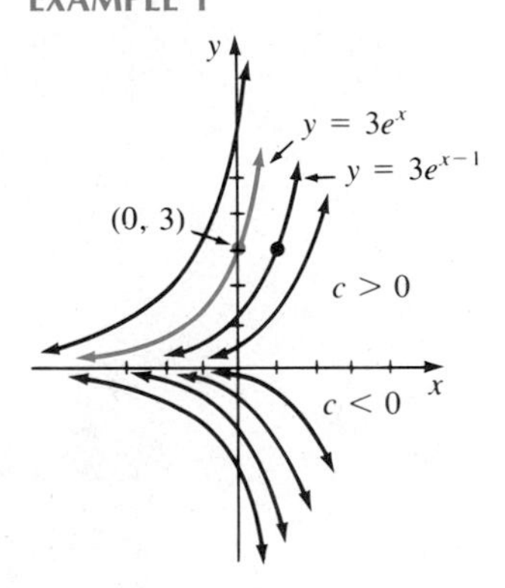

Figure 2.1

We have seen that $y = ce^x$ is a one-parameter family of solutions for $y' = y$ on the interval $(-\infty, \infty)$. If we specify, say, $y(0) = 3$, then substituting $x = 0$, $y = 3$ in the family yields $3 = ce^0 = c$. Thus, as shown in Figure 2.1,

$$y = 3e^x$$

is a solution of the initial-value problem

$$\begin{aligned} y' &= y \\ y(0) &= 3. \end{aligned}$$

Had we demanded that a solution of $y' = y$ pass through the point (1, 3) rather than (0, 3), then $y(1) = 3$ would yield $c = 3e^{-1}$ and so $y = 3e^{x-1}$. The graph of this function is also indicated in Figure 2.1.

Two fundamental questions arise in considering an initial-value problem such as (2):

Does a solution of the problem *exist*?

If a solution exists, is it *unique*, or the only solution of the problem?

Geometrically we are asking in the second question: Of all the solutions of a differential equation (1) that exist on an interval I, is there only one whose graph passes through (x_0, y_0) (see Figure 2.2)?

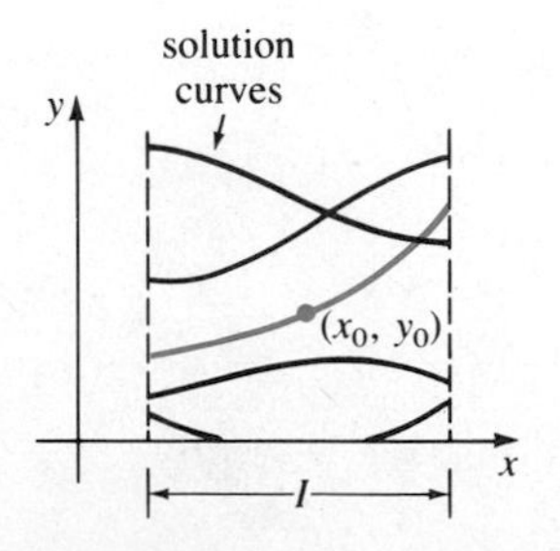

Figure 2.2

* In this text we shall assume that a differential equation $F(x, y, y', \ldots, y^{(n)}) = 0$ can be solved for the highest order derivative: $y^{(n)} = f(x, y, y', \ldots, y^{(n-1)})$. There are exceptions.

As the next example shows, the answer to the second question is sometimes no.

EXAMPLE 2

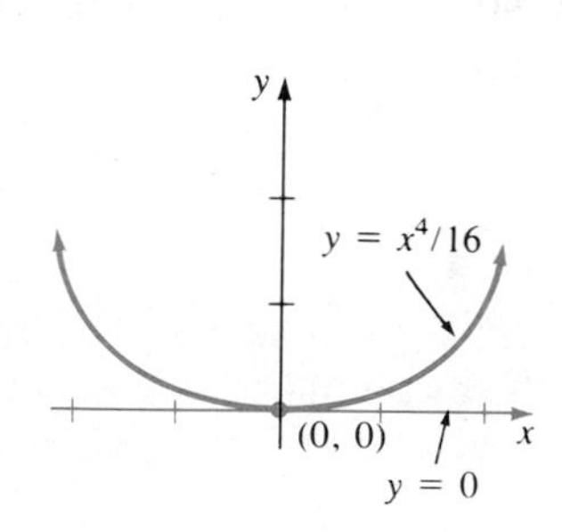

Figure 2.3

Figure 2.3 shows that the initial-value problem

$$\frac{dy}{dx} - xy^{1/2} = 0$$

$$y(0) = 0$$

has at least *two* solutions on the interval $(-\infty, \infty)$. The functions

$$y = 0 \quad \text{and} \quad y = \frac{x^4}{16}$$

satisfy the differential equation and have graphs passing through (0, 0).

It is often desirable to know before tackling an initial-value problem whether a solution exists, and when it does, whether it is the only solution of the problem. The following theorem due to Picard* gives *sufficient* conditions for the **existence of a unique solution** of (2).

THEOREM 2.1 Let R be a rectangular region in the xy-plane defined by $a \le x \le b$, $c \le y \le d$ that contains the point (x_0, y_0) in its interior. If $f(x, y)$ and $\partial f/\partial y$ are continuous on R, then there exists an interval I centered at x_0 and a unique function $y(x)$ defined on I satisfying the initial-value problem (2).

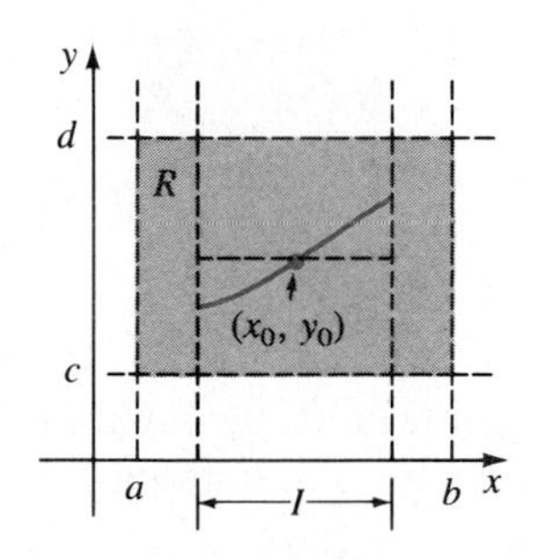

Figure 2.4

The foregoing is one of the most popular existence and uniqueness theorems for first-order differential equations because the criteria of continuity of $f(x, y)$ and $\partial f/\partial y$ are relatively easy to check. In general, it is not always possible to find a specific interval I on which a solution is defined without actually solving the differential equation (see Problem 16). The geometry of Theorem 2.1 is illustrated in Figure 2.4.

---

* **Charles Émile Picard** (1856–1941) Picard was one of the prominent French mathematicians of the latter nineteenth and early twentieth centuries. He made significant contributions to the fields of differential equations and complex variables. In 1899 Picard lectured at Clark University in Worcester, Massachusetts.

EXAMPLE 3

We have already seen in Example 2 that the differential equation

$$\frac{dy}{dx} - xy^{1/2} = 0$$

possesses at least two solutions whose graphs pass through (0, 0). By writing the equation in the form $dy/dx = xy^{1/2}$, we can identify

$$f(x, y) = xy^{1/2} \quad \text{and} \quad \frac{\partial f}{\partial y} = \frac{x}{2y^{1/2}}.$$

We note that both functions are continuous in the upper half plane defined by $y > 0$. We conclude from Theorem 2.1 that through any point (x_0, y_0), $y_0 > 0$ (say, for example, (0, 1)) there is some interval around x_0 on which the given differential equation has a unique solution.

EXAMPLE 4

Theorem 2.1 guarantees that there exists an interval about $x = 0$ on which $y = 3e^x$ is the only solution of the initial-value problem of Example 1:

$$y' = y$$
$$y(0) = 3.$$

This follows from the fact that $f(x, y) = y$ and $\partial f/\partial y = 1$ are continuous throughout the entire xy-plane. It can be further shown that this interval is $(-\infty, \infty)$.

EXAMPLE 5

For

$$\frac{dy}{dx} = x^2 + y^2$$

we observe that $f(x, y) = x^2 + y^2$ and $\partial f/\partial y = 2y$ are continuous throughout the entire xy-plane. Therefore, through any given point (x_0, y_0) there passes one and only one solution of the differential equation.

The reader should be aware of the distinction between a solution existing and exhibiting a solution. Clearly if we find a solution by exhibiting it we can say that it exists, but on the other hand a solution can exist but we may not be able to display it. From the last example we know that a solution of the problem $dy/dx = x^2 + y^2$, $y(0) = 1$ exists on some interval around $x = 0$ and is unique. However, the best we can do in this case is to approximate the solution (see Chapter 9).

Remark: The conditions stated in Theorem 2.1 are *sufficient* but not *necessary*. When $f(x, y)$ and $\partial f/\partial y$ are continuous on a rectangular region R, it must *always* follow that there exists a unique solution of (2) when (x_0, y_0) is a point interior to R. However, if the conditions stated in the hypothesis of the theorem do not hold, then the initial-value problem (2) may still have either (a) no solution, (b) more than one solution, or (c) a unique solution. Furthermore, the continuity condition on $\partial f/\partial y$ can be relaxed somewhat without changing the conclusion of the theorem. This results in a stronger theorem but is, unfortunately, not as easy to apply as Theorem 2.1. Indeed if we are not interested in uniqueness, then a famous theorem due to the Italian mathematician Guiseppe Peano states that the continuity of $f(x, y)$ on R is sufficient to guarantee the existence of at least one solution of $dy/dx = f(x, y)$ through a point (x_0, y_0) interior to R.

Exercises 2.1

Answers to odd-numbered problems begin on page A–2.

In Problems 1–10 determine a region of the xy-plane for which the given differential equation would have a unique solution through a point (x_0, y_0) in the region.

1. $\dfrac{dy}{dx} = y^{2/3}$
2. $\dfrac{dy}{dx} = \sqrt{xy}$
3. $x\dfrac{dy}{dx} = y$
4. $\dfrac{dy}{dx} - y = x$
5. $(4 - y^2)y' = x^2$
6. $(1 + y^3)y' = x^2$
7. $(x^2 + y^2)y' = y^2$
8. $(y - x)y' = y + x$
9. $\dfrac{dy}{dx} = x^3 \cos y$
10. $\dfrac{dy}{dx} = (x - 1)e^{y/(x-1)}$
11. By inspection find at least two solutions of the initial-value problem

$$y' = 3y^{2/3}$$
$$y(0) = 0.$$

12. By inspection find at least two solutions of the initial-value problem

$$x\frac{dy}{dx} = 2y$$
$$y(0) = 0.$$

13. By inspection determine a solution of the nonlinear differential equation $y' = y^3$ satisfying $y(0) = 0$. Is the solution unique?

14. By inspection find a solution of the initial-value problem

$$y' = |y - 1|$$
$$y(0) = 1.$$

State why the conditions of Theorem 2.1 do not hold for this differential equation. Although we shall not prove it, the solution to this initial-value problem is unique.

15. Verify that $y = cx$ is a solution of the differential equation $xy' = y$ for every value of the parameter c. Find at least two solutions of the initial-value problem

$$xy' = y$$
$$y(0) = 0.$$

Observe that the piecewise-defined function

$$y = \begin{cases} 0, & x < 0, \\ x, & x \geq 0 \end{cases}$$

satisfies the condition $y(0) = 0$. Is it a solution of the initial-value problem?

16. **(a)** Consider the differential equation

$$\frac{dy}{dx} = 1 + y^2.$$

Determine a region of the xy-plane for which the equation has a unique solution through a point (x_0, y_0) in the region.

(b) Formally show that $y = \tan x$ satisfies the differential equation and the condition $y(0) = 0$.

(c) Explain why $y = \tan x$ is not a solution of the initial-value problem

$$\frac{dy}{dx} = 1 + y^2$$
$$y(0) = 0$$

on the interval $(-2, 2)$.

(d) Explain why $y = \tan x$ is a solution of the initial-value problem in part (c) on the interval $(-1, 1)$.

In Problems 17–20 determine whether Theorem 2.1 guarantees that the differential equation $y' = \sqrt{y^2 - 9}$ possesses a unique solution through the given point.

17. (1, 4)
18. (5, 3)
19. (2, −3)
20. (−1, 1)

2.2 Separable Variables

Note to the student: In solving a differential equation you will often have to utilize, say, integration by parts, partial fractions, or possibly a substitution. It will be worth a few minutes of your time to review some techniques of integration.

We begin our study of the methodology of solving first-order equations with the simplest of all differential equations.

If $g(x)$ is a given continuous function, then the first-order equation

$$\frac{dy}{dx} = g(x) \tag{1}$$

can be solved by integration. The solution of (1) is

$$y = \int g(x)\,dx + c.$$

EXAMPLE 1

Solve **(a)** $\dfrac{dy}{dx} = 1 + e^{2x}$, **(b)** $\dfrac{dy}{dx} = \sin x$.

SOLUTION **(a)** $y = \int (1 + e^{2x})\,dx = x + \frac{1}{2}e^{2x} + c$

(b) $y = \int \sin x\,dx = -\cos x + c.$

Equation (1), as well as its method of solution, is just a special case of the following.

DEFINITION 2.1 A differential equation of the form

$$\frac{dy}{dx} = \frac{g(x)}{h(y)}$$

is said to be **separable** or to have **separable variables**.

Observe that a separable equation can be written as

$$h(y)\frac{dy}{dx} = g(x). \tag{2}$$

It is seen immediately that (2) reduces to (1) when $h(y) = 1$.

Now if $y = f(x)$ denotes a solution of (2), we must have

$$h(f(x))f'(x) = g(x)$$

and therefore

$$\int h(f(x))f'(x)\,dx = \int g(x)\,dx + c. \tag{3}$$

But $dy = f'(x)\,dx$ so (3) is the same as

$$\int h(y)\,dy = \int g(x)\,dx + c. \tag{4}$$

Method of Solution

Equation (4) indicates the procedure for solving separable differential equations. A one-parameter family of solutions, usually given implicitly, is obtained by integrating both sides of $h(y)\,dy = g(x)\,dx$.

Note: There is no need to use two constants in the integration of a separable equation since

$$\int h(y)\,dy + c_1 = \int g(x)\,dx + c_2$$

$$\int h(y)\,dy = \int g(x)\,dx + c_2 - c_1$$

$$= \int g(x)\,dx + c,$$

where c is completely arbitrary. In many instances throughout the following chapters, we shall not hesitate to relabel constants in a manner which may prove convenient for a given equation. For example, multiples of constants or combinations of constants can sometimes be replaced by one constant.

EXAMPLE 2 Solve $(1 + x)\,dy - y\,dx = 0.$

SOLUTION Dividing by $(1 + x)y$ we can write $dy/y = dx/(1 + x)$ from which it follows

$$\int \frac{dy}{y} = \int \frac{dx}{1 + x}.$$

$$\ln|y| = \ln|1 + x| + c_1$$

$$y = e^{\ln|1 + x| + c_1}$$

$$= e^{\ln|1 + x|} \cdot e^{c_1}$$

$$= (1 + x)e^{c_1}.$$

Relabeling e^{c_1} as c then gives $y = c(1 + x)$.*

ALTERNATIVE SOLUTION Since each integral results in a logarithm, a judicious choice for the constant of integration is $\ln|c|$ rather than c:

$$\ln|y| = \ln|1 + x| + \ln|c|$$

or

$$\ln|y| = \ln|c(1 + x)|$$

so that

$$y = c(1 + x).$$

Even if not *all* the indefinite integrals are logarithms, it may still be advantageous to use $\ln|c|$. However, no firm rule can be given.

EXAMPLE 3 Solve $\dfrac{dy}{dx} = -\dfrac{x}{y}$ subject to $y(4) = 3.$

SOLUTION In Example 4 of Section 1.2 we saw that the differential equation of the family of concentric circles $x^2 + y^2 = c^2$ was $dy/dx = -x/y$. We now

* One might object that since $e^{c_1} > 0$ then necessarily $c > 0$. Because $\ln|y| - \ln|1 + x| = c_1$ implies $|y/(1 + x)| = e^{c_1}$ and $y = \pm e^{c_1}(1 + x)$ we are, strictly speaking, replacing $\pm e^{c_1}$ by c. In other words, $y = c(1 + x)$ is a solution of the differential equation on the interval $(-\infty, \infty)$ *for any real choice of* c. The reader is encouraged to worry about such things. Albeit unfortunate, to save time in solving differential equations many manipulations are carried out without much thought to mathematical rigor. This is what is known in the trade as "proceeding formally."

are in a position of being able to work "backwards." Obviously $y\,dy = -x\,dx$ and so

$$\int y\,dy = -\int x\,dx$$

$$\frac{y^2}{2} = -\frac{x^2}{2} + c_1$$

or
$$x^2 + y^2 = c^2,$$

where the constant $2c_1$ is replaced by c^2.

Now when $x = 4$, $y = 3$ so that $16 + 9 = 25 = c^2$. Thus the initial-value problem determines the solution $x^2 + y^2 = 25$. In view of Theorem 2.1, we can conclude that it is the only circle of the family passing through the point (4, 3) (see Figure 2.5).

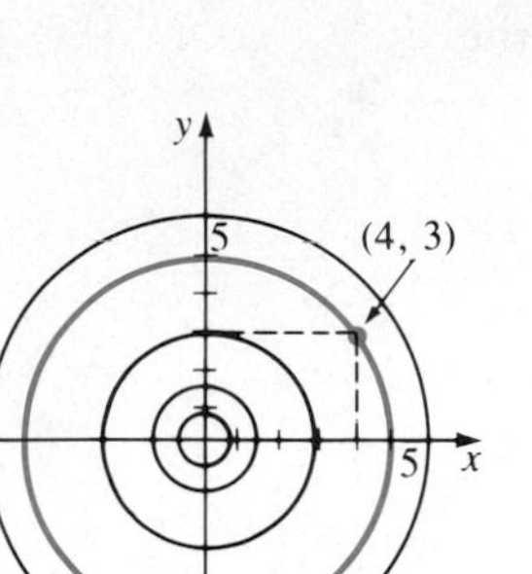

Figure 2.5

EXAMPLE 4

Solve
$$x \sin x e^{-y}\,dx - y\,dy = 0.$$

SOLUTION After multiplying by e^y the equation becomes

$$x \sin x\,dx = ye^y\,dy.$$

Using integration by parts on both sides of the equality gives

$$-x \cos x + \sin x = ye^y - e^y + c.$$

EXAMPLE 5

Solve
$$xy^4\,dx + (y^2 + 2)e^{-3x}\,dy = 0. \tag{5}$$

SOLUTION By multiplying the given equation by e^{3x} and dividing by y^4, we obtain

$$xe^{3x}\,dx + \frac{y^2 + 2}{y^4}\,dy = 0 \qquad \text{or} \qquad xe^{3x}\,dx + (y^{-2} + 2y^{-4})\,dy = 0. \tag{6}$$

Using integration by parts on the first term yields

$$\frac{1}{3}xe^{3x} - \frac{1}{9}e^{3x} - y^{-1} - \frac{2}{3}y^{-3} = c_1.$$

The one-parameter family of solutions can also be written as

$$e^{3x}(3x - 1) = \frac{9}{y} + \frac{6}{y^3} + c, \tag{7}$$

where the constant $9c_1$ is rewritten as c.

Two points are worth mentioning at this time. First, unless it is important or convenient there is no need to try to solve an expression representing a family of solutions for y explicitly in terms of x. Equation (7) shows this task may present more problems than just the drudgery of symbol pushing. As a consequence it is often the case that the interval over which a solution is valid is not apparent. Second, some care should be exercised when separating variables to make certain that divisors are not zero. A constant solution may sometimes get lost in the shuffle of solving the problem. In Example 5 observe that $y = 0$ is a perfectly good solution of equation (5) but is not a member of the set of solutions defined by (7).

EXAMPLE 6

Solve $\dfrac{dy}{dx} = y^2 - 4$ subject to $y(0) = -2$.

SOLUTION We put the equation into the form

$$\frac{dy}{y^2 - 4} = dx \tag{8}$$

and use partial fractions on the left side. We have

$$\left[\frac{-\frac{1}{4}}{y + 2} + \frac{\frac{1}{4}}{y - 2}\right] dy = dx \tag{9}$$

so that

$$-\frac{1}{4}\ln|y + 2| + \frac{1}{4}\ln|y - 2| = x + c_1. \tag{10}$$

Thus

$$\ln\left|\frac{y - 2}{y + 2}\right| = 4x + c_2 \qquad [c_2 = 4c_1]$$

and

$$\frac{y - 2}{y + 2} = ce^{4x},$$

where we have replaced e^{c_2} by c. Finally, we obtain

$$y = 2\frac{1 + ce^{4x}}{1 - ce^{4x}}. \tag{11}$$

Substituting $x = 0$, $y = -2$ leads to the dilemma

$$-2 = 2\frac{1 + c}{1 - c}$$

$$-1 + c = 1 + c \qquad \text{or} \qquad -1 = 1.$$

Let us consider the differential equation a little more carefully. The fact is, the equation

$$\frac{dy}{dx} = (y + 2)(y - 2)$$

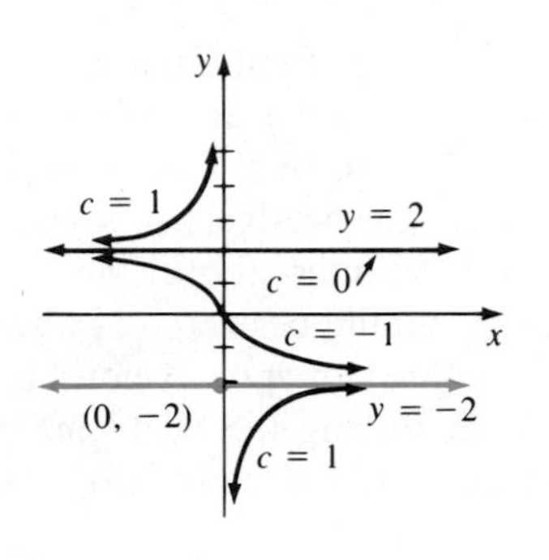

Figure 2.6

is satisfied by two constant functions, namely, $y = -2$ and $y = 2$. Inspection of equations (8), (9), and (10) clearly indicates we must preclude $y = -2$ and $y = 2$ at those steps in our solution. But it is interesting to observe that we can subsequently recover the solution $y = 2$ by setting $c = 0$ in equation (11). However, there is no finite value of c which will ever yield the solution $y = -2$. This latter constant function is the only solution to the original initial-value problem (see Figure 2.6).

If, in Example 6, we had used $\ln |c|$ for the constant of integration, then the form of the one-parameter family of solutions would be

$$y = 2\frac{c + e^{4x}}{c - e^{4x}}. \tag{12}$$

Note that (12) reduces to $y = -2$ when $c = 0$, but now there is no finite value of c that will give the constant solution $y = 2$.

If an initial condition leads to a particular solution by finding a specific value of the parameter c in a family of solutions for a first-order differential equation, it is a natural inclination of most students (and instructors) to relax and be content. In Section 2.1 we saw, however, that a solution of an initial-value problem may not be unique. For example, the problem

$$\frac{dy}{dx} = xy^{1/2}$$
$$y(0) = 0 \tag{13}$$

has at least two solutions, namely, $y = 0$ and $y = x^4/16$. We are now in a position to solve the equation. Separating variables

$$y^{-1/2}\, dy = x\, dx$$

and integrating give

$$2y^{1/2} = \frac{x^2}{2} + c_1 \qquad \text{or} \qquad y = \left(\frac{x^2}{4} + c\right)^2.$$

When $x = 0$, $y = 0$, so necessarily $c = 0$. Therefore $y = x^4/16$. The solution $y = 0$ was lost by dividing by $y^{1/2}$. In addition, the initial-value problem (13) possesses infinitely more solutions, since for any choice of the parameter $a \geq 0$ the piecewise defined function

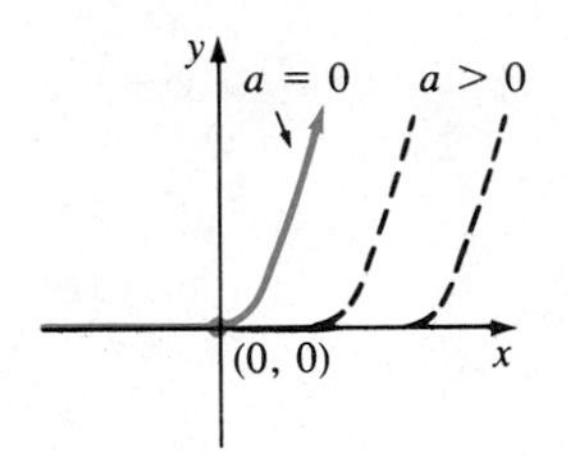

Figure 2.7

$$y = \begin{cases} 0, & x < a \\ \dfrac{(x^2 - a^2)^2}{16}, & x \geq a \end{cases}$$

satisfies both the differential equation and the initial condition (see Figure 2.7).

Exercises 2.2

Answers to odd-numbered problems begin on page A–3.

In Problems 1–40 solve the given differential equation by separation of variables.

1. $\dfrac{dy}{dx} = \sin 5x$

2. $\dfrac{dy}{dx} = (x + 1)^2$

3. $dx + e^{3x}\,dy = 0$

4. $dx - x^2\,dy = 0$

5. $(x + 1)\dfrac{dy}{dx} = x + 6$

6. $e^x \dfrac{dy}{dx} = 2x$

7. $xy' = 4y$

8. $\dfrac{dy}{dx} + 2xy = 0$

9. $\dfrac{dy}{dx} = \dfrac{y^3}{x^2}$

10. $\dfrac{dy}{dx} = \dfrac{y + 1}{x}$

11. $\dfrac{dx}{dy} = \dfrac{x^2 y^2}{1 + x}$

12. $\dfrac{dx}{dy} = \dfrac{1 + 2y^2}{y \sin x}$

13. $\dfrac{dy}{dx} = e^{3x + 2y}$

14. $e^x y \dfrac{dy}{dx} = e^{-y} + e^{-2x - y}$

15. $(4y + yx^2)\,dy - (2x + xy^2)\,dx = 0$

16. $(1 + x^2 + y^2 + x^2 y^2)\,dy = y^2\,dx$

17. $2y(x + 1)\,dy = x\,dx$

18. $x^2 y^2\,dy = (y + 1)\,dx$

19. $y \ln x \dfrac{dx}{dy} = \left(\dfrac{y + 1}{x}\right)^2$

20. $\dfrac{dy}{dx} = \left(\dfrac{2y + 3}{4x + 5}\right)^2$

21. $\dfrac{dS}{dr} = kS$

22. $\dfrac{dQ}{dt} = k(Q - 70)$

23. $\frac{dP}{dt} = P - P^2$

24. $\frac{dN}{dt} + N = Nte^{t+2}$

25. $\sec^2 x\, dy + \csc y\, dx = 0$

26. $\sin 3x\, dx + 2y \cos^3 3x\, dy = 0$

27. $e^y \sin 2x\, dx + \cos x(e^{2y} - y)\, dy = 0$

28. $\sec x\, dy = x \cot y\, dx$

29. $(e^y + 1)^2 e^{-y}\, dx + (e^x + 1)^3 e^{-x}\, dy = 0$

30. $\frac{y}{x}\frac{dy}{dx} = (1 + x^2)^{-1/2}(1 + y^2)^{1/2}$

31. $(y - yx^2)\frac{dy}{dx} = (y + 1)^2$

32. $2\frac{dy}{dx} - \frac{1}{y} = \frac{2x}{y}$

33. $\frac{dy}{dx} = \frac{xy + 3x - y - 3}{xy - 2x + 4y - 8}$

34. $\frac{dy}{dx} = \frac{xy + 2y - x - 2}{xy - 3y + x - 3}$

35. $\frac{dy}{dx} = \sin x(\cos 2y - \cos^2 y)$

36. $\sec y \frac{dy}{dx} + \sin(x - y) = \sin(x + y)$

37. $x\sqrt{1 - y^2}\, dx = dy$

38. $y(4 - x^2)^{1/2}\, dy = (4 + y^2)^{1/2}\, dx$

39. $(e^x + e^{-x})\frac{dy}{dx} = y^2$

40. $(x + \sqrt{x})\frac{dy}{dx} = y + \sqrt{y}$

In Problems 41–48 solve the given differential equation subject to the indicated initial condition.

41. $\sin x(e^{-y} + 1)\, dx = (1 + \cos x)\, dy, \quad y(0) = 0$

42. $(1 + x^4)\, dy + x(1 + 4y^2)\, dx = 0, \quad y(1) = 0$

43. $y\, dy = 4x(y^2 + 1)^{1/2}\, dx, \quad y(0) = 1$

44. $\frac{dy}{dt} + ty = y, \quad y(1) = 3$

45. $\frac{dx}{dy} = 4(x^2 + 1), \quad x\left(\frac{\pi}{4}\right) = 1$

46. $\frac{dy}{dx} = \frac{y^2 - 1}{x^2 - 1}, \quad y(2) = 2$

47. $x^2 y' = y - xy, \quad y(-1) = -1$

48. $y' + 2y = 1, \quad y(0) = \frac{5}{2}$

In Problems 49 and 50 find a solution of the given differential equation that passes through the indicated points.

49. $\dfrac{dy}{dx} - y^2 = -9$ **(a)** $(0, 0)$ **(b)** $(0, 3)$ **(c)** $\left(\dfrac{1}{3}, 1\right)$

50. $x\dfrac{dy}{dx} = y^2 - y$ **(a)** $(0, 1)$ **(b)** $(0, 0)$ **(c)** $\left(\dfrac{1}{2}, \dfrac{1}{2}\right)$

51. Find a singular solution for the equation in Problem 37.

52. Find a singular solution for the equation in Problem 39.

Miscellaneous Problems

A differential equation of the form $dy/dx = f(ax + by + c), b \neq 0$, can always be reduced to an equation with separable variables by means of the substitution $u = ax + by + c$. Use this procedure to solve Problems 53–58.

EXAMPLE 7

Solve
$$\frac{dy}{dx} = \frac{1}{x + y + 1}.$$

SOLUTION Let $u = x + y + 1$ so that $du/dx = 1 + dy/dx$. The given equation then becomes

$$\frac{du}{dx} - 1 = \frac{1}{u} \quad \text{or} \quad \frac{u\,du}{1 + u} = dx.$$

Integrating
$$\left(1 - \frac{1}{1 + u}\right) du = dx$$

gives
$$u - \ln|1 + u| = x + c_1$$
$$x + y + 1 - \ln|x + y + 2| = x + c_1$$
$$y + 1 - \ln|x + y + 2| = c_1 \quad \text{or} \quad x + y + 2 = ce^y,$$

where we have replaced e^{1-c_1} by c.

53. $\dfrac{dy}{dx} = (x + y + 1)^2$

54. $\dfrac{dy}{dx} = \dfrac{1 - x - y}{x + y}$

55. $\dfrac{dy}{dx} = \tan^2(x + y)$

56. $\dfrac{dy}{dx} = \sin(x + y)$

57. $\dfrac{dy}{dx} = 2 + \sqrt{y - 2x + 3}$

58. $\dfrac{dy}{dx} = 1 + e^{y-x+5}$

2.3 Homogeneous Equations

If an equation in the differential form

$$M(x, y)\,dx + N(x, y)\,dy = 0 \tag{1}$$

has the property that

$$M(tx, ty) = t^n M(x, y) \qquad \text{and} \qquad N(tx, ty) = t^n N(x, y),$$

we say it has **homogeneous coefficients** or is a **homogeneous equation**. The important point in the subsequent discussion is the fact that a homogeneous differential equation *can always be reduced to a separable equation* through a simple algebraic substitution. Before pursuing the method of solution for this type of differential equation, let us closely examine the nature of **homogeneous functions**.

DEFINITION 2.2 If $f(tx, ty) = t^n f(x, y)$ for some real number n, then $f(x, y)$ is said to be a homogeneous function of degree n.

EXAMPLE 1

(a) $f(x, y) = x - 3\sqrt{xy} + 5y$

$$\begin{aligned} f(tx, ty) &= (tx) - 3\sqrt{(tx)(ty)} + 5(ty) \\ &= tx - 3\sqrt{t^2xy} + 5ty \\ &= t[x - 3\sqrt{xy} + 5y] = tf(x, y) \end{aligned}$$

The function is homogeneous of degree one.

(b) $f(x, y) = \sqrt{x^3 + y^3}$

$$\begin{aligned} f(tx, ty) &= \sqrt{t^3x^3 + t^3y^3} \\ &= t^{3/2}\sqrt{x^3 + y^3} = t^{3/2}f(x, y) \end{aligned}$$

The function is homogeneous of degree 3/2.

(c) $f(x, y) = x^2 + y^2 + 1$

$$f(tx, ty) = t^2x^2 + t^2y^2 + 1 \neq t^2 f(x, y)$$

since $t^2 f(x, y) = t^2x^2 + t^2y^2 + t^2$. The function is not homogeneous.

(d) $f(x, y) = \dfrac{x}{2y} + 4$

$$f(tx, ty) = \frac{tx}{2ty} + 4$$

$$= \frac{x}{2y} + 4 = t^0 f(x, y)$$

The function is homogeneous of degree zero.

As parts (c) and (d) of Example 1 show, a constant added to a function destroys homogeneity, *unless* the function is homogeneous of degree zero. Also, in many instances a homogeneous function can be recognized by examining the total degree of each term.

EXAMPLE 2

(a) $f(x, y) = 6xy^3 - x^2y^2$

(In $6xy^3$: x is degree 1, y^3 is degree 3, total degree 4. In x^2y^2: x^2 is degree 2, y^2 is degree 2, total degree 4.)

The function is homogeneous of degree 4.

(b) $f(x, y) = x^2 - y$

(x^2 is degree 2; y is degree 1.)

The function is not homogeneous since the degrees of the two terms are different.

If $f(x, y)$ is a homogeneous function of degree n, notice that we can write

$$f(x, y) = x^n f\left(1, \frac{y}{x}\right) \quad \text{and} \quad f(x, y) = y^n f\left(\frac{x}{y}, 1\right), \tag{2}$$

where $f(1, y/x)$ and $f(x/y, 1)$ are both of degree zero.

EXAMPLE 3

We see $f(x, y) = x^2 + 3xy + y^2$ is homogeneous of degree 2. Thus

$$f(x, y) = x^2\left[1 + 3\left(\frac{y}{x}\right) + \left(\frac{y}{x}\right)^2\right] = x^2 f\left(1, \frac{y}{x}\right)$$

$$f(x, y) = y^2\left[\left(\frac{x}{y}\right)^2 + 3\left(\frac{x}{y}\right) + 1\right] = y^2 f\left(\frac{x}{y}, 1\right).$$

Method of Solution

An equation of the form $M(x, y)\,dx + N(x, y)\,dy = 0$, where M and N have the same degree of homogeneity, can be reduced to separable variables by *either* the substitution $y = ux$ or $x = vy$, where u and v are new dependent variables. In particular, if we choose $y = ux$ then $dy = u\,dx + x\,du$. Hence the differential equation (1) becomes

$$M(x, ux)\,dx + N(x, ux)[u\,dx + x\,du] = 0.$$

Now, by the homogeneity property given in (2), we can write

$$x^n M(1, u)\,dx + x^n N(1, u)[u\,dx + x\,du] = 0$$

or

$$[M(1, u) + uN(1, u)]\,dx + xN(1, u)\,du = 0,$$

which gives

$$\frac{dx}{x} + \frac{N(1, u)\,du}{M(1, u) + uN(1, u)} = 0.$$

We hasten to point out that the preceding formula should not be memorized; rather, the *procedure should be worked through each time.* The proof that the substitution $x = vy$ in (1) also leads to a separable equation is left as an exercise.

EXAMPLE 4

Solve $$(x^2 + y^2)\,dx + (x^2 - xy)\,dy = 0.$$

SOLUTION Both $M(x, y)$ and $N(x, y)$ are homogeneous of degree 2. If we let $y = ux$, it follows that

$$(x^2 + u^2x^2)\,dx + (x^2 - ux^2)[u\,dx + x\,du] = 0$$

$$x^2(1 + u)\,dx + x^3(1 - u)\,du = 0$$

$$\frac{1 - u}{1 + u}\,du + \frac{dx}{x} = 0$$

$$\left[-1 + \frac{2}{1 + u}\right]du + \frac{dx}{x} = 0$$

$$-u + 2\ln|1 + u| + \ln|x| + \ln|c| = 0$$

$$-\frac{y}{x} + 2\ln\left|1 + \frac{y}{x}\right| + \ln|x| + \ln|c| = 0.$$

Using the properties of logarithms we can write the preceding solution in the alternative form

$$c(x + y)^2 = xe^{y/x}.$$

EXAMPLE 5

Solve $$(2\sqrt{xy} - y)\,dx - x\,dy = 0.$$

SOLUTION The coefficients $M(x, y)$ and $N(x, y)$ are homogeneous of degree one. If $y = ux$ the differential equation becomes, after simplifying,

$$\frac{du}{2u - 2u^{1/2}} + \frac{dx}{x} = 0.$$

The integral of the first term can be evaluated by the further substitution $t = u^{1/2}$. The result is

$$\frac{dt}{t-1} + \frac{dx}{x} = 0$$

$$\ln|t - 1| + \ln|x| = \ln|c|$$

$$\ln\left|\sqrt{\frac{y}{x}} - 1\right| + \ln|x| = \ln|c|$$

$$x\left(\sqrt{\frac{y}{x}} - 1\right) = c$$

$$\sqrt{xy} - x = c.$$

By now the reader may be asking: When should the substitution $x = vy$ be used? Although it can be used for every homogeneous differential equation, in practice we try $x = vy$ whenever the function $M(x, y)$ is simpler than $N(x, y)$. In solving $(x^2 + y^2)\,dx + (x^2 - xy)\,dy = 0$, for example, there is no appreciable difference between M and N, so either $y = ux$ or $x = vy$ could be used. Also, it could happen that after using one substitution, we may encounter integrals that are difficult or impossible to evaluate in closed form; switching substitutions may result in an easier problem.

EXAMPLE 6

Solve $$2x^3y\,dx + (x^4 + y^4)\,dy = 0.$$

SOLUTION Each coefficient is a homogeneous function of degree four. Since the coefficient of dx is slightly simpler than the coefficient of dy we try $x = vy$. After substituting, the equation

$$2v^3y^4[v\,dy + y\,dv] + (v^4y^4 + y^4)\,dy = 0$$

simplifies to

$$\frac{2v^3\,dv}{3v^4 + 1} + \frac{dy}{y} = 0.$$

Integrating gives

$$\frac{1}{6}\ln(3v^4+1)+\ln|y|=\ln|c_1|$$

or

$$3x^4y^2+y^6=c.$$

Had $y = ux$ been used, then

$$\frac{dx}{x}+\frac{u^4+1}{u^5+3u}\,du=0.$$

The reader is urged to reflect on how to evaluate the integral of the second term in the last equation.

A homogeneous differential equation can always be expressed in the alternative form

$$\frac{dy}{dx}=F\left(\frac{y}{x}\right).$$

To see this suppose we write the equation $M(x, y)\,dx + N(x, y)\,dy = 0$ as $dy/dx = f(x, y)$, where

$$f(x,y)=-\frac{M(x,y)}{N(x,y)}.$$

The function $f(x, y)$ must necessarily be homogeneous of degree zero when M and N are homogeneous of degree n. Using (2) it follows that

$$f(x,y)=-\frac{x^nM\left(1,\frac{y}{x}\right)}{x^nN\left(1,\frac{y}{x}\right)}=-\frac{M\left(1,\frac{y}{x}\right)}{N\left(1,\frac{y}{x}\right)}.$$

The last ratio is recognized as a function of the form $F(y/x)$. We leave it as an exercise to demonstrate that a homogeneous differential equation can also be written as $dy/dx = G(x/y)$ (see Problem 49).

EXAMPLE 7

Solve $$x\frac{dy}{dx}=y+xe^{y/x}$$ subject to $y(1) = 1$.

SOLUTION By writing the equation as

$$\frac{dy}{dx}=\frac{y}{x}+e^{y/x},$$

we see that the function to the right of the equality is homogeneous of degree zero. From the form of this function we are prompted to use $u = y/x$. After differentiating $y = ux$ by the product rule and substituting, we find

$$u + x\frac{du}{dx} = u + e^u$$

$$e^{-u}\,du = dx/x.$$

Hence

$$-e^{-u} + c = \ln|x|$$

$$-e^{-y/x} + c = \ln|x|.$$

Since $y = 1$ when $x = 1$, we get $-e^{-1} + c = 0$ or $c = e^{-1}$. Therefore the solution to the initial-value problem is

$$e^{-1} - e^{-y/x} = \ln|x|.$$

Exercises 2.3

Answers to odd-numbered problems begin on page A–3.

In Problems 1–10 determine whether the given function is homogeneous. If so, state the degree of homogeneity.

1. $x^3 + 2xy^2 - y^4/x$

2. $\sqrt{x + y}(4x + 3y)$

3. $\dfrac{x^3y - x^2y^2}{(x + 8y)^2}$

4. $\dfrac{x}{y^2 + \sqrt{x^4 + y^4}}$

5. $\cos\dfrac{x^2}{x + y}$

6. $\sin\dfrac{x}{x + y}$

7. $\ln x^2 - 2\ln y$

8. $\dfrac{\ln x^3}{\ln y^3}$

9. $(x^{-1} + y^{-1})^2$

10. $(x + y + 1)^2$

In Problems 11–30 solve the given differential equation by using an appropriate substitution.

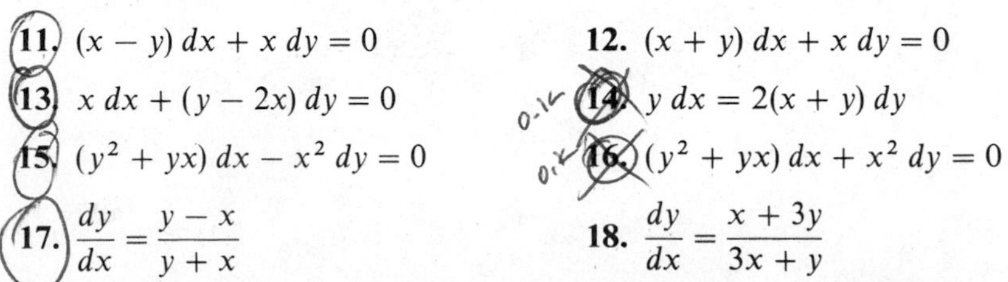

11. $(x - y)\,dx + x\,dy = 0$

12. $(x + y)\,dx + x\,dy = 0$

13. $x\,dx + (y - 2x)\,dy = 0$

14. $y\,dx = 2(x + y)\,dy$

15. $(y^2 + yx)\,dx - x^2\,dy = 0$

16. $(y^2 + yx)\,dx + x^2\,dy = 0$

17. $\dfrac{dy}{dx} = \dfrac{y - x}{y + x}$

18. $\dfrac{dy}{dx} = \dfrac{x + 3y}{3x + y}$

19. $-y\,dx + (x + \sqrt{xy})\,dy = 0$

20. $x\dfrac{dy}{dx} - y = \sqrt{x^2 + y^2}$

21. $2x^2y\,dx = (3x^3 + y^3)\,dy$

22. $(x^4 + y^4)\,dx - 2x^3y\,dy = 0$

23. $\dfrac{dy}{dx} = \dfrac{y}{x} + \dfrac{x}{y}$

24. $\dfrac{dy}{dx} = \dfrac{y}{x} + \dfrac{x^2}{y^2} + 1$

25. $y\dfrac{dx}{dy} = x + 4ye^{-2x/y}$

26. $(x^2e^{-y/x} + y^2)\,dx = xy\,dy$

27. $\left(y + x\cot\dfrac{y}{x}\right)dx - x\,dy = 0$

28. $\dfrac{dy}{dx} = \dfrac{y}{x}\ln\dfrac{y}{x}$

29. $(x^2 + xy - y^2)\,dx + xy\,dy = 0$

30. $(x^2 + xy + 3y^2)\,dx - (x^2 + 2xy)\,dy = 0$

In Problems 31–44 solve the given differential equation subject to the indicated initial conditions.

31. $xy^2\dfrac{dy}{dx} = y^3 - x^3, \quad y(1) = 2$

32. $(x^2 + 2y^2)\,dx = xy\,dy, \quad y(-1) = 1$

33. $2x^2\dfrac{dy}{dx} = 3xy + y^2, \quad y(1) = -2$

34. $xy\,dx - x^2\,dy = y\sqrt{x^2 + y^2}\,dy, \quad y(0) = 1$

35. $(x + ye^{y/x})\,dx - xe^{y/x}\,dy = 0, \quad y(1) = 0$

36. $y\,dx + \left(y\cos\dfrac{x}{y} - x\right)dy = 0, \quad y(0) = 2$

37. $(y^2 + 3xy)\,dx = (4x^2 + xy)\,dy, \quad y(1) = 1$

38. $y^3\,dx = 2x^3\,dy - 2x^2y\,dx \quad y(1) = \sqrt{2}$

39. $(x + \sqrt{xy})\dfrac{dy}{dx} + x - y = x^{-1/2}y^{3/2}, \quad y(1) = 1$

40. $y\,dx + x(\ln x - \ln y - 1)\,dy = 0, \quad y(1) = e$

41. $y^2\,dx + (x^2 + xy + y^2)\,dy = 0, \quad y(0) = 1$

42. $(\sqrt{x} + \sqrt{y})^2\,dx = x\,dy, \quad y(1) = 0$

43. $(x + \sqrt{y^2 - xy})\dfrac{dy}{dx} = y, \quad y(\tfrac{1}{2}) = 1$

44. $\dfrac{dy}{dx} - \dfrac{y}{x} = \cosh\dfrac{y}{x}, \quad y(1) = 0$

Miscellaneous Problems

A different equation of the form

$$\frac{dy}{dx} = f\left(\frac{ax + by + c}{Ax + By + C}\right), \quad aB - bA \neq 0,$$

can always be reduced to a homogeneous equation by means of the substitutions $x = u + h$, $y = v + k$. In Problems 45 and 46 use the indicated substitutions to reduce the equation to one with homogeneous coefficients. Solve.

45. $\dfrac{dy}{dx} = \dfrac{x - y - 3}{x + y - 1}$; $\quad x = u + 2, \quad y = v - 1$

46. $\dfrac{dy}{dx} = \dfrac{x + y - 6}{x - y}$; $\quad x = u + 3, \quad y = v + 3$

47. Suppose $M(x, y)\,dx + N(x, y)\,dy = 0$ is a homogeneous equation. Show that the substitution $x = vy$ reduces the equation to one with separable variables.

48. Suppose $M(x, y)\,dx + N(x, y)\,dy = 0$ is a homogeneous equation. Show that the substitutions $x = r\cos\theta$, $y = r\sin\theta$ reduce the equation to one with separable variables.

49. Suppose $M(x, y)\,dx + N(x, y)\,dy = 0$ is a homogeneous equation. Show that the equation has the alternative form $dy/dx = G(x/y)$.

50. If $f(x, y)$ is a homogeneous function of degree n, show that

$$x\frac{\partial f}{\partial x} + y\frac{\partial f}{\partial y} = nf.$$

2.4 Exact Equations

While the simple equation

$$y\,dx + x\,dy = 0$$

is both separable and homogeneous, we should recognize that it is also equivalent to the differential of the product of x and y—that is,

$$y\,dx + x\,dy = d(xy) = 0.$$

By integrating we immediately obtain the implicit solution $xy = c$.

From calculus you might remember that if $z = f(x, y)$ is a function having continuous first partial derivatives in a region R of the xy-plane, then its *total differential* is

$$dz = \frac{\partial f}{\partial x}\,dx + \frac{\partial f}{\partial y}\,dy. \tag{1}$$

Now if $f(x, y) = c$, it follows from (1) that

$$\frac{\partial f}{\partial x}\,dx + \frac{\partial f}{\partial y}\,dy = 0. \tag{2}$$

In other words, given a family of curves $f(x, y) = c$, we can generate a first-order differential equation by computing the total differential.

EXAMPLE 1

If $x^2 - 5xy + y^3 = c$, then (2) gives

$$(2x - 5y)\,dx + (-5x + 3y^2)\,dy = 0 \qquad \text{or} \qquad \frac{dy}{dx} = \frac{5y - 2x}{-5x + 3y^2}.$$

For our purposes, it is more important to turn the problem around; namely, given an equation such as

$$\frac{dy}{dx} = \frac{5y - 2x}{-5x + 3y^2}, \tag{3}$$

can we identify the equation as being equivalent to the statement

$$d(x^2 - 5xy + y^3) = 0?$$

Notice that equation (3) is neither separable nor homogeneous.

Exact Equations

DEFINITION 2.3 A differential expression

$$M(x, y)\,dx + N(x, y)\,dy$$

is an **exact differential** in a region R of the xy-plane if it corresponds to the total differential of some function $f(x, y)$. An equation

$$M(x, y)\,dx + N(x, y)\,dy = 0$$

is said to be an **exact equation** if the expression on the left side is an exact differential.

EXAMPLE 2

The equation $x^2y^3\,dx + x^3y^2\,dy = 0$ is exact since it is recognized that

$$d(\tfrac{1}{3}x^3y^3) = x^2y^3\,dx + x^3y^2\,dy.$$

The following theorem is a test for an exact differential.

THEOREM 2.2 Let $M(x, y)$ and $N(x, y)$ be continuous and have continuous first partial derivatives in a rectangular region R defined by $a < x < b$, $c < y < d$. Then a necessary and sufficient condition that

$$M(x, y)\,dx + N(x, y)\,dy$$

be an exact differential is

$$\frac{\partial M}{\partial y} = \frac{\partial N}{\partial x}. \tag{4}$$

PROOF OF THE NECESSITY For simplicity let us assume that $M(x, y)$ and $N(x, y)$ have continuous first partial derivatives for all (x, y). Now if the expression $M(x, y)\,dx + N(x, y)\,dy$ is exact, there exists some function f for which

$$M(x, y)\,dx + N(x, y)\,dy = \frac{\partial f}{\partial x}\,dx + \frac{\partial f}{\partial y}\,dy$$

for all (x, y) in R. Therefore

$$M(x, y) = \frac{\partial f}{\partial x}, \qquad N(x, y) = \frac{\partial f}{\partial y},$$

and

$$\frac{\partial M}{\partial y} = \frac{\partial}{\partial y}\left(\frac{\partial f}{\partial x}\right) = \frac{\partial^2 f}{\partial y\,\partial x} = \frac{\partial}{\partial x}\left(\frac{\partial f}{\partial y}\right) = \frac{\partial N}{\partial x}.$$

The equality of the mixed partials is a consequence of the continuity of the first partial derivatives of $M(x, y)$ and $N(x, y)$. ■

The sufficiency part of Theorem 2.2 consists of showing there exists a function f for which $\partial f/\partial x = M(x, y)$ and $\partial f/\partial y = N(x, y)$ whenever (4) holds. The construction of the function f actually reflects a basic procedure for solving exact equations.

Method of Solution

Given the equation

$$M(x, y)\,dx + N(x, y)\,dy = 0, \tag{5}$$

first show

$$\frac{\partial M}{\partial y} = \frac{\partial N}{\partial x}.$$

Then assume that

$$\frac{\partial f}{\partial x} = M(x, y)$$

so we can find f by integrating $M(x, y)$ with respect to x, while holding y constant. We write

$$f(x, y) = \int M(x, y)\,dx + g(y), \tag{6}$$

where the arbitrary function $g(y)$ is the "constant" of integration. Now differentiate (6) with respect to y and assume $\partial f/\partial y = N(x, y)$.

$$\begin{aligned}\frac{\partial f}{\partial y} &= \frac{\partial}{\partial y}\int M(x, y)\,dx + g'(y)\\ &= N(x, y).\end{aligned}$$

This gives

$$g'(y) = N(x, y) - \frac{\partial}{\partial y}\int M(x, y)\,dx. \tag{7}$$

It is important to observe that the expression $N(x, y) - (\partial/\partial y)\int M(x, y)\,dx$ is independent of x since

$$\begin{aligned}\frac{\partial}{\partial x}\left[N(x, y) - \frac{\partial}{\partial y}\int M(x, y)\,dx\right] &= \frac{\partial N}{\partial x} - \frac{\partial}{\partial y}\left(\frac{\partial}{\partial x}\int M(x, y)\,dx\right)\\ &= \frac{\partial N}{\partial x} - \frac{\partial M}{\partial y} = 0.\end{aligned}$$

Finally, integrate (7) with respect to y and substitute the result in (6). The solution of the equation is $f(x, y) = c$.

Note: In the foregoing procedure we could just as well start out with the assumption that $\partial f/\partial y = N(x, y)$. The analogues of (6) and (7) would be, respectively,

$$f(x, y) = \int N(x, y)\,dy + h(x)$$

and

$$h'(x) = M(x, y) - \frac{\partial}{\partial x}\int N(x, y)\,dy.$$

In either case *none of these formulas should be memorized*. Also, when testing an equation for exactness, make sure it is of form (5). Often a differential equation is written $G(x, y)\,dx = H(x, y)\,dy$. In this case write the equation as $G(x, y)\,dx - H(x, y)\,dy = 0$ and then identify $M(x, y) = G(x, y)$ and $N(x, y) = -H(x, y)$.

EXAMPLE 3

Solve
$$2xy\,dx + (x^2 - 1)\,dy = 0.$$

SOLUTION With $M(x, y) = 2xy$ and $N(x, y) = x^2 - 1$ we have

$$\frac{\partial M}{\partial y} = 2x = \frac{\partial N}{\partial x}.$$

Thus the equation is exact and so, by Theorem 2.2, there exists a function $f(x, y)$ such that

$$\frac{\partial f}{\partial x} = 2xy \quad \text{and} \quad \frac{\partial f}{\partial y} = x^2 - 1.$$

From the first of these equations we obtain, after integrating,

$$f(x, y) = x^2y + g(y).$$

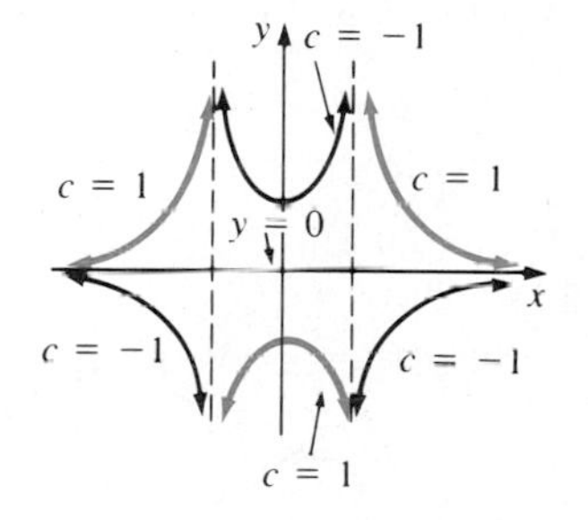

Figure 2.8

Taking the partial derivative of the last expression with respect to y and setting the result equal to $N(x, y)$ give

$$\frac{\partial f}{\partial y} = x^2 + g'(y) = x^2 - 1.$$

It follows that $\quad g'(y) = -1 \quad$ and $\quad g(y) = -y.$

The constant of integration need not be included in the preceding line since the solution is $f(x, y) = c$. Some of the family of curves $x^2y - y = c$ are given in Figure 2.8.

Note: The solution of the equation is *not* $f(x, y) = x^2y - y$. Rather it is $f(x, y) = c$ or $f(x, y) = 0$ if a constant is used in the integration of $g'(y)$. Observe that the equation could also be solved by separation of variables.

EXAMPLE 4

Solve $\quad (e^{2y} - y\cos xy)\,dx + (2xe^{2y} - x\cos xy + 2y)\,dy = 0.$

SOLUTION The equation is neither separable nor homogeneous but is exact since

$$\frac{\partial M}{\partial y} = 2e^{2y} + xy \sin xy - \cos xy = \frac{\partial N}{\partial x}.$$

Hence a function $f(x, y)$ exists for which

$$M(x, y) = \frac{\partial f}{\partial x} \qquad \text{and} \qquad N(x, y) = \frac{\partial f}{\partial y}.$$

Now for variety we shall start with the assumption that $\partial f/\partial y = N(x, y)$;

that is,
$$\frac{\partial f}{\partial y} = 2xe^{2y} - x \cos xy + 2y$$

$$f(x, y) = 2x \int e^{2y}\, dy - x \int \cos xy\, dy + 2 \int y\, dy.$$

Remember, the reason that x can come out in front of the symbol $\int$ is that in the integration with respect to y, x is treated as an ordinary constant. It follows that

$$f(x, y) = xe^{2y} - \sin xy + y^2 + h(x)$$

$$\frac{\partial f}{\partial x} = e^{2y} - y \cos xy + h'(x)$$

$$= e^{2y} - y \cos xy,$$

so that
$$h'(x) = 0 \qquad \text{and} \qquad h(x) = c.$$

Hence a one-parameter family of solutions is given by

$$xe^{2y} - \sin xy + y^2 + c = 0.$$

EXAMPLE 5 Solve
$$(\cos x \sin x - xy^2)\, dx + y(1 - x^2)\, dy = 0$$

subject to $y(0) = 2$.

SOLUTION The equation is exact since

$$\frac{\partial M}{\partial y} = -2xy = \frac{\partial N}{\partial x}.$$

Now
$$\frac{\partial f}{\partial y} = y(1 - x^2)$$

$$f(x, y) = \frac{y^2}{2}(1 - x^2) + h(x)$$

$$\frac{\partial f}{\partial x} = -xy^2 + h'(x) = \cos x \sin x - xy^2.$$

The last equation implies

$$h'(x) = \cos x \sin x$$

$$h(x) = \int (\cos x)(-\sin x \, dx) = -\frac{1}{2}\cos^2 x.$$

Thus

$$\frac{y^2}{2}(1 - x^2) - \frac{1}{2}\cos^2 x = c_1$$

or

$$y^2(1 - x^2) - \cos^2 x = c. \qquad (c = 2c_1)$$

The initial condition $y = 2$ when $x = 0$ demands that $4(1) - \cos^2(0) = c$ or that $c = 3$. Thus a solution of the problem is

$$y^2(1 - x^2) - \cos^2 x = 3.$$

Exercises 2.4

Answers to odd-numbered problems begin on page A–3.

In Problems 1–24 determine whether the given equation is exact. If exact, solve.

1. $(2x - 1)\,dx + (3y + 7)\,dy = 0$

2. $(2x + y)\,dx - (x + 6y)\,dy = 0$

3. $(5x + 4y)\,dx + (4x - 8y^3)\,dy = 0$

4. $(\sin y - y \sin x)\,dx + (\cos x + x \cos y - y)\,dy = 0$

5. $(2y^2x - 3)\,dx + (2yx^2 + 4)\,dy = 0$

6. $\left(2y - \frac{1}{x} + \cos 3x\right)\frac{dy}{dx} + \frac{y}{x^2} - 4x^3 + 3y \sin 3x = 0$

7. $(x + y)(x - y)\,dx + x(x - 2y)\,dy = 0$

8. $\left(1 + \ln x + \frac{y}{x}\right)dx = (1 - \ln x)\,dy$

9. $(y^3 - y^2 \sin x - x)\,dx + (3xy^2 + 2y \cos x)\,dy = 0$

10. $(x^3 + y^3)\,dx + 3xy^2\,dy = 0$

11. $(y \ln y - e^{-xy})\,dx + \left(\frac{1}{y} + x \ln y\right)dy = 0$

12. $\frac{2x}{y}\,dx - \frac{x^2}{y^2}\,dy = 0$

13. $x\frac{dy}{dx} = 2xe^x - y + 6x^2$

14. $(3x^2y + e^y)\,dx + (x^3 + xe^y - 2y)\,dy = 0$

15. $\left(1 - \frac{3}{x} + y\right) dx + \left(1 - \frac{3}{y} + x\right) dy = 0$

16. $(e^y + 2xy \cosh x)y' + xy^2 \sinh x + y^2 \cosh x = 0$

17. $\left(x^2y^3 - \frac{1}{1 + 9x^2}\right) \frac{dx}{dy} + x^3y^2 = 0$

18. $(5y - 2x)y' - 2y = 0$

19. $(\tan x - \sin x \sin y)\,dx + \cos x \cos y\,dy = 0$

20. $(3x \cos 3x + \sin 3x - 3)\,dx + (2y + 5)\,dy = 0$

21. $(1 - 2x^2 - 2y) \frac{dy}{dx} = 4x^3 + 4xy$

22. $(2y \sin x \cos x - y + 2y^2e^{xy^2})\,dx = (x - \sin^2 x - 4xye^{xy^2})\,dy$

23. $(4x^3y - 15x^2 - y)\,dx + (x^4 + 3y^2 - x)\,dy = 0$

24. $\left(\frac{1}{x} + \frac{1}{x^2} - \frac{y}{x^2 + y^2}\right) dx + \left(ye^y + \frac{x}{x^2 + y^2}\right) dy = 0$

In Problems 25–30 solve the given differential equation subject to the indicated initial condition.

25. $(x + y)^2\,dx + (2xy + x^2 - 1)\,dy = 0, \quad y(1) = 1$

26. $(e^x + y)\,dx + (2 + x + ye^y)\,dy = 0, \quad y(0) = 1$

27. $(4y + 2x - 5)\,dx + (6y + 4x - 1)\,dy = 0, \quad y(-1) = 2$

28. $\left(\frac{3y^2 - x^2}{y^5}\right) \frac{dy}{dx} + \frac{x}{2y^4} = 0, \quad y(1) = 1$

29. $(y^2 \cos x - 3x^2y - 2x)\,dx + (2y \sin x - x^3 + \ln y)\,dy = 0, \quad y(0) = e$

30. $\left(\frac{1}{1 + y^2} + \cos x - 2xy\right) \frac{dy}{dx} = y(y + \sin x), \quad y(0) = 1$

In Problems 31–34 find the value of k so that the given differential equation is exact.

31. $(y^3 + kxy^4 - 2x)\,dx + (3xy^2 + 20x^2y^3)\,dy = 0$

32. $(2x - y \sin xy + ky^4)\,dx - (20xy^3 + x \sin xy)\,dy = 0$

33. $(2xy^2 + ye^x)\,dx + (2x^2y + ke^x - 1)\,dy = 0$

34. $(6xy^3 + \cos y)\,dx + (kx^2y^2 - x \sin y)\,dy = 0$

35. Determine a function $M(x, y)$ so that the following differential equation is exact.

$$M(x, y)\,dx + \left(xe^{xy} + 2xy + \frac{1}{x}\right)dy = 0$$

36. Determine a function $N(x, y)$ so that the following differential equation is exact.

$$\left(y^{1/2}x^{-1/2} + \frac{x}{x^2 + y}\right)dx + N(x, y)\,dy = 0$$

Miscellaneous Problems

It is sometimes possible to transform a nonexact differential equation $M(x, y)dx + N(x, y)\,dy = 0$ into an exact equation by multiplying it by an **integrating factor** $\mu(x, y)$. In Problems 37–42 solve the equation by verifying that the given $\mu(x, y)$ is an integrating factor.*

EXAMPLE 6

Solve $(x + y)\,dx + x \ln x\,dy = 0, \qquad \mu(x, y) = \dfrac{1}{x}, \qquad$ on $(0, \infty)$.

SOLUTION Let $M(x, y) = x + y$ and $N(x, y) = x \ln x$ so that $\partial M/\partial y = 1$ and $\partial N/\partial x = 1 + \ln x$. The equation is not exact. However, if we multiply the equation by $\mu(x, y) = 1/x$, we obtain

$$\left(1 + \frac{y}{x}\right)dx + \ln x\,dy = 0.$$

From this latter form we make the identifications:

$$M(x, y) = 1 + y/x, \quad N(x, y) = \ln x, \qquad \partial M/\partial y = 1/x = \partial N/\partial x.$$

Therefore the second differential equation is exact. It follows that

$$\frac{\partial f}{\partial x} = 1 + \frac{y}{x} = M(x, y)$$

$$f(x, y) = x + y \ln x + g(y)$$

$$\frac{\partial f}{\partial y} = 0 + \ln x + g'(y) = \ln x$$

and so $\qquad g'(y) = 0 \qquad$ and $\qquad g(y) = c.$

---

* The two equations $M\,dx + N\,dy = 0$ and $\mu M\,dx + \mu N\,dy = 0$ are not necessarily equivalent in the sense that a solution of one is also a solution of the other. A solution may be lost or gained as a result of the multiplication.

Hence $f(x, y) = x + y \ln x + c$. It is readily verified that

$$x + y \ln x + c = 0$$

is a solution of both equations on $(0, \infty)$.

37. $6xy\,dx + (4y + 9x^2)\,dy = 0, \quad \mu(x, y) = y^2$

38. $-y^2\,dx + (x^2 + xy)\,dy = 0, \quad \mu(x, y) = 1/x^2y$

39. $(-xy \sin x + 2y \cos x)\,dx + 2x \cos x\,dy = 0, \quad \mu(x, y) = xy$

40. $y(x + y + 1)\,dx + (x + 2y)\,dy = 0, \quad \mu(x, y) = e^x$

41. $(2y^2 + 3x)\,dx + 2xy\,dy = 0, \quad \mu(x, y) = x$

42. $(x^2 + 2xy - y^2)\,dx + (y^2 + 2xy - x^2)\,dy = 0, \quad \mu(x, y) = (x + y)^{-2}$

43. Show that any separable first-order differential equation is also exact.

2.5 Linear Equations

In Chapter 1 we defined the general form of a *linear* differential equation of order n to be

$$a_n(x)\frac{d^ny}{dx^n} + a_{n-1}(x)\frac{d^{n-1}y}{dx^{n-1}} + \cdots + a_1(x)\frac{dy}{dx} + a_0(x)y = g(x).$$

We remind the reader that linearity means that all coefficients are functions of x only, and that y and all its derivatives are raised to the first power. Now when $n = 1$ we obtain the linear first-order equation

$$a_1(x)\frac{dy}{dx} + a_0(x)y = g(x).$$

Dividing by $a_1(x)$ gives the more useful form

$$\frac{dy}{dx} + P(x)y = f(x). \tag{1}$$

We seek the solution of (1) on an interval I for which $P(x)$ and $f(x)$ are continuous. In the discussion that follows, we tacitly assume that (1) has a solution.

Integrating Factor

Let us suppose equation (1) is written in the differential form

$$dy + [P(x)y - f(x)]\,dx = 0. \tag{2}$$

Linear equations possess the pleasant property that a function $\mu(x)$ can always be found such that the multiple of (2)

$$\mu(x)\,dy + \mu(x)[P(x)y - f(x)]\,dx = 0 \tag{3}$$

is an exact differential equation. By Theorem 2.2 we know that the left side of equation (3) will be an exact differential if

$$\frac{\partial}{\partial x}\mu(x) = \frac{\partial}{\partial y}\mu(x)[P(x)y - f(x)] \tag{4}$$

or

$$\frac{d\mu}{dx} = \mu P(x).$$

This is a separable equation from which we can determine $\mu(x)$. We have

$$\frac{d\mu}{\mu} = P(x)\,dx$$

$$\ln|\mu| = \int P(x)\,dx \tag{5}$$

so that

$$\mu(x) = e^{\int P(x)\,dx}. \tag{6}$$

The function $\mu(x)$ defined in (6) is called an **integrating factor** for the linear equation. Note that we need not use a constant of integration in (5) since (3) is unaffected by a constant multiple. Also, $\mu(x) \neq 0$ for every x in I and is continuous and differentiable.

It is interesting to observe that equation (3) is still an exact differential equation even when $f(x) = 0$. In fact, $f(x)$ plays no part in determining $\mu(x)$ since we see from (4) that $\partial/\partial y\;\mu(x)f(x) = 0$. Thus both

$$e^{\int P(x)\,dx}\,dy + e^{\int P(x)\,dx}[P(x)y - f(x)]\,dx$$

and

$$e^{\int P(x)\,dx}\,dy + e^{\int P(x)\,dx}P(x)y\,dx$$

are exact differentials. We now write (3) in the form

$$e^{\int P(x)\,dx}\,dy + e^{\int P(x)\,dx}P(x)y\,dx = e^{\int P(x)\,dx}f(x)\,dx$$

and recognize that we can write the equation as

$$d[e^{\int P(x)\,dx}y] = e^{\int P(x)\,dx}f(x)\,dx.$$

Integrating the last equation gives

$$e^{\int P(x)\,dx}y = \int e^{\int P(x)\,dx}f(x)\,dx + c$$

or $$y = e^{-\int P(x)\,dx} \int e^{\int P(x)\,dx} f(x)\,dx + ce^{-\int P(x)\,dx}. \tag{7}$$

In other words, if (1) has a solution, it must be of form (7). Conversely, it is a straightforward matter to verify that (7) constitutes a one-parameter family of solutions of equation (1).

Summary of the Method

No attempt should be made to memorize the formula given in (7). The procedure should be followed each time, so for convenience we summarize the results.

(i) To solve a linear first-order equation first put it into the form (1); that is, make the coefficient of dy/dx unity.

(ii) Identify $P(x)$ and find the integrating factor

$$e^{\int P(x)\,dx}.$$

(iii) Multiply the equation obtained in step (i) by the integrating factor:

$$e^{\int P(x)\,dx} \frac{dy}{dx} + P(x) e^{\int P(x)\,dx} y = e^{\int P(x)\,dx} f(x).$$

(iv) The left side of the equation in step (iii) is the derivative of the integrating factor and the dependent variable y; that is,

$$\frac{d}{dx}\left[e^{\int P(x)\,dx} y\right] = e^{\int P(x)\,dx} f(x).$$

(v) Integrate both sides of the equation found in step (iv).

EXAMPLE 1

Solve $$x\frac{dy}{dx} - 4y = x^6 e^x.$$

SOLUTION Write the equation as

$$\frac{dy}{dx} - \frac{4}{x}y = x^5 e^x \tag{8}$$

and determine the integrating factor

$$e^{-4\int dx/x} = e^{-4\ln|x|} = e^{\ln x^{-4}} = x^{-4}.$$

Here we have used the basic identity $b^{\log_b N} = N$. Now multiply (8) by this term

$$x^{-4}\frac{dy}{dx} - 4x^{-5}y = xe^x \tag{9}$$

and obtain
$$\frac{d}{dx}[x^{-4}y] = xe^x.^* \tag{10}$$

It follows from integration by parts that

$$x^{-4}y = xe^x - e^x + c$$

or
$$y = x^5e^x - x^4e^x + cx^4.$$

EXAMPLE 2 Solve
$$\frac{dy}{dx} - 3y = 0.$$

SOLUTION The equation is already in form (1). Hence the integrating factor is

$$e^{\int(-3)\,dx} = e^{-3x}.$$

Therefore
$$e^{-3x}\frac{dy}{dx} - 3e^{-3x}y = 0$$

$$\frac{d}{dx}[e^{-3x}y] = 0$$

$$e^{-3x}y = c$$

and so
$$y = ce^{3x}.$$

General Solution

If it is assumed that $P(x)$ and $f(x)$ are continuous on an interval I and x_0 is any point in the interval, then it follows from Theorem 2.1 that there exists only one solution of the initial-value problem

$$\frac{dy}{dx} + P(x)y = f(x)$$
$$y(x_0) = y_0. \tag{11}$$

* The reader should perform the indicated differentiations a few times in order to be convinced that all equations, such as (8), (9), and (10), are formally equivalent.

But we saw earlier that (1) possesses a family of solutions and that every solution of the equation on the interval I is of form (7). Thus, obtaining the solution of (11) is a simple matter of finding an appropriate value of c in (7). Consequently, we are justified in calling (7) the **general solution** of the differential equation. In retrospect, the reader should recall that in several instances we found singular solutions of nonlinear equations. This cannot happen in the case of a linear equation when proper attention is paid to solving the equation over a common interval on which $P(x)$ and $f(x)$ are continuous.

EXAMPLE 3

Find the general solution of

$$(x^2 + 9)\frac{dy}{dx} + xy = 0.$$

SOLUTION We write $$\frac{dy}{dx} + \frac{x}{x^2 + 9}y = 0.$$

The function $P(x) = x/(x^2 + 9)$ is continuous on $(-\infty, \infty)$. Now the integrating factor for the equation is

$$e^{\int x\,dx/(x^2+9)} = e^{\frac{1}{2}\int 2x\,dx/(x^2+9)} = e^{\frac{1}{2}\ln(x^2+9)} = \sqrt{x^2 + 9}$$

so that
$$\sqrt{x^2 + 9}\frac{dy}{dx} + \frac{x}{\sqrt{x^2 + 9}}y = 0$$

$$\frac{d}{dx}[\sqrt{x^2 + 9}y] = 0$$

$$\sqrt{x^2 + 9}y = c.$$

Hence the general solution on the interval is

$$y = \frac{c}{\sqrt{x^2 + 9}}.$$

EXAMPLE 4

Solve $$\frac{dy}{dx} + 2xy = x \qquad \text{subject to } y(0) = -3.$$

SOLUTION The functions $P(x) = 2x$ and $f(x) = x$ are continuous on $(-\infty, \infty)$. The integrating factor is

$$e^{2\int x\,dx} = e^{x^2}$$

so that
$$e^{x^2}\frac{dy}{dx} + 2xe^{x^2}y = xe^{x^2}$$

$$\frac{d}{dx}[e^{x^2}y] = xe^{x^2}$$

$$e^{x^2}y = \int xe^{x^2}\,dx$$

$$= \frac{1}{2}e^{x^2} + c.$$

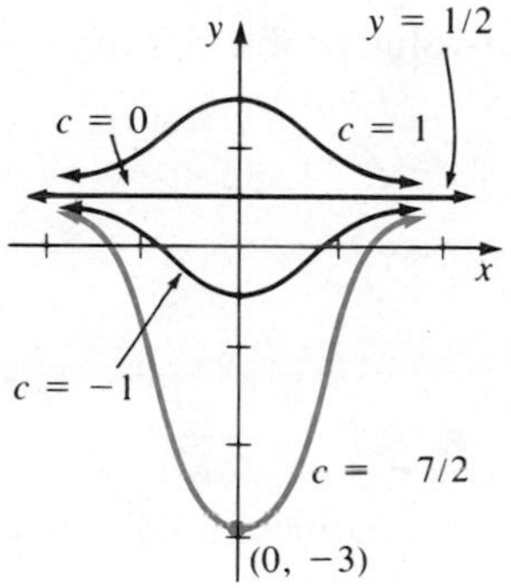

Thus the general solution of the differential equation is

$$y = \frac{1}{2} + ce^{-x^2}.$$

The condition $y(0) = -3$ gives $c = -7/2$ and hence the solution of the initial-value problem on the interval is

$$y = \frac{1}{2} - \frac{7}{2}e^{-x^2}.$$

See Figure 2.9.

y
y = 1/2
c = 0
c = 1
x
c = −1
c = −7/2
(0, −3)

Figure 2.9

EXAMPLE 5

Solve
$$x\frac{dy}{dx} + y = 2x \qquad \text{subject to } y(1) = 0.$$

SOLUTION Write the given equation as

$$\frac{dy}{dx} + \frac{1}{x}y = 2$$

and observe that $P(x) = 1/x$ is continuous on any interval not containing the origin. In view of the initial condition we solve the problem on the interval $(0, \infty)$.

The integrating factor is

$$e^{\int dx/x} = e^{\ln|x|} = x$$

and so

$$\frac{d}{dx}[xy] = 2x$$

gives
$$xy = x^2 + c.$$

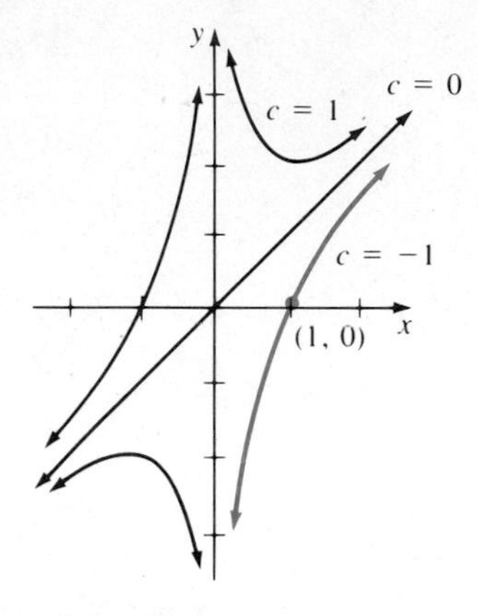

Figure 2.10

The general solution of the equation is

$$y = x + \frac{c}{x}. \tag{12}$$

But $y(1) = 0$ implies $c = -1$. Hence we obtain

$$y = x - \frac{1}{x}, \qquad 0 < x < \infty. \tag{13}$$

Considered as a one-parameter family of curves, the graph of (12) is given in Figure 2.10. The solution (13) of the initial-value problem is indicated by the colored portion of the graph.

EXAMPLE 6

Solve $\dfrac{dy}{dx} = \dfrac{1}{x + y^2}$ subject to $y(-2) = 0$.

SOLUTION The given differential equation is neither separable, homogeneous, exact, nor linear in the variable y. However, if we take the reciprocal, then

$$\frac{dx}{dy} = x + y^2 \qquad \text{or} \qquad \frac{dx}{dy} - x = y^2.$$

This latter equation is *linear in* x, so the corresponding integrating factor is $e^{-\int dy} = e^{-y}$. Therefore, it follows that for $-\infty < y < \infty$,

$$\frac{d}{dy}[e^{-y}x] = y^2e^{-y}$$

$$e^{-y}x = \int y^2e^{-y}\,dy$$

$$= -y^2e^{-y} - 2ye^{-y} - 2e^{-y} + c$$

$$x = -y^2 - 2y - 2 + ce^y$$

When $x = -2$, $y = 0$, we find $c = 0$, and so

$$x = -y^2 - 2y - 2.$$

The next example illustrates a way of solving (1) when the function f is discontinuous.

EXAMPLE 7 Find a continuous solution satisfying

$$\frac{dy}{dx} + y = f(x), \qquad \text{where} \qquad f(x) = \begin{cases} 1, & 0 \le x \le 1 \\ 0, & x > 1 \end{cases}$$

and the initial condition $y(0) = 0$.

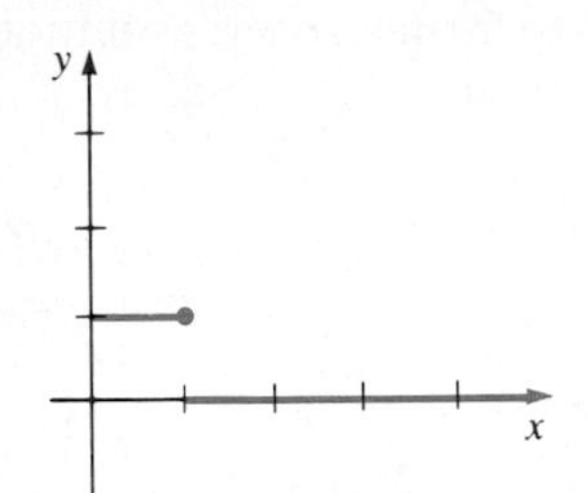

Figure 2.11

SOLUTION From Figure 2.11 we see that f is discontinuous at $x = 1$. Consequently, we solve the problem in two parts. For $0 \le x \le 1$ we have

$$\frac{dy}{dx} + y = 1$$

$$\frac{d}{dx}[e^x y] - e^x$$

$$y = 1 + c_1 e^{-x}.$$

Since $y(0) = 0$, we must have $c_1 = -1$, and therefore

$$y = 1 - e^{-x}, \qquad 0 \le x \le 1.$$

For $x > 1$ we then have

$$\frac{dy}{dx} + y = 0,$$

which leads to

$$y = c_2 e^{-x}.$$

Hence we can write

$$y = \begin{cases} 1 - e^{-x}, & 0 \le x \le 1 \\ c_2 e^{-x}, & x > 1. \end{cases}$$

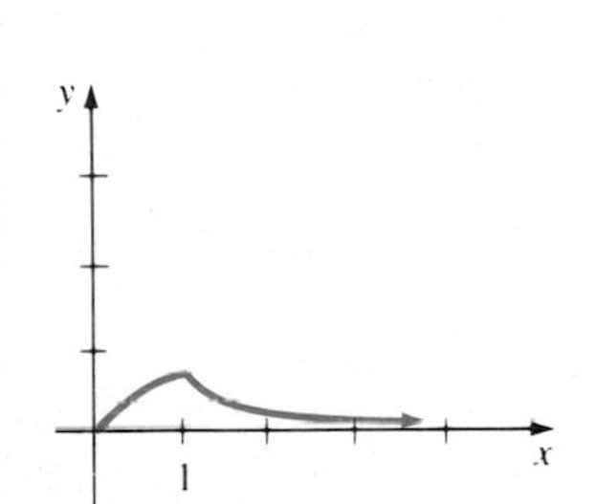

Figure 2.12

Now in order that y be a continuous function, we certainly want $\lim_{x \to 1^+} y(x) = y(1)$. This latter requirement is equivalent to $c_2 e^{-1} = 1 - e^{-1}$ or $c_2 = e - 1$. As Figure 2.12 shows, the function

$$y = \begin{cases} 1 - e^{-x}, & 0 \le x \le 1 \\ (e - 1)e^{-x}, & x > 1 \end{cases}$$

is continuous but not differentiable at $x = 1$.

Remark: Formula (7), representing the general solution of (1), actually consists of the sum of two solutions. We define

$$y = y_c + y_p, \tag{14}$$

where

$$y_c = ce^{-\int P(x)\,dx} \qquad \text{and} \qquad y_p = e^{-\int P(x)\,dx} \int e^{\int P(x)\,dx} f(x)\,dx.$$

The function y_c is readily shown to be the general solution of $y' + P(x)y = 0$, whereas y_p is a particular solution of $y' + P(x)y = f(x)$. As we shall see in Chapter 4, the additivity property of solutions (14) to form a general solution is an intrinsic property of linear equations of any order.

Exercises 2.5

Answers to odd-numbered problems begin on page A-4.

In Problems 1–40 find the general solution of the given differential equation. State an interval on which the general solution is defined.

1. $\dfrac{dy}{dx} = 5y$

2. $\dfrac{dy}{dx} + 2y = 0$

3. $3\dfrac{dy}{dx} + 12y = 4$

4. $x\dfrac{dy}{dx} + 2y = 3$

5. $\dfrac{dy}{dx} + y = e^{3x}$

6. $\dfrac{dy}{dx} = y + e^{x}$

7. $y' + 3x^2y = x^2$

8. $y' + 2xy = x^3$

9. $x^2y' + xy = 1$

10. $y' = 2y + x^2 + 5$

11. $(x + 4y^2)\,dy + 2y\,dx = 0$

12. $\dfrac{dx}{dy} = x + y$

13. $x\,dy = (x \sin x - y)\,dx$

14. $(1 + x^2)\,dy + (xy + x^3 + x)\,dx = 0$

15. $(1 + e^x)\dfrac{dy}{dx} + e^x y = 0$

16. $(1 - x^3)\dfrac{dy}{dx} = 3x^2y$

17. $\cos x\dfrac{dy}{dx} + y \sin x = 1$

18. $\dfrac{dy}{dx} + y \cot x = 2 \cos x$

19. $x\dfrac{dy}{dx} + 4y = x^3 - x$

20. $(1 + x)y' - xy = x + x^2$

21. $x^2y' + x(x + 2)y = e^x$

22. $xy' + (1 + x)y = e^{-x} \sin 2x$

23. $\cos^2 x \sin x\,dy + (y \cos^3 x - 1)\,dx = 0$

24. $(1 - \cos x)\,dy + (2y \sin x - \tan x)\,dx = 0$

25. $y\,dx + (xy + 2x - ye^y)\,dy = 0$

26. $(x^2 + x)\,dy = (x^5 + 3xy + 3y)\,dx$

27. $x\dfrac{dy}{dx} + (3x + 1)y = e^{-3x}$

28. $(x + 1)\dfrac{dy}{dx} + (x + 2)y = 2xe^{-x}$

29. $y\,dx - 4(x + y^6)\,dy = 0$

30. $xy' + 2y = e^x + \ln x$

31. $\dfrac{dy}{dx} + y = \dfrac{1 - e^{-2x}}{e^x + e^{-x}}$

32. $\dfrac{dy}{dx} - y = \sinh x$

33. $y\,dx + (x + 2xy^2 - 2y)\,dy = 0$

34. $y\,dx = (ye^y - 2x)\,dy$

35. $\dfrac{dr}{d\theta} + r\sec\theta = \cos\theta$

36. $\dfrac{dP}{dt} + 2tP = P + 4t - 2$

37. $(x + 2)^2\dfrac{dy}{dx} = 5 - 8y - 4xy$

38. $(x^2 - 1)\dfrac{dy}{dx} + 2y = (x + 1)^2$

39. $y' = (10 - y)\cosh x$

40. $dx = (3e^y - 2x)\,dy$

In Problems 41–54 solve the given differential equation subject to the indicated initial condition.

41. $\dfrac{dy}{dx} + 5y = 20, \quad y(0) = 2$

42. $y' = 2y + x(e^{3x} - e^{2x}), \quad y(0) = 2$

43. $L\dfrac{di}{dt} + Ri = E; \quad L, R,$ and E constants; $\quad i(0) = i_0$

44. $y\dfrac{dx}{dy} - x = 2y^2, \quad y(1) = 5$

45. $y' + (\tan x)y = \cos^2 x, \quad y(0) = -1$

46. $\dfrac{dQ}{dx} = 5x^4 Q, \quad Q(0) = -7$

47. $\dfrac{dT}{dt} = k(T - 50), \quad k$ a constant, $\quad T(0) = 200$

48. $x\,dy + (xy + 2y - 2e^{-x})\,dx = 0, \quad y(1) = 0$

49. $(x + 1)\dfrac{dy}{dx} + y = \ln x, \quad y(1) = 10$

50. $xy' + y = e^x, \quad y(1) = 2$

51. $x(x - 2)y' + 2y = 0, \quad y(3) = 6$

52. $\sin x\dfrac{dy}{dx} + (\cos x)y = 0, \quad y\left(-\dfrac{\pi}{2}\right) = 1$

53. $\dfrac{dy}{dx} = \dfrac{y}{y - x}, \quad y(5) = 2$

54. $\cos^2 x\dfrac{dy}{dx} + y = 1, \quad y(0) = -3$

In Problems 55–58 find a continuous solution satisfying each differential equation and the given initial condition.

55. $\dfrac{dy}{dx} + 2y = f(x), \quad f(x) = \begin{cases} 1, & 0 \le x \le 3 \\ 0, & x > 3 \end{cases} \qquad y(0) = 0$

56. $\dfrac{dy}{dx} + y = f(x), \quad f(x) = \begin{cases} 1, & 0 \le x \le 1 \\ -1, & x > 1 \end{cases} \qquad y(0) = 1$

57. $\dfrac{dy}{dx} + 2xy = f(x), \quad f(x) = \begin{cases} x, & 0 \le x < 1 \\ 0, & x \ge 1 \end{cases} \qquad y(0) = 2$

58. $(1 + x^2)\dfrac{dy}{dx} + 2xy = f(x), \quad f(x) = \begin{cases} x, & 0 \le x < 1 \\ -x, & x \ge 1 \end{cases} \qquad y(0) = 0$

[O] 2.6 Equations of Bernoulli, Ricatti, and Clairaut*

In this section we are not going to study any one particular type of differential equation. Rather, we are going to consider three classical equations that in some instances can be transformed into equations we have already studied.

Bernoulli's Equation

The differential equation

$$\frac{dy}{dx} + P(x)y = f(x)y^n, \tag{1}$$

where n is any real number, is called **Bernoulli's equation**. For $n = 0$ and $n = 1$

---

* **Jakob Bernoulli** (1654–1705) The Bernoullis were a Swiss family of scholars whose contributions to mathematics, physics, astronomy, and history spanned from the sixteenth to the twentieth century. Jakob, the elder of the two sons of the patriarch Jacques Bernoulli, made many contributions to the then-new fields of calculus and probability. Originally the second of the two major divisions of calculus was called *calculus summatorius*. In 1696, at Jakob Bernoulli's suggestion, its name was changed to *calculus integralis*, or, as we know it today, integral calculus.

Jacobo Francesco Ricatti (1676–1754) An Italian count, Ricatti was also a mathematician and philosopher.

Alexis Claude Clairaut (1713–1765) Born in Paris in 1713, Clairaut was a child prodigy who wrote his first book on mathematics at the age of eleven. He was among the first to discover singular solutions of differential equations. Like many mathematicians of his era, Clairaut was also a physicist and astronomer.

equation (1) is linear in y. Now for $y \neq 0$ (1) can be written as

$$y^{-n}\frac{dy}{dx} + P(x)y^{1-n} = f(x). \tag{2}$$

If we let $w = y^{1-n}$, $n \neq 0$, $n \neq 1$, then

$$\frac{dw}{dx} = (1 - n)y^{-n}\frac{dy}{dx}.$$

With these substitutions, (2) simplifies to the linear equation

$$\frac{dw}{dx} + (1 - n)P(x)w = (1 - n)f(x). \tag{3}$$

Solving (3) for w and using $y^{1-n} = w$ lead to a solution of (1).

EXAMPLE 1

Solve
$$\frac{dy}{dx} + \frac{1}{x}y = xy^2.$$

SOLUTION From (1) we identify $P(x) = 1/x$, $f(x) = x$, and $n = 2$. Thus the substitution $w - y^{-1}$ gives

$$\frac{dw}{dx} - \frac{1}{x}w = -x.$$

The integrating factor for this linear equation on, say $(0, \infty)$ is

$$e^{-\int dx/x} = e^{-\ln|x|} = e^{\ln|x|^{-1}} = x^{-1}.$$

Hence
$$\frac{d}{dx}[x^{-1}w] = -1.$$

Integrating this latter form gives

$$x^{-1}w = -x + c \qquad \text{or} \qquad w = -x^2 + cx.$$

Since $w = y^{-1}$, we obtain $y = 1/w$ or

$$y = \frac{1}{-x^2 + cx}.$$

For $n > 0$ note that the trivial solution $y = 0$ is a solution of (1). In Example 1, $y = 0$ is a singular solution of the given equation.

Ricatti's Equation

The nonlinear differential equation

$$\frac{dy}{dx} = P(x) + Q(x)y + R(x)y^2 \tag{4}$$

is called **Ricatti's equation**. If y_1 is a *known* particular solution of (4), then a family of solutions of the equation is given by $y = y_1 + u$, where u is a solution of

$$\frac{du}{dx} - (Q + 2y_1R)u = Ru^2. \tag{5}$$

Since (5) is a Bernoulli equation with $n = 2$, it can in turn be reduced to the linear equation

$$\frac{dw}{dx} + (Q + 2y_1R)w = -R \tag{6}$$

by the substitution $w = u^{-1}$ (see Problems 25 and 26).

As Example 2 illustrates, in many cases a solution of a Ricatti equation cannot be expressed in terms of elementary functions.

EXAMPLE 2

Solve $$\frac{dy}{dx} = 2 - 2xy + y^2.$$

SOLUTION It is easily verified that a particular solution of this equation is $y_1 = 2x$. From (4) we make the identifications $P(x) = 2$, $Q(x) = -2x$, and $R(x) = 1$ and then solve the linear equation (6),

$$\frac{dw}{dx} + (-2x + 4x)w = -1 \quad \text{or} \quad \frac{dw}{dx} + 2xw = -1.$$

The integrating factor for the last equation is e^{x^2}, and so

$$\frac{d}{dx}[e^{x^2}w] = -e^{x^2}.$$

Now the integral $\int_{x_0}^{x} e^{t^2}\,dt$ cannot be expressed in terms of elementary functions.* Thus we write

$$e^{x^2}w = -\int_{x_0}^{x} e^{t^2}\,dt + c \quad \text{or} \quad e^{x^2}\left(\frac{1}{u}\right) = -\int_{x_0}^{x} e^{t^2}\,dt + c$$

* When an integral $\int f(x)\,dx$ cannot be evaluated in terms of elementary functions, it is customary to write $\int_{x_0}^{x} f(t)\,dt$, where x_0 is a constant. When an initial condition is specified, it is imperative that this form be used.

so that
$$u = \frac{e^{x^2}}{c - \int_{x_0}^{x} e^{t^2}\,dt}.$$

A solution of the equation is then $y = 2x + u$.

Clairaut's Equation

It is left as an exercise to show that a solution of **Clairaut's equation**

$$y = xy' + f(y') \tag{7}$$

is the family of straight lines $y = cx + f(c)$, where c is an arbitrary constant. Furthermore (7) may also possess a solution in parametric form:

$$\begin{aligned} x &= -f'(t) \\ y &= f(t) - tf'(t). \end{aligned} \tag{8}$$

This last solution is a singular solution since, if $f''(t) \neq 0$, it cannot be obtained from the family of solutions $y = cx + f(c)$.

EXAMPLE 3

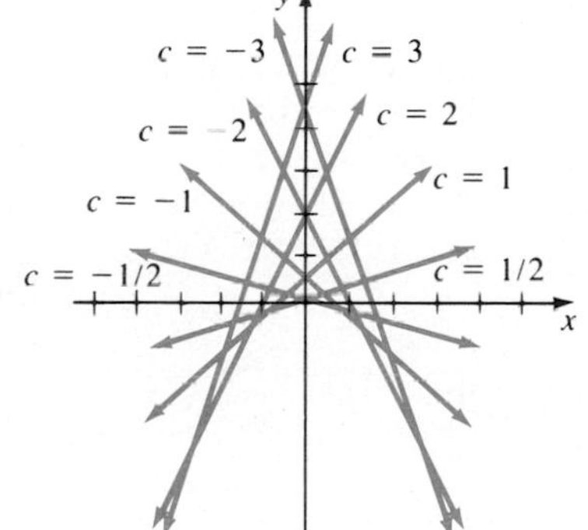

Figure 2.13

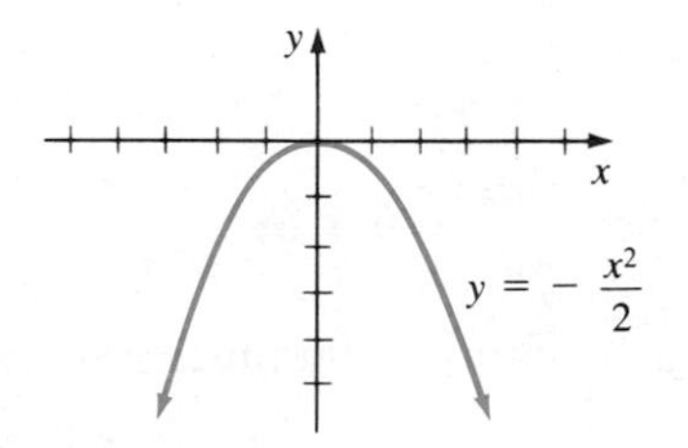

Figure 2.14

Solve
$$y = xy' + \frac{1}{2}(y')^2.$$

SOLUTION We first make the identification $f(y') = (1/2)(y')^2$ so that $f(t) = (1/2)t^2$. It follows from the preceding discussion that a family of solutions is

$$y = cx + \frac{1}{2}c^2. \tag{9}$$

The graph of this family is given in Figure 2.13. Since $f'(t)$, a singular solution is obtained from (8):

$$\begin{aligned} x &= -t \\ y &= \frac{1}{2}t^2 - t \cdot t = -\frac{1}{2}t^2. \end{aligned}$$

After eliminating the parameter, we find this latter solution is the same as

$$y = -\frac{1}{2}x^2.$$

One can readily see that this function is not part of the family (9) (see Figure 2.14).

Exercises 2.6

Answers to odd-numbered problems begin on page A-4.

In Problems 1–6 solve the given Bernoulli equation.

1. $x\dfrac{dy}{dx} + y = \dfrac{1}{y^2}$

2. $\dfrac{dy}{dx} - y = e^x y^2$

3. $\dfrac{dy}{dx} = y(xy^3 - 1)$

4. $x\dfrac{dy}{dx} - (1 + x)y = xy^2$

5. $x^2\dfrac{dy}{dx} + y^2 = xy$

6. $3(1 + x^2)\dfrac{dy}{dx} = 2xy(y^3 - 1)$

In Problems 7–10 solve the given differential equation subject to the indicated initial condition.

7. $x^2\dfrac{dy}{dx} - 2xy = 3y^4, \quad y(1) = \dfrac{1}{2}$

8. $y^{1/2}\dfrac{dy}{dx} + y^{3/2} = 1, \quad y(0) = 4$

9. $xy(1 + xy^2)\dfrac{dy}{dx} = 1, \quad y(1) = 0$

10. $2\dfrac{dy}{dx} = \dfrac{y}{x} - \dfrac{x}{y^2}, \quad y(1) = 1$

In Problems 11–16 solve the given Ricatti equation. y_1 is a known solution of the equation.

11. $\dfrac{dy}{dx} = -2 - y + y^2, \quad y_1 = 2$

12. $\dfrac{dy}{dx} = 1 - x - y + xy^2, \quad y_1 = 1$

13. $\dfrac{dy}{dx} = -\dfrac{4}{x^2} - \dfrac{1}{x}y + y^2, \quad y_1 = \dfrac{2}{x}$

14. $\dfrac{dy}{dx} = 2x^2 + \dfrac{1}{x}y - 2y^2, \quad y_1 = x$

15. $\dfrac{dy}{dx} = e^{2x} + (1 + 2e^x)y + y^2, \quad y_1 = -e^x$

16. $\dfrac{dy}{dx} = \sec^2 x - (\tan x)y + y^2, \quad y_1 = \tan x$

17. Solve $\dfrac{dy}{dx} = 6 + 5y + y^2$.

18. Solve $\dfrac{dy}{dx} = 9 + 6y + y^2$.

In Problems 19–24 solve the given Clairaut equation. Obtain a singular solution.

19. $y = xy' + 1 - \ln y'$

20. $y = xy' + (y')^{-2}$

21. $y = x\dfrac{dy}{dx} - \left(\dfrac{dy}{dx}\right)^3$

22. $y = (x + 4)y' + (y')^2$

23. $xy' - y = e^{y'}$

24. $y - xy' = \ln y'$

Miscellaneous Problems

25. Show that if y_1 is a solution of (4), then a family of solutions of (4) is given by $y = y_1 + u$, where u is a solution of (5).

26. Show that (5) can be reduced to (6) by means of the substitution $w = u^{-1}$.

27. When $R(x) = -1$, the Ricatti equation can be written as $y' + y^2 - Q(x)y - P(x) = 0$. Show that the substitution $y = w'/w$ leads to the linear second-order equation $w'' - Q(x)w' - P(x)w = 0$. (When Q and P are also constants there is little difficulty in solving equations of this type.)

28. An alternative definition of Clairaut's equation is any equation of the form $F(y - xy', y') = 0$. **(a)** Show that a family of solutions of the latter equation is $F(y - cx, c) = 0$. **(b)** Use the result of part (a) to solve

$$(xy' - y)^3 = (y')^2 + 5.$$

29. Show that $y = cx + f(c)$, where c is an arbitrary constant, is a solution of (7).

30. Show that (8) is a solution of (7). [*Hint:* Differentiate both sides of (7) with respect to x and consider two cases. Use parametric differentiation to show

$$\frac{dy}{dx} = \frac{dy/dt}{dx/dt} = t, \quad f''(t) \neq 0.$$

Note that since the slope of $y = cx + f(c)$ is constant, the singular solution cannot be obtained from it.]

[O] 2.7 Substitutions

In the preceding sections, we saw that sometimes a differential equation could be transformed by means of a substitution into a form that could then be solved by one of the standard methods. An equation may look different from any of those that we have just studied, but through a judicious change of variables perhaps an apparently difficult problem may be readily solved. Although we can give no firm rules on what, *if any*, substitution to use, a working axiom might be: Try something! It sometimes pays to be clever.

EXAMPLE 1 The differential equation

$$y(1 + 2xy)\,dx + x(1 - 2xy)\,dy = 0$$

is neither separable, homogeneous, exact, linear, nor Bernoulli. However, if we stare at the equation long enough, we might be prompted to try the substitution

$$u = 2xy \quad \text{or} \quad y = \frac{u}{2x}.$$

Since
$$dy = \frac{x\,du - u\,dx}{2x^2},$$

the equation becomes, after simplifying,

$$2u^2\,dx + (1 - u)x\,du = 0.$$

We recognize the last equation as separable, and so from

$$2\frac{dx}{x} + \frac{1 - u}{u^2}\,du = 0$$

we obtain
$$2\ln|x| - u^{-1} - \ln|u| = c$$

$$\ln\left|\frac{x}{2y}\right| = c + \frac{1}{2xy}$$

$$\frac{x}{2y} = c_1 e^{1/2xy}$$

$$x = 2c_1 y e^{1/2xy},$$

where e^c was replaced by c_1.

Notice that the differential equation in Example 1 possesses the trivial solution $y = 0$ but that this function is not included in the one-parameter family of solutions.

EXAMPLE 2 Solve
$$2xy\frac{dy}{dx} + 2y^2 = 3x - 6.$$

SOLUTION The presence of the term $2y\dfrac{dy}{dx}$ prompts us to try $u = y^2$ since

$$\frac{du}{dx} = 2y\frac{dy}{dx}.$$

Now
$$x\frac{du}{dx} + 2u = 3x - 6$$

has the linear form
$$\frac{du}{dx} + \frac{2}{x}u = 3 - \frac{6}{x},$$

so that multiplication by the integrating factor $e^{\int (2/x)\,dx} = e^{\ln x^2} = x^2$ gives

$$\frac{d}{dx}[x^2u] = 3x^2 - 6x$$

$$x^2u = x^3 - 3x^2 + c \qquad \text{or} \qquad x^2y^2 = x^3 - 3x^2 + c.$$

EXAMPLE 3

Solve
$$x\frac{dy}{dx} - y = \frac{x^3}{y}e^{y/x}.$$

SOLUTION If we let
$$u = \frac{y}{x},$$

the differential equation simplifies to

$$ue^{-u}\,du = dx.$$

Integration by parts then gives

$$-ue^{-u} - e^{-u} = x + c$$

$$u + 1 = (c_1 - x)e^u \qquad (c_1 = -c)$$

$$\frac{y}{x} + 1 = (c_1 - x)e^{y/x}$$

$$y + x = x(c_1 - x)e^{y/x}.$$

Some higher-order differential equations can be reduced to first-order equations by a substitution.

EXAMPLE 4

Solve
$$y'' = 2x(y')^2.$$

SOLUTION If we let $u = y'$ so that $du/dx = y''$, the equation then reduces to a separable form. We have

$$\frac{du}{dx} = 2xu^2 \qquad \text{or} \qquad \frac{du}{u^2} = 2x\,dx$$

$$\int u^{-2}\,du = \int 2x\,dx$$

$$-u^{-1} = x^2 + c_1^2.$$

The constant of integration is written as c_1^2 for convenience. The reason should be obvious in the next few steps. Since $u^{-1} = 1/y'$ it follows that

$$\frac{dy}{dx} = -\frac{1}{x^2 + c_1^2} \quad \text{or} \quad dy = -\frac{dx}{x^2 + c_1^2}$$

$$\int dy = -\int \frac{dx}{x^2 + c_1^2}$$

$$y + c_2 = -\frac{1}{c_1}\tan^{-1}\frac{x}{c_1}.$$

Exercises 2.7

Answers to odd-numbered problems begin on page A-5.

In Problems 1–26 solve the given differential equation by using an appropriate substitution.

1. $xe^{2y}\dfrac{dy}{dx} + e^{2y} = \dfrac{\ln x}{x}$

2. $y' + y\ln y = ye^x$

3. $y\,dx + (1 + ye^x)\,dy = 0$

4. $(2 + e^{-x/y})\,dx + 2\left(1 - \dfrac{x}{y}\right)dy = 0$

5. $\dfrac{dy}{dx} - \dfrac{4}{x}y = 2x^5e^{y/x^4}$

6. $\dfrac{dy}{dx} + x + y + 1 = (x + y)^2e^{3x}$

7. $2yy' + x^2 + y^2 + x = 0$

8. $y' = y + x(y + 1)^2 + 1$

9. $2x\csc 2y\dfrac{dy}{dx} = 2x - \ln(\tan y)$

10. $x^2\dfrac{dy}{dx} + 2xy = x^4y^2 + 1$

11. $x^4y^2y' + x^3y^3 = 2x^3 - 3$

12. $xe^yy' - 2e^y = x^2$

13. $y' + 1 = e^{-(x+y)}\sin x$

14. $\sin y\sinh x\,dx + \cos y\cosh x\,dy = 0$

15. $y\dfrac{dx}{dy} + 2x\ln x = xe^y$

16. $x\sin y\dfrac{dy}{dx} + \cos y = -x^2e^x$

17. $y'' + (y')^2 + 1 = 0$

18. $xy'' = y' + x(y')^2$

19. $xy'' = y' + (y')^3$

20. $x^2y'' + (y')^2 = 0$

21. $y' - xy'' - (y'')^3 = 1$

22. $y'' = 1 + (y')^2$

23. $xy'' - y' = 0$

24. $y'' + (\tan x)y' = 0$

25. $y'' + 2y(y')^3 = 0$

26. $y^2y'' = y'$

$$\left[\textit{Hint: } \text{Let } u = y' \text{ so that } y'' = \frac{du}{dx} = \frac{du}{dy}\frac{dy}{dx} = \frac{du}{dy}u.\right]$$

Miscellaneous Problems

27. In calculus the curvature of a curve whose equation is $y = f(x)$ is defined to be the number

$$\kappa = y''/[1 + (y')^2]^{3/2}.$$

Determine a function for which $\kappa = 1$. [*Hint:* For simplicity ignore constants of integration. Also consider a trigonometric substitution.]

[O] 2.8 Picard's Method

The initial-value problem

$$\begin{aligned} y' &= f(x, y) \\ y(x_0) &= y_0 \end{aligned} \tag{1}$$

first considered in Section 2.1 can be written in an alternative manner. Let f be continuous in a region containing the point (x_0, y_0). By integrating both sides of the differential equation with respect to x, we get

$$y(x) = c + \int_{x_0}^{x} f(t, y(t))\,dt.$$

Now

$$y(x_0) = c + \int_{x_0}^{x_0} f(t, y(t))\,dt = c$$

implies $c = y_0$. Thus

$$y(x) = y_0 + \int_{x_0}^{x} f(t, y(t))\,dt. \tag{2}$$

Conversely, if we start with (2), we can obtain (1). In other words, the integral equation (2) and the initial-value problem (1) are equivalent. We now try to solve (2) by a *method of successive approximations.*

Suppose $y_0(x)$ is an arbitrary continuous function that represents a guess or approximation to the solution of (2). Since $f(x, y_0(x))$ is a known function depending solely on x, it can be integrated. With $y(t)$ replaced by $y_0(t)$, the right-hand side of (2) defines another function, which we write as

$$y_1(x) = y_0 + \int_{x_0}^{x} f(t, y_0(t))\, dt.$$

It is hoped that this new function is a better approximation to the solution. Repeating the procedure, yet another function is given by

$$y_2(x) = y_0 + \int_{x_0}^{x} f(t, y_1(t))\, dt.$$

In this manner we obtain a sequence of functions $y_1(x), y_2(x), y_3(x), \ldots$, whose nth term is defined by the relation

$$y_n(x) = y_0 + \int_{x_0}^{x} f(t, y_{n-1}(t))\, dt \qquad n = 1, 2, 3, \ldots. \tag{3}$$

In the application of (3), it is common practice to choose the initial function as $y_0(x) = y_0$. The repetitive use of formula (3) is known as **Picard's method of iteration.**

EXAMPLE 1

Consider the problem

$$y' = y - 1$$
$$y(0) = 2.$$

Use Picard's method to find the approximations y_1, y_2, y_3, y_4.

SOLUTION By identifying $x_0 = 0$, $y_0(x) = 2$, and $f(t, y_{n-1}(t)) = y_{n-1}(t) - 1$, equation (3) becomes

$$y_n(x) = 2 + \int_0^x (y_{n-1}(t) - 1)\, dt, \qquad n = 1, 2, 3, \ldots.$$

Iterating this last expression then gives

$$y_1(x) = 2 + \int_0^x 1 \cdot dt = 2 + x$$

$$y_2(x) = 2 + \int_0^x (1 + t)\, dt = 2 + x + \frac{x^2}{2}$$

$$y_3(x) = 2 + \int_0^x \left(1 + t + \frac{t^2}{2}\right) dt$$

$$= 2 + x + \frac{x^2}{2} + \frac{x^3}{2 \cdot 3}$$

$$y_4(x) = 2 + \int_0^x \left(1 + t + \frac{t^2}{2} + \frac{t^3}{2 \cdot 3}\right) dt$$
$$= 2 + x + \frac{x^2}{2} + \frac{x^3}{2 \cdot 3} + \frac{x^4}{2 \cdot 3 \cdot 4}.$$

By induction it can be shown in the last example that the nth term of the sequence of approximations is

$$y_n(x) = 2 + x + \frac{x^2}{2!} + \frac{x^3}{3!} + \cdots + \frac{x^n}{n!} = 1 + \sum_{k=0}^{n} \frac{x^k}{k!}.$$

From this latter form we recognize that the *limit* of $y_n(x)$ and $n \to \infty$ is $y(x) = 1 + e^x$. It should come as no surprise to note that the function is an exact solution of the given initial-value problem.

The reader should not be deceived by the relative ease with which the iterants $y_n(x)$ were obtained in the last example. In general, the integration involved in generating each $y_n(x)$ can become complicated very quickly. Nor, for that matter, is it always apparent that the sequence $\{y_n(x)\}$ converges to a nice explicit function. Thus it is fair to ask at this point: Is Picard's method a practical means of solving a first-order equation $y' = f(x, y)$ subject to $y(x_0) = y_0$? In most cases the answer is: No. One might ask further, in the spirit of a scientist/engineer: What *is* it good for? The answer is not bound to please: Picard's method of iteration is a theoretical tool used in the consideration of existence and uniqueness of solutions of differential equations. Under certain conditions on $f(x, y)$ it can be shown that as $n \to \infty$, the sequence $\{y_n(x)\}$ defined by (3) converges to a function $y(x)$ that satisfies the integral equation (2) and hence the initial-value problem (1). Indeed, it is precisely Picard's method of successive approximations that is used in proving Picard's theorem of Section 2.1. However, the proof of Theorem 2.1 uses concepts from advanced calculus and will not be presented here. Our purpose in introducing this topic is twofold: to gain an appreciation for the potential of the procedure and to obtain at least a nodding acquaintance with an iterative technique. In Chapter 9 we shall consider other methods for approximating solutions of differential equations that also utilize iteration.

Exercises 2.8

Answers to odd-numbered problems begin on page A–5.

In Problems 1–6 use Picard's method to find y_1, y_2, y_3, y_4. Determine the limit of the sequence $\{y_n(x)\}$ as $n \to \infty$.

1. $y' = -y, \quad y(0) = 1$

2. $y' = x + y, \quad y(0) = 1$

3. $y' = 2xy, \quad y(0) = 1$

4. $y' + 2xy = x, \quad y(0) = 0$

5. $y' + y^2 = 0, \quad y(0) = 0$

6. $y' = 2e^x - y, \quad y(0) = 1$

7. **(a)** Use Picard's method to find y_1, y_2, y_3 for the problem

$$y' = 1 + y^2, \quad y(0) = 0.$$

(b) Solve the initial-value problem in part (a) by one of the methods of this chapter.

(c) Compare the results of parts (a) and (b).

8. In Picard's method the initial choice $y_0(x) = y_0$ is not necessary. Rework Problem 3 with **(a)** $y_0(x) = k$ a constant and $k \neq 1$, **(b)** $y_0(x) = x$.

CHAPTER 2 SUMMARY

An **initial-value problem** consists of finding a solution of

$$\frac{dy}{dx} = f(x, y)$$

$$y(x_0) = y_0$$

on an interval I containing x_0. If $f(x, y)$ and $\partial f/\partial y$ are continuous in a rectangular region of the xy-plane with (x_0, y_0) in its interior, then we are guaranteed that there exists an interval around x_0 on which the problem has a unique solution.

The method of solution for a first-order differential equation depends on an appropriate classification of the equation. We summarize five cases.

An equation is **separable** if it can be put into the form $h(y)\,dy = g(x)\,dx$. The solution results from integrating both sides of the equation.

If $M(x, y)$ and $N(x, y)$ are **homogeneous functions** of the same degree, then $M(x, y)\,dx + N(x, y)\,dy = 0$ can be reduced to an equation with separable variables by either the substitution $y = ux$ or $x = vy$. The choice of substitution usually depends on which coefficient is simplier.

The differential equation $M(x, y)\,dx + N(x, y)\,dy = 0$ is said to be **exact** if the form $M(x, y)\,dx + N(x, y)\,dy$ is an exact differential. When $M(x, y)$ and $N(x, y)$ are continuous and have continuous first partial derivatives, then $\partial M/\partial y = \partial N/\partial x$ is a necessary and sufficient condition that $M(x, y)\,dx + N(x, y)\,dy$ be exact. This means there exists some function $f(x, y)$ for which $M(x, y) = \partial f/\partial x$ and $N(x, y) = \partial f/\partial y$. The method of solution for an exact equation starts by integrating either of these latter expressions.

If a first-order equation can be put into the form $dy/dx + P(x)y = f(x)$, it is said to be **linear** in the variable y. We solve the equation by first finding the **integrating factor**, $e^{\int P(x)\,dx}$, multiplying both sides of the equation by this factor, and then integrating both sides of

$$\frac{d}{dx}[e^{\int P(x)\,dx}y] = e^{\int P(x)\,dx}f(x).$$

Bernoulli's equation is $dy/dx + P(x)y = f(x)y^n$, where n is any real number. When $n \neq 0$ and $n \neq 1$, Bernoulli's equation can be reduced to a linear equation by the substitution $w = y^{1-n}$.

In certain circumstances a differential equation can be reduced to one of the familiar forms by an appropriate **substitution,** or **change, of variables.** Of course we already know that this is the procedure when solving a homogeneous or a Bernoulli equation. In the general context, no rule on when to use a substitution can be given.

By converting an initial-value problem to an equivalent integral equation, **Picard's method of iteration** is one way of obtaining an approximation to the solution of the problem.

CHAPTER 2 REVIEW EXERCISES

Answers to odd-numbered problems begin on page A-5.

Answer Problems 1–4 without referring back to the text. Fill in the blank or answer true/false.

1. The differential equation $y' = 1/(25 - x^2 - y^2)$ will have a unique solution through any point (x_0, y_0) in the region(s) defined by ______.
2. The initial-value problem $xy' = 3y$, $y(0) = 0$ has the solutions $y = x^3$ and ______.
3. The initial-value problem $y' = y^{1/2}$, $y(0) = 0$ has no solution since $\partial f/\partial y$ is discontinuous on the line $y = 0$. ______
4. There exists an interval centered at 2 on which the unique solution of the initial-value problem $y' = (y - 1)^3$, $y(2) = 1$ is $y = 1$. ______
5. Without solving, classify each of the following equations as to: separable, homogeneous, exact, linear, Bernoulli, Ricatti, or Clairaut.

(a) $\dfrac{dy}{dx} = \dfrac{1}{y - x}$ **(b)** $\dfrac{dy}{dx} = \dfrac{x - y}{x}$

(c) $\left(\dfrac{dy}{dx}\right)^2 + 2y = 2x\dfrac{dy}{dx}$ **(d)** $\dfrac{dy}{dx} = \dfrac{1}{x(x - y)}$

(e) $\dfrac{dy}{dx} = \dfrac{y^2 + y}{x^2 + x}$ **(f)** $\dfrac{dy}{dx} = 4 + 5y + y^2$

(g) $y\,dx = (y - xy^2)\,dy$ **(h)** $x\dfrac{dy}{dx} = ye^{x/y} - x$

(i) $xyy' + y^2 = 2x$

(j) $2xyy' + y^2 = 2x^2$

(k) $y\,dx + x\,dy = 0$

(l) $\left(x^2 + \dfrac{2y}{x}\right) dx = (3 - \ln x^2)\,dy$

(m) $\dfrac{dy}{dx} = \dfrac{x}{y} + \dfrac{y}{x} + 1$

(n) $\dfrac{y}{x^2}\dfrac{dy}{dx} + e^{2x^3 + y^2} = 0$

(o) $y = xy' + (y' - 3)^2$

(p) $y' + 5y^2 = 3x^4 - 2xy$

6. Solve $(y^2 + 1)\,dx = y \sec^2 x\,dy$.

7. Solve $\dfrac{y}{x}\dfrac{dy}{dx} = \dfrac{e^x}{\ln y}$ subject to $y(1) = 1$.

8. Solve $y(\ln x - \ln y)\,dx = (x \ln x - x \ln y - y)\,dy$.

9. Solve $xyy' = 3y^2 + x^2$ subject to $y(-1) = 2$.

10. Solve $(6x + 1)y^2 \dfrac{dy}{dx} + 3x^2 + 2y^3 = 0$.

11. Solve $ye^{xy}\dfrac{dx}{dy} + xe^{xy} = 12y^2$ subject to $y(0) = -1$.

12. Solve $x\,dy + (xy + y - x^2 - 2x)\,dx = 0$.

13. Solve $(x^2 + 4)\dfrac{dy}{dx} = 2x - 8xy$ subject to $y(0) = -1$.

14. Solve $(2x + y)y' = 1$.

15. Solve $x\dfrac{dy}{dx} + 4y = x^4y^2$ subject to $y(1) = 1$.

16. Solve $-xy' + y = (y' + 1)^2$ subject to $y(0) = 0$.

In Problems 17 and 18 solve the given differential equation by means of a substitution.

17. $\dfrac{dy}{dx} + xy^3 \sec \dfrac{1}{y^2} = 0$

18. $y'' = x - y'$

19. Use the Picard method to find the approximations y_1 and y_2 for $y' = x^2 + y^2$, $y(0) = 1$.

20. Solve $y' + 2y = 4$, $y(0) = 3$ by one of the usual methods. Solve the same problem by Picard's method and compare the results.

CHAPTER 3

Applications of First-Order Differential Equations

IMPORTANT TERMS

mathematical model
orthogonal trajectories
exponential growth and decay
half-life
carbon dating
response
transient term
steady-state term
logistic equation
first- and second-order reactions

A differential equation used to describe a physical phenomenon is said to be a **mathematical model**. Mathematical models for phenomena such as radioactive decay, population growth, chemical reactions, cooling of bodies, current in a series circuit, velocity of a falling body, and so on are often first-order differential equations. In this chapter we shall be concerned with solving some of the more commonly occurring linear and nonlinear first-order equations that arise in applications.

3.1 Orthogonal Trajectories

Orthogonal Curves

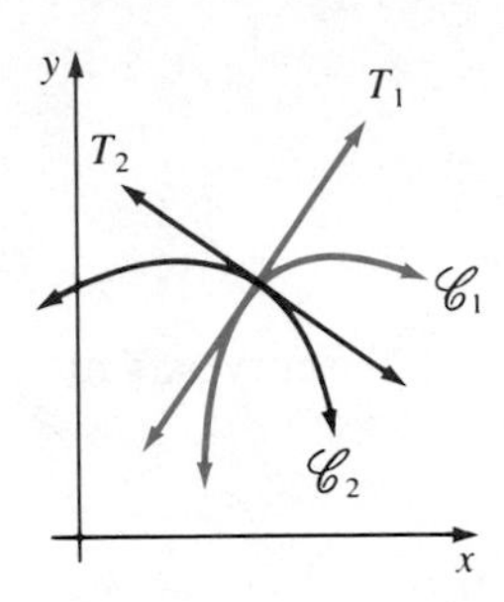

Figure 3.1

Recall from your study of analytic geometry that two lines L_1 and L_2, which are not parallel to the coordinate axes, are perpendicular if and only if their respective slopes satisfy the relationship $m_1m_2 = -1$. For this reason, the graphs of $y = (-1/2)x + 1$ and $y = 2x + 4$ are obviously perpendicular. In general, two curves $\mathscr{C}_1$ and $\mathscr{C}_2$ are said to be **orthogonal** at a point if and only if their tangent lines T_1 and T_2 are perpendicular at the point of intersection (see Figure 3.1). Except for the case when T_1 and T_2 are parallel to the coordinate axes, this means the slopes of the tangents are negative reciprocals of one another.

EXAMPLE 1

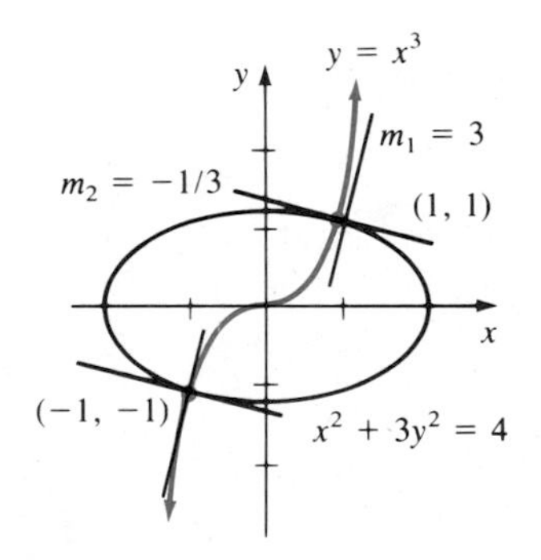

Figure 3.2

Show that the curves $\mathscr{C}_1$ and $\mathscr{C}_2$ defined by $y = x^3$ and $x^2 + 3y^2 = 4$ are orthogonal at their point(s) of intersection.

SOLUTION In Figure 3.2 it is seen that the points of intersection of the graphs are (1, 1) and (−1, −1). Now the slope of the tangent line to $y = x^3$ at any point is $dy/dx = 3x^2$

so that
$$\left.\frac{dy}{dx}\right|_{x=1} = \left.\frac{dy}{dx}\right|_{x=-1} = 3.$$

We use implicit differentiation to obtain dy/dx for the second curve:

$$2x + 6y\frac{dy}{dx} = 0 \qquad \text{or} \qquad \frac{dy}{dx} = -\frac{x}{3y}$$

and therefore,
$$\left.\frac{dy}{dx}\right|_{(1,1)} = \left.\frac{dy}{dx}\right|_{(-1,-1)} = -\frac{1}{3}.$$

Thus, at either (1, 1) or (−1, −1) we have

$$\left(\frac{dy}{dx}\right)_{\mathscr{C}_1} \cdot \left(\frac{dy}{dx}\right)_{\mathscr{C}_2} = -1.$$

It is easy to show that any curve $\mathscr{C}_1$ in the family $y = c_1x^3$, $c_1 \neq 0$, is orthogonal to each curve $\mathscr{C}_2$ in the family $x^2 + 3y^2 = c_2$, $c_2 > 0$. The differential equation of the first family is:

$$\frac{dy}{dx} = 3c_1x^2 \qquad \text{or} \qquad \frac{dy}{dx} = \frac{3y}{x}$$

since $c_1 = y/x^3$. Now implicit differentiation of $x^2 + 3y^2 = c_2$ leads to exactly the same differential equation as for $x^2 + 3y^2 = 4$; namely,

$$\frac{dy}{dx} = -\frac{x}{3y}.$$

Hence, at the point (x, y) on each curve

$$\left(\frac{dy}{dx}\right)_{\mathscr{C}_1} \cdot \left(\frac{dy}{dx}\right)_{\mathscr{C}_2} = \left(\frac{3y}{x}\right)\left(-\frac{x}{3y}\right) = -1.$$

Since the slopes of the tangent lines are negative reciprocals, the curves $\mathscr{C}_1$ and $\mathscr{C}_2$ intersect each other in an orthogonal manner.

This discussion leads to the following definition.

DEFINITION 3.1 When *all* the curves of one family of curves $G(x, y, c_1) = 0$ intersect orthogonally *all* the curves of another family $H(x, y, c_2) = 0$, then the families are said to be **orthogonal trajectories** of each other.

In other words, an orthogonal trajectory is any *one* curve that intersects every curve of another family at right angles.

EXAMPLE 2

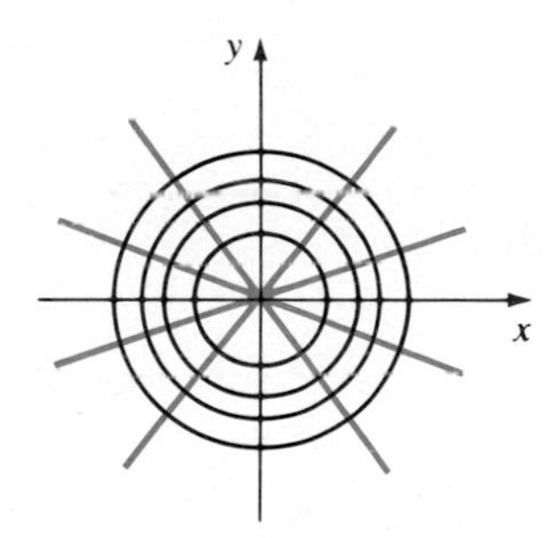

Figure 3.3

(a) The graph of $y = (-1/2)x + 1$ is an orthogonal trajectory of $y = 2x + c_1$. The families $y = (-1/2)x + c_2$ and $y = 2x + c_1$ are orthogonal trajectories.

(b) The graph of $y = 4x^3$ is an orthogonal trajectory of $x^2 + 3y^2 = c_2$. The families $y = c_1x^3$ and $x^2 + 3y^2 = c_2$ are orthogonal trajectories.

(c) In Figure 3.3 it is seen that the family of straight lines $y = c_1x$ through the origin and the family $x^2 + y^2 = c_2$ of concentric circles with center at the origin are orthogonal trajectories.

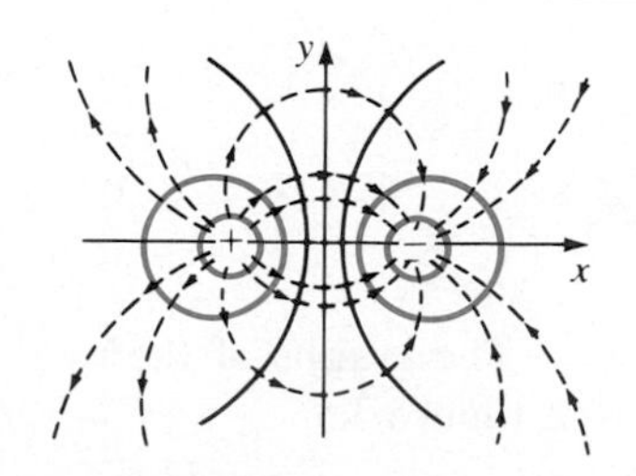

Figure 3.4

Orthogonal trajectories occur naturally in the construction of meteorological maps and in the study of electricity and magnetism. For example, in an electric field around two bodies of opposite charge the lines of force are perpendicular to the equipotential curves (that is, curves along which the potential is constant). The lines of force are indicated in Figure 3.4 by dashed lines.

General Method

To find the orthogonal trajectories of a given family of curves we first find the differential equation

$$\frac{dy}{dx} = f(x, y)$$

that describes the family. The differential equation of the second, and orthogonal, family is then

$$\frac{dy}{dx} = \frac{-1}{f(x, y)}.$$

EXAMPLE 3

Find the orthogonal trajectories of the family of rectangular hyperbolas

$$y = \frac{c_1}{x}.$$

SOLUTION The derivative of $y = c_1/x$ is

$$\frac{dy}{dx} = \frac{-c_1}{x^2}.$$

Replacing c_1 by $c_1 = xy$ yields the differential equation of the given family:

$$\frac{dy}{dx} = -\frac{y}{x}.$$

The differential equation of the orthogonal family is then

$$\frac{dy}{dx} = \frac{-1}{(-y/x)} = \frac{x}{y}.$$

We solve this last equation by separation of variables:

$$y\,dy = x\,dx$$

$$\int y\,dy = \int x\,dx$$

$$\frac{y^2}{2} = \frac{x^2}{2} + c_2' \quad \text{or} \quad y^2 - x^2 = c_2,$$

where, for convenience, we have replaced $2c_2'$ by c_2. The graphs of the two families, for various values of c_1 and c_2, are given in Figure 3.5.

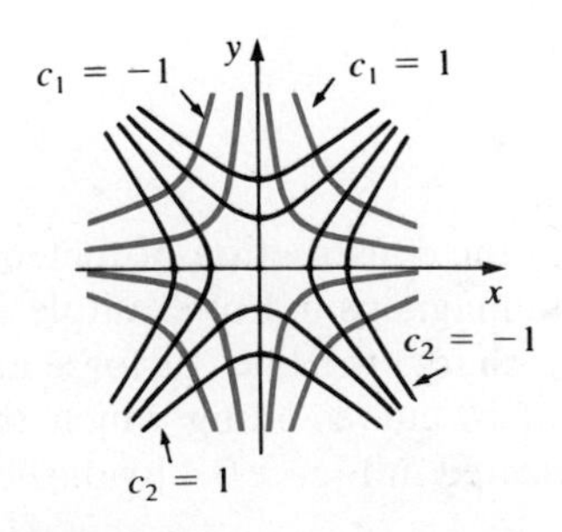

Figure 3.5

EXAMPLE 4 Find the orthogonal trajectories of

$$y = \frac{c_1 x}{1 + x}.$$

SOLUTION From the quotient rule we find

$$\frac{dy}{dx} = \frac{c_1}{(1 + x)^2} \quad \text{or} \quad \frac{dy}{dx} = \frac{y}{x(1 + x)},$$

since $c_1 = y(1 + x)/x$. The differential equation of the orthogonal trajectories is then

$$\frac{dy}{dx} = -\frac{x(1 + x)}{y}.$$

Again, by separating variables, we have:

$$y\,dy = -x(1 + x)\,dx$$

$$\int y\,dy = -\int (x + x^2)\,dx$$

$$\frac{y^2}{2} = -\frac{x^2}{2} - \frac{x^3}{3} + c_2' \quad \text{or} \quad 3y^2 + 3x^2 + 2x^3 = c_2.$$

Polar Curves

In calculus it is shown that for a graph of a polar equation $r = f(\theta)$,

$$r\frac{d\theta}{dr} = \tan\psi,$$

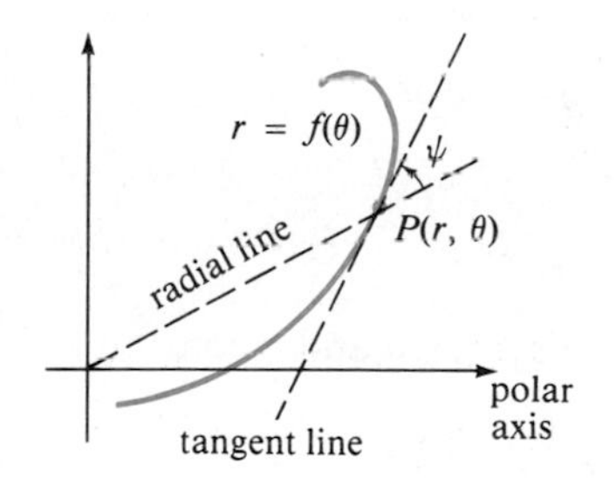

Figure 3.6

where ψ is the positive counterclockwise angle between the radial line and the tangent line (see Figure 3.6). It is left as an exercise to show that two polar curves $r = f_1(\theta)$ and $r = f_2(\theta)$ are orthogonal at a point of intersection if and only if

$$(\tan\psi_1)_{\mathscr{C}_1}(\tan\psi_2)_{\mathscr{C}_2} = -1. \tag{1}$$

EXAMPLE 5 Find the orthogonal trajectories of

$$r = c_1(1 - \sin\theta).$$

SOLUTION For the given curve we can write

$$\frac{dr}{d\theta} = -c_1\cos\theta = \frac{-r\cos\theta}{1 - \sin\theta}$$

so that
$$r\frac{d\theta}{dr} = -\frac{1 - \sin\theta}{\cos\theta} = \tan\psi_1.$$

Thus, by (1), the differential equation of the orthogonal trajectories is

$$r\frac{d\theta}{dr} = \frac{\cos\theta}{1 - \sin\theta} = \tan\psi_2.$$

Separating variables then gives

$$\begin{aligned}\frac{dr}{r} &= \frac{1 - \sin\theta}{\cos\theta}\, d\theta \\ &= (\sec\theta - \tan\theta)\, d\theta\end{aligned}$$

so that
$$\begin{aligned}\ln|r| &= \ln|\sec\theta + \tan\theta| + \ln|\cos\theta| + \ln c_2 \\ &= \ln|c_2(1 + \sin\theta)|.\end{aligned}$$

Hence
$$r = c_2(1 + \sin\theta).$$

Exercises 3.1

Answers to odd-numbered problems begin on page A–6.

In Problems 1–26 find the orthogonal trajectories of the given family of curves.

1. $y = c_1 x$

2. $3x + 4y = c_1$

3. $y = c_1 x^2$

4. $y = (x - c_1)^2$

5. $c_1 x^2 + y^2 = 1$

6. $2x^2 + y^2 = c_1^2$

7. $y = c_1 e^{-x}$

8. $y = e^{c_1 x}$

9. $y^2 = c_1 x^3$

10. $y^a = c_1 x^b$, a and b constants

11. $y = \dfrac{x}{1 + c_1 x}$

12. $y = \dfrac{1 + c_1 x}{1 - c_1 x}$

13. $2x^2 + y^2 = 4c_1 x$

14. $x^2 + y^2 = 2c_1 x$

15. $y^3 + 3x^2 y = c_1$

16. $y^2 - x^2 = c_1 x^3$

17. $y = \dfrac{c_1}{1 + x^2}$

18. $y = \dfrac{1}{c_1 + x}$

19. $4y + x^2 + 1 + c_1 e^{2y} = 0$

20. $y = -x - 1 + c_1 e^x$

21. $y = \dfrac{1}{\ln c_1 x}$

22. $y = \ln(\tan x + c_1)$

23. $\sinh y = c_1 x$

24. $y = c_1 \sin x$

25. $x^{1/3} + y^{1/3} = c_1$

26. $x^a + y^a = c_1, \quad a \neq 2$

27. Find the member of the orthogonal trajectories for $x + y = c_1 e^y$ that passes through (0, 5).

28. Find the member of the orthogonal trajectories for $3xy^2 = 2 + 3c_1 x$ that passes through (0, 10).

In Problems 29–34 find the orthogonal trajectories of the given polar curves.

29. $r = 2c_1 \cos \theta$

30. $r = c_1(1 + \cos \theta)$

31. $r^2 = c_1 \sin 2\theta$

32. $r = \dfrac{c_1}{1 + \cos \theta}$

33. $r = c_1 \sec \theta$

34. $r = c_1 e^\theta$

Miscellaneous Problems

35. A family of curves that intersects a given family of curves at a specified constant angle $\alpha \neq \pi/2$ is said to be an isogonal family. The two families are said to be **isogonal trajectories** of each other. If $dy/dx = f(x, y)$ is the differential equation of the given family, show that the differential equation of the isogonal family is

$$\frac{dy}{dx} = \frac{f(x, y) \pm \tan \alpha}{1 \mp f(x, y) \tan \alpha}.$$

In Problems 36–38 use the results of Problem 35 to find the isogonal family that intersects the one-parameter family of straight lines $y = c_1 x$ at the given angle.

36. $\alpha = 45°$

37. $\alpha = 60°$

38. $\alpha = 30°$

A family of curves can be **self-orthogonal** in the sense that a member of the orthogonal trajectories is also a member of the original family. In Problems 39 and 40 show that the given family of curves is self-orthogonal.

39. parabolas $y^2 = c_1(2x + c_1)$

40. confocal conics $\dfrac{x^2}{c_1 + 1} + \dfrac{y^2}{c_1} = 1$

41. Verify that the orthogonal trajectories of the family of curves given by the parametric equations $x = c_1 e^t \cos t$, $y = c_1 e^t \sin t$ are

$$x = c_2 e^{-t} \cos t, \qquad y = c_2 e^{-t} \sin t.$$

[*Hint:* $dy/dx = (dy/dt)/(dx/dt)$.]

42. Show that two polar curves $r = f_1(\theta)$ and $r = f_2(\theta)$ are orthogonal at a point of intersection if and only if

$$(\tan \psi_1)_{\mathscr{C}_1}(\tan \psi_2)_{\mathscr{C}_2} = -1.$$

3.2 Applications of Linear Equations

3.2.1 Growth and Decay

The initial-value problem

$$\frac{dx}{dt} = kx \qquad x(t_0) = x_0, \tag{1}$$

where k is a constant of proportionality, occurs in many physical theories involving either **growth** or **decay**. For example, in biology it is often observed that the rate at which certain bacteria grow is proportional to the number of bacteria present at any time. Over short intervals of time, the population of small animals, such as rodents, can be predicted fairly accurately by the solution of (1). In physics an initial-value problem such as (1) provides a model for approximating the remaining amount of a substance that is disintegrating, or decaying, through radioactivity. The differential equation in (1) could also determine the temperature of a cooling body. In chemistry the amount of a substance remaining during certain reactions is also described by (1).

The constant of proportionality k in (1) is either positive or negative and can be determined from the solution of the problem using a subsequent measurement of x at a time $t_1 > t_0$.

EXAMPLE 1

A culture initially has N_0 number of bacteria. At $t = 1$ hour the number of bacteria is measured to be $(3/2)N_0$. If the rate of growth is proportional to the number of bacteria present, determine the time necessary for the number of bacteria to triple.

SOLUTION We first solve the differential equation

$$\frac{dN}{dt} = kN, \tag{2}$$

subject to $N(0) = N_0$.

After we have solved the above problem, we then use the empirical condition $N(1) = (3/2)N_0$ to determine the constant of proportionality k.

Now (2) is both separable and linear. When put into the form

$$\frac{dN}{dt} - kN = 0,$$

we can see by inspection that the integrating factor is e^{-kt}. Multiplying both sides of the equation by this term gives immediately

$$\frac{d}{dt}[e^{-kt}N] = 0.$$

Integrating both sides of the last equation yields

$$e^{-kt}N = c \qquad \text{or} \qquad N(t) = ce^{kt}$$

At $t = 0$ it follows that $N_0 = ce^0 = c$ and so $N(t) = N_0e^{kt}$. At $t = 1$ we have

$$\frac{3}{2}N_0 = N_0e^k \qquad \text{or} \qquad e^k = \frac{3}{2},$$

from which we get to four decimal places

$$k = \ln\left(\frac{3}{2}\right) = 0.4055.$$

Thus $$N(t) = N_0e^{0.4055t}$$

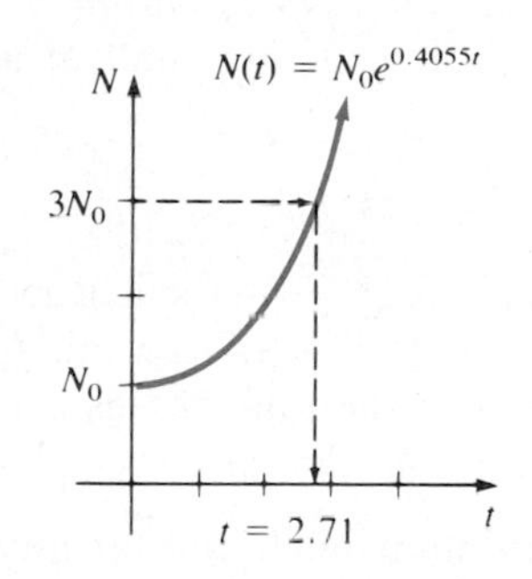

Figure 3.7

To find the time at which the bacteria have tripled we solve

$$3N_0 = N_0e^{0.4055t}$$

for t. It follows that

$$0.4055t = \ln 3$$

$$t = \frac{\ln 3}{0.4055} \approx 2.71 \text{ hours.}$$

See Figure 3.7.

Note: We can write the function $N(t)$ obtained in the preceding example in an alternative form. From the laws of exponents

$$N(t) = N_0(e^k)^t$$
$$= N_0\left(\frac{3}{2}\right)^t$$

since $e^k = 3/2$. This latter solution provides a convenient method for computing $N(t)$ for small positive integral values of t. It also clearly shows the influence of the subsequent experimental observation at $t = 1$ on the solution for all time. We notice, too, that the actual number of bacteria present

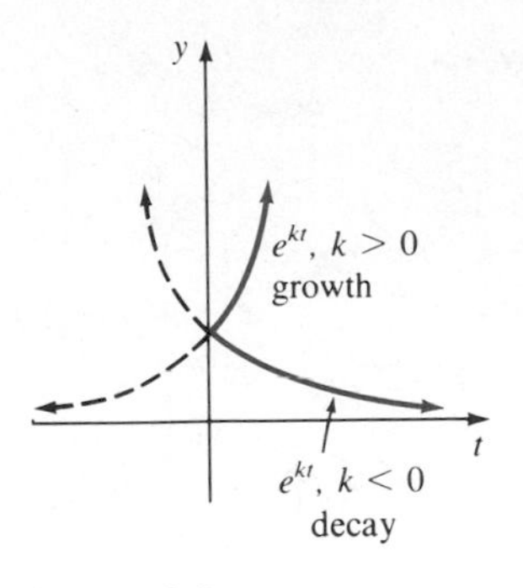

Figure 3.8

at time $t = 0$ is quite irrelevant in finding the time required to triple the number in the culture. The necessary time to triple, say, 100 or 10,000 bacteria is still approximately 2.71 hours.

As shown in Figure 3.8, the exponential function e^{kt} increases as t increases for $k > 0$, and decreases as t increases if $k < 0$. Thus problems describing growth, such as population, bacteria, or even capital, are characterized by a positive value of k, whereas problems involving decay, as in radioactive disintegration, will yield a negative k value.

Half-life

In physics the **half-life** is a measure of the stability of a radioactive substance. The half-life is simply the time it takes for one-half of the atoms in an initial amount A_0 to disintegrate, or transmute, into the atoms of another element. The longer the half-life of a substance, the more stable it is. For example, the half-life of highly radioactive radium, Ra-226, is about 1700 years. In 1700 years one-half of a given quantity of Ra-226 is transmuted into radon, Rn-222. The most commonly occurring uranium isotope, U-238, has a half-life of approximately 4,500,000,000 years. In about 4.5 billion years, one-half of a quantity of U-238 is transmuted into lead, Pb-206.

EXAMPLE 2

A breeder reactor converts the relatively stable uranium 238 into the isotope plutonium 239. After 15 years it is determined that 0.043% of the initial amount A_0 of the plutonium has disintegrated. Find the half-life of this isotope if the rate of disintegration is proportional to the amount remaining.

SOLUTION Let $A(t)$ denote the amount of plutonium remaining at any time. As in Example 1, the solution of the initial value problem

$$\frac{dA}{dt} = kA$$

$$A(0) = A_0$$

is

$$A(t) = A_0 e^{kt}.$$

If 0.043% of the atoms of A_0 have disintegrated, then 99.957% of the substance remains. To find k we must solve

$$0.99957A_0 = A_0 e^{15k}.$$

Therefore

$$e^{15k} = 0.99957$$

$$15k = \ln(0.99957)$$

$$k = \frac{\ln(0.99957)}{15} = -0.00002867.$$

Hence $$A(t) = A_0 e^{-0.00002867t}.$$

Now the half-life is the corresponding value of time for which $A(t) = A_0/2$. Solving for t gives

$$\frac{A_0}{2} = A_0 e^{-0.00002867t} \quad \text{or} \quad \frac{1}{2} = e^{-0.00002867t},$$

$$-0.00002867t = \ln\left(\frac{1}{2}\right) = -\ln 2$$

$$t = \frac{\ln 2}{0.00002867} \approx 24{,}180 \text{ years.}$$

Carbon Dating

About 1950 the chemist Willard Libby devised a method of using radioactive carbon as a means of determining the approximate ages of fossils. The theory of **carbon dating** is based on the fact that the isotope carbon 14 is produced in the atmosphere by the action of cosmic radiation on nitrogen. The ratio of the amount of C-14 to ordinary carbon in the atmosphere appears to be a constant, and as a consequence the proportionate amount of the isotope present in all living organisms is the same as that in the atmosphere. When an organism dies, the absorption of C-14, by either breathing or eating, ceases. Thus, by comparing the proportionate amount of C-14 present, say, in a fossil with the constant ratio found in the atmosphere, it is possible to obtain a reasonable estimation of its age. The method is based upon the knowledge that the half-life of the radioactive C-14 is approximately 5600 years. For his work Libby won the Nobel Prize for chemistry in 1960. Libby's method has been used to date wooden furniture in Egyptian tombs and the woven flax wrappings of the Dead Sea scrolls.

EXAMPLE 3

A fossilized bone is found to contain 1/1000 the original amount of C-14. Determine the age of the fossil.

SOLUTION The starting point is again

$$A(t) = A_0 e^{kt}.$$

When $t = 5600$ years, $A(t) = A_0/2$, from which we can determine the value of k, as follows:

$$\frac{A_0}{2} = A_0 e^{5600k}$$

$$5600k = \ln\left(\frac{1}{2}\right) = -\ln 2$$

$$k = -\frac{\ln 2}{5600} = -0.00012378.$$

Therefore $$A(t) = A_0 e^{-0.00012378t}.$$

When $A(t) = A_0/1000$, we have

$$\frac{A_0}{1000} = A_0 e^{-0.00012378t}$$

so that $$-0.00012378t = \ln\left(\frac{1}{1000}\right) = -\ln 1000$$

$$t = \frac{\ln 1000}{0.00012378} \approx 55{,}800 \text{ years.}$$

Remark: The date found in the last example is really at the border of accuracy for this method. The usual carbon 14 technique is limited to about 9 half-lives of the isotope or about 50,000 years. One reason is that the chemical analysis needed to obtain an accurate measurement of the remaining C-14 becomes somewhat formidable around the point of $A_0/1000$. Also, this analysis demands the destruction of a rather large sample of the specimen. If this measurement is accomplished indirectly, based on the actual radioactivity of the specimen, then it is very difficult to distinguish between the radiation from the fossil and the normal background radiation. But in recent developments, the use of a particle accelerator has enabled scientists to separate the C-14 from the stable C-12 directly. By computing the precise value of the ratio of C-14 to C-12, the accuracy of this method can be extended to 70,000–100,000 years. Other isotopic techniques such as using potassium 40 and argon 40 can give dates of several million years. Nonisotopic methods based on the use of amino acids are also sometimes possible.

3.2.2 Cooling, Circuits, and Chemical Mixtures

Cooling

Newton's law of cooling states that the rate at which the temperature $T(t)$ changes in a cooling body is proportional to the difference between the temperature in the body and the constant temperature T_0 of the surrounding medium—that is,

$$\frac{dT}{dt} = k(T - T_0), \tag{3}$$

where k is a constant of proportionality.

EXAMPLE 4 When a cake is removed from a baking oven, its temperature is measured at 300°F. Three minutes later its temperature is 200°F. How long will it take to cool off to a room temperature of 70°F?

SOLUTION We must solve the initial-value problem

$$\frac{dT}{dt} = k(T - 70) \qquad T(0) = 300 \tag{4}$$

and determine the value of k so that $T(3) = 200$.

Equation (3) is both linear and separable. Separating variables yields

$$\frac{dT}{T - 70} = k\,dt$$

$$\ln|T - 70| = kt + c_1$$

$$T - 70 = c_2 e^{kt}$$

$$T = 70 + c_2 e^{kt}.$$

When $t = 0$, $T = 300$ so that $300 = 70 + c_2$ gives $c_2 = 230$ and, therefore, $T = 70 + 230e^{kt}$.

From $T(3) = 200$ we find

$$e^{3k} = \frac{13}{23} \qquad \text{or} \qquad k = \frac{1}{3}\ln\frac{13}{23} = \quad 0.19018.$$

Thus

$$T(t) = 70 + 230e^{-0.19018t}. \tag{5}$$

Unfortunately (5) furnishes no finite solution to $T(t) = 70$ since $\lim_{t\to\infty} T(t) = 70$. Yet intuitively we expect the cake will assume the room temperature after a reasonably long period of time. How long is long? Of course, we should not be the least bit disturbed by the fact that the model (4) does not quite live up to our physical intuition. Parts (a) and (b) of Figure 3.9 clearly show that the cake will be approximately at room temperature in about one-half hour.

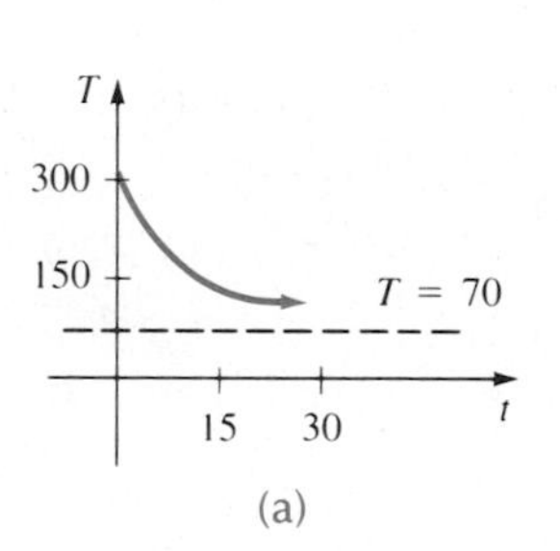

(a)

$T(t)$	t (minutes)
75°	20.1
74°	21.3
73°	22.8
72°	24.9
71°	28.6
70.5°	32.3

(b)

Figure 3.9

L-R Series Circuit

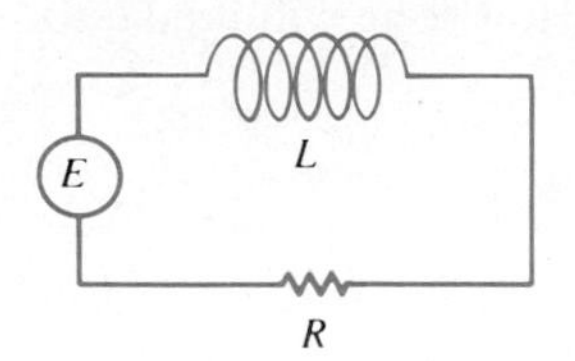

Figure 3.10

In a series circuit containing only a resistor and an inductor, Kirchhoff's second law states that the sum of the voltage drop across the inductor ($L(di/dt)$) and the voltage drop across the resistor (iR) is the same as the impressed voltage ($E(t)$) on the circuit (see Figure 3.10).

Thus we obtain the linear differential equation for the current $i(t)$,

$$L\frac{di}{dt} + Ri = E(t), \tag{6}$$

where L and R are constants known as the inductance and the resistance, respectively. The current $i(t)$ is sometimes called the **response** of the system.

R-C Series Circuit

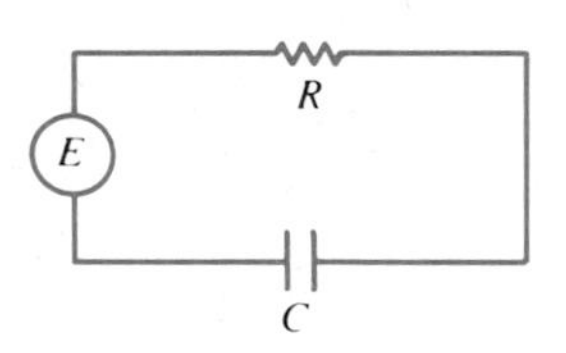

Figure 3.11

The voltage drop across a capacitor with capacitance C is given by $q(t)/C$, where q is the charge on the capacitor. Hence, for the series circuit shown in Figure 3.11, Kirchhoff's second law gives

$$Ri + \frac{1}{C}q = E(t). \tag{7}$$

But current i and charge q are related by $i = dq/dt$, so (7) becomes the linear differential equation

$$R\frac{dq}{dt} + \frac{1}{C}q = E(t). \tag{8}$$

EXAMPLE 5

A 12-volt battery is connected to a series circuit in which the inductance is 1/2 henry and the resistance is 10 ohms. Determine the current i if the initial current is zero.

SOLUTION From (6) we see that we must solve

$$\frac{1}{2}\frac{di}{dt} + 10i = 12$$

subject to $i(0) = 0$. First, multiply the differential equation by 2 and read off the integrating factor e^{20t}. We then obtain

$$\frac{d}{dt}[e^{20t}i] = 24e^{20t}$$

$$e^{20t}i = \frac{24}{20}e^{20t} + c$$

$$i = \frac{6}{5} + ce^{-20t}.$$

Now $i(0) = 0$ implies $0 = 6/5 + c$ or $c = -6/5$. Therefore, the response is

$$i(t) = \frac{6}{5} - \frac{6}{5}e^{-20t}.$$

Transient and Steady-State Terms

From (7) of Section 2.5 we can write down a general solution of (6),

$$i(t) = \frac{e^{-(R/L)t}}{L}\int e^{(R/L)t}E(t)\,dt + ce^{-(R/L)t}. \tag{9}$$

In particular, when $E(t) = E_0$ is a constant, (9) becomes

$$i(t) = \frac{E_0}{R} + ce^{-(R/L)t}. \tag{10}$$

Note that as $t \to \infty$, the second term in equation (10) approaches zero. Such a term is usually called a **transient term**; any remaining terms are called the **steady-state** part of the solution. In this case E_0/R is also called the **steady-state current**; for large values of time it then appears that the current in the circuit is simply governed by Ohm's law ($E = iR$).

Mixture Problem

The mixing of two fluids sometimes gives rise to a linear first-order differential equation. In the next example we consider the mixture of two salt solutions of different concentrations.

EXAMPLE 6

Initially 50 pounds of salt is dissolved in a large tank holding 300 gallons of water. A brine solution is pumped into the tank at a rate of 3 gallons per minute, and the well-stirred solution is then pumped out at the same rate (see Figure 3.12). If the concentration of the solution entering is 2 pounds per gallon, determine the amount of salt in the tank at any time. How much salt is present after 50 minutes? after a long time?

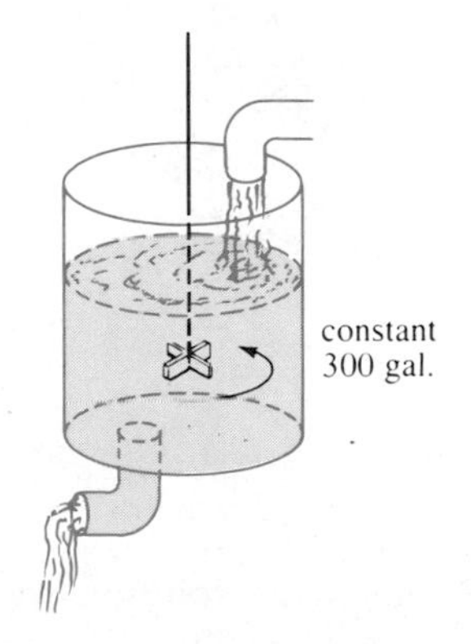

Figure 3.12

SOLUTION Let $A(t)$ be the amount of salt (in pounds) in the tank at any time. For problems of this sort, the net rate at which $A(t)$ changes is given by

$$\frac{dA}{dt} = \begin{pmatrix}\text{rate of}\\ \text{substance entering}\end{pmatrix} - \begin{pmatrix}\text{rate of}\\ \text{substance leaving}\end{pmatrix} = R_1 - R_2. \tag{11}$$

Now the rate at which the salt enters the tank is, in pounds per minute,

$$R_1 = (3\text{ gal/min})\cdot(2\text{ lb/gal}) = 6\text{ lb/min},$$

whereas the rate at which salt is leaving is

$$R_2 = (3\text{ gal/min})\cdot\left(\frac{A}{300}\text{ lb/gal}\right) = \frac{A}{100}\text{ lb/min}.$$

Thus equation (11) becomes

$$\frac{dA}{dt} = 6 - \frac{A}{100}, \tag{12}$$

which we solve subject to the initial condition $A(0) = 50$.

Since the integrating factor is $e^{t/100}$, we can write (12) as

$$\frac{d}{dt}[e^{t/100}A] = 6e^{t/100}$$

and therefore

$$e^{t/100}A = 600e^{t/100} + c$$

$$A = 600 + ce^{-t/100}. \tag{13}$$

When $t = 0$, $A = 50$, so we find that $c = -550$. Finally, we obtain

$$A(t) = 600 - 550e^{-t/100}. \tag{14}$$

At $t = 50$ we find $A(50) = 266.41$ lb. Also, as $t \to \infty$ it is seen from (14) and Figure 3.13 that $A \to 600$. Of course this is what we would expect; over a long period of time the number of pounds of salt in the solution must be

$$(300 \text{ gal})(2 \text{ lb/gal}) = 600 \text{ lb}.$$

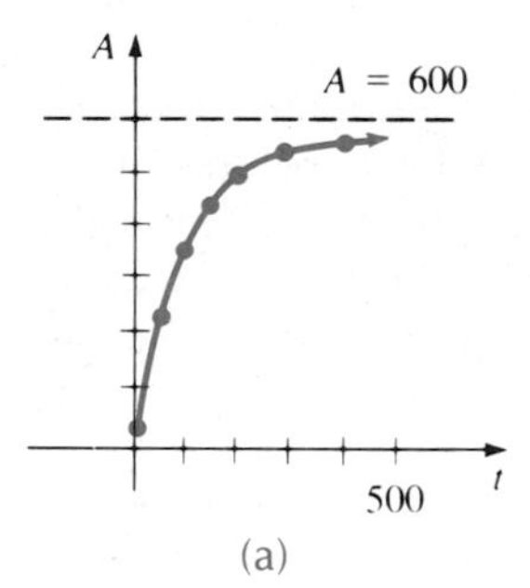

(a)

t(minutes)	A(lbs)
50	266.41
100	397.67
150	477.27
200	525.57
300	572.62
400	589.93

(b)

Figure 3.13

In Example 6 we assumed that the rate at which the solution was pumped in was the same as the rate at which the solution was pumped out. However, this need not be the case; the mixed brine solution could be pumped out at a rate faster or slower than the rate at which the other solution is pumped in. The resulting differential equation in this latter situation is linear with a variable coefficient.

EXAMPLE 7 If the well-stirred solution in Example 6 is pumped out at a slower rate of 2 gallons per minute, then the solution is *accumulating* at a rate of

$$(3 - 2) \text{ gal/min} = 1 \text{ gal/min}.$$

After t minutes there are $300 + t$ gallons of brine in the tank. The rate at which the salt is leaving is then

$$R_2 = (2 \text{ gal/min}) \cdot \left(\frac{A}{300 + t} \text{ lb/gal}\right) = \frac{2A}{300 + t} \text{ lb/min}.$$

Hence equation (11) becomes

$$\frac{dA}{dt} = 6 - \frac{2A}{300 + t} \quad \text{or} \quad \frac{dA}{dt} + \frac{2A}{300 + t} = 6.$$

Finding the integrating factor and solving the last equation give

$$A(t) = 2(300 + t) + c(300 + t)^{-2}.$$

The initial condition $A(0) = 50$ yields $c = -4.95 \times 10^7$ and so

$$A(t) = 2(300 + t) - (4.95 \times 10^7)(300 + t)^{-2}.$$

Exercises 3.2

Answers to odd-numbered problems begin on page A–7.

[3.2.1] 1. The population of a certain community is known to increase at a rate proportional to the number of people present at any time. If the population has doubled in 5 years, how long will it take to triple? to quadruple?

2. Suppose it is known that the population of the community in Problem 1 is 10,000 after 3 years. What was the initial population? What will be the population in 10 years?

3. The population of a town grows at a rate proportional to the population at any time. Its initial population of 500 increases by 15% in 10 years. What will be the population in 30 years?

4. Bacteria in a culture grow at a rate proportional to the number of bacteria present at any time. After 3 hours it is observed that there are 400 bacteria present. After 10 hours there are 2000 bacteria present. What is the initial number of bacteria?

5. The radioactive isotope of lead, Pb-209, decays at a rate proportional to the amount present at any time and has a half-life of 3.3 hours. If 1 gram of lead is present initially, how long will it take for 90% of the lead to decay?

6. Initially there were 100 milligrams of a radioactive substance present. After 6 hours the mass decreased by 3%. If the rate of decay is proportional to the amount of the substance present at any time, find the amount remaining after 24 hours.

7. Determine the half-life of the radioactive substance described in Problem 6.

8. Show that the half-life of a radioactive substance is, in general,

$$t = \frac{(t_2 - t_1)\ln 2}{\ln(A_1/A_2)},$$

where $A_1 = A(t_1)$ and $A_2 = A(t_2)$, $t_1 < t_2$.

9. When a vertical beam of light passes through a transparent substance, the rate which its intensity I decreases is proportional to $I(t)$, where t represents the thickness of the medium in feet. In clear seawater the intensity 3 feet below the surface is 25% of the initial intensity I_0 of the incident beam. What is the intensity of the beam 15 feet below the surface?

10. When interest is compounded continuously, the amount of money S increases at a rate proportional to the amount present at any time: $dS/dt = rS$ where r is the annual rate of interest (see (19) of Section 1.2).

(a) Find the amount of money accrued at the end of 5 years when $5000 is deposited in a savings account drawing $5\frac{3}{4}$% annual interest compounded continuously.

(b) In how many years will the initial sum deposited be doubled?

(c) Use a hand calculator to compare the number obtained in part (a) with the value

$$S = 5000\left(1 + \frac{0.0575}{4}\right)^{5(4)}.$$

This value represents the amount accrued when interest is compounded quarterly.

11. In a piece of burned wood, or charcoal, it was found that 85.5% of the C-14 had decayed. Use the information in Example 3 to determine the approximate age of the wood. (It is precisely this data that archaeologists used to date prehistoric paintings in a cave in Lascaux, France.)

[3.2.2] **12.** A thermometer is taken from an inside room to the outside where the air temperature is 5°F. After 1 minute the thermometer reads 55°F and after 5 minutes the reading is 30°F. What is the initial temperature of the room?

13. A thermometer is removed from a room where the air temperature is 70°F to the outside where the temperature is 10°F. After $\frac{1}{2}$ minute the thermometer reads 50°F. What is the reading at $t = 1$ minute? How long will it take for the thermometer to reach 15°F?

14. Formula (3) also obtains when an object absorbs heat from the surrounding medium. If a small metal bar whose initial temperature is 20°C is dropped into a container of boiling water, how long will it take for the bar to reach 90°C if it is known that its temperature increased 2° in 1 second? How long will it take the bar to reach 98°C?

15. A 30-volt electromotive force is applied to an L-R series circuit in which the inductance is 0.1 henry and the resistance is 50 ohms. Find the current $i(t)$ if $i(0) = 0$. Determine the current as $t \to \infty$.

16. Solve equation (6) under the assumption that $E(t) = E_0 \sin \omega t$ and $i(0) = i_0$.

17. A 100-volt electromotive force is applied to an R-C series circuit in which the resistance is 200 ohms and the capacitance is 10^{-4} farad. Find the charge $q(t)$ on the capacitor if $q(0) = 0$. Find the current $i(t)$.

18. A 200-volt electromotive force is applied to an R-C series circuit in which the resistance is 1000 ohms and the capacitance is 5×10^{-6} farad. Find the charge $q(t)$ on the capacitor if $i(0) = 0.4$. Determine the charge and current at $t = 0.005$ second. Determine the charge as $t \to \infty$.

19. An electromotive force

$$E(t) = \begin{cases} 120, & 0 \le t \le 20 \\ 0, & t > 20 \end{cases}$$

is applied to an L-R series circuit in which the inductance is 20 henry and the resistance is 2 ohms. Find the current $i(t)$ if $i(0) = 0$.

20. Suppose an R-C series circuit has a variable resistor. If the resistance at any time t is given by $R = k_1 + k_2 t$, where $k_1 > 0$ and $k_2 > 0$ are known constants, then (8) becomes

$$(k_1 + k_2 t)\frac{dq}{dt} + \frac{1}{C}q = E(t).$$

Show that if $E(t) = E_0$ and $q(0) = q_0$, then

$$q(t) = E_0 C + (q_0 - E_0 C)\left(\frac{k_1}{k_1 + k_2 t}\right)^{1/Ck_2}.$$

21. A tank contains 200 liters of fluid in which 30 g of salt is dissolved. Brine containing 1 g of salt per liter is then pumped into the tank at a rate of 4 liters per minute; the well-mixed solution is pumped out at the same rate. Find the number of grams of salt $A(t)$ in the tank at any time.

22. Solve Problem 21 assuming pure water is pumped into the tank.

23. A large tank is filled with 500 gallons of pure water. Brine containing 2 lb of salt per gallon is pumped into the tank at a rate of 5 gallons per minute. The well-mixed solution is pumped out at the same rate. Find the number of pounds of salt $A(t)$ in the tank at any time.

24. Solve Problem 23 under the assumption that the solution is pumped out at a faster rate of 10 gallons per minute. When is the tank empty?

25. A large tank is partially filled with 100 gallons of fluid in which 10 lb of salt is dissolved. Brine containing $\frac{1}{2}$ lb of salt per gallon is pumped into the tank at a rate of 6 gallons per minute. The well-mixed solution is then pumped out at a slower rate of 4 gallons per minute. Find the number of pounds of salt in the tank after 30 minutes.

26. Beer containing 6% alcohol per gallon is pumped into a vat that initially contains 400 gallons of beer at 3% alcohol. The rate at which the beer is pumped in is 3 gallons per minute, whereas the mixed liquid is pumped out at a rate of 4 gallons per minute. Find the number of gallons of alcohol $A(t)$ in the tank at any time. What is the percentage of alcohol in the tank after 60 minutes? When is the tank empty?

Miscellaneous Problems

27. The differential equation governing the velocity v of a falling weight w subjected to air resistance proportional to the instantaneous velocity is

$$m\frac{dv}{dt} = mg - kv,$$

where k is a positive constant of proportionality. Solve the equation subject to the initial condition $v(0) = v_0$ and determine the limiting velocity of the weight. If distance s is related to velocity $ds/dt = v$, find an explicit expression for s if it is further known that $s(0) = s_0$.

28. The rate at which a drug disseminates into the bloodstream is governed by the differential equation

$$\frac{dX}{dt} = A - BX,$$

where A and B are positive constants. The function $X(t)$ describes the concentration of the drug in the bloodstream at any time t. Find the limiting value of X as $t \to \infty$. At what time is the concentration one-half this limiting value? Assume that $X(0) = 0$.

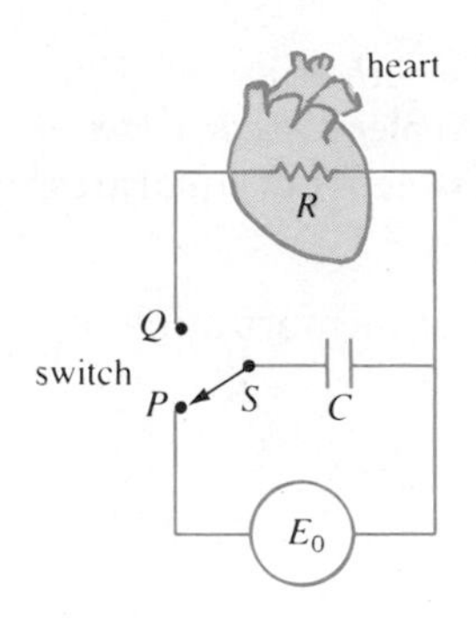

Figure 3.14

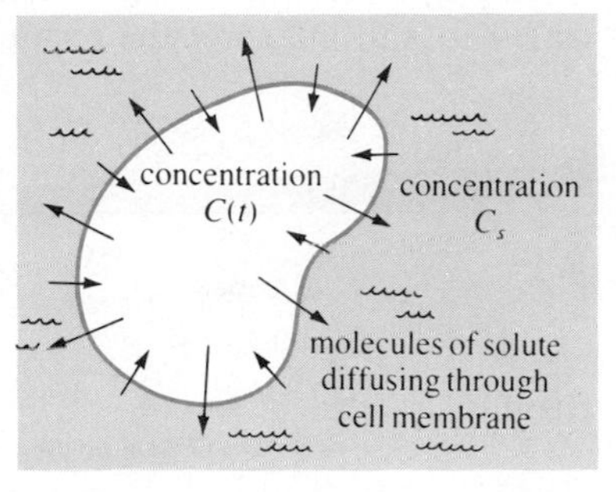

Figure 3.15

29. A heart pacemaker, as shown in Figure 3.14, consists of a battery, a capacitor, and the heart as a resistor. When the switch S is at P the capacitor charges; when S is at Q the capacitor discharges, sending an electrical stimulus to the heart. During this time the voltage E applied to the heart is given by

$$\frac{dE}{dt} = -\frac{1}{RC}E, \quad t_1 < t < t_2,$$

where R and C are constants. Determine $E(t)$ if $E(t_1) = E_0$. (Of course, the opening and closing of the switch is periodic in time, to simulate the natural heartbeat.)

30. Suppose a cell is suspended in a solution containing a solute of constant concentration C_s. Suppose further that the cell has constant volume V and that the area of its permeable membrane is the constant A. By Fick's law the rate of change of its mass m is directly proportional to the area A and the difference $C_s - C(t)$, where $C(t)$ is the concentration of the solute inside the cell at any time t. Find $C(t)$ if $m = VC(t)$ and $C(0) = C_0$ (see Figure 3.15).

31. In one model of the changing population $P(t)$ of a community, it is assumed that

$$\frac{dP}{dt} = \frac{dB}{dt} - \frac{dD}{dt},$$

where dB/dt and dD/dt are the birth and death rates, respectively.

(a) Solve for $P(t)$ if

$$\frac{dB}{dt} = k_1 P \quad \text{and} \quad \frac{dD}{dt} = k_2 P.$$

(b) Analyze the cases $k_1 > k_2$, $k_1 = k_2$, and $k_1 < k_2$.

32. The differential equation

$$\frac{dP}{dt} = (k \cos t)P,$$

where k is a positive constant, is often used as a model of a population that undergoes yearly seasonal fluctuations. Solve for $P(t)$ and graph the solution. Assume $P(0) = P_0$.

33. In polar coordinates the angular momentum of a moving body of mass m is defined to be $L = mr^2 \dfrac{d\theta}{dt}$. Assume that the coordinates of the body are (r_1, θ_1) and (r_2, θ_2) at times $t = a$ and $t = b$, $a < b$, respectively. If L is

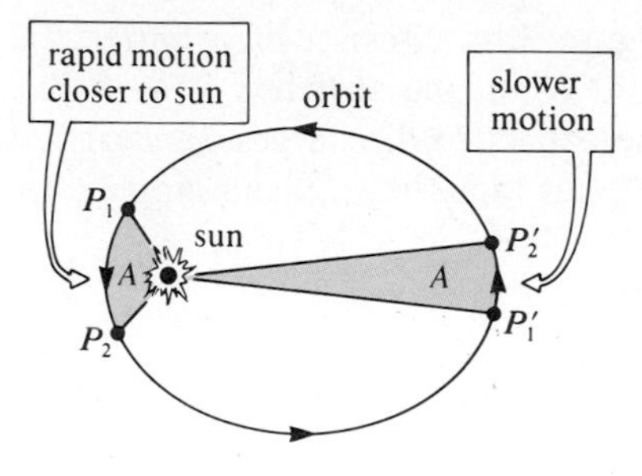

Figure 3.16

constant, show that the area A swept out by r is $A = L(b - a)/2m$. When the sun is taken to be at the origin, this proves **Kepler's second law of planetary motion**: The radius vector joining the sun sweeps out equal areas in equal intervals of time (see Figure 3.16).

34. When forgetfulness is taken into account, the rate of memorization of a subject is given by

$$\frac{dA}{dt} = k_1(M - A) - k_2A,$$

where $k_1 > 0$, $k_2 > 0$, $A(t)$ is the amount of material memorized in time t, M is the total amount to be memorized, and $M - A$ is the amount remaining to be memorized. Solve for $A(t)$ and graph the solution. Assume $A(0) = 0$. Find the limiting value of A as $t \to \infty$ and interpret the result (see (28) of Section 1.2).

3.3 Applications of Nonlinear Equations

We have seen that if a population P is described by

$$\frac{dP}{dt} = kP, \qquad k > 0, \tag{1}$$

then $P(t)$ exhibits unbounded exponential growth. In many instances this differential equation provides an unrealistic model of the growth of a population; that is, what is actually observed differs substantially from what is predicted.

Around 1840 the Belgian mathematician-biologist P. F. Verhulst was concerned with mathematical formulations for predicting the human populations of various countries. One of the equations he studied was

$$\frac{dP}{dt} = P(a - bP), \tag{2}$$

where a and b are positive constants. Equation (2) came to be known as the **logistic equation** and its solution is called the **logistic function** (the graph of which is naturally called a logistic curve).

Equation (1) does not provide a very accurate model for population growth when the population itself is very large. Overcrowded conditions with the resulting detrimental effects on the environment, such as pollution and excessive and competitive demands for food and fuel, can have an inhibitive effect on the population growth. If a, $a > 0$, is a constant average birth rate, let us assume that the average death rate is proportional to the population $P(t)$ at any time. Thus, if $(1/P)\ (dP/dt)$ is the rate of growth per individual in a population, then

$$\frac{1}{P}\frac{dP}{dt} = \begin{pmatrix}\text{average}\\ \text{birth rate}\end{pmatrix} - \begin{pmatrix}\text{average}\\ \text{death rate}\end{pmatrix} = a - bP, \tag{3}$$

where b is a positive constant of proportionality. Cross multiplying (3) by P immediately gives (2).

As we shall now see, the solution of (2) is bounded as $t \to \infty$. If we rewrite (2) as $dP/dt = aP - bP^2$, the term $-bP^2$, $b > 0$, can be interpreted as an "inhibition" or "competition" term. Also, in most applications, the positive constant a is much larger than the constant b.

Logistic curves have proved to be quite accurate in predicting the growth patterns, in a limited space, of certain types of bacteria, protozoa, water fleas (*Daphnia*), and fruit flies (*Drosophila.*). We have already seen equation (2) in the form $dx/dt = kx(n + 1 - x)$, $k > 0$. This differential equation provides a reasonable model for describing the spread of an epidemic brought about by initially introducing an infected individual into a static population. The solution $x(t)$ represents the number of individuals infected with the disease at any time (see Example 14 of Section 1.2). Sociologists, and even business analysts, have borrowed this latter model to study the spread of information and the impact of advertising in certain centers of population.

Solution

One method for solving equation (2) is separation of variables.* By using partial fractions, we can write

$$\frac{dP}{P(a - bP)} = dt$$

$$\left[\frac{1/a}{P} + \frac{b/a}{a - bP}\right] dP = dt$$

$$\frac{1}{a}\ln|P| - \frac{1}{a}\ln|a - bP| = t + c$$

$$\ln\left|\frac{P}{a - bP}\right| = at + ac$$

$$\frac{P}{a - bP} = c_1 e^{at}. \tag{4}$$

* In the form

$$\frac{dP}{dt} - aP = -bP^2$$

you might recognize the logistic equation as a special case of Bernoulli's equation (see Section 2.6).

It follows from the last equation that

$$P(t) = \frac{ac_1e^{at}}{1 + bc_1e^{at}} = \frac{ac_1}{bc_1 + e^{-at}}. \tag{5}$$

Now if we are given the initial condition $P(0) = P_0$, $P_0 \neq a/b$,* equation (4) implies $c_1 = P_0/(a - bP_0)$ and therefore

$$P(t) = \frac{aP_0/(a - bP_0)}{[bP_0/(a - bP_0)] + e^{-at}}$$

or

$$P(t) = \frac{aP_0}{bP_0 + (a - bP_0)e^{-at}}. \tag{6}$$

Graphs of $P(t)$

The basic shape of the graph of the logistic function $P(t)$ can be obtained without too much effort. Although the variable t usually represents time and we are seldom concerned with applications in which $t < 0$, it is nonetheless of some interest to include this interval when displaying the various graphs of P. From (6) we see that, for $t > 0$,

$$P(t) \to \frac{aP_0}{bP_0} = \frac{a}{b} \qquad \text{as } t \to \infty;$$

and for $t < 0$,

$$P(t) \to 0 \qquad \text{as } t \to -\infty.$$

Now differentiating (2) by the product rule gives

$$\begin{aligned}
\frac{d^2P}{dt^2} &= P\left(-b\frac{dP}{dt}\right) + (a - bP)\frac{dP}{dt} \\
&= \frac{dP}{dt}(a - 2bP) \\
&= P(a - bP)(a - 2bP) \\
&= 2b^2P\left(P - \frac{a}{b}\right)\left(P - \frac{a}{2b}\right).
\end{aligned} \tag{7}$$

From calculus recall that the points where $d^2P/dt^2 = 0$ are possible points of inflection, but $P = 0$ and $P = a/b$ can obviously be ruled out. Hence $P = a/2b$ is the only possible ordinate value at which the concavity of the graph can change. For $0 < P < a/2b$ it follows from (7) that $P'' > 0$, and $a/2b < P < a/b$ implies $P'' < 0$. Thus, reading from left to right, the graph changes from concave up to concave down at the point corresponding to $P = a/2b$. When

* Notice that $P = a/b$ is a singular solution of equation (2).

(a)

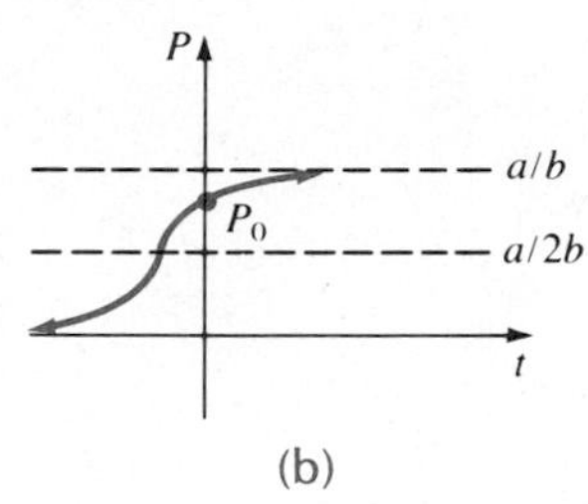

(b)

Figure 3.17

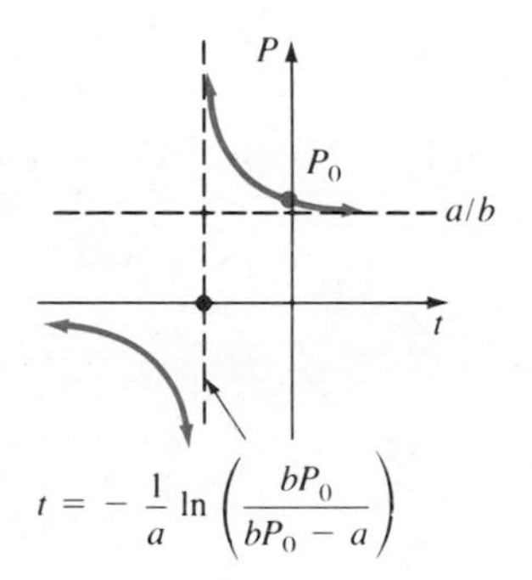

$t = -\frac{1}{a} \ln\left(\frac{bP_0}{bP_0 - a}\right)$

Figure 3.18

the initial value satisfies $0 < P_0 < a/2b$, the graph of $P(t)$ assumes the shape of an S, as we see in Figure 3.17(a). For $a/2b < P_0 < a/b$ the graph is still S-shaped but the point of inflection occurs at a negative value of t, as shown in Figure 3.17(b).

If $P_0 > a/b$, equation (7) shows $P'' > 0$ for all t in the domain of $P(t)$ for which $P > 0$. When $P < 0$, equation (7) implies $P'' < 0$. However, $P = 0$ is not a point of inflection since, whenever $a - bP_0 < 0$, an inspection of (6) reveals a vertical asymptote at

$$t = -\frac{1}{a} \ln\left(\frac{bP_0}{bP_0 - a}\right).$$

The graph of $P(t)$ in this case is given in Figure 3.18.

EXAMPLE 1

Suppose a student carrying a flu virus returns to an isolated college campus of 1000 students. If it is assumed that the rate at which the virus spreads is proportional not only to the number x of infected students but also to the number of students not infected, determine the number of infected students after 6 days if it is further observed that after 4 days $x(4) = 50$.

SOLUTION Assuming that no one leaves the campus throughout the duration of the disease, we must then solve the initial-value problem

$$\frac{dx}{dt} = kx(1000 - x)$$

$$x(0) = 1.$$

By making the identifications $a = 1000k$ and $b = k$, we have immediately from (6) that

$$x(t) = \frac{1000k}{k + 999ke^{-1000kt}}$$

$$= \frac{1000}{1 + 999e^{-1000kt}}. \tag{8}$$

Now, using the information $x(4) = 50$, we determine k from

$$50 = \frac{1000}{1 + 999e^{-4000k}}.$$

We find $$k = \frac{-1}{4000} \ln \frac{19}{999} = -0.0009906.$$

Thus (8) becomes $$x(t) = \frac{1000}{1 + 999e^{-0.9906t}}.$$

Finally $$x(6) = \frac{1000}{1 + 999e^{-5.9436}} = 276 \text{ students}.$$

Additional calculated values of $x(t)$ are given in the table in Figure 3.19.

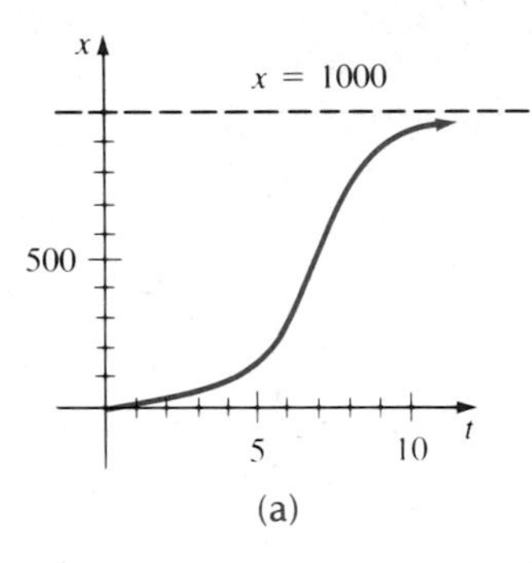

(a)

t days	x (number infected)
4	50 (observed)
5	124
6	276
7	507
8	735
9	882
10	953

(b)

Figure 3.19

Gompertz Curves

A modification of the logistic equation is

$$\frac{dP}{dt} = P(a - b \ln P), \tag{9}$$

where a and b are constants. It is readily shown by separation of variables (see Problem 5) that a solution of (9) is

$$P(t) = e^{a/b}e^{-ce^{-bt}}, \tag{10}$$

where c is an arbitrary constant. We note that when $b > 0$, $P \to e^{a/b}$ as $t \to \infty$, whereas for $b < 0$, $c > 0$, $P \to 0$ as $t \to \infty$. The graph of the function (10), called a **Gompertz curve**, is quite similar to the graph of the logistic function. Figure 3.20 shows two possibilities for the graph of $P(t)$.

Functions such as (10) are encountered, for example, in studies of the growth or decline of certain populations; in actuarial predictions; and in the study of growth of revenue in the sale of a commercial product.

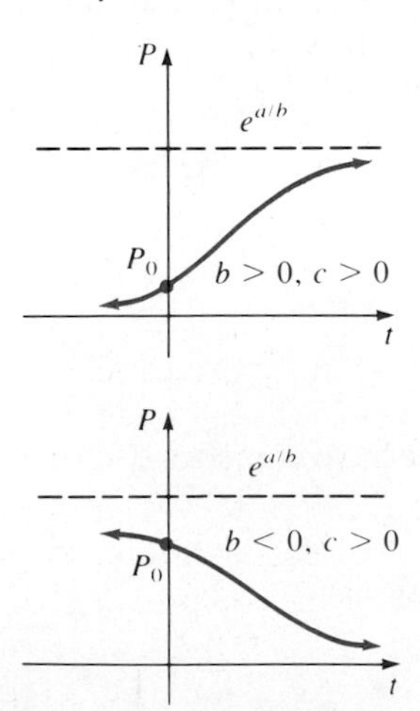

Figure 3.20

Chemical Reactions

The disintegration of a radioactive substance, governed by equation (1) of the preceding section, is said to be a **first-order reaction**. In chemistry a few reactions follow the same empirical law: If the molecules of a substance A decompose into smaller molecules, it is a natural assumption to suppose that the rate at which this decomposition takes place is proportional to the amount of the first substance that has not undergone conversion—that is, if $X(t)$ is the amount of substance A remaining at any time, then

$$\frac{dX}{dt} = kX,$$

where k is negative, since X is decreasing. An example of a first-order chemical reaction is the conversion of t-butyl chloride into t-butyl alcohol

$$(CH_3)_3CCl + NaOH \rightarrow (CH_3)_3COH + NaCl.$$

Only the concentration of the t-butyl chloride controls the rate of reaction.

Now in the reaction

$$CH_3Cl + NaOH \rightarrow CH_3OH + NaCl,$$

for every molecule of methyl chloride one molecule of sodium hydroxide is consumed, thus forming one molecule of methyl alcohol and one molecule of sodium chloride. In this case the rate at which the reaction proceeds is proportional to the product of the remaining concentrations of CH_3Cl and of NaOH. If X denotes the amount of CH_3OH formed and α and β are the given amounts of the first two chemicals A and B, then the instantaneous amounts not converted to chemical C are $\alpha - X$ and $\beta - X$, respectively. Hence the rate of formation of C is given by

$$\frac{dX}{dt} = k(\alpha - X)(\beta - X), \tag{11}$$

where k is a constant of proportionality. A reaction described by equation (11) is said to be of **second-order**.

EXAMPLE 2

A compound C is formed when two chemicals A and B are combined. The resulting reaction between the two chemicals is such that for each gram of A, 4 grams of B are used. It is observed that 30 grams of the compound C are formed in 10 minutes. Determine the amount of C at any time if the rate of the reaction is proportional to the amounts of A and B remaining and if initially there are 50 grams of A and 32 grams of B. How much of the compound C is present at 15 minutes? Interpret the solution as $t \rightarrow \infty$.

SOLUTION Let $X(t)$ denote the number of grams of the compound C present at any time. Clearly $X(0) = 0$ and $X(10) = 30$.

Now for example, if there are 2 grams of compound C, we must have used, say, a grams of A and b grams of B so that

$$a + b = 2 \qquad \text{and} \qquad b = 4a.$$

Thus we must use $a = 2/5 = 2(1/5)$ grams of chemical A and $b = 8/5 = 2(4/5)$ grams of B. In general, for X grams of C we must use

$$\frac{X}{5} \text{ grams of } A \qquad \text{and} \qquad \frac{4}{5}X \text{ grams of } B.$$

The amounts of A and B remaining at any time are then

$$50 - \frac{X}{5} \qquad \text{and} \qquad 32 - \frac{4}{5}X,$$

respectively.

Now we know that the rate at which chemical C is formed satisfies

$$\frac{dX}{dt} \propto \left(50 - \frac{X}{5}\right)\left(32 - \frac{4}{5}X\right).$$

To simplify the subsequent algebra, we factor 1/5 from the first term and 4/5 from the second, and then introduce the constant of proportionality,

$$\frac{dX}{dt} = k(250 - X)(40 - X).$$

By separation of variables and partial fractions, we can write

$$\frac{dX}{(250 - X)(40 - X)} = k\,dt$$

$$-\frac{1/210}{250 - X}\,dX + \frac{1/210}{40 - X}\,dX = k\,dt$$

$$\ln\left|\frac{250 - X}{40 - X}\right| = 210kt + c_1$$

$$\frac{250 - X}{40 - X} = c_2 e^{210kt}. \tag{12}$$

When $t = 0$, $X = 0$, so it follows at this point that $c_2 = 25/4$. Using $X = 30$ at $t = 10$, we find

$$210k = \frac{1}{10}\ln\frac{88}{25} = 0.1258.$$

Using this information, we solve (12) for X

$$X(t) = 1000\,\frac{1 - e^{-0.1258t}}{25 - 4e^{-0.1258t}}. \tag{13}$$

The behavior of X as a function of time is displayed in Figure 3.21. It is clear from the accompanying table and equation (13) that $X \to 40$ as $t \to \infty$. This means there are 40 grams of compound C formed, leaving

$$50 - \tfrac{1}{5}(40) = 42 \text{ grams of chemical } A$$

and

$$32 - \tfrac{4}{5}(40) = 0 \text{ grams of chemical } B.$$

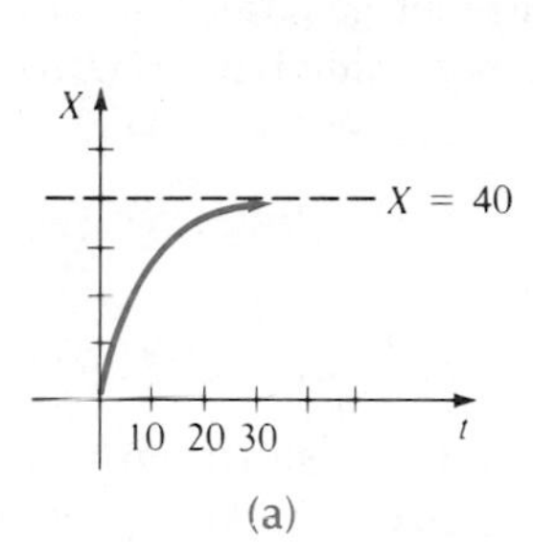

(a)

t (minutes)	X (grams)
10	30 (measured)
15	34.78
20	37.25
25	38.54
30	39.22
35	39.59

(b)

Figure 3.21

Law of Mass Action

The preceding example can be generalized in the following manner. Suppose that a grams of substance A are combined with b grams of substance B. If there are M parts of A and N parts of B formed in the compound, then the amounts of substances A and B remaining at any time are, respectively,

$$a - \frac{M}{M+N}X \quad \text{and} \quad b - \frac{N}{M+N}X.$$

Thus

$$\frac{dX}{dt} \propto \left[a - \frac{M}{M+N}X\right]\left[b - \frac{N}{M+N}X\right]. \tag{14}$$

Proceeding as before, if we factor out $M/(M+N)$ from the first term and $N/(M+N)$ from the second term, the resulting differential equation is the same as (11)

$$\frac{dX}{dt} = k(\alpha - X)(\beta - X), \tag{15}$$

where

$$\alpha = \frac{a(M+N)}{M} \quad \text{and} \quad \beta = \frac{b(M+N)}{N}.$$

Chemists refer to reactions described by equation (15) as the **law of mass action**.

When $\alpha \neq \beta$, it is readily shown (see Problem 9) that a solution of (15) is

$$\frac{1}{\alpha - \beta} \ln \left| \frac{\alpha - X}{\beta - X} \right| = kt + c. \tag{16}$$

Assuming the natural initial condition $X(0) = 0$, equation (16) yields the explicit solution

$$X(t) = \frac{\alpha\beta[1 - e^{(\alpha - \beta)kt}]}{\beta - \alpha e^{(\alpha - \beta)kt}}. \tag{17}$$

Without loss of generality we assume in (17) that $\beta > \alpha$ or $\alpha - \beta < 0$. Since $X(t)$ is an increasing function, we expect $k > 0$, and so it follows immediately from (17) that $X \to \alpha$ as $t \to \infty$.

Exercises 3.3

Answers to odd-numbered problems begin on page A–7.

1. The number of supermarkets $C(t)$ throughout the country that are using a computerized checkout system is described by the initial-value problem

$$\frac{dC}{dt} = C(1 - 0.0005C), \qquad t > 0$$

$$C(0) = 1.$$

How many supermarkets are using the computerized method when $t = 10$? How many companies are estimated to adopt the new procedure over a long period of time?

2. The number of people $N(t)$ in a community who are exposed to a particular advertisement is governed by the logistic equation. Initially $N(0) = 500$, and it is observed that $N(1) = 1000$. If it is predicted that the limiting number of people in the community who will see the advertisement is 50,000, determine $N(t)$ at any time.

3. The population $P(t)$ at any time in a suburb of a large city is governed by the initial-value problem

$$\frac{dP}{dt} = P(10^{-1} - 10^{-7} P)$$

$$P(0) = 5000,$$

where t is measured in months. What is the limiting value of the population? At what time will the population be equal to one-half of this limiting value?

4. Find a solution of the **modified logistic equation**

$$\frac{dP}{dt} = P(a - bP)(1 - cP^{-1}), \qquad a, b, c > 0.$$

5. **(a)** Solve equation (9):

$$\frac{dP}{dt} = P(a - b \ln P).$$

(b) Determine the value of c in equation (10) if $P(0) = P_0$.

6. Assuming $0 < P_0 < e^{a/b}$, and $a > 0$, use equation (9) to find the ordinate of the point of inflection for a Gompertz curve.

7. Two chemicals A and B are combined to form a chemical C. The rate or velocity of the reaction is proportional to the product of the instantaneous amounts of A and B not converted to chemical C. Initially there are 40 grams of A and 50 grams of B, and for each gram of B, 2 grams of A are used. It is observed that 10 grams of C are formed in 5 minutes. How much is formed in 20 minutes? What is the limiting amount of C after a long time? How much of chemicals A and B remains after a long time?

8. Solve the preceding problem if 100 grams of chemical A are present initially. At what time is chemical C half-formed?

9. Obtain a solution of the equation

$$\frac{dX}{dt} = k(\alpha - X)(\beta - X)$$

governing second-order reactions in the two cases $\alpha \neq \beta$ and $\alpha = \beta$.

10. In a third-order chemical reaction the number of grams X of a compound obtained by combining three chemicals is governed by

$$\frac{dX}{dt} = k(\alpha - X)(\beta - X)(\gamma - X).$$

Solve the equation under the assumption $\alpha \neq \beta \neq \gamma$.

EXAMPLE 3

A rocket is shot vertically upward from the ground with an initial velocity v_0 (see Figure 3.22). If the positive direction is taken to be upward, the student might recall from Section 1.2 (Problem 43) that in the absence of air resistance, the differential equation of motion after fuel burnout is

$$\frac{d^2y}{dt^2} = -\frac{k}{y^2},$$

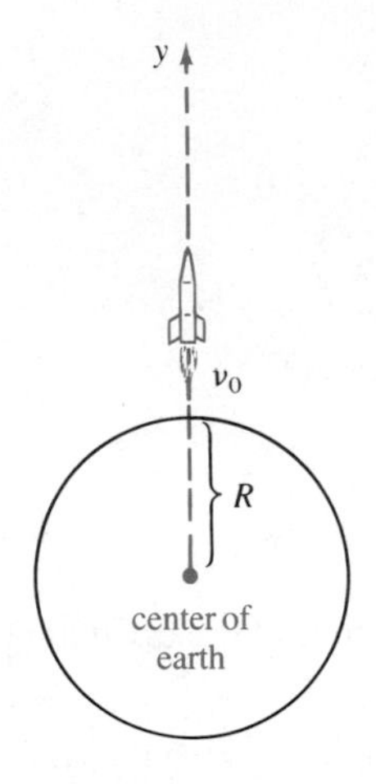

Figure 3.22

where k is a positive constant. Although this is not a first-order equation, we note that if we write the acceleration as

$$\frac{d^2y}{dt^2} = \frac{dv}{dt} = \frac{dv}{dy}\frac{dy}{dt} = v\frac{dv}{dy},$$

then the given equation becomes first-order in v, that is,

$$v\frac{dv}{dy} = -\frac{k}{y^2}.$$

Separating variables and integrating then gives

$$\frac{v^2}{2} = \frac{k}{y} + c.$$

Since $v = v_0$ at $y = R$, we obtain

$$\frac{v^2}{2} = \frac{k}{y} - \frac{k}{R} + \frac{v_0^2}{2}.$$

11. The reader might object that in the preceding example we really have not solved the original equation for y. Actually the solution gives quite a bit of information. If $k = gR^2$, show that the "escape velocity" for a rocket is $v_0 = 25{,}000$ mi/hr. Use a hand calculator and the value $R = 4000$ miles.

12. In the discussion of Section 1.2, we saw that the differential equation describing the shape of a wire of constant density w hanging under its own weight is

$$\frac{d^2y}{dx^2} = \frac{w}{T_1}\sqrt{1 + \left(\frac{dy}{dx}\right)^2},$$

where T_1 is the horizontal tension in the wire at its lowest point. Using the substitution $p = dy/dx$, solve this equation subject to the initial conditions

$$y(0) = 1, \qquad \left.\frac{dy}{dx}\right|_{x=0} = 0.$$

13. An equation similar to that given in the preceding problem is

$$x\frac{d^2y}{dx^2} = \frac{v_1}{v_2}\sqrt{1 + \left(\frac{dy}{dx}\right)^2}.$$

In this case the equation arises in the study of the shape of the path that a pursuer, traveling at a speed v_2, must take in order to intercept a prey traveling at speed v_1. Use the same substitution as in Problem 12 and the initial conditions

$$y(1) = 0, \qquad \left.\frac{dy}{dx}\right|_{x=1} = 0$$

to solve the equation. Consider the two cases $v_1 = v_2$ and $v_1 \neq v_2$.

14. According to **Stefan's law** of radiation, the rate of change of temperature from a body at absolute temperature T is

$$\frac{dT}{dt} = k(T^4 - T_0^4),$$

where T_0 is the absolute temperature of the surrounding medium. Find a solution of this differential equation. It can be shown that when $T - T_0$ is small compared to T_0, this particular equation is closely approximated by Newton's law of cooling (Equation (3), Section 3.2).

15. The height h of water that is flowing through an orifice at the bottom of a cylindrical tank is given by

$$\frac{dh}{dt} = -\frac{A_2}{A_1}\sqrt{2gh}, \qquad g = 32 \text{ ft/sec}^2,$$

where A_1 and A_2 are the cross-sectional areas of the tank and orifice, respectively (see Problem 34, Exercise 1.2). Solve the equation if the initial height of the water is 20 ft and $A_1 = 50 \text{ ft}^2$ and $A_2 = 1/4 \text{ ft}^2$. At what time is the tank empty?

16. The nonlinear differential equation

$$\left(\frac{dr}{dt}\right)^2 = \frac{2\mu}{r} + 2h,$$

where μ and h are nonnegative constants, arises in the study of the two-body problem of celestial mechanics. Here the variable r represents the distance between the two masses. Solve the equation in the two cases $h = 0$ and $h > 0$.

17. Solve the differential equation of the **tractrix**

$$\frac{dy}{dx} = -\frac{y}{\sqrt{s^2 - y^2}}$$

(see Problem 35, Exercise 1.2). Assume that the initial point on the y-axis is (0, 10) and the length of rope is $s = 10$ ft.

18. A body of mass m falling through a viscous medium encounters a resisting force proportional to the square of its instantaneous velocity. In this

situation the differential equation for the velocity $v(t)$ at any time is

$$m\frac{dv}{dt} = mg - kv^2,$$

where k is a positive constant of proportionality. Solve the equation subject to $v(0) = v_0$. What is the limiting velocity of the falling body?

19. The differential equation

$$x\left(\frac{dx}{dy}\right)^2 + 2y\frac{dx}{dy} = x,$$

where $x = x(y)$, occurs in the study of optics. The equation describes the type of plane curve that will reflect all incoming light rays to the same point (see Problem 41, Exercise 1.2). Show that the curve must be a parabola. (*Hint:* Use the substitution $w = x^2$ and then re-examine Section 2.6.]

20. Solve the equation of Problem 19 with the aid of the quadratic formula.

21. The equations of Lotka and Volterra*

$$\frac{dy}{dt} = y(\alpha - \beta x)$$

$$\frac{dx}{dt} = x(-\gamma + \delta y),$$

where α, β, γ, and δ are positive constants, occur in the analysis of the biological balance of two species of animals such as predators and prey (for example, foxes and rabbits.) Here $x(t)$ and $y(t)$ denote the populations of the two species at any time. Although no explicit solutions of the system

* **A.J. Lotka** (1880–1949) Lotka, born in Austria, was an American biomathematician.

Vito Volterra (1860–1940) Born in Ancona, Italy, Vito Volterra showed an early aptitude for mathematics. He studied calculus on his own initiative and investigated problems in gravitation at the age of twelve. Although his education was a constant financial struggle, Volterra quickly attained prominence as a scientist and mathematician. He was also an active politician and was appointed Senator of the Kingdom of Italy in 1905. Volterra became interested in the applications of mathematics to ecology in the mid-1920s and formulated this system of differential equations in an attempt to explain the variations in the fish population in the Mediterranean as a result of predator-prey interactions. (Lotka, working independently, arrived at the same system equations and published the result in 1925 in his text *Elements of Physical Biology*.) Through his research into mathematical models of population, Volterra established the groundwork for a field of mathematics known as integral equations. A man of principle, Volterra refused to sign a loyalty oath to the fascist regime of Benito Mussolini and eventually resigned his chair of mathematics at the University of Rome and all his memberships in Italian scientific societies.

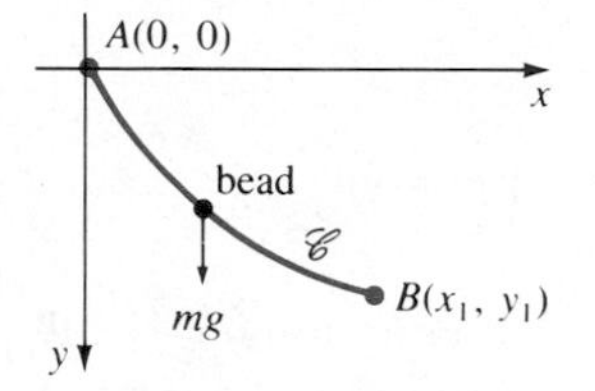

Figure 3.23

exist, solutions can be found relating the two populations at any time. Divide the first equation by the second and solve the resulting nonlinear first-order differential equation.

22. A classical problem in the calculus of variations is to find the shape of a curve $\mathscr{C}$ (see Figure 3.23) such that a bead, under the influence of gravity, will slide from $A(0, 0)$ to $B(x_1, y_1)$ in the least time. It can be shown that the differential equation for the shape of the path is $y[1 + (y')^2] = k$, where k is a constant. First solve for dx in terms of y and dy, and then use the substitution $y = k \sin^2\theta$ to obtain the parametric form of the solution. The curve $\mathscr{C}$ turns out to be a cycloid.

23. The initial-value problem describing the motion of a simple pendulum released from rest from an angle $\theta_0 > 0$ is

$$\frac{d^2\theta}{dt^2} + \frac{g}{l}\sin\theta = 0$$

$$\theta(0) = \theta_0, \qquad \left.\frac{d\theta}{dt}\right|_{t=0} = 0.$$

(a) Obtain the first-order equation

$$\left(\frac{d\theta}{dt}\right)^2 = \frac{2g}{l}(\cos\theta - \cos\theta_0).$$

[*Hint:* Multiply the given equation by $2\, d\theta/dt$.]

(b) Use the equation in part (a) to show that the period of motion is

$$T = 2\sqrt{\frac{2l}{g}}\int_0^{\theta_0} \frac{d\theta}{\sqrt{\cos\theta - \cos\theta_0}}.$$

CHAPTER 3 SUMMARY

If every curve in a one-parameter family of curves $G(x, y, c_1) = 0$ is orthogonal to every curve in a second one-parameter family $H(x, y, c_2) = 0$, we say that the two families are **orthogonal trajectories**. Two curves are orthogonal if their tangent lines are perpendicular at a point of intersection. When given a family, we find its differential equation $dy/dx = f(x, y)$ by differentiating the equation $G(x, y, c_1) = 0$ and eliminating the parameter c_1. The differential equation of the second and orthogonal family is then $dy/dx = -1/f(x, y)$. We solve this latter equation by the methods of Chapter 2.

In the mathematical analysis of population growth, radioactive decay, or chemical mixtures, we often encounter **linear** differential equations such as

$$\frac{dx}{dt} = kx \qquad \text{and} \qquad \frac{dx}{dt} = a + bx$$

or **nonlinear** differential equations such as

$$\frac{dx}{dt} = x(a - bx) \qquad \text{and} \qquad \frac{dx}{dt} = k(\alpha - x)(\beta - x).$$

The student should be able to solve these particular equations without hesitation. It is never a good idea simply to memorize solutions of differential equations.

CHAPTER 3 REVIEW EXERCISES

Answers to odd-numbered problems begin on page A–8.

1. Find the orthogonal trajectories of the family of curves $y(x^3 + c_1) = 3$.
2. Find the orthogonal trajectory to the family $y = 4x + 1 + c_1 e^{4x}$ passing through the point (0, 0).
3. Find the orthogonal trajectories of the family of parabolas opening in the y-direction with vertex at (1, 2.)
4. Show that if a population expands at a rate proportional to the number of people present at any time, then the doubling time of the population is $T = (\ln 2)/k$, where k is the positive growth rate. This is known as the **Law of Malthus.***
5. In March of 1976 the world population reached 4 billion. A popular news magazine has predicted that with an average yearly growth rate of 1.8%, the world population will be 8 billion in 45 years. How does this value compare with that predicted by the model which says that the rate of increase is proportional to the population at any time?
6. Air containing 0.06% carbon dioxide is pumped into a room whose volume is 8000 ft^3. The rate at which the air is pumped in is 2000 ft^3/min, and the circulated air is then pumped out at the same rate. If there is an initial concentration of 0.2% carbon dioxide, determine the subsequent amount in the room at any time. What is the concentration at 10 minutes? What is the steady-state or equilibrium concentration of carbon-dioxide?

---

* **Thomas R. Malthus** (1766–1834) An English clergyman and political economist, Malthus achieved notoriety through his theory published in 1798 that human population grows at a (geometrical) rate that is faster than the (arithmetical) rate of growth of the supply of commodities necessary for life. He predicted famine and wars as a consequence.

7. The populations of two competing species of animals are described by the nonlinear system of first-order differential equations

$$\frac{dx}{dt} = k_1 x(\alpha - x), \qquad \frac{dy}{dt} = k_2 xy.$$

Solve for x and y in terms of t.

8. A projectile is shot vertically into the air with an initial velocity of v_0 ft/sec. Assuming that air resistance is proportional to the square of the instantaneous velocity, the motion is described by the pair of differential equations:

$$m\frac{dv}{dt} = -mg - kv^2, \qquad k > 0,$$

positive y-axis up, origin at ground level so that $v = v_0$ at $y = 0$,

and
$$m\frac{dv}{dt} = mg - kv^2, \qquad k > 0,$$

positive y-axis down, origin at the maximum height so that $v = 0$ at $y = h$. The first and second equations describe the motion of the projectile when rising and falling, respectively. Prove that the impact velocity v_i of the projectile is less than the initial velocity v_0. It can also be shown that the time t_1 needed to attain its maximum height h is less than the time t_2 that it takes to fall from this height (see Figure 3.24).

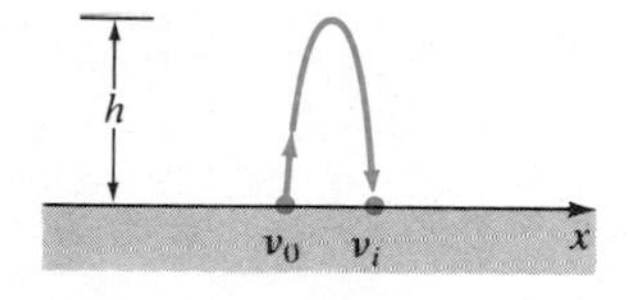

Figure 3.24

CHAPTER 4

Linear Differential Equations of Higher Order

IMPORTANT TERMS

initial-value problem
initial conditions
boundary-value problem
boundary conditions
linear dependence
linear independence
Wronskian
homogeneous equation
nonhomogeneous equation
superposition principle
fundamental set of solutions
general solution
complete solution
particular solution
complementary function
reduction of order
auxiliary equation
characteristic equation
Euler's formula
differential operator
annihilator operator
method of undetermined coefficients
variation of parameters

Except in a few cases in which a first-order differential equation can be solved, in general, nonlinear differential equations cannot be solved in terms of the familiar elementary functions. Consequently, in attempting to solve higher-order differential equations we shall confine our attention to linear equations. Specifically, in this chapter we shall examine linear equations with constant coefficients. We shall see that our ability to solve an nth-order linear differential equation with constant coefficients hinges on our ability to solve an nth-degree polynomial equation.

4.1 Preliminary Theory

We begin the discussion of higher-order differential equations, as we did with first-order equations, with the notion of an initial-value problem. However, we shall confine our attention to linear differential equations.

4.1.1 Initial-Value and Boundary-Value Problems

Initial-Value Problem

For a linear nth-order differential equation, the problem

$$\textit{Solve:}\quad a_n(x)\frac{d^n y}{dx^n} + a_{n-1}(x)\frac{d^{n-1}y}{dx^{n-1}} + \cdots + a_1(x)\frac{dy}{dx} + a_0(x)y = g(x)$$

$$\textit{Subject to:}\quad ty(x_0) = y_0,\quad y'(x_0) = y'_0, \ldots, y^{(n-1)}(x_0) = y_0^{(n-1)}, \tag{1}$$

where $y_0, y'_0, \ldots, y_0^{(n-1)}$ are arbitrary constants, is called an **inital-value problem**. The specified values $y(x_0) = y_0, y'(x_0) = y'_0, \ldots, y^{(n-1)}(x_0) = y_0^{(n-1)}$ are called **initial conditions**. We seek a solution on some interval I containing the point $x = x_0$.

In the case of a linear second-order equation, a solution of the initial-value problem

$$a_2(x)\frac{d^2y}{dx^2} + a_1(x)\frac{dy}{dx} + a_0(x)y = g(x)$$

$$y(x_0) = y_0, \qquad y'(x_0) = y'_0$$

is a function defined on I whose graph passes through (x_0, y_0) such that the slope of the curve at the point is the number y'_0.

The next theorem gives sufficient conditions for the existence of a unique solution to (1).

THEOREM 4.1 Let $a_n(x)$, $a_{n-1}(x), \ldots, a_1(x)$, $a_0(x)$, and $g(x)$ be continuous on an interval I and let $a_n(x) \neq 0$ for every x in this interval. If $x = x_0$ is any point in this interval, then a solution $y(x)$ of the initial-value problem (1) exists on the interval and is unique.

While we are not in a position to prove Theorem 4.1 in its full generality, a demonstration of the *uniqueness* of the solution in the special case

$$a_2y'' + a_1y' + a_0y = g(x)$$

$$y(0) = y_0, \qquad y'(0) = y'_0,$$

where a_2, a_1, and a_0 are positive constants and $g(x)$ is continuous for all x, is given in Appendix I.

EXAMPLE 1

The reader should verify that $y = 3e^{2x} + e^{-2x} - 3x$ is a solution of the initial-value problem

$$y'' - 4y = 12x$$
$$y(0) = 4, \qquad y'(0) = 1.$$

Now the differential equation is linear and the coefficients as well as $g(x) = 12x$ are continuous on any interval containing $x = 0$. We conclude from Theorem 4.1 that the given function is the unique solution.

EXAMPLE 2

The initial-value problem

$$3y''' + 5y'' - y' + 7y = 0$$
$$y(1) = 0, \qquad y'(1) = 0, \qquad y''(1) = 0$$

possesses the trivial solution $y = 0$. Since the third-order equation is linear with constant coefficients, it follows that all the conditions of Theorem 4.1 are fulfilled. Hence $y = 0$ is the *only* solution on any interval containing $x = 1$.

EXAMPLE 3

The function $y = \frac{1}{4}\sin 4x$ is a solution of the initial-value problem

$$y'' + 16y = 0$$
$$y(0) = 0, \qquad y'(0) = 1.$$

It follows from Theorem 4.1 that on any interval containing $x = 0$ the solution is unique.

The requirements in Theorem 4.1 that $a_i(x)$, $i = 0, 1, 2, \ldots, n$ be continuous and $a_n(x) \neq 0$ for every x in I are both important. Specifically, if $a_n(x) = 0$ for some x in the interval then the solution of a linear initial-value problem may not be unique or even exist.

EXAMPLE 4

Verify that the function $y = cx^2 + x + 3$ is a solution of the initial-value problem

$$x^2y'' - 2xy' + 2y = 6$$
$$y(0) = 3, \qquad y'(0) = 1,$$

on the interval $(-\infty, \infty)$ for any choice of the parameter c.

SOLUTION Since $y' = 2cx + 1$ and $y'' = 2c$, it follows that

$$x^2y'' - 2xy' + 2y = x^2(2c) - 2x(2cx + 1) + 2(cx^2 + x + 3)$$
$$= 2cx^2 - 4cx^2 - 2x + 2cx^2 + 2x + 6 = 6.$$

Also $$y(0) = c(0)^2 + 0 + 3 = 3$$

and $$y'(0) = 2c(0) + 1 = 1.$$

Although the differential equation in the foregoing example is linear and the coefficients and $g(x) = 6$ are continuous everywhere, the obvious difficulty is that $a_2(x) = x^2$ is zero at $x = 0$.

Boundary-Value Problem

Another type of problem consists of solving a differential equation of order two or greater in which the dependent variable y, or its derivatives, is specified at *different points*. A problem such as

$$\textit{Solve:} \quad a_2(x)\frac{d^2y}{dx^2} + a_1(x)\frac{dy}{dx} + a_0(x)y = g(x)$$

$$\textit{Subject to:} \quad y(a) = y_0, \qquad y(b) = y_1$$

is called a **two-point boundary-value problem** or simply a **boundary-value problem**. The specified values $y(a) = y_0$ and $y(b) = y_1$ are called **boundary conditions**. For a second-order equation, other pairs of boundary conditions could be

$$y'(a) = y'_0, \qquad y(b) = y_1;$$

$$y(a) = y_0, \qquad y'(b) = y'_1;$$

or $$y'(a) = y'_0, \qquad y'(b) = y'_1,$$

where y_0, y'_0, y_1, and y'_1 denote arbitrary constants.

EXAMPLE 5

For the boundary-value problem

$$x^2y'' - 2xy' + 2y = 6$$

$$y(1) = 0, \qquad y(2) = 3,$$

we seek a function defined on an interval containing $x = 1$ and $x = 2$ that satisfies the differential equation and whose graph passes through the two points (1, 0) and (2, 3).

The next examples show that even when the conditions of Theorem 4.1 are fulfilled, a boundary-value problem may either have (a) several solutions, (b) a unique solution, or (c) no solution at all.

EXAMPLE 6

In Example 6 of Section 1.1, we have seen that a two-parameter family of solutions for the differential equation $y'' + 16y = 0$ is

$$y = c_1 \cos 4x + c_2 \sin 4x.$$

Suppose we now wish to determine that solution of the equation that further satisfies the boundary conditions

$$y(0) = 0, \qquad y\left(\frac{\pi}{2}\right) = 0.$$

Observe that the first condition

$$0 = c_1 \cos 0 + c_2 \sin 0$$

implies $c_1 = 0$ so that $y = c_2 \sin 4x$. But when $x = \pi/2$ we have

$$0 = c_2 \sin 2\pi.$$

Since $\sin 2\pi = 0$ this latter condition is satisfied for any choice of c_2, so it follows that a solution of the problem

$$y'' + 16y = 0$$

$$y(0) = 0, \qquad y\left(\frac{\pi}{2}\right) = 0$$

is the one-parameter family

$$y = c_2 \sin 4x.$$

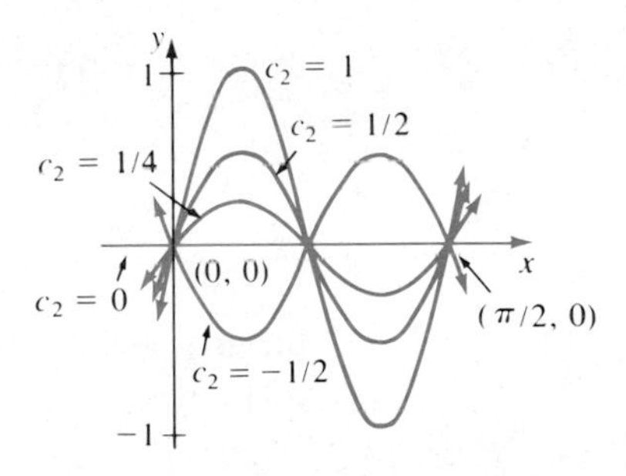

Figure 4.1

As Figure 4.1 shows, there are an infinite number of functions satisfying the differential equation whose graphs pass through the two points (0, 0) and $(\pi/2, 0)$.

If the boundary conditions were $y(0) = 0$ and $y(\pi/8) = 0$, then necessarily c_1 and c_2 would both equal 0. Thus $y = 0$ would be a solution of this new boundary-value problem. In fact, as we shall see later on in this section, it is the only solution.

EXAMPLE 7 The boundary-value problem

$$y'' + 16y = 0$$

$$y(0) = 0, \qquad y\left(\frac{\pi}{2}\right) = 1$$

has no solution in the family $y = c_1 \cos 4x + c_2 \sin 4x$. As in Example 6 the condition $y(0) = 0$ still implies that $c_1 = 0$. Thus $y = c_2 \sin 4x$ so that when $x = \pi/2$ we obtain the contradiction $1 = c_2 \cdot 0 = 0$.

Boundary-value problems are often encountered in the applications of partial differential equations.

4.1.2 Linear Dependence and Linear Independence

The next two concepts are basic to the study of linear differential equations.

DEFINITION 4.1 A set of functions $f_1(x), f_2(x), \ldots, f_n(x)$ is said to be **linearly dependent** on an interval I if there exist constants $c_1, c_2, \ldots, c_n$, not all zero, such that

$$c_1 f_1(x) + c_2 f_2(x) + \cdots + c_n f_n(x) = 0$$

for every x in the interval.

DEFINITION 4.2 A set of functions $f_1(x), f_2(x), \ldots, f_n(x)$ is said to be **linearly independent** on an interval I if it is not linearly dependent on the interval.

In other words, a set of functions is linearly independent on an interval if the only constants for which

$$c_1 f_1(x) + c_2 f_2(x) + \cdots + c_n f_n(x) = 0,$$

for every x in the interval, are $c_1 = c_2 = \cdots = c_n = 0$.

It is easy to understand these definitions in the case of two functions $f_1(x)$ and $f_2(x)$. If the functions are linearly dependent on an interval, then there exist constants c_1 and c_2 that are not both zero such that for every x in the interval

$$c_1 f_1(x) + c_2 f_2(x) = 0.$$

Therefore, if we assume that $c_1 \neq 0$, it follows that

$$f_1(x) = -\frac{c_2}{c_1} f_2(x),$$

that is, *if two functions are linearly dependent, then one is simply a constant multiple of the other*. Conversely, if $f_1(x) = c_2 f_2(x)$ for some constant c_2, then

$$(-1) \cdot f_1(x) + c_2 f_2(x) = 0$$

for every x on some interval. Hence the functions are linearly dependent since at least one of the constants (namely, $c_1 = -1$) is not zero. We conclude that *two functions are linearly independent when neither is a constant multiple of the other* on an interval.

EXAMPLE 8

The functions $f_1(x) = \sin 2x$ and $f_2(x) = \sin x \cos x$ are linearly dependent on the interval $(-\infty, \infty)$ since

$$c_1 \sin 2x + c_2 \sin x \cos x = 0$$

is satisfied for every real x if we choose $c_1 = 1/2$ and $c_2 = -1$. (Recall the trigonometric identity $\sin 2x = 2 \sin x \cos x$.)

EXAMPLE 9

The functions $f_1(x) = x$ and $f_2(x) = |x|$ are linearly independent on the interval $(-\infty, \infty)$. Inspection of Figure 4.2 should convince the reader that neither function is a constant multiple of the other. Thus in order to have $c_1 f_1(x) + c_2 f_2(x) = 0$ for every real x, we must choose $c_1 = 0$ and $c_2 = 0$.

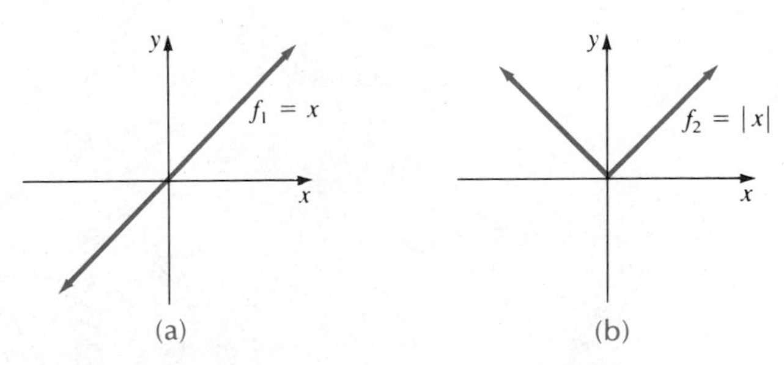

Figure 4.2

In the consideration of linear dependence or linear independence, the interval on which the functions are defined is important. The functions $f_1(x) = x$ and $f_2(x) = |x|$ in Example 9 are linearly dependent on the interval $(0, \infty)$ since

$$c_1 x + c_2|x| = c_1 x + c_2 x = 0$$

is satisfied for any nonzero choice of c_1 and c_2 for which $c_1 = -c_2$.

EXAMPLE 10

The functions $f_1(x) = \cos^2 x$, $f_2(x) = \sin^2 x$, $f_3(x) = \sec^2 x$, $f_4(x) = \tan^2 x$ are linearly dependent on the interval $(-\pi/2, \pi/2)$ since

$$c_1 \cos^2 x + c_2 \sin^2 x + c_3 \sec^2 x + c_4 \tan^2 x = 0,$$

when $c_1 = c_2 = 1$, $c_3 = -1$, $c_4 = 1$. We note that $\cos^2 x + \sin^2 x = 1$ and $1 + \tan^2 x = \sec^2 x$.

A set of functions $f_1(x), f_2(x), \ldots, f_n(x)$ are linearly dependent on an interval if at least one function can be expressed as a nontrivial linear combination of the remaining functions.

EXAMPLE 11

The functions $f_1(x) = \sqrt{x} + 5$, $f_2(x) = \sqrt{x} + 5x$, $f_3(x) = x - 1$, $f_4(x) = x^2$ are linearly dependent on the interval $(0, \infty)$ since

$$f_2(x) = 1 \cdot f_1(x) + 5 \cdot f_3(x) + 0 \cdot f_4(x)$$

for every x in the interval.

Wronskian

The following theorem provides a sufficient condition for the linear independence of n functions on an interval. Each function is assumed to be differentiable at least $n - 1$ times.

THEOREM 4.2 Suppose $f_1(x), f_2(x), \ldots, f_n(x)$ possess at least $n - 1$ derivatives. If the determinant

$$\begin{vmatrix} f_1 & f_2 & \cdots & f_n \\ f'_1 & f'_2 & \cdots & f'_n \\ \vdots & \vdots & & \vdots \\ f_1^{(n-1)} & f_2^{(n-1)} & \cdots & f_n^{(n-1)} \end{vmatrix}$$

is not zero for at least one point in the interval I, then the functions $f_1(x)$, $f_2(x), \ldots, f_n(x)$ are linearly independent on the interval.

The determinant in the preceding theorem is denoted by

$$W(f_1(x), f_2(x), \ldots, f_n(x))$$

and is called the **Wronskian*** of the functions.

PROOF We prove Theorem 4.2 by contradiction for the case when $n = 2$. Assume that $W(f_1(x_0), f_2(x_0)) \neq 0$ for a fixed x_0 in the interval I and that $f_1(x)$ and $f_2(x)$ are linearly dependent on the interval. The fact that the functions are linearly dependent means there exist constants c_1 and c_2, not both zero, for which

$$c_1 f_1(x) + c_2 f_2(x) = 0$$

for every x in I. Differentiating this combination then gives

$$c_1 f'_1(x) + c_2 f'_2(x) = 0.$$

Thus we obtain the system of linear equations

$$\begin{aligned} c_1 f_1(x) + c_2 f_2(x) &= 0. \\ c_1 f'_1(x) + c_2 f'_2(x) &= 0. \end{aligned} \tag{2}$$

But the linear dependence of f_1 and f_2 implies that (2) possesses a nontrivial solution for each x in the interval. Hence

$$W(f_1(x), f_2(x)) = \begin{vmatrix} f_1(x) & f_2(x) \\ f'_1(x) & f'_2(x) \end{vmatrix} = 0$$

for every x in I.† This contradicts the assumption that $W(f_1(x_0), f_2(x_0)) \neq 0$. We conclude that f_1 and f_2 are linearly independent. ■

COROLLARY If $f_1(x), f_2(x), \ldots, f_n(x)$ possess at least $n - 1$ derivatives and are linearly dependent on I then

$$W(f_1(x), f_2(x), \ldots, f_n(x)) = 0$$

for every x in the interval.

---

* **Josef Maria Hoëne Wronski** (1778–1853) Born in Poland, educated in Germany, Wronski lived most of his life in France. More a philosopher than a mathematician, he believed that absolute truth could be attained through mathematics. Wronski's only noteworthy contribution to mathematics was the above determinant. Always an eccentric, he eventually went insane.

† See Appendix IV for a review of determinants.

EXAMPLE 12

The functions $f_1(x) = \sin^2 x$ and $f_2(x) = 1 - \cos 2x$ are linearly dependent on $(-\infty, \infty)$. (Why?) By the preceding corollary, $W(\sin^2 x, 1 - \cos 2x) = 0$ for every real number. To see this we observe

$$\begin{aligned} W(\sin^2 x, 1 - \cos 2x) &= \begin{vmatrix} \sin^2 x & 1 - \cos 2x \\ 2 \sin x \cos x & 2 \sin 2x \end{vmatrix} \\ &= 2 \sin^2 x \sin 2x - 2 \sin x \cos x \\ &\quad + 2 \sin x \cos x \cos 2x \\ &= \sin 2x[2 \sin^2 x - 1 + \cos 2x] \\ &= \sin 2x[2 \sin^2 x - 1 + \cos^2 x - \sin^2 x] \\ &= \sin 2x[\sin^2 x + \cos^2 x - 1] \\ &= 0. \end{aligned}$$

EXAMPLE 13

For $f_1(x) = e^{m_1 x}$, $f_2(x) = e^{m_2 x}$, $m_1 \neq m_2$

$$\begin{aligned} W(e^{m_1 x}, e^{m_2 x}) &= \begin{vmatrix} e^{m_1 x} & e^{m_2 x} \\ m_1 e^{m_1 x} & m_2 e^{m_2 x} \end{vmatrix} \\ &= (m_2 - m_1) e^{(m_1 + m_2)x} \neq 0 \end{aligned}$$

for every real value of x. Thus f_1 and f_2 are linearly independent on any interval of the x-axis.

EXAMPLE 14

If α and β are real numbers, $\beta \neq 0$, then $y_1 = e^{\alpha x} \cos \beta x$ and $y_2 = e^{\alpha x} \sin \beta x$ are linearly independent on any interval of the x-axis since

$$\begin{aligned} W(e^{\alpha x} \cos \beta x, e^{\alpha x} \sin \beta x) &= \begin{vmatrix} e^{\alpha x} \cos \beta x & e^{\alpha x} \sin \beta x \\ -\beta e^{\alpha x} \sin \beta x + \alpha e^{\alpha x} \cos \beta x & \beta e^{\alpha x} \cos \beta x + \alpha e^{\alpha x} \sin \beta x \end{vmatrix} \\ &= \beta e^{2\alpha x}(\cos^2 \beta x + \sin^2 \beta x) \\ &= \beta e^{2\alpha x} \neq 0. \end{aligned}$$

Notice when $\alpha = 0$ we see that $\cos \beta x$ and $\sin \beta x$, $\beta \neq 0$, are also linearly independent on any interval of the x-axis.

EXAMPLE 15

The functions $f_1(x) = e^x$, $f_2(x) = xe^x$, and $f_3(x) = x^2e^x$ are linearly independent on any interval of the x-axis since

$$W(e^x, xe^x, x^2e^x) = \begin{vmatrix} e^x & xe^x & x^2e^x \\ e^x & xe^x + e^x & x^2e^x + 2xe^x \\ e^x & xe^x + 2e^x & x^2e^x + 4xe^x + 2e^x \end{vmatrix} = 2e^{3x}$$

is not zero for any real value of x.

EXAMPLE 16

In Example 9 we saw that $f_1(x) = x$ and $f_2(x) = |x|$ are linearly independent on $(-\infty, \infty)$; however, we cannot compute the Wronskian since f_2 is not differentiable at $x = 0$.

We leave it as an exercise to show that a set of functions $f_1(x), f_2(x), \dots, f_n(x)$ could be linearly independent on some interval and yet have a vanishing Wronskian. In other words, if $W(f_1(x), f_2(x), \dots, f_n(x)) = 0$ for every x in an interval, it does not necessarily mean that the functions are linearly dependent.

4.1.3 Solutions of Linear Equations

Homogeneous Equations

A linear nth-order differential equation of the form

$$a_n(x)\frac{d^ny}{dx^n} + a_{n-1}(x)\frac{d^{n-1}y}{dx^{n-1}} + \cdots + a_1(x)\frac{dy}{dx} + a_0(x)y = 0 \tag{3}$$

is said to be **homogeneous**, whereas

$$a_n(x)\frac{d^ny}{dx^n} + a_{n-1}(x)\frac{d^{n-1}y}{dx^{n-1}} + \cdots + a_1(x)\frac{dy}{dx} + a_0(x)y = g(x), \tag{4}$$

$g(x)$ not identically zero, is said to be **nonhomogeneous**.

The word *homogeneous* in this context does not refer to coefficients that are homogeneous functions (see Section 2.3).

EXAMPLE 17

(a) The equation $$2y'' + 3y' - 5y = 0$$

is a homogeneous linear second-order differential equation.

(b) The equation $$x^3y''' - 2xy'' + 5y' + 6y = e^x$$

is a nonhomogeneous linear third-order ordinary differential equation.

We shall see in the latter part of this section, as well as in the subsequent sections of this chapter, that in order to solve a nonhomogeneous equation (4), we must first solve the associated homogeneous equation (3).

Note: To avoid needless repetition throughout the remainder of this text we shall, as a matter of course, make the following important assumptions when giving definitions and proving theorems about the linear equations (3) and (4). On some common interval I

(i) the coefficients $a_i(x)$, $i = 0, 1, \ldots, n$ are continuous;
(ii) the right-hand member $g(x)$ is continuous;
(iii) and $a_n(x) \neq 0$ for every x in the interval.

Superposition Principle

The following theorem is known as the **superposition principle**.

THEOREM 4.3 Let $y_1, y_2, \ldots, y_k$ be solutions of the homogeneous linear nth-order differential equation (3) on an interval I. Then the linear combination

$$y = c_1 y_1(x) + c_2 y_2(x) + \cdots + c_k y_k(x), \tag{5}$$

where the c_i, $i = 1, 2, \ldots, k$ are arbitrary constants, is also a solution on the interval.

PROOF We prove the case when $n = k = 2$. Let $y_1(x)$ and $y_2(x)$ be solutions of

$$a_2(x)y'' + a_1(x)y' + a_0(x)y = 0.$$

If we define $y = c_1 y_1(x) + c_2 y_2(x)$, then

$$\begin{aligned} &a_2(x)[c_1 y_1'' + c_2 y_2''] + a_1(x)[c_1 y_1' + c_2 y_2'] + a_0(x)[c_1 y_1 + c_2 y_2] \\ &\quad = c_1 \underbrace{[a_2(x)y_1'' + a_1(x)y_1' + a_0(x)y_1]}_{\text{zero}} + c_2 \underbrace{[a_2(x)y_2'' + a_1(x)y_2' + a_0(x)y_2]}_{\text{zero}} \\ &\quad = c_1 \cdot 0 + c_2 \cdot 0 = 0. \end{aligned}$$

■

COROLLARIES **(A)** A constant multiple $y = c_1 y_1(x)$ of a solution $y_1(x)$ of a homogeneous linear differential equation is also a solution. **(B)** A homogeneous linear differential equation always possesses the trivial solution $y = 0$.

The superposition principle defined by (5), and its special case given in Corollary (A), are properties that nonlinear differential equations, in general, do not possess (see Problems 31 and 32).

EXAMPLE 18

The functions $y_1 = x^2$ and $y_2 = x^2 \ln x$

are both solutions of the homogeneous third-order equation

$$x^3y''' - 2xy' + 4y = 0$$

on the interval $(0, \infty)$. By the superposition principle the linear combination

$$y = c_1x^2 + c_2x^2 \ln x$$

is also a solution of the equation on the interval.

EXAMPLE 19

The functions $y_1 = e^x$, $y_2 = e^{2x}$, and $y_3 = e^{3x}$ all satisfy the homogeneous equation

$$\frac{d^3y}{dx^3} - 6\frac{d^2y}{dx^2} + 11\frac{dy}{dx} - 6y = 0$$

on $(-\infty, \infty)$. By Theorem 4.3 another solution is

$$y = c_1e^x + c_2e^{2x} + c_3e^{3x}.$$

EXAMPLE 20

The function $y = x^2$ is a solution of the homogeneous linear equation

$$x^2y'' - 3xy' + 4y = 0$$

on $(0, \infty)$. Hence $y = cx^2$ is also a solution. For various values of c we see that $y = 3x^2$, $y = ex^2$, $y = 0, \ldots$ are all solutions of the equation on the interval.

Linearly Independent Solutions

We are interested in determining when n solutions, $y_1, y_2, \ldots, y_n$, of the **homogeneous** differential equation (3) are linearly independent. Surprisingly the nonvanishing of the Wronskian of a set of n such solutions on an interval I is both necessary and sufficient for linear independence.

THEOREM 4.4 Let $y_1, y_2, \ldots, y_n$ be n solutions of the homogeneous linear nth-order differential equation (3) on an interval I. Then the set of solutions is linearly independent on I if and only if

$$W(y_1, y_2, \ldots, y_n) \neq 0$$

for every x in the interval.

PROOF We prove Theorem 4.4 for the case when $n = 2$. First, if $W(y_1, y_2) \neq 0$ for every x in I, it follows immediately from Theorem 4.2 that y_1 and y_2 are linearly independent. Next, we must show that if y_1 and y_2 are linearly independent solutions of a homogeneous linear second-order differential equation, then $W(y_1, y_2) \neq 0$ for every x in I. To show this, let us suppose y_1 and y_2 are linearly independent and there is some fixed x_0 in I for which $W(y_1(x_0), y_2(x_0)) = 0$. Hence there must exist c_1 and c_2, not both zero, such that

$$\begin{aligned} c_1 y_1(x_0) + c_2 y_2(x_0) &= 0 \\ c_1 y_1'(x_0) + c_2 y_2'(x_0) &= 0. \end{aligned} \tag{6}$$

If we define $$y(x) = c_1 y_1(x) + c_2 y_2(x),$$

then in view of (6), $y(x)$ must also satisfy

$$y(x_0) = 0, \qquad y'(x_0) = 0. \tag{7}$$

But the identically zero function satisfies both the differential equation and the initial conditions (7), and thus by Theorem 4.1 it is the only solution. In other words, $y = 0$ or

$$c_1 y_1(x) + c_2 y_2(x) = 0$$

for every x in I. This contradicts the assumption that y_1 and y_2 are linearly independent on the interval. ■

From the foregoing discussion we conclude that when $y_1, y_2, \ldots, y_n$ are n solutions of (3) on an interval I, either the Wronskian is identically zero or is never zero on the interval.

DEFINITION 4.3 Any set $y_1, y_2, \ldots, y_n$ of n linearly independent solutions of the homogeneous linear nth-order differential equation (3) on an interval I is said to be a **fundamental set of solutions** on the interval.

THEOREM 4.5 Let $y_1, y_2, \ldots, y_n$ be a fundamental set of solutions of the homogeneous linear nth-order differential equation (3) on an interval I. Then for any solution $Y(x)$ of (3) on I, constants $C_1, C_2, \ldots, C_n$ can be found so that

$$Y = C_1 y_1(x) + C_2 y_2(x) + \cdots + C_n y_n(x).$$

PROOF We prove the case when $n = 2$. Let Y be a solution, and let y_1 and y_2 be linearly independent solutions of

$$a_2(x)y'' + a_1(x)y' + a_0(x)y = 0$$

on an interval I. Suppose $x = t$ is a point in this interval for which $W(y_1(t), y_2(t)) \neq 0$. Suppose also that the values of $Y(t)$ and $Y'(t)$ are given by

$$Y(t) = k_1, \qquad Y'(t) = k_2.$$

If we now examine the system of equations

$$C_1 y_1(t) + C_2 y_2(t) = k_1$$
$$C_1 y_1'(t) + C_2 y_2'(t) = k_2,$$

it follows that we can determine C_1 and C_2 uniquely provided the determinant of the coefficients satisfies

$$\begin{vmatrix} y_1(t) & y_2(t) \\ y_1'(t) & y_2'(t) \end{vmatrix} \neq 0.$$

But this latter determinant is simply the Wronskian evaluated at $x = t$, and by assumption $W \neq 0$. If we now define the function

$$G(x) = C_1 y_1(x) + C_2 y_2(x),$$

we then observe:

(i) $G(x)$ satisfies the differential equation since it is the superposition of two known solutions y_1 and y_2,

(ii) $G(x)$ satisfies the initial conditions

$$G(t) = C_1 y_1(t) + C_2 y_2(t) = k_1$$
$$G'(t) = C_1 y_1'(t) + C_2 y_2'(t) = k_2,$$

(iii) $Y(x)$ satisfies the *same* linear equation and the *same* initial conditions.

Since the solution of this linear initial-value problem is unique

(Theorem 4.1), we have $Y(x) = G(x)$, or

$$Y(x) = C_1 y_1(x) + C_2 y_2(x). \qquad \blacksquare$$

The basic question of whether a fundamental set of solutions exists for a linear equation is answered in the next theorem.

THEOREM 4.6 There exists a fundamental set of solutions for the homogeneous linear nth-order differential equation (3) on an interval I.

The proof of this result follows from Theorem 4.1. The justification of Theorem 4.6 in the special case of second-order equations is left as an exercise.

Since we have shown that any solution of (3) is obtained from a linear combination of functions in a fundamental set of solutions, we are able to make the following definition.

DEFINITION 4.4 Let $y_1, y_2, \ldots, y_n$ be a fundamental set of solutions of the homogeneous linear nth-order differential equation (3) on an interval I. The **general solution** of the equation on the interval is defined to be

$$y = c_1 y_1(x) + c_2 y_2(x) + \cdots + c_n y_n(x),$$

where the c_i, $i = 1, 2, \ldots, n$ are arbitrary constants.

Recall, the general solution as defined in Section 1.1 is also called the **complete solution** of the differential equation.

EXAMPLE 21 The second-order equation $y'' - 9y = 0$ possesses two solutions

$$y_1 = e^{3x} \quad \text{and} \quad y_2 = e^{-3x}.$$

Since
$$W(e^{3x}, e^{-3x}) = \begin{vmatrix} e^{3x} & e^{-3x} \\ 3e^{3x} & -3e^{-3x} \end{vmatrix} = -6 \neq 0$$

for every value of x, y_1 and y_2 form a fundamental set of solutions on $(-\infty, \infty)$. The general solution of the differential equation on the interval is

$$y = c_1 e^{3x} + c_2 e^{-3x}.$$

EXAMPLE 22

The reader should verify that the function $y = 4 \sinh 3x - 5e^{-3x}$ also satisfies the differential equation in Example 21. By choosing $c_1 = 2$, $c_2 = -7$ in the general solution $y = c_1e^{3x} + c_2e^{-3x}$, we obtain

$$\begin{aligned} y &= 2e^{3x} - 7e^{-3x} \\ &= 2e^{3x} - 2e^{-3x} - 5e^{-3x} \\ &= 4\left(\frac{e^{3x} - e^{-3x}}{2}\right) - 5e^{-3x} \\ &= 4 \sinh 3x - 5e^{-3x}. \end{aligned}$$

EXAMPLE 23

The functions $y_1 = e^x$, $y_2 = e^{2x}$, and $y_3 = e^{3x}$ satisfy the third-order equation

$$\frac{d^3y}{dx^3} - 6\frac{d^2y}{dx^2} + 11\frac{dy}{dx} - 6y = 0.$$

Since

$$W(e^x, e^{2x}, e^{3x}) = \begin{vmatrix} e^x & e^{2x} & e^{3x} \\ e^x & 2e^{2x} & 3e^{3x} \\ e^x & 4e^{2x} & 9e^{3x} \end{vmatrix} = 2e^{6x} \neq 0$$

for every real value of x, y_1, y_2, and y_3 form a fundamental set of solutions on $(-\infty, \infty)$. We conclude

$$y = c_1e^x + c_2e^{2x} + c_3e^{3x}$$

is the general solution of the differential equation on the interval.

Nonhomogeneous Equations

We now turn our attention to defining the general solution of a **nonhomogeneous** linear equation. Any function y_p, free of arbitrary parameters, that satisfies (4) is said to be a **particular solution** of the equation (sometimes also called a **particular integral**).

EXAMPLE 24

(a) A particular solution of

$$y'' + 9y = 27$$

is $y_p = 3$ since $y_p'' = 0$, and $0 + 9y_p = 9(3) = 27$.

(b) $y_p = x^3 - x$ is a particular solution of

$$x^2y'' + 2xy' - 8y = 4x^3 + 6x$$

since $y'_p = 3x^2 - 1$, $y''_p = 6x$, and

$$x^2y''_p + 2xy'_p - 8y_p = x^2(6x) + 2x(3x^2 - 1) - 8(x^3 - x)$$
$$= 4x^3 + 6x.$$

THEOREM 4.7 Let $y_1, y_2, \ldots, y_k$ be solutions of the homogeneous linear nth-order differential equation (3) on an interval I and let y_p be any solution of the nonhomogeneous equation (4) on the same interval. Then

$$y = c_1y_1(x) + c_2y_2(x) + \cdots + c_ky_k(x) + y_p(x)$$

is also a solution of the nonhomogeneous equation on the interval for any constants $c_1, c_2, \ldots, c_k$.

We can now prove the following analogue of Theorem 4.5 for nonhomogeneous differential equations.

THEOREM 4.8 Let y_p be a given solution of the nonhomogeneous linear nth-order differential equation (4) on an interval I and let $y_1, y_2, \ldots, y_n$ be a fundamental set of solutions of the associated homogeneous equation (3) on the interval. Then for any solution $Y(x)$ of (4) on I, constants $C_1, C_2, \ldots, C_n$ can be found so that

$$Y = C_1y_1(x) + C_2y_2(x) + \cdots + C_ny_n(x) + y_p(x).$$

PROOF We prove the case when $n = 2$. Suppose Y and y_p are both solutions of

$$a_2(x)y'' + a_1(x)y' + a_0(x)y = g(x).$$

If we define a function u by $u(x) = Y(x) - y_p(x)$ then

$$\begin{aligned}
&a_2(x)u'' + a_1(x)u' + a_0(x)u \\
&\quad = a_2(x)[Y'' - y''_p] + a_1(x)[Y' - y'_p] + a_0(x)[Y - y_p] \\
&\quad = a_2(x)Y'' + a_1(x)Y' + a_0(x)Y - [a_2(x)y''_p + a_1(x)y'_p + a_0(x)y_p] \\
&\quad = g(x) - g(x) = 0.
\end{aligned}$$

Therefore, in view of Definition 4.4 and Theorem 4.5, we can write

$$u(x) = C_1y_1(x) + C_2y_2(x)$$

$$Y(x) - y_p(x) = C_1y_1(x) + C_2y_2(x)$$

or

$$Y(x) = C_1y_1(x) + C_2y_2(x) + y_p(x).$$ ■

Thus we come to the last definition of this section.

DEFINITION 4.5 Let y_p be a given solution of the nonhomogeneous linear nth-order differential equation (4) on an interval I and let

$$y_c = c_1y_1(x) + c_2y_2(x) + \cdots + c_ny_n(x)$$

denote the general solution of the associated homogeneous equation (3) on the interval. The **general solution** of the nonhomogeneous equation on the interval is defined to be

$$\begin{aligned} y &= c_1y_1(x) + c_2y_2(x) + \cdots + c_ny_n(x) + y_p(x) \\ &= y_c(x) + y_p(x). \end{aligned}$$

Complementary Function

In Definition 4.5 the linear combination

$$y_c(x) = c_1y_1(x) + c_2y_2(x) + \cdots + c_ny_n(x),$$

which is the general solution of (3), is called the **complementary function** for equation (4). In other words, the general solution of a nonhomogeneous linear differential equation is

$$y = \textit{complementary function} + \textit{any particular solution.}$$

EXAMPLE 25

The function

$$y_p = -\frac{11}{12} - \frac{1}{2}x$$

can be shown to be a particular solution of the nonhomogeneous equation

$$\frac{d^3y}{dx^3} - 6\frac{d^2y}{dx^2} + 11\frac{dy}{dx} - 6y = 3x. \tag{8}$$

In order to write down the general solution of (8), we must also be able to solve the associated homogeneous equation

$$\frac{d^3y}{dx^3} - 6\frac{d^2y}{dx^2} + 11\frac{dy}{dx} - 6y = 0.$$

But in Example 23 we saw that the general solution of this latter equation on the interval $(-\infty, \infty)$ was

$$y_c = c_1e^x + c_2e^{2x} + c_3e^{3x}.$$

Hence the general solution of (8) on the interval is

$$\begin{aligned} y &= y_c + y_p \\ &= c_1e^x + c_2e^{2x} + c_3e^{3x} - \frac{11}{12} - \frac{1}{2}x. \end{aligned}$$

Before we actually start solving homogeneous and nonhomogeneous linear differential equations, we need the one additional bit of theory presented in the next section.

Remark: A physical system that changes with time and whose mathematical model is a differential equation of the form given in (4) is said to be a **linear system**. The function g in (4) is called the **input** or **forcing function**. A solution y of (4) is called the **output**, or **response**, of the system. In a linear system the response of the system to a superposition of inputs is a superposition of outputs (see Problem 51).

Exercises 4.1

Answers to odd-numbered problems begin on page A–8.

[4.1.1] **1.** Given that $$y = c_1e^x + c_2e^{-x}$$

is a two-parameter family of solutions of $y'' - y = 0$ on the interval $(-\infty, \infty)$, find a member of the family satisfying the initial conditions $y(0) = 0$, $y'(0) = 1$.

2. Find a solution of the differential equation in Problem 1 satisfying the boundary conditions $y(0) = 0$, $y(1) = 1$.

3. Given that $$y = c_1e^{4x} + c_2e^{-x}$$

is a two-parameter family of solutions of $y'' - 3y' - 4y = 0$ on the interval $(-\infty, \infty)$, find a member of the family satisfying the initial conditions $y(0) = 1$, $y'(0) = 2$.

4. Given that $$y = c_1 + c_2 \cos x + c_3 \sin x$$

is a three-parameter family of solutions of $y''' + y' = 0$ on the interval $(-\infty, \infty)$, find a member of the family satisfying the initial conditions $y(\pi) = 0$, $y'(\pi) = 2$, $y''(\pi) = -1$.

5. Given that $$y = c_1x + c_2x \ln x$$

is a two-parameter family of solutions of $x^2y'' - xy' + y = 0$ on the interval $(-\infty, \infty)$, find a member of the family satisfying the initial conditions $y(1) = 3$, $y'(1) = -1$.

6. Given that $$y = c_1 + c_2x^2$$

is a two-parameter family of solutions of $xy'' - y' = 0$ on the interval $(-\infty, \infty)$, show that constants c_1 and c_2 cannot be found so that a member of the family satisfies the initial conditions $y(0) = 0$, $y'(0) = 1$. Explain why this does not violate Theorem 4.1.

7. Find two members of the family of solutions of $xy'' - y' = 0$ given in Problem 6 satisfying the initial conditions $y(0) = 0$, $y'(0) = 0$.

8. Find a member of the family of solutions of $xy'' - y' = 0$ given in Problem 6 satisfying the boundary conditions $y(0) = 1$, $y'(1) = 6$. Does Theorem 4.1 guarantee that this solution is unique?

9. Given that $$y = c_1e^x \cos x + c_2e^x \sin x$$

is a two-parameter family of solutions of $y'' - 2y' + 2y = 0$ on the interval $(-\infty, \infty)$, determine whether a member of the family can be found that satisfies the conditions

(a) $y(0) = 1, \quad y'(0) = 0$

(b) $y(0) = 1, \quad y(\pi) = -1$

(c) $y(0) = 1, \quad y(\pi/2) = 1$

(d) $y(0) = 0, \quad y(\pi) = 0.$

10. Given that $$y = c_1x^2 + c_2x^4 + 3$$

is a two-parameter family of solutions of $x^2y'' - 5xy' + 8y = 24$ on the interval $(-\infty, \infty)$, determine whether a member of the family can be found that satisfies the conditions

(a) $y(-1) = 0, \quad y(1) = 4$

(b) $y(0) = 1, \quad y(1) = 2$

(c) $y(0) = 3, \quad y(1) = 0$

(d) $y(1) = 3, \quad y(2) = 15.$

In Problems 11 and 12 find an interval around $x = 0$ for which the given initial-value problem has a unique solution.

11. $(x - 2)y'' + 3y = x, \quad y(0) = 0, \quad y'(0) = 1$

12. $y'' + (\tan x)y = e^x, \quad y(0) = 1, \quad y'(0) = 0$

13. Given that $y = c_1 \cos \lambda x + c_2 \sin \lambda x$ is a family of solutions of the differential equation $y'' + \lambda^2y = 0$, determine the values of the parameter λ for which the boundary-value problem

$$y'' + \lambda^2y = 0, \qquad y(0) = 0, \qquad y(\pi) = 0$$

has nontrivial solutions.

14. Determine the values of the parameter λ for which the boundary-value problem

$$y'' + \lambda^2 y = 0, \qquad y(0) = 0, \qquad y(5) = 0$$

has nontrivial solutions (see Problem 13).

[4.1.2] In Problems 15–22 determine whether the given functions are linearly independent or dependent on $(-\infty, \infty)$.

15. $f_1(x) = x, \quad f_2(x) = x^2, \quad f_3(x) = 4x - 3x^2$

16. $f_1(x) = 0, \quad f_2(x) = x, \quad f_3(x) = e^x$

17. $f_1(x) = 5, \quad f_2(x) = \cos^2 x, \quad f_3(x) = \sin^2 x$

18. $f_1(x) = \cos 2x, \quad f_2(x) = 1, \quad f_3(x) = \cos^2 x$

19. $f_1(x) = x, \quad f_2(x) = x - 1, \quad f_3(x) = x + 3$

20. $f_1(x) = 2 + x, \quad f_2(x) = 2 + |x|$

21. $f_1(x) = 1 + x, \quad f_2(x) = x, \quad f_3(x) = x^2$

22. $f_1(x) = e^x, \quad f_2(x) = e^{-x}, \quad f_3(x) = \sinh x$

In Problems 23–28 show by computing the Wronskian that the given functions are linearly independent on the indicated interval.

23. $x^{1/2}, x^2; \quad (0, \infty)$

24. $1 + x, x^3; \quad (-\infty, \infty)$

25. $\sin x, \csc x; \quad (0, \pi)$

26. $\tan x, \cot x; \quad (0, \pi/2)$

27. $e^x, e^{-x}, e^{4x}; \quad (-\infty, \infty)$

28. $x, x \ln x, x^2 \ln x; \quad (0, \infty)$

29. Observe that for the functions $f_1(x) = 2$ and $f_2(x) = e^x$

$$1 \cdot f_1(0) - 2 \cdot f_2(0) = 0.$$

Does this imply that f_1 and f_2 are linearly dependent on any interval containing $x = 0$?

30. **(a)** Show graphically that $f_1(x) = x^2$ and $f_2(x) = x|x|$ are linearly independent on $(-\infty, \infty)$.

(b) Show that $W(f_1(x), f_2(x)) = 0$ for every real number.

[4.1.3] **31.** **(a)** Verify that $y = 1/x$ is a solution of the nonlinear differential equation $y'' = 2y^3$ on the interval $(0, \infty)$.

(b) Show that the constant multiple $y = c/x$ is not a solution of the equation when $c \neq 0, \pm 1$.

32. **(a)** Verify that $y_1 = 1$ and $y_2 = \ln x$ are solutions of the nonlinear differential equation $y'' + (y')^2 = 0$ on the interval $(0, \infty)$.

(b) Is $y_1 + y_2$ a solution of the equation? Is $c_1 y_1 + c_2 y_2$, c_1 and c_2 arbitrary, a solution of the equation?

In Problems 33–40 verify that the given functions form a fundamental set of solutions of the differential equation on the indicated interval. Form the general solution.

33. $y'' - y' - 12y = 0; \quad e^{-3x}, e^{4x}, (-\infty, \infty)$

34. $y'' - 4y = 0; \quad \cosh 2x, \sinh 2x, (-\infty, \infty)$

35. $y'' - 2y' + 5y = 0; \quad e^x \cos 2x, e^x \sin 2x, (-\infty, \infty)$

36. $4y'' - 4y' + y = 0; \quad e^{x/2}, xe^{x/2}, (-\infty, \infty)$

37. $x^2y'' - 6xy' + 12y = 0; \quad x^3, x^4, (0, \infty)$

38. $x^2y'' + xy' + y = 0; \quad \cos(\ln x), \sin(\ln x), (0, \infty)$

39. $x^3y''' + 6x^2y'' + 4xy' - 4y = 0; \quad x, x^{-2}, x^{-2} \ln x, (0, \infty)$

40. $y^{(4)} + y'' = 0; \quad 1, x, \cos x, \sin x, (-\infty, \infty)$

In Problems 41–44 verify that the given two-parameter family of functions is the general solution of the nonhomogeneous differential equation on the indicated interval.

41. $y'' - 7y' + 10y = 24e^x;$
$y = c_1e^{2x} + c_2e^{5x} + 6e^x, (-\infty, \infty)$

42. $y'' + y = \sec x;$
$y = c_1 \cos x + c_2 \sin x + x \sin x + (\cos x) \ln(\cos x), (-\pi/2, \pi/2)$

43. $y'' - 4y' + 4y = 2e^{2x} + 4x - 12;$
$y = c_1e^{2x} + c_2xe^{2x} + x^2e^{2x} + x - 2, (-\infty, \infty)$

44. $2x^2y'' + 5xy' + y = x^2 - x;$
$y = c_1x^{-1/2} + c_2x^{-1} + \frac{1}{15}x^2 - \frac{1}{6}x, (0, \infty)$

45. **(a)** Verify that $y_1 = x^3$ and $y_2 = |x|^3$ are linearly independent solutions of the differential equation $x^2y'' - 4xy' + 6y = 0$ on $(-\infty, \infty)$.

(b) Show that $W(y_1, y_2) = 0$ for every real number.

(c) Does the result of part (b) violate Theorem 4.4?

(d) Verify that $Y_1 = x^3$ and $Y_2 = x^2$ are also linearly independent solutions of the differential equation on the interval $(-\infty, \infty)$.

(e) Find a solution of the equation satisfying $y(0) = 0, y'(0) = 0$.

(f) By the superposition principle both linear combinations

$$y = c_1y_1 + c_2y_2 \quad \text{and} \quad y = c_1Y_1 + c_2Y_2$$

are solutions of the differential equation. Is one, both, or neither the general solution of the differential equation on $(-\infty, \infty)$?

46. Consider the second-order differential equation

$$a_2(x)y'' + a_1(x)y' + a_0(x)y = 0, \tag{9}$$

where $a_2(x)$, $a_1(x)$, and $a_0(x)$ are continuous on an interval I and $a_2(x) \neq 0$ for every x in the interval. From Theorem 4.1 there exists only one solution y_1 of the equation satisfying $y(x_0) = 1$ and $y'(x_0) = 0$, where x_0 is a point in I. Similarly there exists a unique solution y_2 of the equation satisfying $y(x_0) = 0$ and $y'(x_0) = 1$. Show that y_1 and y_2 form a fundamental set of solutions of the differential equation on the interval I.

Miscellaneous Problems

47. Let y_1 and y_2 be two solutions of (9).

(a) If $W(y_1, y_2)$ is the Wronskian of y_1 and y_2, show that

$$a_2(x)\frac{dW}{dx} + a_1(x)W = 0.$$

(b) Derive **Abel's formula***

$$W = ce^{-\int [a_1(x)/a_2(x)]\,dx},$$

where c is a constant.

(c) Using an alternative form of Abel's formula,

$$W = ce^{-\int_{x_0}^{x} [a_1(t)/a_2(t)]\,dt},$$

for x_0 in I show that

$$W(y_1, y_2) = W(x_0)e^{-\int_{x_0}^{x} [a_1(t)/a_2(t)]\,dt}.$$

(d) Show that if $W(x_0) = 0$, then $W = 0$ for every x in I, whereas if $W(x_0) \neq 0$, then $W \neq 0$ for every x in the interval.

In Problems 48 and 49 use the results of Problem 47.

48. If y_1 and y_2 are two solutions of

$$(1 - x^2)y'' - 2xy' + n(n + 1)y = 0$$

on $(-1, 1)$, show that $W(y_1, y_2) = c/(1 - x^2)$, where c is a constant.

---

* **Niels Henrik Abel** (1802–1829) Abel was a brilliant Norwegian mathematician whose tragic death at age 26 due to tuberculosis was an inestimable loss for mathematics. His greatest achievement was the solution of a problem that baffled mathematicians for centuries: he showed that a general fifth-degree polynomial equation cannot be solved algebraically—that is, in terms of radicals. Abel's contemporary, the Frenchman Evariste Galois, then proved that it was impossible to solve *any* general polynomial equation of degree greater than four in an algebraic manner. Galois is another tragic figure in the history of mathematics; a political activist, he was killed in a duel at the age of 22.

49. In Chapter 6 we shall see that the solutions y_1 and y_2 of $xy'' + y' + xy = 0$, $0 < x < \infty$, are infinite series. Suppose we consider initial conditions

$$y_1(x_0) = k_1 \qquad y_1'(x_0) = k_2$$

and

$$y_2(x_0) = k_3 \qquad y_2'(x_0) = k_4$$

for $x_0 > 0$. Show that

$$W(y_1, y_2) = \frac{(k_1k_4 - k_2k_3)x_0}{x}.$$

50. Suppose f is a differentiable function on an interval I and $f(x) \neq 0$ for every x in the interval. Prove that $f(x)$ and $xf(x)$ are linearly independent on I.

51. Suppose the mathematical model of a linear system is given by

$$a_2(t)\frac{d^2y}{dt^2} + a_1(t)\frac{dy}{dt} + a_0(t) = E(t).$$

If y_1 is a response of the system to an input $E_1(t)$ and y_2 is a response of the same system to an input $E_2(t)$, show that $y_1 + y_2$ is a response of the system to the input $E_1(t) + E_2(t)$.

4.2 Constructing a Second Solution from a Known Solution

Reduction of Order

It is one of the more interesting as well as important facts of life in the study of linear *second-order* differential equations that we can construct a second solution from a *known* solution. Suppose $y_1(x)$ is a nonzero solution of the equation*

$$a_2(x)y'' + a_1(x)y' + a_0(x)y = 0. \tag{1}$$

The process we shall use to find a second solution $y_2(x)$ consists of **reducing the order** of equation (1) to a first-order equation. For example, it is easily verified that $y_1 = e^x$ satisfies the differential equation $y'' - y = 0$. If we try to determine a solution of the form $y = u(x)e^x$, then

$$y' = ue^x + e^x u'$$

* We naturally assume, as we did in the preceding section, that the coefficients in (1) are continuous and $a_2(x) \neq 0$ for every x in some interval I.

$$y'' = ue^x + 2e^x u' + e^x u''$$

and so $$y'' - y = e^x(u'' + 2u') = 0.$$

Since $e^x \neq 0$, this last equation requires that $u'' + 2u' = 0$.

If we let $w = u'$, then the latter equation is recognized as a linear first-order equation $w' + 2w = 0$. Using the integrating factor e^{2x}, we can write

$$\frac{d}{dx}[e^{2x}w] = 0$$

$$w = c_1 e^{-2x} \quad \text{or} \quad u' = c_1 e^{-2x}.$$

Thus $$u = -\frac{c_1}{2}e^{-2x} + c_2$$

and so $$y = u(x)e^x = -\frac{c_1}{2}e^{-x} + c_2 e^x.$$

By picking $c_2 = 0$ and $c_1 = -2$, we obtain the second solution $y_2 = e^{-x}$. Since $W(e^x, e^{-x}) \neq 0$ for every x, the solutions are linearly independent on $(-\infty, \infty)$ and thus the expression for y is actually the general solution of the given equation.

EXAMPLE 1

Given that $y_1 = x^3$ is a solution of $x^2y'' - 6y = 0$. Use reduction of order to find a second solution on the interval $(0, \infty)$.

SOLUTION Define $$y = u(x)x^3$$

so that $$y' = 3x^2u + x^3u'$$

$$y'' = x^3u'' + 6x^2u' + 6xu$$

$$x^2y'' - 6y = x^2(x^3u'' + 6x^2u' + 6xu) - 6ux^3$$

$$= x^5u'' + 6x^4u' = 0$$

provided $u(x)$ is a solution of

$$x^5u'' + 6x^4u' = 0 \quad \text{or} \quad u'' + \frac{6}{x}u' = 0.$$

If $w = u'$, we obtain the linear first-order equation

$$w' + \frac{6}{x}w = 0,$$

which possesses the integrating factor $e^{6\int dx/x} = e^{6\ln x} = x^6$. Now

$$\frac{d}{dx}[x^6w] = 0 \quad \text{gives} \quad x^6w = c_1.$$

Therefore
$$w = u' = \frac{c_1}{x^6}$$
$$u = -\frac{c_1}{5x^5} + c_2$$
$$y = u(x)x^3 = -\frac{c_1}{5x^2} + c_2x^3.$$

Choosing $c_2 = 0$ and $c_1 = -5$ yields the second solution $y_2 = 1/x^2$.

General Case

Suppose we divide by $a_2(x)$ in order to put equation (1) in the standard form
$$y'' + P(x)y' + Q(x)y = 0, \tag{2}$$
where $P(x)$ and $Q(x)$ are continuous on some interval I. Let us suppose further that $y_1(x)$ is a known solution of (2) on I and that $y_1(x) \neq 0$ for every x in the interval. If we define $y = u(x)y_1(x)$, it follows that
$$y' = uy_1' + y_1u'$$
$$y'' = uy_1'' + 2y_1'u' + y_1u''$$
$$y'' + Py' + Qy = u\underbrace{[y_1'' + Py_1' + Qy_1]}_{\text{zero}} + y_1u'' + (2y_1' + Py_1)u' = 0.$$

This implies that we must have
$$y_1u'' + (2y_1' + Py_1)u' = 0$$
or
$$y_1w' + (2y_1' + Py_1)w = 0, \tag{3}$$
where we have let $w = u'$. Observe the equation (3) is both linear and separable. Applying the latter technique, we obtain
$$\frac{dw}{w} + 2\frac{y_1'}{y_1}\,dx + P\,dx = 0$$
$$\ln|w| + 2\ln|y_1| = -\int P\,dx + c$$
$$\ln|wy_1^2| = -\int P\,dx + c$$
$$wy_1^2 = c_1e^{-\int P\,dx}$$
$$w = u' = c_1\frac{e^{-\int P\,dx}}{y_1^2}.$$

Integrating again gives $u = c_1 \int \frac{e^{-\int P\,dx}}{y_1^2}\,dx + c_2$ and therefore

$$y = u(x)y_1(x) = c_1 y_1(x) \int \frac{e^{-\int P(x)\,dx}}{y_1^2(x)}\,dx + c_2 y_1(x).$$

By choosing $c_2 = 0$ and $c_1 = 1$, we find that a second solution of equation (2) is

$$y_2 = y_1(x) \int \frac{e^{-\int P(x)\,dx}}{y_1^2(x)}\,dx. \tag{4}$$

It makes a good review exercise in differentiation to start with formula (4) and actually verify that equation (2) is satisfied.

Now $y_1(x)$ and $y_2(x)$ are linearly independent since

$$W(y_1(x), y_2(x)) = \begin{vmatrix} y_1 & y_1 \int \frac{e^{-\int P\,dx}}{y_1^2}\,dx \\ y_1' & \frac{e^{-\int P\,dx}}{y_1} + y_1' \int \frac{e^{-\int P\,dx}}{y_1^2}\,dx \end{vmatrix} = e^{-\int P\,dx}$$

is not zero on any interval on which $y_1(x)$ is not zero.*

EXAMPLE 2

The function $y_1 = x^2$ is a solution of $x^2y'' - 3xy' + 4y = 0$. Find the general solution on the interval $(0, \infty)$.

SOLUTION Since the equation has the alternative form

$$y'' - \frac{3}{x}y' + \frac{4}{x^2}y = 0,$$

we find from (4)

$$y_2 = x^2 \int \frac{e^{3\int dx/x}}{x^4}\,dx \qquad [e^{3\int dx/x} = e^{\ln x^3} = x^3]$$

$$= x^2 \int \frac{dx}{x} = x^2 \ln x.$$

The general solution on $(0, \infty)$ is given by $y = c_1y_1 + c_2y_2$;

that is, $$y = c_1x^2 + c_2x^2 \ln x.$$

---

* Alternatively, if $y_2 = u(x)y_1$ then $W(y_1, y_2) = u'(y_1)^2 \neq 0$, since $y_1 \neq 0$ for every x on some interval. If $u' = 0$, then $u =$ constant.

EXAMPLE 3 It can be verified that $y_1 = \dfrac{\sin x}{\sqrt{x}}$ is a solution of $x^2y'' + xy' + (x^2 - \frac{1}{4})y = 0$ on $(0, \pi)$. Find a second solution.

SOLUTION First put the equation into the form

$$y'' + \frac{1}{x}y' + \left(1 - \frac{1}{4x^2}\right)y = 0.$$

Then from (4) we have

$$y_2 = \frac{\sin x}{\sqrt{x}} \int \frac{e^{-\int dx/x}}{\left(\dfrac{\sin x}{\sqrt{x}}\right)^2} dx$$

$$= \frac{\sin x}{\sqrt{x}} \int \csc^2 x\, dx \qquad [e^{-\int dx/x} = e^{\ln x^{-1}} = x^{-1}]$$

$$= \frac{\sin x}{\sqrt{x}}(-\cot x) = -\frac{\cos x}{\sqrt{x}}.$$

Since the differential equation is homogeneous, we can disregard the negative sign and take the second solution to be $y_2 = (\cos x)/\sqrt{x}$.

Observe that $y_1(x)$ and $y_2(x)$ of the previous example are linearly independent solutions of the given differential equation on the larger interval $(0, \infty)$.

Exercises 4.2

Answers to odd-numbered problems begin on page A–10.

In Problems 1–30 find a second solution of each differential equation. Assume an appropriate interval of validity.

1. $y'' + 5y' = 0; \quad y_1 = 1$

2. $y'' - y' = 0; \quad y_1 = 1$

3. $y'' - 4y' + 4y = 0; \quad y_1 = e^{2x}$

4. $y'' + 2y' + y = 0; \quad y_1 = xe^{-x}$

5. $y'' + 16y = 0; \quad y_1 = \cos 4x$

6. $y'' + 9y = 0; \quad y_1 = \sin 3x$

7. $y'' - y = 0; \quad y_1 = \cosh x$

8. $y'' - 25y = 0; \quad y_1 = e^{5x}$

9. $9y'' - 12y' + 4y = 0; \quad y_1 = e^{2x/3}$

10. $6y'' + y' - y = 0; \quad y_1 = e^{x/3}$

11. $x^2y'' - 7xy' + 16y = 0; \quad y_1 = x^4$

12. $x^2y'' + 2xy' - 6y = 0; \quad y_1 = x^2$

13. $xy'' + y' = 0; \quad y_1 = \ln x$

14. $4x^2y'' + y = 0; \quad y_1 = x^{1/2} \ln x$

15. $(1 - 2x - x^2)y'' + 2(1 + x)y' - 2y = 0; \quad y_1 = x + 1$

16. $(1 - x^2)y'' - 2xy' = 0; \quad y_1 = 1$

17. $x^2y'' - xy' + 2y = 0; \quad y_1 = x \sin(\ln x)$

18. $x^2y'' - 3xy' + 5y = 0; \quad y_1 = x^2 \cos(\ln x)$

19. $(1 + 2x)y'' + 4xy' - 4y = 0; \quad y_1 = e^{-2x}$

20. $(1 + x)y'' + xy' - y = 0; \quad y_1 = x$

21. $x^2y'' - xy' + y = 0; \quad y_1 = x$

22. $x^2y'' - 20y = 0; \quad y_1 = x^{-4}$

23. $x^2y'' - 5xy' + 9y = 0; \quad y_1 = x^3 \ln x$

24. $x^2y'' + xy' + y = 0; \quad y_1 = \cos(\ln x)$

25. $x^2y'' - 4xy' + 6y = 0; \quad y_1 = x^2 + x^3$

26. $x^2y'' - 7xy' - 20y = 0; \quad y_1 = x^{10}$

27. $(3x + 1)y'' - (9x + 6)y' + 9y = 0; \quad y_1 = e^{3x}$

28. $xy'' - (x + 1)y' + y = 0; \quad y_1 = e^x$

29. $y'' - 3(\tan x)y' = 0; \quad y_1 = 1$

30. $xy'' - (2 + x)y' = 0; \quad y_1 = 1$

In Problems 31–34 use the method of reduction of order to find a solution of the given nonhomogeneous equation. The indicated function $y_1(x)$ is a solution of the associated homogeneous equation. Determine a second solution of the homogeneous equation and a particular solution of the nonhomogeneous equation.

31. $y'' - 4y = 2; \quad y_1 = e^{-2x}$

32. $y'' + y' = 1; \quad y_1 = 1$

33. $y'' - 3y' + 2y = 5e^{3x}; \quad y_1 = e^x$

34. $y'' - 4y' + 3y = x; \quad y_1 = e^x$

Miscellaneous Problems

35. Verify by direct substitution that formula (4) satisfies equation (2).

4.3 Homogeneous Linear Equations with Constant Coefficients

We have seen that the linear first-order equation $dy/dx + ay = 0$, where a is a constant, has the exponential solution $y = c_1e^{-ax}$ on $(-\infty, \infty)$. Therefore, it is natural to seek to determine whether exponential solutions exist on $(-\infty, \infty)$

for higher-order equations such as

$$a_n y^{(n)} + a_{n-1} y^{(n-1)} + \cdots + a_2 y'' + a_1 y' + a_0 y = 0, \tag{1}$$

where the a_i, $i = 0, 1, \ldots, n$ are constants. The surprising fact is that *all* solutions of (1) are exponential functions or constructed out of exponential functions. We shall begin by considering the special case of the second-order equation

$$ay'' + by' + cy = 0. \tag{2}$$

Auxiliary Equation

If we try a solution of the form $y = e^{mx}$, then $y' = me^{mx}$ and $y'' = m^2 e^{mx}$ so that equation (2) becomes

$$am^2 e^{mx} + bme^{mx} + ce^{mx} = 0 \qquad \text{or} \qquad e^{mx}(am^2 + bm + c) = 0.$$

Because e^{mx} is never zero for real values of x, it is apparent that the only way that this exponential function can satisfy the differential equation is to choose m so that it is a root of the quadratic equation

$$am^2 + bm + c = 0. \tag{3}$$

This latter equation is called the **auxiliary equation**, or **characteristic equation**, of the differential equation (2). We shall consider three cases, namely, the solutions of the auxiliary equation corresponding to real distinct roots, real but equal roots, and lastly, complex conjugate roots.

CASE I Under the assumption that the auxiliary equation (3) has two unequal real roots m_1 and m_2, we find two solutions,

$$y_1 = e^{m_1 x} \qquad \text{and} \qquad y_2 = e^{m_2 x}.$$

We have already seen that these functions are linearly independent on $(-\infty, \infty)$ (see page 138) and hence form a fundamental set. It follows that the general solution of (2) on this interval is

$$y = c_1 e^{m_1 x} + c_2 e^{m_2 x}. \tag{4}$$

■

CASE II When $m_1 = m_2$ we necessarily obtain only one exponential solution $y_1 = e^{m_1 x}$. However, it follows immediately from the discussion of Section 4.2 that a second solution is

$$y_2 = e^{m_1 x} \int \frac{e^{-(b/a)x}}{e^{2m_1 x}}\, dx. \tag{5}$$

But from the quadratic formula, we have $m_1 = -b/2a$ since the only way to have $m_1 = m_2$ is to have $b^2 - 4ac = 0$. In view of the fact that

$2m_1 = -b/a$, (5) becomes

$$y_2 = e^{m_1 x} \int \frac{e^{2m_1 x}}{e^{2m_1 x}}\, dx = e^{m_1 x} \int dx$$

$$= xe^{m_1 x}.$$

The general solution of (2) is then

$$y = c_1 e^{m_1 x} + c_2 x e^{m_1 x}. \tag{6}$$

■

CASE III If m_1 and m_2 are complex, then we can write

$$m_1 = \alpha + i\beta \qquad \text{and} \qquad m_2 = \alpha - i\beta,$$

where α and $\beta > 0$ are real and $i^2 = -1$. Formally there is no difference between this case and Case I, and hence

$$y = C_1 e^{(\alpha + i\beta)x} + C_2 e^{(\alpha - i\beta)x}. \tag{7}$$

However, in practice we would prefer to work with real functions instead of complex exponentials. Now we can write (7) in a more practical form by using **Euler's formula***

$$e^{i\theta} = \cos\theta + i\sin\theta,$$

where θ is any real number. From this result we can write

$$e^{i\beta x} = \cos\beta x + i\sin\beta x \quad \text{and} \quad e^{-i\beta x} = \cos\beta x - i\sin\beta x,$$

---

* **Leonhard Euler** (1707–1783) A man with a prodigious memory and phenomenal powers of concentration, Euler had almost universal interests; he was a theologian, physicist, astronomer, linguist, physiologist, classical scholar, and, primarily, a mathematician. Euler is considered to be an example of a true genius of the era. In mathematics he made lasting contributions to algebra, trigonometry, analytic geometry, calculus, calculus of variations, differential equations, complex variables, number theory, and topology. The volume of his mathematical output did not seem to be affected by the distractions of thirteen children or the fact that he was totally blind for the last seventeen years of his life. Euler wrote over 700 papers and 32 books on mathematics and was responsible for introducing many of the symbols (such as e, π, and $i = \sqrt{-1}$) and notations that are still used (such as $f(x)$, Σ, $\sin x$, and $\cos x$.) Euler was born in Basel, Switzerland, on April 15, 1707, and died of a stroke in St. Petersburg on September 18, 1783, while serving in the court of the Russian empress Catherine the Great.

See Appendix V for a review of complex numbers and a derivation of Euler's formula.

where we have used $\cos(-\beta x) = \cos \beta x$ and $\sin(-\beta x) = -\sin \beta x$. Thus (7) becomes

$$\begin{aligned} y &= e^{\alpha x}[C_1 e^{i\beta x} + C_2 e^{-i\beta x}] \\ &= e^{\alpha x}[C_1\{\cos \beta x + i \sin \beta x\} + C_2\{\cos \beta x - i \sin \beta x\}] \\ &= e^{\alpha x}[(C_1 + C_2)\cos \beta x + (C_1 i - C_2 i)\sin \beta x]. \end{aligned}$$

Since $e^{\alpha x}\cos \beta x$ and $e^{\alpha x}\sin \beta x$ themselves form a fundamental set of solutions of the given differential equation on $(-\infty, \infty)$, we can simply relabel $C_1 + C_2$ as c_1 and $C_1 i - C_2 i$ as c_2 and use the superposition principle to write the general solution

$$\begin{aligned} y &= c_1 e^{\alpha x}\cos \beta x + c_2 e^{\alpha x}\sin \beta x \\ &= e^{\alpha x}(c_1 \cos \beta x + c_2 \sin \beta x). \end{aligned} \tag{8}$$

■

EXAMPLE 1

Solve the following differential equations.

(a) $2y'' - 5y' - 3y = 0$

(b) $y'' - 10y' + 25y = 0$

(c) $y'' + y' + y = 0$

SOLUTION **(a)** $2m^2 - 5m - 3 = (2m + 1)(m - 3) = 0$

$$m_1 = -\frac{1}{2}, \qquad m_2 = 3$$

$$y = c_1 e^{-x/2} + c_2 e^{3x}$$

(b) $m^2 - 10m + 25 = (m - 5)^2 = 0$

$$m_1 = m_2 = 5$$

$$y = c_1 e^{5x} + c_2 x e^{5x}$$

(c) $m^2 + m + 1 = 0$

$$m = \frac{-1 \pm \sqrt{-3}}{2}$$

$$m_1 = -\frac{1}{2} + \frac{\sqrt{3}}{2} i, \qquad m_2 = -\frac{1}{2} - \frac{\sqrt{3}}{2} i$$

$$y = e^{-x/2}\left(c_1 \cos \frac{\sqrt{3}}{2} x + c_2 \sin \frac{\sqrt{3}}{2} x\right)$$

EXAMPLE 2

Solve $$y'' - 4y' + 13y = 0$$

subject to $$y(0) = -1, \qquad y'(0) = 2.$$

SOLUTION The roots of the auxiliary equation $m^2 - 4m + 13 = 0$ are $m_1 = 2 + 3i$ and $m_2 = 2 - 3i$ so that

$$y = e^{2x}(c_1 \cos 3x + c_2 \sin 3x).$$

The condition $y(0) = -1$ implies

$$-1 = e^0(c_1 \cos 0 + c_2 \sin 0) = c_1,$$

from which we can write

$$y = e^{2x}(-\cos 3x + c_2 \sin 3x).$$

Differentiating this latter expression and using the second condition give

$$y' = e^{2x}(3 \sin 3x + 3c_2 \cos 3x) + 2e^{2x}(-\cos 3x + c_2 \sin 3x)$$
$$2 = 3c_2 - 2$$

and so $c_2 = 4/3$. Hence

$$y = e^{2x}\left(-\cos 3x + \frac{4}{3} \sin 3x\right).$$

EXAMPLE 3

The two equations $$y'' + k^2 y = 0 \tag{9}$$

and $$y'' - k^2 y = 0 \tag{10}$$

are frequently encountered in the study of applied mathematics. For the former differential equation, the auxiliary equation $m^2 + k^2 = 0$ has the roots $m_1 = ki$ and $m_2 = -ki$. It follows from (8) that the general solution of (9) is

$$y = c_1 \cos kx + c_2 \sin kx. \tag{11}$$

The latter differential equation has the auxiliary equation $m^2 - k^2 = 0$, with real roots $m_1 = k$ and $m_2 = -k$, so that its general solution is

$$y = c_1 e^{kx} + c_2 e^{-kx}. \tag{12}$$

Notice that if we choose $c_1 = c_2 = 1/2$ in (12), then

$$y = \frac{e^{kx} + e^{-kx}}{2} = \cosh kx$$

is also a solution of (10). Furthermore, when $c_1 = 1/2$, $c_2 = -1/2$ then (12) becomes

$$y = \frac{e^{kx} - e^{-kx}}{2} = \sinh kx.$$

Since cosh kx and sinh kx are linearly independent on any interval of the x-axis, they form a fundamental set. Thus an alternative form for the general solution of (10) is

$$y = c_1 \cosh kx + c_2 \sinh kx. \tag{13}$$

Higher-Order Equations

In general, to solve an nth-order differential equation

$$a_n y^{(n)} + a_{n-1} y^{(n-1)} + \cdots + a_2 y'' + a_1 y' + a_0 y = 0, \tag{14}$$

where the a_i, $i = 0, 1, \ldots, n$ are real constants, we must solve an nth-degree polynominal equation

$$a_n m^n + a_{n-1} m^{n-1} + \cdots + a_2 m^2 + a_1 m + a_0 = 0. \tag{15}$$

If all the roots of (15) are real and distinct, then the general solution of (14) is

$$y = c_1 e^{m_1 x} + c_2 e^{m_2 x} + \cdots + c_n e^{m_n x}. \tag{16}$$

It is somewhat harder to summarize the analogues of Cases II and III because the roots of an auxiliary equation of degree greater than two can occur in many combinations. For example, a fifth-degree equation could have five distinct real roots, or three distinct real and two complex roots, or one real and four complex roots, or five real but equal roots, or five real roots but two of them equal, and so on. When m_1 is a root of multiplicity k of an nth-degree auxiliary equation (that is, k roots are equal to m_1), then it can be shown that the linearly independent solutions are

$$e^{m_1 x},\ xe^{m_1 x},\ x^2 e^{m_1 x}, \ldots, x^{k-1} e^{m_1 x}$$

and the general solution must contain the linear combination

$$c_1 e^{m_1 x} + c_2 x e^{m_1 x} + c_3 x^2 e^{m_1 x} + \cdots + c_k x^{k-1} e^{m_1 x}.$$

Lastly, it should be remembered that when the coefficients are real, complex roots of an auxiliary equation will always appear in conjugate pairs. Thus, for example, a cubic polynomial equation can have at most two complex roots.

EXAMPLE 4

Solve
$$y''' + 3y'' - 4y = 0.$$

SOLUTION It should be apparent from inspection of

$$m^3 + 3m^2 - 4 = 0$$

that one root is $m_1 = 1$. Now if we divide* $m^3 + 3m^2 - 4$ by $m - 1$, we find

$$\begin{aligned} m^3 + 3m^2 - 4 &= (m - 1)(m^2 + 4m + 4) \\ &= (m - 1)(m + 2)^2, \end{aligned}$$

and so the other roots are $m_2 = m_3 = -2$. Thus the general solution is

$$y = c_1 e^x + c_2 e^{-2x} + c_3 x e^{-2x}.$$

Of course, the most difficult aspect of solving constant-coefficient equations is finding the roots of auxiliary equations of degree greater than two. As illustrated in the preceding example, one way to solve an equation is to guess† a root m_1 and then divide by $m - m_1$ to obtain the factorization $(m - m_1)Q(m)$. We then try to find the roots of $Q(m)$.

EXAMPLE 5 Solve

$$3y''' - 19y'' + 36y' - 10y = 0.$$

SOLUTION It is easily verified that $m_1 = 1/3$ is one root of

$$3m^3 - 19m^2 + 36m - 10 = 0.$$

By division we find that

$$\begin{aligned} 3m^3 - 19m^2 + 36m - 10 &= \left(m - \frac{1}{3}\right)(3m^2 - 18m + 30) \\ &= (3m - 1)(m^2 - 6m + 10). \end{aligned}$$

---

* Synthetic division provides a quick way of testing for roots. In the above case

$$\begin{array}{rrrrl} 1 & 3 & 0 & -4 & \lfloor +1 \\ & 1 & 4 & 4 & \\ \hline 1 & 4 & 4 & \lfloor 0 = R. & \end{array}$$

If $R \neq 0$ the number tested is not a root. For example, dividing $m + 1$ is equivalent to

$$\begin{array}{rrrrl} 1 & 3 & 0 & -4 & \lfloor -1 \\ & -1 & -2 & 2 & \\ \hline 1 & 2 & -2 & \lfloor -2 = R. & \end{array}$$

Therefore, $m = -1$ is not a root and so $m + 1$ is not a factor.

† If the equation $a_n m^n + \cdots + a_1 m + a_0 = 0$ has a *rational* real root $m_1 = p/q$, where p and q are integers, then q must be a factor of a_n and p must be a factor of a_0. We can test all the possible ratios by synthetic division.

From the quadratic formula we find $m_2 = 3 + i$ and $m_3 = 3 - i$. The general solution is

$$y = c_1 e^{x/3} + e^{3x}(c_2 \cos x + c_3 \sin x).$$

EXAMPLE 6

Solve
$$\frac{d^4 y}{dx^4} + 2\frac{d^2 y}{dx^2} + y = 0.$$

SOLUTION The auxiliary equation

$$m^4 + 2m^2 + 1 = (m^2 + 1)^2 = 0$$

has roots $m_1 = m_3 = i$ and $m_2 = m_4 = -i$. Thus from Case II the solution is

$$y = C_1 e^{ix} + C_2 e^{-ix} + C_3 x e^{ix} + C_4 x e^{-ix}.$$

By Euler's formula the grouping $C_1 e^{ix} + C_2 e^{-ix}$ can be rewritten as

$$c_1 \cos x + c_2 \sin x$$

after a relabeling of constants. Similarly, $x(C_3 e^{ix} + C_4 e^{-ix})$ can be expressed as $x(c_3 \cos x + c_4 \sin x)$. Hence the general solution is

$$y = c_1 \cos x + c_2 \sin x + c_3 x \cos x + c_4 x \sin x.$$

When $m_1 = \alpha + i\beta$ is a complex root of multiplicity k of an auxiliary equation with real coefficients, its complex conjugate $m_2 = \alpha - i\beta$ is also a root of multiplicity k. The general solution of the corresponding differential equation must then contain a linear combination of the linearly independent solutions

$$e^{\alpha x} \cos \beta x,\ x e^{\alpha x} \cos \beta x,\ x^2 e^{\alpha x} \cos \beta x, \ldots, x^{k-1} e^{\alpha x} \cos \beta x,$$

$$e^{\alpha x} \sin \beta x,\ x e^{\alpha x} \sin \beta x,\ x^2 e^{\alpha x} \sin \beta x, \ldots, x^{k-1} e^{\alpha x} \sin \beta x.$$

In Example 6 we identify $k = 2$, $\alpha = 0$, and $\beta = 1$.

Exercises 4.3

Answers to odd-numbered problems begin on page A–10.

In Problems 1–36 find the general solution of the given differential equation.

1. $4y'' + y' = 0$

2. $2y'' - 5y' = 0$

3. $y'' - 36y = 0$

4. $y'' - 8y = 0$

5. $y'' + 9y = 0$

6. $3y'' + y = 0$

7. $y'' - y' - 6y = 0$

8. $y'' - 3y' + 2y = 0$

9. $\dfrac{d^2y}{dx^2} + 8\dfrac{dy}{dx} + 16y = 0$

10. $\dfrac{d^2y}{dx^2} - 10\dfrac{dy}{dx} + 25y = 0$

11. $y'' + 3y' - 5y = 0$

12. $y'' + 4y' - y = 0$

13. $12y'' - 5y' - 2y = 0$

14. $8y'' + 2y' - y = 0$

15. $y'' - 4y' + 5y = 0$

16. $2y'' - 3y' + 4y = 0$

17. $3y'' + 2y' + y = 0$

18. $2y'' + 2y' + y = 0$

19. $y''' - 4y'' - 5y' = 0$

20. $4y''' + 4y'' + y' = 0$

21. $y''' - y = 0$

22. $y''' + 5y'' = 0$

23. $y''' - 5y'' + 3y' + 9y = 0$

24. $y''' + 3y'' - 4y' - 12y = 0$

25. $y''' + y'' - 2y = 0$

26. $y''' - y'' - 4y = 0$

27. $y''' + 3y'' + 3y' + y = 0$

28. $y''' - 6y'' + 12y' - 8y = 0$

29. $\dfrac{d^4y}{dx^4} + \dfrac{d^3y}{dx^3} + \dfrac{d^2y}{dx^2} = 0$

30. $\dfrac{d^4y}{dx^4} - 2\dfrac{d^2y}{dx^2} + y = 0$

31. $16\dfrac{d^4y}{dx^4} + 24\dfrac{d^2y}{dx^2} + 9y = 0$

32. $\dfrac{d^4y}{dx^4} - 7\dfrac{d^2y}{dx^2} - 18y = 0$

33. $\dfrac{d^5y}{dx^5} - 16\dfrac{dy}{dx} = 0$

34. $\dfrac{d^5y}{dx^5} - 2\dfrac{d^4y}{dx^4} + 17\dfrac{d^3y}{dx^3} = 0$

35. $\dfrac{d^5y}{dx^5} + 5\dfrac{d^4y}{dx^4} - 2\dfrac{d^3y}{dx^3} - 10\dfrac{d^2y}{dx^2} + \dfrac{dy}{dx} + 5y = 0$

36. $2\dfrac{d^5y}{dx^5} - 7\dfrac{d^4y}{dx^4} + 12\dfrac{d^3y}{dx^3} + 8\dfrac{d^2y}{dx^2} = 0$

In Problems 37–52 solve the given differential equation subject to the indicated initial conditions.

37. $y'' + 16y = 0, \quad y(0) = 2, \quad y'(0) = -2$

38. $y'' - y = 0, \quad y(0) = y'(0) = 1$

39. $y'' + 6y' + 5y = 0, \quad y(0) = 0, \quad y'(0) = 3$

40. $y'' - 8y' + 17y = 0, \quad y(0) = 4, \quad y'(0) = -1$

41. $2y'' - 2y' + y = 0, \quad y(0) = -1, \quad y'(0) = 0$

42. $y'' - 2y' + y = 0, \quad y(0) = 5, \quad y'(0) = 10$

43. $y'' + y' + 2y = 0, \quad y(0) = y'(0) = 0$

44. $4y'' - 4y' - 3y = 0, \quad y(0) = 1, \quad y'(0) = 5$

45. $y'' - 3y' + 2y = 0, \quad y(1) = 0, \quad y'(1) = 1$

46. $y'' + y = 0, \quad y(\pi/3) = 0, \quad y'(\pi/3) = 2$

47. $y''' + 12y'' + 36y' = 0, \quad y(0) = 0, \quad y'(0) = 1, \quad y''(0) = -7$

48. $y''' + 2y'' - 5y' - 6y = 0, \quad y(0) = y'(0) = 0, \quad y''(0) = 1$

49. $y''' - 8y = 0, \quad y(0) = 0, \quad y'(0) = -1, \quad y''(0) = 0$

50. $\dfrac{d^4y}{dx^4} = 0, \quad y(0) = 2, \quad y'(0) = 3, \quad y''(0) = 4, \quad y'''(0) = 5$

51. $\dfrac{d^4y}{dx^4} - 3\dfrac{d^3y}{dx^3} + 3\dfrac{d^2y}{dx^2} - \dfrac{dy}{dx} = 0, \quad y(0) = y'(0) = 0, \quad y''(0) = y'''(0) = 1$

52. $\dfrac{d^4y}{dx^4} - y = 0, \quad y(0) = y'(0) = y''(0) = 0, \quad y'''(0) = 1$

In Problems 53–56 solve the given differential equation subject to the indicated boundary conditions.

53. $y'' - 10y' + 25y = 0, \quad y(0) = 1, \quad y(1) = 0$

54. $y'' + 4y = 0, \quad y(0) = 0, \quad y(\pi) = 0$

55. $y'' + y = 0, \quad y'(0) = 0, \quad y'\left(\dfrac{\pi}{2}\right) = 2$

56. $y'' - y = 0, \quad y(0) = 1, \quad y'(1) = 0$

57. The roots of an auxiliary equation are $m_1 = 4$, $m_2 = m_3 = -5$. What is the corresponding differential equation?

58. The roots of an auxiliary equation are $m_1 = -\frac{1}{2}$, $m_2 = 3 + i$, $m_3 = 3 - i$. What is the corresponding differential equation?

In Problems 59 and 60 find the general solution of the given equation if it is known that y_1 is a solution.

59. $y''' - 9y'' + 25y' - 17y = 0; \quad y_1 = e^x$

60. $y''' + 6y'' + y' - 34y = 0; \quad y_1 = e^{-4x}\cos x$

In Problems 61–64 determine a linear differential equation with constant coefficients having the given solutions.

61. $4e^{6x}, 3e^{-3x}$

62. $10 \cos 4x, -5 \sin 4x$

63. $3, 2x, -e^{7x}$

64. $8 \sinh 3x, 12 \cosh 3x$

65. Use the fact that

$$i = \left(\frac{\sqrt{2}}{2} + \frac{\sqrt{2}}{2}i\right)^2 \quad \text{and} \quad -i = \left(\frac{\sqrt{2}}{2} - \frac{\sqrt{2}}{2}i\right)^2$$

to solve the differential equation

$$\frac{d^4y}{dx^4} + y = 0.$$

[*Hint:* Write the auxiliary equation $m^4 + 1 = 0$ as $(m^2 + 1)^2 - 2m^2 = 0$. See what happens when you factor.]

4.4 Undetermined Coefficients

4.4.1 Differential Operators

In calculus the symbol D^n is often used to denote the nth-order derivative of a function:

$$D^n y = \frac{d^n y}{dx^n}.$$

Hence a linear differential equation with constant coefficients

$$a_n y^{(n)} + a_{n-1} y^{(n-1)} + \cdots + a_2 y'' + a_1 y' + a_0 y = g(x)$$

can be written as

$$a_n D^n y + a_{n-1} D^{n-1} y + \cdots + a_2 D^2 y + a_1 Dy + a_0 y = g(x)$$

or

$$(a_n D^n + a_{n-1} D^{n-1} + \cdots + a_2 D^2 + a_1 D + a_0) y = g(x).$$

The expression

$$a_n D^n + a_{n-1} D^{n-1} + \cdots + a_2 D^2 + a_1 D + a_0 \tag{1}$$

is called a **linear nth-order differential operator**. Since (1) is polynomial in the symbol D, it is often abbreviated by $P(D)$.

It can be shown that when the a_i, $i = 0, 1, \ldots, n$ are constants

(i) $P(D)$ can possibly be factored into differential operators of lower order.* This is accomplished by treating $P(D)$ as an ordinary polynomial.

(ii) The factors of $P(D)$ commute.

---

* If one is willing to use complex numbers, then a differential operator with constant coefficients can *always* be factored. We are primarily concerned with writing differential equations in operator form with real coefficients.

EXAMPLE 1

(a) The operators $D^2 + D$ and $D^2 - 1$ can be factored as

$$D(D + 1) \quad \text{and} \quad (D + 1)(D - 1),$$

respectively.

(b) The operator $D^2 + 1$ does not factor using real numbers.

(c) The operator $D^2 + 5D + 6$ can be written as $(D + 2)(D + 3)$ or $(D + 3)(D + 2)$. The following example shows that these factors commute.

EXAMPLE 2

If $y = f(x)$ possesses a second derivative then

$$\begin{aligned}(D^2 + 5D + 6)y &= (D + 2)(D + 3)y \\ &= (D + 3)(D + 2)y.\end{aligned}$$

To show this let $w = (D + 3)y = y' + 3y$:

$$\begin{aligned}(D + 2)w &= Dw + 2w \\ &= \frac{d}{dx}[y' + 3y] + 2[y' + 3y] \\ &= y'' + 3y' + 2y' + 6y \\ &= y'' + 5y' + 6y.\end{aligned}$$

Similarly, if we let $w = (D + 2)y = y' + 2y$, then

$$\begin{aligned}(D + 3)w &= Dw + 3w \\ &= \frac{d}{dx}[y' + 2y] + 3[y' + 2y] \\ &= y'' + 2y' + 3y' + 6y \\ &= y'' + 5y' + 6y.\end{aligned}$$

EXAMPLE 3

The operator $D^2 + 4D + 4$ can be written as $(D + 2)(D + 2)$ or $(D + 2)^2$.

Annihilator Operator

Let $y = f(x)$ be a function that possesses at least n derivatives. If

$$(a_n D^n + a_{n-1} D^{n-1} + \cdots + a_1 D + a_0) f(x) = 0,$$

then the differential operator $a_n D^n + a_{n-1} D^{n-1} + \cdots + a_1 D + a_0$ is said to

annihilate f. For example, if $f(x) = k$, a constant, then $Dk = 0$. Also, $D^2x = 0$, $D^3x^2 = 0$, and so on.

> The differential operator D^n annihilates each of the functions
> $$1, x, x^2, \ldots, x^{n-1}. \tag{2}$$

As an immediate consequence of (2) and of the fact that differentiation can be done term-by-term, a polynomial

$$c_0 + c_1x + \cdots + c_{n-1}x^{n-1}$$

can be annihilated by finding an operator that annihilates the highest power of x.

EXAMPLE 4

Find a differential operator that annihilates $1 - 5x^2 + 8x^3$.

SOLUTION From (2) we know that $D^4x^3 = 0$, and so it follows that

$$D^4(1 - 5x^2 + 8x^3) = 0.$$

> The differential operator $(D - \alpha)^n$ annihilates each of the functions
> $$e^{\alpha x}, xe^{\alpha x}, x^2e^{\alpha x}, \ldots, x^{n-1}e^{\alpha x}. \tag{3}$$

To see this, note that the auxiliary equation of the homogeneous equation $(D - \alpha)^n y = 0$ is $(m - \alpha)^n = 0$. Since α is a root of multiplicity n, the general solution is

$$y = c_1e^{\alpha x} + c_2xe^{\alpha x} + \cdots + c_nx^{n-1}e^{\alpha x}. \tag{4}$$

EXAMPLE 5

Find an annihilator operator for **(a)** e^{5x}, **(b)** $4e^{2x} - 6xe^{2x}$.

SOLUTION **(a)** From (3) with $\alpha = 5$ and $n = 1$, we see

$$(D - 5)e^{5x} = 0.$$

(b) From (3) and (4) with $\alpha = 2$ and $n = 2$,

$$(D - 2)^2(4e^{2x} - 6xe^{2x}) = 0.$$

EXAMPLE 6

Find a differential operator that annihilates $e^{-3x} + xe^x$.

SOLUTION From (3),

$$(D + 3)e^{-3x} = 0 \qquad \text{and} \qquad (D - 1)^2 xe^x = 0.$$

The *product* of the two operators $(D + 3)(D - 1)^2$ will annihilate the given linear combination. Since this may not be obvious, a verification is in order:

$$\begin{aligned}(D + 3)(D - 1)^2(e^{-3x} + xe^x) &= (D + 3)[(D - 1)^2 e^{-3x} + (D - 1)^2 xe^x] \\ &= (D + 3)[16e^{-3x} + 0] \\ &= 16(D + 3)e^{-3x} = 0.\end{aligned}$$

When α and β are real numbers, the quadratic formula reveals that $[m^2 - 2\alpha m + (\alpha^2 + \beta^2)]^n = 0$ has complex roots $\alpha + i\beta$, $\alpha - i\beta$, both of multiplicity n. From the discussion of Section 4.3, we have the next result.

> The differential operator $[D^2 - 2\alpha D + (\alpha^2 + \beta^2)]^n$ annihilates each of the functions
>
> $$e^{\alpha x} \cos \beta x, xe^{\alpha x} \cos \beta x, x^2 e^{\alpha x} \cos \beta x, \ldots, x^{n-1} e^{\alpha x} \cos \beta x,$$
> $$e^{\alpha x} \sin \beta x, xe^{\alpha x} \sin \beta x, x^2 e^{\alpha x} \sin \beta x, \ldots, x^{n-1} e^{\alpha x} \sin \beta x. \tag{5}$$

EXAMPLE 7

From (5), with $\alpha = -1$, $\beta = 2$, and $n = 1$, we see that

$$(D^2 + 2D + 5)e^{-x} \cos 2x = 0 \qquad \text{and} \qquad (D^2 + 2D + 5)e^{-x} \sin 2x = 0.$$

EXAMPLE 8

From (5), with $\alpha = 0$, $\beta = 1$, and $n = 2$, it is seen that the differential operator $(D^2 + 1)^2$ or $D^4 + 2D^2 + 1$ will annihilate $\cos x$, $x \cos x$, $\sin x$, and $x \sin x$. Moreover, $(D^2 + 1)^2$ will annihilate any linear combination of these functions.

When $\alpha = 0$ and $n = 1$, a special case of (5) is

$$(D^2 + \beta^2)\begin{cases}\cos \beta x \\ \sin \beta x\end{cases} = 0. \tag{6}$$

EXAMPLE 9 Find a differential operator that annihilates $7 - x + 6 \sin 3x$.

SOLUTION From (2) and (6) we have, respectively,

$$D^2(7 - x) = 0 \quad \text{and} \quad (D^2 + 9) \sin 3x = 0.$$

Hence the operator $D^2(D^2 + 9)$ will annihilate the given linear combination.

The concept of an annihilator operator will be of immediate importance in the discussion that follows.

4.4.2 Solving a Nonhomogeneous Linear Equation

To obtain the general solution of a nonhomogeneous differential equation with constant coefficients we must do two things: find the complementary function y_c and then find *any* particular solution y_p of the nonhomogeneous equation. Recall from the discussion of Section 4.1 that a particular solution is any function, free of arbitrary constants, that satisfies the equation identically. The general solution of a nonhomogeneous equation is the sum of y_c and y_p.

Method of Undetermined Coefficients

If $P(D)$ denotes the differential operator (1), then a nonhomogeneous linear differential equation with constant coefficients can be written simply as

$$P(D)y = g(x). \tag{7}$$

When $g(x)$ is

(i) a constant k,
(ii) a polynomial in x,
(iii) an exponential function $e^{\alpha x}$,
(iv) $\sin \beta x$, $\cos \beta x$.

or finite sums and products of these functions, it is always possible to find another differential operator $P_1(D)$ that annihilates $g(x)$. Applying $P_1(D)$ to (7) yields

$$P_1(D)P(D)y = P_1(D)g(x) = 0.$$

By solving the homogeneous equation $P_1(D)P(D)y = 0$, it is possible to discover the *form* of a particular solution y_p for the nonhomogeneous equation (7).

The next several examples illustrate the so-called **method of undetermined coefficients** for finding y_p. The general solution of each equation is defined on the interval $(-\infty, \infty)$.

EXAMPLE 10 Solve
$$\frac{d^2y}{dx^2} + 3\frac{dy}{dx} + 2y = 4x^2. \tag{8}$$

SOLUTION ***Step 1.*** We first solve the homogeneous equation

$$\frac{d^2y}{dx^2} + 3\frac{dy}{dx} + 2y = 0.$$

From the auxiliary equation $m^2 + 3m + 2 = (m + 1)(m + 2) = 0$, we find the complementary function

$$y_c = c_1e^{-x} + c_2e^{-2x}.$$

Step 2. In view of (2), (8) can be rendered homogeneous by taking three derivatives of each side of the equation. In other words,

$$D^3(D^2 + 3D + 2)y = 4D^3x^2 = 0, \tag{9}$$

since $D^3x^2 = 0$. The auxiliary equation of (9) is

$$m^3(m^2 + 3m + 2) = 0 \qquad \text{or} \qquad m^3(m + 1)(m + 2) = 0,$$

and so its general solution must be

$$y = \underbrace{c_1e^{-x} + c_2e^{-2x}}_{y_c} + \underbrace{c_3 + c_4x + c_5x^2}_{y_p}. \tag{10}$$

We can then argue that every solution of equation (8) should also be a solution of equation (9). Since the complementary function $y_c = c_1e^{-x} + c_2e^{-2x}$ appears as part of the solution of (9), the remaining terms in (10) must be the basic structure of y_p:

$$y_p = A + Bx + Cx^2, \tag{11}$$

where we have replaced c_3, c_4, and c_5 by A, B, and C, respectively. For (11) to be a particular solution of (8), it is necessary to find specific coefficients A, B, and C. Differentiating (11) and substituting in (8) gives

$$\begin{aligned} y_p'' + 3y_p' + 2y_p &= 2A + 3B + 2C + (2B + 6C)x + 2Cx^2 \\ &= 4x^2. \end{aligned}$$

By equating coefficients in the last identity, we obtain the system of equations

$$\begin{aligned} 2A + 3B + 2C &= 0 \\ 2B + 6C &= 0 \\ 2C &= 4. \end{aligned} \tag{12}$$

Soving (12) gives $A = 7$, $B = -6$, and $C = 2$. Thus $y_p = 7 - 6x + 2x^2$.

Step 3. The general solution of (8) is $y = y_c + y_p$, or

$$y = c_1 e^{-x} + c_2 e^{-2x} + 7 - 6x + 2x^2.$$

EXAMPLE 11

Solve
$$y'' - 3y' = 8e^{3x} + 4 \sin x. \tag{13}$$

SOLUTION ***Step 1.*** The auxiliary equation for the homogeneous equation $y'' - 3y' = 0$ is $m(m - 3) = 0$ and so

$$y_c = c_1 + c_2 e^{3x}.$$

Step 2. Now since $(D - 3)e^{3x} = 0$ and $(D^2 + 1) \sin x = 0$, we apply the differential operator $(D - 3)(D^2 + 1)$ to both sides of (13):

$$(D - 3)(D^2 + 1)(D^2 - 3D)y = 0. \tag{14}$$

The auxiliary equation of (14) is

$$(m - 3)(m^2 + 1)(m^2 - 3m) = 0 \qquad \text{or} \qquad m(m - 3)^2(m^2 + 1) = 0.$$

Thus
$$y = \underbrace{c_1 + c_2 e^{3x}}_{y_c} + \underbrace{c_3 x e^{3x} + c_4 \cos x + c_5 \sin x}_{y_p}. \tag{15}$$

After excluding the linear combination of terms in (15) corresponding to y_c, we arrive at the form of y_p:

$$y_p = Axe^{3x} + B \cos x + C \sin x.$$

Substituting y_p in (13) and simplifying yield

$$\begin{aligned} y_p'' - 3y_p' &= 3Ae^{3x} + (-B - 3C) \cos x + (3B - C) \sin x \\ &= 8e^{3x} + 4 \sin x. \end{aligned}$$

Equating coefficients gives

$$\begin{aligned} 3A &= 8 \\ -B - 3C &= 0 \\ 3B - C &= 4. \end{aligned}$$

We find $A = 8/3$, $B = 6/5$, $C = -2/5$, and consequently,

$$y_p = \frac{8}{3} x e^{3x} + \frac{6}{5} \cos x - \frac{2}{5} \sin x.$$

Step 3. The general solution of (13) is then

$$y = c_1 + c_2 e^{3x} + \frac{8}{3} x e^{3x} + \frac{6}{5} \cos x - \frac{2}{5} \sin x.$$

EXAMPLE 12

Solve
$$y'' + 8y = 5x + 2e^{-x}. \tag{16}$$

SOLUTION From (2) and (3) we know that $D^2 x = 0$ and $(D + 1)e^{-x} = 0$, respectively. Hence we apply $D^2(D + 1)$ to (16):

$$D^2(D + 1)(D^2 + 8)y = 0.$$

It is seen that

$$y = \underbrace{c_1 \cos 2\sqrt{2}\,x + c_2 \sin 2\sqrt{2}\,x}_{y_c} + \underbrace{c_3 + c_4 x + c_5 e^{-x}}_{y_p}$$

and so
$$y_p = A + Bx + Ce^{-x}.$$

Substituting y_p into (16) yields

$$\begin{aligned} y_p'' + 8y_p &= 8A + 8Bx + 9Ce^{-x} \\ &= 5x + 2e^{-x}. \end{aligned}$$

This implies $A = 0$, $B = 5/8$, and $C = 2/9$ so that the general solution of (16) is

$$y = c_1 \cos 2\sqrt{2}\,x + c_2 \sin 2\sqrt{2}\,x + \frac{5}{8} x + \frac{2}{9} e^{-x}.$$

EXAMPLE 13

Solve
$$y'' + y = x \cos x - \cos x. \tag{17}$$

SOLUTION In Example 8 we saw that $x \cos x$ and $\cos x$ are annihilated by the operator $(D^2 + 1)^2$. Thus

$$(D^2 + 1)^2(D^2 + 1)y = 0 \qquad \text{or} \qquad (D^2 + 1)^3 y = 0.$$

Since i and $-i$ are both complex roots of multiplicity 3 of the auxiliary equation of the last differential equation, we conclude

$$y = \underbrace{c_1 \cos x + c_2 \sin x}_{y_c} + \underbrace{c_3 x \cos x + c_4 x \sin x + c_5 x^2 \cos x + c_6 x^2 \sin x}_{y_p}.$$

Substitute

$$y_p = Ax \cos x + Bx \sin x + Cx^2 \cos x + Ex^2 \sin x$$

into (17) and simplify:

$$y_p'' + y_p = 4Ex\cos x - 4Cx\sin x + (2B + 2C)\cos x + (-2A + 2E)\sin x$$
$$= x\cos x - \cos x.$$

Equating coefficients gives the equations

$$\begin{aligned} 4E &= 1 \\ -4C &= 0 \\ 2B + 2C &= -1 \\ -2A + 2E &= 0, \end{aligned}$$

from which we find $E = 1/4$, $C = 0$, $B = -1/2$, and $A = 1/4$. Hence the general solution of (17) is

$$y = c_1\cos x + c_2\sin x + \frac{1}{4}x\cos x - \frac{1}{2}x\sin x + \frac{1}{4}x^2\sin x.$$

EXAMPLE 14

Determine the form of a particular solution for

$$y'' - 2y' + y = 10e^{-2x}\cos x. \tag{18}$$

SOLUTION From (5), with $\alpha = -2$, $\beta = 1$, and $n = 1$, we know that

$$(D^2 + 4D + 5)e^{-2x}\cos x = 0.$$

Applying the operator $D^2 + 4D + 5$ to (18) gives

$$(D^2 + 4D + 5)(D^2 - 2D + 1)y = 0. \tag{19}$$

Since the roots of the auxiliary equation of (19) are $-2 - i$, $-2 + i$, 1, 1,

$$y = \underbrace{c_1e^x + c_2xe^x}_{y_c} + \underbrace{c_3e^{-2x}\cos x + c_4e^{-2x}\sin x}_{y_p}.$$

Hence a particular solution of (18) can be found having the form

$$y_p = Ae^{-2x}\cos x + Be^{-2x}\sin x.$$

EXAMPLE 15

Determine the form of a particular solution for

$$y''' - 4y'' + 4y' = 5x^2 - 6x + 4x^2e^{2x} + 3e^{5x}. \tag{20}$$

SOLUTION Observe that

$$D^3(5x^2 - 6x) = 0, \qquad (D-2)^3x^2e^{2x} = 0, \qquad \text{and} \qquad (D-5)e^{5x} = 0.$$

Therefore $D^3(D-2)^3(D-5)$ applied to (20) gives

$$D^3(D-2)^3(D-5)(D^3 - 4D^2 + 4D)y = 0$$

or

$$D^4(D-2)^5(D-5)y = 0.$$

Hence

$$y = c_1 + c_2x + c_3x^2 + c_4x^3 + c_5e^{2x} + c_6xe^{2x} + c_7x^2e^{2x} + c_8x^3e^{2x} + c_9x^4e^{2x} + c_{10}e^{5x} \quad (21)$$

Since the linear combination $c_1 + c_5e^{2x} + c_6xe^{2x}$ can be taken as the complementary function of (20), we see from (21) that the terms indicated in color give the form of a particular solution of the differential equation:

$$y_p = Ax + Bx^2 + Cx^3 + Ex^2e^{2x} + Fx^3e^{2x} + Gx^4e^{2x} + He^{5x}.$$

Summary of the Method

For your convenience the method of undetermined coefficients is summarized here.

The differential equation $P(D)y = g(x)$ has constant coefficients and the function $g(x)$ consists of finite sums and products of constants, polynomials, exponential functions $e^{\alpha x}$, sines, and cosines.

(i) Find the complementary solution y_c for the homogeneous equation $P(D)y = 0$.

(ii) Operate on both sides of the nonhomogeneous equation $P(D)y = g(x)$ with a differential operator $P_1(D)$ that annihilates the function $g(x)$.

(iii) Find the general solution of the homogeneous differential equation $P_1(D)P(D)y = 0$.

(iv) Delete all those terms from the solution in step (iii) that are duplicated in the complementary solution y_c found in step (i). Form a linear combination y_p of the terms that remain. This is the form of a particular solution of $P(D)y = g(x)$.

(v) Substitute y_p found in step (iv) into $P(D)y = g(x)$. Match coefficients of the various functions on each side of the equality and solve the resulting system of equations for the unknown coefficients in y_p.

(vi) With the particular solution found in step (v), form the general solution $y = y_c + y_p$ of the given differential equation.

Exercises 4.4

Answers to odd-numbered problems begin on page A–11.

[4.4.1] In Problems 1–10, if possible, factor the given differential operator.

1. $9D^2 - 4$

2. $D^2 - 5$

3. $D^2 - 4D - 12$

4. $2D^2 - 3D - 2$

5. $D^3 + 10D^2 + 25D$

6. $D^3 + 4D$

7. $D^3 + 2D^2 - 13D + 10$

8. $D^3 + 4D^2 + 3D$

9. $D^4 + 8D$

10. $D^4 - 8D^2 + 16$

In Problems 11–22, find a differential operator that annihilates the given function.

11. $1 + 6x - 2x^3$

12. $x^3(1 - 5x)$

13. $1 + 7e^{2x}$

14. $x + 3xe^{6x}$

15. $\cos 2x$

16. $1 + \sin x$

17. $13x + 9x^2 - \sin 4x$

18. $8x - \sin x + 10 \cos 5x$

19. $e^{-x} + 2xe^x - x^2e^x$

20. $(2 - e^x)^2$

21. $3 + e^x \cos 2x$

22. $e^{-x} \sin x - e^{2x} \cos x$

[4.4.2] In Problems 23–54 solve the given differential equation by undetermined coefficients.

23. $y'' - 9y = 54$

24. $2y'' - 7y' + 5y = -29$

25. $y'' + y' = 3$

26. $y''' + 2y'' + y' = 10$

27. $y'' + 4y' + 4y = 2x + 6$

28. $y'' + 3y' = 4x - 5$

29. $y''' + y'' = 8x^2$

30. $y'' - 2y' + y = x^3 + 4x$

31. $y'' - y' - 12y = e^{4x}$

32. $y'' + 2y' + 2y = 5e^{6x}$

33. $y'' - 2y' - 3y = 4e^x - 9$

34. $y'' + 6y' + 8y = 3e^{-2x} + 2x$

35. $y'' + 25y = 6 \sin x$

36. $y'' + 4y = 4 \cos x + 3 \sin x - 8$

37. $y'' + 6y' + 9y = -xe^{4x}$

38. $y'' + 3y' - 10y = x(e^x + 1)$

39. $y'' - y = x^2e^x + 5$

40. $y'' + 2y' + y = x^2e^{-x}$

41. $y'' - 2y' + 5y = e^x \sin x$

42. $y'' + y' + \frac{1}{4}y = e^x(\sin 3x - \cos 3x)$

43. $y'' + 25y = 20 \sin 5x$

44. $y'' + y = 4 \cos x - \sin x$

45. $y'' + y' + y = x \sin x$

46. $y'' + 4y = \cos^2 x$

47. $y''' + 8y'' = -6x^2 + 9x + 2$

48. $y''' - y'' + y' - y = xe^x - e^{-x} + 7$

49. $y''' - 3y'' + 3y' - y = e^x - x + 16$

50. $2y''' - 3y'' - 3y' + 2y = (e^x + e^{-x})^2$

51. $y^{(4)} - 2y''' + y'' = e^x + 1$

52. $y^{(4)} - 4y'' = 5x^2 - e^{2x}$

53. $16y^{(4)} - y = e^{x/2}$

54. $y^{(4)} - 5y'' + 4y = 2 \cosh x - 6$

In Problems 55–62 solve the given differential equation subject to the indicated initial conditions.

55. $y'' - 64y = 16; \quad y(0) = 1, y'(0) = 0$

56. $y'' + y' = x; \quad y(0) = 1, y'(0) = 0$

57. $y'' - 5y' = x - 2; \quad y(0) = 0, y'(0) = 2$

58. $y'' + 5y' - 6y = 10e^{2x}; \quad y(0) = 1, y'(0) = 1$

59. $y'' + y = 8 \cos 2x - 4 \sin x; \quad y(\pi/2) = -1, y'(\pi/2) = 0$

60. $y''' - 2y'' + y' = xe^x + 5; \quad y(0) = 2, y'(0) = 2, y''(0) = -1$

61. $y'' - 4y' + 8y = x^3; \quad y(0) = 2, y'(0) = 4$

62. $y^{(4)} - y''' = x + e^x; \quad y(0) = 0, y'(0) = 0, y''(0) = 0, y'''(0) = 0$

In Problems 63 and 64 determine the form of a particular solution for the given differential equation.

63. $y'' - y = e^x(2 + 3x \cos 2x)$

64. $y'' + y' = 9 - e^{-x} + x^2 \sin x$

Miscellaneous Problems

65. Determine whether the operator $(xD - 1)(D + 4)$ is the same as the operator $(D + 4)(xD - 1)$.

66. Prove that the differential equation

$$a_n y^{(n)} + a_{n-1} y^{(n-1)} + \cdots + a_1 y' + a_0 y = k,$$

k a constant, $a_0 \neq 0$, has the particular solution $y_p = k/a_0$.

4.5 Variation of Parameters

Linear First-Order Equation Revisited

In Chapter 2 we have seen that the general solution of the linear first-order different equation

$$\frac{dy}{dx} + P(x)y = f(x), \tag{1}$$

where $P(x)$ and $f(x)$ are continuous on an interval I, is

$$y = e^{-\int P(x)\,dx} \int e^{\int P(x)\,dx} f(x)\,dx + c_1 e^{-\int P(x)\,dx}. \tag{2}$$

Now (2) has the form

$$y = y_c + y_p,$$

where

$$y_c = c_1 e^{-\int P(x)\,dx}$$

is a solution of

$$\frac{dy}{dx} + P(x)y = 0 \tag{3}$$

and

$$y_p = e^{-\int P(x)\,dx} \int e^{\int P(x)\,dx} f(x)\,dx \tag{4}$$

is a particular solution of (1). As a means of motivating an additional method for solving nonhomogeneous linear equations of higher order, we shall rederive (4) by a method known as **variation of parameters**. The basic procedure is esssentially that employed in Section 4.2.

Suppose y_1 is a known solution of (3), that is,

$$\frac{dy_1}{dx} + P(x)y_1 = 0.$$

We have already proved in Section 2.5 that $y_1 = e^{-\int P(x)\,dx}$ is a solution, and since the differential equation is linear, its general solution is $y = c_1 y_1(x)$. Variation of parameters consists of finding a function u_1 such that

$$y_p = u_1(x)y_1(x)$$

is a particular solution of (1). In other words, we replace the parameter c_1 by a variable u_1.

Substituting $y_p = u_1 y_1$ into (1) gives

$$\frac{d}{dx}[u_1 y_1] + P(x)u_1 y_1 = f(x)$$

$$u_1 \frac{dy_1}{dx} + y_1 \frac{du_1}{dx} + P(x)u_1 y_1 = f(x)$$

$$\underbrace{u_1\left[\frac{dy_1}{dx} + P(x)y_1\right]}_{\text{zero}} + y_1 \frac{du_1}{dx} = f(x)$$

so that

$$y_1 \frac{du_1}{dx} = f(x).$$

By separating variables, we find

$$du_1 = \frac{f(x)}{y_1(x)}\,dx \qquad \text{and} \qquad u_1 = \int \frac{f(x)}{y_1(x)}\,dx,$$

from which it follows that

$$y = u_1 y_1 = y_1 \int \frac{f(x)}{y_1(x)}\,dx.$$

From the definition of y_1, we see that the last result is identical to (4).

Second-Order Equations

To adapt the foregoing procedure to a linear second-order differential equation

$$a_2(x)y'' + a_1(x)y' + a_0(x)y = g(x), \tag{5}$$

we put (5) in the standard form

$$y'' + P(x)y' + Q(x)y = f(x) \tag{6}$$

by dividing through by $a_2(x)$. Here we assume that $P(x)$, $Q(x)$, and $f(x)$ are continuous on some interval I. Equation (6) is the analogue of (1). As we know, when $P(x)$ and $Q(x)$ are constants, there is absolutely no difficulty in writing down y_c.

Suppose y_1 and y_2 form a fundamental set of solutions on I of the associated homogeneous form of (6)—that is,

$$y_1'' + P(x)y_1' + Q(x)y_1 = 0 \qquad \text{and} \qquad y_2'' + P(x)y_2' + Q(x)y_2 = 0.$$

Now we ask: Can two functions u_1 and u_2 be found so that

$$y_p = u_1(x)y_1(x) + u_2(x)y_2(x)$$

is a particular solution of (1)? Notice that our assumption for y_p is the same as $y_c = c_1y_1 + c_2y_2$, but we have replaced c_1 and c_2 by the "variable parameters" u_1 and u_2. Using the product rule to differentiate y_p gives

$$y_p' = u_1y_1' + y_1u_1' + u_2y_2' + y_2u_2'. \tag{7}$$

If we make the further demand that u_1 and u_2 be functions for which

$$y_1u_1' + y_2u_2' = 0, \tag{8}$$

then (7) becomes $\quad y_p' = u_1y_1' + u_2y_2'.$

Continuing, we find $\quad y_p'' = u_1y_1'' + y_1'u_1' + u_2y_2'' + y_2'u_2'$

and hence

$$y_p'' + Py_p' + Qy_p = u_1y_1'' + y_1'u_1' + u_2y_2'' + y_2'u_2' + Pu_1y_1' + Pu_2y_2' + Qu_1y_1 + Qu_2y_2$$
$$= u_1\underbrace{[y_1'' + Py_1' + Qy_1]}_{\text{zero}} + u_2\underbrace{[y_2'' + Py_2' + Qy_2]}_{\text{zero}} + y_1'u_1' + y_2'u_2' = f(x).$$

In other words, u_1 and u_2 must be functions that also satisfy the condition

$$y_1'u_1' + y_2'u_2' = f(x). \tag{9}$$

Equations (8) and (9) constitute a linear system of equations for determining the derivatives u_1' and u_2'. By Cramer's rule*, the solution of

$$y_1u_1' + y_2u_2' = 0$$
$$y_1'u_1' + y_2'u_2' = f(x)$$

can be expressed in terms of determinants.

$$u_1' = \frac{W_1}{W} \quad \text{and} \quad u_2' = \frac{W_2}{W}, \tag{10}$$

where $W_1 = \begin{vmatrix} 0 & y_2 \\ f(x) & y_2' \end{vmatrix} = -y_2f(x), \quad W_2 = \begin{vmatrix} y & 0 \\ y_1' & f(x) \end{vmatrix} = y_1f(x), \quad (11)$

and $$W = \begin{vmatrix} y_1 & y_2 \\ y_1' & y_2' \end{vmatrix}.$$

The determinant W is recognized as the Wronskian of y_1 and y_2. By linear independence of y_1 and y_2 on I, we know that $W(y_1(x), y_2(x)) \neq 0$ for every x in the interval.

Summary of the Method

Usually it is not a good idea to memorize formulas in lieu of understanding a procedure. However, the foregoing procedure is too long and complicated to use each time we wish to solve a differential equation. In this case it is more efficient to simply use the formulas in (10). Thus to solve $a_2y'' + a_1y' + a_0y = g(x)$, first find the complementary function $y_c = c_1y_1 + c_2y_2$, and then compute the Wronskian

$$W = \begin{vmatrix} y_1 & y_2 \\ y_1' & y_2' \end{vmatrix}.$$

By dividing by a_2, we put the equation into the form $y'' + Py' + Qy = f(x)$ to determine $f(x)$. We find u_1 and u_2 by integrating

$$u_1' = -\frac{y_2f(x)}{W} \quad \text{and} \quad u_2' = \frac{y_1f(x)}{W}. \tag{12}$$

* See Appendix IV for a review of Cramer's rule.

Finally, form the particular solution

$$y_p = u_1 y_1 + u_2 y_2.$$

EXAMPLE 1 Solve

$$y'' - 4y' + 4y = (x + 1)e^{2x}.$$

SOLUTION Since the auxiliary equation is $m^2 - 4m + 4 = (m - 2)^2 = 0$, we have

$$y_c = c_1 e^{2x} + c_2 x e^{2x}.$$

Identifying $y_1 = e^{2x}$ and $y_2 = xe^{2x}$, we next compute the Wronskian:

$$W(e^{2x}, xe^{2x}) = \begin{vmatrix} e^{2x} & xe^{2x} \\ 2e^{2x} & 2xe^{2x} + e^{2x} \end{vmatrix}$$
$$= e^{4x}.$$

Since the given differential equation is already in form (6) (that is, the coefficient of y'' is 1), we identify $f(x) = (x + 1)e^{2x}$. From (12) we obtain

$$u_1' = -\frac{xe^{2x}(x + 1)e^{2x}}{e^{4x}}$$
$$= -x^2 - x$$

and

$$u_2' = \frac{e^{2x}(x + 1)e^{2x}}{e^{4x}}$$
$$= x + 1.$$

It follows that

$$u_1 = -\frac{x^3}{3} - \frac{x^2}{2} \quad \text{and} \quad u_2 = \frac{x^2}{2} + x.$$

Therefore

$$y_p = \left(-\frac{x^3}{3} - \frac{x^2}{2}\right)e^{2x} + \left(\frac{x^2}{2} + x\right)xe^{2x}$$
$$= \left(\frac{x^3}{6} + \frac{x^2}{2}\right)e^{2x}.$$

Hence

$$y = y_c + y_p = c_1 e^{2x} + c_2 x e^{2x} + \left(\frac{x^3}{6} + \frac{x^2}{2}\right)e^{x}.$$

EXAMPLE 2 Solve

$$4y'' + 36y = \csc 3x.$$

SOLUTION We first put the equation in the standard form (6) by dividing by 4:

$$y'' + 9y = \frac{1}{4} \csc 3x.$$

Since the roots of the auxiliary equation $m^2 + 9 = 0$ are $m_1 = 3i$ and $m_2 = -3i$, we have

$$y_c = c_1 \cos 3x + c_2 \sin 3x$$

and

$$W(\cos 3x, \sin 3x) = \begin{vmatrix} \cos 3x & \sin 3x \\ -3 \sin 3x & 3 \cos 3x \end{vmatrix} = 3.$$

Therefore

$$u_1' = -\frac{(\sin 3x)(\frac{1}{4} \csc 3x)}{3} = -\frac{1}{12}$$

$$u_2' = \frac{(\cos 3x)(\frac{1}{4} \csc 3x)}{3} = \frac{1}{12} \frac{\cos 3x}{\sin 3x},$$

from which we obtain

$$u_1 = -\frac{1}{12} x \quad \text{and} \quad u_2 = \frac{1}{36} \ln |\sin 3x|$$

and

$$y_p = -\frac{1}{12} x \cos 3x + \frac{1}{36} (\sin 3x) \ln |\sin 3x|.$$

Thus we have

$$y = y_c + y_p = c_1 \cos 3x + c_2 \sin 3x - \frac{1}{12} x \cos 3x + \frac{1}{36} (\sin 3x) \ln |\sin 3x|. \tag{13}$$

Equation (13) represents the general solution of the differential equation on, say, $(0, \pi/6)$.

Constants of Integration

When computing the indefinite integrals of u_1' and u_2', we need not introduce any constants. This is because

$$\begin{aligned} y = y_c + y_p &= c_1 y_1 + c_2 y_2 + (u_1 + a_1) y_1 + (u_2 + b_1) y_2 \\ &= (c_1 + a_1) y_1 + (c_2 + b_1) y_2 + u_1 y_1 + u_2 y_2 \\ &= C_1 y_1 + C_2 y_2 + u_1 y_1 + u_2 y_2. \end{aligned}$$

EXAMPLE 3 Solve

$$y'' - y = \frac{1}{x}.$$

SOLUTION The auxiliary equation $m^2 - 1 = 0$ yields $m_1 = -1$ and $m_2 = 1$. Therefore

$$y_c = c_1 e^x + c_2 e^{-x}$$

$$W(e^x, e^{-x}) = \begin{vmatrix} e^x & e^{-x} \\ e^x & -e^{-x} \end{vmatrix} = -2$$

$$u_1' = -\frac{e^{-x}(1/x)}{-2}, \qquad u_1 = \frac{1}{2}\int_{x_0}^{x} \frac{e^{-t}}{t}\,dt$$

$$u_2' = \frac{e^x(1/x)}{-2}, \qquad u_2 = -\frac{1}{2}\int_{x_0}^{x} \frac{e^{t}}{t}\,dt.$$

It is well known that the integrals defining u_1 and u_2 cannot be expressed in terms of elementary functions. Hence we write

$$y_p = \frac{1}{2}e^x \int_{x_0}^{x} \frac{e^{-t}}{t}\,dt - \frac{1}{2}e^{-x}\int_{x_0}^{x} \frac{e^{t}}{t}\,dt,$$

and so

$$y = y_c + y_p = c_1 e^x + c_2 e^{-x} + \frac{1}{2}e^x \int_{x_0}^{x} \frac{e^{-t}}{t}\,dt - \frac{1}{2}e^{-x}\int_{x_0}^{x} \frac{e^{t}}{t}\,dt.$$

In Example 3 we can integrate on any interval $x_0 \le t \le x$ not containing the origin.

nth-Order Equations

The method that we have just examined for nonhomogeneous second-order differential equations can be generalized to nth-order linear equations that have been put into the form

$$y^{(n)} + P_{n-1}(x)y^{(n-1)} + \cdots + P_1(x)y' + P_0(x)y = f(x). \tag{14}$$

If $y = c_1y_1 + c_2y_2 + \cdots + c_ny_n$ is the complementary function for (14), then it is straightforward, although tedious, to show that substituting

$$y_p = u_1(x)y_1(x) + u_2(x)y_2(x) + \cdots + u_n(x)y_n(x)$$

into the differential equation leads to a system of n linear equations

$$\begin{aligned} y_1u_1' + \quad y_2u_2' + \cdots + \quad y_nu_n' &= 0 \\ y_1'u_1' + \quad y_2'u_2' + \cdots + \quad y_n'u_n' &= 0 \\ &\vdots \\ y_1^{(n-1)}u_1' + y_2^{(n-1)}u_2' + \cdots + y_n^{(n-1)}u_n' &= f(x) \end{aligned}$$

for determining the u_k', $k = 1, 2, \ldots, n$. In this case, Cramer's rule gives

$$u_k' = \frac{W_k}{W}, \qquad k = 1, 2, \ldots, n,$$

where W is the Wronskian of $y_1, y_2, \ldots, y_n$, and W_k is the determinant obtained by replacing the kth column of the Wronskian by the column

$$\begin{matrix} 0 \\ 0 \\ \vdots \\ 0 \\ f(x). \end{matrix}$$

When $n = 2$ we get (10) and (11).

Remark: Variation of parameters has a distinct advantage over the method of undetermined coefficients in that it will *always* yield a particular solution y_p, provided the related homogeneous equation can be solved. The present method is not limited to a function $f(x)$, which is a combination of the four functions (i)–(iv) on page 172. Also, variation of parameters, unlike undetermined coefficients, is applicable to differential equations with variable coefficients.

In the problems that follow do not hesitate to simplify the form of y_p. Depending on how the antiderivatives of u_1' and u_2' are found, you may not obtain the same y_p as given in the answer section. For example, in Problem 3, both $y_p = \frac{1}{2}\sin x - \frac{1}{2}x \cos x$ and $y_p = \frac{1}{4}\sin x - \frac{1}{2}x \cos x$ are valid answers. In either case, the general solution $y = y_c + y_p$ simplifies to $y = c_1 \cos x + c_2 \sin x - \frac{1}{2}x \cos x$. Why?

Exercises 4.5

Answers to odd-numbered problems begin on page A-11.

In Problems 1–24 solve each differential equation by variation of parameters. State an interval on which the general solution is defined.

1. $y'' + y = \sec x$

2. $y'' + y = \tan x$

3. $y'' + y = \sin x$

4. $y'' + y = \sec x \tan x$

5. $y'' + y = \cos^2 x$

6. $y'' + y = \sec^2 x$

7. $y'' - y = \cosh x$

8. $y'' - y = \sinh 2x$

9. $y'' - 4y = e^{2x}/x$

10. $y'' - 9y = 9x/e^{3x}$

11. $y'' + 3y' + 2y = 1/(1 + e^x)$

12. $y'' - 3y' + 2y = e^{3x}/(1 + e^x)$

13. $y'' + 3y' + 2y = \sin e^x$

14. $y'' - 2y' + y = e^x \arctan x$

15. $y'' - 2y' + y = e^x/(1 + x^2)$

16. $y'' - 2y' + 2y = e^x \sec x$

17. $y'' + 2y' + y = e^{-x} \ln x$

18. $y'' + 10y' + 25y = e^{-10x}/x^2$

19. $3y'' - 6y' + 30y = e^x \tan 3x$

20. $4y'' - 4y' + y = e^{x/2}\sqrt{1 - x^2}$

21. $y''' + y' = \tan x$

22. $y''' + 4y' = \sec 2x$

23. $y''' - 2y'' - y' + 2y = e^{3x}$

24. $2y''' - 6y'' = x^2$

In Problems 25–28 solve each differential equation by variations of parameters subject to the initial conditions $y(0) = 1$, $y'(0) = 0$.

25. $4y'' - y = xe^{x/2}$

26. $2y'' + y' - y = x + 1$

27. $y'' + 2y' - 8y = 2e^{-2x} - e^{-x}$

28. $y'' - 4y' + 4y = (12x^2 - 6x)e^{2x}$

29. Given that $y_1 = x$ and $y_2 = x \ln x$, form a fundamental set of solutions of $x^2y'' - xy' + y = 0$ on $(0, \infty)$. Find the general solution of

$$x^2y'' - xy' + y = 4x \ln x.$$

30. Given that $y_1 = x^2$ and $y_2 = x^3$, form a fundamental set of solutions of $x^2y'' - 4xy' + 6y = 0$ on $(0, \infty)$. Find the general solution of

$$x^2y'' - 4xy' + 6y = 1/x.$$

31. Given that $y_1 = x^{-1/2} \cos x$ and $y_2 = x^{-1/2} \sin x$, form a fundamental set of solutions of $x^2y'' + xy' + (x^2 - \frac{1}{4})y = 0$ on $(0, \infty)$. Find the general solution of

$$x^2y'' + xy' + \left(x^2 - \frac{1}{4}\right)y = x^{3/2}.$$

32. Given that $y_1 = \cos(\ln x)$ and $y_2 = \sin(\ln x)$ are known linearly independent solutions of $x^2y'' + xy' + y = 0$ on $(0, \infty)$.

(a) Find a particular solution of

$$x^2y'' + xy' + y = \sec(\ln x).$$

(b) Given the general solution of the equation and state an interval of validity. [*Hint:* It is *not* $(0, \infty)$. Why?]

Miscellaneous Problems

33. Solve the third-order equation

$$y''' - y' = xe^x.$$

[*Hint:* $(d/dx)(y'' - y) = y''' - y'$, integrate and then use variation of parameters.]

34. Solve $y''' + 4y' = \sin x \cos x$ by the method outlined in Problem 33.

CHAPTER 4 SUMMARY

We summarize the important results of this chapter for **linear second-order** differential equations.

The equation

$$a_2(x)y'' + a_1(x)y' + a_0(x)y = 0 \tag{1}$$

is said to be **homogeneous**, whereas

$$a_2(x)y'' + a_1(x)y' + a_0(x)y = g(x), \tag{2}$$

$g(x)$ not identically zero, is **nonhomogeneous**. In the consideration of the linear equations (1) and (2), we assume that $a_2(x)$, $a_1(x)$, $a_0(x)$, and $g(x)$ are continuous on an interval I and that $a_2(x) \neq 0$ for every x in the interval. Under these assumptions there exists a unique solution of (2) satisfying the **initial conditions** $y(x_0) = y_0$, $y'(x_0) = y'_0$, where x_0 is a point in I.

The **Wronskian** of two differentiable functions $f_1(x)$ and $f_2(x)$ is the determinant

$$W(f_1(x), f_2(x)) = \begin{vmatrix} f_1(x) & f_2(x) \\ f'_1(x) & f'_2(x) \end{vmatrix}.$$

When $W \neq 0$ for at least one point in an interval, the functions are **linearly independent** on the interval. If the functions are **linearly dependent** on the interval, then $W = 0$ for every x in the interval.

In solving the homogeneous equation (1), we want linearly independent solutions. A necessary and sufficient condition that two solutions y_1 and y_2 are linearly independent on I is that $W(y_1, y_2) \neq 0$ for every x in I. We say y_1 and y_2 form a **fundamental set** on I when they are linearly independent solutions of (1) on the interval. For *any* two solutions y_1 and y_2, the **superposition principle** states that the linear combination $c_1y_1 + c_2y_2$ is also a solution of (1). When y_1 and y_2 form a fundamental set, the function $y = c_1y_1 + c_2y_2$ is called the **general solution** of (1). The **general solution** of (2) is $y = y_c + y_p$, where y_c is the **complementary function**, or general solution, of (1), and y_p is any **particular solution** of (2).

To solve (1) in the case $ay'' + by' + cy = 0$, a, b, and c constants, we first solve the **auxiliary equation** $am^2 + bm + c = 0$. There are three forms of the general solution depending on the three possible ways in which the roots of the auxiliary equation can occur.

Roots		General Solution
1. m_1 and m_2:	real and distinct	$y = c_1e^{m_1x} + c_2e^{m_2x}$
2. m_1 and m_2:	real but $m_1 = m_2$	$y = c_1e^{m_1x} + c_2xe^{m_1x}$
3. m_1 and m_2:	complex	
$m_1 = \alpha + i\beta$,	$m_2 = \alpha - i\beta$	$y = e^{\alpha x}(c_1 \cos \beta x + c_2 \sin \beta x)$

To solve a nonhomogeneous differential equation we use either the method of **undetermined coefficients** or **variation of parameters** to find a particular solution y_p. The former procedure is limited to differential equations $ay'' + by' + cy = g(x)$, where a, b, and c are constants and $g(x)$ is a constant, a polynomial, $e^{\alpha x}$, $\cos \beta x$, $\sin \beta x$, or finite sums and products of these functions.

CHAPTER 4 REVIEW EXERCISES

Answers to odd-numbered problems begin on page A–12.

Answer Problems 1–10 without referring back to the text. Fill in the blank or answer true/false. In some cases there can be more than one correct answer.

1. The only solution to $y'' + x^2y = 0$, $y(0) = 0$, $y'(0) = 0$ is ______.

2. If two differentiable functions $f_1(x)$ and $f_2(x)$ are linearly independent on an interval, then $W(f_1(x), f_2(x)) \neq 0$ for at least one point in the interval. ______

3. Two functions $f_1(x)$ and $f_2(x)$ are linearly independent on an interval if one is not a constant multiple of the other. ______

4. The functions $f_1(x) = x^2$, $f_2(x) = 1 - x^2$, and $f_3(x) = 2 + x^2$ are linearly ______ on the interval $(-\infty, \infty)$.

5. The functions $f_1(x) = x^2$ and $f_2(x) = x|x|$ are linearly independent on the interval ______ whereas they are linearly dependent on the interval ______.

6. Two solutions y_1 and y_2 of $y'' + y' + y = 0$ are linearly dependent if $W(y_1, y_2) = 0$ for every real value of x. ______

7. A constant multiple of a solution of a differential equation is also a solution. ______

8. A fundamental set of two solutions of $(x - 2)y'' + y = 0$ exists on any interval not containing the point ______.

9. Using the method of undetermined coefficients, the assumed form of the particular solution y_p for $y'' - y = 1 + e^x$ is ______.

10. A differential operator that annihilates $e^{2x}(x + \sin x)$ is ______.

In Problems 11 and 12 find a second solution for the differential equation given that $y_1(x)$ is a known solution.

11. $y'' + 4y = 0; \quad y_1 = \cos 2x$

12. $xy'' - 2(x + 1)y' + (x + 2)y = 0; \quad y_1 = e^x$

In Problems 13–18 find the general solution of each differential equation.

13. $y'' - 2y' - 2y = 0$

14. $2y'' + 2y' + 3y = 0$

15. $y''' + 10y'' + 25y' = 0$

16. $2y''' + 9y'' + 12y' + 5y = 0$

17. $3y''' + 10y'' + 15y' + 4y = 0$

18. $2\dfrac{d^4y}{dx^4} + 3\dfrac{d^3y}{dx^3} + 2\dfrac{d^2y}{dx^2} + 6\dfrac{dy}{dx} - 4y = 0$

In Problems 19–21 solve each differential equation by the method of undetermined coefficients.

19. $y'' - 3y' + 5y = 4x^3 - 2x$

20. $y'' - 2y' + y = x^2e^x$

21. $y''' - 5y'' + 6y' = 2\sin x + 8$

22. Solve $y''' - y'' = 6$ subject to $y(0) = 0$, $y'(0) = 1$, $y''(0) = 4$.

In Problems 23 and 24 solve each differential equation by the method of variation of parameters.

23. $y'' - 2y' + 2y = e^x \tan x$

24. $y'' - y = 2e^x/(e^x + e^{-x})$

25. Solve $y'' + y = \sec^3 x$ subject to $y(0) = 1$, $y'(0) = 1/2$.

CHAPTER 5

Applications of Second-Order Differential Equations: Vibrational Models

IMPORTANT TERMS

free motion
simple harmonic motion
free undamped motion
period
frequency
equation of motion
amplitude
phase angle
free damped motion
overdamping
critical damping
underdamping
forced motion
transient term
transient solution
steady-state solution
pure resonance
resonance
electrical vibrations
overdamped circuit
critically damped circuit
underdamped circuit
steady-state current
reactance
impedance

A single differential equation can serve as a mathematical model for diverse physical phenomena. The simple linear second-order differential equation $ay'' + by' + cy = f(t)$ appears in the mathematical analysis of many problems in physics, engineering, chemistry, and even biology. A substantial number of these problems deal with classical vibrational phenomena. In this chapter we shall focus on one such application that involves this differential equation: the motion of a mass on a spring. Our goal, of course, is not to study all possible applications but to acquaint the student with the mathematical procedures that are common to these problems.

5.1 Simple Harmonic Motion

Hooke's Law

Suppose, as in Figure 5.1(b), a mass m_1 is attached to a flexible spring suspended from a rigid support. When m_1 is replaced with a different mass m_2, the amount of stretch, or elongation, of the spring will of course be different.

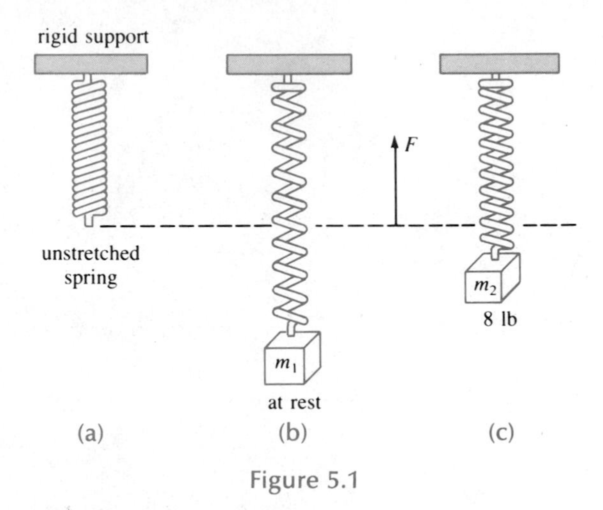

Figure 5.1

By Hooke's law,* the spring itself exerts a restoring force F opposite to the direction of elongation and proportional to the amount of elongation s. Simply stated, $F = ks$, where k is a constant of proportionality. Although masses with different weights stretch a spring by different amounts, the spring is essentially characterized by the number k. For example, if a mass weighing 10 lb stretches a spring by 1/2 ft, then

$$10 = k\left(\frac{1}{2}\right) \qquad \text{implies} \qquad k = 20 \text{ lb/ft}.$$

Necessarily then a mass weighing 8 lb stretches the same spring 2/5 ft.

Newton's Second Law

After a mass m is attached to a spring, it will stretch the spring by an amount s and attain a position of equilibrium at which its weight W is balanced by the

* **Robert Hooke** (1635–1703) An English physicist and inventor, Hooke published this law in 1658. The idea of attaching a spring to a balance wheel, causing oscillatory motion that enabled a clock to mark units of time, is usually attributed to Hooke. The concept of the balance spring led to the invention of the pocket watch by Christian Huygens in 1674. Hooke accused Huygens of stealing his invention. Irascible and contentious, Hooke charged many of his colleagues, notably Isaac Newton, with plagiarism.

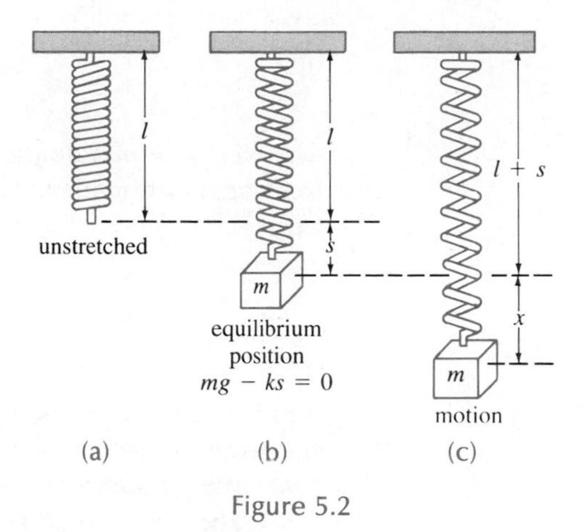

Figure 5.2

restoring force ks. Recall from Section 1.2 that weight is defined by

$$W = mg,$$

where the mass is measured in slugs, kilograms, or grams and $g = 32 \text{ ft/sec}^2$, 9.8 m/sec^2, or 980 cm/sec^2, respectively. As indicated in Figure 5.2(b), the condition of equilibrium is $mg = ks$ or $mg - ks = 0$. If the mass is now displaced by an amount x from its equilibrium position and released, the net force F in this dynamic case is given by **Newton's second law of motion** $F = ma$, where a is the acceleration d^2x/dt^2. Assuming that there are no retarding forces acting on the system and assuming that the mass vibrates free of other external influencing forces—**free motion**—we can then equate F to the resultant force of the weight and the restoring force:

$$m\frac{d^2x}{dt^2} = -k(s + x) + mg$$

$$= -kx + \underbrace{mg - ks}_{\text{zero}} = -kx. \tag{1}$$

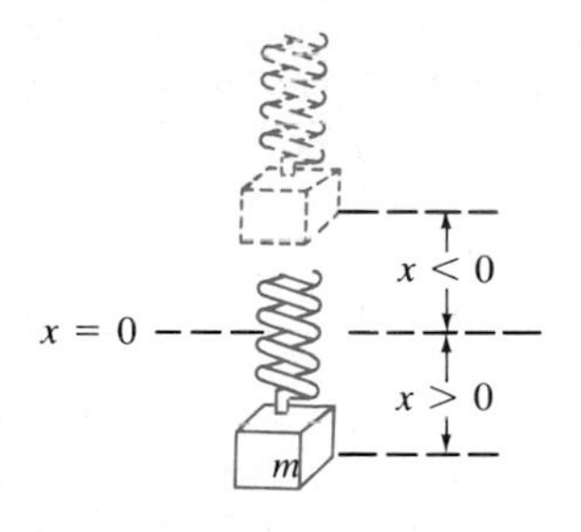

Figure 5.3

The negative sign in (1) indicates that the restoring force of the spring acts opposite to the direction of motion. Furthermore, we shall adopt the convention that displacements measured *below* the equilibrium position are positive (see Figure 5.3).

Differential Equation of Free Undamped Motion

By dividing (1) by the mass m, we obtain the second-order differential equation

$$\frac{d^2x}{dt^2} + \frac{k}{m}x = 0 \tag{2}$$

or

$$\frac{d^2x}{dt^2} + \omega^2 x = 0, \tag{3}$$

where $\omega^2 = k/m$. Equation (3) is said to describe **simple harmonic motion**, or **free undamped motion**. There are two obvious initial conditions associated with (3):

$$x(0) = \alpha, \qquad \left.\frac{dx}{dt}\right|_{t=0} = \beta \tag{4}$$

representing the amount of initial displacement and the initial velocity, respectively. For example, if $\alpha > 0$, $\beta < 0$, the mass would start from a point *below* the equilibrium position with an imparted *upward* velocity. If $\alpha < 0$, $\beta = 0$, the mass would be released from *rest* from a point $|\alpha|$ units *above* the equilibrium position, and so on.

Solution and Equation of Motion

To solve equation (3) we note that the solutions of the auxiliary equation $m^2 + \omega^2 = 0$ are the complex numbers

$$m_1 = \omega i, \qquad m_2 = -\omega i.$$

Thus from (8) of Section 4.3, we find the general solution of (3) to be

$$x(t) = c_1 \cos \omega t + c_2 \sin \omega t. \tag{5}$$

The **period** of free vibrations described by (5) is $T = 2\pi/\omega$ and the **frequency** is $f = 1/T = \omega/2\pi$.* For example, for $x(t) = 2 \cos 3t - 4 \sin 3t$, the period is $2\pi/3$ and the frequency is $3/2\pi$. The former number means that the graph of $x(t)$ repeats every $2\pi/3$ units; the latter number means that there are 3 cycles of the graph every 2π units, or equivalently, the mass undergoes $3/2\pi$ complete vibrations per unit time. In addition, it can be shown that the period $2\pi/\omega$ is the time interval between two successive maxima of $x(t)$.† Finally, when the initial conditions (4) are used to determine the constants c_1 and c_2 in (5), we say that the resulting particular solution is the **equation of motion**.

---

* Sometimes the number ω is called the *circular frequency* of vibrations. For free undamped motion the numbers $2\pi/\omega$ and $\omega/2\pi$ are also referred to as the *natural period* and *natural frequency*, respectively.

† Keep in mind that a maximum of $x(t)$ is a positive displacement corresponding to the mass attaining a maximum distance *below* the equilibrium position, whereas a minimum of $x(t)$ is a negative displacement corresponding to the mass attaining a maximum height *above* the equilibrium position. We shall refer to either case as an *extreme displacement* of the mass.

Solve and interpret the initial-value problem

$$\frac{d^2x}{dt^2} + 16x = 0$$

$$x(0) = 10, \qquad \left.\frac{dx}{dt}\right|_{t=0} = 0.$$

SOLUTION The problem is equivalent to pulling a mass on a spring down 10 units below the equilibrium position, holding it until $t = 0$, and then releasing it from rest. Applying the initial conditions to the solution

$$x(t) = c_1 \cos 4t + c_2 \sin 4t$$

gives

$$x(0) = 10 = c_1 \cdot 1 + c_2 \cdot 0$$

so that $c_1 = 10$, and hence

$$x(t) = 10 \cos 4t + c_2 \sin 4t$$

$$\frac{dx}{dt} = -40 \sin 4t + 4c_2 \cos 4t$$

$$\left.\frac{dx}{dt}\right|_{t=0} = 0 = 4c_2 \cdot 1.$$

The latter equation implies $c_2 = 0$, so therefore the equation of motion is $x(t) = 10 \cos 4t$.

The solution clearly shows that once the system is set in motion, it stays in motion with the mass bouncing back and forth 10 units on either side of the equilibrium position $x = 0$. As shown in Figure 5.4(b), the period of oscillation is $2\pi/4 = \pi/2$ sec.

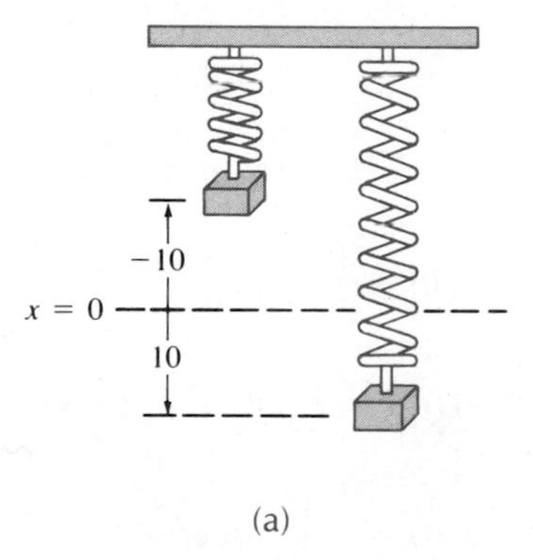

(a)

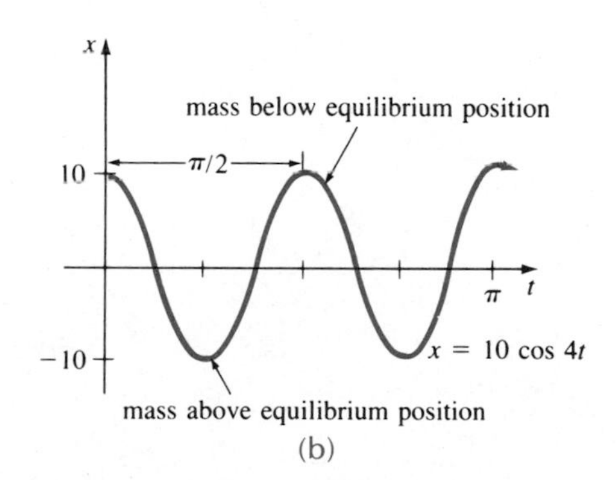

(b)

Figure 5.4

EXAMPLE 2

A mass weighing 2 lb stretches a spring 6 inches. At $t = 0$ the mass is released from a point 8 inches below the equilibrium position with an upward velocity of 4/3 ft/sec. Determine the function $x(t)$ that describes the subsequent free motion.

SOLUTION Since we are using the engineering system of units, the measurements given in terms of inches must be converted to feet: 6 inches = 1/2 foot; 8 inches = 2/3 foot. In addition we must convert the units of weight given in pounds to units of mass. From $m = W/g$ we have

$$m = \frac{2}{32} = \frac{1}{16} \text{ slug.}$$

Also, from Hooke's law,

$$2 = k\left(\frac{1}{2}\right) \quad \text{implies} \quad k = 4 \text{ lb/ft.}$$

Hence the analogues of (1) and (2) are, respectively,

$$\frac{1}{16}\frac{d^2x}{dt^2} = -4x \quad \text{and} \quad \frac{d^2x}{dt^2} + 64x = 0.$$

The initial displacement and initial velocity are given by

$$x(0) = \frac{2}{3}, \quad \left.\frac{dx}{dt}\right|_{t=0} = -\frac{4}{3},$$

where the negative sign in the last condition is a consequence of the fact that the mass is given an initial velocity in the negative or upward direction.

Now $\omega^2 = 64$ or $\omega = 8$ so that the general solution of the differential equation is

$$x(t) = c_1 \cos 8t + c_2 \sin 8t. \tag{6}$$

Applying the initial conditions to (6), we obtain

$$x(0) = \frac{2}{3} = c_1 \cdot 1 + c_2 \cdot 0, \quad \left(c_1 = \frac{2}{3}\right)$$

and

$$x(t) = \frac{2}{3} \cos 8t + c_2 \sin 8t$$

$$x'(t) = -\frac{16}{3} \sin 8t + 8c_2 \cos 8t$$

$$x'(0) = -\frac{4}{3} = -\frac{16}{3} \cdot 0 + 8c_2 \cdot 1, \quad \left(c_2 = -\frac{1}{6}\right).$$

Thus the equation of motion is

$$x(t) = \frac{2}{3}\cos 8t - \frac{1}{6}\sin 8t. \tag{7}$$

Note: The distinction between weight and mass, regrettably, is often blurred. Thus one often speaks of both the motion of a mass on a spring and the motion of a weight on a spring.

Alternative Form of $x(t)$

When $c_1 \neq 0$ and $c_2 \neq 0$, the actual **amplitude** A of free vibrations is not obvious from inspection of equation (5). For example, although the mass in Example 2 is initially displaced 2/3 ft beyond the equilibrium position, the amplitude of vibrations is a number larger than 2/3. Hence it is often convenient to convert a solution of form (5) to the simpler form

$$x(t) = A\sin(\omega t + \phi), \tag{8}$$

where

$$A = \sqrt{c_1^2 + c_2^2}$$

and ϕ is a **phase angle** defined by

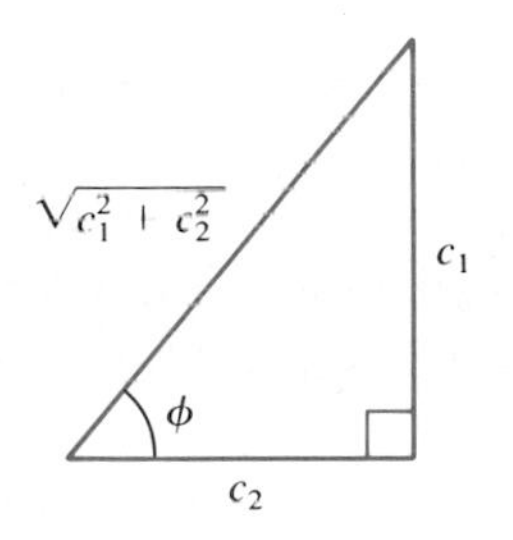

Figure 5.5

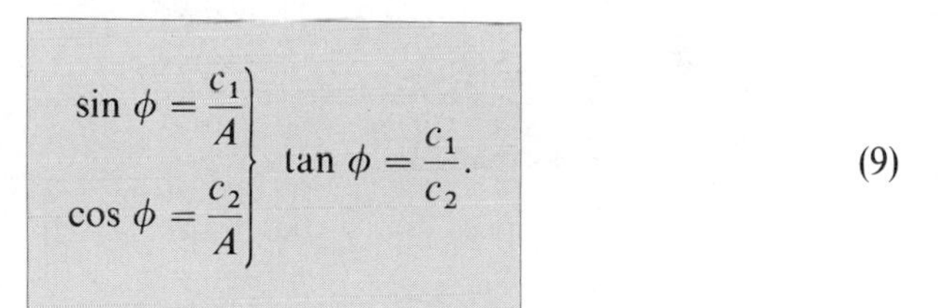

$$\left.\begin{aligned} \sin\phi &= \frac{c_1}{A} \\ \cos\phi &= \frac{c_2}{A} \end{aligned}\right\} \tan\phi = \frac{c_1}{c_2}. \tag{9}$$

To verify this we expand (8) by the addition formula for the sine function:

$$A\sin\omega t\cos\phi + A\cos\omega t\sin\phi = (A\sin\phi)\cos\omega t + (A\cos\phi)\sin\omega t. \tag{10}$$

It follows from Figure 5.5 that if ϕ is defined by

$$\sin\phi = \frac{c_1}{\sqrt{c_1^2 + c_2^2}} = \frac{c_1}{A}, \qquad \cos\phi = \frac{c_2}{\sqrt{c_1^2 + c_2^2}} = \frac{c_2}{A}$$

then (10) becomes

$$A\frac{c_1}{A}\cos\omega t + A\frac{c_2}{A}\sin\omega t = c_1\cos\omega t + c_2\sin\omega t = x(t).$$

EXAMPLE 3

In view of the foregoing discussion, we can write the solution (7) in Example 2

$$x(t) = \frac{2}{3}\cos 8t - \frac{1}{6}\sin 8t \qquad \text{alternatively as} \qquad x(t) = A\sin(8t + \phi).$$

The amplitude is given by

$$A = \sqrt{\left(\frac{2}{3}\right)^2 + \left(-\frac{1}{6}\right)^2} = \frac{\sqrt{17}}{6} \approx 0.69 \text{ ft.}$$

One should exercise some care when finding the phase angle ϕ defined by (9). In this case

$$\tan\phi = \frac{2/3}{-1/6} = -4$$

and a scientific hand calculator would give

$$\tan^{-1}(-4) = -1.326 \text{ radians.*}$$

But this angle is located in the fourth quadrant and therefore contradicts the fact that $\sin\phi > 0$ and $\cos\phi < 0$ (recall, $c_1 > 0$ and $c_2 < 0$). Hence we must take ϕ to be the second quadrant angle

$$\phi = \pi + (-1.326) = 1.816 \text{ radians.}$$

Thus we have

$$x(t) = \frac{\sqrt{17}}{6}\sin(8t + 1.816). \tag{11}$$

Form (8) is very useful since it is easy to find the values of time for which the graph of $x(t)$ crosses the positive t-axis (the line $x = 0$). We observe that $\sin(\omega t + \phi) = 0$ when

$$\omega t + \phi = n\pi,$$

where n is a nonnegative integer.

EXAMPLE 4

For motion described by $x(t) = (\sqrt{17}/6)\sin(8t + 1.816)$, find the first value of time for which the mass passes through the equilibrium position heading downward.

* The range of the inverse tangent is $-\pi/2 < \tan^{-1}x < \pi/2$.

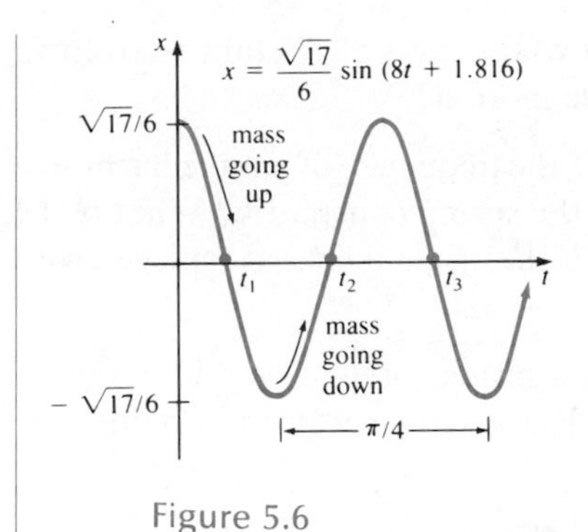

Figure 5.6

SOLUTION The values $t_1, t_2, t_3, \ldots$, for which $\sin(8t + 1.816) = 0$ are determined from

$$8t_1 + 1.816 = \pi, \qquad 8t_2 + 1.816 = 2\pi, \qquad 8t_3 + 1.816 = 3\pi, \ldots.$$

We find $t_1 = 0.166$, $t_2 = 0.558$, $t_3 = 0.951, \ldots$, respectively.

Inspection of Figure 5.6 shows that the mass passes through $x = 0$ heading downward (namely, toward $x > 0$) the first time at $t_2 = 0.558$ sec.

Exercises 5.1

Answers to odd-numbered problems begin on page A–12.

In Problems 1 and 2 state in words a possible physical interpretation of the given problem.

1. $\frac{4}{32}x'' + 3x = 0$

$x(0) = -3, \quad x'(0) = -2$

2. $\frac{1}{16}x'' + 4x = 0$

$x(0) = 0.7, \quad x'(0) = 0$

In Problems 3–8 write the solution of the given initial-value problem in form (8).

3. $x'' + 25x = 0$

$x(0) = -2, \quad x'(0) = 10$

4. $\frac{1}{2}x'' + 8x = 0$

$x(0) = 1, \quad x'(0) = -2$

5. $x'' + 2x = 0$

$x(0) = -1, \quad x'(0) = -2\sqrt{2}.$

6. $\frac{1}{4}x'' + 16x = 0$

$x(0) = 4, \quad x'(0) = 16$

7. $0.1x'' + 10x = 0$

$x(0) = 1, \quad x'(0) = 1$

8. $x'' + x = 0$

$x(0) = -4, \quad x'(0) = 3$

9. The period of free undamped oscillations of a mass on a spring is $\pi/4$ sec. If the spring constant is 16 lb/ft, what is the numerical value of the weight?

10. A spring is suspended from a ceiling. When a mass weighing 60 lb is attached, the spring is stretched 1/2 ft. The mass is removed and a person grabs the end of the spring and proceeds to bounce up and down with a period of 1 sec. How much does the person weigh?

11. A 4-lb weight is attached to a spring whose spring constant is 16 lb/ft. What is the period of simple harmonic motion?

12. A 20-kg mass is attached to a spring. If the frequency of simple harmonic motion is $2/\pi$ vibrations/sec, what is the spring constant k? What is the frequency of simple harmonic motion if the original mass is replaced with an 80-kg mass?

13. A 24-lb weight, attached to the end of a spring, stretches it 4 in. Find the equation of motion if the weight is released from rest from a point 3 in. above the equilibrium position.

14. Determine the equation of motion if the weight in Problem 13 is released from the equilibrium position with an initial downward velocity of 2 ft/sec.

15. A 20-lb weight stretches a spring 6 in. If the weight is released from rest 6 in. below the equilibrium position,

(a) find the position of the weight at $t = \pi/12, \pi/8, \pi/6, \pi/4, 9\pi/32$ sec.

(b) What is the velocity of the weight when $t = 3\pi/16$ sec? In which direction is the weight heading at this instant?

(c) At what times does the weight pass through the equilibrium position?

16. A force of 400 newtons stretches a spring 2 m. A mass of 50 kg is attached to the end of the spring and released from the equilibrium position with an upward velocity of 10 m/sec. Find the equation of motion.

17. Another spring whose constant is 20 nt/m is suspended from the same rigid support but parallel to the spring-mass system in Problem 16. A mass of 20 kg is attached to the second spring and both masses are released from the equilibrium position with an upward velocity of 10 m/sec.

(a) Which mass exhibits the greater amplitude of motion?

(b) Which mass is moving faster at $t = \pi/4$ sec? at $\pi/2$ sec?

(c) At what times are the two masses in the same position? Where are the masses at these times? In which directions are they moving?

18. A 32-lb weight stretches a spring 2 ft. Determine the amplitude and period of motion if the weight is released 1 ft above the equilibrium position with an initial upward velocity of 2 ft/sec. How many complete vibrations will the weight have completed at the end of 4π sec?

19. An 8-lb weight attached to a spring exhibits simple harmonic motion. Determine the equation of motion if the spring constant is 1 lb/ft and if the weight is released 6 in. below the equilibrium position with a downward velocity of 3/2 ft/sec. Express the solution in form (8).

20. A mass weighing 10 lb stretches a spring 1/4 ft. This mass is removed and replaced with a mass of 1.6 slugs, which is released 1/3 ft above the equilibrium position with a downward velocity of 5/4 ft/sec. Express the solution in form (8). At what times does the mass attain a displacement below the equilibrium position numerically equal to one-half the amplitude?

21. A 64-lb weight attached to the end of a spring stretches it 0.32 ft. From a position 8 in. above the equilibrium position the weight is given a downward velocity of 5 ft/sec.

(a) Find the equation of motion.

(b) What is the amplitude and period of motion?

(c) How many complete vibrations will the weight have completed at the end of 3π sec?

(d) At what time does the weight pass through the equilibrium position heading downward for the second time?

(e) At what time does the weight attain its extreme displacement on either side of the equilibrium position?

(f) What is the position of the weight at $t = 3$ sec?

(g) What is the instantaneous velocity at $t = 3$ sec?

(h) What is the acceleration at $t = 3$ sec?

(i) What is the instantaneous velocity at the times when the weight passes through the equilibrium position?

(j) At what times is the weight 5 in. below the equilibrium position?

(k) At what times is the weight 5 in. below the equilibrium position heading in the upward direction?

22. A mass of 1 slug is suspended from a spring whose characteristic spring constant is 9 lb/ft. Initially the mass starts from a point 1 ft above the equilibrium position with an upward velocity of $\sqrt{3}$ ft/sec. Find the times for which the mass is heading downward at a velocity of 3 ft/sec.

k_1 k_2

20 lb

Figure 5.7

23. Under some circumstances when two parallel springs, with constants k_1 and k_2, support a single weight W, the **effective spring constant** of the system is given by $k = 4k_1k_2/(k_1 + k_2)$*. A 20-lb weight stretches one spring 6 in. and another spring 2 in. The springs are attached to a common rigid support and then to a metal plate. As shown in Figure 5.7, the 20-lb weight is attached to the center of the plate in the double spring arrangement. Determine the effective spring constant of this system. Find

* If the two springs have the same natural length and if the plate is guided either by rails or by an external force so that the extensions of both springs are always the same, then the effective spring constant is simply $k = k_1 + k_2$.

the equation of motion if the weight is released from the equilibrium position with a downward velocity of 2 ft/sec.

24. A certain weight stretches one spring 1/3 ft and another spring 1/2 ft. The two springs are attached to a common rigid support in a manner indicated in Problem 23 and Figure 5.7. The first weight is set aside, and an 8-lb weight is attached to the double spring arrangement and the system is set in motion. If the period of motion is $\pi/15$ sec, determine the numerical value of the first weight.

Miscellaneous Problems

25. If x_0 and v_0 are the initial position and velocity, respectively, of a weight exhibiting simple harmonic motion, show that the amplitude of vibrations is

$$A = \sqrt{x_0^2 + \left(\frac{v_0}{\omega}\right)^2}.$$

26. Show that any linear combination $x(t) = c_1 \cos \omega t + c_2 \sin \omega t$ can also be written in the form

$$x(t) = A \cos(\omega t + \phi), \qquad \text{where} \qquad A = \sqrt{c_1^2 + c_2^2}$$

and

$$\sin \phi = -\frac{c_2}{A}, \qquad \cos \phi = \frac{c_1}{A}.$$

27. Express the solution of Problem 3 in the form of the cosine function given in Problem 26.

28. Show that when a weight attached to a spring exhibits simple harmonic motion, the maximum value of the speed (that is, $|v(t)|$) occurs when the weight is passing through the equilibrium position.

29. A weight attached to a spring exhibits simple harmonic motion. Show that the maximum acceleration of the weight occurs at an extreme displacement and has the magnitude $4\pi^2 A/T^2$, where A is the amplitude and T is the period of free vibrations.

30. Use (8) to prove that the time interval between two successive maxima of $x(t)$ is $2\pi/\omega$.

5.2 Damped Motion

The discussion of free harmonic motion is somewhat unrealistic since the motion described by equation (2) of Section 5.1 assumes that no retarding forces are acting on the moving mass. Unless the mass is suspended in a perfect vacuum, there will be at least a resisting force due to the surrounding medium.

(a)

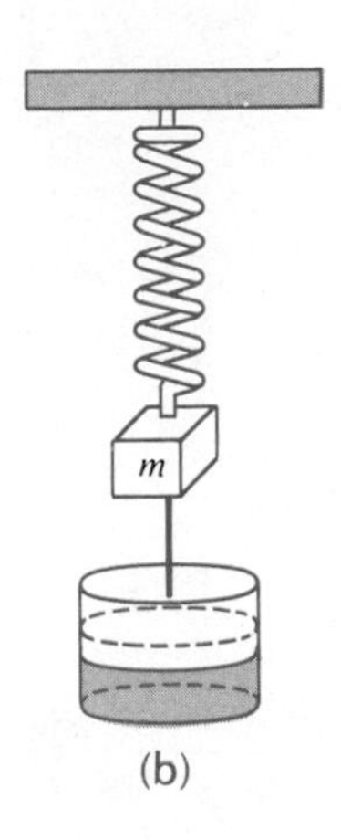

(b)

Figure 5.8

For example, as Figure 5.8 shows, the mass m could be suspended in a viscous medium or connected to a dashpot damping device.

Differential Equation of Motion with Damping

In the study of mechanics, damping forces acting on a body are considered to be proportional to a power of the instantaneous velocity. In particular, we shall assume throughout the subsequent discussion that this force is given by a constant multiple dx/dt.* When no other external forces are impressed on the system, it follows from Newton's second law that

$$m\frac{d^2x}{dt^2} = -kx - \beta\frac{dx}{dt}, \tag{1}$$

where β is a positive *damping constant* and the negative sign is a consequence of the fact that the damping force acts in a direction opposite to the motion.

Dividing (1) by the mass m, the differential equation of **free damped motion** then is

$$\frac{d^2x}{dt^2} + \frac{\beta}{m}\frac{dx}{dt} + \frac{k}{m}x = 0 \tag{2}$$

or

$$\frac{d^2x}{dt^2} + 2\lambda\frac{dx}{dt} + \omega^2 x = 0. \tag{3}$$

---

* In many instances, such as problems in hydrodynamics, the damping force is proportional to $(dx/dt)^2$.

In equation (3) we make the identifications

$$2\lambda = \frac{\beta}{m}, \qquad \omega^2 = \frac{k}{m}. \tag{4}$$

The symbol 2λ is used only for algebraic convenience since the auxiliary equation is $m^2 + 2\lambda m + \omega^2 = 0$ and the corresponding roots are then

$$m_1 = -\lambda + \sqrt{\lambda^2 - \omega^2}, \qquad m_2 = -\lambda - \sqrt{\lambda^2 - \omega^2}.$$

We can now distinguish three possible cases depending on the algebraic sign of $\lambda^2 - \omega^2$. Since each solution will contain the *damping factor* $e^{-\lambda t}$, $\lambda > 0$, the displacements of the mass will become negligible for large time.

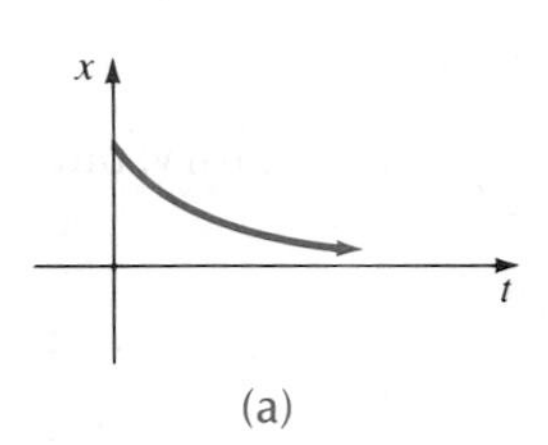

(a)

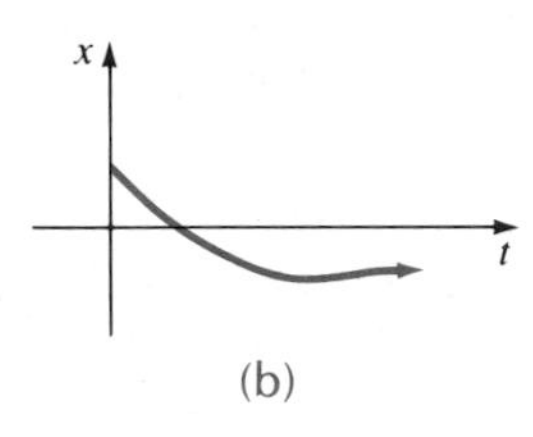

(b)

Figure 5.9

CASE I $\lambda^2 - \omega^2 > 0$. In this situation the system is said to be **overdamped** since the damping coefficient β is large when compared to the spring constant k. The corresponding solution of (3) is

$$x(t) = c_1 e^{m_1 t} + c_2 e^{m_2 t}$$

or

$$x(t) = e^{-\lambda t}\left(c_1 e^{\sqrt{\lambda^2 - \omega^2}\, t} + c_2 e^{-\sqrt{\lambda^2 - \omega^2}\, t}\right). \tag{5}$$

This equation represents a smooth and nonoscillatory motion. Figure 5.9 shows two possible graphs of $x(t)$. ■

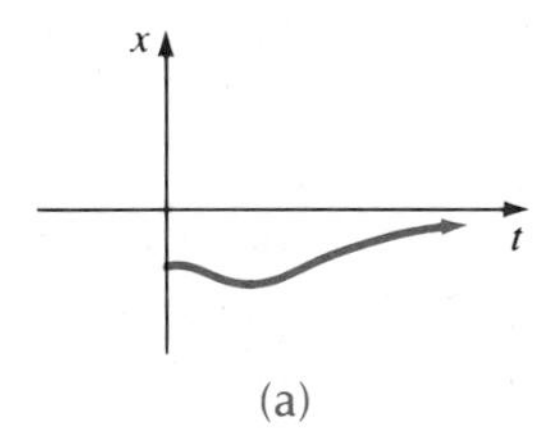

(a)

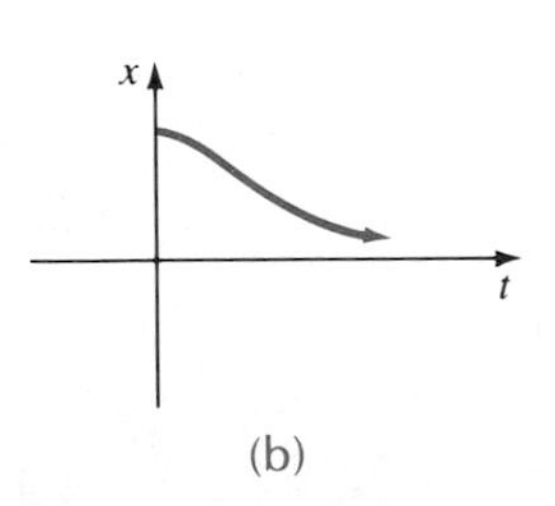

(b)

Figure 5.10

CASE II $\lambda^2 - \omega^2 = 0$. The system is said to be **critically damped** since any slight decrease in the damping force would result in oscillatory motion. The general solution of (3) is

$$x(t) = c_1 e^{m_1 t} + c_2 t e^{m_1 t}$$

or

$$x(t) = e^{-\lambda t}(c_1 + c_2 t). \tag{6}$$

Some graphs of typical motion are given in Figure 5.10. Notice that the motion is quite similar to that of an overdamped system. It is also apparent from (6) that the mass can pass through the equilibrium position at most one time.* ■

CASE III $\lambda^2 - \omega^2 < 0$. In this case the system is said to be **underdamped** since the damping coefficient is small compared to the spring constant.

* An examination of the derivatives of (5) and (6) would show that these functions can have at most one relative maximum or one relative minimum for $t > 0$.

The roots m_1 and m_2 are now complex:

$$m_1 = -\lambda + \sqrt{\omega^2 - \lambda^2}\,i, \qquad m_2 = -\lambda - \sqrt{\omega^2 - \lambda^2}\,i$$

and so the general solution of equation (3) is

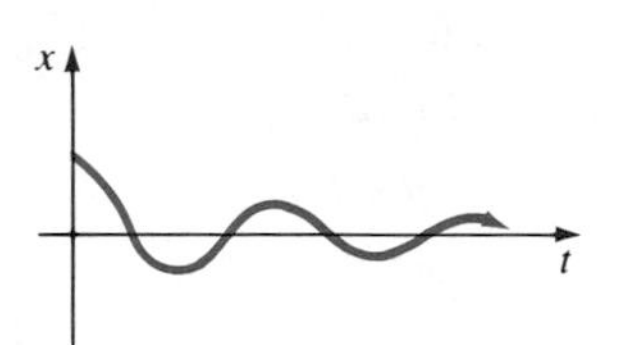

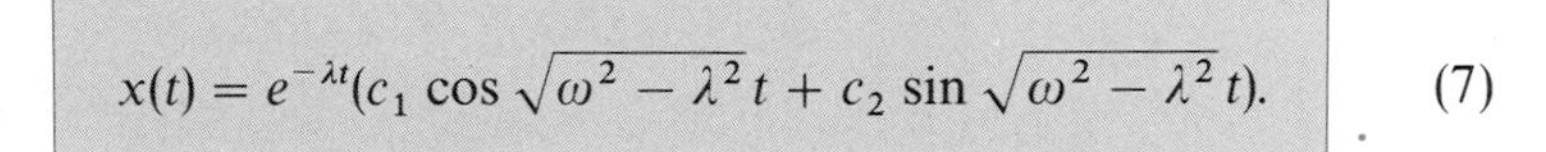

$$x(t) = e^{-\lambda t}(c_1 \cos \sqrt{\omega^2 - \lambda^2}\,t + c_2 \sin \sqrt{\omega^2 - \lambda^2}\,t). \tag{7}$$

Figure 5.11

As indicated in Figure 5.11, the motion described by (7) is oscillatory; but because of the coefficient $e^{-\lambda t}$, the amplitudes of vibration $\to 0$ as $t \to \infty$. ■

EXAMPLE 1

It is readily verified that the solution of the initial-value problem

$$\frac{d^2x}{dt^2} + 5\frac{dx}{dt} + 4x = 0$$

$$x(0) = 1, \qquad \left.\frac{dx}{dt}\right|_{t=0} = 1$$

is

$$x(t) = \frac{5}{3}e^{-t} - \frac{2}{3}e^{-4t}. \tag{8}$$

The problem can be interpreted as representing the overdamped motion of a mass on a spring. The mass starts from a position 1 unit *below* the equilibrium position with a *downward* velocity of 1 ft/sec.

To graph $x(t)$ we find the value of t for which the function has an extremum, that is, the value of time for which the first derivative (velocity) is zero. Differentiating (8) gives

$$x'(t) = -\frac{5}{3}e^{-t} + \frac{8}{3}e^{-4t}$$

so that $x'(t) = 0$ implies

$$e^{3t} = \frac{8}{5} \qquad \text{or} \qquad t = \frac{1}{3}\ln\frac{8}{5} = 0.157.$$

It follows from the first derivative test, as well as our physical intuition, that $x(0.157) = 1.069$ ft is actually a maximum. In other words, the mass attains an extreme displacement of 1.069 ft below the equilibrium position.

We should also check to see whether the graph crosses the t-axis, that is, whether the mass passes through the equilibrium position. This cannot happen in this instance since the equation $x(t) = 0$, or $e^{3t} = 2/5$, has the physically irrelevant solution $t = (1/3)\ln(2/5) = -0.305$.

The graph of $x(t)$, along with some other pertinent data, is given in Figure 5.12.

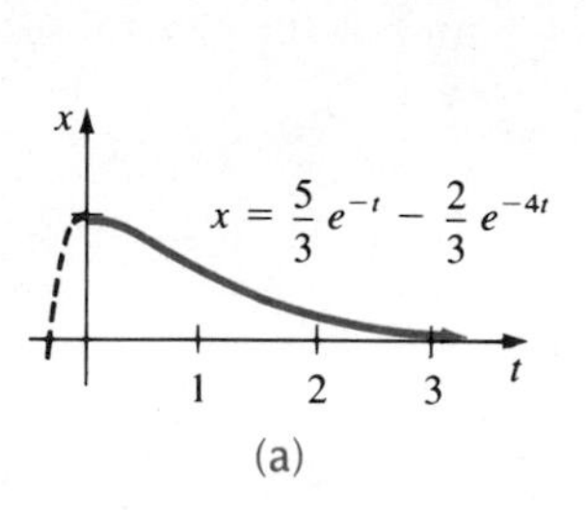

(a)

t	$x(t)$
1	0.601
1.5	0.370
2	0.225
2.5	0.137
3	0.083

(b)

Figure 5.12

EXAMPLE 2

An 8-lb weight stretches a spring 2 ft. Assuming a damping force numerically equal to two times the instantaneous velocity acts on the system, determine the equation of motion if the weight is released from the equilibrium position with an upward velocity of 3 ft/sec.

SOLUTION From Hooke's law we have

$$8 = k(2), \qquad k = 4 \text{ lb/ft}$$

and from $m = W/g$,

$$m = \frac{8}{32} = \frac{1}{4} \text{ slug.}$$

Thus the differential equation of motion is

$$\frac{1}{4}\frac{d^2x}{dt^2} = -4x - 2\frac{dx}{dt} \quad \text{or} \quad \frac{d^2x}{dt^2} + 8\frac{dx}{dt} + 16x = 0. \tag{9}$$

The initial conditions are

$$x(0) = 0, \qquad \left.\frac{dx}{dt}\right|_{t=0} = -3.$$

Now the auxiliary equation for (9) is

$$m^2 + 8m + 16 = (m + 4)^2 = 0$$

so that $m_1 = m_2 = -4$. Hence the system is critically damped and

$$x(t) = c_1e^{-4t} + c_2te^{-4t}. \tag{10}$$

The initial condition $x(0) = 0$ immediately demands that $c_1 = 0$, whereas using $x'(0) = -3$ gives $c_2 = -3$. Thus the equation of motion is

$$x(t) = -3te^{-4t}. \tag{11}$$

To graph $x(t)$ we proceed as in Example 1:

$$\begin{aligned} x'(t) &= -3(-4te^{-4t} + e^{-4t}) \\ &= -3e^{-4t}(1 - 4t). \end{aligned}$$

Clearly $x'(t) = 0$ when $t = 1/4$. The corresponding extreme displacement is

$$x\left(\frac{1}{4}\right) = -3\left(\frac{1}{4}\right)e^{-1} = -0.276 \text{ ft}.$$

As shown in Figure 5.13, we interpret this value to mean that the weight reaches a maximum height of 0.276 ft above the equilibrium position.

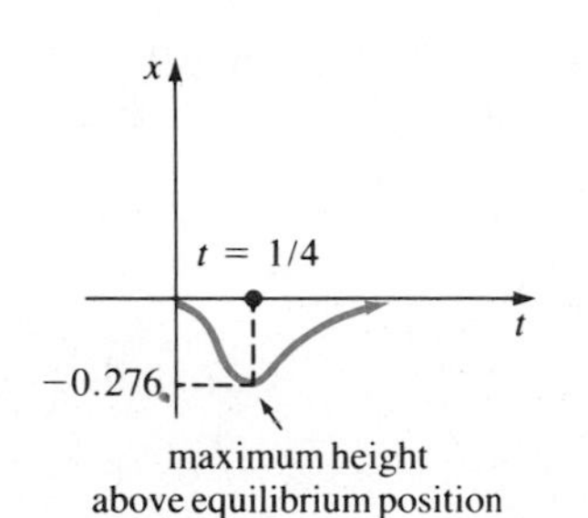

Figure 5.13

EXAMPLE 3

A 16-lb weight is attached to a 5-ft-long spring. At equilibrium the spring measures 8.2 ft. If the weight is pushed up and released from rest at a point 2 ft above the equilibrium position, find the displacements $x(t)$ if it is further known that the surrounding medium offers a resistance numerically equal to the instantaneous velocity.

SOLUTION The elongation of the spring after the weight is attached is $8.2 - 5 = 3.2$ ft so it follows from Hooke's law that

$$16 = k(3.2), \qquad k = 5 \text{ lb/ft}.$$

In addition,

$$m = \frac{16}{32} = \frac{1}{2} \text{ slug},$$

so that the differential equation is given by

$$\frac{1}{2}\frac{d^2x}{dt^2} = -5x - \frac{dx}{dt} \qquad \text{or} \qquad \frac{d^2x}{dt^2} + 2\frac{dx}{dt} + 10x = 0. \tag{12}$$

This latter equation is solved subject to the conditions

$$x(0) = -2, \qquad \left.\frac{dx}{dt}\right|_{t=0} = 0.$$

Proceeding, we find that the roots of $m^2 + 2m + 10 = 0$ are $m_1 = -1 + 3i$ and $m_2 = -1 - 3i$, which then implies the system is underdamped and

$$x(t) = e^{-t}(c_1 \cos 3t + c_2 \sin 3t) \tag{13}$$

Now $x(0) = -2 = c_1$

$$x(t) = e^{-t}(-2 \cos 3t + c_2 \sin 3t)$$

$$x'(t) = e^{-t}(6 \sin 3t + 3c_2 \cos 3t) - e^{-t}(-2 \cos 3t + c_2 \sin 3t)$$

$$x'(0) = 0 = 3c_2 + 2,$$

which gives $c_2 = -2/3$. Thus we finally obtain

$$x(t) = e^{-t}\left(-2 \cos 3t - \frac{2}{3} \sin 3t\right). \tag{14}$$

Alternative Form of the Solution

In a manner identical to the procedure used in Section 5.1, we can write any solution

$$x(t) = e^{-\lambda t}(c_1 \cos \sqrt{\omega^2 - \lambda^2}\, t + c_2 \sin \sqrt{\omega^2 - \lambda^2}\, t)$$

in the alternative form

$$x(t) = Ae^{-\lambda t} \sin(\sqrt{\omega^2 - \lambda^2}\, t + \phi), \tag{15}$$

where $A = \sqrt{c_1^2 + c_2^2}$ and the phase angle ϕ is determined from the equations

$$\sin \phi = \frac{c_1}{A}, \qquad \cos \phi = \frac{c_2}{A}, \qquad \tan \phi = \frac{c_1}{c_2}.$$

The coefficient $Ae^{-\lambda t}$ is sometimes called the **damped amplitude** of vibrations. Because (15) is not a periodic function, the number $2\pi/\sqrt{\omega^2 - \lambda^2}$ is called the **quasi period** and $\sqrt{\omega^2 - \lambda^2}/2\pi$ is the **quasi frequency**. The quasi period is the time interval between two successive maxima of $x(t)$.

To graph an equation such as (15), we first find the intercepts $t_1, t_2, \ldots, t_k, \ldots$; that is, for some integer n we must solve

$$\sqrt{\omega^2 - \lambda^2}\, t + \phi = n\pi$$

for t. It follows that

$$t = \frac{n\pi - \phi}{\sqrt{\omega^2 - \lambda^2}}. \tag{16}$$

In addition we note that $|x(t)| \leq Ae^{-\lambda t}$ since

$$|\sin(\sqrt{\omega^2 - \lambda^2}\, t + \phi)| \leq 1.$$

Indeed, the graph of (15) touches the graphs of $\pm Ae^{-\lambda t}$ at the values t_1^*, $t_2^*, \ldots, t_k^*, \ldots$ for which

$$\sin(\sqrt{\omega^2 + \lambda^2}\, t + \phi) = \pm 1.$$

This means $\sqrt{\omega^2 - \lambda^2}\, t + \phi$ must be an odd multiple of $\pi/2$,

$$\sqrt{\omega^2 - \lambda^2}\, t + \phi = (2n + 1)\frac{\pi}{2}$$

$$t = \frac{(2n + 1)\pi/2 - \phi}{\sqrt{\omega^2 - \lambda^2}} \tag{17}$$

For example, if we are asked to graph $x(t) = e^{-0.5t} \sin(2t - \pi/3)$, we find the intercepts on the *positive* t-axis by solving

$$2t_1 - \frac{\pi}{3} = 0, \qquad 2t_2 - \frac{\pi}{3} = \pi, \qquad 2t_3 - \frac{\pi}{3} = 2\pi, \ldots,$$

which gives, respectively,

$$t_1 = \frac{\pi}{6}, \qquad t_2 = \frac{4\pi}{6}, \qquad t_3 = \frac{7\pi}{6}, \ldots.$$

Notice that even though $x(t)$ is not periodic, the difference between the successive roots is $t_k - t_{k-1} = \pi/2$ units or one-half the quasi period of $2\pi/2 = \pi$ sec. Also $\sin(2t - \pi/3) = \pm 1$ at the solutions of

$$2t_1^* - \frac{\pi}{3} = \frac{\pi}{2}, \qquad 2t_2^* - \frac{\pi}{3} = \frac{3\pi}{2}, \qquad 2t_3^* - \frac{\pi}{3} = \frac{5\pi}{2}, \ldots$$

or

$$t_1^* = \frac{5\pi}{12}, \qquad t_2^* = \frac{11\pi}{12}, \qquad t_3^* = \frac{17\pi}{12}, \ldots.$$

It is readily shown that the difference between the successive t_k^* is also $\pi/2$.[†] The graph of $x(t)$ is given in Figure 5.14.

[†] We note that the values of t for which the graph of $x(t)$ touches the exponential graphs are *not* the values for which the function attains its relative extrema.

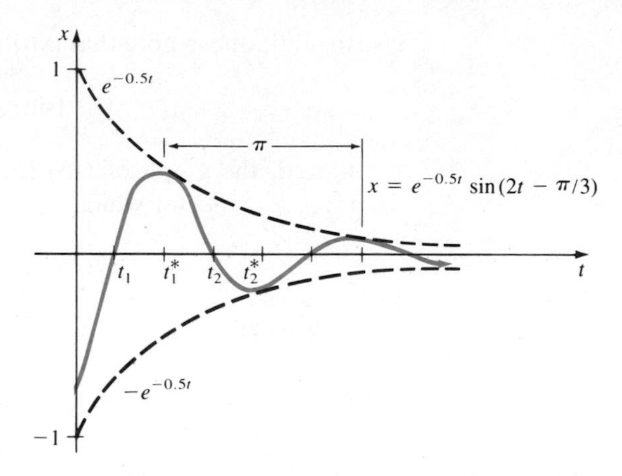

Figure 5.14

EXAMPLE 4

Using (15), we can write the solution of the preceding example

$$\frac{d^2x}{dt^2} + 2\frac{dx}{dt} + 10x = 0$$

$$x(0) = -2, \qquad \left.\frac{dx}{dt}\right|_{t=0} = 0$$

in the form $x(t) = Ae^{-t}\sin(3t + \phi)$.

Now from (14) we have $c_1 = -2$, $c_2 = -2/3$ so that

$$A = \sqrt{4 + \frac{4}{9}} = \frac{2}{3}\sqrt{10}$$

$$\tan\phi = \frac{-2}{-2/3} = 3 \qquad \text{and} \qquad \tan^{-1}(3) = 1.249 \text{ radians.}$$

But since $\sin\phi < 0$ and $\cos\phi < 0$, we take ϕ to be the third quadrant angle $\phi = \pi + 1.249 = 4.391$ radians. Hence

$$x(t) = \frac{2}{3}\sqrt{10}e^{-t}\sin(3t + 4.391).$$

The graph of this function is given in Figure 5.15. The values of t_k and t_k^* given in the accompanying table are the intercepts and the points at which the graph of $x(t)$ touches the graphs of $\pm(2/3)\sqrt{10}e^{-t}$, respectively. In this example the quasi period is $2\pi/3$ sec and so the difference between the successive t_k (and the successive t_k^*) is $\pi/3$ units.

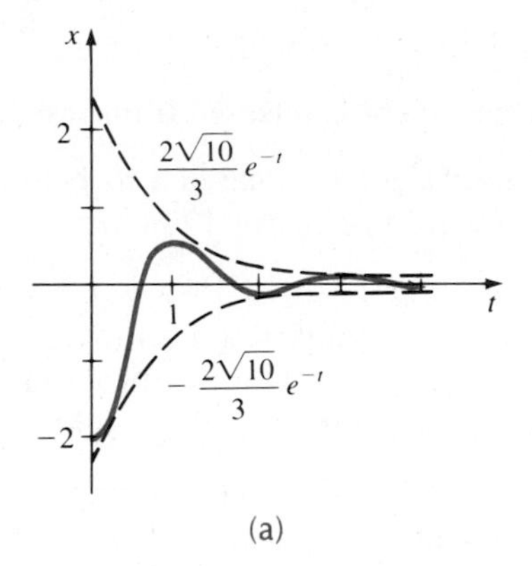

k	t_k	t_k^*	$x(t_k^*)$
1	0.631	1.154	0.665
2	1.678	2.202	−0.233
3	2.725	3.249	0.082
4	3.772	4.296	−0.029

(b)

Figure 5.15

Exercises 5.2

Answers to odd-numbered problems begin on page A–12.

In Problems 1 and 2 give a possible physical interpretation of the given initial-value problem.

1. $\frac{1}{16}x'' + 2x' + x = 0$

$x(0) = 0, \quad \left.\frac{dx}{dt}\right|_{t=0} = -1.5$

2. $\frac{16}{32}x'' + x' + 2x = 0$

$x(0) = -2, \quad x'(0) = 1$

3. A 4-lb weight is attached to a spring whose constant is 2 lb/ft. The medium offers a resistance to the motion of the weight numerically equal to the instantaneous velocity. If the weight is released from a point 1 ft above the equilibrium position with a downward velocity of 8 ft/sec, determine the time that the weight passes through the equilibrium position. Find the time for which the weight attains its extreme displacement from the equilibrium position. What is the position of the weight at this instant?

4. A 4-ft spring measures 8 ft long after an 8-lb weight is attached to it. The medium through which the weight moves offers a resistance numerically equal to $\sqrt{2}$ times the instantaneous velocity. Find the equation of motion if the weight is released from the equilibrium position with a downward velocity of 5 ft/sec. Find the time for which the weight attains its extreme displacement from the equilibrium position. What is the position of the weight at this instant?

5. A 1-kg mass is attached to a spring whose constant is 16 nt/m and the entire system is then submerged in a liquid that imparts a damping force numerically equal to 10 times the instantaneous velocity. Determine the equations of motion if

(a) the weight is released from rest 1 m below the equilibrium position;

(b) the weight is released 1 m below the equilibrium position with an upward velocity of 12 m/sec.

6. In parts (a) and (b) of Problem 5, determine whether the weight passes through the equilibrium position. In each case find the time at which the weight attains its extreme displacement from the equilibrium position. What is the position of the weight at this instant?

7. A force of 2 lb stretches a spring 1 ft. A 3.2-lb weight is attached to the spring and the system is then immersed in a medium that imparts a damping force numerically equal to 0.4 times the instantaneous velocity.

(a) Find the equation of motion if the weight is released from rest 1 ft above the equilibrium position.

(b) Express the equation of motion in the form given in (15).

(c) Find the first time for which the weight passes through the equilibrium position heading upward.

8. After a 10-lb weight is attached to a 5-ft spring, the spring measures 7 ft long. The 10-lb weight is removed and replaced with an 8-lb weight and the entire system is placed in a medium offering a resistance numerically equal to the instantaneous velocity.

(a) Find the equation of motion if the weight is released 1/2 ft below the equilibrium position with a downward velocity of 1 ft/sec.

(b) Express the equation of motion in the form given in (15).

(c) Find the times for which the weight passes through the equilibrium position heading downward.

(d) Graph the equation of motion.

9. A 10-lb weight attached to a spring stretches it 2 ft. The weight is attached to a dashpot damping device that offers a resistance numerically equal to $\beta(\beta > 0)$ times the instantaneous velocity. Determine the values of the damping constant β so that the subsequent motion is **(a)** overdamped, **(b)** critically damped, and **(c)** underdamped.

10. A 24-lb weight stretches a spring 4 ft. The subsequent motion takes place in a medium offering a resistance numerically equal to $\beta(\beta > 0)$ times the instantaneous velocity. If the weight starts from the equilibrium position with an upward velocity of 2 ft/sec, show that if $\beta > 3\sqrt{2}$, the equation of motion is

$$x(t) = \frac{-3}{\sqrt{\beta^2 - 18}} e^{-2\beta t/3} \sinh \frac{2}{3} \sqrt{\beta^2 - 18}\, t.$$

11. A mass of 40 g stretches a spring 10 cm. A damping device imparts a resistance to motion numerically equal to 560 times the instantaneous velocity. Find the equation of motion if the mass is released from the equilibrium position with a downward velocity of 2 cm/sec.

12. Find the equation of motion for the mass in Problem 11 if the damping constant is doubled.

13. A mass of 1 slug is attached to a spring whose constant is 9 lb/ft. The medium offers a resistance to the motion numerically equal to 6 times the instantaneous velocity. The mass is released from a point 8 in. above the equilibrium position with a downward velocity of v_0 ft/sec. Determine the values of v_0 such that the mass will subsequently pass through the equilibrium position.

14. The quasi period of an underdamped, vibrating 1-slug mass on a spring is $\pi/2$ sec. If the spring constant is 25 lb/ft, find the damping constant β.

Miscellaneous Problems

15. In the case of underdamped motion show that the difference in times between two successive positive maxima of the equation of motion is $2\pi/\sqrt{\omega^2 - \lambda^2}$.

16. Use (16) to show that the time interval between successive intercepts of (15) is one-half the quasi period.

17. Use (17) to show that the time interval between successive values of t for which the graph of (15) touches the graphs of $\pm Ae^{-\lambda t}$ is one-half the quasi period.

18. Use equation (17) to show that the intercepts of the graph of $x(t) = Ae^{-\lambda t}\sin(\sqrt{\omega^2 - \lambda^2}\,t + \phi)$ are halfway between the values of t for which the graph of $x(t)$ touches the graphs of $\pm Ae^{-\lambda t}$. The values of t for which $x(t)$ is a maximum or minimum are not located halfway between the intercepts of the graph of $x(t)$. Verify this last statement by considering the function $x(t) = e^{-t}\sin(t + \pi/4)$.

19. In the case of underdamped motion show that the ratio between two consecutive maximum (or minimum) displacements x_n and x_{n+2} is the constant

$$x_n/x_{n+2} = e^{2\pi\lambda/\sqrt{\omega^2 - \lambda^2}}.$$

The number $\delta = \ln(x_n/x_{n+2}) = 2\pi\lambda/\sqrt{\omega^2 - \lambda^2}$ is called the **logarithmic decrement**.

20. The logarithmic decrement defined in Problem 19 is an indicator of the rate at which the motion is damped out.

(a) Describe the motion of an underdamped system if δ is a very small positive number.

(b) Compute the logarithmic decrement for the motion described in Problem 8.

5.3 Forced Motion

With Damping

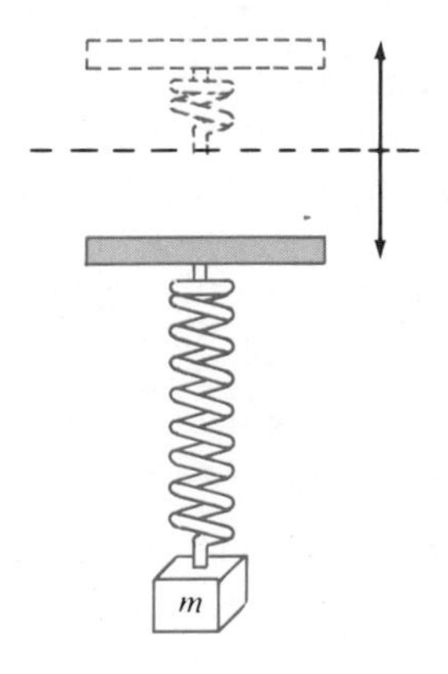

Figure 5.16

Suppose we now take into consideration an external force $f(t)$ acting on a vibrating mass on a spring. For example, $f(t)$ could represent a driving force causing an oscillatory vertical motion of the support of the spring (see Figure 5.16). The inclusion of $f(t)$ in the formulation of Newton's second law gives

$$m\frac{d^2x}{dt^2} = -kx - \beta\frac{dx}{dt} + f(t), \tag{1}$$

$$\frac{d^2x}{dt^2} + \frac{\beta}{m}\frac{dx}{dt} + \frac{k}{m}x = \frac{f(t)}{m}, \tag{2}$$

or

$$\frac{d^2x}{dt^2} + 2\lambda\frac{dx}{dt} + \omega^2 x = F(t), \tag{3}$$

where $F(t) = f(t)/m$ and, as in the preceding section, $2\lambda = \beta/m$, $\omega^2 = k/m$. To solve the latter nonhomogeneous equation we can employ either the method of undetermined coefficients or variation of parameters.

EXAMPLE 1

Interpret and solve the initial-value problem

$$\frac{1}{5}\frac{d^2x}{dt^2} + 1.2\frac{dx}{dt} + 2x = 5\cos 4t \tag{4}$$

$$x(0) = \frac{1}{2}, \qquad \left.\frac{dx}{dt}\right|_{t=0} = 0.$$

SOLUTION We can interpret the problem to represent a vibrational system consisting of a mass ($m = 1/5$ slug or kilogram) attached to a spring ($k = 2$ lb/ft or nt/m). The mass is released from rest 1/2 unit (foot or meter) below the equilibrium position. The motion is damped ($\beta = 1.2$) and is being driven by an external periodic ($T = \pi/2$ sec) force beginning at $t = 0$. Intuitively we would expect that even with damping the system will remain in motion until such time as the forcing function is "turned off," in which case the amplitudes would diminish. However, as the problem is given, $f(t) = 5\cos 4t$ will remain "on" forever.

We first multiply (4) by 5 and solve the homogeneous equation

$$\frac{dx^2}{dt^2} + 6\frac{dx}{dt} + 10x = 0$$

by the usual methods. Since $m_1 = -3 + i$, $m_2 = -3 - i$, it follows that

$$x_c(t) = e^{-3t}(c_1 \cos t + c_2 \sin t).$$

Using the method of undetermined coefficients, we assume a particular solution of the form $x_p(t) = A \cos 4t + B \sin 4t$. Now

$$x'_p = -4A \sin 4t + 4B \cos 4t$$
$$x''_p = -16A \cos 4t - 16B \sin 4t$$

so that

$$\begin{aligned} x''_p + 6x'_p + 10x_p &= -16A \cos 4t - 16B \sin 4t - 24A \sin 4t \\ &\quad + 24B \cos 4t + 10A \cos 4t + 10B \sin 4t \\ &= (-6A + 24B) \cos 4t + (-24A - 6B) \sin 4t \\ &= 25 \cos 4t. \end{aligned}$$

The resulting system of equations

$$-6A + 24B = 25$$
$$-24A - 6B = 0$$

yields $A = -25/102$ and $B = 50/51$. It follows that

$$x(t) = e^{-3t}(c_1 \cos t + c_2 \sin t) - \frac{25}{102}\cos 4t + \frac{50}{51}\sin 4t. \tag{5}$$

When we set $t = 0$ in the above equation, we immediately obtain $c_1 = 38/51$. By differentiating the expression and then setting $t = 0$, we also find that $c_2 = -86/51$. Therefore the equation of motion is

$$x(t) = e^{-3t}\left(\frac{38}{51}\cos t - \frac{86}{51}\sin t\right) - \frac{25}{102}\cos 4t + \frac{50}{51}\sin 4t. \tag{6}$$

Transient and Steady-State Terms

Notice that the complementary function

$$x_c(t) = e^{-3t}\left(\frac{38}{51}\cos t - \frac{86}{51}\sin t\right)$$

in the preceding example possesses the property that

$$\lim_{t \to \infty} x_c(t) = 0.$$

Since $x_c(t)$ becomes negligible (namely, $\to 0$) as $t \to \infty$, it is said to be a **transient term**, or **transient solution**. Thus for large time the displacements of the weight in the preceding problem are closely approximated by the particular solution $x_p(t)$. This latter function is also called the **steady-state solution**. When F is a periodic function, such as $F(t) = F_0 \sin \gamma t$ or $F(t) = F_0 \cos \gamma t$, the general solution of (3) consists of

$$x(t) = \textit{transient} + \textit{steady-state}.$$

EXAMPLE 2 The solution to the initial-value problem

$$\frac{d^2x}{dt^2} + 2\frac{dx}{dt} + 2x = 4 \cos t + 2 \sin t$$

$$x(0) = 0, \qquad \left.\frac{dx}{dt}\right|_{t=0} = 3$$

is readily shown to be

$$x = x_c + x_p = \underbrace{e^{-t} \sin t}_{\text{transient}} + \underbrace{2 \sin t}_{\text{steady-state}}.$$

Inspection of Figure 5.17 shows that the effect of the transient term on the solution is, in this case, negligible for about $t > 2\pi$.

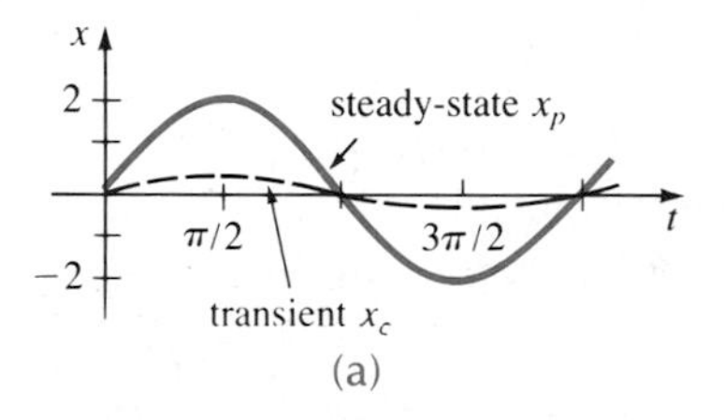

(a)

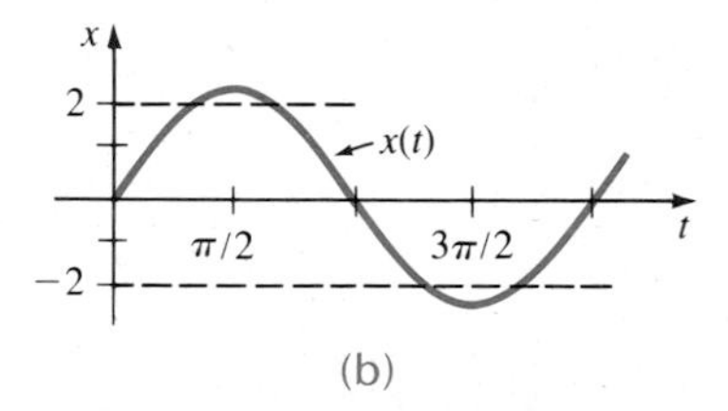

(b)

Figure 5.17

Without Damping

In the absence of a damping force, there will be no transient term in the solution of a problem. Also, we shall see that a periodic impressed force with a frequency near, or the same as, the frequency of free undamped vibrations can cause a severe problem in any oscillatory mechanical system.

EXAMPLE 3 Solve the initial-value problem

$$\frac{d^2x}{dt^2} + \omega^2 x = F_0 \sin \gamma t, \qquad F_0 = \text{constant}, \tag{7}$$

$$x(0) = 0, \qquad \left.\frac{dx}{dt}\right|_{t=0} = 0.$$

SOLUTION The complementary function is $x_c(t) = c_1 \cos \omega t + c_2 \sin \omega t$. To obtain a particular solution we assume $x_p(t) = A \cos \gamma t + B \sin \gamma t$, so that

$$\begin{aligned} x_p' &= -A\gamma \sin \gamma t + B\gamma \cos \gamma t \\ x_p'' &= -A\gamma^2 \cos \gamma t - B\gamma^2 \sin \gamma t \\ x_p'' + \omega^2 x_p &= A(\omega^2 - \gamma^2) \cos \gamma t + B(\omega^2 - \gamma^2) \sin \gamma t \\ &= F_0 \sin \gamma t. \end{aligned}$$

It follows that

$$A = 0 \quad \text{and} \quad B = \frac{F_0}{\omega^2 - \gamma^2} \qquad (\gamma \neq \omega).$$

Therefore $$x_p(t) = \frac{F_0}{\omega^2 - \gamma^2} \sin \gamma t.$$

Applying the given initial conditions to the general solution

$$x(t) = c_1 \cos \omega t + c_2 \sin \omega t + \frac{F_0}{\omega^2 - \gamma^2} \sin \gamma t$$

yields $c_1 = 0$ and $c_2 = -\gamma F_0/\omega(\omega^2 - \gamma^2)$. Thus the solution is

$$x(t) = \frac{F_0}{\omega(\omega^2 - \gamma^2)}(-\gamma \sin \omega t + \omega \sin \gamma t), \qquad \gamma \neq \omega. \tag{8}$$

Pure Resonance

Although equation (8) is not defined for $\gamma = \omega$, it is interesting to observe that its limiting value as $\gamma \to \omega$ can be obtained by applying L'Hôpital's rule. This limiting process is analogous to "tuning in" the frequency of the driving force $(\gamma/2\pi)$ to the frequency of free vibrations $(\omega/2\pi)$. Intuitively we expect that over a length of time we should be able to substantially increase the amplitudes of vibration.* For $\gamma = \omega$ we define the solution to be

$$x(t) = \lim_{\gamma \to \omega} F_0 \frac{-\gamma \sin \omega t + \omega \sin \gamma t}{\omega(\omega^2 - \gamma^2)}$$

$$= F_0 \lim_{\gamma \to \omega} \frac{\dfrac{d}{d\gamma}(-\gamma \sin \omega t + \omega \sin \gamma t)}{\dfrac{d}{d\gamma}(\omega^3 - \omega\gamma^2)}$$

$$= F_0 \lim_{\gamma \to \omega} \frac{-\sin \omega t + \omega t \cos \gamma t}{-2\omega\gamma}$$

$$= F_0 \frac{-\sin \omega t + \omega t \cos \omega t}{-2\omega^2}$$

$$= \frac{F_0}{2\omega^2} \sin \omega t - \frac{F_0}{2\omega} t \cos \omega t. \tag{9}$$

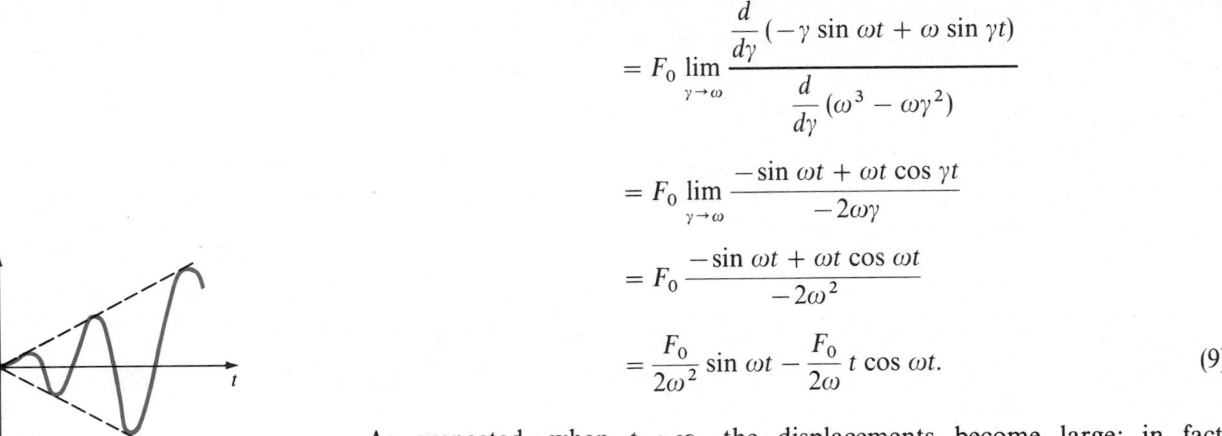

Figure 5.18

As suspected, when $t \to \infty$, the displacements become large; in fact, $|x(t_n)| \to \infty$, when $t_n = n\pi/\omega$, $n = 1, 2, \ldots$. The phenomenon we have just described is known as **pure resonance**. The graph given in Figure 5.18 shows typical motion in this case.

In conclusion it should be noted that there is no actual need to use a limiting process on (8) to obtain the solution for $\gamma = \omega$. Alternatively, equation (9) follows by solving the initial-value problem

$$\frac{d^2x}{dt^2} + \omega^2 x = F_0 \sin \omega t$$

$$x(0) = 0, \qquad \left.\frac{dx}{dt}\right|_{t=0} = 0$$

directly by conventional methods.

Remark: If a mechanical system were actually described by a function such as (9) of this section, it would necessarily fail. Large oscillations of a weight on a spring would eventually force the spring beyond its elastic limit. One might argue too that the resonating model presented in Figure 5.18 is completely unrealistic since it ignores the retarding effects of ever-present damping forces. Although it is true that pure resonance cannot occur when the smallest amount of damping is taken into consideration, large and equally destructive amplitudes of vibration (although bounded as $t \to \infty$) can occur.

* Forgetting about damping effects of shock absorbers, the situation is roughly equivalent to a number of passengers jumping up and down in the back of a bus in time with the natural vertical motion caused by equally spaced faults (such as cracks) in the road. Theoretically these passengers could upset the bus—assuming they are not kicked off first.

If you have ever looked out a window while in flight, you have probably observed that wings on an airplane are not perfectly rigid. A reasonable amount of flex or flutter is not only tolerated but necessary to prevent the wing from snapping like a piece of peppermint stick candy. In late 1959 and early 1960 two commercial plane crashes involving a relatively new model of propjet occurred, illustrating the destructive effects of large mechanical oscillations.

The unusual aspect of these crashes was that they both happened while the planes were in mid-flight. Barring midair collisions, the safest period during any flight is when the plane has attained its cruising altitude. It is well known that a plane is most vulnerable to an accident when it is least maneuverable, namely, during either take-off or landing. So, having two planes simply fall out of the sky was not only a tragedy but an embarrassment to the aircraft industry and a thoroughly puzzling problem to aerodynamic engineers. In crashes of this sort, a structural failure of some kind is immediately suspected. After a massive technical investigation, the problem was eventually traced in each case to an outboard engine and engine housing. Roughly, it was determined that when each plane surpassed a critical speed of approximately 400 mph, a propeller and engine began to wobble, causing a gyroscopic force, which could not be quelled or damped by the engine housing. This external vibrational force was then transferred to the already oscillating wing. This, in itself, need not have been destructively dangerous since aircraft wings are designed to withstand the stress of unusual and excessive forces. (In fact the particular wing in question was so incredibly strong that test engineers and pilots who were deliberately trying to snap a wing under every conceivable flight condition failed to do so.) But, unfortunately, after a short period of time during which the engine wobbled rapidly, the frequency of the impressed force actually slowed to a point at which it approached and finally coincided with the maximum frequency of wing flutter (around 3 cycles per second). The resulting resonance situation finally accomplished what the test engineers could not do; namely, the amplitudes of wing flutter became large enough to snap the wing (see Figure 5.19).

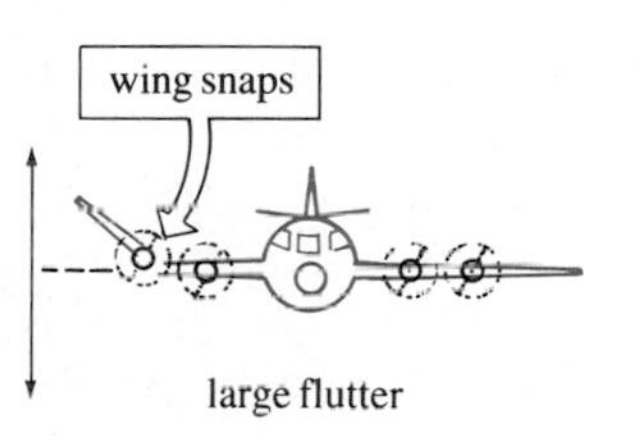

Figure 5.19

The problem was solved in two steps. All models of this particular plane were required to fly at speeds substantially below 400 mph until each plane could be modified by considerably strengthening (or stiffening) the engine housings. A strengthened engine housing was shown to be able to impart a damping effect capable of preventing the critical resonance phenomenon even in the unlikely event of a subsequent engine wobble.*

You may be aware that soldiers usually do not march in step across bridges. The reason for breaking stride is simply to avoid any possibility of resonance occurring between the natural vibrations inherent in the bridge's

* For a fascinating nontechnical account of the investigation see Robert J. Serling, *Loud and Clear*, New York: Dell, 1970, Chapter 5.

Figure 5.20 Courtesy of Wide World Photos/AP

structure and the frequency of the external force of a multitude of feet stomping in unison on the bridge.

Bridges are good examples of vibrating mechanical systems, which are constantly being subjected to external forces, from people walking on them, cars and trucks driving on them, water pushing against their foundations, and wind blowing against their superstructures. On November 7, 1940, the Tacoma Narrows Bridge at Puget Sound in the state of Washington collapsed. However, the crash came as no surprise since "Galloping Gertie," as the bridge was called by local residents, was famous for a vertical undulating motion of its roadway, which gave many motorists a very exciting crossing. On November 7, only four months after its grand opening, the amplitudes of these undulations became so large that the bridge failed and a substantial portion was sent splashing into the water below. In the investigation that followed, it was found that a poorly designed superstructure caused the wind blowing across it to vortex in a periodic manner. When the frequency of this periodic force approached the natural frequency of the bridge, large upheavals of the road resulted. In a word, the bridge was another victim of the destructive effect of mechanical resonance. Since this disaster developed over a matter of months, there was sufficient opportunity to record on film the strange and frightening phenomenon of a bucking and heaving bridge and its ultimate collapse (see Figure 5.20).*

Figure 5.21 © 1988 Memtek Products

Acoustic vibrations can be equally as destructive as large mechanical vibrations. In recent television commercials, jazz singers have inflicted destruction on the lowly wine glass (see Figure 5.21). Sounds from organs and piccolos have been known to crack windows.

> "As the horns blew, the people began to shout. When they heard the signal horn, they raised a tremendous shout. The wall collapsed. . . ." *Joshua* 6:20

---

* National Committee for Fluid Mechanics Films, Educational Services, Inc., Watertown, Mass. See also, *American Society of Civil Engineers: Proceedings*, "Failure of the Tacoma Narrows Bridge," Vol. 69, pp. 1555–85, Dec. 1943.

Did the power of acoustic resonance cause the walls of Jericho to tumble down? This is the conjecture of some contemporary scholars.

The phenomenon of resonance is not always destructive, however. For example, it is resonance of an electrical circuit that enables a radio to be tuned to a specific station.

Exercises 5.3

Answers to odd-numbered problems begin on page A–13.

1. A 16-lb weight stretches a spring 8/3 ft. Initially the weight starts from rest 2 ft below the equilibrium position and the subsequent motion takes place in a medium that offers a damping force numerically equal to 1/2 the instantaneous velocity. Find the equation of motion if the weight is driven by an external force equal to $f(t) = 10 \cos 3t$.

2. A mass of 1 slug is attached to a spring whose constant is 5 lb/ft. Initially the mass is released 1 ft below the equilibrium position with a downward velocity of 5 ft/sec, and the subsequent motion takes place in a medium that offers a damping force numerically equal to 2 times the instantaneous velocity.

   (a) Find the equation of motion if the mass is driven by an external force equal to $f(t) = 12 \cos 2t + 3 \sin 2t$.

   (b) Graph the transient and steady-state solutions on the same coordinate axes.

   (c) Graph the equation of motion.

3. A mass of 1 slug when attached to a spring stretches it 2 ft and then comes to rest in the equilibrium position. Starting at $t = 0$ an external force equal to $f(t) = 8 \sin 4t$ is applied to the system. Find the equation of motion if the surrounding medium offers a damping force numerically equal to 8 times the instantaneous velocity.

4. In Problem 3 determine the equation of motion if the external force is $f(t) = e^{-t} \sin 4t$. Analyze the displacements for $t \to \infty$.

5. When a mass of 2 kilograms is attached to a spring whose constant is 32 nt/m it comes to rest in the equilibrium position. Starting at $t = 0$ a force equal to $f(t) = 68e^{-2t} \cos 4t$ is applied to the system. Find the equation of motion in the absence of damping.

6. In Problem 5 write the equation of motion in the form $x(t) = A \sin(\omega t + \phi) + Be^{-2t} \sin(4t + \theta)$. What is the amplitude of vibrations after a very long time?

7. A mass m is attached to the end of a spring whose constant is k. After the mass reaches equilibrium, its support begins to oscillate vertically about

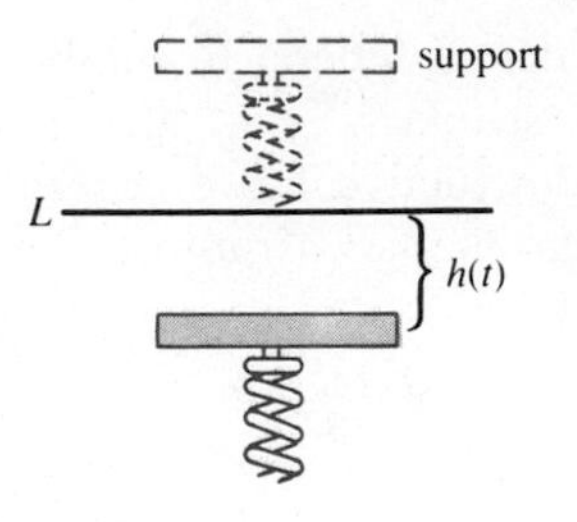

Figure 5.22

a horizontal line L according to a formula $h(t)$. The value of h represents the distance in feet measured from L (see Figure 5.22). Determine the differential equation of motion if the entire system moves through a medium offering a damping force numerically equal to $\beta(dx/dt)$.

8. Solve the differential equation of the preceding problem if the spring is stretched 4 ft by a weight of 16 lb, and $\beta = 2$, $h(t) = 5\cos t$, $x(0) = x'(0) = 0$.

9. A mass of 100 g is attached to a spring whose constant is 1600 dynes/cm. After the mass reaches equilibrium, its support oscillates according to the formula $h(t) = \sin 8t$, where h represents displacement from its original position (see Problem 7 and Figure 5.22).

(a) In the absence of damping, determine the equation of motion if the mass starts from rest from the equilibrium position.

(b) At what times does the mass pass through the equilibrium position?

(c) At what times does the mass attain its extreme displacements?

(d) What are the maximum and minimum displacements?

(e) Graph the equation of motion.

10. In the case of underdamped vibrations, show that the general solution of the differential equation

$$\frac{d^2x}{dt^2} + 2\lambda\frac{dx}{dt} + \omega^2 x = F_0 \sin \gamma t$$

is

$$x(t) = Ae^{-\lambda t}\sin(\sqrt{\omega^2 - \lambda^2}\,t + \phi) + \frac{F_0}{\sqrt{(\omega^2 - \gamma^2)^2 + 4\lambda^2\gamma^2}}\sin(\gamma t + \theta),$$

where $A = \sqrt{c_1^2 + c_2^2}$ and the phase angles ϕ and θ are, respectively, defined by

$$\sin\phi = \frac{c_1}{A}, \qquad \cos\phi = \frac{c_2}{A}$$

$$\sin\theta = \frac{-2\lambda\gamma}{\sqrt{(\omega^2 - \gamma^2)^2 + 4\lambda^2\gamma^2}}, \qquad \cos\theta = \frac{\omega^2 - \gamma^2}{\sqrt{(\omega^2 - \gamma^2)^2 + 4\lambda^2\gamma^2}}.$$

EXAMPLE 4

Inspection of Problem 10 shows that $x_c(t)$ is transient when damping is present and hence for large values of time the solution is closely approximated by the steady-state solution

$$x_p(t) = g(\gamma)\sin(\gamma t + \theta),$$

where we define

$$g(\gamma) = \frac{F_0}{\sqrt{(\omega^2 - \gamma^2)^2 + 4\lambda^2\gamma^2}}. \tag{10}$$

Although the amplitude of x_p is bounded as $t \to \infty$, it is easily shown that the maximum oscillations will occur at the value $\gamma_1 = \sqrt{\omega^2 - 2\lambda^2}$ (see Problem 11). Thus when the frequency of the external force is $\sqrt{\omega^2 - 2\lambda^2}/2\pi$, the system is said to be in **resonance**.

In the specific case $k = 4$, $m = 1$, $F_0 = 2$, $g(\gamma)$ becomes

$$g(\gamma) = \frac{2}{\sqrt{(4 - \gamma^2)^2 + \beta^2\gamma^2}}. \tag{11}$$

Figure 5.23(a) shows the graph of (11) for various values of the damping coefficients β. This family of graphs is called the **resonance curve** of the system. Observe the behavior of the amplitudes $g(\gamma)$ as $\beta \to 0$, that is, as the system approaches pure resonance.

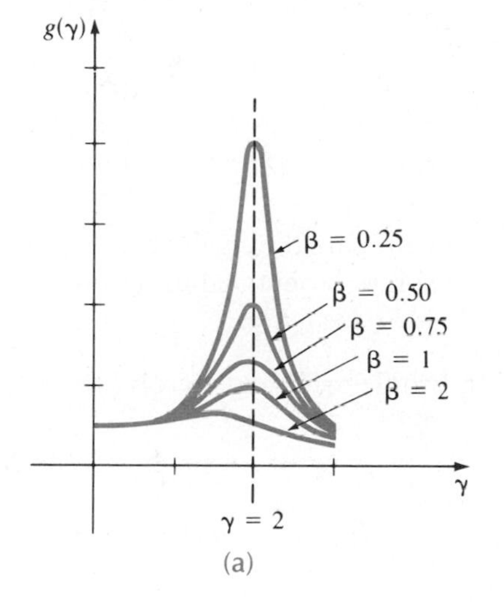

(a)

β	γ_1	$g(\gamma_1)$
2	1.41	0.58
1	1.87	1.03
0.75	1.93	1.36
0.50	1.97	2.02
0.25	1.99	4.01

(b)

Figure 5.23

11. **(a)** Prove that $g(\gamma)$ given in (10) of Example 4 has a maximum value at $\gamma_1 = \sqrt{\omega^2 - 2\lambda^2}$. [*Hint:* Differentiate with respect to γ.]

(b) What is the maximum value of $g(\gamma)$ at resonance?

12. **(a)** If $k = 3$ lb/ft and $m = 1$ slug, use the information in Example 4 to show that the system is underdamped when the damping coefficient

β satisfies $0 < \beta < 2\sqrt{3}$ but that resonance can occur only if $0 < \beta < \sqrt{6}$.

(b) Construct the resonance curve of the system when $F_0 = 3$.

13. A mass of 1/2 slug is suspended on a spring whose constant is 6 lb/ft. The system is set in motion in a medium offering a damping force numerically equal to twice the instantaneous velocity. Find the steady-state solution if an external force $f(t) = 40 \sin 2t$ is applied to the system starting at $t = 0$. Write this solution in the form of a constant multiple of $\sin(2t + \theta)$.

14. Verify that the mechanical system described in Problem 13 is in resonance. Show that the amplitude of the steady-state solution is the maximum value of $g(\gamma)$ described in Problem 11.

15. (a) Show that the solution of the initial-value problem

$$\frac{d^2x}{dt^2} + \omega^2 x = F_0 \cos \gamma t$$

$$x(0) = 0, \qquad \left.\frac{dx}{dt}\right|_{t=0} = 0$$

is
$$x(t) = \frac{F_0}{\omega^2 - \gamma^2}(\cos \gamma t - \cos \omega t).$$

(b) Evaluate
$$\lim_{\gamma \to \omega} \frac{F_0}{\omega^2 - \lambda^2}(\cos \gamma t - \cos \omega t).$$

16. Compare the result obtained in part (b) of the preceding problem with the solution obtained using variation of parameters when the external force is $F_0 \cos \omega t$.

In Problems 17 and 18 solve the given initial-value problem.

17. $\dfrac{d^2x}{dt^2} + 4x = -5 \sin 2t + 3 \cos 2t$

$x(0) = -1, \qquad \left.\dfrac{dx}{dt}\right|_{t=0} = 1$

18. $\dfrac{d^2x}{dt^2} + 9x = 5 \sin 3t$

$x(0) = 2, \qquad \left.\dfrac{dx}{dt}\right|_{t=0} = 0$

19. (a) Show that $x(t)$ given in part (a) of Problem 15 can be written in the form

$$x(t) = \frac{-2F_0}{\omega^2 - \gamma^2} \sin \frac{1}{2}(\gamma - \omega)t \sin \frac{1}{2}(\gamma + \omega)t.$$

(b) If we define $\varepsilon = \frac{1}{2}(\gamma - \omega)$, show that when ε is small an *approximate* solution is

$$x(t) = \frac{F_0}{2\varepsilon\gamma} \sin \varepsilon t \sin \gamma t.$$

(c) Evaluate

$$\lim_{\varepsilon \to 0} \frac{F_0}{2\varepsilon\gamma} \sin \varepsilon t \sin \gamma t.$$

EXAMPLE 5

In part (b) of the preceding problem, when ε is small, the frequency $\gamma/2\pi$ of the impressed force is close to the frequency $\omega/2\pi$ of free vibrations. When this occurs, the motion is as indicated in Figure 5.24. Oscillations of this kind are called *beats* and are due to the fact that the frequency of sin εt is quite small in comparison to the frequency of sin γt. The dashed curves, or *envelope* of the graph of $x(t)$, are obtained from the graphs of $\pm(F_0/2\varepsilon\gamma) \sin \varepsilon t$,

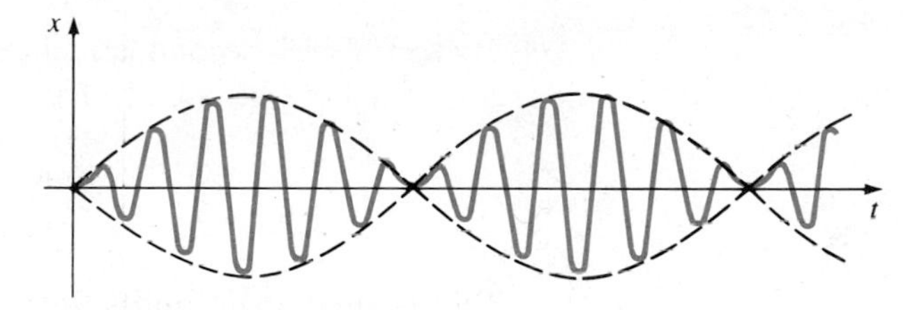

Figure 5.24

20. Show that the solution of

$$\frac{d^2x}{dt^2} + 25x = 10 \cos 7t$$

$$x(0) = 0, \qquad \left.\frac{dx}{dt}\right|_{t=0} = 0$$

is $x(t) = \frac{5}{6} \sin t \sin 6t$.

5.4 Electric Circuits and Other Analogous Systems

L-R-C Series Circuits

As mentioned in the introduction to this chapter, many different physical systems can be described by a linear second-order differential equation similar to the differential equation of forced motion with damping:

$$m\frac{d^2x}{dt^2} + \beta\frac{dx}{dt} + kx = f(t). \tag{1}$$

If $i(t)$ denotes current in an ***L-R-C* series electrical circuit** shown in Figure 5.25(a), then the voltage drops across the inductor, resistor, and capacitor are shown in Figure 5.25(b).

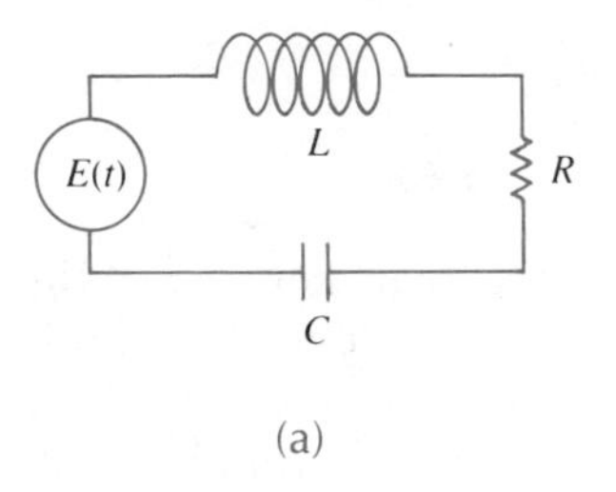

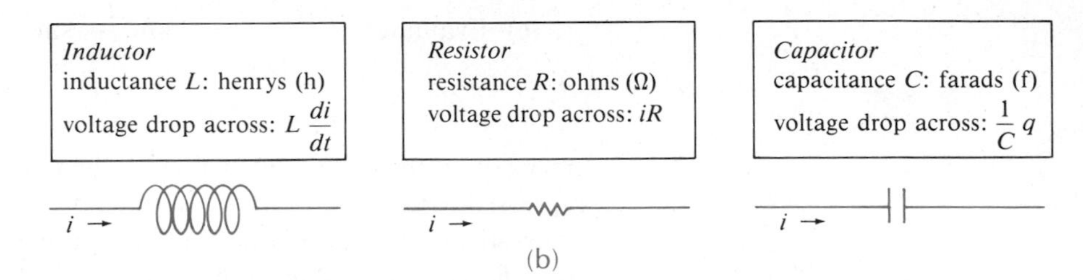

Figure 5.25

By Kirchhoff's second law, the sum of these voltages equals the voltage $E(t)$ impressed on the circuit; that is,

$$L\frac{di}{dt} + Ri + \frac{1}{C}q = E(t). \tag{2}$$

But the charge $q(t)$ on the capacitor is related to the current $i(t)$ by $i = dq/dt$ and so (2) becomes the linear second-order differential equation

$$L\frac{d^2q}{dt^2} + R\frac{dq}{dt} + \frac{1}{C}q = E(t). \tag{3}$$

The nomenclature used in the analysis of circuits is similar to that used to describe spring-mass systems.

If $E(t) = 0$, the **electrical vibrations** of the circuit are said to be *free*. Since the auxiliary equation for (3) is $Lm^2 + Rm + \frac{1}{C} = 0$, there will three forms of the solution when $R \neq 0$, depending on the value of the discriminant $R^2 - 4L/C$. We say that the circuit is

	overdamped if	$R^2 - 4L/C > 0$,
	critically damped if	$R^2 - 4L/C = 0$,
and	**underdamped** if	$R^2 - 4L/C < 0$.

In each of these three cases, the general solution of (3) contains the factor $e^{-Rt/2L}$ and so $q(t) \to 0$ as $t \to \infty$. In the underdamped case when $q(0) = q_0$, the charge on the capacitor will oscillate as it decays; in other words, the capacitor is charging and discharging as $t \to \infty$. When $E(t) = 0$ and $R = 0$, the circuit is said to be undamped and the electrical vibrations do not approach zero as t increases without bound; the response of the circuit is **simple harmonic**.

EXAMPLE 1

Consider an L-C series circuit in which $E(t) = 0$. Determine the charge $q(t)$ on the capacitor for $t > 0$ if its initial charge is q_0 and if initially there is no current flowing in the circuit.

SOLUTION In an L-C circuit there is no resistor, so from (3) we obtain

$$L\frac{d^2q}{dt^2} + \frac{1}{C}q = 0.$$

The initial conditions are $q(0) = q_0$ and $i(0) = 0$. Since $q'(t) = i(t)$ the latter condition is the same as $q'(0) = 0$. The general solution of the differential equation is

$$q(t) = c_1 \cos\frac{1}{\sqrt{LC}}t + c_2 \sin\frac{1}{\sqrt{LC}}t.$$

Now the initial conditions imply $c_1 = q_0$ and $c_2 = 0$, so that

$$q(t) = q_0 \cos\frac{1}{\sqrt{LC}}t.$$

In Example 1 if we want the current in the circuit, we use $i(t) = q'(t)$:

$$i(t) = -\frac{q_0}{\sqrt{LC}}\sin\frac{1}{\sqrt{LC}}t.$$

EXAMPLE 2

Find the charge $q(t)$ on the capacitor in an L-R-C series circuit when $L = 0.25$ henry, $R = 10$ ohms, $C = 0.001$ farad, $E(t) = 0$, $q(0) = q_0$ coulombs, and $i(0) = 0$.

SOLUTION Since $1/C = 1000$, equation (3) becomes

$$\frac{1}{4}q'' + 10q' + 1000q = 0 \qquad \text{or} \qquad q'' + 40q' + 4000q = 0.$$

Solving this homogeneous equation in the usual manner, we find that the circuit is underdamped and

$$q(t) = e^{-20t}(c_1 \cos 60t + c_2 \sin 60t).$$

Applying the initial conditions, we find $c_1 = q_0$ and $c_2 = q_0/3$. Thus the solution is given by

$$q(t) = q_0 e^{-20t}\left(\cos 60t + \frac{1}{3}\sin 60t\right).$$

The solution in Example 2 can be written as a single sine function using the method discussed in Section 5.2. From (15) of that section, we find

$$q(t) = \frac{q_0\sqrt{10}}{3} e^{-20t} \sin(60t + 1.249).$$

When there is an impressed voltage $E(t)$ on the circuit, the electrical vibrations are said to be *forced*. Note in Example 1 that the free electrical vibrations are simple harmonic with period $2\pi/1/\sqrt{LC} = 2\pi\sqrt{LC}$ and frequency $1/2\pi\sqrt{LC}$. If a periodic voltage $E(t)$ with the same frequency were impressed on the circuit, the system would be in **resonance**. In the case when $R \neq 0$, the complementary function $q_c(t)$ of (3) is called a **transient solution**. If $E(t)$ is periodic or a constant, then the particular solution $q_p(t)$ of (3) is a **steady-state solution**.

EXAMPLE 3

Find the steady-state solution $q_p(t)$ and the **steady-state current** in an L-R-C series circuit when the impressed voltage is $E(t) = E_0 \sin \gamma t$.

SOLUTION The steady-state solution $q_p(t)$ is a particular solution of the differential equation

$$L\frac{d^2q}{dt^2} + R\frac{dq}{dt} + \frac{1}{C}q = E_0 \sin \gamma t.$$

Using the method of undetermined coefficients, we assume a particular solution of the form

$$q_p(t) = A \sin \gamma t + B \cos \gamma t. \tag{4}$$

Substituting (4) into the differential equation, simplifying, and equating coefficients give

$$A = \frac{E_0(L\gamma - 1/C\gamma)}{-\gamma\left[L^2\gamma^2 - \frac{2L}{C} + \frac{1}{C^2\gamma^2} + R^2\right]}$$

and

$$B = \frac{E_0 R}{-\gamma\left[L^2\gamma^2 - \frac{2L}{C} + \frac{1}{C^2\gamma^2} + R^2\right]}$$

It is convenient to express A and B in terms of some new symbols. If

$$X = L\gamma - \frac{1}{C\gamma} \quad \text{then} \quad X^2 = L^2\gamma^2 - \frac{2L}{C} + \frac{1}{C^2\gamma^2}$$

and

$$Z = \sqrt{X^2 + R^2} \quad \text{then} \quad Z^2 = L^2\gamma^2 - \frac{2L}{C} + \frac{1}{C^2\gamma^2} + R^2.$$

Therefore $A = E_0X/(-\gamma Z^2)$ and $B = E_0R/(-\gamma Z^2)$ and so the steady-state charge is

$$q_p(t) = -\frac{E_0X}{\gamma Z^2}\sin\gamma t - \frac{E_0R}{\gamma Z^2}\cos\gamma t.$$

Now the steady-state current is given by $i_p(t) = q_p'(t)$:

$$i_p(t) = \frac{E_0}{Z}\left(\frac{R}{Z}\sin\gamma t - \frac{X}{Z}\cos\gamma t\right). \tag{5}$$

The quantities $X = L\gamma - \dfrac{1}{C\gamma}$ and $Z = \sqrt{X^2 + R^2}$ defined in Example 3 are called, respectively, the **reactance** and **impedance** of the circuit. Both the reactance and the impedance are measured in ohms.

Twisted Shaft

The differential equation governing the torsional motion of a weight suspended from the end of an elastic shaft is

$$I\frac{d^2\theta}{dt^2} + c\frac{d\theta}{dt} + k\theta = T(t). \tag{6}$$

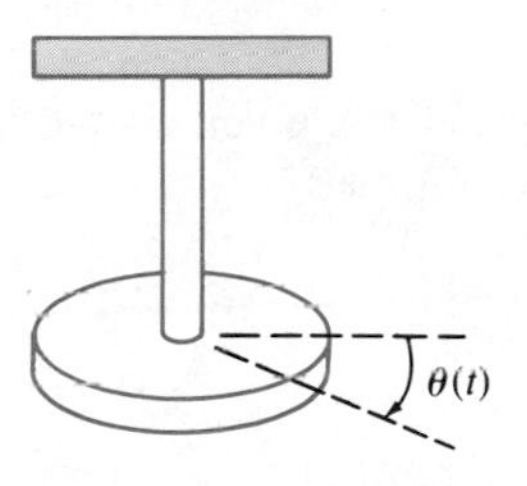

Figure 5.26

As shown in Figure 5.26, the function $\theta(t)$ represents the amount of twist of the weight at any time.

By comparing equations (3) and (6) with (1), we see that, with the exception of terminology, there is absolutely no difference between the mathematics of vibrating springs, simple series circuits, and torsional vibrations. The following table gives a comparison of the analogous parts of these three kinds of systems.

Mechanical	Series Electrical	Torsional
m (mass)	L (inductance)	I (moment of inertia)
β (damping)	R (resistance)	c (damping)
k (spring constant)	$\dfrac{1}{C}$ (reciprocal of capacitance—called elastance)	k (elastic shaft constant)
$f(t)$ (applied force)	$E(t)$ (impressed voltage)	$T(t)$ (applied torque)

Simple Pendulum

In Example 11 of Section 1.2 it was seen that the angular displacements θ of a simple pendulum are described by the nonlinear second-order equation

$$\frac{d^2\theta}{dt^2} + \frac{g}{l}\sin\theta = 0,$$

where l is the length of the pendulum rod. For small displacements $\sin\theta$ is replaced by θ, and the resulting differential equation

$$\frac{d^2\theta}{dt^2} + \frac{g}{l}\theta = 0 \tag{7}$$

indicates that the pendulum exhibits simple harmonic motion. Inspection of the solution (7) reveals that the period of small oscillations is given by the familiar formula from physics $T = 2\pi\sqrt{l/g}$.

Exercises 5.4

Answers to odd-numbered problems begin on page A–14.

In Problems 1 and 2 find the charge on the capacitor and the current in the given L-C series circuit. Assume $q(0) = 0$ and $i(0) = 0$.

1. $L = 1$ henry, $C = \dfrac{1}{16}$ farad, $E(t) = 60$ volts

2. $L = 5$ henry, $C = 0.01$ farad, $E(t) = 20t$ volts

In Problems 3 and 4 without solving (3) determine whether the given L-R-C series circuit is overdamped, critically damped, or underdamped.

3. $L = 3$ henry, $R = 10$ ohms, $C = 0.1$ farad

4. $L = 1$ henry, $R = 20$ ohms, $C = 0.01$ farad

5. Find the charge on the capacitor in an L-R-C series circuit at $t = 0.01$ sec when $L = 0.05$ henry, $R = 2$ ohms, $C = 0.01$ farad, $E(t) = 0$, $q(0) = 5$ coulombs, and $i(0) = 0$. Determine the first time at which the charge on the capacitor is equal to zero.

6. Find the charge on the capacitor in an L-R-C series circuit when $L = 1/4$ henry, $R = 20$ ohms, $C = 1/300$ farad, $E(t) = 0$, $q(0) = 4$ coulombs, and $i(0) = 0$. Is the charge on the capacitor ever equal to zero?

In Problems 7 and 8 find the charge on the capacitor and the current in the given L-R-C series circuit. Find the maximum charge on the capacitor.

7. $L = 5/3$ henry, $R = 10$ ohms, $C = 1/30$ farad, $E(t) = 300$ volts, $q(0) = 0$, $i(0) = 0$.

8. $L = 1$ henry, $R = 100$ ohms, $C = 0.0004$ farad, $E(t) = 30$ volts, $q(0) = 0$, $i(0) = 2$ amp.

9. Find the steady-state charge and the steady-state current in an L-R-C series circuit when $L = 1$ henry, $R = 2$ ohms, $C = 0.25$ farad, and $E(t) = 50\cos t$ volts.

10. Show that the amplitude of the steady-state current in the L-R-C series circuit in Example 3 is given by E_0/Z, where Z is the impedance of the circuit.

11. Show that the steady-state current in an L-R-C series circuit when $L = 1/2$ henry, $R = 20$ ohms, $C = 0.001$ farad, and $E(t) = 100 \sin 60t$ is given by $i_p(t) = (4.160) \sin (60t - 0.588)$. [*Hint:* Use Problem 10.]

12. Find the steady-state current in an L-R-C series circuit when $L = 1/2$ henry, $R = 20$ ohms, $C = 0.001$ farad, and $E(t) = 100 \sin 60t + 200 \cos 40t$ volts.

13. Find the charge on the capacitor in an L-R-C series circuit when $L = 1/2$ henry, $R = 10$ ohms, $C = 0.01$ farad, $E(t) = 150$ volts, $q(0) = 1$ coulomb, and $i(0) = 0$. What is the charge on the capacitor after a long time?

14. Show that if L, R, C, and E_0 are constant, then the amplitude of the steady-state current in Example 3 is a maximum when $\gamma = 1/\sqrt{LC}$. What is the maximum amplitude?

15. Show that if L, R, E_0, and γ are constant, then the amplitude of the steady-state current in Example 3 is a maximum when the capacitance is $C = 1/L\gamma^2$.

16. Find the charge on the capacitor and current in an L-C circuit when $L = 0.1$ henry, $C = 0.1$ farad, $E(t) = 100 \sin \gamma t$ volts, $q(0) = 0$, and $i(0) = 0$.

17. Find the charge on the capacitor and current in an L-C circuit when $E(t) = E_0 \cos \gamma t$ volts, $q(0) = q_0$ coulombs, and $i(0) = i_0$ amps.

18. In Problem 17 find the current when the circuit is in resonance.

19. Find the equation of motion describing the small displacements $\theta(t)$ of a simple pendulum of length 2 ft released at $t = 0$ with a displacement of $1/2$ radians to the right of the vertical and angular velocity of $2\sqrt{3}$ ft/sec to the right. What are the amplitude, period, and frequency of motion?

20. In Problem 19 at what times does the pendulum pass through its equilibrium position? At what times does the pendulum attain its extreme angular displacements on either side of its equilibrium position?

CHAPTER 5 SUMMARY

When a mass is attached to a spring it stretches to a position where the restoring force ks of the spring is balanced by the weight mg. Any subsequent motion is then measured x units (feet in the engineering system) above or below this **equilibrium position**. When the mass is above the equilibrium position, we adopt the convention that $x < 0$; whereas, when the mass is below the equilibrium position, we take $x > 0$.

The differential equation of motion is obtained by equating Newton's second law $F = ma = m(d^2x/dt^2)$ with the net force acting on the mass at any time. We distinguish three cases.

CASE I The equation

$$m\frac{d^2x}{dt^2} = -kx \qquad \text{or} \qquad \frac{d^2x}{dt^2} + \omega^2 x = 0, \tag{1}$$

where $\omega^2 = k/m$, describes the motion under the assumptions that no damping force and no external impressed forces are acting on the system. The solution of (1) is $x(t) = c_1 \cos \omega t + c_2 \sin \omega t$ and the mass is said to exhibit **simple harmonic motion**. The constants c_1 and c_2 are determined by the initial position $x(0)$ and the initial velocity $x'(0)$ of the mass. ■

CASE II When a damping force is present, the differential equation becomes

$$m\frac{d^2x}{dt^2} = -kx - \beta\frac{dx}{dt} \qquad \text{or} \qquad \frac{d^2x}{dt^2} + 2\lambda\frac{dx}{dt} + \omega^2 x = 0, \tag{2}$$

where $\beta > 0$, $2\lambda = \beta/m$, and $\omega^2 = k/m$. The resulting motion is said to be **overdamped**, **critically damped**, or **underdamped** accordingly as $\lambda^2 - \omega^2 > 0$, $\lambda^2 - \omega^2 = 0$, or $\lambda^2 - \omega^2 < 0$. The respective solutions of (2) are then

$$x(t) = c_1 e^{m_1 t} + c_2 e^{m_2 t},$$

where $m_1 = -\lambda + \sqrt{\lambda^2 - \omega^2}$, $m_2 = -\lambda - \sqrt{\lambda^2 - \omega^2}$;

$$x(t) = c_1 e^{m_1 t} + c_2 t e^{m_1 t},$$

where $m_1 = -\lambda$; and

$$x(t) = e^{-\lambda t}(c_1 \cos \sqrt{\omega^2 - \lambda^2}\, t + c_2 \sin \sqrt{\omega^2 - \lambda^2}\, t).$$

In each case the damping force is responsible for the displacements becoming negligible for large time, that is, $x \to 0$ as $t \to \infty$. ■

The motion described in Cases I and II is said to be **free motion**.

CASE III When an external force is impressed on the system for $t > 0$, the differential equation becomes

$$m\frac{d^2x}{dt^2} = -kx - \beta\frac{dx}{dt} + f(t) \qquad \text{or} \qquad \frac{d^2x}{dt^2} + 2\lambda\frac{dx}{dt} + \omega^2 x = F(t), \tag{3}$$

where λ and ω^2 are defined in Case II. The solution of the nonhomogeneous equation (3) is $x(t) = x_c + x_p$.

Since the complementary function x_c always contains the factor $e^{-\lambda t}$, it will be **transient**, that is, $x_c \to 0$ as $t \to \infty$. If $F(t)$ is periodic, then x_p will be a **steady-state solution**. ■

In the absence of a damping force, an impressed periodic force can cause the amplitudes of vibration to become very large. If the frequency of the external force is the same as the frequency $\omega/2\pi$ of free vibrations, we say that the system is in a state of **pure resonance**. In this case the amplitudes of vibrations become unbounded as $t \to \infty$. In the presence of a damping force, amplitudes of oscillatory motion are always bounded. However, large and potentially destructive amplitudes can occur.

When a series circuit containing an inductor, resistor, and capacitor is driven by an electromotive force $E(t)$, the resulting differential equations for the charge $q(t)$ or the current $i(t)$ are quite similar to equation (3). Hence the analysis of such circuits is the same as outlined above.

CHAPTER 5 REVIEW EXERCISES

Answers to odd-numbered problems begin on page A–14.

Answer Problems 1–9 without referring back to the text. Fill in the blank or answer true/false.

1. If a 10-lb weight stretches a spring 2.5 ft, a 32-lb weight will stretch it ______ ft.

2. The period of simple harmonic motion of an 8-lb weight attached to a spring whose constant is 6.25 lb/ft is ______ sec.

3. The differential equation of a weight on a spring is $x'' + 16x = 0$. If the weight is released at $t = 0$ from 1 m above the equilibrium position with a downward velocity of 3 m/sec, the amplitude of vibrations is ______ m.

4. Pure resonance cannot take place in the presence of a damping force. ______

5. In the presence of damping, the displacements of a weight on a spring will always approach zero as $t \to \infty$. ______

6. A weight on a spring whose motion is critically damped can possibly pass through the equilibrium position twice. ______

7. At critical damping any increase in damping will result in an ______ system.

8. In describing simple harmonic motion by $x = (\sqrt{2}/2)\sin(2t + \phi)$, the phase angle ϕ is ______ when $x(0) = -1/2$ and $x'(0) = 1$.

9. A 16-lb weight attached to a spring exhibits simple harmonic motion. If the frequency of oscillations is $3/2\pi$ vibrations/sec, the spring constant is ______.

10. A 12-lb weight stretches a spring 2 ft. The weight is released from a point 1 ft below the equilibrium position with an upward velocity of 4 ft/sec.

(a) Find the equation describing the resulting simple harmonic motion.

(b) What is the amplitude, period, and frequency of motion?

(c) At what times does the weight return to the point 1 ft below the equilibrium position?

(d) At what times does the weight pass through the equilibrium position moving upward? moving downward?

(e) What is the velocity of weight at $t = 3\pi/16$?

(f) At what times is the velocity zero?

11. A force of 2 lb stretches a spring 1 ft. With one end held fixed, an 8-lb weight is attached to the other end and the system lies on a table that imparts a frictional force numerically equal to 3/2 times the instantaneous velocity. Initially the weight is displaced 4 in. above the equilibrium position and released from rest. Find the equation of motion if the motion takes place along a horizontal straight line that is taken as the x-axis.

12. A 32-lb weight stretches a spring 6 in. The weight moves through a medium offering a damping force numerically equal to β times the instantaneous velocity. Determine the values of β for which the system will exhibit oscillatory motion.

13. A spring with constant $k = 2$ is suspended in a liquid that offers a damping force numerically equal to 4 times the instantaneous velocity. If a mass m is suspended from the spring, determine the values of m for which the subsequent free motion is nonoscillatory.

14. The vertical motion of a weight attached to a spring is described by the initial-value problem

$$\frac{1}{4}\frac{d^2x}{dt^2} + \frac{dx}{dt} + x = 0$$

$$x(0) = 4, \qquad x'(0) = 2.$$

Determine the maximum vertical displacement.

15. A 4-lb weight stretches a spring 18 in. A periodic force equal to $f(t) = \cos \gamma t + \sin \gamma t$ is impressed on the system starting at $t = 0$. In the absence of a damping force, for what value of γ will the system be in a state of pure resonance?

16. Find a particular solution for $\frac{d^2x}{dt^2} + 2\lambda \frac{dx}{dt} + \omega^2 x = A$, where A is a constant force.

17. A 4-lb weight is suspended from a spring whose constant is 3 lb/ft. The entire system is immersed in a fluid offering a damping force numerically equal to the instantaneous velocity. Beginning at $t = 0$, an external force equal to $f(t) = e^{-t}$ is impressed on the system. Determine the equation of motion if the weight is released from rest at a point 2 ft below the equilibrium position.

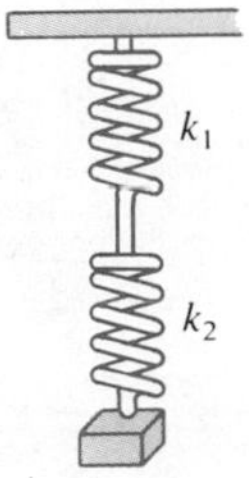

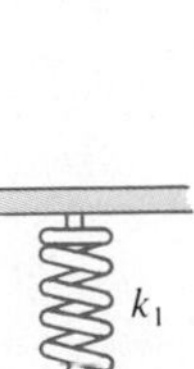

Figure 5.27

18. A weight of W lb stretches one spring 1/2 ft and stretches a different spring 1/4 ft. If the two springs are attached in series, the effective spring constant k is given by $1/k = 1/k_1 + 1/k_2$. The weight W is then attached to the double spring, as shown in Figure 5.27. Assume that the motion is free and that there is no damping force present. Determine the equation of motion if the weight is released at a point 1 ft below the equilibrium position with a downward velocity of 2/3 ft/sec. Show that the maximum speed of the weight is $(2/3)\sqrt{3g + 1}$.

19. A series circuit contains an inductance of $L = 1$ henry, a capacitance of $C = 10^{-4}$ farad, and an electromotive force of $E(t) = 100 \sin 50t$ volts. Initially the charge q and current i are zero.

(a) Find the equation for the charge at any time.

(b) Find the equation for the current at any time.

(c) Find the times for which the charge on the capacitor is zero.

20. Show that the current $i(t)$ in an L-R-C series circuit satisfies the differential equation

$$L \frac{d^2i}{dt^2} + R \frac{di}{dt} + \frac{1}{C} i = E'(t),$$

where $E'(t)$ denotes the derivative of $E(t)$.

CHAPTER 6

Differential Equations with Variable Coefficients

The same ease with which we solved differential equations with constant coefficients does not generally carry over to equations with variable coefficients. In fact we cannot expect to be able to express the solutions of even a simple linear equation such as $y'' - 2xy = 0$ in terms of the usual sines, cosines, logarithms, exponentials, and other elementary functions. Although it is easily verified that

$$(1 - x^2)y'' - 2xy' + 2y = 0$$

and

$$x^2y'' + xy' + (x^2 - \tfrac{1}{4})y = 0$$

have elementary solutions

$$y = x \qquad \text{and} \qquad y = \frac{\sin x}{\sqrt{x}},$$

respectively, the best that we can usually expect from equations of this sort is an *infinite series solution.* On the other hand, we shall begin this chapter with one important type of equation with variable coefficients whose general solution can always be written in terms of elementary functions.

IMPORTANT TERMS

- Cauchy-Euler equation
- equidimensional equation
- auxiliary equation
- power series solution
- ordinary point
- analytic
- singular point
- regular singular point
- irregular singular point
- method of Frobenius
- indicial equation
- indicial roots
- Bessel's equation
- Legendre's equation
- Bessel functions of the first kind
- Bessel function of the second kind
- parametric Bessel equation
- Legendre polynomials

6.1 Cauchy-Euler Equation

Any differential equation of the form

$$a_n x^n \frac{d^n y}{dx^n} + a_{n-1} x^{n-1} \frac{d^{n-1} y}{dx^{n-1}} + \cdots + a_1 x \frac{dy}{dx} + a_0 y = g(x),$$

where $a_n, a_{n-1}, \ldots, a_0$ are constants, is said to be a **Cauchy-Euler equation***, or **equidimensional equation**. The obvious characteristic of this type of equation is that the polynomial coefficients x^k match the order of differentiation in the terms $d^k y/dx^k$ for $k = 1, 2, \ldots, n$.

For the sake of discussion, we shall confine our attention to solving the homogeneous second-order equation

$$ax^2 \frac{d^2 y}{dx^2} + bx \frac{dy}{dx} + cy = 0.$$

The solution of higher-order equations follows analogously. Also, we can solve the nonhomogeneous equation

$$ax^2 \frac{d^2 y}{dx^2} + bx \frac{dy}{dx} + cy = g(x)$$

by variation of parameters once we have determined the complementary function $y_c(x)$.

Note: The coefficient of d^2y/dx^2 is zero at $x = 0$. Hence in order to guarantee that the fundamental results of Theorem 4.1 are applicable to the Cauchy-Euler equation, we shall confine our attention to finding the general solution on the interval $(0, \infty)$. Solutions on the interval $(-\infty, 0)$ can be obtained by substituting $t = -x$ in the differential equation.

---

* **Augustin-Louis Cauchy** (1789–1857) Born during a period of upheaval in French history, Augustin-Louis Cauchy was destined to initiate a revolution of his own—in mathematics. For many original contributions, but especially for his efforts in clarifying mathematical obscurities and for his incessant demand for satisfactory definitions and rigorous proofs of theorems, Cauchy is often called "the father of modern analysis." A prolific writer whose output was surpassed by only a few, Cauchy produced nearly 800 papers in astronomy, physics, and mathematics. It was he who developed the concept of convergence of an infinite series and the theory of functions of a complex variable. The same mind that was always open and inquiring in science and mathematics was narrow and unquestioning in many other areas. Outspoken and arrogant, Cauchy was also passionate on political and religious issues. His stands on these issues often alienated him from his colleagues.

Method of Solution

We try a solution of the form $y = x^m$, where m is to be determined. The first and second derivatives are, respectively,

$$\frac{dy}{dx} = mx^{m-1} \quad \text{and} \quad \frac{d^2y}{dx^2} = m(m-1)x^{m-2}.$$

Consequently the differential equation becomes

$$\begin{aligned} ax^2\frac{d^2y}{dx^2} + bx\frac{dy}{dx} + cy &= ax^2 \cdot m(m-1)x^{m-2} + bx \cdot mx^{m-1} + cx^m \\ &= am(m-1)x^m + bmx^m + cx^m \\ &= x^m(am(m-1) + bm + c). \end{aligned}$$

Thus $y = x^m$ will be a solution of the differential equation whenever m is a solution of the **auxiliary equation**

$$am(m-1) + bm + c = 0 \quad \text{or} \quad am^2 + (b-a)m + c = 0. \tag{1}$$

There are three different cases to be considered, depending on whether the roots of this quadratic equation are real and distinct, real and equal, or complex conjugates.

CASE I Let m_1 and m_2 be the real roots of (1) such that $m_1 \neq m_2$.

Then $$y_1 = x^{m_1} \quad \text{and} \quad y_2 = x^{m_2}$$

form a fundamental set of solutions. Hence the general solution is

$$y = c_1x^{m_1} + c_2x^{m_2}. \tag{2}$$

■

EXAMPLE 1

Solve $$x^2\frac{d^2y}{dx^2} - 2x\frac{dy}{dx} - 4y = 0.$$

SOLUTION Rather than just memorizing equation (1), it is preferable to assume $y = x^m$ as the solution a few times in order to understand the origin and the difference between this new form of the auxiliary equation and that obtained in Chapter 4. Differentiate twice

$$\frac{dy}{dx} = mx^{m-1}, \quad \frac{d^2y}{dx^2} = m(m-1)x^{m-2}$$

and substitute back into the differential equation:

$$x^2 \frac{d^2y}{dx^2} - 2x\frac{dy}{dx} - 4y = x^2 \cdot m(m-1)x^{m-2} - 2x \cdot mx^{m-1} - 4x^m$$

$$= x^m(m(m-1) - 2m - 4)$$

$$= x^m(m^2 - 3m - 4) = 0$$

if $m^2 - 3m - 4 = 0$. Now $(m+1)(m-4) = 0$ implies $m_1 = -1$, $m_2 = 4$

so that $$y = c_1x^{-1} + c_2x^4.$$

CASE II If $m_1 = m_2$, then we obtain only one solution, namely, $y = x^{m_1}$. When the roots of the quadratic equation $am^2 + (b-a)m + c = 0$ are equal, the discriminant of the coefficients is necessarily zero. It follows from the quadratic formula that the root must be $m_1 = -(b-a)/2a$.

Now we can construct a second solution y_2, using (4) of Section 4.2. We first write the Cauchy-Euler equation in the form

$$\frac{d^2y}{dx^2} + \frac{b}{ax}\frac{dy}{dx} + \frac{c}{ax^2}y = 0$$

and make the identification $P(x) = b/ax$. Thus

$$y_2 = x^{m_1}\int \frac{e^{-\int (b/ax)\,dx}}{(x^{m_1})^2}\,dx$$

$$= x^{m_1}\int \frac{e^{-(b/a)\ln x}}{x^{2m_1}}\,dx$$

$$= x^{m_1}\int x^{-b/a} \cdot x^{-2m_1}\,dx \qquad [e^{-(b/a)\ln x} = e^{\ln x^{-b/a}} = x^{-b/a}]$$

$$= x^{m_1}\int x^{-b/a} \cdot x^{(b-a)/a}\,dx \qquad \left[2m_1 = -\frac{b-a}{a}\right]$$

$$= x^{m_1}\int \frac{dx}{x}$$

$$= x^{m_1}\ln x.$$

The general solution is then

$$y = c_1x^{m_1} + c_2x^{m_1}\ln x. \tag{3}$$

■

EXAMPLE 2 Solve
$$4x^2\frac{d^2y}{dx^2} + 8x\frac{dy}{dx} + y = 0.$$

SOLUTION The substitution $y = x^m$ yields

$$4x^2\frac{d^2y}{dx^2} + 8x\frac{dy}{dx} + y = x^m(4m(m-1) + 8m + 1)$$
$$= x^m(4m^2 + 4m + 1) = 0,$$

when
$$4m^2 + 4m + 1 = 0 \qquad \text{or} \qquad (2m+1)^2 = 0.$$

Since $m_1 = -1/2$, the general solution is

$$y = c_1x^{-1/2} + c_2x^{-1/2}\ln x.$$

For higher-order equations, if m_1 is a root of multiplicity k, then it can be shown that

$$x^{m_1},\ x^{m_1}\ln x,\ x^{m_1}(\ln x)^2, \ldots,\ x^{m_1}(\ln x)^{k-1}$$

are k linearly independent solutions.

CASE III If m_1 and m_2 are complex conjugates,

$$m_1 = \alpha + i\beta, \qquad m_2 = \alpha - i\beta,$$

where α and $\beta > 0$ are real, then a solution is

$$y = C_1x^{\alpha+i\beta} + C_2x^{\alpha-i\beta}.$$

But, as in the case of equations with constant coefficients, when the roots of the auxiliary equation are complex, we wish to write the solution in terms of real functions only. We note the identity

$$x^{i\beta} = (e^{\ln x})^{i\beta} = e^{i\beta\ln x},$$

which, by Euler's formula, is the same as

$$x^{i\beta} = \cos(\beta\ln x) + i\sin(\beta\ln x).$$

Therefore

$$\begin{aligned} y = C_1x^{\alpha+i\beta} + C_2x^{\alpha-i\beta} &= x^\alpha[C_1x^{i\beta} + C_2x^{-i\beta}] \\ &= x^\alpha[C_1\{\cos(\beta\ln x) + i\sin(\beta\ln x)\} + C_2\{\cos(\beta\ln x) - i\sin(\beta\ln x)\}] \\ &= x^\alpha[(C_1 + C_2)\cos(\beta\ln x) + (C_1i - C_2i)\sin(\beta\ln x)]. \end{aligned}$$

On the interval $(0, \infty)$ it can be verified that

$$y_1 = x^\alpha \cos(\beta \ln x) \qquad \text{and} \qquad y_2 = x^\alpha \sin(\beta \ln x)$$

constitute a fundamental set of solutions of the differential equation. It follows that the general solution is

$$y = x^\alpha[c_1 \cos(\beta \ln x) + c_2 \sin(\beta \ln x)]. \tag{4}$$

■

EXAMPLE 3

Solve
$$x^2 \frac{d^2y}{dx^2} + 3x \frac{dy}{dx} + 3y = 0.$$

SOLUTION We have

$$x^2 \frac{d^2y}{dx^2} + 3x \frac{dy}{dx} + 3y = x^m(m(m-1) + 3m + 3)$$
$$= x^m(m^2 + 2m + 3) = 0,$$

when $m^2 + 2m + 3 = 0$.

From the quadratic formula we find $m_1 = -1 + \sqrt{2}i$ and $m_2 = -1 - \sqrt{2}i$. If we make the identifications $\alpha = -1$ and $\beta = \sqrt{2}$, we see from (4) that the general solution is

$$y = x^{-1}[c_1 \cos(\sqrt{2} \ln x) + c_2 \sin(\sqrt{2} \ln x)].$$

EXAMPLE 4

Solve the third-order Cauchy-Euler equation

$$x^3 \frac{d^3y}{dx^3} + 5x^2 \frac{d^2y}{dx^2} + 7x \frac{dy}{dx} + 8y = 0.$$

SOLUTION The first three derivatives of $y = x^m$ are

$$\frac{dy}{dx} = mx^{m-1}, \qquad \frac{d^2y}{dx^2} = m(m-1)x^{m-2}, \qquad \frac{d^3y}{dx^3} = m(m-1)(m-2)x^{m-3},$$

so that the given differential equation becomes

$$x^3 \frac{d^3y}{dx^3} + 5x^2 \frac{d^2y}{dx^2} + 7x \frac{dy}{dx} + 8y$$
$$= x^3 m(m-1)(m-2)x^{m-3} + 5x^2 m(m-1)x^{m-2} + 7xmx^{m-1} + 8x^m$$
$$= x^m(m(m-1)(m-2) + 5m(m-1) + 7m + 8)$$
$$= x^m(m^3 + 2m^2 + 4m + 8).$$

In this case we see that $y = x^m$ will be a solution of the differential equation, provided m is a root of the cubic equation

$$m^3 + 2m^2 + 4m + 8 = 0 \qquad \text{or} \qquad (m + 2)(m^2 + 4) = 0.$$

The roots are: $m_1 = -2$, $m_2 = 2i$, $m_3 = -2i$. Hence the general solution is

$$y = c_1 x^{-2} + c_2 \cos(2 \ln x) + c_3 \sin(2 \ln x).$$

EXAMPLE 5

Solve the nonhomogeneous equation

$$x^2 y'' - 3xy' + 3y = 2x^4 e^x.$$

SOLUTION The substitution $y = x^m$ leads to the auxiliary equation

$$m(m - 1) - 3m + 3 = 0 \qquad \text{or} \qquad (m - 1)(m - 3) = 0.$$

Thus $$y_c = c_1 x + c_2 x^3.$$

Before using variation of parameters, recall that the formulas $u_1' = -y_2 f(x)/W$ and $u_2' = y_1 f(x)/W$ were derived under the assumption that the differential equation has been put into the form $y'' + P(x)y' + Q(x)y = f(x)$. Therefore we divide the given equation by x^2 and then make the identification that $f(x) = 2x^2 e^x$.

Now $$W = \begin{vmatrix} x & x^3 \\ 1 & 3x^2 \end{vmatrix} = 3x^3 - x^3 = 2x^3$$ so that

$$u_1' = -\frac{x^3(2x^2 e^x)}{2x^3} = -x^2 e^x \qquad \text{and} \qquad u_2' = \frac{x(2x^2 e^x)}{2x^3} = e^x.$$

The integral of the latter function is immediate, but in the case of u_1' we integrate by parts twice. The results are

$$u_1 = -x^2 e^x + 2xe^x - 2e^x \qquad \text{and} \qquad u_2 = e^x.$$

Hence $$\begin{aligned} y_p &= u_1 y_1 + u_2 y_2 \\ &= (-x^2 e^x + 2xe^x - 2e^x)x + e^x x^3 = 2x^2 e^x - 2xe^x. \end{aligned}$$

Finally we have $$\begin{aligned} y &= y_c + y_p \\ &= c_1 x + c_2 x^3 + 2x^2 e^x - 2xe^x. \end{aligned}$$

Alternative Method of Solution

Any Cauchy-Euler differential equation can be reduced to an equation with constant coefficients by means of the substitution $x = e^t$. The next example illustrates this method.

EXAMPLE 6 Solve $$x^2\frac{d^2y}{dx^2} - x\frac{dy}{dx} + y = \ln x.$$

SOLUTION With the substitution $x = e^t$ or $t = \ln x$, it follows from the chain rule that

$$\frac{dy}{dx} = \frac{dy}{dt}\frac{dt}{dx} = \frac{1}{x}\frac{dy}{dt}$$

$$\frac{d^2y}{dx^2} = \frac{1}{x}\frac{d}{dx}\left(\frac{dy}{dt}\right) + \frac{dy}{dt}\left(-\frac{1}{x^2}\right)$$

$$= \frac{1}{x}\left(\frac{d^2y}{dt^2}\frac{1}{x}\right) + \frac{dy}{dt}\left(-\frac{1}{x^2}\right) = \frac{1}{x^2}\left(\frac{d^2y}{dt^2} - \frac{dy}{dt}\right).$$

Substituting in the given differential equation and simplifying yield

$$\frac{d^2y}{dt^2} - 2\frac{dy}{dt} + y = t.$$

Since this last equation has constant coefficients, its auxiliary equation is $m^2 - 2m + 1 = 0$ or $(m - 1)^2 = 0$. Thus we obtain $y_c = c_1e^t + c_2te^t$.

By undetermined coefficients we try a particular solution of the form $y_p = A + Bt$. This assumption leads to $-2B + A + Bt = t$ so that $A = 2$ and $B = 1$. Using $y = y_c + y_p$, we get

$$y = c_1e^t + c_2te^t + 2 + t;$$

and so the general solution of the original differential equation on the interval $(0, \infty)$ is

$$y = c_1x + c_2x\ln x + 2 + \ln x.$$

Exercises 6.1

Answers to odd-numbered problems begin on page A–14.

In Problems 1–22 solve the given differential equation.

1. $x^2y'' - 2y = 0$

2. $4x^2y'' + y = 0$

3. $xy'' + y' = 0$

4. $xy'' - y' = 0$

5. $x^2y'' + xy' + 4y = 0$

6. $x^2y'' + 5xy' + 3y = 0$

7. $x^2y'' - 3xy' - 2y = 0$

8. $x^2y'' + 3xy' - 4y = 0$

9. $25x^2y'' + 25xy' + y = 0$

10. $4x^2y'' + 4xy' - y = 0$

11. $x^2y'' + 5xy' + 4y = 0$

12. $x^2y'' + 8xy' + 6y = 0$

13. $x^2y'' - xy' + 2y = 0$

14. $x^2y'' - 7xy' + 41y = 0$

15. $3x^2y'' + 6xy' + y = 0$

16. $2x^2y'' + xy' + y = 0$

17. $x^3y''' - 6y = 0$

18. $x^3y''' + xy' - y = 0$

19. $x^3\dfrac{d^3y}{dx^3} - 2x^2\dfrac{d^2y}{dx^2} - 2x\dfrac{dy}{dx} + 8y = 0$

20. $x^3\dfrac{d^3y}{dx^3} - 2x^2\dfrac{d^2y}{dx^2} + 4x\dfrac{dy}{dx} - 4y = 0$

21. $x\dfrac{d^4y}{dx^4} + 6\dfrac{d^3y}{dx^3} = 0$

22. $x^4\dfrac{d^4y}{dx^4} + 6x^3\dfrac{d^3y}{dx^3} + 9x^2\dfrac{d^2y}{dx^2} + 3x\dfrac{dy}{dx} + y = 0$

In Problems 23–26 solve the given differential equation subject to the indicated initial conditions.

23. $x^2y'' + 3xy' = 0, \quad y(1) = 0, \quad y'(1) = 4$

24. $x^2y'' - 5xy' + 8y = 0, \quad y(2) = 32, \quad y'(2) = 0$

25. $x^2y'' + xy' + y = 0, \quad y(1) = 1, \quad y'(1) = 2$

26. $x^2y'' - 3xy' + 4y = 0, \quad y(1) = 5, \quad y'(1) = 3$

In Problems 27 and 28 solve the given differential equation subject to the indicated initial conditions. [*Hint:* Let $t = -x$.]

27. $4x^2y'' + y = 0, \quad y(-1) = 2, \quad y'(-1) = 4$

28. $x^2y'' - 4xy' + 6y = 0, \quad y(-2) = 8, \quad y'(-2) = 0$

Solve Problems 29–34 by variation of parameters.

29. $xy'' + y' = x$

30. $xy'' - 4y' = x^4$

31. $2x^2y'' + 5xy' + y = x^2 - x$

32. $x^2y'' - 2xy' + 2y = x^4e^x$

33. $x^2y'' - xy' + y = 2x$

34. $x^2y'' - 2xy' + 2y = x^3 \ln x$

In Problems 35–40 solve the given differential equation by means of the substitution $x = e^t$.

35. $x^2\dfrac{d^2y}{dx^2} + 10x\dfrac{dy}{dx} + 8y = x^2$

36. $x^2y'' - 4xy' + 6y = \ln x^2$

37. $x^2y'' - 3xy' + 13y = 4 + 3x$

38. $2x^2y'' - 3xy' - 3y = 1 + 2x + x^2$

39. $x^2y'' + 9xy' - 20y = 5/x^3$

40. $x^3\dfrac{d^3y}{dx^3} - 3x^2\dfrac{d^2y}{dx^2} + 6x\dfrac{dy}{dx} - 6y = 3 + \ln x^3$

Miscellaneous Problems

In Problems 41–43 solve the given differential equation.

41. $(x-1)^2 \dfrac{d^2y}{dx^2} - 2(x-1)\dfrac{dy}{dx} - 4y = 0$ [*Hint:* Let $t = x - 1$.]

42. $(3x+4)^2 y'' + 10(3x+4)y' + 9y = 0$

43. $(x+2)^2 y'' + (x+2)y' + y = 0$

44. Show that a solution of the Cauchy-Euler equation $r^2R'' + rR' - n^2R = 0$ satisfying $R(b) = 0$ is

(a) $R = C[(b/r)^n - (r/b)^n]$ for $n = 1, 2, 3, \ldots$, C a constant,

(b) $R = C\ln(r/b)$ for $n = 0$, C a constant.

6.2 Power Series Solutions

6.2.1 Procedure

Review

In the discussion that follows, we shall be concerned with the use of *power series*. The reader should recall the following facts from calculus.

- A power series in $(x - a)$ is an infinite series of the form $\Sigma_{n=0}^{\infty} c_n(x-a)^n$.
- Every power series has an *interval of convergence*. The interval of convergence is the set of all numbers for which the series converges.
- A power series *converges absolutely* at a number x_1 if $\Sigma_{n=0}^{\infty} |c_n||x_1 - a|^n$ converges.
- Every interval of convergence has a *radius of convergence* R.
- A power series converges absolutely for $|x - a| < R$ and diverges for $|x - a| > R$. When $R = 0$, the interval of convergence consists of the single number a; when $R = \infty$, the power series converges for all numbers x.
- The radius of convergence is usually obtained from the *ratio test:**

$$\lim_{n\to\infty}\left|\frac{c_{n+1}}{c_n}\right||x-a| = L.$$

The series will converge absolutely for those values of x for which $L < 1$.

* The *root test* is sometimes applicable.

- If R is not 0 or ∞ then the interval of convergence *may* include the endpoints $a - R$ and $a + R$.
- If $\Sigma_{n=0}^{\infty} c_n(x - a)^n = 0$ for all numbers x in the interval of convergence, then $c_n = 0$ for all values of n.
- A power series represents a continuous function within its interval of convergence.
- A power series can be differentiated termwise within its interval of convergence.
- A power series can be integrated termwise within its interval of convergence.
- Two power series with a common interval of convergence can be added term by term.

Solution of a Differential Equation

We have already seen in Section 1.1 that the function $y = e^{x^2}$ is an explicit solution of the linear first-order differential equation

$$\frac{dy}{dx} - 2xy = 0. \tag{1}$$

But e^x has the well-known power series representation

$$e^x = \sum_{n=0}^{\infty} \frac{x^n}{n!} \tag{2}$$

so that the solution of (1) can then be written as

$$y = e^{x^2} = \sum_{n=0}^{\infty} \frac{x^{2n}}{n!}. \tag{3}$$

Both series (2) and (3) converge for all real values of x.

In other words, knowing the solution in advance, we were able to find an infinite series solution of the differential equation. We now propose to obtain the **power series solution** of (1) directly; the method of attack is similar to the technique of undetermined coefficients.

If we assume that a solution exists in the form of power series in x,

$$y = \sum_{n=0}^{\infty} c_n x^n, \tag{4}$$

we pose the question: Can we determine coefficients c_n for which (4) converges to a function satisfying (1)? Formal* term-by-term differentiation of (4) gives

$$\frac{dy}{dx} = \sum_{n=0}^{\infty} nc_n x^{n-1} = \sum_{n=1}^{\infty} nc_n x^{n-1}.$$

* At this point we do not know the interval of convergence.

Note that the first term in the first series corresponding to $n = 0$ is zero. Now using this last result and assumption (4), we find

$$\frac{dy}{dx} - 2xy = \sum_{n=1}^{\infty} nc_n x^{n-1} - \sum_{n=0}^{\infty} 2c_n x^{n+1}. \tag{5}$$

We would like to add the two series in (5), but in order to do this both summation indices must start at the same value. In addition it is desirable for the numerical values of the powers of x to be "in phase" in each summation—that is, if one series starts with a multiple of, say, x to the first power, then we want the other series to start with this same power. To this end we write (5) as

$$\frac{dy}{dx} - 2xy = 1 \cdot c_1 x^0 + \sum_{n=2}^{\infty} nc_n x^{n-1} - \sum_{n=0}^{\infty} 2c_n x^{n+1} \tag{6}$$

and let $k = n - 1$ in the first series and $k = n + 1$ in the second.* The right side of equation (6) then becomes

$$c_1 + \sum_{k=1}^{\infty} (k + 1)c_{k+1}x^k - \sum_{k=1}^{\infty} 2c_{k-1}x^k.$$

By adding the series termwise it follows that

$$\frac{dy}{dx} - 2xy = c_1 + \sum_{k=1}^{\infty} [(k + 1)c_{k+1} - 2c_{k-1}]x^k = 0. \tag{7}$$

Hence in order to have (7) identically zero it is necessary that the coefficients satisfy

$$c_1 = 0 \quad \text{and} \quad (k + 1)c_{k+1} - 2c_{k-1} = 0,\ k = 1, 2, 3 \ldots. \tag{8}$$

Equation (8) provides a *recurrence relation* that determines the c_k. Since $k + 1 \neq 0$ for all of the indicated values of k, we can write (8) as

$$c_{k+1} = \frac{2c_{k-1}}{k + 1}. \tag{9}$$

Iteration of this last formula then gives

$$k = 1, \quad c_2 = \frac{2}{2}c_0 = c_0$$

$$k = 2, \quad c_3 = \frac{2}{3}c_1 = 0$$

---

* Recall that the summation index is a "dummy" variable. The fact that $k = n - 1$ in one case and $k = n + 1$ in the other should cause no confusion if you keep in mind that it is the *value* of the summation index that is important. In both cases k takes on the successive values $1, 2, 3, \ldots$ for $n = 2, 3, 4, \ldots$ (for $k = n - 1$) and $n = 0, 1, 2, \ldots$ (for $k = n + 1$), respectively.

$$k = 3, \quad c_4 = \frac{2}{4}c_2 = \frac{1}{2}c_0 = \frac{1}{2!}c_0$$

$$k = 4, \quad c_5 = \frac{2}{5}c_3 = 0$$

$$k = 5, \quad c_6 = \frac{2}{6}c_4 = \frac{1}{3 \cdot 2!}c_0 = \frac{1}{3!}c_0$$

$$k = 6, \quad c_7 = \frac{2}{7}c_5 = 0$$

$$k = 7, \quad c_8 = \frac{2}{8}c_6 = \frac{1}{4 \cdot 3!}c_0 = \frac{1}{4!}c_0$$

and so on. Thus from the original assumption (4), we find

$$y = \sum_{n=0}^{\infty} c_n x^n = c_0 + c_1 x + c_2 x^2 + c_3 x^3 + c_4 x^4 + c_5 x^5 + \cdots$$

$$= c_0 + 0 + c_0 x^2 + 0 + \frac{1}{2!}c_0 x^4 + 0 + \frac{1}{3!}c_0 x^6 + 0 + \cdots$$

$$= c_0\left[1 + x^2 + \frac{1}{2!}x^4 + \frac{1}{3!}x^6 + \cdots\right] = c_0 \sum_{n=0}^{\infty} \frac{x^{2n}}{n!}. \tag{10}$$

Since the iteration of (9) leaves c_0 completely undetermined, we have in fact found the general solution of (1).

6.2.2 Solutions Around Ordinary Points

Ordinary and Singular Points

Suppose the linear second-order differential equation

$$a_2(x)y'' + a_1(x)y' + a_0(x)y = 0 \tag{11}$$

is put into the form

$$y'' + P(x)y' + Q(x)y = 0, \tag{12}$$

where $P(x) = a_1(x)/a_2(x)$ and $Q(x) = a_0(x)/a_2(x)$. We make the following definition.

DEFINITION 6.1 A point $x = x_0$ is said to be an **ordinary point** of equation (11) if both $P(x)$ and $Q(x)$ are **analytic** at x_0, that is, both $P(x)$ and $Q(x)$ have a power series in $(x - x_0)$ with a positive radius of convergence. A point that is not an ordinary point is said to be a **singular point** of the equation.

EXAMPLE 1

Every finite value of x is an ordinary point of

$$y'' + (e^x)y' + (\sin x)y = 0.$$

In particular we see that $x = 0$ is an ordinary point since

$$e^x = 1 + \frac{x}{1!} + \frac{x^2}{2!} + \cdots \quad \text{and} \quad \sin x = x - \frac{x^3}{3!} + \frac{x^5}{5!} + \cdots$$

converge for all finite values of x.

EXAMPLE 2

The differential equation $xy'' + (\sin x)y = 0$ has an ordinary point at $x = 0$ since it can be shown that $Q(x) = (\sin x)/x$ possesses the power series expansion

$$Q(x) = 1 - \frac{x^2}{3!} + \frac{x^4}{5!} - \frac{x^6}{7!} + \cdots$$

that converges for all finite values of x.

EXAMPLE 3

The differential equation $y'' + (\ln x)y = 0$ has a singular point at $x = 0$ because $Q(x) = \ln x$ possesses no power series in x.

Polynomial Coefficients

Primarily we shall be concerned with the case when (11) has *polynomial* coefficients. As a consequence of Definition 6.1, we note that when $a_2(x)$, $a_1(x)$, and $a_0(x)$ are polynomials with *no common factors*, a point $x = x_0$ is

(i) an ordinary point if $a_2(x_0) \neq 0$,

(ii) a singular point if $a_2(x_0) = 0$.

EXAMPLE 4

(a) The singular points of the equation $(x^2 - 1)y'' + 2xy' + 6y = 0$ are the solutions of $x^2 - 1 = 0$ or $x = \pm 1$. All other finite values of x are ordinary points.

(b) Singular points need not be real numbers. The equation $(x^2 + 1)y'' + xy' - y = 0$ has singular points at the solutions of $x^2 + 1 = 0$, namely, $x = \pm i$. All other finite values of x, real or complex, are ordinary points.

EXAMPLE 5

The Cauchy-Euler equation $ax^2y'' + bxy' + cy = 0$, where a, b, and c are constants, has a singular point at $x = 0$. All other finite values of x, real or complex, are ordinary points.

In the remaining discussion of this section, our goal is to find power series solutions about ordinary points for differential equations of type (11) in which the coefficients are polynomials.

Note: For our purposes ordinary points and singular points will always be finite. It is possible for a differential equation to have, say, a singular point at infinity (see Exercises 6.3).

We state the following theorem without proof.

THEOREM 6.1 If $x = x_0$ is an ordinary point of equation (11), we can always find two distinct power series solutions of the form

$$y = \sum_{n=0}^{\infty} c_n(x - x_0)^n.$$

A series solution will converge at least for $|x - x_0| < R_1$, where R_1 is the distance to the closest singular point.*

To solve a linear second-order equation such as (11) we find two different sets of coefficients c_n so that we have two linearly independent power series $y_1(x)$ and $y_2(x)$, both expanded about the same ordinary point $x = x_0$. On a common interval of convergence not containing the origin, the general solution of the equation is $y = C_1y_1(x) + C_2y_2(x)$. The procedure used to solve a second-order equation is the same as that used in solving $y' - 2xy = 0$; namely, we assume a solution $y = \Sigma_{n=0}^{\infty} c_n(x - x_0)^n$ and then determine the c_n. In fact we shall find that $C_1 = c_0$ and $C_2 = c_1$, where c_0 and c_1 are arbitrary.

Note: For the sake of simplicity, we shall assume an ordinary point is always located at $x = 0$, since, if not, the substitution $t = x - x_0$ translates the value $x = x_0$ to $t = 0$.

---

* For example, polynomial solutions could exist, in which case the solution would be valid for all finite values of x.

EXAMPLE 6 Solve $$y'' - 2xy = 0.$$

SOLUTION We see that $x = 0$ is an ordinary point of the equation. Since there are no finite singular points, Theorem 6.1 guarantees two solutions of the form $y = \sum_{n=0}^{\infty} c_n x^n$ convergent for $|x| < \infty$. Proceeding, we write

$$y' = \sum_{n=0}^{\infty} nc_n x^{n-1} = \sum_{n=1}^{\infty} nc_n x^{n-1}$$

$$y'' = \sum_{n=1}^{\infty} n(n-1)c_n x^{n-2} = \sum_{n=2}^{\infty} n(n-1)c_n x^{n-2},$$

where we have used the fact that the first term in each series, corresponding to $n = 0$ and $n = 1$, respectively, is zero. Therefore

$$y'' - 2xy = \sum_{n=2}^{\infty} n(n-1)c_n x^{n-2} - \sum_{n=0}^{\infty} 2c_n x^{n+1}$$

$$= 2 \cdot 1c_2 x^0 + \underbrace{\sum_{n=3}^{\infty} n(n-1)c_n x^{n-2} - \sum_{n=0}^{\infty} 2c_n x^{n+1}}_{\text{both series start with } x^1.}$$

Letting $k = n - 2$ in the first series and $k = n + 1$ in the second gives

$$y'' - 2xy = 2c_2 + \sum_{k=1}^{\infty} (k+2)(k+1)c_{k+2}x^k - \sum_{k=1}^{\infty} 2c_{k-1}x^k$$

$$= 2c_2 + \sum_{k=1}^{\infty} [(k+2)(k+1)c_{k+2} - 2c_{k-1}]x^k = 0.$$

We must then have

$$2c_2 = 0 \qquad \text{and} \qquad (k+2)(k+1)c_{k+2} - 2c_{k-1} = 0.$$

The last expression is the same as

$$c_{k+2} = \frac{2c_{k-1}}{(k+2)(k+1)}, \qquad k = 1, 2, 3 \dots.$$

Iterating,

$$c_3 = \frac{2c_0}{3 \cdot 2}$$

$$c_4 = \frac{2c_1}{4 \cdot 3}$$

$$c_5 = \frac{2c_2}{5 \cdot 4} = 0$$

$$c_6 = \frac{2c_3}{6 \cdot 5} = \frac{2^2}{6 \cdot 5 \cdot 3 \cdot 2} c_0$$

$$c_7 = \frac{2c_4}{7 \cdot 6} = \frac{2^2}{7 \cdot 6 \cdot 4 \cdot 3} c_1$$

$$c_8 = \frac{2c_5}{8 \cdot 7} = 0$$

$$c_9 = \frac{2c_6}{9 \cdot 8} = \frac{2^3}{9 \cdot 8 \cdot 6 \cdot 5 \cdot 3 \cdot 2} c_0$$

$$c_{10} = \frac{2c_7}{10 \cdot 9} = \frac{2^3}{10 \cdot 9 \cdot 7 \cdot 6 \cdot 4 \cdot 3} c_1$$

$$c_{11} = \frac{2c_8}{11 \cdot 10} = 0$$

and so on. It should be apparent that both c_0 and c_1 are arbitrary. Now

$$\begin{aligned} y &= c_0 + c_1 x + c_2 x^2 + c_3 x^3 + c_4 x^4 + c_5 x^5 + c_6 x^6 + c_7 x^7 + c_8 x^8 \\ &\quad + c_9 x^9 + c_{10} x^{10} + c_{11} x^{11} + \cdots \\ &= c_0 + c_1 x + 0 + \frac{2}{3 \cdot 2} c_0 x^3 + \frac{2}{4 \cdot 3} c_1 x^4 + 0 + \frac{2^2}{6 \cdot 5 \cdot 3 \cdot 2} c_0 x^6 \\ &\quad + \frac{2^2}{7 \cdot 6 \cdot 4 \cdot 3} c_1 x^7 + 0 + \frac{2^3}{9 \cdot 8 \cdot 6 \cdot 5 \cdot 3 \cdot 2} c_0 x^9 \\ &\quad + \frac{2^3}{10 \cdot 9 \cdot 7 \cdot 6 \cdot 4 \cdot 3} c_1 x^{10} + 0 + \cdots \\ &= c_0 \left[1 + \frac{2}{3 \cdot 2} x^3 + \frac{2^2}{6 \cdot 5 \cdot 3 \cdot 2} x^6 + \frac{2^3}{9 \cdot 8 \cdot 6 \cdot 5 \cdot 3 \cdot 2} x^9 + \cdots \right] \\ &\quad + c_1 \left[x + \frac{2}{4 \cdot 3} x^4 + \frac{2^2}{7 \cdot 6 \cdot 4 \cdot 3} x^7 + \frac{2^3}{10 \cdot 9 \cdot 7 \cdot 6 \cdot 4 \cdot 3} x^{10} + \cdots \right]. \end{aligned}$$

Although the pattern of the coefficients in the preceding example should be clear, it is sometimes useful to write the solutions in terms of summation notation. By using the properties of the factorial, we can write

$$y_1(x) = c_0 \left[1 + \sum_{k=1}^{\infty} \frac{2^k [1 \cdot 4 \cdot 7 \cdots (3k-2)]}{(3k)!} x^{3k} \right]$$

and

$$y_2(x) = c_1 \left[x + \sum_{k=1}^{\infty} \frac{2^k [2 \cdot 5 \cdot 8 \cdots (3k-1)]}{(3k+1)!} x^{3k+1} \right].$$

In this form the ratio test can be used to show that each series converges for $|x| < \infty$.

EXAMPLE 7 Solve
$$(x^2+1)y''+xy'-y=0.$$

SOLUTION Since the singular points are $x=\pm i$, a power series solution will converge at least for $|x|<1$.* The assumption $y=\sum_{n=0}^{\infty} c_n x^n$ leads to

$$\begin{aligned}
(x^2+1)\sum_{n=2}^{\infty} n(n-1)c_n x^{n-2} &+ x\sum_{n-1}^{\infty} nc_n x^{n-1} - \sum_{n=0}^{\infty} c_n x^n \\
&= \sum_{n=2}^{\infty} n(n-1)c_n x^n + \sum_{n=2}^{\infty} n(n-1)c_n x^{n-2} \\
&\quad + \sum_{n=1}^{\infty} nc_n x^n - \sum_{n=0}^{\infty} c_n x^n \\
&= 2c_2x^0 - c_0x^0 + 6c_3x + c_1x - c_1x + \underbrace{\sum_{n=2}^{\infty} n(n-1)c_n x^n}_{k=n} \\
&\quad + \underbrace{\sum_{n=4}^{\infty} n(n-1)c_n x^{n-2}}_{k=n-2} + \underbrace{\sum_{n=2}^{\infty} nc_n x^n}_{k=n} - \underbrace{\sum_{n=2}^{\infty} c_n x^n}_{k=n} \\
&= 2c_2 - c_0 + 6c_3x \\
&\quad + \sum_{k=2}^{\infty} [k(k-1)c_k + (k+2)(k+1)c_{k+2} + kc_k - c_k]x^k \\
&= 2c_2 - c_0 + 6c_3x \\
&\quad + \sum_{k=2}^{\infty} [(k+1)(k-1)c_k + (k+2)(k+1)c_{k+2}]x^k = 0.
\end{aligned}$$

Thus
$$2c_2 - c_0 = 0$$
$$c_3 = 0$$
$$(k+1)(k-1)c_k + (k+2)(k+1)c_{k+2} = 0$$

or, after dividing by $(k+2)(k+1)$,

$$c_3 = 0$$
$$c_2 = \frac{1}{2}c_0$$
$$c_{k+2} = \frac{1-k}{k+2}c_k, \qquad k = 2, 3, 4\dots.$$

---

* The **modulus** or magnitude of the complex number $x=i$ is $|x|=1$. If $x=a+bi$ is a singular point then $|x|=\sqrt{a^2+b^2}$. See Appendix V.

Iteration of the last formula gives

$$c_4 = -\frac{1}{4}c_2 = -\frac{1}{2\cdot 4}c_0 = -\frac{1}{2^2 2!}c_0$$

$$c_5 = -\frac{2}{5}c_3 = 0$$

$$c_6 = -\frac{3}{6}c_4 = \frac{3}{2\cdot 4\cdot 6}c_0 = \frac{1\cdot 3}{2^3 3!}c_0$$

$$c_7 = -\frac{4}{7}c_5 = 0$$

$$c_8 = -\frac{5}{8}c_6 = -\frac{3\cdot 5}{2\cdot 4\cdot 6\cdot 8}c_0 = -\frac{1\cdot 3\cdot 5}{2^4 4!}c_0$$

$$c_9 = -\frac{6}{9}c_7 = 0$$

$$c_{10} = -\frac{7}{10}c_8 = \frac{3\cdot 5\cdot 7}{2\cdot 4\cdot 6\cdot 8\cdot 10}c_0 = \frac{1\cdot 3\cdot 5\cdot 7}{2^5 5!}c_0$$

and so on. Therefore

$$\begin{aligned} y &= c_0 + c_1 x + c_2 x^2 + c_3 x^3 + c_4 x^4 + c_5 x^5 + c_6 x^6 + c_7 x^7 + c_8 x^8 + \cdots \\ &= c_1 x + c_0\left[1 + \frac{1}{2}x^2 - \frac{1}{2^2 2!}x^4 + \frac{1\cdot 3}{2^3 3!}x^6 - \frac{1\cdot 3\cdot 5}{2^4 4!}x^8 + \frac{1\cdot 3\cdot 5\cdot 7}{2^5 5!}x^{10} + \cdots\right]. \end{aligned}$$

The solutions are

$$y_1(x) = c_0\left[1 + \frac{1}{2}x^2 + \sum_{n=2}^{\infty}(-1)^{n-1}\frac{1\cdot 3\cdot 5\cdots(2n-3)}{2^n n!}x^{2n}\right], \quad |x| < 1,$$

$$y_2(x) = c_1 x.$$

EXAMPLE 8

If we seek a solution $y = \Sigma_{n=0}^{\infty} c_n x^n$ for the equation

$$y'' - (1 + x)y = 0,$$

we obtain $c_2 = c_0/2$ and the three-term recurrence relation

$$c_{k+2} = \frac{c_k + c_{k-1}}{(k+1)(k+2)}, \qquad k = 1, 2, 3 \ldots.$$

To simplify the iteration we can first choose $c_0 \neq 0$, $c_1 = 0$; this will yield one solution. The other solution follows from next choosing $c_0 = 0$, $c_1 \neq 0$. With the first assumption we find

$$c_2 = \frac{1}{2} c_0$$

$$c_3 = \frac{c_1 + c_0}{2 \cdot 3} = \frac{c_0}{2 \cdot 3} = \frac{1}{6} c_0$$

$$c_4 = \frac{c_2 + c_1}{3 \cdot 4} = \frac{c_0}{2 \cdot 3 \cdot 4} = \frac{1}{24} c_0$$

$$c_5 = \frac{c_3 + c_2}{4 \cdot 5} = \frac{c_0}{4 \cdot 5}\left[\frac{1}{2 \cdot 3} + \frac{1}{2}\right] = \frac{1}{30} c_0$$

and so on. Thus one solution is

$$y_1(x) = c_0\left[1 + \frac{1}{2}x^2 + \frac{1}{6}x^3 + \frac{1}{24}x^4 + \frac{1}{30}x^5 + \cdots\right].$$

Similarly if we choose $c_0 = 0$ then

$$c_2 = 0$$

$$c_3 = \frac{c_1 + c_0}{2 \cdot 3} = \frac{c_1}{2 \cdot 3} = \frac{1}{6} c_1$$

$$c_4 = \frac{c_2 + c_1}{3 \cdot 4} = \frac{c_1}{3 \cdot 4} = \frac{1}{12} c_1$$

$$c_5 = \frac{c_3 + c_2}{4 \cdot 5} = \frac{c_1}{2 \cdot 3 \cdot 4 \cdot 5} = \frac{1}{120} c_1$$

and so on. Hence another solution is

$$y_2(x) = c_1\left[x + \frac{1}{6}x^3 + \frac{1}{12}x^4 + \frac{1}{120}x^5 + \cdots\right].$$

Each series converges for all finite values of x.

Exercises 6.2

Answers to odd-numbered problems begin on page A–15.

[6.2.1] In Problems 1–10 solve each differential equation in the manner of the previous chapters and then compare the results with the solutions obtained by assuming a power series solution $y = \sum_{n=0}^{\infty} c_n x^n$.

1. $y' + y = 0$

2. $y' = 2y$

3. $y' - x^2 y = 0$

4. $y' + x^3 y = 0$

5. $(1 - x)y' - y = 0$

6. $(1 + x)y' - 2y = 0$

7. $y'' + y = 0$

8. $y'' - y = 0$

9. $y'' = y'$

10. $2y'' + y' = 0$

[6.2.2] In Problems 11–24 for each differential equation find two linearly independent power series solutions about the ordinary point $x = 0$.

11. $y'' = xy$

12. $y'' + x^2y = 0$

13. $y'' - 2xy' + y = 0$

14. $y'' - xy' + 2y = 0$

15. $y'' + x^2y' + xy = 0$

16. $y'' + 2xy' + 2y = 0$

17. $(x - 1)y'' + y' = 0$

18. $(x + 2)y'' + xy' - y = 0$

19. $(x^2 - 1)y'' + 4xy' + 2y = 0$

20. $(x^2 + 1)y'' - 6y = 0$

21. $(x^2 + 2)y'' + 3xy' - y = 0$

22. $(x^2 - 1)y'' + xy' - y = 0$

23. $y'' - (x + 1)y' - y = 0$

24. $y'' - xy' - (x + 2)y = 0$

In Problems 25–28 use the power series method to solve the given differential equation subject to the indicated initial conditions.

25. $(x - 1)y'' - xy' + y = 0, \quad y(0) = -2, \quad y'(0) = 6$

26. $(x + 1)y'' - (2 - x)y' + y = 0, \quad y(0) = 2, \quad y'(0) = -1$

27. $y'' - 2xy' + 8y = 0, \quad y(0) = 3, \quad y'(0) = 0$

28. $(x^2 + 1)y'' + 2xy' = 0, \quad y(0) = 0, \quad y'(0) = 1$

Miscellaneous Problems

The series method of this section can be used when the coefficients are not polynomials. In Problems 29–32 find two power series solutions about the ordinary point $x = 0$.

EXAMPLE 9

$$y'' + (\cos x)y = 0.$$

SOLUTION Since $\cos x = 1 - \dfrac{x^2}{2!} + \dfrac{x^4}{4!} - \dfrac{x^6}{6!} + \cdots$, it is seen that $x = 0$ is an ordinary point. Thus the assumption $y = \sum_{n=0}^{\infty} c_n x^n$ leads to

$$\begin{aligned} y'' + (\cos x)y &= \sum_{n=2}^{\infty} n(n-1)c_n x^{n-2} + \left(1 - \frac{x^2}{2!} + \frac{x^4}{4!} + \cdots\right)\sum_{n=0}^{\infty} c_n x^n \\ &= (2c_2 + 6c_3x + 12c_4x^2 + 20c_5x^3 + \cdots) \\ &\quad + \left(1 - \frac{x^2}{2} + \frac{x^4}{24} + \cdots\right)(c_0 + c_1x + c_2x^2 + c_3x^3 + \cdots) \\ &= 2c_2 + c_0 + (6c_3 + c_1)x + \left(12c_4 + c_2 - \frac{1}{2}c_0\right)x^2 + \left(20c_5 + c_3 - \frac{1}{2}c_1\right)x^3 + \cdots. \end{aligned}$$

Since the last line is to be identically zero, we must have

$$2c_2 + c_0 = 0$$

$$6c_3 + c_1 = 0$$

$$12c_4 + c_2 - \frac{1}{2}c_0 = 0$$

$$20c_5 + c_3 - \frac{1}{2}c_1 = 0$$

and so on. Since c_0 and c_1 are arbitrary we find

$$y_1(x) = c_0\left[1 - \frac{1}{2}x^2 + \frac{1}{12}x^4 + \cdots\right] \quad \text{and} \quad y_2(x) = c_1\left[x - \frac{1}{6}x^3 + \frac{1}{30}x^5 + \cdots\right].$$

Both series converge for all finite values of x.

29. $y'' + (\sin x)y = 0$

30. $xy'' + (\sin x)y = 0$
[*Hint:* See page 250.]

31. $y'' + e^{-x}y = 0$

32. $y'' + e^x y' - y = 0$

In Problems 33 and 34 use the power series method to solve the nonhomogeneous equations.

33. $y'' - xy = 1$

34. $y'' - 4xy' - 4y = e^x$

6.3 Solutions Around Singular Points

6.3.1 Regular Singular Points; Method of Frobenius—Case I

We saw in the preceding section that there is no basic problem in finding a power series solution of

$$a_2(x)y'' + a_1(x)y' + a_0(x)y = 0 \tag{1}$$

around an ordinary point $x = x_0$. However, when $x = x_0$ is a singular point it is not always possible to find a solution of the form $y = \Sigma_{n=0}^{\infty} c_n(x - x_0)^n$; it turns out that we *may* be able to find a solution of the form $y = \Sigma_{n=0}^{\infty} c_n(x - x_0)^{n+r}$, where r is a constant that must be determined.

Regular and Irregular Singular Points

Singular points are further classified as either regular or irregular. To define these concepts we again put (1) into the standard form

$$y'' + P(x)y' + Q(x)y = 0. \tag{2}$$

DEFINITION 6.2 A singular point $x = x_0$ of equation (1) is said to be a **regular singular point** if both $(x - x_0)P(x)$ and $(x - x_0)^2Q(x)$ are analytic at x_0, that is, both $(x - x_0)P(x)$ and $(x - x_0)^2Q(x)$ have a power series in $(x - x_0)$ with a positive radius of convergence. A singular point that is not regular is said to be an **irregular singular point** of the equation.

Polynomial Coefficients

In the case in which coefficients in (1) are polynomials with no common factors, Definition 6.2 is equivalent to the following.

> Let $a_2(x_0) = 0$. Form $P(x)$ and $Q(x)$ by reducing $a_1(x)/a_2(x)$ and $a_0(x)/a_2(x)$ to lowest terms, respectively. If the factor $(x - x_0)$ appears *at most* to the first power in the denominator of $P(x)$ and *at most* to the second power in the denominator of $Q(x)$, then $x = x_0$ is a regular singular point.

EXAMPLE 1

It should be clear that $x = -2$ and $x = 2$ are singular points of the equation

$$(x^2 - 4)^2y'' + (x - 2)y' + y = 0.$$

Dividing the equation by $(x^2 - 4)^2 = (x - 2)^2(x + 2)^2$, we find that

$$P(x) = \frac{1}{(x - 2)(x + 2)^2} \quad \text{and} \quad Q(x) = \frac{1}{(x - 2)^2(x + 2)^2}.$$

We now test $P(x)$ and $Q(x)$ at each singular point.

In order that $x = -2$ be a regular singular point, the factor $x + 2$ can appear at most to the first power in the denominator of $P(x)$, and can appear at most to the second power in the denominator of $Q(x)$. Inspection of $P(x)$ and $Q(x)$ shows that the first condition does not obtain, and so we conclude that $x = -2$ is an irregular singular point.

In order that $x = 2$ be a regular singular point, the factor $x - 2$ can appear at most to the first power in the denominator of $P(x)$ and can appear at most to the second power in the denominator of $Q(x)$. Further inspection of $P(x)$ and $Q(x)$ shows that both these conditions are satisfied, so $x = 2$ is a regular singular point.

EXAMPLE 2

Both $x = 0$ and $x = -1$ are singular points of the differential equation

$$x^2(x + 1)^2y'' + (x^2 - 1)y' + 2y = 0.$$

Inspection of

$$P(x) = \frac{x-1}{x^2(x+1)} \quad \text{and} \quad Q(x) = \frac{2}{x^2(x+1)^2}$$

shows that $x = 0$ is an irregular singular point since $(x - 0)$ appears to the second power in the denominator of $P(x)$. Note, however, that $x = -1$ is a regular singular point.

EXAMPLE 3

(a) $x = 1$ and $x = -1$ are regular singular points of

$$(1 - x^2)y'' - 2xy' + 30y = 0.$$

(b) $x = 0$ is an irregular singular point of

$$x^3y'' - 2xy' + 5y = 0$$

since

$$Q(x) = 5/x^3.$$

(c) $x = 0$ is a regular singular point of

$$xy'' - 2xy' + 5y = 0$$

since

$$P(x) = -2 \quad \text{and} \quad Q(x) = 5/x.$$

In part **(c)** of the preceding example, notice that $(x - 0)$ and $(x - 0)^2$ do not even appear in the denominators of $P(x)$ and $Q(x)$. Remember these factors can appear at most in this fashion. For a singular point $x = x_0$, any nonnegative power of $(x - x_0)$ less than one (namely, zero) and nonnegative power less than two (namely, zero and one) in the denominators of $P(x)$ and $Q(x)$, respectively, imply x_0 is a regular singular point.

Also, recall that singular points can be complex numbers. It should be apparent that both $x = 3i$ and $x = -3i$ are regular singular points of the equation $(x^2 + 9)y'' - 3xy' + (1 - x)y = 0$ since

$$P(x) = \frac{-3x}{(x-3i)(x+3i)} \quad \text{and} \quad Q(x) = \frac{1-x}{(x-3i)(x+3i)}.$$

EXAMPLE 4

From our discussion of the Cauchy-Euler equation in Section 6.1, we can show that $y_1 = x^2$ and $y_2 = x^2 \ln x$ are solutions of the equation $x^2y'' - 3xy' + 4y = 0$ on the interval $(0, \infty)$. If the procedure of Theorem 6.1 is attempted at the regular singular point $x = 0$ (that is, an assumed solution of the form $y = \Sigma_{n=0}^{\infty} c_n x^n$) we would succeed in obtaining only the solution

$y_1 = x^2$. The fact that we would not obtain the second solution is not really surprising since $\ln x$ does not possess a Taylor series expansion about $x = 0$; it follows that $y_2 = x^2 \ln x$ does not have a power series in x.

EXAMPLE 5

The differential equation $6x^2y'' + 5xy' + (x^2 - 1)y = 0$ has a regular singular point at $x = 0$ but does not possess *any* solution of the form $y = \sum_{n=0}^{\infty} c_n x^n$. By the procedure that we shall now consider it can be shown, however, that there exist two series solutions of the form

$$y = \sum_{n=0}^{\infty} c_n x^{n+1/2} \quad \text{and} \quad y = \sum_{n=0}^{\infty} c_n x^{n-1/3}.$$

Method of Frobenius

To solve a differential equation such as (1) about a regular singular point we employ the following theorem called the **method of Frobenius.***

THEOREM 6.2 If $x = x_0$ is a regular singular point of equation (1) then there exists at least one series solution of the form

$$y = (x - x_0)^r \sum_{n=0}^{\infty} c_n(x - x_0)^n = \sum_{n=0}^{\infty} c_n(x - x_0)^{n+r} \tag{3}$$

where the number r is a constant which must be determined. The series will converge at least on some interval $0 < x - x_0 < R$.

As in the preceding section, for the sake of simplicity we shall always assume that $x_0 = 0$.

EXAMPLE 6

Since $x = 0$ is a regular singular point of the differential equation

$$3xy'' + y' - y = 0, \tag{4}$$

* Ferdinand Georg Frobenius (1848–1917) Although the basic idea of this series method can be traced back to Euler, the German mathematician Ferdinand Frobenius was the first to prove the result, which he published in 1878. Frobenius made many contributions to the field of analysis, but his name appears more in texts on abstract algebra than in texts on differential equations. His most significant contributions to mathematics were in the field of group theory.

we try a solution of the form

$$y = \sum_{n=0}^{\infty} c_n x^{n+r}.$$

Now

$$y' = \sum_{n=0}^{\infty} (n + r)c_n x^{n+r-1}$$

$$y'' = \sum_{n=0}^{\infty} (n + r)(n + r - 1)c_n x^{n+r-2}$$

so that

$$3xy'' + y' - y = 3 \sum_{n=0}^{\infty} (n + r)(n + r - 1)c_n x^{n+r-1} + \sum_{n=0}^{\infty} (n + r)c_n x^{n+r-1} - \sum_{n=0}^{\infty} c_n x^{n+r}$$

$$= \sum_{n=0}^{\infty} (n + r)(3n + 3r - 2)c_n x^{n+r-1} - \sum_{n=0}^{\infty} c_n x^{n+r}$$

$$= x^r\left[r(3r - 2)c_0 x^{-1} + \underbrace{\sum_{n=1}^{\infty} (n + r)(3n + r - 2)c_n x^{n-1}}_{k = n - 1} - \underbrace{\sum_{n=0}^{\infty} c_n x^n}_{k = n}\right]$$

$$= x^r\left[r(3r - 2)c_0 x^{-1} + \sum_{k=0}^{\infty} [(k + r + 1)(3k + 3r + 1)c_{k+1} - c_k]x^k\right] = 0$$

which implies

$$r(3r - 2)c_0 = 0$$

$$(k + r + 1)(3k + 3r + 1)c_{k+1} - c_k = 0, \qquad k = 0, 1, 2. \ldots \tag{5}$$

Since nothing is gained by taking $c_0 = 0$, we must then have

$$r(3r - 2) = 0 \tag{6}$$

and

$$c_{k+1} = \frac{c_k}{(k + r + 1)(3k + 3r + 1)}, \qquad k = 0, 1, 2. \ldots \tag{7}$$

The two values of r that satisfy (6), $r_1 = 2/3$ and $r_2 = 0$, when substituted in (7), give two different recurrence relations:

$$r_1 = \frac{2}{3}, \qquad c_{k+1} = \frac{c_k}{(3k + 5)(k + 1)}, \qquad k = 0, 1, 2, \ldots, \tag{8}$$

and

$$r_2 = 0, \qquad c_{k+1} = \frac{c_k}{(k + 1)(3k + 1)}, \qquad k = 0, 1, 2, \ldots. \tag{9}$$

Iteration of (8) gives

$$c_1 = \frac{c_0}{5 \cdot 1}$$

$$c_2 = \frac{c_1}{8 \cdot 2} = \frac{c_0}{2!5 \cdot 8}$$

$$c_3 = \frac{c_2}{11 \cdot 3} = \frac{c_0}{3!5 \cdot 8 \cdot 11}$$

$$c_4 = \frac{c_3}{14 \cdot 4} = \frac{c_0}{4!5 \cdot 8 \cdot 11 \cdot 14}$$

$$\vdots$$

$$c_n = \frac{c_0}{n!5 \cdot 8 \cdot 11 \cdots (3n + 2)}, \qquad n = 1, 2, 3 \ldots,$$

whereas iteration of (9) yields

$$c_1 = \frac{c_0}{1 \cdot 1}$$

$$c_2 = \frac{c_1}{2 \cdot 4} = \frac{c_0}{2!1 \cdot 4}$$

$$c_3 = \frac{c_2}{3 \cdot 7} = \frac{c_0}{3!1 \cdot 4 \cdot 7}$$

$$c_4 = \frac{c_3}{4 \cdot 10} = \frac{c_0}{4!1 \cdot 4 \cdot 7 \cdot 10}$$

$$\vdots$$

$$c_n = \frac{c_0}{n!1 \cdot 4 \cdot 7 \cdots (3n - 2)}, \qquad n = 1, 2, 3 \ldots.$$

Thus we obtain two series solutions

$$y_1 = c_0 x^{2/3}\left[1 + \sum_{n=1}^{\infty} \frac{1}{n!5 \cdot 8 \cdot 11 \cdots (3n + 2)} x^n\right] \tag{10}$$

and

$$y_2 = c_0 x^0\left[1 + \sum_{n=1}^{\infty} \frac{1}{n!1 \cdot 4 \cdot 7 \cdots (3n - 2)} x^n\right]. \tag{11}$$

By the ratio test it can be demonstrated that both (10) and (11) converge for all finite values of x. Also it should be clear from the form of (10) and (11) that neither series is a constant multiple of the other and, therefore, $y_1(x)$ and $y_2(x)$ are linearly independent solutions on the x-axis. Hence by the superposition principle

$$y = C_1 y_1(x) + C_2 y_2(x) = C_1\left[x^{2/3} + \sum_{n=1}^{\infty} \frac{1}{n!5 \cdot 8 \cdot 11 \cdots (3n + 2)} x^{n+2/3}\right]$$
$$+ C_2\left[1 + \sum_{n=1}^{\infty} \frac{1}{n!1 \cdot 4 \cdot 7 \cdots (3n - 2)} x^n\right], \qquad |x| < \infty$$

is another solution of (4). On any interval not containing the origin, this combination represents the general solution of the differential equation.

Although Example 6 illustrates the general procedure for using the method of Frobenius, we hasten to point out that we may not always be able to find two solutions so readily or for that matter find two solutions that are infinite series consisting entirely of powers of x.

Indicial Equation

Equation (6) is called the **indicial equation** of the problem, and the values $r_1 = 2/3$ and $r_2 = 0$ are called the **indicial roots**, or **exponents**, of the singularity. In general, if $x = 0$ is a regular singular point of (1), then the functions $xP(x)$ and $x^2Q(x)$ obtained from (2) are analytic at 0; that is, the expansions

$$\begin{aligned} xP(x) &= p_0 + p_1x + p_2x^2 + \cdots \\ x^2Q(x) &= q_0 + q_1x + q_2x^2 + \cdots \end{aligned} \tag{12}$$

are valid on intervals that have a positive radius of convergence. After substituting $y = \sum_{n=0}^{\infty} c_n x^{n+r}$ in (1) or (2) and simplifying, the indicial equation is a quadratic equation in r that results from equating the *total coefficient of the lowest power of x equal to zero.* It is left as an exercise to show that the general indicial equation is

$$r(r-1) + p_0r + q_0 = 0. \tag{13}$$

We then solve the latter equation for the two values of the exponents and substitute these values into a recurrence relation such as (7). Theorem 6.2 guarantees that at least one solution of the assumed series form can be found.

EXAMPLE 7

The differential equation

$$xy'' + 3y' - y = 0 \tag{14}$$

has a regular singular point at $x = 0$. The method of Frobenius yields

$$xy'' + 3y' - y = x^r\left[r(r+2)c_0x^{-1} + \sum_{k=0}^{\infty}[(k+r+1)(k+r+3)c_{k+1} - c_k]x^k\right] = 0$$

so that the indicial equation and exponents and $r(r+2) = 0$, and $r_1 = 0$, $r_2 = -2$, respectively.

Since

$$(k+r+1)(k+r+3)c_{k+1} - c_k = 0, \qquad k = 0, 1, 2\ldots, \tag{15}$$

it follows that when $r_1 = 0$

$$c_{k+1} = \frac{c_k}{(k+1)(k+3)}$$

$$c_1 = \frac{c_0}{1 \cdot 3}$$

$$c_2 = \frac{c_1}{2 \cdot 4} = \frac{2c_0}{2!4!}$$

$$c_3 = \frac{c_2}{3 \cdot 5} = \frac{2c_0}{3!5!}$$

$$c_4 = \frac{c_3}{4 \cdot 6} = \frac{2c_0}{4!6!}$$

$$\vdots$$

$$c_n = \frac{2c_0}{n!(n+2)!}, \qquad n = 1, 2, 3, \ldots.$$

Thus one series solution is

$$\begin{aligned} y_1 &= c_0 x^0 \left[1 + \sum_{n=1}^{\infty} \frac{2}{n!(n+2)!} x^n \right] \\ &= c_0 \sum_{n=0}^{\infty} \frac{2}{n!(n+2)!} x^n, \qquad |x| < \infty. \end{aligned} \tag{16}$$

Now when $r_2 = -2$, (15) becomes

$$(k-1)(k+1)c_{k+1} - c_k = 0, \tag{17}$$

but note here that we *do not divide* by $(k-1)(k+1)$ immediately since this term is zero for $k = 1$. However, we use the recurrence relation (17) for the cases $k = 0$ and $k = 1$:

$$-1 \cdot 1c_1 - c_0 = 0 \qquad \text{and} \qquad 0 \cdot 2c_2 - c_1 = 0.$$

The latter equation implies that $c_1 = 0$ and so the former equation implies $c_0 = 0$. Continuing, we find

$$c_{k+1} = \frac{c_k}{(k-1)(k+1)}, \qquad k = 2, 3, \ldots$$

and so

$$c_3 = \frac{c_2}{1 \cdot 3}$$

$$c_4 = \frac{c_3}{2 \cdot 4} = \frac{2c_2}{2!4!}$$

$$c_5 = \frac{c_4}{3 \cdot 5} = \frac{2c_2}{3!5!}$$

$$\vdots$$

$$c_n = \frac{2c_2}{(n-2)!n!}, \qquad n = 2, 3, 4, \ldots.$$

Thus
$$y_2 = c_2 x^{-2} \sum_{n=2}^{\infty} \frac{2}{(n-2)!n!} x^n. \tag{18}$$

However, close inspection of (18) reveals that y_2 is simply a constant multiple of (16). To see this, let $k = n - 2$ in (18). We conclude that the method of Frobenius gives only one series solution of (14).

Cases of Indicial Roots

When using the method of Frobenius, we usually distinguish three cases corresponding to the nature of the indicial roots. For the sake of discussion let us suppose that r_1 and r_2 are the *real* solutions of the indicial equation and that, when appropriate, r_1 *denotes the largest root.*

CASE I If r_1 and r_2 are distinct and *do not* differ by an integer, then there exist two linearly independent solutions of equation (1) of the form

$$y_1 = \sum_{n=0}^{\infty} c_n x^{n+r_1}, \qquad c_0 \neq 0, \tag{19a}$$

$$y_2 = \sum_{n=0}^{\infty} b_n x^{n+r_2}, \qquad b_0 \neq 0. \tag{19b}$$

■

CASE I Roots Not Differing by an Integer

EXAMPLE 8

Solve
$$2xy'' + (1 + x)y' + y = 0. \tag{20}$$

SOLUTION If $y = \sum_{n=0}^{\infty} c_n x^{n+r}$ then

$$2xy'' + (1 + x)y' + y = 2 \sum_{n=0}^{\infty} (n + r)(n + r - 1)c_n x^{n+r-1} + \sum_{n=0}^{\infty} (n + r)c_n x^{n+r-1}$$
$$+ \sum_{n=0}^{\infty} (n + r)c_n x^{n+r} + \sum_{n=0}^{\infty} c_n x^{n+r}$$
$$= \sum_{n=0}^{\infty} (n + r)(2n + 2r - 1)c_n x^{n+r-1} + \sum_{n=0}^{\infty} (n + r + 1)c_n x^{n+r}$$
$$= x^r \Bigg[r(2r - 1)c_0 x^{-1} + \underbrace{\sum_{n=1}^{\infty} (n + r)(2n + 2r - 1)c_n x^{n-1}}_{k=n-1} + \underbrace{\sum_{n=0}^{\infty} (n + r + 1)c_n x^n}_{k=n} \Bigg]$$
$$= x^r \Bigg[r(2r - 1)c_0 x^{-1} + \sum_{k=0}^{\infty} [(k + r + 1)(2k + 2r + 1)c_{k+1} + (k + r + 1)c_k]x^k \Bigg] = 0,$$

which implies
$$r(2r - 1) = 0 \tag{21}$$
$$(k + r + 1)(2k + 2r + 1)c_{k+1} + (k + r + 1)c_k = 0, \qquad k = 0, 1, 2. \ldots \tag{22}$$

For $r_1 = 1/2$, we can divide by $k + 3/2$ in (22) to obtain

$$c_{k+1} = \frac{-c_k}{2(k+1)}$$

$$c_1 = \frac{-c_0}{2 \cdot 1}$$

$$c_2 = \frac{-c_1}{2 \cdot 2} = \frac{c_0}{2^2 \cdot 2!}$$

$$c_3 = \frac{-c_2}{2 \cdot 3} = \frac{-c_0}{2^3 \cdot 3!}$$

$$\vdots$$

$$c_n = \frac{(-1)^n c_0}{2^n n!}, \qquad n = 1, 2, 3. \ldots$$

Thus we have

$$y_1 = c_0 x^{1/2}\left[1 + \sum_{n=0}^{\infty} \frac{(-1)^n}{2^n n!} x^n\right]$$

$$= c_0 \sum_{n=0}^{\infty} \frac{(-1)^n}{2^n 2!} x^{n+1/2}, \tag{23}$$

which converges for $x \geq 0$. As given, the series is not meaningful for $x < 0$ because of the presence of $x^{1/2}$.

Now for $r_2 = 0$, (22) becomes

$$c_{k+1} = \frac{-c_k}{2k+1}$$

$$c_1 = \frac{-c_0}{1}$$

$$c_2 = \frac{-c_1}{3} = \frac{c_0}{1 \cdot 3}$$

$$c_3 = \frac{-c_2}{5} = \frac{-c_0}{1 \cdot 3 \cdot 5}$$

$$c_4 = \frac{-c_3}{7} = \frac{c_0}{1 \cdot 3 \cdot 5 \cdot 7}$$

$$\vdots$$

$$c_n = \frac{(-1)^n c_0}{1 \cdot 3 \cdot 5 \cdot 7 \cdots (2n-1)}, \qquad n = 1, 2, 3. \ldots$$

We conclude that a second solution to (20) is

$$y_2 = c_0\left[1 + \sum_{n=1}^{\infty} \frac{(-1)^n}{1 \cdot 3 \cdot 5 \cdot 7 \cdots (2n-1)} x^n\right], \qquad |x| < \infty. \tag{24}$$

On the interval $(0, \infty)$, the general solution is $y = C_1 y_1(x) + C_2 y_2(x)$.

6.3.2 Method of Frobenius—Cases II and III

When the roots of the indicial equation differ by a positive integer, we may or may not be able to find two solutions of (1) having form (3). If not, then one solution corresponding to the smaller root contains a logarithmic term. When the exponents are equal, a second solution will *always* contain a logarithm. This latter situation is analogous to the solutions of the Cauchy-Euler differential equation when the roots of the auxiliary equation are equal. We have the next two cases.

CASE II If $r_1 - r_2 = N$, where N is a positive integer, then there exist two linearly independent solutions of equation (1) of the form

$$y_1 = \sum_{n=0}^{\infty} c_n x^{n+r_1}, \qquad c_0 \neq 0 \tag{25a}$$

$$y_2 = Cy_1(x)\ln x + \sum_{n=0}^{\infty} b_n x^{n+r_2}, \qquad b_0 \neq 0 \tag{25b}$$

where C is a constant that could be zero. ■

CASE III If $r_1 = r_2$, there always exist two linearly independent solutions of equation (1) of the form

$$y_1 = \sum_{n=0}^{\infty} c_n x^{n+r_1}, \qquad c_0 \neq 0 \tag{26a}$$

$$y_2 = y_1(x)\ln x + \sum_{n=1}^{\infty} b_n x^{n+r_1}. \tag{26b}$$

■

CASE II Roots Differing by a Positive Interger

EXAMPLE 9 Solve

$$xy'' + (x-6)y' - 3y = 0. \tag{27}$$

SOLUTION The assumption $y = \sum_{n=0}^{\infty} c_n x^{n+r}$ leads to

$$xy'' + (x-6)y' - 3y$$

$$= \sum_{n=0}^{\infty}(n+r)(n+r-1)c_n x^{n+r-1} - 6\sum_{n=0}^{\infty}(n+r)c_n x^{n+r-1} + \sum_{n=0}^{\infty}(n+r)c_n x^{n+r} - 3\sum_{n=0}^{\infty} c_n x^{n+r}$$

$$= x^r\Bigg[r(r-7)c_0x^{-1} + \underbrace{\sum_{n=1}^{\infty}(n+r)(n+r-7)c_n x^{n-1}}_{k=n-1} + \underbrace{\sum_{n=0}^{\infty}(n+r-3)c_n x^n}_{k=n}\Bigg]$$

$$= x^r\Bigg[r(r-7)c_0x^{-1} + \sum_{k=0}^{\infty}[(k+r+1)(k+r-6)c_{k+1} + (k+r-3)c_k]x^k\Bigg] = 0.$$

Thus $r(r-7)=0$ so that $r_1=7$, $r_2=0$, $r_1-r_2=7$, and

$$(k+r+1)(k+r-6)c_{k+1}+(k+r-3)c_k=0, \qquad k=0,1,2.\ldots \tag{28}$$

For the smaller root $r_2=0$, (28) becomes

$$(k+1)(k-6)c_{k+1}+(k-3)c_k=0. \tag{29}$$

Since $k-6=0$ when $k=6$, we do not divide by this term until $k>6$. We find

$$\begin{aligned}
&1\cdot(-6)c_1+(-3)c_0=0\\
&2\cdot(-5)c_2+(-2)c_1=0\\
&3\cdot(-4)c_3+(-1)c_2=0\\
&\left.\begin{aligned}
&4\cdot(-3)c_4+0\cdot c_3=0\\
&5\cdot(-2)c_5+1\cdot c_4=0\\
&6\cdot(-1)c_6+2\cdot c_5=0\\
&7\cdot 0c_7+3\cdot c_6=0
\end{aligned}\right\}
\end{aligned}$$

implies $c_4=c_5=c_6=0$ but c_0 and c_7 can be chosen arbitrarily.

Hence

$$\begin{aligned}
c_1&=-\frac{1}{2}c_0\\
c_2&=-\frac{1}{5}c_1=\frac{1}{10}c_0\\
c_3&=-\frac{1}{12}c_2=-\frac{1}{120}c_0
\end{aligned} \tag{30}$$

and for $k\geq 7$

$$\begin{aligned}
c_{k+1}&=\frac{-(k-3)c_k}{(k+1)(k-6)}\\
c_8&=\frac{-4}{8\cdot 1}c_7\\
c_9&=\frac{-5}{9\cdot 2}c_8=\frac{4\cdot 5}{2!8\cdot 9}c_7\\
c_{10}&=\frac{-6}{10\cdot 3}c_9=\frac{-4\cdot 5\cdot 6}{3!8\cdot 9\cdot 10}c_7\\
&\vdots\\
c_n&=\frac{(-1)^{n+1}4\cdot 5\cdot 6\cdots(n-4)}{(n-7)!8\cdot 9\cdot 10\cdots n}c_7, \qquad n=8,9,10.\ldots
\end{aligned} \tag{31}$$

If we choose $c_7=0$ and $c_0\neq 0$, we obtain the polynomial solution

$$y_1=c_0\left[1-\frac{1}{2}x+\frac{1}{10}x^2-\frac{1}{120}x^3\right], \tag{32}$$

but when $c_7 \neq 0$ and $c_0 = 0$, it follows that a second, though infinite series, solution is

$$y_2 = c_7\left[x^7 + \sum_{n=8}^{\infty} \frac{(-1)^{n+1}4 \cdot 5 \cdot 6 \cdots (n-4)}{(n-7)!8 \cdot 9 \cdot 10 \cdots n} x^n\right]$$

$$= c_7\left[x^7 + \sum_{k=1}^{\infty} \frac{(-1)^k 4 \cdot 5 \cdot 6 \cdots (k+3)}{k!8 \cdot 9 \cdot 10 \cdots (k+7)} x^{k+7}\right], \qquad |x| < \infty \tag{33}$$

Finally, the general solution of (27) on the interval $(0, \infty)$ is

$$y = C_1 y_1(x) + C_2 y_2(x)$$

$$= C_1\left[1 - \frac{1}{2}x + \frac{1}{10}x^2 - \frac{1}{120}x^3\right] + C_2\left[x^7 + \sum_{k=1}^{\infty} \frac{(-1)^k 4 \cdot 5 \cdot 6 \cdots (k+3)}{k!8 \cdot 9 \cdot 10 \cdots (k+7)} x^{k+7}\right].$$

It is interesting to observe that in the preceding example the larger root $r_1 = 7$ was not used. Had we done so, we would have obtained a series solution of the form*

$$y = \sum_{n=0}^{\infty} c_n x^{n+7}, \tag{34}$$

where the c_n are defined by (28) with $r_1 = 7$:

$$c_{k+1} = \frac{-(k-4)}{(k+8)(k+1)} c_k, \qquad k = 0, 1, 2. \ldots$$

Iteration of this latter recurrence relation then would yield only *one* solution, namely, the solution given by (33) (with c_0 playing the part of c_7).

When the roots of the indicial equation differ by a positive integer, the second solution *may* contain a logarithm. In practice this is something we do not know in advance but which is determined after we have found the indicial roots and have carefully examined the recurrence relation that defines the coefficients c_n. As the foregoing example shows, we just may be lucky enough to find two solutions that involve only powers of x. On the other hand, if we fail to find a second series-type solution, we can always use the fact that

$$y_2 = y_1(x) \int \frac{e^{-\int P(x)\,dx}}{y_1^2(x)} dx \tag{35}$$

is also a solution of the equation $y'' + P(x)y' + Q(x)y = 0$, whenever y_1 is a known solution (see Section 4.2).

---

* Observe that both (33) and (34) start with the power x^7. In Case II it is always a good idea to work with the smaller root first.

EXAMPLE 10 Find the general solution of $xy'' + 3y' - y = 0$.

SOLUTION Recall from Example 7 that the method of Frobenius provides only one solution to this equation, namely,

$$y_1 = \sum_{n=0}^{\infty} \frac{2}{n!(n+2)!} x^n$$
$$= 1 + \frac{1}{3}x + \frac{1}{24}x^2 + \frac{1}{360}x^3 + \cdots. \tag{36}$$

From (35) we obtain a second solution

$$y_2 = y_1(x)\int \frac{e^{-\int(3/x)\,dx}}{y_1^2(x)}\,dx = y_1(x)\int \frac{dx}{x^3\left[1 + \frac{1}{3}x + \frac{1}{24}x^2 + \frac{1}{360}x^3 + \cdots\right]^2}$$

$$= y_1(x)\int \frac{dx}{x^3\left[1 + \frac{2}{3}x + \frac{7}{36}x^2 + \frac{1}{30}x^3 + \cdots\right]} \qquad \text{(squaring)}$$

$$= y_1(x)\int \frac{1}{x^3}\left[1 - \frac{2}{3}x + \frac{1}{4}x^2 - \frac{19}{270}x^3 + \cdots\right]dx \qquad \text{(long division)}$$

$$= y_1(x)\int \left[\frac{1}{x^3} - \frac{2}{3x^2} + \frac{1}{4x} - \frac{19}{270} + \cdots\right]dx$$

$$= y_1(x)\left[-\frac{1}{2x^2} + \frac{2}{3x} + \frac{1}{4}\ln x - \frac{19}{270}x + \cdots\right]$$

or

$$y_2 = \frac{1}{4}y_1(x)\ln x + y_1(x)\left[-\frac{1}{2x^2} + \frac{2}{3x} - \frac{19}{270}x + \cdots\right]. \tag{37}$$

Hence on the interval $(0, \infty)$, the general solution is

$$y = C_1 y_1(x) + C_2\left[\frac{1}{4}y_1(x)\ln x + y_1(x)\left(-\frac{1}{2x^2} + \frac{2}{3x} - \frac{19}{270}x + \cdots\right)\right], \tag{38}$$

where $y_1(x)$ is defined by (36).

Alternative Procedure

There are several alternative procedures to formula (35) when the method of Frobenius fails to provide a second series solution. Although the next method is somewhat tedious, it is nonetheless straightforward. The basic idea is to assume a solution either of the form (25b) or (26b) and determine coefficients b_n in terms of the coefficients c_n that define the known solution $y_1(x)$.

EXAMPLE 11

The smaller of the two indicial roots for the equation $xy'' + 3y' - y = 0$ is $r_2 = -2$. From (25b) we now assume a second solution

$$y_2 = y_1 \ln x + \sum_{n=0}^{\infty} b_n x^{n-2}, \tag{39}$$

where

$$y_1 = \sum_{n=0}^{\infty} \frac{2}{n!(n+2)!} x^n. \tag{40}$$

Differentiation of (39) gives

$$y_2' = \frac{y_1}{x} + y_1' \ln x + \sum_{n=0}^{\infty} (n-2) b_n x^{n-3}$$

$$y_2'' = -\frac{y_1}{x^2} + \frac{2y_1'}{x} + y_1'' \ln x + \sum_{n=0}^{\infty} (n-2)(n-3) b_n x^{n-4}$$

so that

$$\begin{aligned} xy_2'' + 3y_2' - y_2 &= \ln x \underbrace{[xy_1'' + 3y_1' - y_1]}_{\text{zero}} + 2y_1' + \frac{2y_1}{x} + \sum_{n=0}^{\infty} (n-2)(n-3) b_n x^{n-3} \\ &\quad + 3\sum_{n=0}^{\infty} (n-2) b_n x^{n-3} - \sum_{n=0}^{\infty} b_n x^{n-2} \\ &= 2y_1' + \frac{2y_1}{x} + \sum_{n=0}^{\infty} (n-2) n b_n x^{n-3} - \sum_{n=0}^{\infty} b_n x^{n-2}, \end{aligned} \tag{41}$$

where we have combined the first two summations and have used the fact that $xy_1'' + 3y_1' - y_1 = 0$.

By differentiating (40), we can write (41) as

$$\begin{aligned} &\sum_{n=0}^{\infty} \frac{4n}{n!(n+2)!} x^{n-1} + \sum_{n=0}^{\infty} \frac{4}{n!(n+2)!} x^{n-1} + \sum_{n=0}^{\infty} (n-2) n b_n x^{n-3} - \sum_{n=0}^{\infty} b_n x^{n-2} \\ &= 0(-2) b_0 x^{-3} + (-b_0 - b_1) x^{-2} + \underbrace{\sum_{n=0}^{\infty} \frac{4(n+1)}{n!(n+2)!} x^{n-1}}_{k=n} + \underbrace{\sum_{n=2}^{\infty} (n-2) n b_n x^{n-3}}_{k=n-2} - \underbrace{\sum_{n=1}^{\infty} b_n x^{n-2}}_{k=n-1} \\ &= -(b_0 + b_1) x^{-2} + \sum_{k=0}^{\infty} \left[\frac{4(k+1)}{k!(k+2)!} + k(k+2) b_{k+2} - b_{k+1} \right] x^{k-1}. \end{aligned} \tag{42}$$

Setting (42) equal to zero then gives $b_1 = -b_0$ and

$$\frac{4(k+1)}{k!(k+2)!} + k(k+2) b_{k+2} - b_{k+1} = 0, \qquad \text{for } k = 0, 1, 2, \ldots. \tag{43}$$

When $k = 0$ in (43), we have $2 + 0 \cdot 2b_2 - b_1 = 0$ so that $b_1 = 2, b_0 = -2$, but b_2 is arbitrary.

Rewriting (43) as

$$b_{k+2} = \frac{b_{k+1}}{k(k+2)} - \frac{4(k+1)}{k!(k+2)!k(k+2)} \tag{44}$$

and evaluating for $k = 1, 2, \ldots$, give

$$b_3 = \frac{b_2}{3} - \frac{4}{9}$$

$$b_4 = \frac{1}{8}b_3 - \frac{1}{32} = \frac{1}{24}b_2 - \frac{25}{288}$$

and so on. Thus we can finally write

$$\begin{aligned} y_2 &= y_1 \ln x + b_0 x^{-2} + b_1 x^{-1} + b_2 + b_3 x + \cdots \\ &= y_1 \ln x - 2x^{-2} + 2x^{-1} + b_2 + \left(\frac{b_2}{3} - \frac{4}{9}\right)x + \cdots, \end{aligned} \tag{45}$$

where b_2 is arbitrary.

Equivalent Solutions

At this point you may be wondering whether (37) and (45) are really equivalent. If we choose $C_2 = 4$ in (38), then

$$\begin{aligned} y_2 &= y_1 \ln x + y_1\left(-\frac{2}{x^2} + \frac{8}{3x} - \frac{38}{135}x + \cdots\right) \\ &= y_1 \ln x + \left(1 + \frac{1}{3}x + \frac{1}{24}x^2 + \frac{1}{360}x^3 + \cdots\right) \\ &\quad \times \left(-\frac{2}{x^2} + \frac{8}{3x} - \frac{38}{135}x + \cdots\right) \\ &= y_1 \ln x - 2x^{-2} + 2x^{-1} + \frac{29}{36} - \frac{19}{108}x + \cdots, \end{aligned} \tag{46}$$

which is precisely what we obtain from (45) if b_2 is chosen as 29/36.

CASE III Equal Indicial Roots

EXAMPLE 12

Find the general solution of

$$xy'' + y' - 4y = 0 \tag{47}$$

SOLUTION The assumption $y = \sum_{n=0}^{\infty} c_n x^{n+r}$ leads to

$$xy'' + y' - 4y = \sum_{n=0}^{\infty} (n+r)(n+r-1)c_n x^{n+r-1} + \sum_{n=0}^{\infty} (n+r)c_n x^{n+r-1} - 4\sum_{n=0}^{\infty} c_n x^{n+r}$$

$$= \sum_{n=0}^{\infty} (n+r)^2 c_n x^{n+r-1} - 4\sum_{n=0}^{\infty} c_n x^{n+r}$$

$$= x^r\Bigg[r^2 c_0 x^{-1} + \underbrace{\sum_{n=1}^{\infty} (n+r)^2 c_n x^{n-1}}_{k=n-1} - \underbrace{4\sum_{n=0}^{\infty} c_n x^n}_{k=n}\Bigg]$$

$$= x^r\Bigg[r^2 c_0 x^{-1} + \sum_{k=0}^{\infty} [(k+r+1)^2 c_{k+1} - 4c_k]x^k\Bigg] = 0.$$

Therefore $r^2 = 0$, $r_1 = r_2 = 0$, and

$$(k+r+1)^2 c_{k+1} - 4c_k = 0, \qquad k = 0, 1, 2, \ldots. \tag{48}$$

Clearly the root $r_1 = 0$ will only yield one solution corresponding to the coefficients defined by the iteration of

$$c_{k+1} = \frac{4c_k}{(k+1)^2}, \qquad k = 0, 1, 2, \ldots.$$

The result is

$$y_1 = c_0 \sum_{n=0}^{\infty} \frac{4^n}{(n!)^2} x^n, \qquad |x| < \infty. \tag{49}$$

To obtain the second linearly independent solution we set $c_0 = 1$ in (49) and then use (35):

$$y_2 = y_1(x)\int \frac{e^{-\int(1/x)\,dx}}{y_1^2(x)}\,dx = y_1(x)\int \frac{dx}{x\left[1 + 4x + 4x^2 + \frac{16}{9}x^3 + \cdots\right]^2}$$

$$= y_1(x)\int \frac{dx}{x\left[1 + 8x + 24x^2 + \frac{16}{9}x^3 + \cdots\right]}$$

$$= y_1(x)\int \frac{1}{x}\left[1 - 8x + 40x^2 - \frac{1472}{9}x^3 + \cdots\right]dx$$

$$= y_1(x)\int \left[\frac{1}{x} - 8 + 40x - \frac{1472}{9}x^2 + \cdots\right]dx$$

$$= y_1(x)\left[\ln x - 8x + 20x^2 - \frac{1472}{27}x^3 + \cdots\right]. \tag{50}$$

Thus on the interval $(0, \infty)$ the general solution of (47) is

$$y = C_1 y_1(x) + C_2\left[y_1(x)\ln x + y_1(x)\left(-8x + 20x^2 - \frac{1472}{27}x^3 + \cdots\right)\right], \quad (51)$$

where $y_1(x)$ is defined by (49).

As in Case II we can also determine $y_2(x)$ of Example 11 directly from assumption (26b).

Remark: We purposely have not considered two further complications when solving a differential equation such as (1) about a point x_0 for which $a_2(x_0) = 0$. When using (3), it is quite possible that the roots of the indicial equation could turn out to be complex numbers. When the exponents r_1 and r_2 are complex, the statement $r_1 > r_2$ is meaningless and must be replaced with $\text{Re}(r_1) > \text{Re}(r_2)$ (for example, if $r = \alpha + i\beta$, then $\text{Re}(r) = \alpha$). In particular, when the indicial equation has real coefficients, the complex roots will be a conjugate pair $r_1 = \alpha + i\beta$, $r_2 = \alpha - i\beta$ and $r_1 - r_2 = 2i\beta \neq$ integer. Thus for $x_0 = 0$ there will always exist two solutions

$$y_1 = \sum_{n=0}^{\infty} c_n x^{n+r_1} \quad \text{and} \quad y_2 = \sum_{n=0}^{\infty} b_n x^{n+r_2}.$$

Unfortunately both solutions will give complex values of y for each real choice of x. This latter difficulty can be surmounted by the superposition principle. Since a combination of solutions is also a solution to the differential equation, we could form appropriate combinations of $y_1(x)$ and $y_2(x)$ to yield real solutions (see Case III of the solution of the Cauchy-Euler equation).

Lastly, if $x = 0$ is an irregular singular point, it should be noted that we may not be able to find *any* solution of the form $y = \sum_{n=0}^{\infty} c_n x^{n+r}$.

Exercises 6.3

Answers to odd-numbered problems begin on page A–15.

[6.3.1] In Problems 1–10 determine the singular points of each differential equation. Classify each singular point as regular or irregular.

1. $x^3y'' + 4x^2y' + 3y = 0$

2. $xy'' - (x + 3)^{-2}y = 0$

3. $(x^2 - 9)^2y'' + (x + 3)y' + 2y = 0$

4. $y'' - \dfrac{1}{x}y' + \dfrac{1}{(x-1)^3}y = 0$

5. $(x^3 + 4x)y'' - 2xy' + 6y = 0$

6. $x^2(x-5)^2y'' + 4xy' + (x^2-25)y = 0$

7. $(x^2+x-6)y'' + (x+3)y' + (x-2)y = 0$

8. $x(x^2+1)^2y'' + y = 0$

9. $x^3(x^2-25)(x-2)^2y'' + 3x(x-2)y' + 7(x+5)y = 0$

10. $(x^3-2x^2-3x)^2y'' + x(x-3)^2y' - (x+1)y = 0$

In Problems 11–22 show that the indicial roots do not differ by an integer. Use the method of Frobenius to obtain two linearly independent series solutions about the regular singular point $x_0 = 0$. Form the general solution on $(0, \infty)$.

11. $2xy'' - y' + 2y = 0$

12. $2xy'' + 5y' + xy = 0$

13. $4xy'' + \frac{1}{2}y' + y = 0$

14. $2x^2y'' - xy' + (x^2+1)y = 0$

15. $3xy'' + (2-x)y' - y = 0$

16. $x^2y'' - \left(x - \frac{2}{9}\right)y = 0$

17. $2xy'' - (3+2x)y' + y = 0$

18. $x^2y'' + xy' + \left(x^2 - \frac{4}{9}\right)y = 0$

19. $9x^2y'' + 9x^2y' + 2y = 0$

20. $2x^2y'' + 3xy' + (2x-1)y = 0$

21. $2x^2y'' - x(x-1)y' - y = 0$

22. $x(x-2)y'' + y' - 2y = 0$

[6.3.2] In Problems 23–34 show that the indicial roots differ by an integer. Use the method of Frobenius to obtain two linearly independent series solutions about the regular singular point $x_0 = 0$. Form the general solution on $(0, \infty)$.

23. $xy'' + 2y' - xy = 0$

24. $x^2y'' + xy' + \left(x^2 - \frac{1}{4}\right)y = 0$

25. $x(x-1)y'' + 3y' - 2y = 0$

26. $y'' + \frac{3}{x}y' - 2y = 0$

27. $xy'' + (1-x)y' - y = 0$

28. $xy'' + y = 0$

29. $xy'' + y' + y = 0$

30. $xy'' - xy' + y = 0$

31. $x^2y'' + x(x-1)y' + y = 0$

32. $xy'' + y' - 4xy = 0$

33. $xy'' + (x-1)y' - 2y = 0$

34. $xy'' - y' + x^3y = 0$

In Problems 35 and 36 note that $x_0 = 0$ is an irregular singular point of each equation. In each case determine whether the method of Frobenius will yield a solution.

35. $x^3y'' + y = 0$

36. $x^2y'' - y' + y = 0$

Miscellaneous Problems

A differential equation is said to have a singular point at ∞ if, after the substitution $w = 1/x$, the resulting equation has a singular point at $w = 0$. In

Problems 37–40 determine whether the given equation has a singular point at ∞, and if so, state whether it is regular or irregular.

EXAMPLE 13

The equation $y'' + xy = 0$ has no singular points in the finite plane. Using the substitution $w = 1/x$, it follows from the chain rule that

$$\frac{dy}{dx} = \frac{dy}{dw}\frac{dw}{dx} = -\frac{1}{x^2}\frac{dy}{dw} = -w^2\frac{dy}{dw},$$

and

$$\frac{d^2y}{dx^2} = -w^2\frac{d}{dx}\left[\frac{dy}{dw}\right] - \frac{dy}{dw}\frac{d}{dx}[w^2] = -w^2\frac{d}{dw}\left(\frac{dy}{dw}\right)\frac{dw}{dx} - \frac{dy}{dw}\left(2w\frac{dw}{dx}\right) = w^4\frac{d^2y}{dw^2} + 2w^3\frac{dy}{dw}$$

so that the original equation transforms into

$$w^4\frac{d^2y}{dw^2} + 2w^3\frac{dy}{dw} + \frac{1}{w}y = 0.$$

Inspection of

$$P(w) = \frac{2}{w} \quad \text{and} \quad Q(w) = \frac{1}{w^5}$$

indicates that $w = 0$ is an irregular singular point. Hence ∞ is an irregular singular point.

37. $x^2y'' - 4y = 0$

38. $(1 - x)y'' + xy' - y = 0$

39. $x^3y'' + 2x^2y' + 3y = 0$

40. $x^2y'' + (2x + 1)y' + 5y = 0$

41. Solve the Cauchy-Euler equation

$$x^2y'' + 3xy' - 8y = 0$$

by the method of Frobenius.

42. If $x = 0$ is a regular singular point, use (12) in (2) to show that (13) is the indicial equation obtained from the method of Frobenius.

43. Use (13) to find the indicial equation and exponents of

$$x^2y'' + \left(\frac{5}{3}x + x^2\right)y' - \frac{1}{3}y = 0.$$

6.4 Two Special Equations

The two equations

$$x^2y'' + xy' + (x^2 - \nu^2)y = 0 \tag{1}$$

$$(1 - x^2)y'' - 2xy' + n(n + 1)y = 0 \tag{2}$$

occur frequently in advanced studies in applied mathematics, physics, and engineering. They are called **Bessel's equation** and **Legendre's equation**, respectively.* In solving (1), we shall assume $\nu \geq 0$; whereas in (2) we shall consider only the case when n is a nonnegative integer. Since we seek series solutions of each equation about $x = 0$, we observe that the origin is a regular singular point of Bessel's equation, but it is an ordinary point of Legendre's equation.

6.4.1 Solution of Bessel's Equation

If we assume $y = \sum_{n=0}^{\infty} c_n x^{n+r}$, then

$$\begin{aligned} x^2y'' + xy' + (x^2 - \nu^2)y &= \sum_{n=0}^{\infty} c_n(n + r)(n + r - 1)x^{n+r} + \sum_{n=0}^{\infty} c_n(n + r)x^{n+r} + \sum_{n=0}^{\infty} c_n x^{n+r+2} \\ &\quad - \nu^2 \sum_{n=0}^{\infty} c_n x^{n+r} \\ &= c_0(r^2 - r + r - \nu^2)x^r + x^r \sum_{n=1}^{\infty} c_n[(n + r)(n + r - 1) \\ &\quad + (n + r) - \nu^2]x^n + x^r \sum_{n=0}^{\infty} c_n x^{n+2} \\ &= c_0(r^2 - \nu^2)x^r + x^r \sum_{n=1}^{\infty} c_n[(n + r)^2 - \nu^2]x^n + x^r \sum_{n=0}^{\infty} c_n x^{n+2}. \end{aligned} \tag{3}$$

---

* **Friedrich Wilhelm Bessel** (1784–1846) Bessel was a German astronomer who in 1838 was the first to measure the distance to a star (61 Cygni). In 1840 he predicted the existence of a planetary mass beyond the orbit of Uranus. The planet Neptune was discovered six years later. Bessel was also the first person to calculate the orbit of Halley's comet. Although Bessel certainly studied equation (1) in his work on planetary motion, the differential equation and its solution were probably discovered by Daniel Bernoulli in his research on determining the displacements of an oscillating chain.

Adrien Marie Legendre (1752–1833) A French mathematician, Legendre is best remembered for spending almost forty years of his life studying and calculating elliptic integrals. However, the particular polynomial solutions of the equation that bears his name were encountered in his studies of gravitation.

From (3) we see that the indicial equation is $r^2 - \nu^2 = 0$, so that the indicial roots are $r_1 = \nu$ and $r_2 = -\nu$. When $r_1 = \nu$, (3) becomes

$$x^\nu \sum_{n=1}^{\infty} c_n n(n+2\nu)x^n + x^\nu \sum_{n=0}^{\infty} c_n x^{n+2}$$

$$= x^\nu \Bigg[(1+2\nu)c_1 x + \underbrace{\sum_{n=2}^{\infty} c_n n(n+2\nu)x^n}_{k=n-2} + \underbrace{\sum_{n=0}^{\infty} c_n x^{n+2}}_{k=n} \Bigg]$$

$$= x^\nu \Bigg[(1+2\nu)c_1 x + \sum_{k=0}^{\infty} [(k+2)(k+2+2\nu)c_{k+2} + c_k]x^{k+2} \Bigg] = 0.$$

Therefore by the usual argument, we can write

$$(1+2\nu)c_1 = 0$$

$$(k+2)(k+2+2\nu)c_{k+2} + c_k = 0$$

or
$$c_{k+2} = \frac{-c_k}{(k+2)(k+2+2\nu)}, \qquad k = 0, 1, 2 \ldots. \tag{4}$$

The choice $c_1 = 0$ in (4) implies $c_3 = c_5 = c_7 = \cdots = 0$, so for $k = 0, 2, 4, \ldots,$ we find, after letting $k + 2 = 2n$, $n = 1, 2, 3, \ldots,$ that

$$c_{2n} = -\frac{c_{2n-2}}{2^2 n(n+\nu)}. \tag{5}$$

Thus
$$c_2 = -\frac{c_0}{2^2 \cdot 1 \cdot (1+\nu)}$$

$$c_4 = -\frac{c_2}{2^2 \cdot 2(2+\nu)} = \frac{c_0}{2^4 \cdot 1 \cdot 2(1+\nu)(2+\nu)}$$

$$c_6 = -\frac{c_4}{2^2 \cdot 3(3+\nu)} = -\frac{c_0}{2^6 \cdot 1 \cdot 2 \cdot 3(1+\nu)(2+\nu)(3+\nu)}$$

$$\vdots$$

$$c_{2n} = \frac{(-1)^n c_0}{2^{2n} n!(1+\nu)(2+\nu)\cdots(n+\nu)}, \qquad n = 1, 2, 3 \ldots. \tag{6}$$

It is standard practice to choose c_0 to be a specific value, namely,

$$c_0 = \frac{1}{2^\nu \Gamma(1+\nu)},$$

where $\Gamma(1+\nu)$ is the Gamma function (see Appendix II). Since this latter function possesses the convenient property $\Gamma(1+\alpha) = \alpha\Gamma(\alpha)$, we can reduce the indicated product in the denominator of (6) to one term. For example,

$$\Gamma(1 + \nu + 1) = (1 + \nu)\Gamma(1 + \nu)$$
$$\Gamma(1 + \nu + 2) = (2 + \nu)\Gamma(2 + \nu) = (2 + \nu)(1 + \nu)\Gamma(1 + \nu).$$

Hence we can write (6) as

$$c_{2n} = \frac{(-1)^n}{2^{2n+\nu}n!(1 + \nu)(2 + \nu)\cdots(n + \nu)\Gamma(1 + \nu)} = \frac{(-1)^n}{2^{2n+\nu}n!\Gamma(1 + \nu + n)}, \qquad n = 0, 1, 2 \ldots.$$

It follows that one solution is

$$y = \sum_{n=0}^{\infty} c_{2n}x^{2n+\nu} = \sum_{n=0}^{\infty} \frac{(-1)^n}{n!\Gamma(1 + \nu + n)}\left(\frac{x}{2}\right)^{2n+\nu}.$$

If $\nu \geq 0$, the series will converge at least on the interval $[0, \infty)$.

Bessel Functions of the First Kind

The foregoing series solution is usually denoted by $J_\nu(x)$:

$$J_\nu(x) = \sum_{n=0}^{\infty} \frac{(-1)^n}{n!\Gamma(1 + \nu + n)}\left(\frac{x}{2}\right)^{2n+\nu}. \tag{7}$$

Also, for the second exponent $r_2 = -\nu$, we obtain, in exactly the same manner,

$$J_{-\nu}(x) = \sum_{n=0}^{\infty} \frac{(-1)^n}{n!\Gamma(1 - \nu + n)}\left(\frac{x}{2}\right)^{2n-\nu}. \tag{8}$$

The functions $J_\nu(x)$ and $J_{-\nu}(x)$ are called **Bessel functions of the first kind** of order ν and $-\nu$, respectively. Depending on the value of ν, (8) may contain negative powers of x and hence converges on $(0, \infty)$.*

Now some care must be taken in writing the general solution of (1). When $\nu = 0$ it is apparent that (7) and (8) are the same. If $\nu > 0$ and $r_1 - r_2 = \nu - (-\nu) = 2\nu$ is not a positive integer, it follows from Case I of Section 6.3 that $J_\nu(x)$ and $J_{-\nu}(x)$ are linearly independent solutions of (1) on $(0, \infty)$, and so the general solution of the interval would be $y = c_1 J_\nu(x) + c_2 J_{-\nu}(x)$. But we also know from Case II of Section 6.3 that when $r_1 - r_2 = 2\nu$ is a positive integer, a second series solution of (1) *may* exist. In this second case we distinguish two possibilities. When $\nu = m =$ positive integer, $J_{-m}(x)$ defined by (8) and $J_m(x)$ are not linearly independent solutions. It can be shown that J_{-m} is a constant multiple of J_m (see Property (i) below). In addition, $r_1 - r_2 = 2\nu$

* By replacing x by $|x|$, the series given in (7) and (8) converge for $0 < |x| < \infty$.

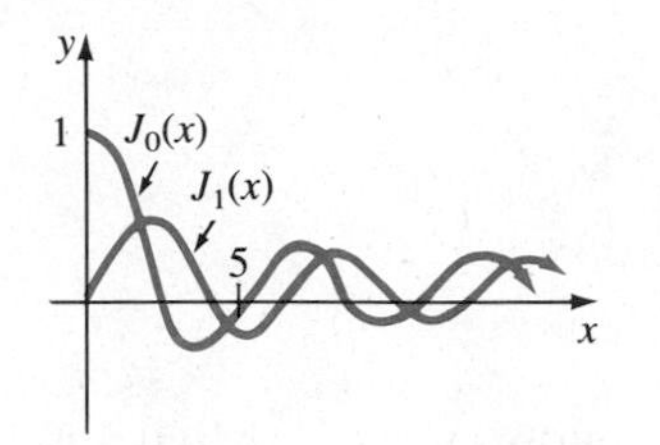

Figure 6.1

can be a positive integer when ν is half an odd positive integer. It can be shown in this latter event that $J_\nu(x)$ and $J_{-\nu}(x)$ are linearly independent. In other words, the general solution of (1) on $(0, \infty)$ is

$$y = c_1 J_\nu(x) + c_2 J_{-\nu}(x), \ \nu \neq \text{integer}. \tag{9}$$

The graphs of $y = J_0(x)$ and $y = J_1(x)$ are given in Figure 6.1. Observe that the graphs of J_0 and J_1 resemble damped cosine and sine graphs, respectively.*

EXAMPLE 1

Find the general solution of the equation

$$x^2 y'' + xy' + (x^2 - \tfrac{1}{4})y = 0$$

on $(0, \infty)$.

SOLUTION We identify $\nu^2 = 1/4$ and so $\nu = 1/2$. From (9) we see that the general solution of the differential equation is

$$y = c_1 J_{1/2}(x) + c_2 J_{-1/2}(x).$$

Bessel Functions of the Second Kind

If $\nu \neq$ integer, the function defined by the linear combination

$$Y_\nu(x) = \frac{\cos \nu\pi \, J_\nu(x) - J_{-\nu}(x)}{\sin \nu\pi} \tag{10}$$

and the function $J_\nu(x)$ are linearly independent solutions of (1). Thus another form of the general solution of (1) is $y = c_1 J_\nu(x) + c_2 Y_\nu(x)$, provided $\nu \neq$ integer. As $\nu \to m$, m an integer, (10) has the indeterminate form 0/0. However, it can be shown by L'Hôpital's rule that $\lim_{\nu \to m} Y_\nu(x)$ exists. Moreover, the function

$$Y_m(x) = \lim_{\nu \to m} Y_\nu(x)$$

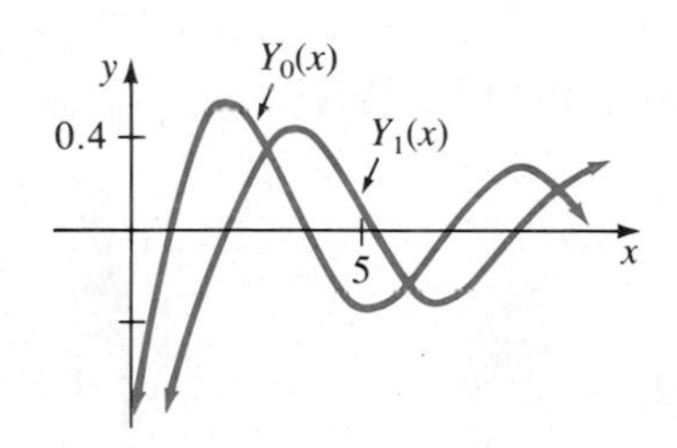

Figure 6.2

and $J_m(x)$ are linearly independent solutions of $x^2 y'' + xy' + (x^2 - m^2)y = 0$. Hence for *any* value of ν the general solution of (1) on $(0, \infty)$ can be written as

$$y = c_1 J_\nu(x) + c_2 Y_\nu(x). \tag{11}$$

$Y_\nu(x)$ is sometimes called **Neumann's function**†; more commonly, $Y_\nu(x)$ is called the **Bessel function of the second kind** of order ν. Figure 6.2 shows the graphs of $Y_0(x)$ and $Y_1(x)$.

---

* Bessel functions belong to a class of functions that are called "almost-periodic."

† The function in (10) is also denoted by $N_\nu(x)$ in honor of the German mathematician C. Neumann (1832–1925), who investigated its properties.

EXAMPLE 2 Find the general solution of the equation

$$x^2y'' + xy' + (x^2 - 9)y = 0$$

on $(0, \infty)$.

SOLUTION Since $\nu^2 = 9$, we identify $\nu = 3$. It follows from (11) that the general solution of the differential equation is

$$y = c_1 J_3(x) + c_2 Y_3(x).$$

Parametric Bessel Equation

By replacing x by λx in (1) and using the Chain Rule, we obtain an alternative form of Bessel's equation known as the **parametric Bessel equation**:

$$x^2y'' + xy' + (\lambda^2x^2 - \nu^2)y = 0. \tag{12}$$

The general solution of (12) is

$$y = c_1 J_\nu(\lambda x) + c_2 Y_\nu(\lambda x). \tag{13}$$

Properties

We list below a few of the more useful properties of Bessel functions of order m, $m = 0, 1, 2, \ldots$.

(i) $J_{-m}(x) = (-1)^m J_m(x)$
(ii) $J_m(-x) = (-1)^m J_m(x)$
(iii) $J_m(0) = 0, \quad m > 0$
(iv) $J_0(0) = 1$
(v) $\lim_{x \to 0^+} Y_m(x) = -\infty$

Note that property (ii) indicates $J_m(x)$ is an even function if m is an even integer and an odd function if m is an odd integer. The graphs of $Y_0(x)$ and $Y_1(x)$ in Figure 6.2 illustrate property (v): $Y_m(x)$ is unbounded at the origin.

6.4.2 Solution of Legendre's Equation

Since $x = 0$ is an ordinary point of equation (2), we assume a solution of the form $y = \sum_{k=0}^{\infty} c_k x^k$. Therefore

$$(1 - x^2)y'' - 2xy' + n(n+1)y = (1 - x^2)\sum_{k=0}^{\infty} c_k k(k-1)x^{k-2} - 2\sum_{k=0}^{\infty} c_k k x^k + n(n+1)\sum_{k=0}^{\infty} c_k x^k$$

$$= \sum_{k=2}^{\infty} c_k k(k-1)x^{k-2} - \sum_{k=2}^{\infty} c_k k(k-1)x^k - 2\sum_{k=1}^{\infty} c_k k x^k + n(n+1)\sum_{k=0}^{\infty} c_k x^k$$

$$= [n(n+1)c_0 + 2c_2]x^0 + [n(n+1)c_1 - 2c_1 + 6c_3]x$$

$$+ \underbrace{\sum_{k=4}^{\infty} c_k k(k-1)x^{k-2}}_{j=k-2} - \underbrace{\sum_{k=2}^{\infty} c_k k(k-1)x^k}_{j=k}$$

$$- 2\underbrace{\sum_{k=2}^{\infty} c_k k x^k}_{j=k} + n(n+1)\underbrace{\sum_{k=2}^{\infty} c_k x^k}_{j=k}$$

$$= [n(n+1)c_0 + 2c_2] + [(n-1)(n+2)c_1 + 6c_3]x$$

$$+ \sum_{j=2}^{\infty} [(j+2)(j+1)c_{j+2} + (n-j)(n+j+1)c_j]x^j = 0$$

implies that

$$n(n+1)c_0 + 2c_2 = 0$$

$$(n-1)(n+2)c_1 + 6c_3 = 0$$

$$(j+2)(j+1)c_{j+2} + (n-j)(n+j+1)c_j = 0$$

or

$$c_2 = -\frac{n(n+1)}{2!}c_0$$

$$c_3 = -\frac{(n-1)(n+2)}{3!}c_1$$

$$c_{j+2} = -\frac{(n-j)(n+j+1)}{(j+2)(j+1)}c_j, \qquad j = 2, 3, 4 \ldots. \tag{14}$$

Iterating (14) gives

$$c_4 = -\frac{(n-2)(n+3)}{4 \cdot 3}c_2 = \frac{(n-2)n(n+1)(n+3)}{4!}c_0$$

$$c_5 = -\frac{(n-3)(n+4)}{5 \cdot 4}c_3 = \frac{(n-3)(n-1)(n+2)(n+4)}{5!}c_1$$

$$c_6 = -\frac{(n-4)(n+5)}{6 \cdot 5}c_4 = -\frac{(n-4)(n-2)n(n+1)(n+3)(n+5)}{6!}c_0$$

$$c_7 = -\frac{(n-5)(n+6)}{7 \cdot 6}c_5$$

$$= -\frac{(n-5)(n-3)(n-1)(n+2)(n+4)(n+6)}{7!}c_1$$

and so on. Thus for at least $|x| < 1$, we obtain two linearly independent power series solutions

$$y_1(x) = c_0\left[1 - \frac{n(n+1)}{2!}x^2 + \frac{(n-2)n(n+1)(n+3)}{4!}x^4 - \frac{(n-4)(n-2)n(n+1)(n+3)(n+5)}{6!}x^6 + \cdots\right]$$

$$y_2(x) = c_1\left[x - \frac{(n-1)(n+2)}{3!}x^3 + \frac{(n-3)(n-1)(n+2)(n+4)}{5!}x^5 - \frac{(n-5)(n-3)(n-1)(n+2)(n+4)(n+6)}{7!}x^7 + \cdots\right]. \tag{15}$$

Notice that if n is an even integer, the first series terminates, whereas $y_2(x)$ is an infinite series. For example, if $n = 4$, then

$$y_1(x) = c_0\left[1 - \frac{4\cdot 5}{2!}x_2 + \frac{2\cdot 4\cdot 5\cdot 7}{4!}x^4\right] = c_0\left[1 - 10x^2 + \frac{35}{3}x^4\right].$$

Similarly, when n is an odd integer, the series for $y_2(x)$ terminates with x^n, that is, *when n is a nonnegative integer we obtain an nth degree polynomial solution* of Legendre's equation.

Since we know that a constant multiple of a solution of Legendre's equation is also a solution, it is traditional to choose specific values for c_0 and c_1, depending on whether n is an even or odd positive integer, respectively. For $n = 0$ we choose $c_0 = 1$ and for $n = 2, 4, 6, \ldots,$

$$c_0 = (-1)^{n/2}\frac{1\cdot 3\cdots(n-1)}{2\cdot 4\cdots n},$$

whereas for $n = 1$ we choose $c_1 = 1$ and for $n = 3, 5, 7, \ldots,$

$$c_1 = (-1)^{(n-1)/2}\frac{1\cdot 3\cdots n}{2\cdot 4\cdots(n-1)}.$$

For example, when $n = 4$, we have

$$\begin{aligned} y_1(x) &= (-1)^{4/2}\frac{1\cdot 3}{2\cdot 4}\left[1 - 10x^2 + \frac{35}{3}x^4\right] \\ &= \frac{3}{8} - \frac{30}{8}x^2 + \frac{35}{8}x^4 \\ &= \frac{1}{8}(35x^4 - 30x^2 + 3). \end{aligned}$$

Legendre Polynomials

These specific nth degree polynomial solutions are called **Legendre polynomials** and are denoted by $P_n(x)$. From the series for $y_1(x)$ and $y_2(x)$ and from the above choices of c_0 and c_1, we find that the first several Legendre polynomials are

$$P_0(x) = 1, \qquad P_1(x) = x,$$
$$P_2(x) = \frac{1}{2}(3x^2 - 1), \qquad P_3(x) = \frac{1}{2}(5x^3 - 3x), \tag{16}$$
$$P_4(x) = \frac{1}{8}(35x^4 - 30x^2 + 3), \qquad P_5(x) = \frac{1}{8}(63x^5 - 70x^3 + 15x).$$

Remember, $P_0(x), P_1(x), P_2(x), P_3(x), \ldots$ are, in turn, particular solutions of the differential equations

$$\begin{aligned} n &= 0, & (1 - x^2)y'' - 2xy' &= 0 \\ n &= 1, & (1 - x^2)y'' - 2xy' + 2y &= 0 \\ n &= 2, & (1 - x^2)y'' - 2xy' + 6y &= 0 \\ n &= 3, & (1 - x^2)y'' - 2xy' + 12y &= 0 \\ & & \vdots & \end{aligned} \tag{17}$$

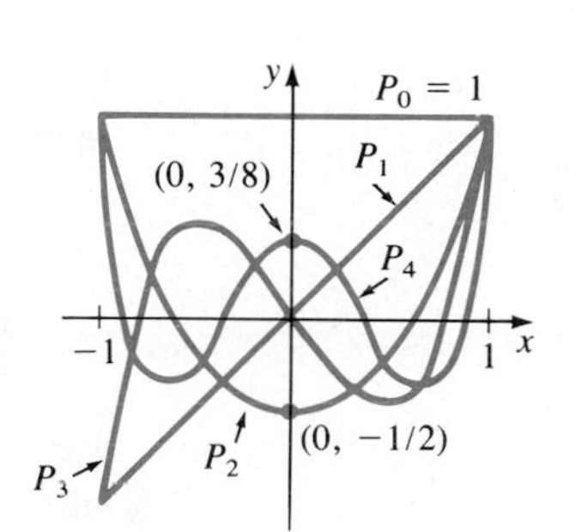

Figure 6.3

The graphs of the first four Legendre polynomials on the interval $-1 \le x \le 1$ are given in Figure 6.3.

Properties

The following properties of the Legendre polynomials are apparent in (16) and Figure 6.3:

(i) $P_n(-x) = (-1)^n P_n(x)$
(ii) $P_n(1) = 1$
(iii) $P_n(-1) = (-1)^n$
(iv) $P_n(0) = 0, \quad n = 1, 3, 5, \ldots$
(v) $P_n'(0) = 0, \quad n = 0, 2, 4, \ldots.$

Property (i) indicates that $P_n(x)$ is an even or odd function according to whether n is even or odd.

Exercises 6.4

Answers to odd-numbered problems begin on page A–17.

[6.4.1] In Problems 1–8 find the general solution of the given differential equation on $(0, \infty)$.

1. $x^2y'' + xy' + \left(x^2 - \frac{1}{9}\right)y = 0$

2. $x^2y'' + xy' + (x^2 - 1)y = 0$

3. $4x^2y'' + 4xy' + (4x^2 - 25)y = 0$

4. $16x^2y'' + 16xy' + (16x^2 - 1)y = 0$

5. $xy'' + y' + xy = 0$

6. $\dfrac{d}{dx}[xy'] + \left(x - \dfrac{4}{x}\right)y = 0$

7. $x^2y'' + xy' + (9x^2 - 4)y = 0$

8. $x^2y'' + xy' + \left(36x^2 - \dfrac{1}{4}\right)y = 0$

9. Use the change of variables $y = x^{-1/2}v(x)$ to find the general solution of the equation

$$x^2y'' + 2xy' + \lambda^2x^2y = 0, \qquad x > 0.$$

10. Verify that the differential equation

$$xy'' + (1 - 2n)y' + xy = 0, \qquad x > 0$$

possesses the particular solution $y = x^nJ_n(x)$.

11. Verify that the differential equation

$$xy'' + (1 + 2n)y' + xy = 0, \qquad x > 0$$

possesses the particular solution $y = x^{-n}J_n(x)$.

12. Verify that the differential equation

$$x^2y'' + \left(\lambda^2x^2 - v^2 + \frac{1}{4}\right)y = 0, \qquad x > 0$$

possesses the particular solution $y = \sqrt{x}\,J_v(\lambda x)$, $\lambda > 0$.

In Problems 13–18 use the results of Problems 10, 11, and 12 to find a particular solution of the given differential equation on $(0, \infty)$.

13. $y'' + y = 0$

14. $xy'' - y' + xy = 0$

15. $xy'' + 3y' + xy = 0$

16. $4x^2y'' + (16x^2 + 1)y = 0$

17. $x^2y'' + (x^2 - 2)y = 0$

18. $xy'' - 5y' + xy = 0$

Recurrence formulas are very important in the study of Bessel functions. In Problems 19–23 derive the given formula.

EXAMPLE 3

$$xJ'_v(x) = vJ_v(x) - xJ_{v+1}(x)$$

SOLUTION $$J_v(x) = \sum_{n=0}^{\infty} \frac{(-1)^n}{n!\Gamma(1 + v + n)}\left(\frac{x}{2}\right)^{2n+v}$$

$$xJ'_v(x) = \sum_{n=0}^{\infty} \frac{(-1)^n(2n + v)}{n!\Gamma(1 + v + n)}\left(\frac{x}{2}\right)^{2n+v}$$

$$= \nu \sum_{n=0}^{\infty} \frac{(-1)^n}{n!\Gamma(1+\nu+n)} \left(\frac{x}{2}\right)^{2n+\nu}$$

$$+ 2 \sum_{n=0}^{\infty} \frac{(-1)^n n}{n!\Gamma(1+\nu+n)} \left(\frac{x}{2}\right)^{2n+\nu}$$

$$= \nu J_\nu(x) + x \underbrace{\sum_{n=1}^{\infty} \frac{(-1)^n}{(n-1)!\Gamma(1+\nu+n)} \left(\frac{x}{2}\right)^{2n+\nu-1}}_{k=n-1}$$

$$= \nu J_\nu(x) - x \sum_{k=0}^{\infty} \frac{(-1)^k}{k!\Gamma(2+\nu+k)} \left(\frac{x}{2}\right)^{2k+\nu+1}$$

$$= \nu J_\nu(x) - x J_{\nu+1}(x).$$

The expression $xJ'_\nu(x) = \nu J_\nu(x) - xJ_{\nu+1}(x)$ is called a **differential recurrence relation**.

19. $xJ'_\nu(x) = -\nu J_\nu(x) + xJ_{\nu-1}(x)$ [*Hint:* $2n + \nu = 2(n + \nu) - \nu$.]

20. $2J'_\nu(x) = J_{\nu-1}(x) - J_{\nu+1}(x)$

21. $2\nu J_\nu(x) = xJ_{\nu+1}(x) + xJ_{\nu-1}(x)$

22. $\dfrac{d}{dx}[x^\nu J_\nu(x)] = x^\nu J_{\nu-1}(x)$

23. $\dfrac{d}{dx}[x^{-\nu} J_\nu(x)] = -x^{-\nu} J_{\nu+1}(x)$

In Problems 24–26 use Problems 19–23 to obtain the given result.

24. $J'_0(x) = J_{-1}(x) = -J_1(x)$

25. $\displaystyle\int_0^x rJ_0(r)\,dr = xJ_1(x)$

26. $\displaystyle\int x^3 J_0(x)\,dx = x^3 J_1(x) - 2x^2 J_2(x) + c$

EXAMPLE 4

Find an alternative expression for $J_{1/2}(x)$. Use the fact that $\Gamma(\frac{1}{2}) = \sqrt{\pi}$.

SOLUTION With $\nu = 1/2$, we have from (7)

$$J_{1/2}(x) = \sum_{n=0}^{\infty} \frac{(-1)^n}{n!\Gamma(1+\frac{1}{2}+n)} \left(\frac{x}{2}\right)^{2n+\frac{1}{2}}$$

Now

$$n = 0, \quad \Gamma\left(1+\frac{1}{2}\right) = \frac{1}{2}\Gamma\left(\frac{1}{2}\right) = \frac{1}{2}\sqrt{\pi}$$

$$n = 1, \quad \Gamma\left(1+\frac{3}{2}\right) = \frac{3}{2}\Gamma\left(\frac{3}{2}\right) = \frac{3}{2^2}\sqrt{\pi}$$

$$n = 2, \quad \Gamma\left(1+\frac{5}{2}\right) = \frac{5}{2}\Gamma\left(\frac{5}{2}\right) = \frac{5\cdot 3}{2^3}\sqrt{\pi} = \frac{5\cdot 4\cdot 3\cdot 2\cdot 1}{2^3 4\cdot 2}\sqrt{\pi} = \frac{5!}{2^5 2!}\sqrt{\pi}$$

$$n = 3, \quad \Gamma\left(1 + \frac{7}{2}\right) = \frac{7}{2}\Gamma\left(\frac{7}{2}\right) = \frac{7 \cdot 5!}{2^6 2!}\sqrt{\pi} = \frac{7 \cdot 6 \cdot 5!}{2^6 \cdot 6 \cdot 2!}\sqrt{\pi} = \frac{7!}{2^7 3!}\sqrt{\pi}.$$

In general,

$$\Gamma\left(1 + \frac{1}{2} + n\right) = \frac{(2n+1)!}{2^{2n+1} n!}\sqrt{\pi}.$$

Hence,

$$J_{1/2}(x) = \sum_{n=0}^{\infty} \frac{(-1)^n}{n! \dfrac{(2n+1)!\sqrt{\pi}}{2^{2n+1} n!}} \left(\frac{x}{2}\right)^{2n+\frac{1}{2}}$$

$$= \sqrt{\frac{2}{\pi x}} \sum_{n=0}^{\infty} \frac{(-1)^n}{(2n+1)!} x^{2n+1}$$

$$= \sqrt{\frac{2}{\pi x}} \sin x.$$

27. Express $J_{-1/2}(x)$ in terms of $\cos x$ and a power of x.

When ν = half an odd integer, $J_\nu(x)$ can be expressed in terms of $\sin x$, $\cos x$, and powers of x. Such Bessel functions are usually called **spherical Bessel functions**. In Problems 28–33 use the results of Problems 21 and 27 and Example 4 to find an alternative expression for the given function.

28. $J_{3/2}(x)$

29. $J_{-3/2}(x)$

30. $J_{5/2}(x)$

31. $J_{-5/2}(x)$

32. $J_{7/2}(x)$

33. $J_{-7/2}(x)$

34. Show that $i^{-\nu} J_\nu(ix)$, $i^2 = -1$, is a real function. The function defined by $I_\nu(x) = i^{-\nu} J_\nu(ix)$ is called a **modified Bessel function of the first kind** of order ν.

35. Find the general solution of the differential equation

$$x^2 y'' + xy' - (x^2 + \nu^2)y = 0, \quad x > 0, \quad \nu \neq \text{integer}.$$

[*Hint:* $i^2 x^2 = -x^2$.]

36. If $y_1 = J_0(x)$ is one solution of the zero-order Bessel equation verify that another solution is

$$y_2 = J_0(x) \ln x + \frac{x^2}{4} - \frac{3x^4}{128} + \frac{11x^6}{13{,}824} - \cdots.$$

37. Use (8) with $\nu = m$, where m is a positive integer, and the fact that $1/\Gamma(N) = 0$, where N is a negative integer, to show that

$$J_{-m}(x) = (-1)^m J_m(x).$$

38. Use (7) with $\nu = m$, where m is a nonnegative integer, to show that

$$J_m(-x) = (-1)^m J_m(x).$$

[6.4.2] **39.** **(a)** Use the explicit solutions $y_1(x)$ and $y_2(x)$ of Legendre's equation and the appropriate choices of c_0 and c_1 to find the Legendre polynomials $P_6(x)$ and $P_7(x)$.

(b) Write the explicit differential equations for which $P_6(x)$ and $P_7(x)$ are particular solutions.

40. Show that Legendre's equation has an alternative form

$$\frac{d}{dx}\left[(1 - x^2)\frac{dy}{dx}\right] + n(n + 1)y = 0.$$

41. Show that the equation

$$\sin\theta \frac{d^2y}{d\theta^2} + \cos\theta \frac{dy}{d\theta} + n(n + 1)(\sin\theta)y = 0$$

can be transformed in Legendre's equation by means of the substitution $x = \cos\theta$.

42. The general Legendre polynomial can be written as

$$P_n(x) = \sum_{k=0}^{[n/2]} \frac{(-1)^k(2n - 2k)!}{2^n k!(n - k)!(n - 2k)!} x^{n-2k},$$

where $[n/2]$ is the greatest integer not greater than $n/2$. Verify the results for $n = 0, 1, 2, 3, 4, 5$.

43. Use the binomial theorem to formally show that

$$(1 - 2xt + t^2)^{-1/2} = \sum_{n=0}^{\infty} P_n(x)t^n.$$

The expression $(1 - 2xt + t^2)^{-1/2}$ is called a **generating function** for the Legendre polynomials.

44. Use Problem 43 to show that $P_n(1) = 1$ and $P_n(-1) = (-1)^n$.

EXAMPLE 5

Differentiating the generating function given in Problem 43 with respect to t gives

$$(1 - 2xt + t^2)^{-3/2}(x - t) = \sum_{n=0}^{\infty} nP_n(x)t^{n-1} = \sum_{n=1}^{\infty} nP_n(x)t^{n-1}$$

so that after multiplying by $1 - 2xt + t^2$, we have

$$(x - t)(1 - 2xt + t^2)^{-1/2} = (1 - 2xt + t^2) \sum_{n=1}^{\infty} nP_n(x)t^{n-1}$$

or

$$(x - t) \sum_{n=0}^{\infty} P_n(x)t^n = (1 - 2xt + t^2) \sum_{n=1}^{\infty} nP_n(x)t^{n-1}. \tag{18}$$

Multiply out and rewrite (18) as

$$\sum_{n=0}^{\infty} xP_n(x)t^n - \sum_{n=0}^{\infty} P_n(x)t^{n+1} - \sum_{n=1}^{\infty} nP_n(x)t^{n-1} + 2x \sum_{n=1}^{\infty} nP_n(x)t^n - \sum_{n=1}^{\infty} nP_n(x)t^{n+1} = 0$$

or

$$x + x^2t + \sum_{n=2}^{\infty} xP_n(x)t^n - t - \sum_{n=1}^{\infty} P_n(x)t^{n+1} - x - 2\left(\frac{3x^2 - 1}{2}\right)t$$

$$- \sum_{n=3}^{\infty} nP_n(x)t^{n-1} + 2x^2t + 2x \sum_{r=2}^{\infty} nP_2(x)t^n - \sum_{n=1}^{\infty} nP_nt^{n+1} = 0.$$

Observing the appropriate cancellations, simplifying, and changing the summation indices give

$$\sum_{k=2}^{\infty} [-(k + 1)P_{k+1}(x) + (2k + 1)xP_k(x) - kP_{k-1}(x)]t^k = 0.$$

Equating the total coefficient of t^k to be zero gives the three-term recurrence relation

$$(k + 1)P_{k+1}(x) - (2k + 1)xP_k(x) + kP_{k-1}(x) = 0, \quad k = 2, 3, 4 \ldots.$$

This formula is also valid when $k = 1$.

45. Use the recurrence relation in Example 5 and the facts that $P_0(x) = 1$, $P_1(x) = x$ to generate the next three Legendre polynomials.

46. The Legendre polynomials are also generated by **Rodrigues' formula***

$$P_n(x) = \frac{1}{2^n n!} \frac{d^n}{dx^n} (x^2 - 1)^n.$$

Verify the results for $n = 0, 1, 2, 3$.

* **Olinde Rodrigues** (1794–1851) Rodrigues was a French banker and amateur mathematician. In mathematics he is remembered solely for the discovery of this one formula in 1816. In politics he is remembered as the financial backer and disciple of Count de Saint-Simon, the founder of French socialism.

47. Use the explicit Legendre polynomials $P_0(x)$, $P_1(x)$, $P_2(x)$, and $P_3(x)$ to evaluate $\int_{-1}^{1} P_n^2(x)\,dx$ for $n = 0, 1, 2, 3$. Generalize the results.

48. Use the explicit Legendre polynomials $P_0(x)$, $P_1(x)$, $P_2(x)$, and $P_3(x)$ to evaluate $\int_{-1}^{1} P_n(x)P_m(x)\,dx$ for $n \neq m$. Generalize the results.

49. We know that $y_1 = x$ is a solution of Legendre's equation when $n = 1$, $(1 - x^2)y'' - 2xy' + 2y = 0$. Show that a second linearly independent solution on the interval $-1 < x < 1$ is

$$y_2 = \frac{x}{2}\ln\left(\frac{1+x}{1-x}\right) - 1.$$

Miscellaneous Problems

50. Show that between two consecutive positive roots of $J_1(x)$ there exists a root of $J_0(x)$. [*Hint:* Look up Rolle's theorem and then inspect Problem 24.]

CHAPTER 6 SUMMARY

The remarkable characteristic of a Cauchy-Euler equation is the fact that even though it is a differential equation with variable coefficients, it can be solved in terms of elementary functions. **A second-order Cauchy-Euler equation** is any differential equation of the form $ax^2y'' + bxy' + cy = g(x)$, where a, b, and c are constants. To solve the homogeneous equation we try a solution of the form $y = x^m$ and this in turn leads to an algebraic **auxiliary equation** $am(m - 1) + bm + c = 0$. Accordingly, when the roots are real and distinct, real and equal, or complex conjugates, the general solutions on the interval $(0, \infty)$ are, respectively,

$$y = c_1x^{m_1} + c_2x^{m_2}$$

$$y = c_1x^{m_1} + c_2x^{m_1}\ln x$$

and

$$y = x^\alpha[c_1\cos(\beta\ln x) + c_2\sin(\beta\ln x)].$$

We say that $x = 0$ is an **ordinary point** of the linear second-order differential equation $a_2(x)y'' + a_1(x)y' + a_0(x)y = 0$ when $a_2(0) \neq 0$ and $a_2(x)$, $a_1(x)$, $a_0(x)$ are polynomials having no common factors. Every solution has the form of a power series in x, $y = \sum_{n=0}^{\infty} c_nx^n$. To find the coefficients c_n we substitute the basic assumption into the differential equation, and after appropriate algebraic manipulations we determine a **recurrence relation** by equating to zero the combined total coefficient of x^k. Iteration of the recurrence relation yields two distinct sets of coefficients, one set containing the arbitrary coefficient c_0 and the other containing c_1. Using each set of coefficients, we form two linearly independent solutions $y_1(x)$ and $y_2(x)$. A solution is valid at least on an interval defined by $|x| < R_1$, where R_1 is the distance to the closest singular point of the equation. If $a_2(0) = 0$, then $x = 0$ is a **singular point**. Singular points are classified as either **regular** or **irregular**. To determine whether $x = 0$ is a regular singular point, we examine the denominators of the rational functions P and Q that result when the equation is put into the form $y'' + P(x)y' + Q(x)y = 0$. It is understood that $a_1(x)/a_2(x)$ and $a_0(x)/a_2(x)$ are reduced to lowest terms. If x appears *at most* to the first power in the denominator of $P(x)$ and *at most* to the second power in the denominator of $Q(x)$, then $x = 0$ is a regular singular point. Around the regular singular point $x = 0$, the **method of Frobenius** guarantees that there exists *at least one* solution of the form $y = \sum_{n=0}^{\infty} c_n x^{n+r}$. The exponent r is a root of a quadratic **indicial equation**. When the indicial roots r_1 and r_2 $(r_1 > r_2)$ satisfy $r_1 - r_2 \neq$ an integer, then we can always find *two* linearly independent solutions of the assumed form. When $r_1 - r_2 =$ a positive integer, then we could *possibly* find two solutions; but when $r_1 - r_2 = 0$, or $r_1 = r_2$, we can find only one solution of the form $y = \sum_{n=0}^{\infty} x^{n+r}$.

Bessel's equation $x^2y'' + xy' + (x^2 - \nu^2)y = 0$ has a regular singular point at $x = 0$, whereas $x = 0$ is an ordinary point of **Legendre's equation** $(1 - x^2)y'' - 2xy' + n(n + 1)y = 0$. The latter equation possesses a polynomial solution when n is a nonnegative integer.

CHAPTER 6 REVIEW EXERCISES

Answers to odd-numbered problems begin on page A–18.

In Problems 1–4 solve the given Cauchy-Euler equation.

1. $6x^2y'' + 5xy' - y = 0$

2. $2x^3y''' + 19x^2y'' + 39xy' + 9y = 0$

3. $x^2y'' - 4xy' + 6y = 2x^4 + x^2$ **4.** $x^2y'' - xy' + y = x^3$

5. Specify the ordinary points of $(x^3 - 8)y'' - 2xy' + y = 0$. ______

6. Specify the singular points of $(x^4 - 16)y'' + 2y = 0$. ______

In Problems 7–10 specify the regular and irregular singular points of the given differential equation.

7. $(x^3 - 10x^2 + 25x)y'' + y' = 0$ RSP ______ ISP ______

8. $(x^3 - 10x^2 + 25x)y'' + y = 0$ RSP ______ ISP ______

9. $x^2(x^2 - 9)^2y'' - (x^2 - 9)y' + xy = 0$ RSP ______ ISP ______

10. $x(x^2 + 1)^3y'' + y' - 8xy = 0$ RSP ______ ISP ______

In Problems 11 and 12 specify an interval around $x = 0$ for which a power series solution of the given differential equation will converge.

11. $y'' - xy' + 6y = 0$ ______

12. $(x^2 - 4)y'' - 2xy' + 9y = 0$ ______

In Problems 13–16 for each differential equation find two power series solutions about the ordinary point $x = 0$.

13. $y'' + xy = 0$

14. $y'' - 4y = 0$

15. $(x - 1)y'' + 3y = 0$

16. $y'' - x^2y' + xy = 0$

In Problems 17–22 find two linearly independent solutions of each equation.

17. $2x^2y'' + xy' - (x + 1)y = 0$

18. $2xy'' + y' + y = 0$

19. $x(1 - x)y'' - 2y' + y = 0$

20. $x^2y'' - xy' + (x^2 + 1)y = 0$

21. $xy'' - (2x - 1)y' + (x - 1)y = 0$

22. $x^2y'' - x^2y' + (x^2 - 2)y = 0$

23. Without referring to Section 6.4, use the method of Frobenius to obtain a solution of the Bessel equation for $\nu = 0$: $xy'' + y' + xy = 0$.

CHAPTER 7

Laplace Transform

IMPORTANT TERMS

linear operation
Laplace transform
linear transform
piecewise continuous
exponential order
inverse Laplace transform
first translation theorem
unit step function
second translation theorem
convolution
convolution theorem
Volterra integral equation
integrodifferential equation
Dirac delta function

In Sections 7.1 and 7.3 of this chapter, we shall examine the definition and properties of the **Laplace transform**. We shall see in Section 7.4 that the Laplace transform applied to a linear nth-order differential equation with constant coefficients transforms it into an algebraic equation involving $y(0), y'(0), y''(0), \ldots, y^{(n-1)}(0)$. As a consequence, the Laplace transform is well suited to solving certain kinds of initial-value problems.

7.1 Laplace Transform

In elementary calculus you have studied the operations of differentiation and integration; recall that

$$\frac{d}{dx}[\alpha f(x) + \beta g(x)] = \alpha \frac{d}{dx} f(x) + \beta \frac{d}{dx} g(x)$$
$$\int [\alpha f(x) + \beta g(x)]\, dx = \alpha \int f(x)\, dx + \beta \int g(x)\, dx \tag{1}$$

for any real constants α and β. Any operation having the property illustrated in (1) is said to be a **linear operation**. A definite integral of a sum can, of course, be written as the sum of the integrals,

$$\int_a^b [\alpha f(x) + \beta g(x)]\, dx = \alpha \int_a^b f(x)\, dx + \beta \int_a^b g(x)\, dx,$$

provided each integral exists. Hence definite integration is a linear operation.

Basic Definition

The foregoing linear operations of differentiation and integration *transform* a function into another expression. For example,

$$\frac{d}{dx} x^2 = 2x, \qquad \int x^2\, dx = \frac{x^3}{3} + c, \qquad \int_0^3 x^2\, dx = 9.$$

Specifically we are concerned with an improper integral that transforms a function $f(t)$ into a function of a parameter s.

DEFINITION 7.1 Let $f(t)$ be defined for $t \geq 0$; then the integral

$$\int_0^\infty e^{-st} f(t)\, dt = \lim_{b \to \infty} \int_0^b e^{-st} f(t)\, dt \tag{2}$$

is said to be the **Laplace transform*** of f, provided the limit exists.

* **Pierre Simon Marquis de Laplace** (1749–1827) A noted mathematician, physicist, and astronomer, Laplace was called by some of his enthusiastic contemporaries the "Newton of France." Although Laplace made use of the integral transform (2) in his work in probability theory, it is likely that the integral was first discovered by Euler. Laplace's noted treatises were *Mécanique Céleste* and *Théorie Analytique des Probabilités.* Born into a poor farming family, Laplace became a friend of Napoleon but was elevated to the nobility by Louis XVIII after the restoration.

Symbolically the Laplace transform of f is denoted by $\mathscr{L}\{f(t)\}$, and since the answer depends on s, we write

$$\mathscr{L}\{f(t)\} = F(s).$$

EXAMPLE 1 Evaluate $\mathscr{L}\{1\}$.

SOLUTION

$$\begin{aligned}\mathscr{L}\{1\} &= \int_0^\infty e^{-st}(1)\,dt = \lim_{b\to\infty}\int_0^b e^{-st}\,dt \\ &= \lim_{b\to\infty}\frac{-e^{-st}}{s}\Big|_0^b \\ &= \lim_{b\to\infty}\frac{-e^{-sb}+1}{s} \\ &= \frac{1}{s} \quad \text{provided } s > 0.\end{aligned}$$

The use of the limit sign becomes somewhat tedious, so we shall adopt the notation $|_0^\infty$ as a shorthand to writing $\lim_{b\to\infty}(\;)|_0^b$.

For example,

$$\begin{aligned}\mathscr{L}\{1\} &= \int_0^\infty e^{-st}\,dt \\ &= \frac{-e^{-st}}{s}\Big|_0^\infty = \frac{1}{s}, \qquad s > 0,\end{aligned}$$

where it is understood that at the upper limit we mean $e^{-st} \to 0$ as $t \to \infty$ for $s > 0$.

$\mathscr{L}$ a Linear Transform

For a sum of functions we can write

$$\int_0^\infty e^{-st}[\alpha f(t) + \beta g(t)]\,dt = \alpha\int_0^\infty e^{-st}f(t)\,dt + \beta\int_0^\infty e^{-st}g(t)\,dt,$$

whenever both integrals converge. Hence it follows that

$$\begin{aligned}\mathscr{L}\{\alpha f(t) + \beta g(t)\} &= \alpha\mathscr{L}\{f(t)\} + \beta\mathscr{L}\{g(t)\} \\ &= \alpha F(s) + \beta G(s). \qquad (3)\end{aligned}$$

Because of the property given in (3), $\mathscr{L}$ is said to be a **linear transform**, or **linear operator**.

Sufficient Conditions for Existence of $\mathscr{L}\{f(t)\}$

The integral that defines the Laplace transform does not have to converge. For example, neither $\mathscr{L}\{1/t\}$ nor $\mathscr{L}\{e^{t^2}\}$ exists. Sufficient conditions that will guarantee the existence of $\mathscr{L}\{f(t)\}$ are that f be piecewise continuous on $[0, \infty)$ and that f be of exponential order for $t > T$. Recall a function f is **piecewise continuous** on $[0, \infty)$ if, in any interval $0 \le a \le t \le b$, there are at most a finite number of points t_k, $k = 1, 2, \ldots, n$ $(t_{k-1} < t_k)$, at which f has

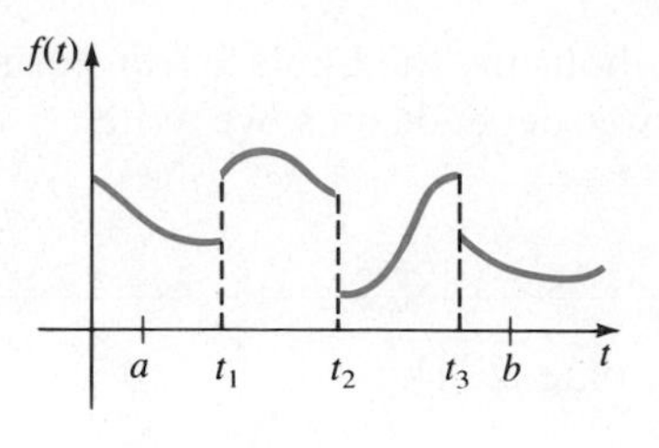

Figure 7.1

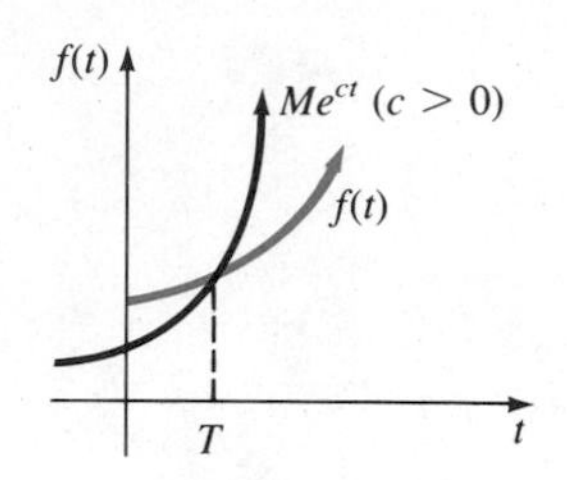

Figure 7.2

finite discontinuities and is continuous on each open interval $t_{k-1} < t < t_k$ (see Figure 7.1). A function f is said to be of **exponential order** if there exist numbers c, $M > 0$, and $T > 0$, such that $|f(t)| \le Me^{ct}$ for $t > T$. If, say, f is an increasing function, then the condition $|f(t)| \le Me^{ct}$, $t > T$, simply states that its graph on the interval (T, ∞) does not grow faster than the graph of Me^{ct}, where c is a positive constant (see Figure 7.2). For example, $f(t) = t$, $f(t) = e^{-t}$, and $f(t) = 2\cos t$ are all of exponential order for $t > 0$ since we have, respectively,

$$|t| \le e^t, \qquad |e^{-t}| \le e^t, \qquad |2\cos t| \le 2e^t.$$

A comparison of the graphs on the interval $(0, \infty)$ is given in Figure 7.3.

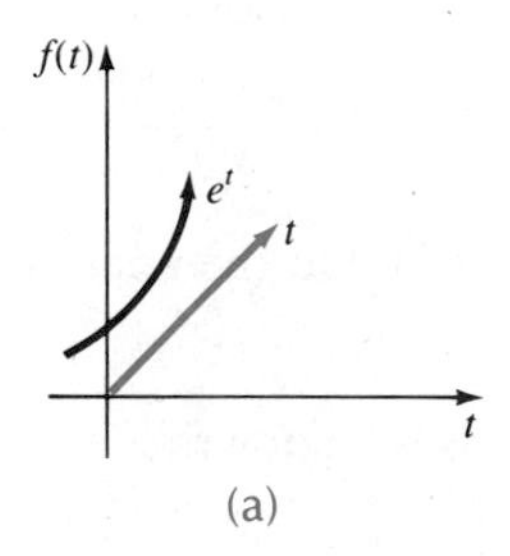

(a)

f(t)
e^t
e^-t
t

(b)

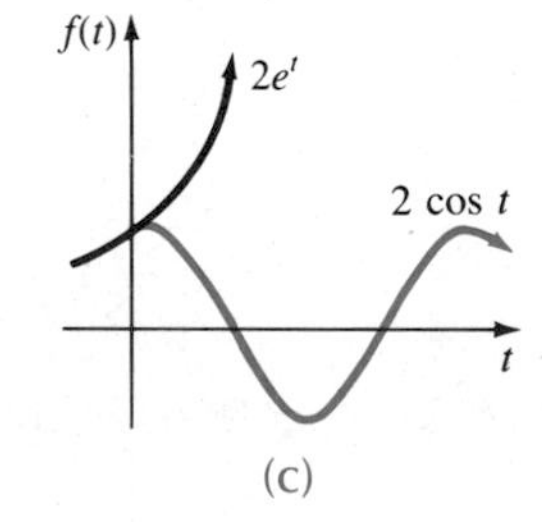

(c)

Figure 7.3

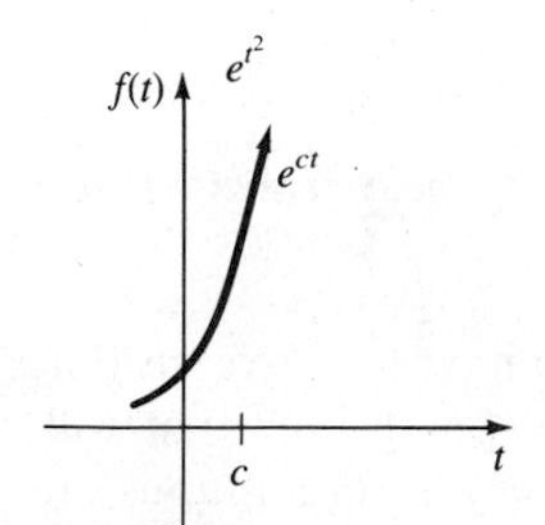

A function such as $f(t) = e^{t^2}$ is not of exponential order since, as shown in Figure 7.4, its graph grows faster than any positive linear power of e for $t > c > 0$.

A positive integral power of t is always of exponential order since for $c > 0$

$$|t^n| \le Me^{ct} \quad \text{or} \quad \left|\frac{t^n}{e^{ct}}\right| \le M \quad \text{for } t > T$$

is equivalent to showing that $\lim_{t\to\infty} t^n/e^{ct}$ is a finite limit for $n = 1, 2, 3 \ldots$. The result follows by n applications of L'Hôpital's rule.

THEOREM 7.1 Let $f(t)$ be piecewise continuous on the interval $[0, \infty)$ and of exponential order for $t > T$; then $\mathscr{L}\{f(t)\}$ exists for $s > c$.

PROOF
$$\mathscr{L}\{f(t)\} = \int_0^\infty e^{-st}f(t)\,dt$$
$$= \int_0^T e^{-st}f(t)\,dt + \int_T^\infty e^{-st}f(t)\,dt = I_1 + I_2.$$

Now I_1 exists since it can be written as a sum of integrals over intervals for which f is continuous. I_2 exists since

$$|I_2| \le \int_T^\infty |e^{-st}f(t)|\,dt \le M\int_T^\infty e^{-st}e^{ct}\,dt$$
$$= M\int_T^\infty e^{-(s-c)t}\,dt = -M\frac{e^{-(s-c)t}}{s-c}\Big|_T^\infty$$
$$= M\frac{e^{-(s-c)T}}{s-c} \qquad \text{for } s > c.$$

■

Throughout this entire chapter we shall be concerned only with functions that are both piecewise continuous and of exponential order. We note, however, that these conditions are sufficient but not necessary for the existence of a Laplace transform. The function $f(t) = t^{-1/2}$ is not piecewise continuous on the interval $[0, \infty)$, but its Laplace transform exists (see Problem 40).

EXAMPLE 2

Evaluate $\mathscr{L}\{t\}$.

SOLUTION From Definition 7.1 we have

$$\mathscr{L}\{t\} = \int_0^\infty e^{-st}t\,dt.$$

Integrating by parts and using the fact that $\lim_{t\to\infty} te^{-st} = 0$ for $s > 0$, we obtain

$$\mathscr{L}\{t\} = \frac{-te^{-st}}{s}\Big|_0^\infty + \frac{1}{s}\int_0^\infty e^{-st}\,dt$$
$$= \frac{1}{s}\mathscr{L}\{1\} = \frac{1}{s}\left(\frac{1}{s}\right)$$
$$= \frac{1}{s^2}, \qquad s > 0.$$

EXAMPLE 3

Evaluate $\mathscr{L}\{e^{-3t}\}$.

SOLUTION From Definition 7.1

$$\begin{aligned}\mathscr{L}\{e^{-3t}\} &= \int_0^\infty e^{-st}e^{-3t}\,dt \\ &= \int_0^\infty e^{-(s+3)t}\,dt \\ &= \frac{-e^{-(s+3)t}}{s+3}\bigg|_0^\infty \\ &= \frac{1}{s+3}, \qquad s > -3.\end{aligned}$$

The last result follows from the fact that $\lim_{t\to\infty} e^{-(s+3)t} = 0$ for $s + 3 > 0$ or $s > -3$.

EXAMPLE 4

Evaluate $\mathscr{L}\{\sin 2t\}$.

SOLUTION From Definition 7.1 and integration by parts

$$\begin{aligned}\mathscr{L}\{\sin 2t\} &= \int_0^\infty e^{-st}\sin 2t\,dt \\ &= \frac{-e^{-st}\sin 2t}{s}\bigg|_0^\infty + \frac{2}{s}\int_0^\infty e^{-st}\cos 2t\,dt \\ &= \frac{2}{s}\int_0^\infty e^{-st}\cos 2t\,dt, \qquad s > 0 \\ &= \frac{2}{s}\left[\frac{-e^{-st}\cos 2t}{s}\bigg|_0^\infty - \frac{2}{s}\int_0^\infty e^{-st}\sin 2t\,dt\right] \\ &= \frac{2}{s^2} - \frac{4}{s^2}\mathscr{L}\{\sin 2t\}, \qquad s > 0.\end{aligned}$$

Now solve for $\mathscr{L}\{\sin 2t\}$:

$$\left[1 + \frac{4}{s^2}\right]\mathscr{L}\{\sin 2t\} = \frac{2}{s^2}$$

$$\mathscr{L}\{\sin 2t\} = \frac{2}{s^2 + 4}, \qquad s > 0.$$

EXAMPLE 5

Evaluate $\mathscr{L}\{3t - 5 \sin 2t\}$.

SOLUTION From Examples 2 and 4 and the linearity property of the Laplace transform we can write

$$\begin{aligned} \mathscr{L}\{3t - 5 \sin 2t\} &= 3\mathscr{L}\{t\} - 5\mathscr{L}\{\sin 2t\} \\ &= 3 \cdot \frac{1}{s^2} - 5 \cdot \frac{2}{s^2 + 4} \\ &= \frac{-7s^2 + 12}{s^2(s^2 + 4)}, \qquad s > 0. \end{aligned}$$

EXAMPLE 6

Evaluate **(a)** $\mathscr{L}\{te^{-2t}\}$, **(b)** $\mathscr{L}\{t^2e^{-2t}\}$.

SOLUTION **(a)** From Definition 7.1 and integration by parts

$$\begin{aligned} \mathscr{L}\{te^{-2t}\} &= \int_0^\infty e^{-st}(te^{-2t})\, dt \\ &= \int_0^\infty te^{-(s+2)t}\, dt \\ &= \frac{-te^{-(s+2)t}}{s+2}\bigg|_0^\infty + \frac{1}{s+2}\int_0^\infty e^{-(s+2)t}\, dt \\ &= \frac{-e^{-(s+2)t}}{(s+2)^2}\bigg|_0^\infty \qquad (s > -2) \\ &= \frac{1}{(s+2)^2}, \qquad (s > -2). \end{aligned}$$

(b) Again, integration by parts gives

$$\begin{aligned} \mathscr{L}\{t^2e^{-2t}\} &= \frac{-t^2e^{-(s+2)t}}{s+2}\bigg|_0^\infty + \frac{2}{s+2}\int_0^\infty te^{-(s+2)t}\, dt \\ &= \frac{2}{s+2}\int_0^\infty e^{-st}(te^{-2t})\, dt, \qquad (s > -2) \\ &= \frac{2}{s+2}\mathscr{L}\{te^{-2t}\} = \frac{2}{s+2}\left[\frac{1}{(s+2)^2}\right] \qquad \text{[from part (a)]} \\ &= \frac{2}{(s+2)^3}, \qquad (s > -2). \end{aligned}$$

EXAMPLE 7

Evaluate $\mathscr{L}\{f(t)\}$ for $f(t) = \begin{cases} 0, & 0 \le t < 3 \\ 2, & t \ge 3. \end{cases}$

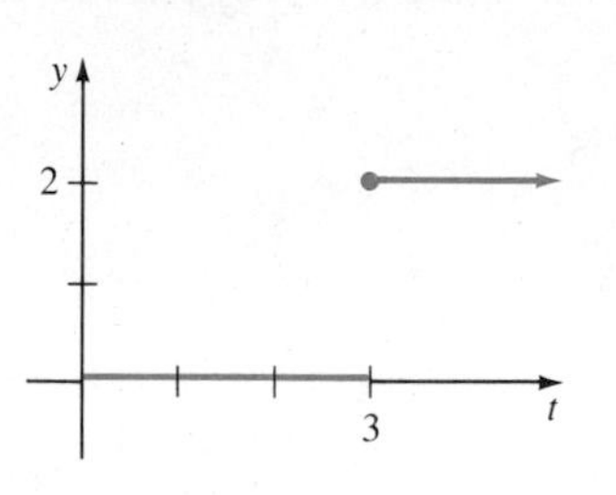

Figure 7.5

SOLUTION This piecewise continuous function is shown in Figure 7.5. From Definition 7.1 we have

$$\begin{aligned}\mathscr{L}\{f(t)\} &= \int_0^\infty e^{-st}f(t)\,dt \\ &= \int_0^3 e^{-st}f(t)\,dt + \int_3^\infty e^{-st}f(t)\,dt \\ &= \int_0^3 e^{-st}(0)\,dt + \int_3^\infty e^{-st}(2)\,dt \\ &= -\frac{2e^{-st}}{s}\bigg|_3^\infty \\ &= \frac{2e^{-3s}}{s}, \qquad s > 0.\end{aligned}$$

We state the generalization of some of the preceding examples by means of the next theorem. From this point on we shall also refrain from stating any restrictions on s; it is understood that s is sufficiently restricted to guarantee the convergence of the appropriate Laplace transform.

THEOREM 7.2

(a) $\mathscr{L}\{1\} = \dfrac{1}{s}$

(b) $\mathscr{L}\{t^n\} = \dfrac{n!}{s^{n+1}}, \qquad n = 1, 2, 3, \ldots$

(c) $\mathscr{L}\{e^{at}\} = \dfrac{1}{s - a}$

(d) $\mathscr{L}\{\sin kt\} = \dfrac{k}{s^2 + k^2}$

(e) $\mathscr{L}\{\cos kt\} = \dfrac{s}{s^2 + k^2}$

(f) $\mathscr{L}\{\sinh kt\} = \dfrac{k}{s^2 - k^2}$

(g) $\mathscr{L}\{\cosh kt\} = \dfrac{s}{s^2 - k^2}.$

Part (b) of Theorem 7.2 can be justified in the following manner: Integration by parts yields

$$\begin{aligned}
\mathscr{L}\{t^n\} &= \int_0^\infty e^{-st} t^n \, dt \\
&= -\frac{1}{s} e^{-st} t^n \Big|_0^\infty + \frac{n}{s} \int_0^\infty e^{-st} t^{n-1} \, dt \\
&= \frac{n}{s} \int_0^\infty e^{-st} t^{n-1} \, dt
\end{aligned}$$

or

$$\mathscr{L}\{t^n\} = \frac{n}{s} \mathscr{L}\{t^{n-1}\}, \qquad n = 1, 2, 3, \ldots.$$

Now $\mathscr{L}\{1\} = 1/s$, so it follows by iteration that

$$\begin{aligned}
\mathscr{L}\{t\} &= \frac{1}{s} \mathscr{L}\{1\} = \frac{1}{s^2} \\
\mathscr{L}\{t^2\} &= \frac{2}{s} \mathscr{L}\{t\} = \frac{2}{s}\left(\frac{1}{s^2}\right) = \frac{2}{s^3} \\
\mathscr{L}\{t^3\} &= \frac{3}{s} \mathscr{L}\{t^2\} = \frac{3}{s}\left(\frac{2}{s^3}\right) = \frac{3!}{s^4}.
\end{aligned}$$

In general it seems reasonable that

$$\mathscr{L}\{t^n\} = \frac{n}{s} \mathscr{L}\{t^{n-1}\} = \frac{n}{s}\left[\frac{(n-1)!}{s^n}\right] = \frac{n!}{s^{n+1}}.^{*}$$

The justifications of parts (f) and (g) of Theorem 7.2 are left to the student (see Problems 29 and 30).

EXAMPLE 8

Evaluate $\mathscr{L}\{\sin^2 t\}$.

SOLUTION With the aid of a trigonometric identity and parts (a) and (e) of Theorem 7.2, we obtain

$$\begin{aligned}
\mathscr{L}\{\sin^2 t\} &= \mathscr{L}\left\{\frac{1 - \cos 2t}{2}\right\} = \frac{1}{2} \mathscr{L}\{1\} - \frac{1}{2} \mathscr{L}\{\cos 2t\} \\
&= \frac{1}{2} \cdot \frac{1}{s} - \frac{1}{2} \cdot \frac{s}{s^2 + 4} = \frac{2}{s(s^2 + 4)}.
\end{aligned}$$

---

* A rigorous proof requires mathematical induction.

Exercises 7.1

Answers to odd-numbered problems begin on page A–19.

In Problems 1–14 use Definition 7.1 to find $\mathscr{L}\{f(t)\}$.

1. $f(t) = \begin{cases} -1, & 0 \le t < 1 \\ 1, & t \ge 1 \end{cases}$

2. $f(t) = \begin{cases} 4, & 0 \le t < 2 \\ 0, & t \ge 2 \end{cases}$

3. $f(t) = \begin{cases} t, & 0 \le t < 1 \\ 1, & t \ge 1 \end{cases}$

4. $f(t) = \begin{cases} 2t+1, & 0 \le t < 1 \\ 0, & t \ge 1 \end{cases}$

5. $f(t) = \begin{cases} \sin t, & 0 \le t < \pi \\ 0, & t \ge \pi \end{cases}$

6. $f(t) = \begin{cases} 0, & 0 \le t < \pi/2 \\ \cos t, & t \ge \pi/2 \end{cases}$

7. $f(t) = e^{t+7}$

8. $f(t) = e^{-2t-5}$

9. $f(t) = te^{4t}$

10. $f(t) = t^2 e^{3t}$

11. $f(t) = e^{-t} \sin t$

12. $f(t) = e^{t} \cos t$

13. $f(t) = t \cos t$

14. $f(t) = t \sin t$

In Problems 15–38 use Theorem 7.2 to find $\mathscr{L}\{f(t)\}$.

15. $f(t) = 2t^4$

16. $f(t) = t^5$

17. $f(t) = 4t - 10$

18. $f(t) = 7t + 3$

19. $f(t) = t^2 + 6t - 3$

20. $f(t) = -4t^2 + 16t + 9$

21. $f(t) = (t+1)^3$

22. $f(t) = (2t-1)^3$

23. $f(t) = 1 + e^{4t}$

24. $f(t) = t^2 - e^{-9t} + 5$

25. $f(t) = (1 + e^{2t})^2$

26. $f(t) = (e^{t} - e^{-t})^2$

27. $f(t) = 4t^2 - 5 \sin 3t$

28. $f(t) = \cos 5t + \sin 2t$

29. $f(t) = \sinh kt$

30. $f(t) = \cosh kt$

31. $f(t) = e^{t} \sinh t$

32. $f(t) = e^{-t} \cosh t$

33. $f(t) = \sin 2t \cos 2t$

34. $f(t) = \cos^2 t$

35. $f(t) = \cos t \cos 2t$ [*Hint:* Examine $\cos(t_1 \pm t_2)$.]

36. $f(t) = \sin t \sin 2t$

37. $f(t) = \sin t \cos 2t$ [*Hint:* Examine $\sin(t_1 \pm t_2)$.]

38. $f(t) = \sin^3 t$ [*Hint:* $\sin^3 t = \sin t \sin^2 t$.]

39. The **gamma function** is defined by the integral

$$\Gamma(\alpha) = \int_0^\infty t^{\alpha-1} e^{-t}\, dt, \qquad \alpha > 0$$

(see Appendix II). Show that $\mathscr{L}\{t^\alpha\} = \dfrac{\Gamma(\alpha+1)}{s^{\alpha+1}}, \quad \alpha > -1.$

In Problems 40–42 use the result of Problem 39 to find $\mathscr{L}\{f(t)\}$.

40. $f(t) = t^{-1/2}$ **41.** $f(t) = t^{1/2}$ **42.** $f(t) = t^{3/2}$

Miscellaneous Problems

43. Show that the function $f(t) = 1/t^2$ does not possess a Laplace transform. $\left[\textit{Hint: } \mathscr{L}\{f(t)\} = \int_0^1 e^{-st}f(t)\,dt + \int_1^\infty e^{-st}f(t)\,dt.\right.$ Use the definition of an improper integral to show $\left.\int_0^1 e^{-st}f(t)\,dt \text{ does not exist.}\right]$

44. Show that if the functions f and g are of exponential order for $t > T$, then the product fg is of exponential order for $t > T$.

7.2 Inverse Transform

By using the integral definition of the Laplace transform of a function f, we determine another function F, that is, a function of the transform parameter s. We have denoted this symbolically by $\mathscr{L}\{f(t)\} = F(s)$.

We now turn the problem around; namely, given $F(s)$, find the function $f(t)$ corresponding to this transform. We say $f(t)$ is the **inverse Laplace transform** of $F(s)$ and write

$$f(t) = \mathscr{L}^{-1}\{F(s)\}.$$

The analogue of Theorem 7.2 for the inverse transform is the following:

THEOREM 7.3 **(a)** $1 = \mathscr{L}^{-1}\left\{\frac{1}{s}\right\}$

(b) $t^n = \mathscr{L}^{-1}\left\{\frac{n!}{s^{n+1}}\right\}, \quad n = 1, 2, 3, \ldots$

(c) $e^{at} = \mathscr{L}^{-1}\left\{\frac{1}{s-a}\right\}$

(d) $\sin kt = \mathscr{L}^{-1}\left\{\frac{k}{s^2+k^2}\right\}$

(e) $\cos kt = \mathscr{L}^{-1}\left\{\frac{s}{s^2+k^2}\right\}$

(f) $\sinh kt = \mathscr{L}^{-1}\left\{\frac{k}{s^2-k^2}\right\}$

(g) $\cosh kt = \mathscr{L}^{-1}\left\{\frac{s}{s^2-k^2}\right\}.$

$\mathscr{L}^{-1}$ a Linear Transform

We shall assume that the inverse Laplace transform is itself a linear transform,* that is, for constants α and β

$$\mathscr{L}^{-1}\{\alpha F(s) + \beta G(s)\} = \alpha\mathscr{L}^{-1}\{F(s)\} + \beta\mathscr{L}^{-1}\{G(s)\},$$

where F and G are the transforms of some functions f and g.

The inverse Laplace transform of a function $F(s)$ *may* not be unique. It is possible that $\mathscr{L}\{f_1(t)\} = \mathscr{L}\{f_2(t)\}$ and yet $f_1 \neq f_2$ (see Problems 29 and 30). For our purposes this is not as bad as it appears. If f_1 and f_2 are piecewise continuous on $[0, \infty)$ and of exponential order for $t > 0$ and if $\mathscr{L}\{f_1(t)\} = \mathscr{L}\{f_2(t)\}$, then it can be proved that the functions f_1 and f_2 are *essentially* the same, that is, they can differ only at points of discontinuity. However, if f_1 and f_2 are continuous on $[0, \infty)$ and $\mathscr{L}\{f_1(t)\} = \mathscr{L}\{f_2(t)\}$, then $f_1 = f_2$ on the interval.

EXAMPLE 1

Evaluate $\mathscr{L}^{-1}\left\{\dfrac{1}{s^5}\right\}$.

SOLUTION To match the form given in part (b) of Theorem 7.3, we identify $n = 4$ and then multiply and divide by 4!. It follows that

$$\mathscr{L}^{-1}\left\{\frac{1}{s^5}\right\} = \frac{1}{4!}\mathscr{L}^{-1}\left\{\frac{4!}{s^5}\right\}$$
$$= \frac{1}{24}t^4.$$

EXAMPLE 2

Evaluate $\mathscr{L}^{-1}\left\{\dfrac{1}{s^2 + 64}\right\}$.

SOLUTION Identifying $k^2 = 64$, we multiply and divide by 8 and use part (d) of Theorem 7.3,

$$\mathscr{L}^{-1}\left\{\frac{1}{s^2 + 64}\right\} = \frac{1}{8}\mathscr{L}^{-1}\left\{\frac{8}{s^2 + 64}\right\}$$
$$= \frac{1}{8}\sin 8t.$$

* The inverse Laplace transform is actually another integral. However, evaluation of this integral demands the use of complex variables, which is beyond the scope of this text.

EXAMPLE 3

Evaluate $\mathscr{L}^{-1}\left\{\frac{3s+5}{s^2+7}\right\}$.

SOLUTION Use termwise division and the linearity property of the inverse transform. From parts (e) and (d) of Theorem 7.3, we have

$$\mathscr{L}^{-1}\left\{\frac{3s+5}{s^2+7}\right\} = 3\mathscr{L}^{-1}\left\{\frac{s}{s^2+7}\right\} + \frac{5}{\sqrt{7}}\mathscr{L}^{-1}\left\{\frac{\sqrt{7}}{s^2+7}\right\}$$

$$= 3\cos\sqrt{7}\,t + \frac{5}{\sqrt{7}}\sin\sqrt{7}\,t.$$

The use of **partial fractions** is very important in finding inverse Laplace transforms. Here we review three basic cases of that theory. For example, the denominators of

(i) $F(s) = \dfrac{1}{(s-1)(s+2)(s+4)}$

(ii) $F(s) = \dfrac{s+1}{s^2(s+2)^3}$

(iii) $F(s) = \dfrac{3s-2}{s^3(s^2+4)}$

contain, respectively, only distinct linear factors, repeated linear factors, and a quadratic factor that is irreducible.*

EXAMPLE 4

Evaluate $\mathscr{L}^{-1}\left\{\frac{1}{(s-1)(s+2)(s+4)}\right\}$.

SOLUTION There exist unique constants A, B, and C so that

$$\frac{1}{(s-1)(s+2)(s+4)} = \frac{A}{s-1} + \frac{B}{s+2} + \frac{C}{s+4}$$

$$= \frac{A(s+2)(s+4) + B(s-1)(s+4) + C(s-1)(s+2)}{(s-1)(s+2)(s+4)}.$$

Since the numerators are identical, we must then have

$$1 = A(s+2)(s+4) + B(s-1)(s+4) + C(s-1)(s+2).$$

* Irreducible means that the quadratic factor has no real zeros.

By comparing coefficients of powers of s on both sides of the equality, we know that the last equation is equivalent to a system of three equations in the three unknowns A, B, and C. However, you might recall the following shortcut for determining these unknowns. If we set $s = 1$, $s = -2$, and $s = -4$, the zeros of the common denominator $(s - 1)(s + 2)(s + 4)$, we obtain, in turn,

$$\begin{aligned} 1 &= A(3)(5), & A &= 1/15, \\ 1 &= B(-3)(2), & B &= -1/6, \\ 1 &= C(-5)(-2), & C &= 1/10. \end{aligned}$$

Hence we can write

$$\frac{1}{(s-1)(s+2)(s+4)} = \frac{1/15}{s-1} - \frac{1/6}{s+2} + \frac{1/10}{s+4}$$

and thus, from part (c) of Theorem 7.3,

$$\begin{aligned} \mathscr{L}^{-1}\left\{\frac{1}{(s-1)(s+2)(s+4)}\right\} &= \frac{1}{15}\mathscr{L}^{-1}\left\{\frac{1}{s-1}\right\} - \frac{1}{6}\mathscr{L}^{-1}\left\{\frac{1}{s+2}\right\} + \frac{1}{10}\mathscr{L}^{-1}\left\{\frac{1}{s+4}\right\} \\ &= \frac{1}{15}e^{t} - \frac{1}{6}e^{-2t} + \frac{1}{10}e^{-4t}. \end{aligned}$$

EXAMPLE 5 Evaluate $\mathscr{L}^{-1}\left\{\dfrac{s+1}{s^2(s+2)^3}\right\}$.

SOLUTION Assume

$$\frac{s+1}{s^2(s+2)^3} = \frac{A}{s} + \frac{B}{s^2} + \frac{C}{s+2} + \frac{D}{(s+2)^2} + \frac{E}{(s+2)^3}$$

so that

$$s + 1 = As(s+2)^3 + B(s+2)^3 + Cs^2(s+2)^2 + Ds^2(s+2) + Es^2.$$

Setting $s = 0$ and $s = -2$ gives $B = 1/8$ and $E = -1/4$, respectively. By equating coefficients of s^4, s^3, and s, we obtain

$$\begin{aligned} 0 &= A + C \\ 0 &= 6A + B + 4C + D \\ 1 &= 8A + 12B, \end{aligned}$$

from which it follows that $A = -1/16$, $C = 1/16$, $D = 0$. Hence from parts (a), (b), and (c) of Theorem 7.3

$$\mathscr{L}^{-1}\left\{\frac{s+1}{s^2(s+2)^3}\right\} = \mathscr{L}^{-1}\left\{-\frac{1/16}{s} + \frac{1/8}{s^2} + \frac{1/16}{s+2} - \frac{1/4}{(s+2)^3}\right\}$$

$$= -\frac{1}{16}\mathscr{L}^{-1}\left\{\frac{1}{s}\right\} + \frac{1}{8}\mathscr{L}^{-1}\left\{\frac{1}{s^2}\right\} + \frac{1}{16}\mathscr{L}^{-1}\left\{\frac{1}{s+2}\right\} - \frac{1}{8}\mathscr{L}^{-1}\left\{\frac{2}{(s+2)^3}\right\}$$

$$= -\frac{1}{16} + \frac{1}{8}t + \frac{1}{16}e^{-2t} - \frac{1}{8}t^2e^{-2t}.$$

Here we have also used $\mathscr{L}^{-1}\{2/(s+2)^3\} = t^2e^{-2t}$ from Example 6 of Section 7.1.

EXAMPLE 6

Evaluate $\mathscr{L}^{-1}\left\{\dfrac{3s-2}{s^3(s^2+4)}\right\}$.

SOLUTION Assume

$$\frac{3s-2}{s^3(s^2+4)} = \frac{A}{s} + \frac{B}{s^2} + \frac{C}{s^3} + \frac{Ds+E}{s^2+4}$$

so that

$$3s - 2 = As^2(s^2+4) + Bs(s^2+4) + C(s^2+4) + (Ds+E)s^3.$$

Setting $s = 0$ gives immediately $C = -1/2$. Now the coefficients of s^4, s^3, s^2, and s are, respectively,

$$0 = A + D, \qquad 0 = B + E, \qquad 0 = 4A + C, \qquad 3 = 4B,$$

from which we obtain $B = 3/4$, $E = -3/4$, $A = 1/8$, $D = -1/8$. Therefore from parts (a), (b), (e), and (d) of Theorem 7.3

$$\mathscr{L}^{-1}\left\{\frac{3s-2}{s^3(s^2+4)}\right\} = \mathscr{L}^{-1}\left\{\frac{1/8}{s} + \frac{3/4}{s^2} - \frac{1/2}{s^3} + \frac{-s/8 - 3/4}{s^2+4}\right\}$$

$$= \frac{1}{8}\mathscr{L}^{-1}\left\{\frac{1}{s}\right\} + \frac{3}{4}\mathscr{L}^{-1}\left\{\frac{1}{s^2}\right\} - \frac{1}{4}\mathscr{L}^{-1}\left\{\frac{2}{s^3}\right\}$$

$$- \frac{1}{8}\mathscr{L}^{-1}\left\{\frac{s}{s^2+4}\right\} - \frac{3}{8}\mathscr{L}^{-1}\left\{\frac{2}{s^2+4}\right\}$$

$$= \frac{1}{8} + \frac{3}{4}t - \frac{1}{4}t^2 - \frac{1}{8}\cos 2t - \frac{3}{8}\sin 2t.$$

Not every arbitrary function of s is a Laplace transform of a piecewise function of exponential order.

THEOREM 7.4 Let $f(t)$ be piecewise continuous on $[0, \infty)$ and of exponential order for $t > T$; then

$$\lim_{s\to\infty} \mathscr{L}\{f(t)\} = 0.$$

PROOF Since $f(t)$ is piecewise continuous on $0 \le t \le T$, it is necessarily bounded on this interval:

$$|f(t)| \le M_1 = M_1 e^{0t}.$$

Also

$$|f(t)| \le M_2 e^{\gamma t}$$

for $t > T$. If M denotes the maximum of $\{M_1, M_2\}$ and c denotes the maximum of $\{0, \gamma\}$, then

$$\begin{aligned} |\mathscr{L}\{f(t)\}| &\le \int_0^\infty e^{-st}|f(t)|\, dt \\ &\le M \int_0^\infty e^{-st} \cdot e^{ct}\, dt \\ &= -M \frac{e^{-(s-c)t}}{s-c}\bigg|_0^\infty \\ &= \frac{M}{s-c} \qquad \text{for } s > c. \end{aligned}$$

As $s \to \infty$, we have $|\mathscr{L}\{f(t)\}| \to 0$ and so $\mathscr{L}\{f(t)\} \to 0$. ■

EXAMPLE 7

The functions

$$F_1(s) = s^2 \qquad \text{and} \qquad F_2(s) = \frac{s}{s+1}$$

are not the Laplace transforms of piecewise continuous functions of exponential order since

$$F_1(s) \not\to 0 \qquad \text{and} \qquad F_2(s) \not\to 0,$$

as $s \to \infty$. We say that $\mathscr{L}^{-1}\{F_1(s)\}$ and $\mathscr{L}^{-1}\{F_2(s)\}$ do not exist.

Exercises 7.2

Answers to odd-numbered problems begin on page A–19.

In Problems 1–28 use Theorem 7.3 to find the given inverse transform.

1. $\mathscr{L}^{-1}\left\{\dfrac{1}{s^3}\right\}$

2. $\mathscr{L}^{-1}\left\{\dfrac{1}{s^4}\right\}$

3. $\mathscr{L}^{-1}\left\{\dfrac{(s+1)^3}{s^4}\right\}$

4. $\mathscr{L}^{-1}\left\{\dfrac{(s+2)^2}{s^3}\right\}$

5. $\mathscr{L}^{-1}\left\{\dfrac{1}{s^2} - \dfrac{1}{s} + \dfrac{1}{s-2}\right\}$

6. $\mathscr{L}^{-1}\left\{\dfrac{4}{s} + \dfrac{6}{s^5} - \dfrac{1}{s+8}\right\}$

7. $\mathscr{L}^{-1}\left\{\dfrac{1}{4s+1}\right\}$

8. $\mathscr{L}^{-1}\left\{\dfrac{1}{5s-2}\right\}$

9. $\mathscr{L}^{-1}\left\{\dfrac{5}{s^2+49}\right\}$

10. $\mathscr{L}^{-1}\left\{\dfrac{10s}{s^2+16}\right\}$

11. $\mathscr{L}^{-1}\left\{\dfrac{4s}{4s^2+1}\right\}$

12. $\mathscr{L}^{-1}\left\{\dfrac{1}{4s^2+1}\right\}$

13. $\mathscr{L}^{-1}\left\{\dfrac{1}{s^2-16}\right\}$

14. $\mathscr{L}^{-1}\left\{\dfrac{10s}{s^2-25}\right\}$

15. $\mathscr{L}^{-1}\left\{\dfrac{2s-6}{s^2+9}\right\}$

16. $\mathscr{L}^{-1}\left\{\dfrac{s+1}{s^2+2}\right\}$

17. $\mathscr{L}^{-1}\left\{\dfrac{1}{s^2+3s}\right\}$

18. $\mathscr{L}^{-1}\left\{\dfrac{s+1}{s^2-4s}\right\}$

19. $\mathscr{L}^{-1}\left\{\dfrac{s}{s^2+2s-3}\right\}$

20. $\mathscr{L}^{-1}\left\{\dfrac{1}{s^2+s-20}\right\}$

21. $\mathscr{L}^{-1}\left\{\dfrac{2s+4}{(s-2)(s^2+4s+3)}\right\}$

22. $\mathscr{L}^{-1}\left\{\dfrac{s+1}{(s^2-4s)(s+5)}\right\}$

23. $\mathscr{L}^{-1}\left\{\dfrac{1}{s^2(s^2+4)}\right\}$

24. $\mathscr{L}^{-1}\left\{\dfrac{s-1}{s^2(s^2+1)}\right\}$

25. $\mathscr{L}^{-1}\left\{\dfrac{s}{(s^2+4)(s+2)}\right\}$

26. $\mathscr{L}^{-1}\left\{\dfrac{1}{s^4-9}\right\}$

27. $\mathscr{L}^{-1}\left\{\dfrac{1}{(s^2+1)(s^2+4)}\right\}$

28. $\mathscr{L}^{-1}\left\{\dfrac{6s+3}{(s^2+1)(s^2+4)}\right\}$

The inverse Laplace transform may not be unique. In Problems 29 and 30 evaluate $\mathscr{L}\{f(t)\}$.

29. $f(t) = \begin{cases} 1, & t \geq 0, t \neq 1, t \neq 2 \\ 3, & t = 1 \\ 4, & t = 2 \end{cases}$

30. $f(t) = \begin{cases} e^{3t}, & t \geq 0, t \neq 5 \\ 1, & t = 5 \end{cases}$

7.3 Operational Properties

7.3.1 Translation Theorems and Derivatives of a Transform

It is not convenient to use Definition 7.1 each time we wish to find the Laplace transform of a function $f(t)$. For example, the integration by parts involved in evaluating, say, $\mathscr{L}\{e^t t^2 \sin 3t)$ is formidable to say the least. In the discussion that follows, we present several labor-saving theorems; and these, in turn, enable us to build up a more extensive list of transforms without the necessity of using the definition of the Laplace transform. Indeed, we shall see that evaluating transforms such as $\mathscr{L}\{e^{4t} \cos 6t\}$, $\mathscr{L}\{t^3 \sin 2t\}$, and $\mathscr{L}\{t^{10}e^{-t}\}$ is fairly straightforward, provided we know $\mathscr{L}\{\cos 6t\}$, $\mathscr{L}\{\sin 2t\}$, and $\mathscr{L}\{t^{10}\}$, respectively. Though extensive tables can be constructed, and we have included a table in Appendix III, it is nonetheless a good idea to know the Laplace transforms of basic functions such as t^n, e^{at}, $\sin kt$, $\cos kt$, $\sinh kt$, and $\cosh kt$.

First Translation Theorem

THEOREM 7.5 If a is any real number then

$$\mathscr{L}\{e^{at}f(t)\} = F(s-a),$$

where $F(s) = \mathscr{L}\{f(t)\}$.

PROOF The proof is immediate, since by Definition 7.1

$$\begin{aligned}\mathscr{L}\{e^{at}f(t)\} &= \int_0^\infty e^{-st}e^{at}f(t)\,dt \\ &= \int_0^\infty e^{-(s-a)t}f(t)\,dt \\ &= F(s-a).\end{aligned}$$ ■

Thus if we already know $\mathscr{L}\{f(t)\} = F(s)$, we can compute $\mathscr{L}\{e^{at}f(t)\}$ with no additional effort other than translating, or shifting, $F(s)$ to $F(s-a)$. For emphasis it is also sometimes useful to employ the symbolism

$$\mathscr{L}\{e^{at}f(t)\} = \mathscr{L}\{f(t)\}_{s \to s-a}.$$

Theorem 7.5 is known as the **first translation theorem**.

EXAMPLE 1

Evaluate **(a)** $\mathscr{L}\{e^{5t}t^3\}$, **(b)** $\mathscr{L}\{e^{-2t}\cos 4t\}$.

SOLUTION The results follow from Theorem 7.5.

(a) $\mathscr{L}\{e^{5t}t^3\} = \mathscr{L}\{t^3\}_{s\to s-5}$

$$= \left.\frac{3!}{s^4}\right|_{s\to s-5} = \frac{6}{(s-5)^4}.$$

(b) $\mathscr{L}\{e^{-2t}\cos 4t\} = \mathscr{L}\{\cos 4t\}_{s\to s+2}$ [*Note:* $a = -2$, so $s - a = s - (-2) = s + 2$.]

$$= \left.\frac{s}{s^2+16}\right|_{s\to s+2} = \frac{s+2}{(s+2)^2+16}.$$

The inverse form of Theorem 7.5 can be written

$$\begin{aligned} e^{at}f(t) &= \mathscr{L}^{-1}\{F(s-a)\} \\ &= \mathscr{L}^{-1}\{F(s)|_{s\to s-a}\}, \end{aligned} \tag{1}$$

where $f(t) = \mathscr{L}^{-1}\{F(s)\}$.

EXAMPLE 2

Evaluate $\mathscr{L}^{-1}\left\{\dfrac{s}{s^2+6s+11}\right\}$.

SOLUTION

$$\mathscr{L}^{-1}\left\{\frac{s}{s^2+6s+11}\right\} = \mathscr{L}^{-1}\left\{\frac{s}{(s+3)^2+2}\right\} \quad \text{[completion of square]}$$

$$= \mathscr{L}^{-1}\left\{\frac{s+3-3}{(s+3)^2+2}\right\} \quad \text{[adding zero in the numerator]}$$

$$= \mathscr{L}^{-1}\left\{\frac{s+3}{(s+3)^2+2} - \frac{3}{(s+3)^2+2}\right\} \quad \text{[termwise division]}$$

$$= \mathscr{L}^{-1}\left\{\frac{s+3}{(s+3)^2+2}\right\} - 3\mathscr{L}^{-1}\left\{\frac{1}{(s+3)^2+2}\right\}$$

$$= \mathscr{L}^{-1}\left\{\left.\frac{s}{s^2+2}\right|_{s\to s+3}\right\} - \frac{3}{\sqrt{2}}\mathscr{L}^{-1}\left\{\left.\frac{\sqrt{2}}{s^2+2}\right|_{s\to s+3}\right\}$$

$$= e^{-3t}\cos\sqrt{2}t - \frac{3}{\sqrt{2}}e^{-3t}\sin\sqrt{2}t. \quad \text{[from (1) and Theorem 7.3]}$$

EXAMPLE 3

Evaluate $\mathscr{L}^{-1}\left\{\dfrac{1}{(s-1)^3}+\dfrac{1}{s^2+2s-8}\right\}$.

SOLUTION

$$\begin{aligned}\mathscr{L}^{-1}\left\{\frac{1}{(s-1)^3}+\frac{1}{s^2+2s-8}\right\} &= \mathscr{L}^{-1}\left\{\frac{1}{(s-1)^3}+\frac{1}{(s+1)^2-9}\right\}\\ &= \frac{1}{2!}\mathscr{L}^{-1}\left\{\frac{2!}{(s-1)^3}\right\}+\frac{1}{3}\mathscr{L}^{-1}\left\{\frac{3}{(s+1)^2-9}\right\}\\ &= \frac{1}{2!}\mathscr{L}^{-1}\left\{\frac{2!}{s^3}\bigg|_{s\to s-1}\right\}+\frac{1}{3}\mathscr{L}^{-1}\left\{\frac{3}{s^2-9}\bigg|_{s\to s+1}\right\}\\ &= \frac{1}{2}e^t t^2+\frac{1}{3}e^{-t}\sinh 3t.\end{aligned}$$

Unit Step Function

In engineering one frequently encounters functions that can be either "on" or "off." For example, an external force acting on a mechanical system or a voltage impressed on a circuit can be turned off after a period of time. It is thus convenient to define a special function called the **unit step function**.

DEFINITION 7.2 The function $\mathscr{U}(t-a)$ is defined to be

$$\mathscr{U}(t-a)=\begin{cases}0, & 0\le t<a\\ 1, & t\ge a.\end{cases}$$

Notice that we define $\mathscr{U}(t-a)$ only on the nonnegative t-axis since this is all that we are concerned with in the study of the Laplace transform. In a broader sense, $\mathscr{U}(t-a)=0$ for $t<a$.

EXAMPLE 4

Graph **(a)** $\mathscr{U}(t)$, **(b)** $\mathscr{U}(t-2)$.

SOLUTION **(a)** $\mathscr{U}(t)=1, \quad t\ge 0$ **(b)** $\mathscr{U}(t-2)=\begin{cases}0, & 0\le t<2\\ 1, & t\ge 2.\end{cases}$

The respective graphs are given in Figure 7.6.

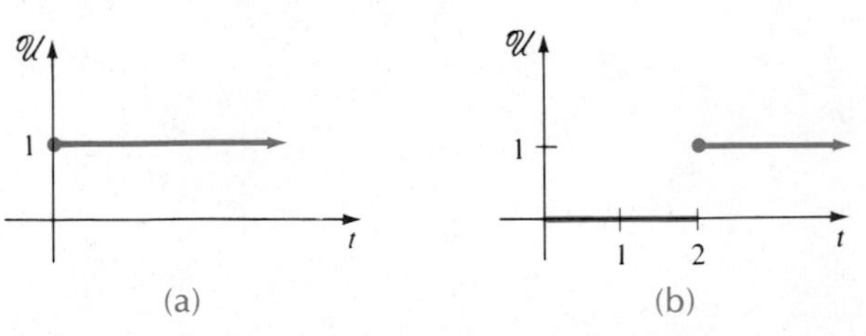

Figure 7.6

When combined with other functions defined for $t \geq 0$, the unit step function "turns off" a portion of their graphs. For example, Figure 7.7 illustrates the graph $y = f(t)$, where

$$f(t) = \sin t\,\mathscr{U}(t - 2\pi), \qquad t \geq 0.$$

$$= \begin{cases} 0, & 0 \leq t < 2\pi \\ \sin t, & t \geq 2\pi. \end{cases}$$

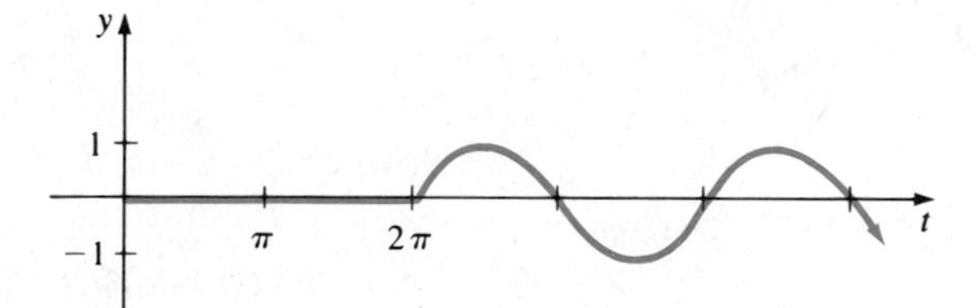

Figure 7.7

EXAMPLE 5

Consider the function $y = f(t)$ defined by $f(t) = t^3$. Compare the graphs of

(a) $f(t) = t^3$, **(b)** $f(t) = t^3, t \geq 0$,

(c) $f(t - 2), t \geq 0$, **(d)** $f(t - 2)\mathscr{U}(t - 2), t \geq 0$.

SOLUTION The respective graphs are given in Figure 7.8.

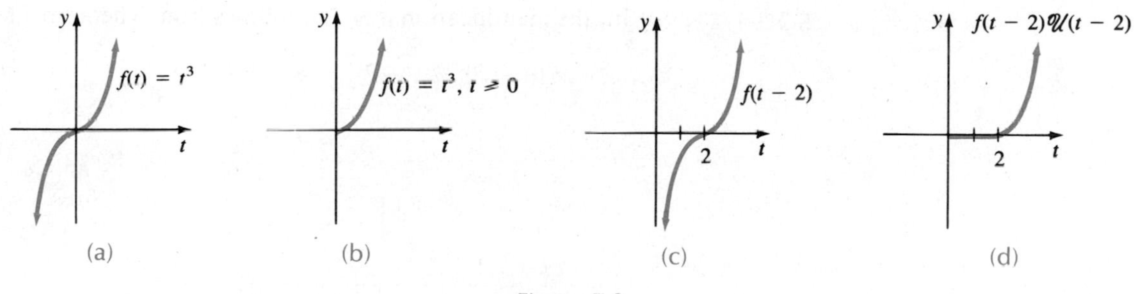

Figure 7.8

We have already seen in Theorem 7.5 that an exponential multiple of $f(t)$ results in a shift or translation of the transform $F(s)$. In the next theorem we see that whenever $F(s)$ is multiplied by an appropriate exponential function, not only is the graph of $f(t)$ translated, but a portion of the graph is turned off as well.

Second Translation Theorem

THEOREM 7.6 If $a > 0$, then

$$\mathscr{L}\{f(t - a)\mathscr{U}(t - a)\} = e^{-as}\mathscr{L}\{f(t)\} = e^{-as}F(s).$$

PROOF From Definition 7.1

$$\mathscr{L}\{f(t-a)\mathscr{U}(t-a)\} = \int_0^\infty e^{-st} f(t-a)\mathscr{U}(t-a)\,dt$$

$$= \int_0^a e^{-st} f(t-a)\underbrace{\mathscr{U}(t-a)}_{\substack{\text{zero for} \\ 0 \le t < a}}\,dt + \int_a^\infty e^{-st} f(t-a)\underbrace{\mathscr{U}(t-a)}_{\substack{\text{one for} \\ t \ge a}}\,dt$$

$$= \int_a^\infty e^{-st} f(t-a)\,dt.$$

Now let $v = t - a$, $dv = dt$; then

$$\mathscr{L}\{f(t-a)\mathscr{U}(t-a)\} = \int_0^\infty e^{-s(v+a)} f(v)\,dv$$

$$= e^{-as}\int_0^\infty e^{-sv} f(v)\,dv = e^{-as}\mathscr{L}\{f(t)\}. \quad \blacksquare$$

Theorem 7.6 is known as the **second translation theorem**.

EXAMPLE 6

Evaluate $\mathscr{L}\{(t-2)^3\mathscr{U}(t-2)\}$.

SOLUTION With the identification $a = 2$, it follows from Theorem 7.6 that

$$\mathscr{L}\{(t-2)^3\mathscr{U}(t-2)\} = e^{-2s}\mathscr{L}\{t^3\}$$

$$= e^{-2s}\frac{3!}{s^4}$$

$$= \frac{6}{s^4}e^{-2s}.$$

EXAMPLE 7

Evaluate $\mathscr{L}\{\mathscr{U}(t-5)\}$.

SOLUTION Making the identifications $f(t) = 1$ and $a = 5$ in Theorem 7.6, we have

$$\mathscr{L}\{\mathscr{U}(t-5)\} = e^{-5s}\mathscr{L}\{1\}$$

$$= \frac{e^{-5s}}{s}.$$

EXAMPLE 8

Find the Laplace transform of the function shown in Figure 7.9.

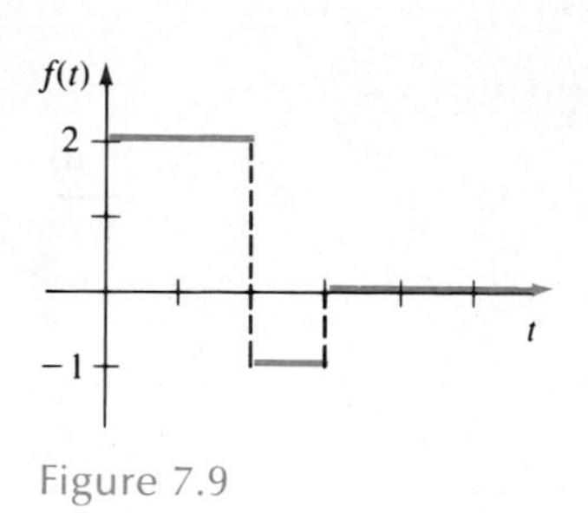

Figure 7.9

SOLUTION With the aid of the unit step function, we can write

$$f(t) = 2 - 3\mathscr{U}(t-2) + \mathscr{U}(t-3).$$

Hence from Theorem 7.6 it follows that

$$\begin{aligned}\mathscr{L}\{f(t)\} &= \mathscr{L}\{2\} - 3\mathscr{L}\{\mathscr{U}(t-2)\} + \mathscr{L}\{\mathscr{U}(t-3)\}\\ &= \frac{2}{s} - 3\frac{e^{-2s}}{s} + \frac{e^{-3s}}{s}.\end{aligned}$$

EXAMPLE 9

Evaluate $\mathscr{L}\{\sin t\,\mathscr{U}(t-2\pi)\}$.

SOLUTION With $a = 2\pi$ we have from Theorem 7.6

$$\begin{aligned}\mathscr{L}\{\sin t\,\mathscr{U}(t-2\pi)\} &= \mathscr{L}\{\sin(t-2\pi)\mathscr{U}(t-2\pi)\} && [\sin t \text{ has period } 2\pi]\\ &= e^{-2\pi s}\mathscr{L}\{\sin t\}\\ &= \frac{e^{-2\pi s}}{s^2+1}.\end{aligned}$$

EXAMPLE 10

Find the Laplace transform of the function shown in Figure 7.10.

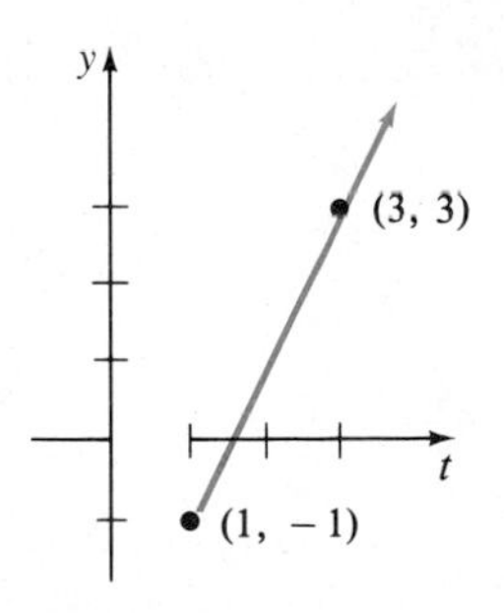

Figure 7.10

SOLUTION An equation of the straight line is found to be $y = 2t - 3$. To "turn off" this graph on the interval $0 \le t < 1$ we form $(2t-3)\mathscr{U}(t-1)$. Now Theorem 7.6 is not immediately applicable in the evaluation of $\mathscr{L}\{(2t-3)\mathscr{U}(t-1)\}$, since the function being transformed lacks the precise form $f(t-a)\mathscr{U}(t-a)$. However, with the little "trick"

$$2t - 3 = 2(t-1) - 1,$$

we can identify $f(t-1) = 2(t-1) - 1$ and consequently $f(t) = 2t - 1$. We are now in a position to apply the second translation theorem:

$$\begin{aligned}\mathscr{L}\{(2t-3)\mathscr{U}(t-1)\} &= \mathscr{L}\{(2(t-1)-1)\mathscr{U}(t-1)\}\\ &= e^{-s}\mathscr{L}\{2t-1\}\\ &= e^{-s}\left(\frac{2}{s^2} - \frac{1}{s}\right).\end{aligned}$$

The inverse form of Theorem 7.6 is

$$f(t-a)\mathscr{U}(t-a) = \mathscr{L}^{-1}\{e^{-as}F(s)\}, \tag{2}$$

where $a > 0$ and $f(t) = \mathscr{L}^{-1}\{F(s)\}$.

EXAMPLE 11 Evaluate $\mathscr{L}^{-1}\left\{\dfrac{e^{-\pi s/2}}{s^2+9}\right\}$.

SOLUTION We identify $a = \dfrac{\pi}{2}$ and $f(t) = \mathscr{L}^{-1}\left\{\dfrac{1}{s^2+9}\right\} = \dfrac{1}{3}\sin 3t$. Thus from (2)

$$\begin{aligned}\mathscr{L}^{-1}\left\{\frac{e^{-\pi s/2}}{s^2+9}\right\} &= \frac{1}{3}\mathscr{L}^{-1}\left\{\frac{3}{s^2+9}\right\}_{t\to t-\pi/2}\mathscr{U}\left(t-\frac{\pi}{2}\right)\\ &= \frac{1}{3}\sin 3\left(t-\frac{\pi}{2}\right)\mathscr{U}\left(t-\frac{\pi}{2}\right)\\ &= \frac{1}{3}\cos 3t\,\mathscr{U}\left(t-\frac{\pi}{2}\right).\end{aligned}$$

If $F(s) = \mathscr{L}\{f(t)\}$ and if we assume that interchanging of differentiation and integration is possible, then

$$\begin{aligned}\frac{d}{ds}F(s) &= \frac{d}{ds}\int_0^\infty e^{-st}f(t)\,dt\\ &= \int_0^\infty \frac{\partial}{\partial s}[e^{-st}f(t)\,dt]\\ &= -\int_0^\infty e^{-st}tf(t)\,dt\\ &= -\mathscr{L}\{tf(t)\};\end{aligned}$$

that is, $\quad \mathscr{L}\{tf(t)\} = -\dfrac{d}{ds}\mathscr{L}\{f(t)\}.$

Similarly, $\quad \mathscr{L}\{t^2f(t)\} = \mathscr{L}\{t\cdot tf(t)\} = -\dfrac{d}{ds}\mathscr{L}\{tf(t)\}$

$$= -\frac{d}{ds}\left(-\frac{d}{ds}\mathscr{L}\{f(t)\}\right) = \frac{d^2}{ds^2}\mathscr{L}\{f(t)\}.$$

The preceding two cases suggest the following result.

Derivatives of a Transform

THEOREM 7.7 For $n = 1, 2, 3, \ldots$

$$\mathscr{L}\{t^n f(t)\} = (-1)^n \frac{d^n}{ds^n} \mathscr{L}\{f(t)\} = (-1)^n \frac{d^n}{ds^n} F(s),$$

where $F(s) = \mathscr{L}\{f(t)\}$.

EXAMPLE 12

Evaluate **(a)** $\mathscr{L}\{te^{3t}\}$, **(b)** $\mathscr{L}\{t \sin kt\}$, **(c)** $\mathscr{L}\{t^2 \sin kt\}$, **(d)** $\mathscr{L}\{te^{-t} \cos t\}$.

SOLUTION The results follow from Theorem 7.7.

(a) $$\mathscr{L}\{te^{3t}\} = -\frac{d}{ds}\mathscr{L}\{e^{3t}\} = -\frac{d}{ds}\left(\frac{1}{s-3}\right) = \frac{1}{(s-3)^2}$$

(b) $$\mathscr{L}\{t \sin kt\} = -\frac{d}{ds}\mathscr{L}\{\sin kt\} = -\frac{d}{ds}\left(\frac{k}{s^2+k^2}\right) = \frac{2ks}{(s^2+k^2)^2}$$

(c) $$\mathscr{L}\{t^2 \sin kt\} = \frac{d^2}{ds^2}\mathscr{L}\{\sin kt\}$$

$$= -\frac{d}{ds}\mathscr{L}\{t \sin kt\}$$

$$= -\frac{d}{ds}\left[\frac{2ks}{(s^2+k^2)^2}\right] \quad \text{[from part (b)]}$$

$$= -\frac{(s^2+k^2)^2 2k - 8ks^2(s^2+k^2)}{(s^2+k^2)^4}$$

$$= \frac{6ks^2 - 2k^3}{(s^2+k^2)^3}$$

(d) $$\mathscr{L}\{te^{-t}\cos t\} = -\frac{d}{ds}\mathscr{L}\{e^{-t}\cos t\}$$
$$= -\frac{d}{ds}\mathscr{L}\{\cos t\}_{s\to s+1}$$
$$= -\frac{d}{ds}\left[\frac{s+1}{(s+1)^2+1}\right]$$
$$= \frac{(s+1)^2-1}{[(s+1)^2+1]^2}$$

Of course, part (a) of Example 12 can also be obtained from Theorem 7.5.

7.3.2 Transforms of Derivatives and Integrals

Our goal is to use the Laplace transform to solve certain kinds of differential equations. To that end we need to evaluate quantities such as $\mathscr{L}\{dy/dt\}$ and $\mathscr{L}\{d^2y/dt^2\}$. For example, if f' is continuous for $t \geq 0$ then

$$\begin{aligned}
\mathscr{L}\{f'(t)\} &= \int_0^\infty e^{-st}f'(t)\,dt \\
&= e^{-st}f(t)\Big|_0^\infty + s\int_0^\infty e^{-st}f(t)\,dt \\
&= -f(0) + s\mathscr{L}\{f(t)\} \\
&= sF(s) - f(0).
\end{aligned}$$

Here we have assumed that $e^{-st}f(t) \to 0$ as $t \to \infty$. Similarly,

$$\begin{aligned}
\mathscr{L}\{f''(t)\} &= \int_0^\infty e^{-st}f''(t)\,dt \\
&= e^{-st}f'(t)\Big|_0^\infty + s\int_0^\infty e^{-st}f'(t)\,dt \\
&= -f'(0) + s\mathscr{L}\{f'(t)\} \\
&= s[sF(s) - f(0)] - f'(0) \\
&= s^2F(s) - sf(0) - f'(0).
\end{aligned}$$

We state the general case in the next theorem. The proof is omitted.

Transform of a Derivative

THEOREM 7.8 If $f(t), f'(t), \ldots, f^{(n-1)}(t)$ are continuous on $[0, \infty)$ and are of exponential order and if $f^{(n)}(t)$ is piecewise continuous on $[0, \infty)$, then

$$\mathscr{L}\{f^{(n)}(t)\} = s^nF(s) - s^{n-1}f(0) - s^{n-2}f'(0) - \cdots - f^{(n-1)}(0),$$

where $F(s) = \mathscr{L}\{f(t)\}$.

EXAMPLE 13

$$\begin{aligned}
\mathscr{L}\{kt\cos kt + \sin kt\} &= \mathscr{L}\left\{\frac{d}{dt}(t\sin kt)\right\} \\
&= s\mathscr{L}\{t\sin kt\} && \text{[Theorem 7.8]} \\
&= s\left[-\frac{d}{ds}\mathscr{L}\{\sin kt\}\right] && \text{[Theorem 7.7]} \\
&= s\left[\frac{2ks}{(s^2+k^2)^2}\right] \\
&= \frac{2ks^2}{(s^2+k^2)^2}.
\end{aligned}$$

Convolution

If functions f and g are piecewise continuous on $[0, \infty)$, then the **convolution** of f and g, denoted by $f * g$, is given by the integral

$$f * g = \int_0^t f(\tau)g(t-\tau)\,d\tau.$$

EXAMPLE 14

The convolution of $f(t) = e^t$ and $g(t) = \sin t$ is

$$e^t * \sin t = \int_0^t e^\tau \sin(t-\tau)\,d\tau \tag{3}$$

$$= \frac{1}{2}(-\sin t - \cos t + e^t). \tag{4}$$

Convolution Theorem

It is possible to find the Laplace transform of the convolution of two functions, such as (3), without actually evaluating the integral as we did in (4). The result that follows is known as the **convolution theorem**.

> **THEOREM 7.9** Let $f(t)$ and $g(t)$ be piecewise continuous on $[0, \infty)$ and of exponential order; then
>
> $$\mathscr{L}\{f * g\} = \mathscr{L}\{f(t)\}\mathscr{L}\{g(t)\} = F(s)G(s).$$

PROOF Let

$$F(s) = \mathscr{L}\{f(t)\} = \int_0^\infty e^{-s\tau} f(\tau)\, d\tau$$

$$G(s) = \mathscr{L}\{g(t)\} = \int_0^\infty e^{-s\beta} g(\beta)\, d\beta.$$

Proceeding formally, we have

$$\begin{aligned} F(s)G(s) &= \left(\int_0^\infty e^{-s\tau} f(\tau)\, d\tau\right)\left(\int_0^\infty e^{-s\beta} g(\beta)\, d\beta\right) \\ &= \int_0^\infty \int_0^\infty e^{-s(\tau+\beta)} f(\tau) g(\beta)\, d\tau\, d\beta \\ &= \int_0^\infty f(\tau)\, d\tau \int_0^\infty e^{-s(\tau+\beta)} g(\beta)\, d\beta. \end{aligned}$$

Holding τ fixed, we let $t = \tau + \beta$, $dt = d\beta$ so that

$$F(s)G(s) = \int_0^\infty f(\tau)\, d\tau \int_\tau^\infty e^{-st} g(t - \tau)\, dt.$$

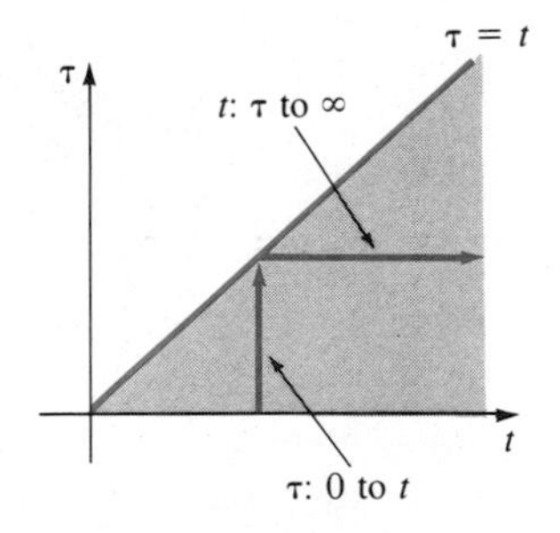

Figure 7.11

In the $t\tau$ plane we are integrating over the shaded region in Figure 7.11. Since f and g are piecewise continuous on $[0, \infty)$ and of exponential order, it is possible to interchange the order of integration:

$$\begin{aligned} F(s)G(s) &= \int_0^\infty e^{-st}\, dt \int_0^t f(\tau) g(t - \tau)\, d\tau \\ &= \int_0^\infty e^{-st} \left\{ \int_0^t f(\tau) g(t - \tau)\, d\tau \right\} dt \\ &= \mathscr{L}\{f * g\}. \end{aligned}$$

■

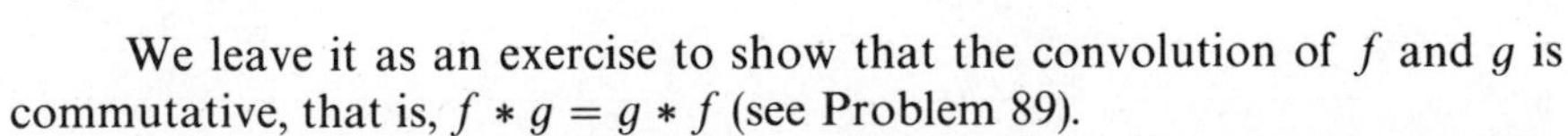

We leave it as an exercise to show that the convolution of f and g is commutative, that is, $f * g = g * f$ (see Problem 89).

When $g(t) = 1$ and $G(s) = 1/s$, the convolution theorem implies that the Laplace transform of the integral of a function f is

$$\mathscr{L}\left\{\int_0^t f(\tau)\, d\tau\right\} = \frac{F(s)}{s}. \tag{5}$$

EXAMPLE 15

Evaluate $\mathscr{L}\left\{\int_0^t e^\tau \sin(t-\tau)\, d\tau\right\}$.

SOLUTION With $f(t) = e^t$ and $g(t) = \sin t$, we have from Theorem 7.9

$$\mathscr{L}\left\{\int_0^t e^\tau \sin(t-\tau)\, d\tau\right\} = \mathscr{L}\{e^t\} \cdot \mathscr{L}\{\sin t\}$$

$$= \frac{1}{s-1} \cdot \frac{1}{s^2+1}$$

$$= \frac{1}{(s-1)(s^2+1)}.$$

The convolution theorem is sometimes useful in finding the inverse Laplace transform of a product of two Laplace transforms. From Theorem 7.9 we have

$$f * g = \mathscr{L}^{-1}\{F(s)G(s)\}. \tag{6}$$

EXAMPLE 16

Evaluate $\mathscr{L}^{-1}\left\{\dfrac{1}{(s-1)(s+4)}\right\}$.

SOLUTION Admittedly partial fractions could be used, but if we identify

$$F(s) = \frac{1}{s-1} \quad \text{and} \quad G(s) = \frac{1}{s+4},$$

then $\mathscr{L}^{-1}\{F(s)\} = f(t) = e^t$ and $\mathscr{L}^{-1}\{G(s)\} = g(t) = e^{-4t}$.

Hence from (6) we obtain

$$\mathscr{L}^{-1}\left\{\frac{1}{(s-1)(s+4)}\right\} = \int_0^t f(\tau)g(t-\tau)\, d\tau = \int_0^t e^\tau e^{-4(t-\tau)}\, d\tau$$

$$= e^{-4t}\int_0^t e^{5\tau}\, d\tau$$

$$= e^{-4t}\frac{1}{5}e^{5\tau}\Big|_0^t$$

$$= \frac{1}{5}e^t - \frac{1}{5}e^{-4t}.$$

EXAMPLE 17 Evaluate $\mathscr{L}^{-1}\left\{\dfrac{1}{(s^2+k^2)^2}\right\}$.

SOLUTION Let
$$F(s) = G(s) = \frac{1}{s^2+k^2}$$

so that
$$f(t) = g(t) = \frac{1}{k}\mathscr{L}^{-1}\left\{\frac{k}{s^2+k^2}\right\} = \frac{1}{k}\sin kt.$$

Thus
$$\mathscr{L}^{-1}\left\{\frac{1}{(s^2+k^2)^2}\right\} = \frac{1}{k^2}\int_0^t \sin k\tau \sin k(t-\tau)\,d\tau.$$

Now recall from trigonometry that

$$\cos(A+B) = \cos A\cos B - \sin A \sin B$$

and
$$\cos(A-B) = \cos A\cos B + \sin A \sin B.$$

Subtracting the first from the second gives the identity

$$\sin A \sin B = \frac{1}{2}[\cos(A-B) - \cos(A+B)].$$

If we set $A = k\tau$ and $B = k(t-\tau)$, it follows that

$$\begin{aligned}\mathscr{L}^{-1}\left\{\frac{1}{(s^2+k^2)^2}\right\} &= \frac{1}{2k^2}\int_0^t [\cos k(2\tau - t) - \cos kt]\,d\tau\\ &= \frac{1}{2k^2}\left[\frac{1}{2k}\sin k(2\tau-t) - \tau\cos kt\right]_0^t\\ &= \frac{\sin kt - kt\cos kt}{2k^3}.\end{aligned}$$

7.3.3 Transform of a Periodic Function

If a periodic function has period T, $T > 0$, then $f(t+T) = f(t)$. The Laplace transform of a periodic function can be obtained by an integration over one period.

THEOREM 7.10 Let $f(t)$ be piecewise continuous on $[0, \infty)$ and of exponential order. If $f(t)$ is periodic with period T, then

$$\mathscr{L}\{f(t)\} = \frac{1}{1-e^{-sT}}\int_0^T e^{-st}f(t)\,dt. \qquad (7)$$

PROOF Write the Laplace transform as

$$\mathscr{L}\{f(t)\} = \int_0^T e^{-st} f(t)\,dt + \int_T^\infty e^{-st} f(t)\,dt. \tag{8}$$

By letting $t = u + T$, the last integral in (8) becomes

$$\begin{aligned}\int_T^\infty e^{-st} f(t)\,dt &= \int_0^\infty e^{-s(u+T)} f(u+T)\,du \\ &= e^{-sT}\int_0^\infty e^{-su} f(u)\,du \\ &= e^{-sT}\mathscr{L}\{f(t)\}.\end{aligned}$$

Hence (8) is

$$\mathscr{L}\{f(t)\} = \int_0^T e^{-st} f(t)\,dt + e^{-sT}\mathscr{L}\{f(t)\}.$$

Solving for $\mathscr{L}\{f(t)\}$ yields

$$\mathscr{L}\{f(t)\} = \frac{1}{1 - e^{-sT}}\int_0^T e^{-st} f(t)\,dt.$$

■

EXAMPLE 18

Find the Laplace transform of the periodic function shown in Figure 7.12.

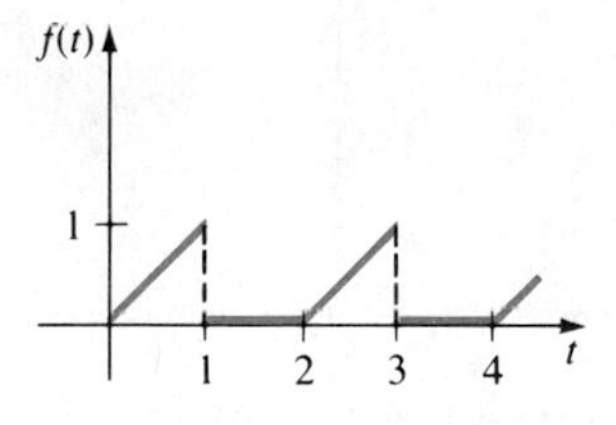

Figure 7.12

SOLUTION On the interval $0 \le t < 2$ the function can be defined by

$$f(t) = \begin{cases} t, & 0 \le t < 1 \\ 0, & 1 \le t < 2 \end{cases}$$

and outside the interval by $f(t + 2) = f(t)$. Identifying $T = 2$, we use (7) and integration by parts to obtain

$$\begin{aligned}\mathscr{L}\{f(t)\} &= \frac{1}{1 - e^{-2s}}\int_0^2 e^{-st} f(t)\,dt \\ &= \frac{1}{1 - e^{-2s}}\left[\int_0^1 e^{-st} t\,dt + \int_1^2 e^{-st} 0\,dt\right] \\ &= \frac{1}{1 - e^{-2s}}\left[-\frac{e^{-s}}{s} + \frac{1 - e^{-s}}{s^2}\right] \\ &= \frac{1 - (s+1)e^{-s}}{s^2(1 - e^{-2s})}.\end{aligned}$$

Although we already know the results, $\mathscr{L}\{\sin t\}$ and $\mathscr{L}\{\cos t\}$ can also be obtained from (7).

Exercises 7.3

Answers to odd-numbered problems begin on page A–19.

[7.3.1] In Problems 1–44 find either $F(s)$ or $f(t)$ as indicated.

1. $\mathscr{L}\{te^{10t}\}$

2. $\mathscr{L}\{te^{-6t}\}$

3. $\mathscr{L}\{t^3e^{-2t}\}$

4. $\mathscr{L}\{t^{10}e^{-7t}\}$

5. $\mathscr{L}\{e^t \sin 3t\}$

6. $\mathscr{L}\{e^{-2t} \cos 4t\}$

7. $\mathscr{L}\{e^{5t} \sinh 3t\}$

8. $\mathscr{L}\left\{\dfrac{\cosh t}{e^t}\right\}$

9. $\mathscr{L}\{t(e^t + e^{2t})^2\}$

10. $\mathscr{L}\{e^{2t}(t - 1)^2\}$

11. $\mathscr{L}\{e^{-t} \sin^2 t\}$

12. $\mathscr{L}\{e^t \cos^2 3t\}$

13. $\mathscr{L}^{-1}\left\{\dfrac{1}{(s + 2)^3}\right\}$

14. $\mathscr{L}^{-1}\left\{\dfrac{1}{(s - 1)^4}\right\}$

15. $\mathscr{L}^{-1}\left\{\dfrac{1}{s^2 - 6s + 10}\right\}$

16. $\mathscr{L}^{-1}\left\{\dfrac{1}{s^2 + 2s + 5}\right\}$

17. $\mathscr{L}^{-1}\left\{\dfrac{s}{s^2 + 4s + 5}\right\}$

18. $\mathscr{L}^{-1}\left\{\dfrac{2s + 5}{s^2 + 6s + 34}\right\}$

19. $\mathscr{L}^{-1}\left\{\dfrac{s}{(s + 1)^2}\right\}$

20. $\mathscr{L}^{-1}\left\{\dfrac{5s}{(s - 2)^2}\right\}$

21. $\mathscr{L}^{-1}\left\{\dfrac{2s - 1}{s^2(s + 1)^3}\right\}$

22. $\mathscr{L}^{-1}\left\{\dfrac{(s + 1)^2}{(s + 2)^4}\right\}$

23. $\mathscr{L}\{(t - 1)\mathscr{U}(t - 1)\}$

24. $\mathscr{L}\{e^{2-t}\mathscr{U}(t - 2)\}$

25. $\mathscr{L}\{t\mathscr{U}(t - 2)\}$

26. $\mathscr{L}\{(3t + 1)\mathscr{U}(t - 3)\}$

27. $\mathscr{L}\{\cos 2t\mathscr{U}(t - \pi)\}$

28. $\mathscr{L}\left\{\sin t\mathscr{U}\left(t - \dfrac{\pi}{2}\right)\right\}$

29. $\mathscr{L}\{(t - 1)^3e^{t-1}\mathscr{U}(t - 1)\}$

30. $\mathscr{L}\{te^{t-5}\mathscr{U}(t - 5)\}$

31. $\mathscr{L}^{-1}\left\{\dfrac{e^{-2s}}{s^3}\right\}$

32. $\mathscr{L}^{-1}\left\{\dfrac{(1 + e^{-2s})^2}{s + 2}\right\}$

33. $\mathscr{L}^{-1}\left\{\dfrac{e^{-\pi s}}{s^2 + 1}\right\}$

34. $\mathscr{L}^{-1}\left\{\dfrac{se^{-\pi s/2}}{s^2 + 4}\right\}$

35. $\mathscr{L}^{-1}\left\{\dfrac{e^{-s}}{s(s+1)}\right\}$

36. $\mathscr{L}^{-1}\left\{\dfrac{e^{-2s}}{s^2(s-1)}\right\}$

37. $\mathscr{L}\{t\cos 2t\}$

38. $\mathscr{L}\{t\sinh 3t\}$

39. $\mathscr{L}\{t^2\sinh t\}$

40. $\mathscr{L}\{t^2\cos t\}$

41. $\mathscr{L}\{te^{2t}\sin 6t\}$

42. $\mathscr{L}\{te^{-3t}\cos 3t\}$

43. $\mathscr{L}^{-1}\left\{\dfrac{s}{(s^2+1)^2}\right\}$

44. $\mathscr{L}^{-1}\left\{\dfrac{s+1}{(s^2+2s+2)^2}\right\}$

In Problems 45–52 write each function in terms of unit step functions. Find the Laplace transform of the given function.

45. $f(t) = \begin{cases} 2, & 0 \le t < 3 \\ -2, & t \ge 3 \end{cases}$

46. $f(t) = \begin{cases} 1, & 0 \le t < 4 \\ 0, & 4 \le t < 5 \\ 1, & t \ge 5 \end{cases}$

47. $f(t) = \begin{cases} 0, & 0 \le t < 1 \\ t^2, & t \ge 1 \end{cases}$

48. $f(t) = \begin{cases} 0, & 0 \le t < \dfrac{3\pi}{2} \\ \sin t, & t \ge \dfrac{3\pi}{2} \end{cases}$

49. $f(t) = \begin{cases} t, & 0 \le t < 2 \\ 0, & t \ge 2 \end{cases}$

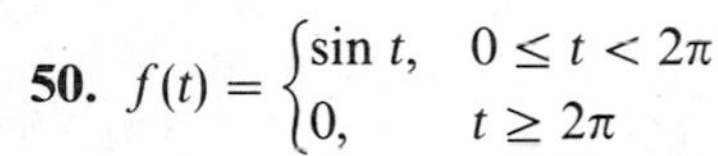

50. $f(t) = \begin{cases} \sin t, & 0 \le t < 2\pi \\ 0, & t \ge 2\pi \end{cases}$

51.

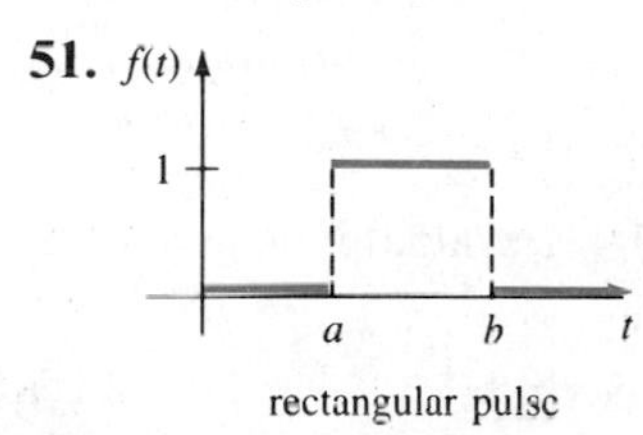

Figure 7.13

52.

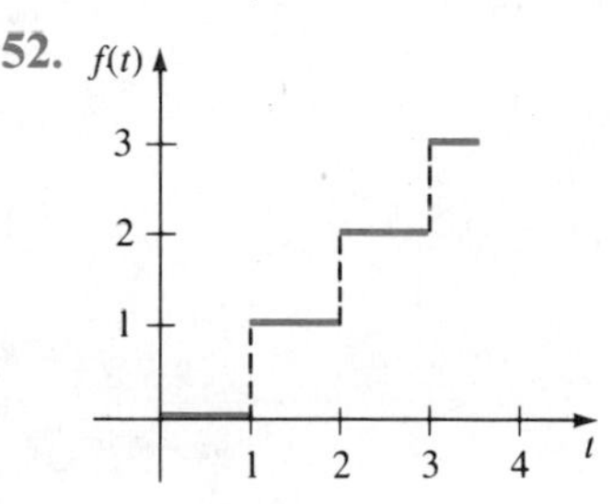

Figure 7.14

In Problems 53 and 54 sketch the graph of the given function.

53. $f(t) = \mathscr{L}^{-1}\left\{\dfrac{1}{s^2} - \dfrac{e^{-s}}{s^2}\right\}$

54. $f(t) = \mathscr{L}^{-1}\left\{\dfrac{2}{s} - \dfrac{3e^{-s}}{s^2} + \dfrac{5e^{-2s}}{s^2}\right\}$

In Problems 55–58 use Theorem 7.7 in the form ($n = 1$)

$$f(t) = -\frac{1}{t}\,\mathscr{L}^{-1}\left\{\frac{d}{ds}\,F(s)\right\}$$

to evaluate the given inverse Laplace transform.

EXAMPLE 19 Evaluate $\mathscr{L}^{-1}\left\{\tan^{-1}\frac{1}{s}\right\}$.

SOLUTION

$$\begin{aligned} f(t) &= -\frac{1}{t}\mathscr{L}^{-1}\left\{\frac{d}{ds}\tan^{-1}\frac{1}{s}\right\} \\ &= -\frac{1}{t}\mathscr{L}^{-1}\left\{\frac{1}{1+(1/s)^2}\cdot(-s^{-2})\right\} \\ &= \frac{1}{t}\mathscr{L}^{-1}\left\{\frac{1}{s^2+1}\right\} \\ &= \frac{\sin t}{t}. \end{aligned}$$

55. $\mathscr{L}^{-1}\left\{\ln\frac{s-3}{s+1}\right\}$

56. $\mathscr{L}^{-1}\left\{\ln\frac{s^2+1}{s^2+4}\right\}$

57. $\mathscr{L}^{-1}\left\{\frac{\pi}{2}-\tan^{-1}\frac{s}{2}\right\}$

58. $\mathscr{L}^{-1}\left\{\frac{1}{s}-\cot^{-1}\frac{4}{s}\right\}$

[7.3.2] **59.** Use Theorem 7.8 to evaluate $\mathscr{L}\{e^t\}$.

60. Use Theorem 7.8 to evaluate $\mathscr{L}\{\cos^2 t\}$. [*Hint:* If $f(t) = \cos^2 t$ then $f'(t) = -\sin 2t$.]

In Problems 61–72 evaluate the given Laplace transform.

61. $\mathscr{L}\left\{\int_0^t e^{-\tau}\cos\tau\,d\tau\right\}$

62. $\mathscr{L}\left\{\int_0^t \tau\sin\tau\,d\tau\right\}$

63. $\mathscr{L}\left\{\int_0^t \tau e^{t-\tau}\,d\tau\right\}$

64. $\mathscr{L}\left\{\int_0^t \sin\tau\cos(t-\tau)\,d\tau\right\}$

65. $\mathscr{L}\left\{t\int_0^t \sin\tau\,d\tau\right\}$

66. $\mathscr{L}\left\{t\int_0^t \tau e^{-\tau}\,d\tau\right\}$

67. $\mathscr{L}\{1 * t^3\}$

68. $\mathscr{L}\{1 * e^{-2t}\}$

69. $\mathscr{L}\{t^2 * t^4\}$

70. $\mathscr{L}\{t^2 * te^t\}$

71. $\mathscr{L}\{e^{-t} * e^t\cos t\}$

72. $\mathscr{L}\{e^{2t} * \sin t\}$

In Problems 73 and 74 suppose $\mathscr{L}^{-1}\{F(s)\} = f(t)$. Find the inverse Laplace transform of the given function.

73. $\frac{1}{s+5}F(s)$

74. $\frac{s}{s^2+4}F(s)$

In Problems 75–80 use (6) to find $f(t)$.

75. $\mathscr{L}^{-1}\left\{\dfrac{1}{s(s+1)}\right\}$

76. $\mathscr{L}^{-1}\left\{\dfrac{1}{s(s^2+1)}\right\}$

77. $\mathscr{L}^{-1}\left\{\dfrac{1}{(s+1)(s-2)}\right\}$

78. $\mathscr{L}^{-1}\left\{\dfrac{1}{(s+1)^2}\right\}$

79. $\mathscr{L}^{-1}\left\{\dfrac{s}{(s^2+4)^2}\right\}$

80. $\mathscr{L}^{-1}\left\{\dfrac{1}{(s^2+4s+5)^2}\right\}$

[7.3.3] In Problems 81–88 use Theorem 7.10 to find the Laplace transform of the given periodic function.

81.

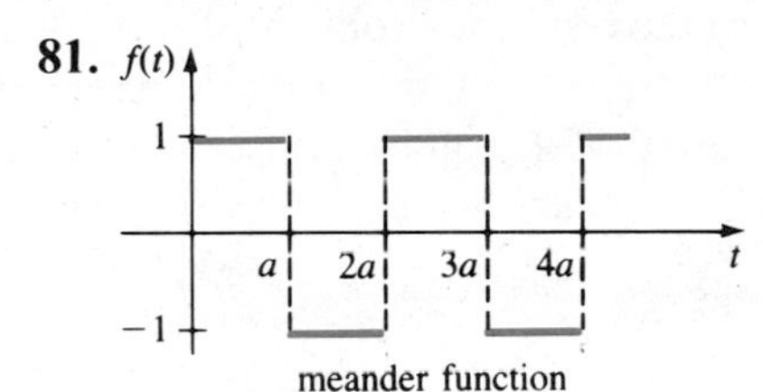

Figure 7.15

82.

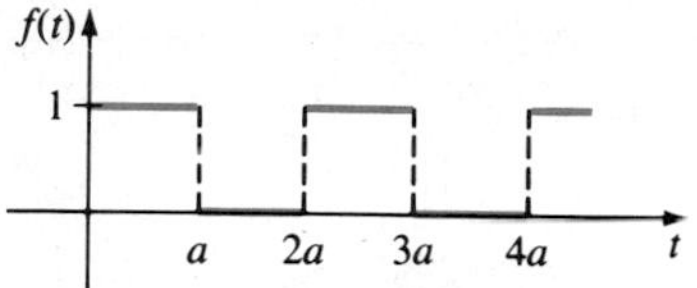

f(t)
1
a 2a 3a 4a t
square wave

Figure 7.16

83.

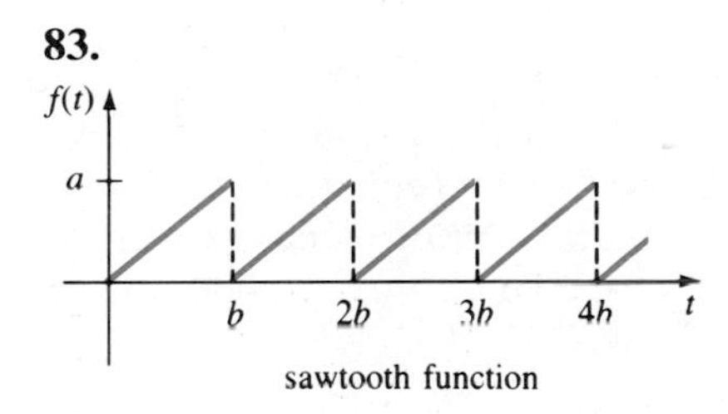

Figure 7.17

84.

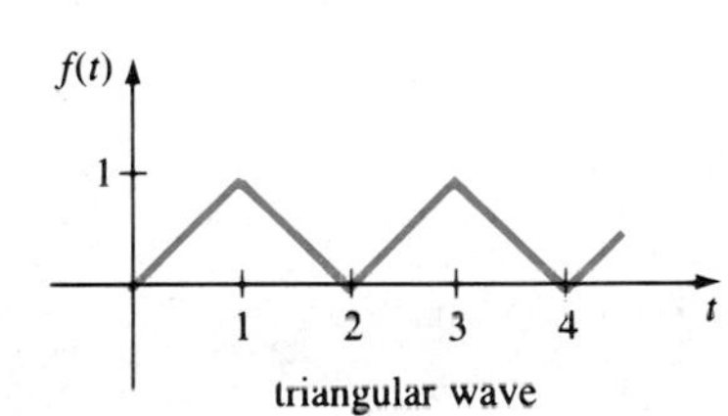

Figure 7.18

85.

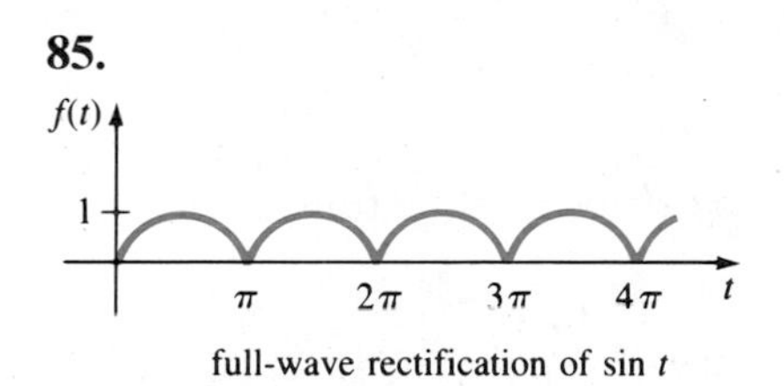

Figure 7.19

86.

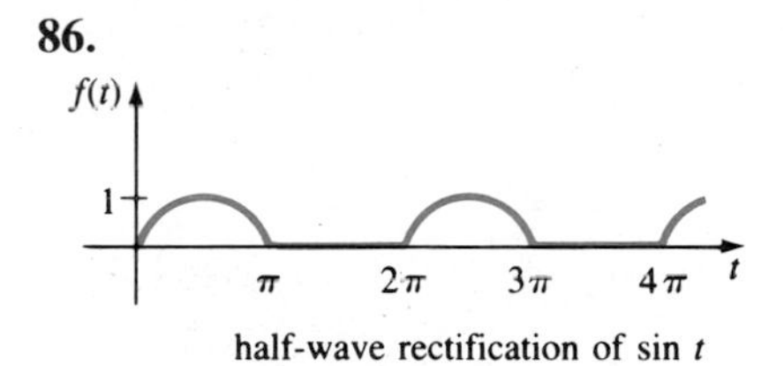

Figure 7.20

87. $f(t) = \sin t$
$f(t + 2\pi) = f(t)$

88. $f(t) = \cos t$
$f(t + 2\pi) = f(t)$

Miscellaneous Problems

89. Prove the commutative property of the convolution integral $f * g = g * f$.

90. Prove the distributive property of the convolution integral $f * (g + h) = f * g + f * h$.

91. Show that if $F(s) = \mathscr{L}\{f(t)\}$, then

$$\mathscr{L}\{f(t)\cosh at\} = \frac{1}{2}[F(s-a) + F(s+a)].$$

92. Use the result of Problem 91 to find $\mathscr{L}\{\sin kt \cosh at\}$.

93. Show that if $a > 0$, then $\mathscr{L}\{f(at)\} = (1/a)\,F(s/a)$.

94. Show that if f is piecewise continuous and of exponential order and $\lim_{t\to 0^+} f(t)/t$ exists, then $\mathscr{L}\{f(t)/t\} = \int_s^\infty F(s)\,ds$, where $F(s) = \mathscr{L}\{f(t)\}$.

95. Use the result of Problem 94 to evaluate $\mathscr{L}\left\{\dfrac{e^t - e^{-t}}{t}\right\}$.

96. Show that if f is piecewise continuous and of exponential order, then $\mathscr{L}\{\int_a^t f(\tau)\,d\tau\} = F(s)/s + (1/s)\int_a^0 f(\tau)\,d\tau$, where $\mathscr{L}\{f(t)\} = F(s)$.*

7.4 Applications

Since $\mathscr{L}\{y^{(n)}(t)\}$, $n > 1$, depends on $y(t)$ and its $n-1$ derivatives evaluated at $t = 0$, the Laplace transform is ideally suited to initial-value problems for linear differential equations with constant coefficients. This kind of differential equation can be reduced to an algebraic equation in the transformed function $Y(s)$. To see this, consider the initial-value problem

$$a_n\frac{d^ny}{dt^n} + a_{n-1}\frac{d^{n-1}y}{dt^{n-1}} + \cdots + a_1\frac{dy}{dt} + a_0 y = g(t)$$

$$y(0) = y_0,\quad y'(0) = y'_0, \ldots,\quad y^{(n-1)}(0) = y_0^{(n-1)},$$

where a_i, $i = 0, 1, \ldots, n$ and $y_0, y'_0, \ldots, y_0^{(n-1)}$ are constants. By the linearity property of the Laplace transform, we can write

$$a_n\mathscr{L}\left\{\frac{d^ny}{dt^n}\right\} + a_{n-1}\mathscr{L}\left\{\frac{d^{n-1}y}{dt^{n-1}}\right\} + \cdots + a_0\mathscr{L}\{y\} = \mathscr{L}\{g(t)\}. \tag{1}$$

Using Theorem 7.8, (1) becomes

* It can be shown that when f is piecewise continuous on $[0, \infty)$ and of exponential order, then $\int_a^t f(\tau)\,d\tau$ is also of exponential order.

$$a_n[s^n Y(s) - s^{n-1}y(0) - \cdots - y^{(n-1)}(0)] + a_{n-1}[s^{n-1}Y(s) - s^{n-2}y(0) - \cdots - y^{(n-2)}(0)] + \cdots + a_0 Y(s) = G(s)$$

or

$$[a_n s^n + a_{n-1}s^{n-1} + \cdots + a_0]Y(s) = a_n[s^{n-1}y_0 + \cdots + y_0^{(n-1)}] + a_{n-1}[s^{n-2}y_0 + \cdots + y_0^{(n-2)}] + \cdots + G(s), \qquad (2)$$

where $Y(s) = \mathscr{L}\{y(t)\}$ and $G(s) = \mathscr{L}\{g(t)\}$. By solving (2) for $Y(s)$, we find $y(t)$ by determining the inverse transform

$$y(t) = \mathscr{L}^{-1}\{Y(s)\}.$$

The procedure is outlined in Figure 7.21.

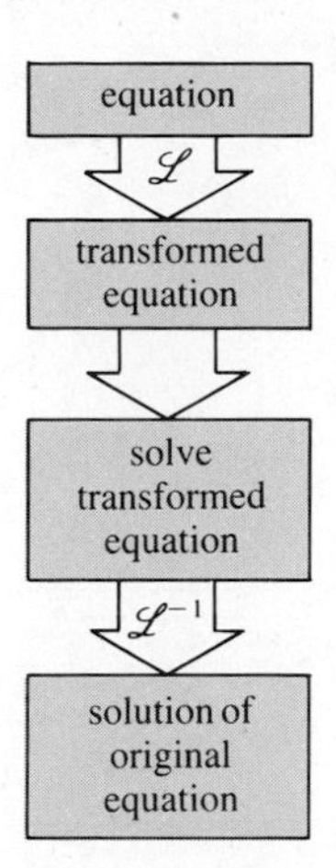

Figure 7.21

EXAMPLE 1 Solve $\dfrac{dy}{dt} - 3y = e^{2t}$ subject to $y(0) = 1$.

SOLUTION We first take the transform of each member of the given differential equation,

$$\mathscr{L}\left\{\frac{dy}{dt}\right\} - 3\mathscr{L}\{y\} = \mathscr{L}\{e^{2t}\}.$$

We then use $\mathscr{L}\{dy/dt\} = sY(s) - y(0) = sY(s) - 1$ and $\mathscr{L}\{e^{2t}\} = 1/(s-2)$.

Solving $\quad sY(s) - 1 - 3Y(s) = \dfrac{1}{s-2} \quad$ for $Y(s)$

gives $\quad Y(s) = \dfrac{s-1}{(s-2)(s-3)}.$

By partial fractions:

$$\frac{s-1}{(s-2)(s-3)} = \frac{A}{s-2} + \frac{B}{s-3},$$

which yields $\quad s - 1 = A(s-3) + B(s-2).$

Setting $s = 2$ and $s = 3$ in the last equation, we obtain $A = -1$ and $B = 2$, respectively. Consequently

$$Y(s) = \frac{-1}{s-2} + \frac{2}{s-3}$$

and $$y(t) = -\mathscr{L}^{-1}\left\{\frac{1}{s-2}\right\} + 2\mathscr{L}^{-1}\left\{\frac{1}{s-3}\right\}.$$

From part (c) of Theorem 7.3 it follows that

$$y(t) = -e^{2t} + 2e^{3t}.$$

EXAMPLE 2

Solve $$y'' - 6y' + 9y = t^2e^{3t}$$

subject to $$y(0) = 2, \qquad y'(0) = 6.$$

SOLUTION $$\mathscr{L}\{y''\} - 6\mathscr{L}\{y'\} + 9\mathscr{L}\{y\} = \mathscr{L}\{t^2e^{3t}\}$$

$$\underbrace{s^2Y(s) - sy(0) - y'(0)}_{\mathscr{L}\{y''\}} - 6\underbrace{[sY(s) - y(0)]}_{\mathscr{L}\{y'\}} + 9\underbrace{Y(s)}_{\mathscr{L}\{y\}} = \underbrace{\frac{2}{(s-3)^3}}_{\mathscr{L}\{t^2e^{3t}\}}.$$

Using the initial conditions and simplifying give

$$(s^2 - 6s + 9)Y(s) = 2s - 6 + \frac{2}{(s-3)^3}$$

$$(s-3)^2Y(s) = 2(s-3) + \frac{2}{(s-3)^3}$$

$$Y(s) = \frac{2}{s-3} + \frac{2}{(s-3)^5}$$

and so $$y(t) = 2\mathscr{L}^{-1}\left\{\frac{1}{s-3}\right\} + \frac{2}{4!}\mathscr{L}^{-1}\left\{\frac{4!}{(s-3)^5}\right\}.$$

Recall from the first translation theorem that

$$\mathscr{L}^{-1}\left\{\frac{4!}{s^5}\bigg|_{s\to s-3}\right\} = t^4e^{3t}.$$

Hence we have $$y(t) = 2e^{3t} + \frac{1}{12}t^4e^{3t}.$$

EXAMPLE 3

Solve $$y'' + 4y' + 6y = 1 + e^{-t}$$

subject to $$y(0) = 0, \qquad y'(0) = 0.$$

SOLUTION $\mathscr{L}\{y''\} + 4\mathscr{L}\{y'\} + 6\mathscr{L}\{y\} = \mathscr{L}\{1\} + \mathscr{L}\{e^{-t}\}$

$$s^2Y(s) - sy(0) - y'(0) + 4[sY(s) - y(0)] + 6Y(s) = \frac{1}{s} + \frac{1}{s+1}$$

$$(s^2 + 4s + 6)Y(s) = \frac{2s+1}{s(s+1)}$$

$$Y(s) = \frac{2s+1}{s(s+1)(s^2+4s+6)}.$$

By partial fractions:

$$\frac{2s+1}{s(s+1)(s^2+4s+6)} = \frac{A}{s} + \frac{B}{s+1} + \frac{Cs+D}{s^2+4s+6},$$

which implies

$$2s + 1 = A(s+1)(s^2+4s+6) + Bs(s^2+4s+6) + (Cs+D)s(s+1).$$

Setting $s = 0$ and $s = -1$ gives, respectively, $A = 1/6$ and $B = 1/3$. Equating the coefficients of s^3 and s gives

$$A + B + C = 0$$

$$10A + 6B + D = 2,$$

so it follows that $C = -1/2$ and $D = -5/3$. Thus

$$Y(s) = \frac{1/6}{s} + \frac{1/3}{s+1} + \frac{-s/2 - 5/3}{s^2+4s+6}$$

$$= \frac{1/6}{s} + \frac{1/3}{s+1} + \frac{(-1/2)(s+2) - 2/3}{(s+2)^2+2}$$

$$= \frac{1/6}{s} + \frac{1/3}{s+1} - \frac{1}{2}\frac{s+2}{(s+2)^2+2} - \frac{2}{3}\frac{1}{(s+2)^2+2}.$$

Finally, from parts (a) and (c) of Theorem 7.3 and the first translation theorem, we obtain

$$y(t) = \frac{1}{6}\mathscr{L}^{-1}\left\{\frac{1}{s}\right\} + \frac{1}{3}\mathscr{L}^{-1}\left\{\frac{1}{s+1}\right\} - \frac{1}{2}\mathscr{L}^{-1}\left\{\frac{s+2}{(s+2)^2+2}\right\}$$

$$- \frac{2}{3\sqrt{2}}\mathscr{L}^{-1}\left\{\frac{\sqrt{2}}{(s+2)^2+2}\right\}$$

$$= \frac{1}{6} + \frac{1}{3}e^{-t} - \frac{1}{2}e^{-2t}\cos\sqrt{2}t - \frac{\sqrt{2}}{3}e^{-2t}\sin\sqrt{2}t.$$

EXAMPLE 4 Solve

$$x'' + 16x = \cos 4t$$

subject to

$$x(0) = 0, \quad x'(0) = 1.$$

SOLUTION Recall that this initial value problem could describe the forced, undamped, and resonant motion of a mass on a spring. The mass starts with an initial velocity of one foot per second in the downward direction from the equilibrium position. Now one could readily solve this problem by, say, variation of parameters, but the use of Laplace transforms obviates the necessity of determining the constants that would naturally occur in the general solution $x = x_c(t) + x_p(t)$.

Transforming the equation gives

$$(s^2 + 16)X(s) = 1 + \frac{s}{s^2 + 16}$$

$$X(s) = \frac{1}{s^2 + 16} + \frac{s}{(s^2 + 16)^2}.$$

With the aid of part (d) of Theorem 7.3 and Theorem 7.7, we find

$$\begin{aligned} x(t) &= \frac{1}{4}\mathscr{L}^{-1}\left\{\frac{4}{s^2 + 16}\right\} + \frac{1}{8}\mathscr{L}^{-1}\left\{\frac{8s}{(s^2 + 16)^2}\right\} \\ &= \frac{1}{4}\sin 4t + \frac{1}{8}t\sin 4t. \end{aligned}$$

EXAMPLE 5 Solve

$$x'' + 16x = f(t), \quad \text{where}$$

$$f(t) = \begin{cases} \cos 4t, & 0 \le t < \pi \\ 0, & t \ge \pi, \end{cases} \quad \text{and} \quad x(0) = 0,\ x'(0) = 1.$$

SOLUTION The function $f(t)$ can be interpreted as an external force that is acting on a mechanical system only for a short period of time and then is removed (see Figure 7.22). Although this problem could be solved by conventional means, the procedure is not at all convenient when $f(t)$ is defined in a piecewise manner. Using the periodicity of the cosine, we can write

$$f(t) = \cos 4t - \cos 4t\,\mathscr{U}(t - \pi) = \cos 4t - \cos 4(t - \pi)\mathscr{U}(t - \pi).$$

With the aid of the second translation theorem it follows that

$$\mathscr{L}\{x''\} + 16\mathscr{L}\{x\} = \mathscr{L}\{f(t)\}$$

$$s^2X(s) - sx(0) - x'(0) + 16X(s) = \frac{s}{s^2 + 16} - \frac{s}{s^2 + 16}e^{-\pi s}$$

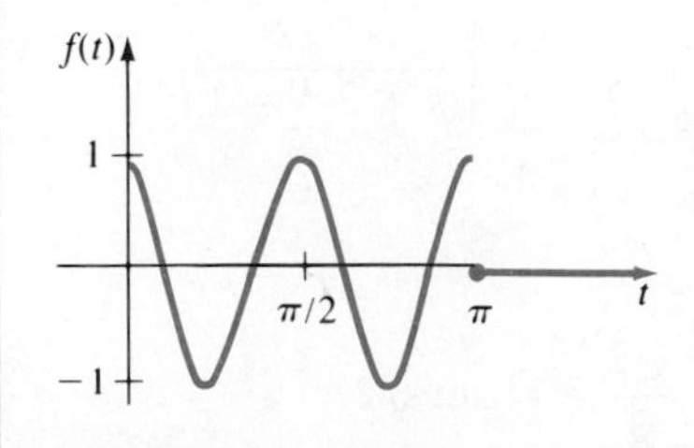

Figure 7.22

$$(s^2 + 16)X(s) = 1 + \frac{s}{s^2 + 16} - \frac{s}{s^2 + 16} e^{-\pi s}$$

$$X(s) = \frac{1}{s^2 + 16} + \frac{s}{(s^2 + 16)^2} - \frac{s}{(s^2 + 16)^2} e^{-\pi s}.$$

From Example 4 and (2) of Section 7.3, we obtain

$$x(t) = \frac{1}{4} \mathscr{L}^{-1}\left\{\frac{4}{s^2 + 16}\right\} + \frac{1}{8} \mathscr{L}^{-1}\left\{\frac{8s}{(s^2 + 16)^2}\right\} - \frac{1}{8} \mathscr{L}^{-1}\left\{\frac{8s}{(s^2 + 16)^2} e^{-\pi s}\right\}$$

$$= \frac{1}{4} \sin 4t + \frac{1}{8} t \sin 4t - \frac{1}{8} (t - \pi) \sin 4(t - \pi)\mathscr{U}(t - \pi).$$

The foregoing solution is the same as

$$x(t) = \begin{cases} \dfrac{1}{4} \sin 4t + \dfrac{1}{8} t \sin 4t, & 0 \le t < \pi \\ \dfrac{2 + \pi}{8} \sin 4t, & t \ge \pi. \end{cases}$$

Observe from the graph of $x(t)$ in Figure 7.23 that the amplitudes of vibration become steady as soon as the external force is turned off.

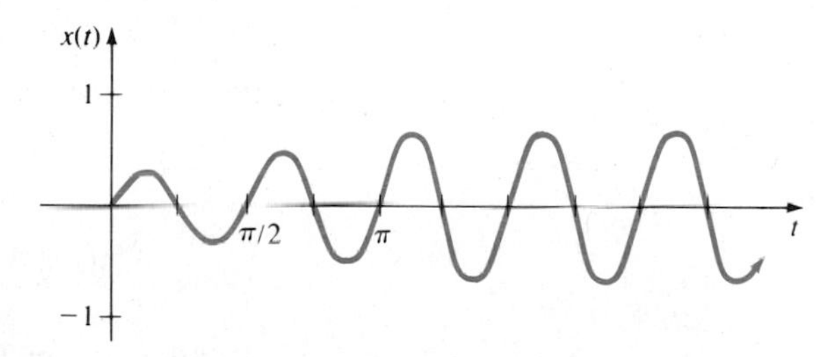

Figure 7.23

EXAMPLE 6 Solve

$$y'' + 2y' + y = f(t), \quad \text{where}$$

$$f(t) = \mathscr{U}(t - 1) - 2\mathscr{U}(t - 2) + \mathscr{U}(t - 3) \quad \text{and} \quad y(0) = 0,\ y'(0) = 0.$$

SOLUTION By the second translation theorem and simplification, the transform of the differential equation is

$$(s + 1)^2 Y(s) = \frac{e^{-s}}{s} - 2\frac{e^{-2s}}{s} + \frac{e^{-3s}}{s}$$

or $$Y(s) = \frac{e^{-s}}{s(s+1)^2} - 2\frac{e^{-2s}}{s(s+1)^2} + \frac{e^{-3s}}{s(s+1)^2}.$$

With the aid of partial fractions, the last equation becomes

$$Y(s) = \left[\frac{1}{s} - \frac{1}{s+1} - \frac{1}{(s+1)^2}\right]e^{-s} - 2\left[\frac{1}{s} - \frac{1}{s+1} - \frac{1}{(s+1)^2}\right]e^{-2s}$$
$$+ \left[\frac{1}{s} - \frac{1}{s+1} - \frac{1}{(s+1)^2}\right]e^{-3s}.$$

Again employing the inverse form of the second translation theorem, we find

$$y(t) = [1 - e^{-(t-1)} - (t-1)e^{-(t-1)}]\mathscr{U}(t-1) - 2[1 - e^{-(t-2)} - (t-2)e^{-(t-2)}]\mathscr{U}(t-2)$$
$$+ [1 - e^{-(t-3)} - (t-3)e^{-(t-3)}]\mathscr{U}(t-3).$$

Volterra Integral Equation

The convolution theorem is useful in solving other types of equations in which an unknown function appears under an integral sign. In the next example we solve a **Volterra integral equation**

$$f(t) = g(t) + \int_0^t f(\tau)h(t-\tau)\,d\tau$$

for $f(t)$. The functions $g(t)$ and $h(t)$ are known.

EXAMPLE 7

Solve $$f(t) = 3t^2 - e^{-t} - \int_0^t f(\tau)e^{t-\tau}\,d\tau$$ for $f(t)$.

SOLUTION It follows from Theorem 7.9 that

$$\mathscr{L}\{f(t)\} = 3\mathscr{L}\{t^2\} - \mathscr{L}\{e^{-t}\} - \mathscr{L}\{f(t)\}\mathscr{L}\{e^{t}\}$$

$$F(s) = 3\cdot\frac{2}{s^3} - \frac{1}{s+1} - F(s)\cdot\frac{1}{s-1}$$

$$\left[1 + \frac{1}{s-1}\right]F(s) = \frac{6}{s^3} - \frac{1}{s+1}$$

$$\frac{s}{s-1}F(s) = \frac{6}{s^3} - \frac{1}{s+1}$$

$$F(s) = \frac{6(s-1)}{s^4} - \frac{s-1}{s(s+1)} \qquad \text{[termwise division and partial fractions]}$$

$$= \frac{6}{s^3} - \frac{6}{s^4} + \frac{1}{s} - \frac{2}{s+1}.$$

Therefore

$$f(t) = 3\mathscr{L}^{-1}\left\{\frac{2!}{s^3}\right\} - \mathscr{L}^{-1}\left\{\frac{3!}{s^4}\right\} + \mathscr{L}^{-1}\left\{\frac{1}{s}\right\} - 2\mathscr{L}^{-1}\left\{\frac{1}{s+1}\right\}$$
$$= 3t^2 - t^3 + 1 - 2e^{-t}.$$

Integrodifferential Equation

In a single loop or series circuit, Kirchhoff's second law states that the sum of the voltage drops across an inductor, resistor, and capacitor is equal to the impressed voltage $E(t)$. Now it is known (see Section 1.2) that

$$\text{the voltage drop across the inductor} = L\frac{di}{dt},$$

$$\text{the voltage drop across the resistor} = Ri(t),$$

and $$\text{the voltage drop across the capacitor} = \frac{1}{C}\int_0^t i(\tau)\,d\tau,$$

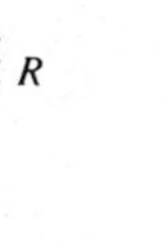

Figure 7.24

where $i(t)$ is the current and L, R, and C are constants. It follows that the current in a circuit, such as that shown in Figure 7.24, is governed by the **integrodifferential equation**

$$L\frac{di}{dt} + Ri + \frac{1}{C}\int_0^t i(\tau)\,d\tau = E(t). \tag{3}$$

EXAMPLE 8

Determine the current $i(t)$ in a single loop L-R-C circuit when $L = 0.1$ henry, $R = 20$ ohms, $C = 10^{-3}$ farads, $i(0) = 0$, and if the impressed voltage $E(t)$ is as given in Figure 7.25.

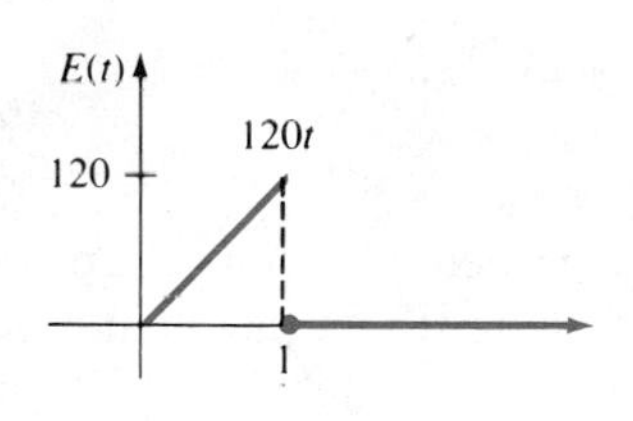

Figure 7.25

SOLUTION Since the voltage is off for $t \geq 1$, we can write

$$E(t) = 120t - 120t\mathscr{U}(t-1). \tag{4}$$

But in order to use the second translation theorem we must rewrite (4) as

$$E(t) = 120t - 120(t-1)\mathscr{U}(t-1) - 120\mathscr{U}(t-1).$$

Equation (3) then becomes

$$0.1\frac{di}{dt} + 20i + 10^3\int_0^t i(\tau)\,d\tau = 120t - 120(t-1)\mathscr{U}(t-1) - 120\mathscr{U}(t-1). \tag{5}$$

Now recall from Theorem 7.9 that

$$\mathscr{L}\left\{\int_0^t i(\tau)\,d\tau\right\} = I(s)/s,$$

where $I(s) = \mathscr{L}\{i(t)\}$. Thus the transform of equation (5) is

$$0.1\, sI(s) + 20I(s) + 10^3 \frac{I(s)}{s} = 120\left[\frac{1}{s^2} - \frac{1}{s^2}e^{-s} - \frac{1}{s}e^{-s}\right],$$

or, after multiplying by $10s$,

$$(s+100)^2 I(s) = 1200\left[\frac{1}{s} - \frac{1}{s}e^{-s} - e^{-s}\right]$$

$$I(s) = 1200\left[\frac{1}{s(s+100)^2} - \frac{1}{s(s+100)^2}e^{-s} - \frac{1}{(s+100)^2}e^{-s}\right].$$

By partial fractions we can write

$$I(s) = 1200\left[\frac{1/10{,}000}{s} - \frac{1/10{,}000}{s+100} - \frac{1/100}{(s+100)^2} - \frac{1/10{,}000}{s}e^{-s} + \frac{1/10{,}000}{s+100}e^{-s} + \frac{1/100}{(s+100)^2}e^{-s} - \frac{1}{(s+100)^2}e^{-s}\right].$$

Employing the inverse form of the second translation theorem, we obtain

$$i(t) = \frac{3}{25}[1 - \mathscr{U}(t-1)] - \frac{3}{25}[e^{-100t} - e^{-100(t-1)}\mathscr{U}(t-1)]$$
$$-12te^{-100t} - 1188(t-1)e^{-100(t-1)}\mathscr{U}(t-1).$$

Beams

The static deflection $y(x)$ of a uniform beam of length L carrying a load $w(x)$ per unit length is found from the fourth-order differential equation

$$EI\frac{d^4y}{dx^4} = w(x), \tag{6}$$

where E is Young's modulus of elasticity and I is a moment of inertia of a cross section of the beam. To apply the Laplace transform to (6) we tacitly assume that $w(x)$ and $y(x)$ are defined on $(0, \infty)$ rather than on $(0, L)$.

EXAMPLE 9

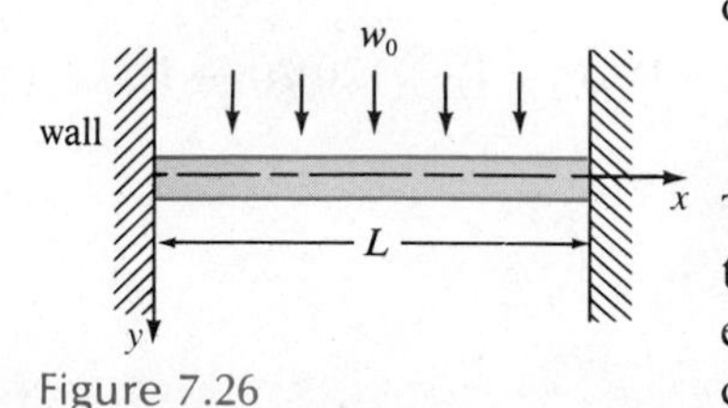

Figure 7.26

A beam of length L is clamped at both ends (see Figure 7.26). In this case the deflection $y(x)$ must satisfy (6) and the conditions

$$y(0) = 0,\ y(L) = 0,\ y'(0) = 0, \text{ and } y'(L) = 0.$$

The first two conditions indicate that there is no vertical deflection at the ends; the last two conditions mean that the x-axis is horizontal (zero slope) at the ends. Find the deflection of the beam when a constant load w_0 is uniformly distributed along its length, that is, when $w(x) = w_0$, $0 < x < L$.

SOLUTION Transforming (6) with respect to the variable x gives

$$EI(s^4Y(s) - s^3y(0) - s^2y'(0) - sy''(0) - y'''(0)) = \frac{w_0}{s}$$

$$s^4Y(s) - sy''(0) - y'''(0) = \frac{w_0}{EIs}$$

If we let $c_1 = y''(0)$ and $c_2 = y'''(0)$, then

$$Y(s) = \frac{c_1}{s^3} + \frac{c_2}{s^4} + \frac{w_0}{EIs^5}$$

and consequently
$$y(x) = \frac{c_1}{2!}\mathscr{L}^{-1}\left\{\frac{2!}{s^3}\right\} + \frac{c_2}{3!}\mathscr{L}^{-1}\left\{\frac{3!}{s^4}\right\} + \frac{w_0}{4!EI}\mathscr{L}^{-1}\left\{\frac{4!}{s^5}\right\}$$
$$= \frac{c_1}{2}x^2 + \frac{c_2}{6}x^3 + \frac{w_0}{24EI}x^4.$$

Applying the given conditions $y(L) = 0$ and $y'(L) = 0$ to the last equation yields the system

$$\frac{c_1}{2}L^2 + \frac{c_2}{6}L^3 + \frac{w_0}{24EI}L^4 = 0$$

$$c_1L + \frac{c_2}{2}L^2 + \frac{w_0}{6EI}L^3 = 0.$$

Solving, we find $c_1 = w_0L^2/12EI$ and $c_2 = -w_0L/2EI$. Thus the deflection is given by

$$y(x) = \frac{w_0L^2}{24EI}x^2 - \frac{w_0L}{12EI}x^3 + \frac{w_0}{24EI}x^4 = \frac{w_0}{24EI}x^2(x - L)^2$$

Exercises 7.4

Answers to odd-numbered problems begin on page A–20.

A table of the transforms of some basic functions is given in Appendix III.

In Problems 1–26 use the Laplace transform to solve the given differential equation subject to the indicated initial conditions. Where appropriate, write f in terms of unit step functions.

1. $\dfrac{dy}{dt} - y = 1, \quad y(0) = 0$

2. $\dfrac{dy}{dt} + 2y = t, \quad y(0) = -1$

3. $y' + 4y = e^{-4t}, \quad y(0) = 2$

4. $y' - y = \sin t, \quad y(0) = 0$

5. $y'' + 5y' + 4y = 0, \quad y(0) = 1, \quad y'(0) = 0$

6. $y'' - 6y' + 13y = 0, \quad y(0) = 0, \quad y'(0) = -3$

7. $y'' - 6y' + 9y = t, \quad y(0) = 0, \quad y'(0) = 1$

8. $y'' - 4y' + 4y = t^3, \quad y(0) = 1, \quad y'(0) = 0$

9. $y'' - 4y' + 4y = t^3e^{2t}, \quad y(0) = 0, \quad y'(0) = 0$

10. $y'' - 2y' + 5y = 1 + t, \quad y(0) = 0, \quad y'(0) = 4$

11. $y'' + y = \sin t, \quad y(0) = 1, \quad y'(0) = -1$

12. $y'' + 16y = 1, \quad y(0) = 1, \quad y'(0) = 2$

13. $y'' - y' = e^t \cos t, \quad y(0) = 0, \quad y'(0) = 0$

14. $y'' - 2y' = e^t \sinh t, \quad y(0) = 0, \quad y'(0) = 0$

15. $2y''' + 3y'' - 3y' - 2y = e^{-t}, \quad y(0) = 0, \quad y'(0) = 0, \quad y''(0) = 1$

16. $y''' + 2y'' - y' - 2y = \sin 3t, \quad y(0) = 0, \quad y'(0) = 0, \quad y''(0) = 1$

17. $y^{(4)} - y = 0, \quad y(0) = 1, \quad y'(0) = 0, \quad y''(0) = -1, \quad y'''(0) = 0$

18. $y^{(4)} - y = t, \quad y(0) = 0, \quad y'(0) = 0, \quad y''(0) = 0, \quad y'''(0) = 0$

19. $y' + y = f(t)$, where $f(t) = \begin{cases} 0, & 0 \le t < 1 \\ 5, & t \ge 1 \end{cases} \quad y(0) = 0$

20. $y' + y = f(t)$, where $f(t) = \begin{cases} 1, & 0 \le t < 1 \\ -1 & t \ge 1 \end{cases} \quad y(0) = 0$

21. $y' + 2y = f(t)$, where $f(t) = \begin{cases} t, & 0 \le t < 1 \\ 0, & t \ge 1 \end{cases} \quad y(0) = 0$

22. $y'' + 4y = f(t)$, where $f(t) = \begin{cases} 1, & 0 \le t < 1 \\ 0, & t \ge 1 \end{cases} \quad y(0) = 0, \quad y'(0) = -1$

23. $y'' + 4y = \sin t\,\mathcal{U}(t - 2\pi), \quad y(0) = 1, \quad y'(0) = 0$

24. $y'' - 5y' + 6y = \mathcal{U}(t - 1), \quad y(0) = 0, \quad y'(0) = 1$

25. $y'' + y = f(t)$, where $f(t) = \begin{cases} 0, & 0 \le t < \pi \\ 1, & \pi \le t < 2\pi \\ 0, & t \ge 2\pi \end{cases} \quad y(0) = 0, \quad y'(0) = 1$

26. $y'' + 4y' + 3y = 1 - \mathcal{U}(t - 2) - \mathcal{U}(t - 4) + \mathcal{U}(t - 6),$
$y(0) = 0, \quad y'(0) = 0$

In Problems 27 and 28 use the Laplace transform to solve the given differential equation subject to the indicated boundary conditions.

27. $y'' + 2y' + y = 0, \quad y'(0) = 2, \quad y(1) = 2$

28. $y'' - 9y' + 20y = 1, \quad y(0) = 0, \quad y'(1) = 0$

In Problems 29–38 use the Laplace transform to solve the given integral equation or integrodifferential equation.

29. $f(t) + \int_0^t (t - \tau) f(\tau)\, d\tau = t$

30. $f(t) = 2t - 4\int_0^t \sin \tau f(t - \tau)\, d\tau$

31. $f(t) = te^t + \int_0^t \tau f(t - \tau)\, d\tau$

32. $f(t) + 2\int_0^t f(\tau) \cos (t - \tau)\, d\tau = 4e^{-t} + \sin t$

33. $f(t) + \int_0^t f(\tau)\, d\tau = 1$

34. $f(t) = \cos t + \int_0^t e^{-\tau} f(t - \tau)\, d\tau$

35. $f(t) = 1 + t - \frac{8}{3}\int_0^t (\tau - t)^3 f(\tau)\, d\tau$

36. $t - 2f(t) = \int_0^t (e^\tau - e^{-\tau}) f(t - \tau)\, d\tau$

37. $y'(t) = 1 - \sin t - \int_0^t y(\tau)\, d\tau, \quad y(0) = 0$

38. $\frac{dy}{dt} + 6y(t) + 9\int_0^t y(\tau)\, d\tau = 1, \quad y(0) = 0$

39. Use equation (3) to determine the current $i(t)$ in a single loop L-R-C circuit when $L = 0.005$ henry, $R = 1$ ohms, $C = 0.02$ farads, $E(t) = 100[1 - \mathscr{U}(t - 1)]$ volts, and $i(0) = 0$.

40. Solve Problem 39 when $E(t) = 100[t - (t - 1)\mathscr{U}(t - 1)]$.

41. Recall that the differential equation for the charge $q(t)$ on the capacitor in an R-C series circuit is

$$R\frac{dq}{dt} + \frac{1}{C}q = E(t),$$

where $E(t)$ is the impressed voltage (see Section 3.2). Use the Laplace transform to determine the charge $q(t)$ when $q(0) = 0$ and $E(t) = E_0 e^{-kt}$, $k > 0$. Consider two cases: $k \neq 1/RC$ and $k = 1/RC$.

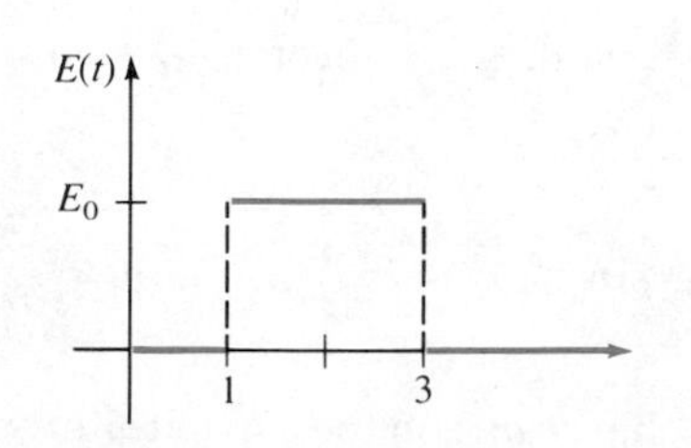

Figure 7.27

42. Use the Laplace transform to determine the charge $q(t)$ on the capacitor in an R-C series circuit when $q(0) = 0$, $R = 50$ ohms, $C = 0.01$ farad, and $E(t)$ is given in Figure 7.27.

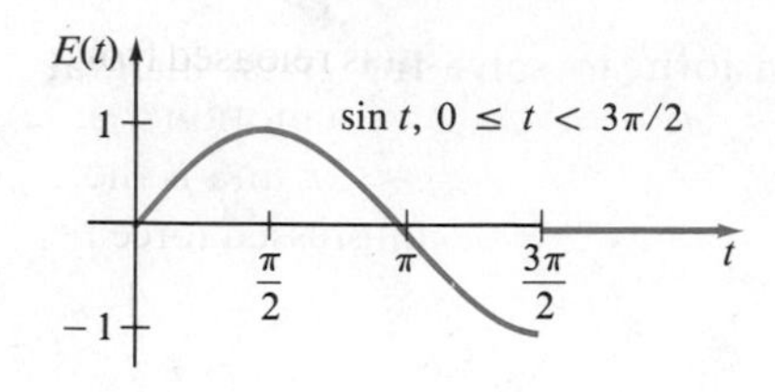

Figure 7.28

43. The differential equation for the current $i(t)$ in an L-R series circuit is

$$L\frac{di}{dt} + Ri = E(t), \tag{7}$$

where $E(t)$ is the impressed voltage. Use the Laplace transform to determine the current $i(t)$ when $i(0) = 0$, $L = 1$ henry, $R = 10$ ohms, and $E(t)$ is given in Figure 7.28.

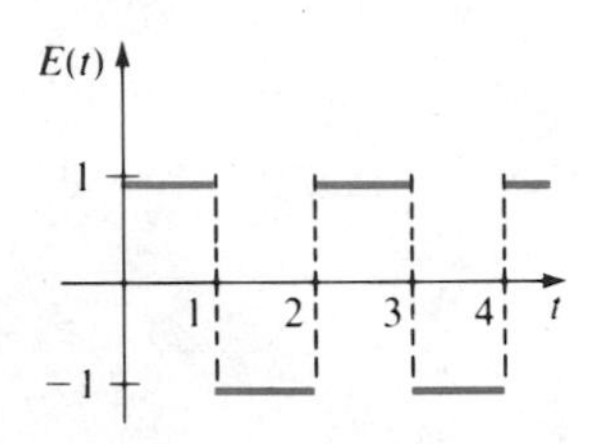

Figure 7.29

44. Solve equation (7) subject to $i(0) = 0$, where $E(t)$ is given in Figure 7.29. [*Hint:* See Problem 81, Exercises 7.3. Also note that $1/(1 - X) = 1 + X + X^2 + \ldots,\ |X| < 1$.]

45. Solve equation (7) subject to $i(0) = 0$, where $E(t)$ is given in Figure 7.30. Specify the solution for $0 \le t < 2$. [*Hint:* See Problem 83, Exercises 7.3.]

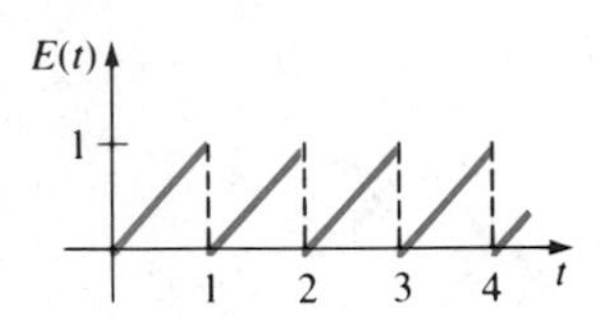

Figure 7.30

46. Recall that the differential equation for the instantaneous charge $q(t)$ on the capacitor in an L-R-C series circuit is given by

$$L\frac{d^2q}{dt^2} + R\frac{dq}{dt} + \frac{1}{C}q = E(t) \tag{8}$$

(see Section 5.4). Use the Laplace transform to determine $q(t)$ when $L = 1$ henry, $R = 20$ ohms, $C = 0.005$ farad, $E(t) = 150$ volts, $t > 0$, and $q(0) = 0$, $i(0) = 0$. What is the current $i(t)$? What is the charge $q(t)$ if the same constant voltage is turned off for $t \ge 2$?

47. Determine the charge $q(t)$ and current $i(t)$ for a series circuit in which $L = 1$ henry, $R = 20$ ohms, $C = 0.01$ farad, $E(t) = 120 \sin 10t$ volts, $q(0) = 0$, and $i(0) = 0$. What is the steady-state current?

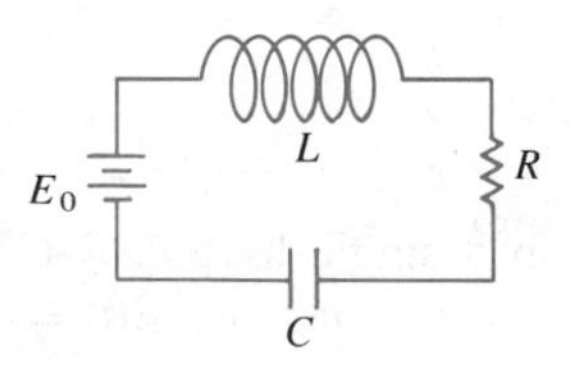

Figure 7.31

48. Consider the battery of constant voltage E_0 that charges the capacitor shown in Figure 7.31. If we divide by L and define $\lambda = R/2L$ and $\omega^2 = 1/LC$, then (8) becomes

$$\frac{d^2q}{dt^2} + 2\lambda\frac{dq}{dt} + \omega^2 q = \frac{E_0}{L}.$$

Use the Laplace transform to show that the solution of this equation, subject to $q(0) = 0$ and $i(0) = 0$, is

$$q(t) = \begin{cases} E_0C[1 - e^{-\lambda t}(\cosh\sqrt{\lambda^2 - \omega^2}\,t + \dfrac{\lambda}{\sqrt{\lambda^2 - \omega^2}}\sinh\sqrt{\lambda^2 - \omega^2}\,t)], & \lambda > \omega \\ E_0C[1 - e^{-\lambda t}(1 + \lambda t)], & \lambda = \omega \\ E_0C[1 - e^{-\lambda t}(\cos\sqrt{\omega^2 - \lambda^2}\,t + \dfrac{\lambda}{\sqrt{\omega^2 - \lambda^2}}\sin\sqrt{\omega^2 - \lambda^2}\,t)], & \lambda < \omega \end{cases}$$

49. Use the Laplace transform to determine the charge $q(t)$ on the capacitor in an L-C series circuit when $q(0) = 0$, $i(0) = 0$, and $E(t) = E_0e^{-kt}$, $k > 0$.

50. Suppose a 32-lb weight stretches a spring 2 ft. If the weight is released from rest at the equilibrium position, determine the equation of a motion if an impressed force $f(t) = \sin t$ acts on the system for $0 \le t < 2\pi$ and is then removed. Ignore any damping forces. [*Hint:* Write the impressed force in terms of the unit step function.]

51. A 4-lb weight stretches a spring 2 ft. The weight is released from rest 18 in. above the equilibrium position, and the resulting motion takes place in a medium offering a damping force numerically equal to 7/8 times the instantaneous velocity. Use the Laplace transform to determine the equation of motion.

52. A 16-lb weight is attached to a spring whose constant is $k = 4.5$ lb/ft. Beginning at $t = 0$, a force equal to $f(t) = 4 \sin 3t + 2 \cos 3t$ acts on the system. Assuming that no damping forces are present, use the Laplace transform to find the equation of motion if the weight is released from rest from the equilibrium position.

53. For a cantilever beam clamped at its left end ($x = 0$) and free at its right end ($x = L$), the deflection $y(x)$ must satisfy (6) and

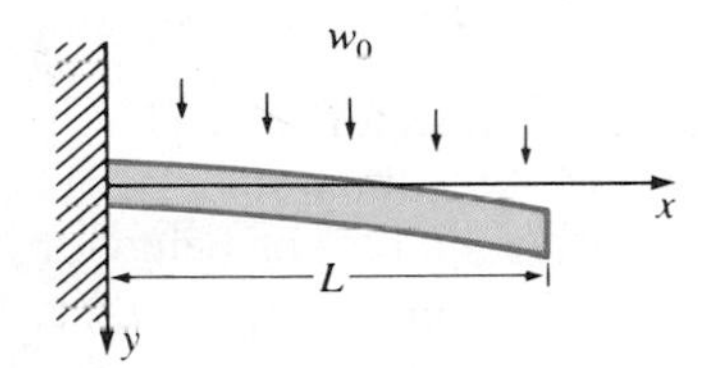

Figure 7.32

$$y(0) = 0, \quad y'(0) = 0, \quad y''(L) = 0, \quad y'''(L) = 0. \tag{9}$$

The first two conditions state that the deflection and slope are zero at $x = 0$, and the last two conditions state that the bending moment and shear force are zero at $x = L$. Use the Laplace transform to solve (6) subject to (9) when a constant load w_0 is uniformly distributed along the length of the beam (see Figure 7.32).

54. Solve Problem 53 when the load is given by

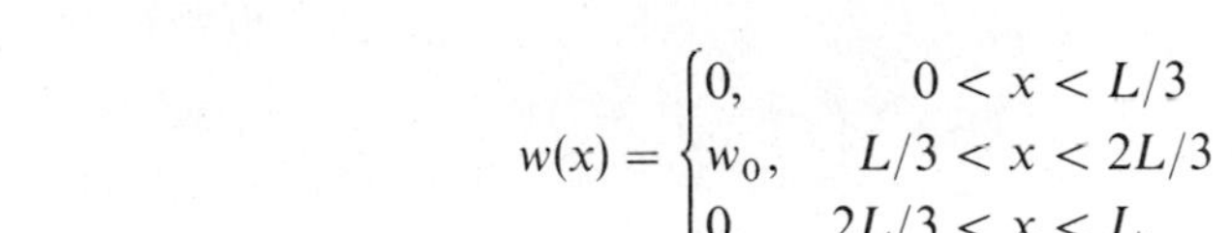

$$w(x) = \begin{cases} 0, & 0 < x < L/3 \\ w_0, & L/3 < x < 2L/3 \\ 0, & 2L/3 < x < L. \end{cases}$$

Write $w(x)$ in terms of unit step functions.

55. Solve Problem 53 when the load is given by

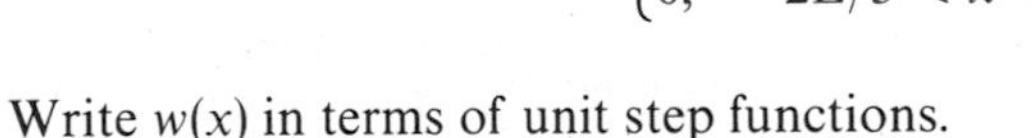

$$w(x) = \begin{cases} w_0, & 0 < x < L/2 \\ 0, & L/2 < x < L. \end{cases}$$

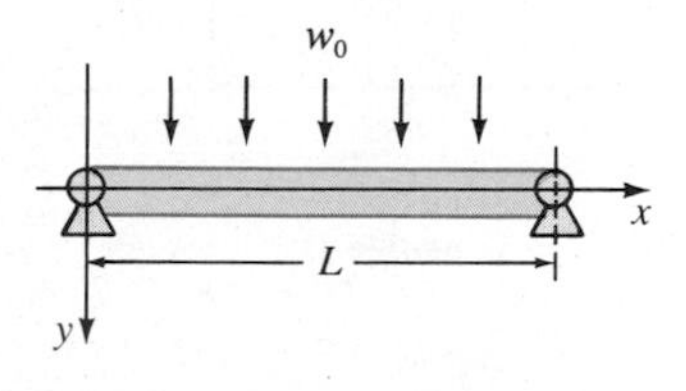

Figure 7.33

56. The static deflection $y(x)$ of a beam that is hinged at both ends must satisfy the differential equation (6) and the conditions

$$y(0) = 0, \quad y''(0) = 0, \quad y(L) = 0, \quad y''(L) = 0. \tag{10}$$

Use the Laplace transform to solve (6) subject to (10) when $w(x) = w_0$, $0 < x < L$ (see Figure 7.33).

Miscellaneous Problems

In Problems 57 and 58 use the Laplace transform and Theorem 7.7 to find a solution of the given equation.

57. $ty'' - y' = t^2, \quad y(0) = 0$

58. $ty'' + 2ty' + 2y = 0, \quad y(0) = 0$

[O] 7.5 Dirac Delta Function

Unit Impulse

Mechanical systems are often acted upon by an external force (or an emf in an electrical circuit) of large magnitude that acts only for a very short period of time. For example, a vibrating airplane wing could be struck by a bolt of lightning, a mass on a spring could be given a sharp blow by a ball peen hammer, or a ball initially at rest could be sent soaring into the air when struck violently by a golf club. The function

$$\delta_a(t - t_0) = \begin{cases} \dfrac{1}{2a}, & t_0 - a < t < t_0 + a \\ 0, & t \le t_0 - a, \quad \text{or} \quad t \ge t_0 + a, \end{cases} \tag{1}$$

shown in Figure 7.34(a), could serve as a mathematical model for such a force. For a small value of a, $\delta_a(t - t_0)$ is essentially a constant function of large magnitude that is "on" for just a short period of time around t_0. The behavior of $\delta_a(t - t_0)$ as $a \to 0$ is illustrated in Figure 7.34(b). The function $\delta_a(t - t_0)$ is called a **unit impulse** since it possesses the integration property

$$\int_{-\infty}^{\infty} \delta_a(t - t_0)\, dt = 1.$$

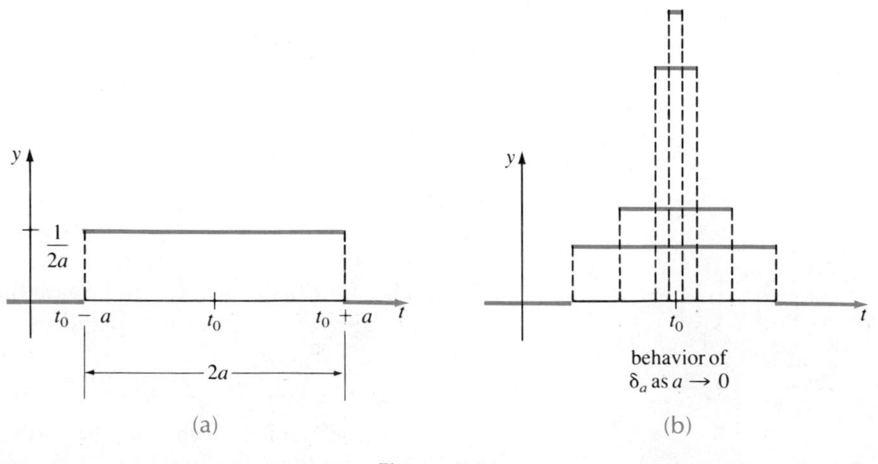

Figure 7.34

Dirac Delta Function

In practice it is convenient to work with another type of unit impulse, a "function" that approximates $\delta_a(t - t_0)$ and is defined by the limit

$$\delta(t - t_0) = \lim_{a \to 0} \delta_a(t - t_0). \tag{2}$$

The latter expression, which is not a function at all, can be characterized by the two properties

(i) $\delta(t - t_0) = \begin{cases} \infty, & t = t_0 \\ 0, & t \neq t_0 \end{cases}$, and

(ii) $\int_{-\infty}^{\infty} \delta(t - t_0)\, dt = 1.$

The expression $\delta(t - t_0)$ is called the **Dirac delta function.***

It is possible to obtain the Laplace transform of $\delta(t - t_0)$ by the formal assumption

$$\mathscr{L}\{\delta(t - t_0)\} = \lim_{a \to 0} \mathscr{L}\{\delta_a(t - t_0)\}.$$

We leave it as an exercise to show that

$$\mathscr{L}\{\delta_a(t - t_0)\} = e^{-st_0}\left(\frac{e^{sa} - e^{-sa}}{2sa}\right). \tag{3}$$

Since (3) has the indeterminate form 0/0 as $a \to 0$, we apply L'Hôpital's rule:

$$\lim_{a \to 0} \mathscr{L}\{\delta_a(t - t_0)\} = e^{-st_0} \lim_{a \to 0}\left(\frac{e^{sa} + e^{-sa}}{2}\right) = e^{-st_0}.$$

Thus we define

$$\mathscr{L}\{\delta(t - t_0)\} = e^{-st_0}. \tag{4}$$

Now when $t_0 = 0$, it seems plausible to conclude from (4) that

$$\mathscr{L}\{\delta(t)\} = 1.$$

---

* Paul Adrian Maurice Dirac (1902–1984) The delta function was the invention of the contemporary British physicist P. A. M. Dirac. Along with Max Planck, Werner Heisenberg, Erwin Schrödinger, and Albert Einstein, Dirac was one of the founding fathers, in the era 1900–1930, of a new way of describing the behavior of atoms, molecules, and elementary particles called *quantum mechanics*. For their pioneering work in this field, Dirac and Schrödinger shared the 1933 Nobel Prize in physics. The Dirac delta function was used extensively throughout his 1932 classic treatise *The Principles of Quantum Mechanics*.

The last result emphasizes the fact that $\delta(t)$ is no ordinary function since we expect from Theorem 7.4 that $\mathscr{L}\{f(t)\} \to 0$ as $s \to \infty$.

EXAMPLE 1

Solve
$$y'' + y = \delta(t - 2\pi)$$
subject to **(a)** $y(0) = 1, y'(0) = 0$; **(b)** $y(0) = 0, y'(0) = 0$.
The two initial-value problems could serve as models for describing the motion of a mass on a spring moving in a medium in which damping is negligible. At $t = 2\pi$ sec the mass is given a sharp blow. In (a) the mass is released from rest 1 unit below the equilibrium position. In (b) the mass is at rest in the equilibrium position.

SOLUTION **(a)** From (4) the Laplace transform of the differential equation is

$$s^2Y(s) - s + Y(s) = e^{-2\pi s}$$

or

$$Y(s) = \frac{s}{s^2 + 1} + \frac{e^{-2\pi s}}{s^2 + 1}.$$

Utilizing the second translation theorem, we find

$$y(t) = \cos t + \sin(t - 2\pi)\mathscr{U}(t - 2\pi).$$

Since $\sin(t - 2\pi) = \sin t$, the foregoing solution can be written

$$y(t) = \begin{cases} \cos t, & 0 \le t < 2\pi \\ \cos t + \sin t, & t \ge 2\pi. \end{cases} \tag{5}$$

(b) In this case the transform of the equation is simply

$$Y(s) = \frac{e^{-2\pi s}}{s^2 + 1}$$

and so

$$\begin{aligned} y(t) &= \sin(t - 2\pi)\mathscr{U}(t - 2\pi) \\ &= \begin{cases} 0 & 0 \le t < 2\pi \\ \sin t, & t > 2\pi. \end{cases} \end{aligned} \tag{6}$$

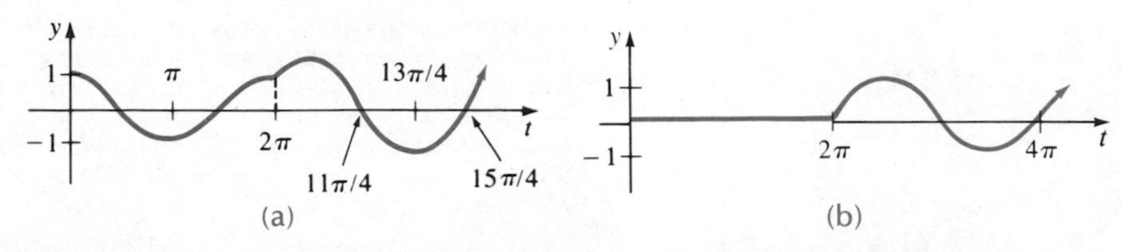

Figure 7.35

In Figure 7.35(a) we see from the graph of (5) that the mass is exhibiting simple harmonic motion until it is struck at $t = 2\pi$. The influence of the unit impulse is to increase the amplitude of vibration to $\sqrt{2}$ for $t > 2\pi$. The graph of (6) in Figure 7.35(b) shows, as we would expect from the initial conditions in (b), that the mass exhibits no motion until it is struck at $t = 2\pi$.

Remark: If $\delta(t - t_0)$ were a function in the usual sense, then property (i) would imply $\int_{-\infty}^{\infty} \delta(t - t_0)\, dt = 0$ rather than $\int_{-\infty}^{\infty} \delta(t - t_0)\, dt = 1$. Since the Dirac delta function did not "behave" as an ordinary function, even though its users produced correct results, it was met initially with great scorn by mathematicians. However, in the 1940s Dirac's controversial function was put on a rigorous footing by the French mathematician Laurent Schwartz in his book *La Théorie de distribution* and this, in turn, led to an entirely new branch of mathematics known as the *theory of distributions*, or *generalized functions.* In the modern theory of generalized functions, (2) is not an accepted definition of $\delta(t - t_0)$, nor does one speak of a function whose values are either ∞ or 0. Although this theory is much beyond the level of this text, it suffices for our purposes to say that the Dirac delta function is defined in terms of its effect or action on other functions. To see this let us suppose that f is a continuous function on $(-\infty, \infty)$. Then, by the mean value theorem for integrals, it follows that

$$\int_{-\infty}^{\infty} f(t)\delta_a(t - t_0)\, dt = \int_{t_0 - a}^{t_0 + a} f(t)\left(\frac{1}{2a}\right) dt$$

$$= \frac{1}{2a}(2af(c)) = f(c),$$

where c is some number in the interval $t_0 - a < t < t_0 + a$. As $a \to 0$ we must have $c \to t_0$ so that

$$\int_{-\infty}^{\infty} f(t)\delta(t - t_0)\, dt = \lim_{a \to 0} \int_{-\infty}^{\infty} f(t)\delta_a(t - t_0)\, dt = \lim_{a \to 0} f(c)$$

implies

$$\int_{-\infty}^{\infty} f(t)\delta(t - t_0)\, dt = f(t_0). \tag{7}$$

Although we have used the intuitive definition (2) to arrive at (7), the result is nevertheless valid and can be obtained in a rigorous fashion. The result in (7) can be taken as the *definition* of $\delta(t - t_0)$. It is known as the *sifting property* since $\delta(t - t_0)$ has the effect of sifting the value $f(t_0)$ from the values of f. Note that property (ii) on page 345 is consistent with (7) when $f(t) = 1$, $-\infty < t < \infty$. The integral operation (7) that corresponds a number $f(t_0)$ with a function f leads to the notion of a *linear functional.* We stop at this point and urge the curious reader to consult an advanced text.*

---

* See M. J. Lighthill, *Introduction to Fourier Analysis and Generalized Functions* (New York: Cambridge University Press, 1958).

Exercises 7.5

Answers to odd-numbered problems begin on page A–21.

In Problems 1–12 use the Laplace transform to solve the given differential equation subject to the indicated initial conditions.

1. $y' - 3y = \delta(t-2), \quad y(0) = 0$

2. $y' + y = \delta(t-1), \quad y(0) = 2$

3. $y'' + y = \delta(t-2\pi), \quad y(0) = 0, \quad y'(0) = 1$

4. $y'' + 16y = \delta(t-2\pi), \quad y(0) = 0, \quad y'(0) = 0$

5. $y'' + y = \delta\left(t - \frac{\pi}{2}\right) + \delta\left(t - \frac{3\pi}{2}\right), \quad y(0) = 0, \quad y'(0) = 0$

6. $y'' + y = \delta(t-2\pi) + \delta(t-4\pi), \quad y(0) = 1, \quad y'(0) = 0$

7. $y'' + 2y' = \delta(t-1), \quad y(0) = 0, \quad y'(0) = 1$

8. $y'' - 2y' = 1 + \delta(t-2), \quad y(0) = 0, \quad y'(0) = 1$

9. $y'' + 4y' + 5y = \delta(t-2\pi), \quad y(0) = 0, \quad y'(0) = 0$

10. $y'' + 2y' + y = \delta(t-1), \quad y(0) = 0, \quad y'(0) = 0$

11. $y'' + 4y' + 13y = \delta(t-\pi) + \delta(t-3\pi), \quad y(0) = 1, \quad y'(0) = 0$

12. $y'' - 7y' + 6y = e^t + \delta(t-2) + \delta(t-4), \quad y(0) = 0, \quad y'(0) = 0$

13. A uniform beam of length L carries a concentrated load P_0 at $x = L/2$. The beam is clamped at its left end and is free at its right end. Use the Laplace transform to determine the deflection $y(x)$ from

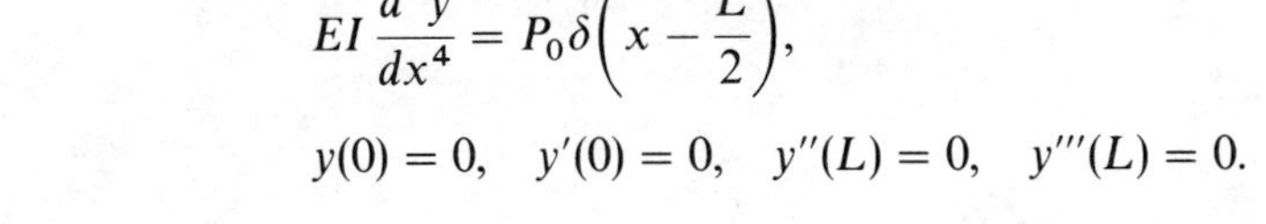

$$EI\frac{d^4y}{dx^4} = P_0\delta\left(x - \frac{L}{2}\right),$$

$$y(0) = 0, \quad y'(0) = 0, \quad y''(L) = 0, \quad y'''(L) = 0.$$

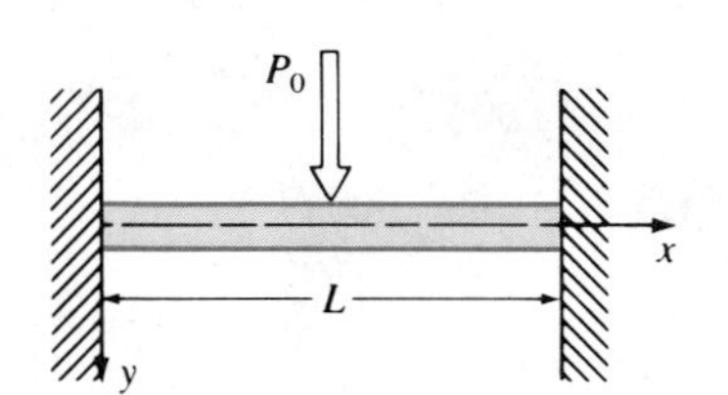

Figure 7.36

14. Solve the differential equation in Problem 13 subject to $y(0) = 0, y'(0) = 0, y(L) = 0, y'(L) = 0$. In this case the beam is clamped at both ends (see Figure 7.36).

Miscellaneous Problems

15. Use the definition of the Laplace transform and (1) to obtain (3).

16. Use (7) to obtain (4).

17. Use the Laplace transform and (7) to solve

$$y'' + 2y' + 2y = \cos t\,\delta(t-3\pi)$$

subject to $y(0) = 1$ and $y'(0) = -1$.

18. To emphasize the unusual nature of the Dirac delta function, show that the "solution" of the initial-value problem $y'' + \omega^2 y = \delta(t)$, $y(0) = 0$, $y'(0) = 0$ does *not* satisfy the initial condition $y'(0) = 0$.

19. Solve the initial-value problem

$$L\frac{di}{dt} + Ri = \delta(t), \qquad i(0) = 0,$$

where L and R are constants. Does the solution satisfy the condition at $t = 0$?

20. If $\delta'(t - t_0)$ is the derivative of the Dirac delta function, then it is known that $\mathscr{L}\{\delta'(t - t_0)\} = se^{-st_0}$, $t_0 \geq 0$. Use this result to solve $y' + 5y = \delta'(t)$ subject to $y(0) = 0$.

CHAPTER 7 SUMMARY

The **Laplace transform** of a function $f(t)$, $t \geq 0$, is defined by the integral

$$\mathscr{L}\{f(t)\} = \int_0^\infty e^{-st} f(t)\, dt = F(s).$$

The parameter s is usually restricted in such a manner that convergence of the integral is guaranteed. When applied to a linear differential equation with constant coefficients such as $ay'' + by' + cy = g(t)$ there results an algebraic equation

$$a[s^2Y(s) - sy(0) - y'(0)] + b[sY(s) - y(0)] + cY(s) = G(s),$$

which depends on the initial conditions $y(0)$ and $y'(0)$. When these values are known, we determine $y(t)$ by evaluating $y(t) = \mathscr{L}^{-1}\{Y(s)\}$.

CHAPTER 7 REVIEW EXERCISES

Answers to odd-numbered problems begin on page A–21.

In Problems 1 and 2 use the definition of the Laplace transform to find $\mathscr{L}\{f(t)\}$.

1. $f(t) = \begin{cases} t, & 0 \leq t < 1 \\ 2 - t, & t \geq 1 \end{cases}$

2. $f(t) = \begin{cases} 0, & 0 \leq t < 2 \\ 1, & 2 \leq t < 4 \\ 0, & t \geq 4 \end{cases}$

In Problems 3–22 fill in the blanks or answer true/false.

3. If f is not piecewise continuous on $[0, \infty)$, then $\mathscr{L}\{f(t)\}$ will not exist. ______

4. The function $f(t) = (e^t)^{10}$ is not of exponential order. ______

5. $F(s) = \dfrac{s^2}{s^2 + 4}$ is not the Laplace transform of a function that is piecewise continuous and of exponential order. ______

6. If $\mathscr{L}\{f(t)\} = F(s)$ and $\mathscr{L}\{g(t)\} = G(s)$ then $\mathscr{L}^{-1}\{F(s)G(s)\} = f(t)g(t)$. ______

7. $\mathscr{L}\{e^{-7t}\} =$ ______

8. $\mathscr{L}\{te^{-7t}\} =$ ______

9. $\mathscr{L}\{\sin 2t\} =$ ______

10. $\mathscr{L}\{e^{-3t}\sin 2t\} =$ ______

11. $\mathscr{L}\{t \sin 2t\} =$ ______

12. $\mathscr{L}\{\sin 2t\,\mathscr{U}(t-\pi)\} =$ ______

13. $\mathscr{L}^{-1}\left\{\dfrac{20}{s^6}\right\} =$ ______

14. $\mathscr{L}^{-1}\left\{\dfrac{1}{4s+1}\right\} =$ ______

15. $\mathscr{L}^{-1}\left\{\dfrac{1}{(s-5)^3}\right\} =$ ______

16. $\mathscr{L}^{-1}\left\{\dfrac{1}{s^2-5}\right\} =$ ______

17. $\mathscr{L}^{-1}\left\{\dfrac{s}{s^2-10s+29}\right\} =$ ______

18. $\mathscr{L}^{-1}\left\{\dfrac{e^{-5s}}{s^2}\right\} =$ ______

19. $\mathscr{L}\{e^{-5t}\}$ exists for $s >$ ______.

20. If $\mathscr{L}\{f(t)\} = F(s)$ then $\mathscr{L}\{te^{8t}f(t)\} =$ ______.

21. If $\mathscr{L}\{f(t)\} = F(s)$ and $k > 0$ then $\mathscr{L}\{e^{a(t-k)}f(t-k)\mathscr{U}(t-k)\} =$ ______.

22. $1 * 1 =$ ______.

In Problems 23–26, **(a)** express f in terms of unit step functions, **(b)** find $\mathscr{L}\{f(t)\}$, **(c)** find $\mathscr{L}\{e^t f(t)\}$.

23.

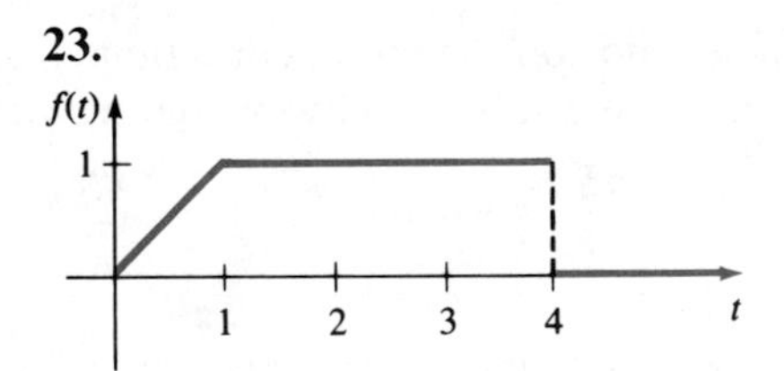

Figure 7.37

24.

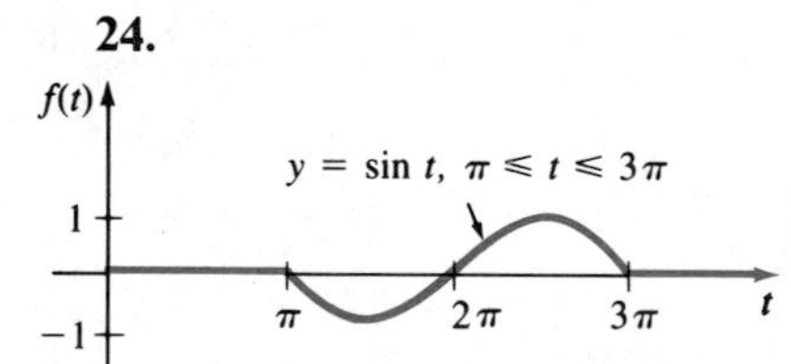

Figure 7.38

25.

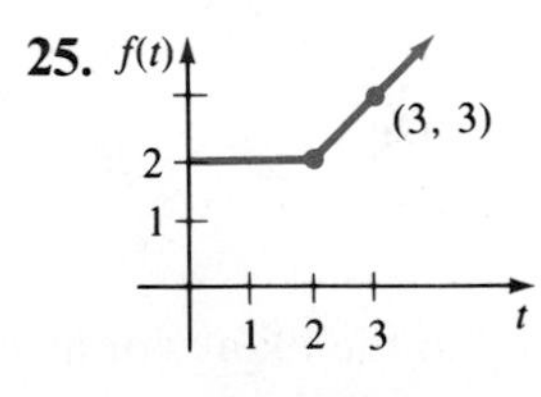

Figure 7.39

26.

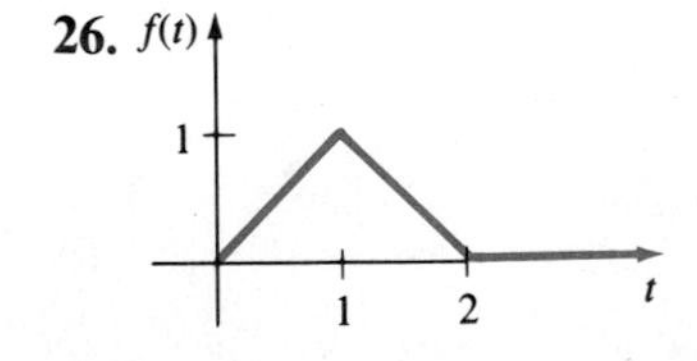

Figure 7.40

In Problems 27–31 use the Laplace transform to solve the given equation.

27. $y'' - 2y' + y = e^t, \quad y(0) = 0, \quad y'(0) = 5$

28. $y'' - 8y' + 20y = te^t, \quad y(0) = 0, \quad y'(0) = 0$

29. $y' - 5y = f(t)$, where $f(t) = \begin{cases} t^2, & 0 \le t < 1 \\ 0, & t \ge 1 \end{cases}$ $\quad y(0) = 1.$

30. $f(t) = 1 - 2\int_0^t e^{-3\tau} f(t - \tau)\, d\tau$

31. $y'(t) = \cos t + \int_0^t y(\tau) \cos(t - \tau)\, d\tau, \quad y(0) = 1$

32. A series circuit contains an inductor, a resistor, and a capacitor for which $L = 1/2$ henry, $R = 10$ ohms, and $C = 0.01$ farad, respectively. The voltage

$$E(t) = \begin{cases} 10, & 0 \le t < 5 \\ 0, & t \ge 5 \end{cases}$$

is applied to the circuit. Determine the instantaneous charge $q(t)$ on the capacitor for $t > 0$ if $q(0) = 0$ and $q'(0) = 0$.

33. The current $i(t)$ in an R-C series circuit can be determined from the integral equation

$$Ri + \frac{1}{C}\int_0^t i(\tau)\, d\tau = E(t),$$

where $E(t)$ is the impressed voltage. Determine $i(t)$ when $R = 10$ ohms, $C = 0.5$ farads, and $E(t) = 2(t^2 + t)$.

CHAPTER 8

Systems of Linear Differential Equations

IMPORTANT TERMS

systematic elimination
normal form
homogeneous system
nonhomogeneous system
degenerate system
matrix
multiplicative inverse
nonsingular matrix
singular matrix
augmented matrix
elementary row operations
Gauss-Jordan elimination method
reduced row-echelon form
eigenvalue
eigenvector
characteristic equation
solution vector
superposition principle
linear dependence
linear independence
Wronskian
fundamental set of solutions
general solution
fundamental matrix
undetermined coefficients
variation of parameters
matrix exponential

In Chapter 5 we saw that the vibrations of a mass attached to a spring could be described by one relatively simple differential equation. Were we to attach two such springs, we would then need two coupled, or **simultaneous**, differential equations to represent the motion. In this chapter we shall confine our attention to the solution of systems of simultaneous linear differential equations, or simply **linear systems**, in which all coefficients are constants.

8.1 Operator Method

Simultaneous ordinary differential equations involve two or more equations containing derivatives of two or more unknown functions of a single independent variable. If x, y, and z are functions of the variable t, then

$$4\frac{d^2x}{dt^2} = -5x + y \qquad\qquad x' - 3x + y' + z' = 5$$
$$\text{and} \qquad x' - y' + 2z' = t^2$$
$$2\frac{d^2y}{dt^2} = 3x - y \qquad\qquad x + y' - 6z' = t - 1$$

are two examples of systems of simultaneous differential equations.

Solution of a System

A **solution** of a system of differential equations is a set of differentiable functions $x = f(t)$, $y = g(t)$, $z = h(t)$, and so on, that satisfies each equation of the system on some interval I.

Systematic Elimination

The first technique that we shall consider for solving systems is based on the fundamental principle of **systematic algebraic elimination** of variables. We shall see that the analogue of multiplying an algebraic equation by a constant is operating on a differential equation with some combination of derivatives. Recall that a linear differential equation

$$a_n y^{(n)} + a_{n-1} y^{(n-1)} + \cdots + a_1 y' + a_0 y = g(t),$$

where the a_i, $i = 0, 1, \ldots, n$ are constants, can be written as

$$(a_n D^n + a_{n-1} D^{n-1} + \cdots + a_1 D + a_0)y = g(t).$$

EXAMPLE 1

Write the system of differential equations

$$x'' + 2x' + y'' = x + 3y + \sin t$$
$$x' + y' = -4x + 2y + e^{-t}$$

in operator notation.

SOLUTION Rewrite the given system as

$$x'' + 2x' - x + y'' - 3y = \sin t \qquad\qquad (D^2 + 2D - 1)x + (D^2 - 3)y = \sin t$$
$$\text{so that}$$
$$x' + 4x + y' - 2y = e^{-t} \qquad\qquad (D + 4)x + (D - 2)y = e^{-t}.$$

Method of Solution

Consider the simple system of linear first-order equations

$$\begin{aligned} Dy &= 2x \\ Dx &= 3y, \end{aligned} \tag{1}$$

or equivalently,

$$\begin{aligned} 2x - Dy &= 0 \\ Dx - 3y &= 0. \end{aligned} \tag{2}$$

Operating on the first equation in (2) by D while multiplying the second by 2 and then subtracting will eliminate x from the system. It follows that

$$-D^2y + 6y = 0 \qquad \text{or} \qquad D^2y - 6y = 0.$$

Since the roots of the auxiliary equation are $m_1 = \sqrt{6}$ and $m_2 = -\sqrt{6}$, we obtain

$$y(t) = c_1 e^{\sqrt{6}t} + c_2 e^{-\sqrt{6}t}. \tag{3}$$

Whereas multiplying the first equation by -3 while operating on the second by D and then adding give the differential equation for x, $D^2x - 6x = 0$. It follows immediately that

$$x(t) = c_3 e^{\sqrt{6}t} + c_4 e^{-\sqrt{6}t}. \tag{4}$$

Now (3) and (4) do not satisfy the system (1) for every choice of c_1, c_2, c_3, and c_4. Substituting $x(t)$ and $x(t)$ into the first equation of the original system (1) gives, after simplifying,

$$(\sqrt{6}\,c_1 - 2c_3)e^{\sqrt{6}t} + (-\sqrt{6}\,c_2 - 2c_4)e^{-\sqrt{6}t} = 0.$$

Since the latter expression is to be zero for all values of t, we must have

$$\sqrt{6}\,c_1 - 2c_3 = 0 \qquad \text{and} \qquad -\sqrt{6}\,c_2 - 2c_4 = 0$$

or

$$c_3 = \frac{\sqrt{6}}{2}c_1, \qquad c_4 = -\frac{\sqrt{6}}{2}c_2. \tag{5}$$

Hence we conclude that a solution of the system must be

$$\begin{aligned} x(t) &= \frac{\sqrt{6}}{2}c_1 e^{\sqrt{6}t} - \frac{\sqrt{6}}{2}c_2 e^{-\sqrt{6}t} \\ y(t) &= c_1 e^{\sqrt{6}t} + c_2 e^{-\sqrt{6}t}. \end{aligned}$$

The reader is urged to substitute (3) and (4) into the second equation of (1) and verify that the same relationship (5) between the constants obtains.

EXAMPLE 2 Solve

$$\begin{aligned} Dx + (D + 2)y &= 0 \\ (D - 3)x - \qquad 2y &= 0. \end{aligned} \tag{6}$$

SOLUTION Operating on the first equation by $D - 3$ and on the second by D and subtracting eliminate x from the system. It follows that the differential equation for y is

$$[(D - 3)(D + 2) + 2D]y = 0 \qquad \text{or} \qquad (D^2 + D - 6)y = 0.$$

Since the characteristic equation of this last differential equation is $m^2 + m - 6 = (m - 2)(m + 3) = 0$, we obtain the solution

$$y(t) = c_1 e^{2t} + c_2 e^{-3t}. \tag{7}$$

Eliminating y in a similar manner yields $(D^2 + D - 6)x = 0$, from which we find

$$x(t) = c_3 e^{2t} + c_4 e^{-3t}. \tag{8}$$

As we noted in the foregoing discussion, a solution of (6) does not contain four independent constants since the system itself puts a constraint on the actual number that can be chosen arbitrarily. Substituting (7) and (8) into the first equation of (6) gives

$$(4c_1 + 2c_3)e^{2t} + (-c_2 - 3c_4)e^{-3t} = 0$$

and so

$$4c_1 + 2c_3 = 0 \qquad \text{and} \qquad -c_2 - 3c_4 = 0.$$

Therefore

$$c_3 = -2c_1 \qquad \text{and} \qquad c_4 = -\tfrac{1}{3}c_2.$$

Accordingly, a solution of the system is

$$\begin{aligned} x(t) &= -2c_1 e^{2t} - \tfrac{1}{3}c_2 e^{-3t} \\ y(t) &= \quad c_1 e^{2t} + \ c_2 e^{-3t}. \end{aligned}$$

Since we could just as easily solve for c_3 and c_4 in terms of c_1 and c_2, the solution in Example 2 can be written in the alternative form

$$\begin{aligned} x(t) &= c_3 e^{2t} + c_4 e^{-3t} \\ y(t) &= -\tfrac{1}{2}c_3 e^{2t} - 3c_4 e^{-3t}. \end{aligned}$$

Also, it sometimes pays to keep one's eyes open when solving systems. Had we solved for x first, then y could be found, along with the relationship between the constants, by simply using the last equation of (6):

$$y = \frac{1}{2}(Dx - 3x) = \frac{1}{2}[2c_3e^{2t} - 3c_4e^{-3t} - 3c_3e^{2t} - 3c_4e^{-3t}]$$

or $$y = -\frac{1}{2}c_3e^{2t} - 3c_4e^{-3t}.$$

EXAMPLE 3

Solve
$$\begin{aligned} x' - 4x + y'' &= t^2 \\ x' + x + y' &= 0. \end{aligned} \tag{9}$$

SOLUTION First write the system in differential operator notation

$$\begin{aligned} (D - 4)x + D^2y &= t^2 \\ (D + 1)x + Dy &= 0. \end{aligned} \tag{10}$$

Then, by eliminating x, we obtain

$$[(D + 1)D^2 - (D - 4)D]y = (D + 1)t^2 - (D - 4)0$$

or $$(D^3 + 4D)y = t^2 + 2t.$$

Since the roots of the auxiliary equation $m(m^2 + 4) = 0$ are $m_1 = 0$, $m_2 = 2i$, and $m_3 = -2i$, the complementary function is

$$y_c = c_1 + c_2 \cos 2t + c_3 \sin 2t.$$

To determine the particular solution y_p we use undetermined coefficients by assuming $y_p = At^3 + Bt^2 + Ct$. Therefore

$$y_p' = 3At^2 + 2Bt + C, \qquad y_p'' = 6At + 2B, \qquad y_p''' = 6A,$$
$$y_p''' + 4y_p' = 12At^2 + 8Bt + 6A + 4C = t^2 + 2t.$$

The last equality implies

$$12A = 1, \qquad 8B = 2, \qquad 6A + 4C = 0$$

and hence $A = \frac{1}{12}$, $B = \frac{1}{4}$, $C = -\frac{1}{8}$. Thus

$$\begin{aligned} y &= y_c + y_p \\ &= c_1 + c_2 \cos 2t + c_3 \sin 2t + \frac{1}{12}t^3 + \frac{1}{4}t^2 - \frac{1}{8}t. \end{aligned} \tag{11}$$

Eliminating y from the system (10) leads to

$$[(D - 4) - D(D + 1)]x = t^2 \qquad \text{or} \qquad (D^2 + 4)x = -t^2.$$

It should be obvious that

$$x_c = c_4 \cos 2t + c_5 \sin 2t$$

and that undetermined coefficients can be applied to obtain a particular solution of the form $x_p = At^2 + Bt + C$. In this case the usual differentiations and algebra yield

$$x_p = -\frac{1}{4}t^2 + \frac{1}{8}$$

and so $$x = x_c + x_p = c_4 \cos 2t + c_5 \sin 2t - \frac{1}{4}t^2 + \frac{1}{8}. \tag{12}$$

Now c_4 and c_5 can be expressed in terms of c_2 and c_3 by substituting (11) and (12) into either equation of (9). By using the second equation, we find, after combining terms,

$$(c_5 - 2c_4 - 2c_2) \sin 2t + (2c_5 + c_4 + 2c_3) \cos 2t = 0$$

so that $$c_5 - 2c_4 - 2c_2 = 0 \quad \text{and} \quad 2c_5 + c_4 + 2c_3 = 0.$$

Solving for c_4 and c_5 in terms of c_2 and c_3 gives

$$c_4 = -\frac{1}{5}(4c_2 + 2c_3) \quad \text{and} \quad c_5 = \frac{1}{5}(2c_2 - 4c_3).$$

Finally a solution of (9) is found to be

$$x(t) = -\frac{1}{5}(4c_2 + 2c_3) \cos 2t + \frac{1}{5}(2c_2 - 4c_3) \sin 2t - \frac{1}{4}t^2 + \frac{1}{8}$$

$$y(t) = c_1 + c_2 \cos 2t + c_3 \sin 2t + \frac{1}{12}t^3 + \frac{1}{4}t^2 - \frac{1}{8}t.$$

Use of Determinants

Symbolically, if L_1, L_2, L_3, and L_4 denote differential operators with constant coefficients, then a system of linear differential equations in two variables x and y can be written as

$$\begin{aligned} L_1x + L_2y &= g_1(t) \\ L_3x + L_4y &= g_2(t). \end{aligned} \tag{13}$$

Eliminating variables, as we would for algebraic equations, leads to

$$(L_1L_4 - L_2L_3)x = f_1(t) \quad \text{and} \quad (L_1L_4 - L_2L_3)y = f_2(t), \tag{14}$$

where

$$f_1(t) = L_4g_1(t) - L_2g_2(t) \quad \text{and} \quad f_2(t) = L_1g_2(t) - L_3g_1(t).$$

Formally the results in (14) can be written in terms of determinants similar to that used in Cramer's rule:

$$\begin{vmatrix} L_1 & L_2 \\ L_3 & L_4 \end{vmatrix} x = \begin{vmatrix} g_1 & L_2 \\ g_2 & L_4 \end{vmatrix} \quad \text{and} \quad \begin{vmatrix} L_1 & L_2 \\ L_3 & L_4 \end{vmatrix} y = \begin{vmatrix} L_1 & g_1 \\ L_3 & g_2 \end{vmatrix}. \tag{15}$$

The left-hand determinant in each equation in (15) can be expanded in the usual algebraic sense, the result then operating on the functions $x(t)$ and $y(t)$. However, some care should be exercised in the expansion of the right-hand determinants in (15). We must expand these determinants in the sense of the internal differential operators actually operating upon the functions $g_1(t)$ and $g_2(t)$.

If
$$\begin{vmatrix} L_1 & L_2 \\ L_3 & L_4 \end{vmatrix} \neq 0$$

in (15) and is a differential operator of order n, then

- The system (13) can be decoupled into two nth-order differential equations in x and y.
- The characteristic equation and hence the complementary function of each of these differential equations is the same.
- Since x and y both contain n constants, there is a total of $2n$ constants appearing.
- The total number of *independent* constants in the solution of the system is n.

If
$$\begin{vmatrix} L_1 & L_2 \\ L_3 & L_4 \end{vmatrix} = 0$$

in (13), then the system may have a solution containing any number of independent constants or may have no solution at all. Similar remarks hold for systems larger than indicated in (13).

EXAMPLE 4

Solve
$$\begin{aligned} x' &= 3x - y - 1 \\ y' &= x + y + 4e^t. \end{aligned} \tag{16}$$

SOLUTION Write the system in terms of differential operators:

$$\begin{aligned} (D-3)x + \quad\quad y &= -1 \\ -x + (D-1)y &= 4e^t \end{aligned}$$

and then use determinants:

$$\begin{vmatrix} D-3 & 1 \\ -1 & D-1 \end{vmatrix} x = \begin{vmatrix} -1 & 1 \\ 4e^t & D-1 \end{vmatrix}$$

$$\begin{vmatrix} D-3 & 1 \\ -1 & D-1 \end{vmatrix} y = \begin{vmatrix} D-3 & -1 \\ -1 & 4e^t \end{vmatrix}.$$

After expanding, we find that

$$(D-2)^2 x = 1 - 4e^t$$
$$(D-2)^2 y = -1 - 8e^t.$$

By the usual methods it follows that

$$x = x_c + x_p = c_1 e^{2t} + c_2 te^{2t} + \frac{1}{4} - 4e^t \tag{17}$$

$$y = y_c + y_p = c_3 e^{2t} + c_4 te^{2t} - \frac{1}{4} - 8e^t. \tag{18}$$

Substituting (17) and (18) into the second equation of (16) gives

$$(c_3 - c_1 + c_4)e^{2t} + (c_4 - c_2)te^{2t} = 0,$$

which then implies

$$c_4 = c_2 \quad \text{and} \quad c_3 = c_1 - c_4 = c_1 - c_2.$$

Thus a solution of (16) is

$$x(t) = c_1 e^{2t} + c_2 te^{2t} + \frac{1}{4} - 4e^t$$

$$y(t) = (c_1 - c_2)e^{2t} + c_2 te^{2t} - \frac{1}{4} - 8e^t.$$

EXAMPLE 5 Given the system

$$\begin{aligned} Dx + \qquad\qquad\quad Dz &= t^2 \\ 2x + D^2 y \qquad\qquad &= e^t \\ -2Dx - \quad 2y + (D+1)z &= 0 \end{aligned}$$

find the differential equation for the variable y.

SOLUTION By determinants we can write

$$\begin{vmatrix} D & 0 & D \\ 2 & D^2 & 0 \\ -2D & -2 & D+1 \end{vmatrix} y = \begin{vmatrix} D & t^2 & D \\ 2 & e^t & 0 \\ -2D & 0 & D+1 \end{vmatrix}$$

In turn, expanding each determinant by cofactors of the first row gives

$$\left(D\begin{vmatrix} D^2 & 0 \\ -2 & D+1 \end{vmatrix} + D\begin{vmatrix} 2 & D^2 \\ -2D & -2 \end{vmatrix}\right)y = D\begin{vmatrix} e^t & 0 \\ 0 & D+1 \end{vmatrix}$$

$$-\begin{vmatrix} 2 & 0 \\ -2D & D+1 \end{vmatrix}t^2 + D\begin{vmatrix} 2 & e^t \\ -2D & 0 \end{vmatrix}$$

or

$$D(3D^3 + D^2 - 4)y = 4e^t - 2t^2 - 4t.$$

Again we remind the reader that the D symbol on the left-hand side is to be treated as an algebraic quantity, but this is not the case on the right-hand side.

Exercises 8.1

Answers to odd-numbered problems begin on page A–21.

In Problems 1–22 solve, if possible, the given system of differential equations by either systematic elimination or determinants.

1. $\dfrac{dx}{dt} = 2x - y$

$\dfrac{dy}{dt} = x$

2. $\dfrac{dx}{dt} = 4x + 7y$

$\dfrac{dy}{dt} = x - 2y$

3. $\dfrac{dx}{dt} = -y + t$

$\dfrac{dy}{dt} = x - t$

4. $\dfrac{dx}{dt} - 4y = 1$

$x + \dfrac{dy}{dt} = 2$

5. $(D^2 + 5)x - 2y = 0$

$-2x + (D^2 + 2)y = 0$

6. $(D + 1)x + (D - 1)y = 2$

$3x + (D + 2)y = -1$

7. $\dfrac{d^2x}{dt^2} = 4y + e^t$

$\dfrac{d^2y}{dt^2} = 4x - e^t$

8. $\dfrac{d^2x}{dt^2} + \dfrac{dy}{dt} = -5x$

$\dfrac{dx}{dt} + \dfrac{dy}{dt} = -x + 4y$

9. $Dx + D^2y = e^{3t}$

$(D + 1)x + (D - 1)y = 4e^{3t}$

10. $D^2x - Dy = t$

$(D + 3)x + (D + 3)y = 2$

11. $(D^2 - 1)x - y = 0$

$(D - 1)x + Dy = 0$

12. $(2D^2 - D - 1)x - (2D + 1)y = 1$

$(D - 1)x + Dy = -1$

13. $2\dfrac{dx}{dt} - 5x + \dfrac{dy}{dt} = e^t$

$\dfrac{dx}{dt} - x + \dfrac{dy}{dt} = 5e^t$

14. $\dfrac{dx}{dt} + \dfrac{dy}{dt} = e^t$

$-\dfrac{d^2x}{dt^2} + \dfrac{dx}{dt} + x + y = 0$

15. $(D-1)x + (D^2+1)y = 1$

$(D^2-1)x + (D+1)y = 2$

16. $D^2x - 2(D^2+D)y = \sin t$

$x + Dy = 0$

17. $Dx = y$

$Dy = z$

$Dz = x$

18. $Dx + z = e^t$

$(D-1)x + Dy + Dz = 0$

$x + 2y + Dz = e^t$

19. $\dfrac{dx}{dt} - 6y = 0$

$x - \dfrac{dy}{dt} + z = 0$

$x + y - \dfrac{dz}{dt} = 0$

20. $\dfrac{dx}{dt} = -x + z$

$\dfrac{dy}{dt} = -y + z$

$\dfrac{dz}{dt} = -x + y$

21. $2Dx + (D-1)y = t$

$Dx + Dy = t^2$

22. $Dx - 2Dy = t^2$

$(D+1)x - 2(D+1)y = 1$

In Problems 23 and 24 solve the given system subject to the indicated initial conditions.

23. $\dfrac{dx}{dt} = -5x - y$

$\dfrac{dy}{dt} = 4x - y,$

$x(1) = 0,\ y(1) = 1$

24. $\dfrac{dx}{dt} = y - 1$

$\dfrac{dy}{dt} = -3x + 2y,$

$x(0) = 0,\ y(0) = 0$

Miscellaneous Problems

25. Determine, if possible, a system of differential equations having

$$x(t) = c_1 + c_2e^{2t}$$

$$y(t) = -c_1 + c_2e^{2t}$$

as its solution.

8.2 Laplace Transform Method

When initial conditions are specified, the Laplace transform will reduce a system of linear differential equations with constant coefficients to a set of simultaneous algebraic equations in the transformed functions.

EXAMPLE 1 Solve

$$\begin{aligned} 2x' + y' - y &= t \\ x' + y' \quad &= t^2 \end{aligned} \tag{1}$$

subject to $x(0) = 1$, $y(0) = 0$.

SOLUTION If $X(s) = \mathscr{L}\{x(t)\}$ and $Y(s) = \mathscr{L}\{y(t)\}$, then after transforming each equation, we obtain

$$2[sX(s) - x(0)] + sY(s) - y(0) - Y(s) = \frac{1}{s^2}$$

$$sX(s) - x(0) \; + sY(s) - y(0) \qquad = \frac{2}{s^3}$$

or

$$\begin{aligned} 2sX(s) + (s-1)Y(s) &= 2 + \frac{1}{s^2} \\ sX(s) + \qquad sY(s) &= 1 + \frac{2}{s^3}. \end{aligned} \tag{2}$$

Multiplying the second equation of (2) by 2 and subtracting yield

$$(-s - 1)Y(s) = \frac{1}{s^2} - \frac{4}{s^3}$$

$$Y(s) = \frac{4 - s}{s^3(s+1)}. \tag{3}$$

Now by partial fractions

$$\frac{4 - s}{s^3(s+1)} = \frac{A}{s} + \frac{B}{s^2} + \frac{C}{s^3} + \frac{D}{s+1}$$

so that

$$4 - s = As^2(s+1) + Bs(s+1) + C(s+1) + Ds^3.$$

Setting $s = 0$ and $s = -1$ in the last line gives $C = 4$ and $D = -5$, respectively, whereas equating the coefficients of s^3 and s^2 on each side of the equality yields

$$A + D = 0 \qquad \text{and} \qquad A + B = 0.$$

It follows that $A = 5$, $B = -5$. Thus (3) becomes

$$Y(s) = \frac{5}{s} - \frac{5}{s^2} + \frac{4}{s^3} - \frac{5}{s+1}$$

and so

$$y(t) = 5\mathscr{L}^{-1}\left\{\frac{1}{s}\right\} - 5\mathscr{L}^{-1}\left\{\frac{1}{s^2}\right\} + 2\mathscr{L}^{-1}\left\{\frac{2!}{s^3}\right\} - 5\mathscr{L}^{-1}\left\{\frac{1}{s+1}\right\}$$

$$= 5 - 5t + 2t^2 - 5e^{-t}.$$

By the second equation of (2),

$$X(s) = -Y(s) + \frac{1}{s} + \frac{2}{s^4},$$

from which it follows that

$$x(t) = -\mathscr{L}^{-1}\{Y(s)\} + \mathscr{L}^{-1}\left\{\frac{1}{s}\right\} + \frac{2}{3!}\mathscr{L}^{-1}\left\{\frac{3!}{s^4}\right\}$$

$$= -4 + 5t - 2t^2 + \frac{1}{3}t^3 + 5e^{-t}.$$

Hence we conclude that the solution of the given system (1) is

$$x(t) = -4 + 5t - 2t^2 + \frac{1}{3}t^3 + 5e^{-t}$$

$$y(t) = 5 - 5t + 2t^2 - 5e^{-t}. \tag{4}$$

Applications

Let us turn now to some elementary applications involving systems of differential equations. The solutions of the problems that we shall consider can be obtained either by the method of the preceding section or through the use of the Laplace transform.

Coupled Springs

Two masses m_1 and m_2 are connected to two springs A and B of negligible mass having spring constants k_1 and k_2, respectively. In turn, the two springs are attached as shown in Figure 8.1. Let $x_1(t)$ and $x_2(t)$ denote the vertical displacements of the masses from their equilibrium positions. When the system is in motion, spring B is subject to both an elongation and a compression; hence its net elongation is $x_2 - x_1$. Therefore it follows from Hooke's law that springs A and B exert forces

$$-k_1x_1 \qquad \text{and} \qquad k_2(x_2 - x_1),$$

respectively, on m_1. If no external force is impressed on the system and if no damping force is present, then the net force on m_1 is

$$-k_1x_1 + k_2(x_2 - x_1).$$

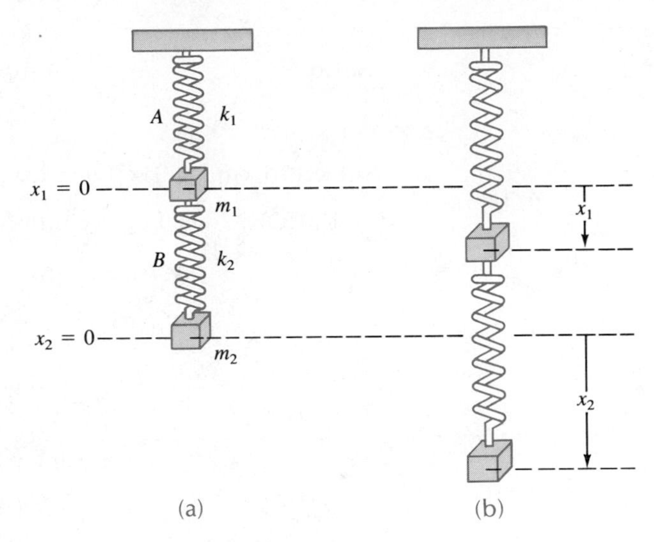

Figure 8.1

By Newton's second law we can write

$$m_1 \frac{d^2x_1}{dt^2} = -k_1x_1 + k_2(x_2 - x_1).$$

Similarly, the net force exerted on mass m_2 is due solely to the net elongation of B, that is,

$$-k_2(x_2 - x_1).$$

Thus it follows that

$$m_2 \frac{d^2x_2}{dt^2} = -k_2(x_2 - x_1).$$

In other words, the motion of the coupled system is represented by the system of simultaneous second-order differential equations

$$\begin{aligned} m_1x_1'' &= -k_1x_1 + k_2(x_2 - x_1) \\ m_2x_2'' &= -k_2(x_2 - x_1). \end{aligned} \tag{5}$$

In the next example we shall solve (5) under the assumptions that

$$k_1 = 6, \qquad k_2 = 4, \qquad m_1 = 1, \qquad m_2 = 1$$

and that the masses start from their equilibrium positions with opposite unit velocities.

EXAMPLE 2 Solve

$$\begin{aligned} x_1'' + 10x_1 \quad - 4x_2 &= 0 \\ -4x_1 + x_2'' + 4x_2 &= 0 \end{aligned} \tag{6}$$

subject to $x_1(0) = 0$, $x_1'(0) = 1$, $x_2(0) = 0$, $x_2'(0) = -1$.

SOLUTION The Laplace transform of each equation is

$$s^2X_1(s) - sx_1(0) - x_1'(0) + 10X_1(s) - 4X_2(s) = 0$$
$$-4X_1(s) + s^2X_2(s) - sx_2(0) - x_2'(0) + 4X_2(s) = 0,$$

where $X_1(s) = \mathscr{L}\{x_1(t)\}$ and $X_2(s) = \mathscr{L}\{x_2(t)\}$. The preceding system is the same as

$$\begin{aligned} (s^2 + 10)X_1(s) - \quad 4X_2(s) &= 1 \\ -4X_1(s) + (s^2 + 4)X_2(s) &= -1. \end{aligned} \tag{7}$$

Eliminating X_2 gives

$$X_1(s) = \frac{s^2}{(s^2 + 2)(s^2 + 12)}.$$

By partial fractions we can write

$$\frac{s^2}{(s^2 + 2)(s^2 + 12)} = \frac{As + B}{s^2 + 2} + \frac{Cs + D}{s^2 + 12}$$

and $$s^2 = (As + B)(s^2 + 12) + (Cs + D)(s^2 + 2).$$

Comparing coefficients of s on each side of the last equality gives

$$A + C = 0$$
$$B + D = 1$$
$$12A + 2C = 0$$
$$12B + 2D = 0$$

so that $$A = 0, \quad C = 0, \quad B = -\frac{1}{5}, \quad D = \frac{6}{5}.$$

Hence $$X_1(s) = -\frac{1/5}{s^2 + 2} + \frac{6/5}{s^2 + 12}$$

and therefore

$$\begin{aligned} x_1(t) &= -\frac{1}{5\sqrt{2}}\mathscr{L}^{-1}\left\{\frac{\sqrt{2}}{s^2 + 2}\right\} + \frac{6}{5\sqrt{12}}\mathscr{L}^{-1}\left\{\frac{\sqrt{12}}{s^2 + 12}\right\} \\ &= -\frac{\sqrt{2}}{10}\sin\sqrt{2}t + \frac{\sqrt{3}}{5}\sin 2\sqrt{3}t. \end{aligned}$$

From the first equation of (7) it follows that

$$X_2(s) = -\frac{s^2 + 6}{(s^2 + 2)(s^2 + 12)}.$$

Proceeding as before with partial fractions, we obtain

$$X_2(s) = -\frac{2/5}{s^2 + 2} - \frac{3/5}{s^2 + 12}$$

and
$$x_2(t) = -\frac{2}{5\sqrt{2}}\mathscr{L}^{-1}\left\{\frac{\sqrt{2}}{s^2 + 2}\right\} - \frac{3}{5\sqrt{12}}\mathscr{L}^{-1}\left\{\frac{\sqrt{12}}{s^2 + 12}\right\}$$
$$= -\frac{\sqrt{2}}{5}\sin\sqrt{2}t - \frac{\sqrt{3}}{10}\sin 2\sqrt{3}t.$$

Finally the solution to the given system (6) is

$$\begin{aligned} x_1(t) &= -\frac{\sqrt{2}}{10}\sin\sqrt{2}t + \frac{\sqrt{3}}{5}\sin 2\sqrt{3}t \\ x_2(t) &= -\frac{\sqrt{2}}{5}\sin\sqrt{2}t - \frac{\sqrt{3}}{10}\sin 2\sqrt{3}t. \end{aligned} \tag{8}$$

Networks

An electrical network having more than one loop also gives rise to simultaneous differential equations. As shown in Figure 8.2, the current $i_1(t)$ splits in the directions shown at point B_1 called a *branch point* of the network. By Kirchhoff's first law we can write

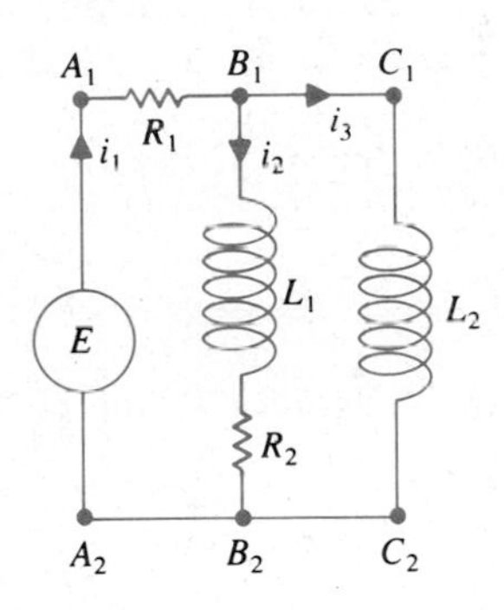

Figure 8.2

$$i_1(t) = i_2(t) + i_3(t). \tag{9}$$

In addition we can also apply **Kirchhoff's second law** to each loop. For loop $A_1B_1B_2A_2A_1$, summing the voltage drops across each part of the loop gives

$$E(t) = i_1R_1 + L_1\frac{di_2}{dt} + i_2R_2. \tag{10}$$

Similarly, for loop $A_1B_1C_1C_2B_2A_2A_1$ we find

$$E(t) = i_1R_1 + L_2\frac{di_3}{dt}. \tag{11}$$

Using (9) to eliminate i_1 in (10) and (11) yields two first-order equations for the currents $i_2(t)$ and $i_3(t)$

$$\begin{aligned} L_1\frac{di_2}{dt} + (R_1 + R_2)i_2 + R_1i_3 &= E(t) \\ L_2\frac{di_3}{dt} + \qquad R_1i_2 + R_1i_3 &= E(t). \end{aligned} \tag{12}$$

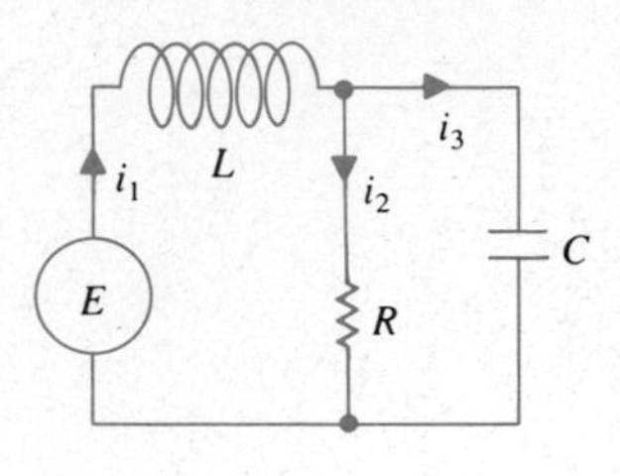

Figure 8.3

Given the natural initial conditions $i_2(0) = 0$, $i_3(0) = 0$, the system (12) is amenable to solution by the Laplace transform.

We leave it as an exercise (see Problem 18) to show that the system of differential equations describing the currents $i_1(t)$ and $i_2(t)$ in the network containing a resistor, an inductor, and a capacitor shown in Figure 8.3 is

$$\begin{aligned} L\frac{di_1}{dt} + Ri_2 \quad &= E(t) \\ RC\frac{di_2}{dt} + i_2 - i_1 &= 0. \end{aligned} \tag{13}$$

EXAMPLE 3

Solve the system (13) under the conditions $E = 60$ volts, $L = 1$ henry, $R = 50$ ohms, $C = 10^{-4}$ farads, and i_1 and i_2 are initially zero.

SOLUTION We must solve

$$\begin{aligned} \frac{di_1}{dt} + 50i_2 \quad &= 60 \\ 50(10^{-4})\frac{di_2}{dt} + i_2 - i_1 &= 0 \end{aligned}$$

subject to $i_1(0) = 0$, $i_2(0) = 0$.

Applying the Laplace transform to each equation of the system and simplifying give

$$\begin{aligned} sI_1(s) + \qquad\qquad 50I_2(s) &= \frac{60}{s} \\ -200I_1(s) + (s + 200)I_2(s) &= 0, \end{aligned}$$

where $I_1(s) = \mathscr{L}\{i_1(t)\}$ and $I_2(s) = \mathscr{L}\{i_2(t)\}$. Solving the system for I_1 and I_2 yields

$$I_1(s) = \frac{60s + 12{,}000}{s(s + 100)^2}$$

$$I_2(s) = \frac{12{,}000}{s(s + 100)^2}.$$

By partial fractions we can write

$$I_1(s) = \frac{6/5}{s} - \frac{6/5}{s + 100} - \frac{60}{(s + 100)^2}$$

$$I_2(s) = \frac{6/5}{s} - \frac{6/5}{s + 100} - \frac{120}{(s + 100)^2},$$

from which it follows that

$$i_1(t) = \frac{6}{5} - \frac{6}{5}e^{-100t} - 60te^{-100t}$$

$$i_2(t) = \frac{6}{5} - \frac{6}{5}e^{-100t} - 120te^{-100t}$$

Note that both $i_1(t)$ and $i_2(t)$ in the preceding example tend toward the value $E/R = 6/5$ as $t \to \infty$. Furthermore since the current through the capacitor is $i_3(t) = i_1(t) - i_2(t) = 60te^{-100t}$, we observe $i_3(t) \to 0$ as $t \to \infty$.

Exercises 8.2

Answers to odd-numbered problems begin on page A–22.

In Problems 1–12 use the Laplace transform to solve the given system of differential equations.

1. $\dfrac{dx}{dt} = -x + y$

$\dfrac{dy}{dt} = 2x,$

$x(0) = 0, \quad y(0) = 1$

2. $\dfrac{dx}{dt} = 2y + e^t$

$\dfrac{dy}{dt} = 8x - t,$

$x(0) = 1, \quad y(0) = 1$

3. $\dfrac{dx}{dt} = x - 2y$

$\dfrac{dy}{dt} = 5x - y,$

$x(0) = -1, \quad y(0) = 2$

4. $\dfrac{dx}{dt} + 3x + \dfrac{dy}{dt} = 1$

$\dfrac{dx}{dt} - x + \dfrac{dy}{dt} - y = e^t,$

$x(0) = 0, \quad y(0) = 0$

5. $2\dfrac{dx}{dt} + \dfrac{dy}{dt} - 2x = 1$

$\dfrac{dx}{dt} + \dfrac{dy}{dt} - 3x - 3y = 2,$

$x(0) = 0, \quad y(0) = 0$

6. $\dfrac{dx}{dt} + x - \dfrac{dy}{dt} + y = 0$

$\dfrac{dx}{dt} + \dfrac{dy}{dt} + 2y = 0$

$x(0) = 0, \quad y(0) = 1$

7. $\dfrac{d^2x}{dt^2} + x - y = 0$

$\dfrac{d^2y}{dt^2} + y - x = 0,$

$x(0) = 0, \quad x'(0) = -2,$

$y(0) = 0, \quad y'(0) = 1$

8. $\dfrac{d^2x}{dt^2} + \dfrac{dx}{dt} + \dfrac{dy}{dt} = 0$

$\dfrac{d^2y}{dt^2} + \dfrac{dy}{dt} - 4\dfrac{dx}{dt} = 0,$

$x(0) = 1, \quad x'(0) = 0,$

$y(0) = -1, \quad y'(0) = 5$

9. $\dfrac{d^2x}{dt^2} + \dfrac{d^2y}{dt^2} = t^2$

$\dfrac{d^2x}{dt^2} - \dfrac{d^2y}{dt^2} = 4t,$

$x(0) = 8, \quad x'(0) = 0,$

$y(0) = 0, \quad y'(0) = 0$

10. $\dfrac{dx}{dt} - 4x + \dfrac{d^3y}{dt^3} = 6 \sin t$

$\dfrac{dx}{dt} + 2x - 2\dfrac{d^3y}{dt^3} = 0,$

$x(0) = 0, \quad y(0) = 0,$

$y'(0) = 0, \quad y''(0) = 0$

11. $\dfrac{d^2x}{dt^2} + 3\dfrac{dy}{dt} + 3y = 0$

$\dfrac{d^2x}{dt^2} + 3y = te^{-t},$

$x(0) = 0, \quad x'(0) = 2, \quad y(0) = 0$

12. $\dfrac{dx}{dt} = 4x - 2y + 2\mathscr{U}(t-1)$

$\dfrac{dy}{dt} = 3x - y + \mathscr{U}(t-1),$

$x(0) = 0, \quad y(0) = \frac{1}{2}$

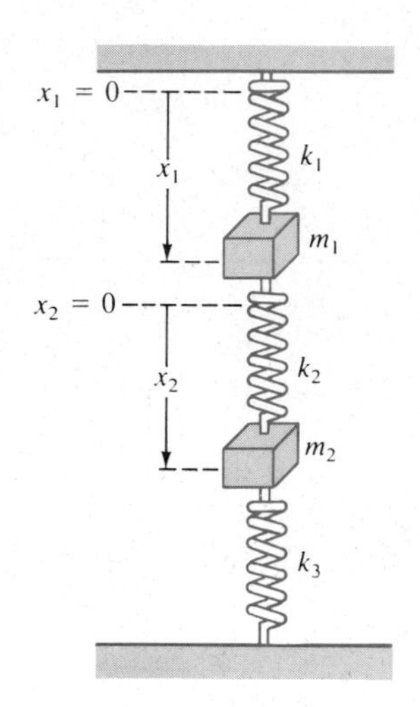

Figure 8.4

13. Solve system (5) when

$$k_1 = 3, \quad k_2 = 2, \quad m_1 = 1, \quad m_2 = 1$$

and

$$x_1(0) = 0, \quad x_1'(0) = 1$$

$$x_2(0) = 1, \quad x_2'(0) = 0.$$

14. Derive the system of differential equations describing the straight line vertical motion of the coupled springs shown in Figure 8.4. Use the Laplace transform to solve the system when $k_1 = 1$, $k_2 = 1$, $k_3 = 1$, $m_1 = 1$, $m_2 = 1$, and $x_1(0) = 0$, $x_1'(0) = -1$, $x_2(0) = 0$, $x_2'(0) = 1$.

15. (a) Show that the system of differential equations for the currents $i_2(t)$ and $i_3(t)$ in the electrical network shown in Figure 8.5 is

$$L_1 \frac{di_2}{dt} + Ri_2 + Ri_3 = E(t)$$

$$L_2 \frac{di_3}{dt} + Ri_2 + Ri_3 = E(t).$$

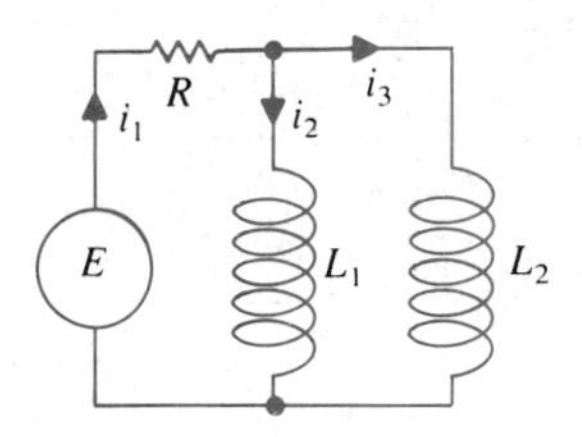

Figure 8.5

(b) Solve the system in part (a) if $R = 5$ ohms, $L_1 = 0.01$ henry, $L_2 = 0.0125$ henry, $E = 100$ volts, $i_2(0) = 0$, and $i_3(0) = 0$.

(c) Determine the current $i_1(t)$.

16. (a) Show that the system of differential equations for the currents $i_2(t)$ and $i_3(t)$ in the electrical network shown in Figure 8.6 is

$$L\frac{di_2}{dt} + L\frac{di_3}{dt} + R_1 i_2 = E(t)$$

$$-R_1\frac{di_2}{dt} + R_2\frac{di_3}{dt} + \frac{1}{C} i_3 = 0.$$

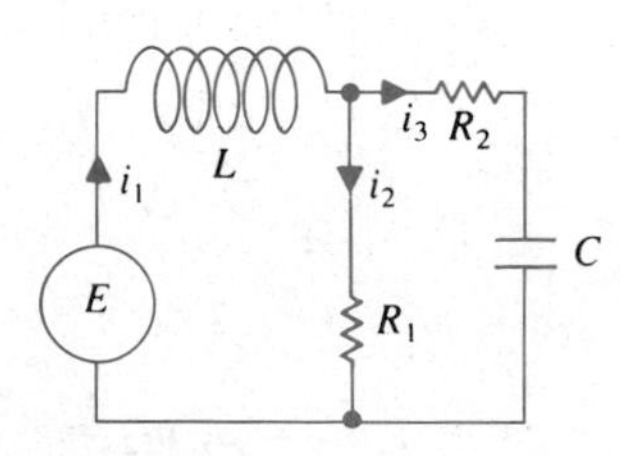

Figure 8.6

(b) Solve the system in part (a) if $R_1 = 10$ ohms, $R_2 = 5$ ohms, $L =$ 1 henry, $C = 0.2$ farads,

$$E(t) = \begin{cases} 120, & 0 \leq t < 2, \\ 0, & t \geq 2, \end{cases}$$

$i_2(0) = 0$, and $i_3(0) = 0$.

(c) Determine the current $i_1(t)$.

17. Solve the system given in (12) when $R_1 = 6$ ohms, $R_2 = 5$ ohms, $L_1 =$ 1 henry, $L_2 = 1$ henry, $E(t) = 50 \sin t$ volts.

18. Derive the system of equations (13).

19. Solve (13) when $E = 60$ volts, $L = 1/2$ henry, $R = 50$ ohms, $C = 10^{-4}$ farads, $i_1(0) = 0$, $i_2(0) = 0$.

20. Solve (13) when $E = 60$ volts, $L = 2$ henry, $R = 50$ ohms, $C = 10^{-4}$ farads, $i_1(0) = 0$, $i_2(0) = 0$.

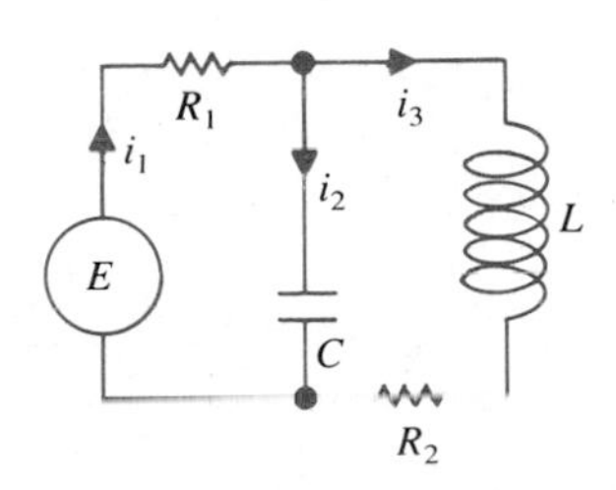

Figure 8.7

21. (a) Show that the system of differential equations for the charge on the capacitor $q(t)$ and the current $i_3(t)$ in the electrical network shown in Figure 8.7 is

$$R_1 \frac{dq}{dt} + \frac{1}{C} q + R_1 i_3 = E(t)$$

$$L \frac{di_3}{dt} + R_2 i_3 - \frac{1}{C} q = 0.$$

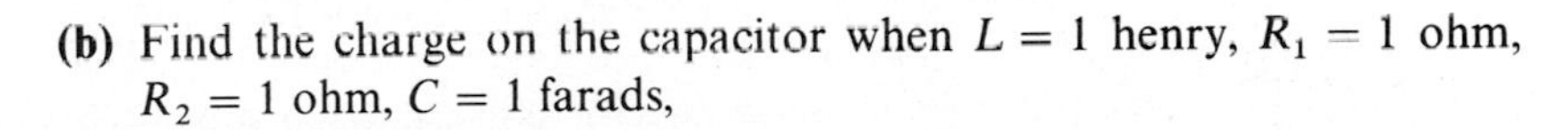

(b) Find the charge on the capacitor when $L = 1$ henry, $R_1 = 1$ ohm, $R_2 = 1$ ohm, $C = 1$ farads,

$$E(t) = \begin{cases} 0, & 0 < t < 1, \\ 50e^{-t}, & t \geq 1, \end{cases}$$

$i_3(0) = 0$ and $q(0) = 0$.

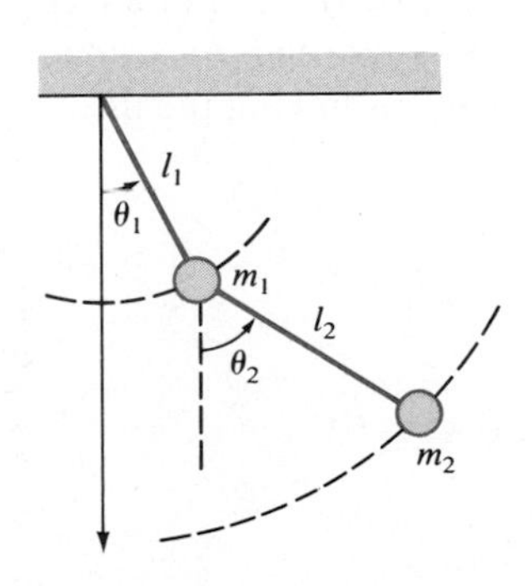

Figure 8.8

22. A double pendulum oscillates in a vertical plane under the influence of gravity (see Figure 8.8). For small displacements $\theta_1(t)$ and $\theta_2(t)$, it can be shown that the differential equations of motion are

$$(m_1 + m_2)l_1^2\theta_1'' + m_2 l_1 l_2 \theta_2'' + (m_1 + m_2)l_1 g\theta_1 = 0$$

$$m_2 l_2^2 \theta_2'' + m_2 l_1 l_2 \theta_1'' + m_2 l_2 g \theta_2 = 0.$$

Use the Laplace transform to solve the system when $m_1 = 3$, $m_2 = 1$, $l_1 = l_2 = 16$, $\theta_1(0) = 1$, $\theta_2(0) = -1$, $\theta_1'(0) = 0$, $\theta_2'(0) = 0$.

8.3 Systems of Linear First-Order Equations

In the preceding two sections we dealt with linear systems that were of the form

$$\begin{aligned}
P_{11}(D)x_1 + P_{12}(D)x_2 + \cdots + P_{1n}(D)x_n &= b_1(t)\\
P_{21}(D)x_1 + P_{22}(D)x_2 + \cdots + P_{2n}(D)x_n &= b_2(t)\\
\vdots \qquad\qquad\qquad \vdots \qquad & \qquad \vdots\\
P_{n1}(D)x_1 + P_{n2}(D)x_2 + \cdots + P_{nn}(D)x_n &= b_n(t),
\end{aligned} \tag{1}$$

where the P_{ij} were polynomials in the differential operator D. However, the study of systems of *first-order* differential equations

$$\begin{aligned}
\frac{dx_1}{dt} &= g_1(t, x_1, x_2, \ldots, x_n)\\
\frac{dx_2}{dt} &= g_2(t, x_1, x_2, \ldots, x_n)\\
&\vdots\\
\frac{dx_n}{dt} &= g_n(t, x_1, x_2, \ldots, x_n)
\end{aligned} \tag{2}$$

is particularly important in advanced mathematics since every nth-order differential equation

$$y^{(n)} = F(t, y, y', \ldots, y^{(n-1)}),$$

as well as most systems of differential equations, can be reduced to form (2).

Linear Normal Form

Of course, a system such as (2) need not be linear and need not have constant coefficients. Consequently, the system may not be readily solvable, if at all. In the remaining sections of this chapter we shall be interested only in a particular, but important, case of (2), namely, those systems having the linear **normal**, or **canonical**, form

$$\begin{aligned}
\frac{dx_1}{dt} &= a_{11}(t)x_1 + a_{12}(t)x_2 + \cdots + a_{1n}(t)x_n + f_1(t)\\
\frac{dx_2}{dt} &= a_{21}(t)x_1 + a_{22}(t)x_2 + \cdots + a_{2n}(t)x_n + f_2(t)\\
\vdots \quad & \qquad\qquad\qquad \vdots \qquad\qquad\qquad \vdots\\
\frac{dx_n}{dt} &= a_{n1}(t)x_1 + a_{n2}(t)x_2 + \cdots + a_{nn}(t)x_n + f_n(t),
\end{aligned} \tag{3}$$

where the coefficients a_{ij} and the f_i are functions continuous on a common interval I. When $f_i(t) = 0$, $i = 1, 2, \ldots, n$, the system (3) is said to be **homogeneous**; otherwise it is called **nonhomogeneous**.

We shall now show that every linear nth-order differential equation can be reduced to a linear system having the normal form (3).

Equation to a System

Suppose a linear nth-order differential equation is first written as

$$\frac{d^n y}{dt^n} = -\frac{a_0}{a_n} y - \frac{a_1}{a_n} y' - \cdots - \frac{a_{n-1}}{a_n} y^{(n-1)} + f(t). \tag{4}$$

If we then introduce the variables

$$y = x_1, \quad y' = x_2, \quad y'' = x_3, \ldots, \quad y^{(n-1)}(x) = x_n, \tag{5}$$

it follows that $y' = x_1' = x_2$, $y'' = x_2' = x_3, \ldots, y^{(n-1)} = x_{n-1}' = x_n$, and $y^{(n)} = x_n'$. Hence from (4) and (5) we obtain

$$\begin{aligned} x_1' &= x_2 \\ x_2' &= x_3 \\ x_3' &= x_4 \\ &\vdots \\ x_{n-1}' &= x_n \\ x_n' &= -\frac{a_0}{a_n} x_1 - \frac{a_1}{a_n} x_2 - \cdots - \frac{a_{n-1}}{a_n} x_n + f(t). \end{aligned} \tag{6}$$

Inspection of (6) reveals that it has the same form as (3).

EXAMPLE 1

Reduce the third-order equation

$$2y''' - 6y'' + 4y' + y = \sin t$$

to the normal form (3).

SOLUTION Write the differential equation as

$$y''' = -\frac{1}{2} y - 2y' + 3y'' + \frac{1}{2} \sin t$$

and then let $y = x_1$, $y' = x_2$, $y'' = x_3$. Since

$$\begin{aligned} x_1' &= y' = x_2 \\ x_2' &= y'' = x_3 \\ x_3' &= y''', \end{aligned}$$

we find

$$x_1' = x_2$$
$$x_2' = x_3$$
$$x_3' = -\frac{1}{2}x_1 - 2x_2 + 3x_3 + \frac{1}{2}\sin t.$$

Systems Reduced to Normal Form

Using a procedure similar to that just outlined, we can reduce *most* systems of the linear form (1) to the linear normal form (3). To accomplish this it is necessary to first solve the system for the highest-order derivative of each dependent variable. As we shall see, this may not always be possible.

EXAMPLE 2

Reduce

$$\begin{aligned}(D^2 - D + 5)x + \quad 2D^2y &= e^t \\ -2x + (D^2 + 2)y &= 3t^2\end{aligned}$$

to the normal form (3).

SOLUTION Write the system as

$$\begin{aligned}D^2x + 2D^2y &= e^t - 5x + Dx \\ D^2y &= 3t^2 + 2x - 2y\end{aligned}$$

and then eliminate D^2y by multiplying the second equation by 2 and subtracting. We have

$$D^2x = e^t - 6t^2 - 9x + 4y + Dx.$$

Since the second equation of the system already expresses the highest-order derivative of y in terms of the remaining functions, we are now in a position to introduce new variables. If we let

$$Dx = u \qquad \text{and} \qquad Dy = v$$

the expressions for D^2x and D^2y become, respectively,

$$\begin{aligned}Du &= e^t - 6t^2 - 9x + 4y + u \\ Dv &= 3t^2 + 2x - 2y.\end{aligned}$$

Thus the original system can be written in the normal form

$$\begin{aligned}Dx &= u \\ Dy &= v \\ Du &= -9x + 4y + u + e^t - 6t^2 \\ Dv &= \quad 2x - 2y + 3t^2.\end{aligned}$$

Degenerate Systems

Those systems of differential equations of form (1) that cannot be reduced to a linear system in normal form are said to be **degenerate**. For example, it is a straightforward matter to show that it is impossible to solve the system

$$
\begin{aligned}
(D + 1)x + (D + 1)y &= 0 \\
2Dx + (2D + 1)y &= 0
\end{aligned} \tag{7}
$$

for the highest derivative of each variable, and hence the system is degenerate.*

The reader may be wondering why anyone would want to convert a single differential equation to a system of equations, or for that matter a system of differential equations to an even larger system. While we are not in a position to completely justify their importance, suffice it to say that these procedures are more than a theoretical exercise. There are times when it is actually desirable to work with a system rather than with one equation. In the numerical analysis of differential equations, almost all computational algorithms are established for first-order equations. Since these algorithms can be generalized directly to systems, to compute numerically, say, a second-order equation, we could reduce it to a system of two first-order equations (see Chapter 9).

A linear system such as (3) also arises naturally in some physical applications. The following example illustrates a homogeneous system in two dependent variables.

EXAMPLE 3

Tank A contains 50 gallons of water in which 25 pounds of salt are dissolved. A second tank, B, contains 50 gallons of pure water. Liquid is pumped in and out of the tanks at rates shown in Figure 8.9. Derive the differential equations that describe the number of pounds $x_1(t)$ and $x_2(t)$ of salt at any time in tanks A and B, respectively.

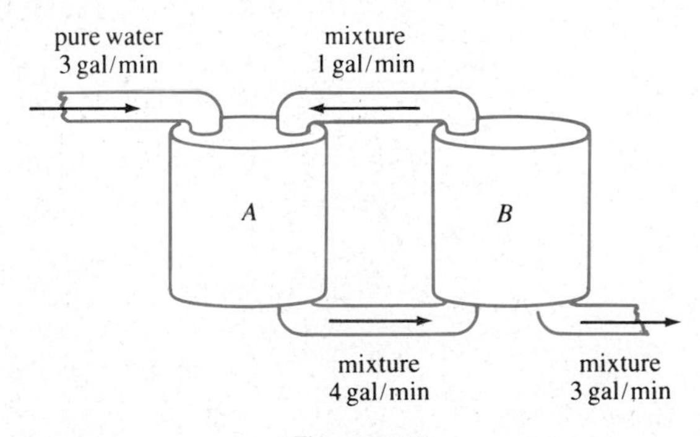

Figure 8.9

* This does *not* mean that the system does not have a solution (see Problem 21).

SOLUTION By an analysis similar to that used in Section 3.2, we see that the net rate of change in $x_1(t)$ in lb/min is

$$\frac{dx_1}{dt} = \overbrace{(3 \text{ gal/min}) \cdot (0 \text{ lb/gal}) + (1 \text{ gal/min}) \cdot \left(\frac{x_2}{50} \text{ lb/gal}\right)}^{\text{input}} - \overbrace{(4 \text{ gal/min}) \cdot \left(\frac{x_1}{50} \text{ lb/gal}\right)}^{\text{output}}$$

$$= -\frac{2}{25}x_1 + \frac{1}{50}x_2.$$

In addition we find the net rate of change in $x_2(t)$ is

$$\frac{dx_2}{dt} = 4 \cdot \frac{x_1}{50} - 3 \cdot \frac{x_2}{50} - 1 \cdot \frac{x_2}{50}$$

$$= \frac{2}{25}x_1 - \frac{2}{25}x_2.$$

Thus we obtain the first-order system

$$\begin{aligned} \frac{dx_1}{dt} &= -\frac{2}{25}x_1 + \frac{1}{50}x_2 \\ \frac{dx_2}{dt} &= \frac{2}{25}x_1 - \frac{2}{25}x_2. \end{aligned} \tag{8}$$

Observe that the foregoing system is accompanied by the initial conditions $x_1(0) = 25$, $x_2(0) = 0$.

It is left as an exercise to solve (8) by the Laplace transform.

Exercises 8.3

Answers to odd-numbered problems begin on page A–22.

In Problems 1–8 rewrite the given differential equation as a system in normal form (3).

1. $y'' - 3y' + 4y = \sin 3t$

2. $2\frac{d^2y}{dt^2} + 4\frac{dy}{dt} - 5y = 0$

3. $y''' - 3y'' + 6y' - 10y = t^2 + 1$

4. $4y''' + y = e^t$

5. $\frac{d^4y}{dt^4} - 2\frac{d^2y}{dt^2} + 4\frac{dy}{dt} + y = t$

6. $2\frac{d^4y}{dt^4} + \frac{d^3y}{dt^3} - 8y = 10$

7. $(t + 1)y'' = ty$

8. $t^2y'' + ty' + (t^2 - 4)y = 0$

In Problems 9–16 rewrite, if possible, the given system in the normal form (3).

9. $x' + 4x - y' = 7t$
$x' + y' - 2y = 3t$

10. $x'' + y' = 1$
$x'' + y' = -1$

11. $(D - 1)x - Dy = t^2$
$x + Dy = 5t - 2$

12. $x'' - 2y'' = \sin t$
$x'' + y'' = \cos t$

13. $(2D + 1)x - 2Dy = 4$
$Dx - Dy = e^t$

14. $m_1 x_1'' = -k_1 x_1 + k_2(x_2 - x_1)$
$m_2 x_2'' = -k_2(x_2 - x_1)$

15. $\dfrac{d^3x}{dt^3} = 4x - 3\dfrac{d^2x}{dt^2} + 4\dfrac{dy}{dt}$
$\dfrac{d^2y}{dt^2} = 10t^2 - 4\dfrac{dx}{dt} + 3\dfrac{dy}{dt}$

16. $D^2x + Dy = 4t$
$-D^2x + (D + 1)y = 6t^2 + 10$

17. Use the Laplace transform to solve system (8) subject to $x_1(0) = 25$ and $x_2(0) = 0$.

18. Consider two tanks A and B with liquid being pumped in and out at the same rates as given in Example 3. What is the system of differential equations if, instead of pure water, a brine solution containing 2 lb of salt per gallon is pumped into tank A?

19. Using the information given in Figure 8.10, derive the system of differential equations describing the number of pounds of salt x_1, x_2, and x_3 at any time in tanks A, B, and C, respectively.

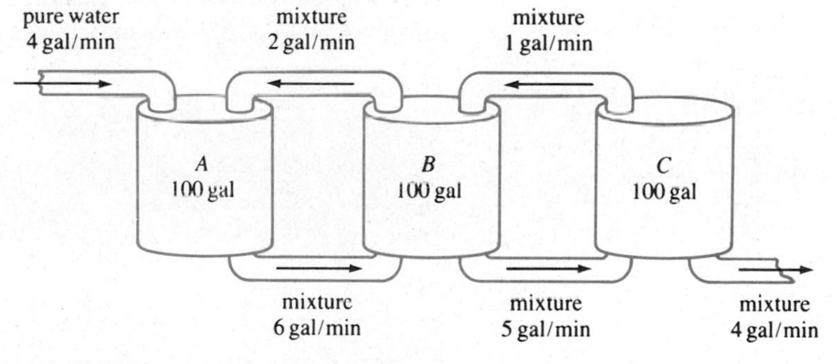

Figure 8.10

Miscellaneous Problems

20. Consider the first-order system

$$(a_1 D - b_1)x + (a_2 D - b_2)y = 0$$
$$(a_3 D - b_3)x + (a_4 D - b_4)y = 0,$$

where the a_i are nonzero constants. Determine a condition on the a_i such that the system is degenerate.

21. Verify that the degenerate system (7) possesses the solution $x(t) = c_1 e^{-t}$, $y(t) = -2c_1 e^{-t}$.

8.4 Introduction to Matrices

8.4.1 Basic Definitions and Theory

Matrix

Before examining a systematic procedure for solving linear first-order systems in normal form, we need the new and useful concept of a **matrix**.

DEFINITION 8.1 A **matrix A** is any rectangular array of numbers or functions:

$$\mathbf{A} = \begin{pmatrix} a_{11} & a_{12} & \cdots & a_{1n} \\ a_{21} & a_{22} & \cdots & a_{2n} \\ \vdots & & & \vdots \\ a_{m1} & a_{m2} & \cdots & a_{mn} \end{pmatrix}. \tag{1}$$

If a matrix has m rows and n columns, we say that its **size** is m by n, (written $m \times n$). An $n \times n$ matrix is called a **square** matrix of order n.

The element, or entry, in the ith row and jth column of an $m \times n$ matrix $\mathbf{A}$ is written a_{ij}. An $m \times n$ matrix $\mathbf{A}$ is then abbreviated as $\mathbf{A} = (a_{ij})_{m \times n}$ or simply $\mathbf{A} = (a_{ij})$. A 1×1 matrix is simply one constant or function.

Equality

DEFINITION 8.2 Two $m \times n$ matrices $\mathbf{A}$ and $\mathbf{B}$ are **equal** if $a_{ij} = b_{ij}$ for each i and j.

Vectors

DEFINITION 8.3 A **column matrix X** is any matrix having n rows and one column:

$$\mathbf{X} = \begin{pmatrix} b_{11} \\ b_{21} \\ \vdots \\ b_{n1} \end{pmatrix} = (b_{i1})_{n \times 1}.$$

A column matrix is also called a **column vector** or simply a **vector**.

Multiples of a Matrix

DEFINITION 8.4 A **multiple** of a matrix $\mathbf{A}$ is defined to be

$$k\mathbf{A} = \begin{pmatrix} ka_{11} & ka_{12} & \cdots & ka_{1n} \\ ka_{21} & ka_{22} & \cdots & ka_{2n} \\ \vdots & & & \vdots \\ ka_{m1} & ka_{m2} & \cdots & ka_{mn} \end{pmatrix} = (ka_{ij})_{m \times n},$$

where k is a constant or a function.

EXAMPLE 1

(a)
$$5\begin{pmatrix} 2 & -3 \\ 4 & -1 \\ \frac{1}{5} & 6 \end{pmatrix} = \begin{pmatrix} 10 & -15 \\ 20 & -5 \\ 1 & 30 \end{pmatrix}$$

(b)
$$e^t\begin{pmatrix} 1 \\ -2 \\ 4 \end{pmatrix} = \begin{pmatrix} e^t \\ -2e^t \\ 4e^t \end{pmatrix}$$

We note in passing that for any matrix $\mathbf{A}$ the product $k\mathbf{A}$ is the same as $\mathbf{A}k$. For example,

$$e^{-3t}\begin{pmatrix} 2 \\ 5 \end{pmatrix} = \begin{pmatrix} 2e^{-3t} \\ 5e^{-3t} \end{pmatrix} = \begin{pmatrix} 2 \\ 5 \end{pmatrix}e^{-3t}.$$

Addition of Matrices

DEFINITION 8.5 The **sum** of two $m \times n$ matrices $\mathbf{A}$ and $\mathbf{B}$ is defined to be the matrix

$$\mathbf{A} + \mathbf{B} = (a_{ij} + b_{ij})_{m \times n}.$$

In other words when adding two matrices of the same size, we add the corresponding elements.

EXAMPLE 2 The sum of

$$\mathbf{A} = \begin{pmatrix} 2 & -1 & 3 \\ 0 & 4 & 6 \\ -6 & 10 & -5 \end{pmatrix} \quad \text{and} \quad \mathbf{B} = \begin{pmatrix} 4 & 7 & -8 \\ 9 & 3 & 5 \\ 1 & -1 & 2 \end{pmatrix}$$

is
$$\mathbf{A}+\mathbf{B}=\begin{pmatrix} 2+4 & -1+7 & 3+(-8) \\ 0+9 & 4+3 & 6+5 \\ -6+1 & 10+(-1) & -5+2 \end{pmatrix}$$
$$=\begin{pmatrix} 6 & 6 & -5 \\ 9 & 7 & 11 \\ -5 & 9 & -3 \end{pmatrix}.$$

EXAMPLE 3

The single matrix
$$\begin{pmatrix} 3t^2-2e^t \\ t^2+7t \\ 5t \end{pmatrix}$$

can be written as the sum of three column vectors

$$\begin{pmatrix} 3t^2-2e^t \\ t^2+7t \\ 5t \end{pmatrix}=\begin{pmatrix} 3t^2 \\ t^2 \\ 0 \end{pmatrix}+\begin{pmatrix} 0 \\ 7t \\ 5t \end{pmatrix}+\begin{pmatrix} -2e^t \\ 0 \\ 0 \end{pmatrix}$$
$$=\begin{pmatrix} 3 \\ 1 \\ 0 \end{pmatrix}t^2+\begin{pmatrix} 0 \\ 7 \\ 5 \end{pmatrix}t+\begin{pmatrix} -2 \\ 0 \\ 0 \end{pmatrix}e^t.$$

The **difference** of two $m \times n$ matrices is defined in the usual manner: $\mathbf{A}-\mathbf{B}=\mathbf{A}+(-\mathbf{B})$ where $-\mathbf{B}=(-1)\mathbf{B}$.

DEFINITION 8.6 Let **A** be a matrix having m rows and n columns and **B** be a matrix having n rows and p columns. We define the **product AB** to be the $m \times p$ matrix

$$\mathbf{AB}=\begin{pmatrix} a_{11} & a_{12} & \cdots & a_{1n} \\ a_{21} & a_{22} & \cdots & a_{2n} \\ \vdots & & & \vdots \\ a_{m1} & a_{m2} & \cdots & a_{mn} \end{pmatrix}\begin{pmatrix} b_{11} & b_{12} & \cdots & b_{1p} \\ b_{21} & b_{22} & \cdots & b_{2p} \\ \vdots & & & \vdots \\ b_{n1} & b_{n2} & \cdots & b_{np} \end{pmatrix}$$
$$=\begin{pmatrix} a_{11}b_{11}+a_{12}b_{21}+\cdots+a_{1n}b_{n1} & \cdots & a_{11}b_{1p}+a_{12}b_{2p}+\cdots+a_{1n}b_{np} \\ a_{21}b_{11}+a_{22}b_{21}+\cdots+a_{2n}b_{n1} & \cdots & a_{21}b_{1p}+a_{22}b_{2p}+\cdots+a_{2n}b_{np} \\ \vdots & & \vdots \\ a_{m1}b_{11}+a_{m2}b_{21}+\cdots+a_{mn}b_{n1} & \cdots & a_{m1}b_{1p}+a_{m2}b_{2p}+\cdots+a_{mn}b_{np} \end{pmatrix}$$
$$=\left(\sum_{k=1}^{n} a_{ik}b_{kj}\right)_{m\times p}$$

Note carefully in Definition 8.6 that the product $\mathbf{AB} = \mathbf{C}$ is defined only when the number of columns in the matrix $\mathbf{A}$ is the same as the number of rows in $\mathbf{B}$. The size of the product can be determined from

$$\mathbf{A}_{m \times n}\mathbf{B}_{n \times p} = \mathbf{C}_{m \times p}.$$

Also, you might recognize that the entries in, say, the ith row of the final matrix $\mathbf{AB}$ are formed by using the component definition of the inner or dot product of the ith row of $\mathbf{A}$ with each of the columns of $\mathbf{B}$.

EXAMPLE 4

(a) For $\mathbf{A} = \begin{pmatrix} 4 & 7 \\ 3 & 5 \end{pmatrix}$ and $\mathbf{B} = \begin{pmatrix} 9 & -2 \\ 6 & 8 \end{pmatrix}$

$$\mathbf{AB} = \begin{pmatrix} 4 \cdot 9 + 7 \cdot 6 & 4 \cdot (-2) + 7 \cdot 8 \\ 3 \cdot 9 + 5 \cdot 6 & 3 \cdot (-2) + 5 \cdot 8 \end{pmatrix} = \begin{pmatrix} 78 & 48 \\ 57 & 34 \end{pmatrix}.$$

(b) For $\mathbf{A} = \begin{pmatrix} 5 & 8 \\ 1 & 0 \\ 2 & 7 \end{pmatrix}$ and $\mathbf{B} = \begin{pmatrix} -4 & -3 \\ 2 & 0 \end{pmatrix}$

$$\mathbf{AB} = \begin{pmatrix} 5 \cdot (-4) + 8 \cdot 2 & 5 \cdot (-3) + 8 \cdot 0 \\ 1 \cdot (-4) + 0 \cdot 2 & 1 \cdot (-3) + 0 \cdot 0 \\ 2 \cdot (-4) + 7 \cdot 2 & 2 \cdot (-3) + 7 \cdot 0 \end{pmatrix} = \begin{pmatrix} -4 & -15 \\ -4 & -3 \\ 6 & -6 \end{pmatrix}.$$

In general, matrix multiplication is not commutative. That is, $\mathbf{AB} \neq \mathbf{BA}$. Observe that in part (a) of Example 4

$$\mathbf{BA} = \begin{pmatrix} 30 & 53 \\ 48 & 82 \end{pmatrix},$$

whereas in part (b) the product $\mathbf{BA}$ is *not defined* since Definition 8.6 requires that the first matrix (in this case $\mathbf{B}$) have the same number of columns as the second matrix has rows.

We are particularly interested in the product of a square matrix and a column vector.

EXAMPLE 5

(a)
$$\begin{pmatrix} 2 & -1 & 3 \\ 0 & 4 & 5 \\ 1 & -7 & 9 \end{pmatrix} \begin{pmatrix} -3 \\ 6 \\ 4 \end{pmatrix} = \begin{pmatrix} 2 \cdot (-3) + (-1) \cdot 6 + 3 \cdot 4 \\ 0 \cdot (-3) + 4 \cdot 6 \quad + 5 \cdot 4 \\ 1 \cdot (-3) + (-7) \cdot 6 + 9 \cdot 4 \end{pmatrix} = \begin{pmatrix} 0 \\ 44 \\ -9 \end{pmatrix}$$

(b)
$$\begin{pmatrix} -4 & 2 \\ 3 & 8 \end{pmatrix} \begin{pmatrix} x \\ y \end{pmatrix} = \begin{pmatrix} -4x + 2y \\ 3x + 8y \end{pmatrix}$$

Multiplicative Identity

For a given positive integer n, the $n \times n$ matrix

$$\mathbf{I} = \begin{pmatrix} 1 & 0 & 0 & \cdots & 0 \\ 0 & 1 & 0 & \cdots & 0 \\ \vdots & & & & \vdots \\ 0 & 0 & 0 & \cdots & 1 \end{pmatrix}$$

is called the **multiplicative identity matrix**. It follows from Definition 8.6 that for any $n \times n$ matrix $\mathbf{A}$

$$\mathbf{AI} = \mathbf{IA} = \mathbf{A}.$$

Also, it is readily verified that if $\mathbf{X}$ is an $n \times 1$ column matrix $\mathbf{IX} = \mathbf{X}$.

Zero Matrix

A matrix consisting of all zero entries is called a **zero matrix** and is denoted by $\mathbf{0}$. For example,

$$\mathbf{0} = \begin{pmatrix} 0 \\ 0 \end{pmatrix}, \quad \mathbf{0} = \begin{pmatrix} 0 & 0 \\ 0 & 0 \end{pmatrix}, \quad \mathbf{0} = \begin{pmatrix} 0 & 0 \\ 0 & 0 \\ 0 & 0 \end{pmatrix},$$

and so on. If $\mathbf{A}$ and $\mathbf{0}$ are $m \times n$ matrices, then

$$\mathbf{A} + \mathbf{0} = \mathbf{0} + \mathbf{A} = \mathbf{A}.$$

Associative Law

Although we shall not prove it, matrix multiplication is **associative**. If $\mathbf{A}$ is an $m \times p$ matrix, $\mathbf{B}$ a $p \times r$ matrix, and $\mathbf{C}$ an $r \times n$ matrix, then

$$\mathbf{A}(\mathbf{BC}) = (\mathbf{AB})\mathbf{C}$$

is an $m \times n$ matrix.

Distributive Law

If $\mathbf{B}$ and $\mathbf{C}$ are $r \times n$ matrices and $\mathbf{A}$ is an $m \times r$ matrix, then the **distributive law** is

$$\mathbf{A}(\mathbf{B} + \mathbf{C}) = \mathbf{AB} + \mathbf{AC}.$$

Furthermore, if the product $(\mathbf{B} + \mathbf{C})\mathbf{A}$ is defined, then

$$(\mathbf{B} + \mathbf{C})\mathbf{A} = \mathbf{BA} + \mathbf{CA}.$$

Determinant of a Matrix

Associated with every *square* matrix $\mathbf{A}$ of constants, there is a number called the **determinant of the matrix**, which is denoted by det $\mathbf{A}$ or $|\mathbf{A}|$.

EXAMPLE 6

For

$$\mathbf{A} = \begin{pmatrix} 3 & 6 & 2 \\ 2 & 5 & 1 \\ -1 & 2 & 4 \end{pmatrix}$$

we expand det **A** by cofactors. Using the first row, we obtain

$$\det \mathbf{A} = \begin{vmatrix} 3 & 6 & 2 \\ 2 & 5 & 1 \\ -1 & 2 & 4 \end{vmatrix} = 3\begin{vmatrix} 5 & 1 \\ 2 & 4 \end{vmatrix} - 6\begin{vmatrix} 2 & 1 \\ -1 & 4 \end{vmatrix} + 2\begin{vmatrix} 2 & 5 \\ -1 & 2 \end{vmatrix}$$

$$= 3(20 - 2) - 6(8 + 1) + 2(4 + 5) = 18.$$

DEFINITION 8.7 The **transpose** of the $m \times n$ matrix (1) is the $n \times m$ matrix $\mathbf{A}^T$ given by

$$\mathbf{A}^T = \begin{pmatrix} a_{11} & a_{21} & \cdots & a_{m1} \\ a_{12} & a_{22} & \cdots & a_{m2} \\ \vdots & & & \vdots \\ a_{1n} & a_{2n} & \cdots & a_{mn} \end{pmatrix}.$$

In other words the rows of a matrix **A** become the columns of its transpose $\mathbf{A}^T$.

EXAMPLE 7

(a) The transpose of matrix **A** in Example 6 is

$$\mathbf{A}^T = \begin{pmatrix} 3 & 2 & -1 \\ 6 & 5 & 2 \\ 2 & 1 & 4 \end{pmatrix}.$$

(b) If $\mathbf{X} = \begin{pmatrix} 5 \\ 0 \\ 3 \end{pmatrix}$, then $\mathbf{X}^T = (5 \quad 0 \quad 3)$.

Multiplicative Inverse

DEFINITION 8.8 Let **A** be an $n \times n$ matrix. If there exists an $n \times n$ matrix **B** such that

$$\mathbf{AB} = \mathbf{BA} = \mathbf{I},$$

where **I** is the multiplicative identity, then **B** is said to be the **multiplicative inverse of A** and is denoted by $\mathbf{B} = \mathbf{A}^{-1}$.

DEFINITION 8.9 Let $\mathbf{A}$ be an $n \times n$ matrix. If $\det \mathbf{A} \neq 0$ then $\mathbf{A}$ is said to be **nonsingular**. If $\det \mathbf{A} = 0$, then $\mathbf{A}$ is said to be **singular**.

The following gives a necessary and sufficient condition for a square matrix to have a multiplicative inverse.

THEOREM 8.1 An $n \times n$ matrix $\mathbf{A}$ has a multiplicative inverse $\mathbf{A}^{-1}$ if and only if $\mathbf{A}$ is nonsingular.

The following theorem gives one way of finding the multiplicative inverse for a nonsingular matrix.

THEOREM 8.2 Let $\mathbf{A}$ be an $n \times n$ nonsingular matrix and let $C_{ij} = (-1)^{i+j}M_{ij}$, where M_{ij} is the determinant of the $(n-1) \times (n-1)$ matrix obtained by deleting the ith row and jth column from $\mathbf{A}$. Then

$$\mathbf{A}^{-1} = \frac{1}{\det \mathbf{A}}(C_{ij})^T. \tag{2}$$

Each C_{ij} in Theorem 8.2 is simply the **cofactor** (signed minor) of the corresponding entry a_{ij} in $\mathbf{A}$. Note that the transpose is utilized in formula (2).

For future reference we observe in the case of a 2×2 nonsingular matrix

$$\mathbf{A} = \begin{pmatrix} a_{11} & a_{12} \\ a_{21} & a_{22} \end{pmatrix},$$

that $C_{11} = a_{22}$, $C_{12} = -a_{21}$, $C_{21} = -a_{12}$, and $C_{22} = a_{11}$.

Thus

$$\mathbf{A}^{-1} = \frac{1}{\det \mathbf{A}}\begin{pmatrix} a_{22} & -a_{21} \\ -a_{12} & a_{11} \end{pmatrix}^T = \frac{1}{\det \mathbf{A}}\begin{pmatrix} a_{22} & -a_{12} \\ -a_{21} & a_{11} \end{pmatrix}. \tag{3}$$

For a 3×3 nonsingular matrix

$$\mathbf{A} = \begin{pmatrix} a_{11} & a_{12} & a_{13} \\ a_{21} & a_{22} & a_{23} \\ a_{31} & a_{32} & a_{33} \end{pmatrix},$$

$$C_{11} = \begin{vmatrix} a_{22} & a_{23} \\ a_{32} & a_{33} \end{vmatrix}, \quad C_{12} = -\begin{vmatrix} a_{21} & a_{23} \\ a_{31} & a_{33} \end{vmatrix}, \quad C_{13} = \begin{vmatrix} a_{21} & a_{22} \\ a_{31} & a_{32} \end{vmatrix},$$

and so on. Carrying out the transposition gives

$$\mathbf{A}^{-1} = \frac{1}{\det \mathbf{A}} \begin{pmatrix} C_{11} & C_{21} & C_{31} \\ C_{12} & C_{22} & C_{32} \\ C_{13} & C_{23} & C_{33} \end{pmatrix}. \tag{4}$$

EXAMPLE 8

Find the multiplicative inverse for $\mathbf{A} = \begin{pmatrix} 1 & 4 \\ 2 & 10 \end{pmatrix}$.

SOLUTION Since $\det \mathbf{A} = 10 - 8 = 2 \neq 0$, $\mathbf{A}$ is nonsingular. It follows from Theorem 8.1 that $\mathbf{A}^{-1}$ exists. From (3) we find

$$\mathbf{A}^{-1} = \frac{1}{2}\begin{pmatrix} 10 & -4 \\ -2 & 1 \end{pmatrix} = \begin{pmatrix} 5 & -2 \\ -1 & \frac{1}{2} \end{pmatrix}.$$

Check:

$$\mathbf{AA}^{-1} = \begin{pmatrix} 1 & 4 \\ 2 & 10 \end{pmatrix}\begin{pmatrix} 5 & -2 \\ -1 & \frac{1}{2} \end{pmatrix} = \begin{pmatrix} 5-4 & -2+2 \\ 10-10 & -4+5 \end{pmatrix} = \begin{pmatrix} 1 & 0 \\ 0 & 1 \end{pmatrix}.$$

$$\mathbf{A}^{-1}\mathbf{A} = \begin{pmatrix} 5 & -2 \\ -1 & \frac{1}{2} \end{pmatrix}\begin{pmatrix} 1 & 4 \\ 2 & 10 \end{pmatrix} = \begin{pmatrix} 5-4 & 20-20 \\ -1+1 & -4+5 \end{pmatrix} = \begin{pmatrix} 1 & 0 \\ 0 & 1 \end{pmatrix}.$$

EXAMPLE 9

The matrix $\mathbf{A} = \begin{pmatrix} 2 & 2 \\ 3 & 3 \end{pmatrix}$ is singular since $\det \mathbf{A} = 2(3) - 2(3) = 0$. We conclude that $\mathbf{A}^{-1}$ does not exist.

EXAMPLE 10

Find the multiplicative inverse for

$$\mathbf{A} = \begin{pmatrix} 2 & 2 & 0 \\ -2 & 1 & 1 \\ 3 & 0 & 1 \end{pmatrix}.$$

SOLUTION Since $\det \mathbf{A} = 12 \neq 0$ the given matrix is nonsingular. The cofactors corresponding to the entries in each row of $\det \mathbf{A}$ are

$$C_{11} = \begin{vmatrix} 1 & 1 \\ 0 & 1 \end{vmatrix} = 1 \qquad C_{12} = -\begin{vmatrix} -2 & 1 \\ 3 & 1 \end{vmatrix} = 5 \qquad C_{13} = \begin{vmatrix} -2 & 1 \\ 3 & 0 \end{vmatrix} = -3$$

$$C_{21} = -\begin{vmatrix} 2 & 0 \\ 0 & 1 \end{vmatrix} = -2 \qquad C_{22} = \begin{vmatrix} 2 & 0 \\ 3 & 1 \end{vmatrix} = 2 \qquad C_{23} = -\begin{vmatrix} 2 & 2 \\ 3 & 0 \end{vmatrix} = 6$$

$$C_{31} = \begin{vmatrix} 2 & 0 \\ 1 & 1 \end{vmatrix} = 2 \qquad C_{32} = -\begin{vmatrix} 2 & 0 \\ -2 & 1 \end{vmatrix} = -2 \qquad C_{33} = \begin{vmatrix} 2 & 2 \\ -2 & 1 \end{vmatrix} = 6$$

It follows from (4) that

$$\mathbf{A}^{-1} = \frac{1}{12}\begin{pmatrix} 1 & -2 & 2 \\ 5 & 2 & -2 \\ -3 & 6 & 6 \end{pmatrix} = \begin{pmatrix} 1/12 & -1/6 & 1/6 \\ 5/12 & 1/6 & -1/6 \\ -1/4 & 1/2 & 1/2 \end{pmatrix}.$$

The reader is urged to verify that $\mathbf{A}^{-1}\mathbf{A} = \mathbf{A}\mathbf{A}^{-1} = \mathbf{I}$.

Formula (2) presents obvious difficulties for nonsingular matrices larger than 3×3. For example, to apply (2) to a 4×4 matrix we would have to calculate *sixteen* 3×3 determinants.* In the case of a large matrix, there are more efficient ways of finding $\mathbf{A}^{-1}$. The curious reader is referred to any text in linear algebra.

Since our goal is to apply the concept of a matrix to systems of linear differential equations in normal form, we need the following definitions.

Derivative of a Matrix of Functions

DEFINITION 8.10 If $\mathbf{A}(t) = (a_{ij}(t))_{m \times n}$ is a matrix whose entries are functions differentiable on a common interval then

$$\frac{d\mathbf{A}}{dt} = \left(\frac{d}{dt}a_{ij}\right)_{m \times n}$$

Integral of a Matrix of Functions

DEFINITION 8.11 If $\mathbf{A}(t) = (a_{ij}(t))_{m \times n}$ is a matrix whose entries are functions continuous on a common interval containing t and t_0, then

$$\int_{t_0}^{t} \mathbf{A}(s)\, ds = \left(\int_{t_0}^{t} a_{ij}(s)\, ds\right)_{m \times n}$$

* Strictly speaking a determinant is a number, but it is sometimes convenient to refer to a determinant as if it were an array.

To differentiate (integrate) a matrix of functions we simply differentiate (integrate) each entry. The derivative of a matrix is also denoted by $\mathbf{A}'(t)$.

EXAMPLE 11

$$\text{If} \quad \mathbf{X}(t) = \begin{pmatrix} \sin 2t \\ e^{3t} \\ 8t - 1 \end{pmatrix}, \quad \text{then} \quad \mathbf{X}'(t) = \begin{pmatrix} \dfrac{d}{dt} \sin 2t \\ \dfrac{d}{dt} e^{3t} \\ \dfrac{d}{dt} (8t - 1) \end{pmatrix} = \begin{pmatrix} 2 \cos 2t \\ 3e^{3t} \\ 8 \end{pmatrix}$$

$$\text{and} \quad \int_0^t \mathbf{X}(s)\, ds = \begin{pmatrix} \displaystyle\int_0^t \sin 2s\, ds \\ \displaystyle\int_0^t e^{3s}\, ds \\ \displaystyle\int_0^t (8s - 1)\, ds \end{pmatrix} = \begin{pmatrix} -\dfrac{1}{2} \cos 2t + \dfrac{1}{2} \\ \dfrac{1}{3} e^{3t} - \dfrac{1}{3} \\ 4t^2 - t \end{pmatrix}.$$

8.4.2 Gauss-Jordan Elimination

In preparation for Section 8.6, we need to know more about solving algebraic systems of n linear equations in n unknowns

$$\begin{aligned} a_{11}x_1 + a_{12}x_2 + \cdots + a_{1n}x_n &= b_1 \\ a_{21}x_1 + a_{22}x_2 + \cdots + a_{2n}x_n &= b_2 \\ \vdots \qquad\qquad\qquad & \quad \vdots \\ a_{n1}x_1 + a_{n2}x_2 + \cdots + a_{nn}x_n &= b_n. \end{aligned} \tag{5}$$

If $\mathbf{A}$ denotes the matrix of coefficients in (5), we know that Cramer's rule (see Appendix IV) could be used to solve the system whenever $\det \mathbf{A} \neq 0$. However, that rule requires a herculean effort if $\mathbf{A}$ is larger than 3×3. The procedure that we shall now consider has the distinct advantage of being not only an efficient way of handling large systems but also a means of solving consistent systems (5) in which $\det \mathbf{A} = 0$ and a means of solving m linear equations in n unknowns.

DEFINITION 8.12 The **augmented matrix** of the system (5) is the $n \times (n + 1)$ matrix

$$\left(\begin{array}{cccc|c} a_{11} & a_{12} & \cdots & a_{1n} & b_1 \\ a_{21} & a_{22} & \cdots & a_{2n} & b_2 \\ \vdots & & & & \vdots \\ a_{n1} & a_{n2} & \cdots & a_{nn} & b_n \end{array}\right).$$

If $\mathbf{B}$ is the column matrix of the b_i, $i = 1, 2, \ldots, n$, the augmented matrix of (5) is denoted by $(\mathbf{A} \mid \mathbf{B})$.

Elementary Row Operations

Recall from algebra that we can transform an algebraic system of equations into an equivalent system (that is, one having the same solution) by multiplying an equation by a nonzero constant, interchanging the position of an equation in a system, and adding a nonzero constant multiple of an equation to another equation. These operations on equations in a system are, in turn, equivalent to **elementary row operations** on an augmented matrix:

(i) Multiply a row by a nonzero constant.

(ii) Interchange any two rows.

(iii) Add a nonzero constant multiple of one row to any other row.

Gauss-Jordan Elimination

To solve a system such as (5) we shall use the so-called **Gauss-Jordan elimination method***. Starting with the augmented matrix of the system, we carry out a succession of elementary row operations until we arrive at an augmented matrix in **reduced row-echelon form**:

(i) The first nonzero entry in a nonzero row is a 1.

(ii) The remaining entries in a column containing a first entry 1 are all zeros.

---

* **Karl Friedrich Gauss** (1777–1855) Gauss was the first of a new breed of precise and demanding mathematicians—the "rigorists." As a child, Gauss was a prodigy in mathematics. As an adult, he often remarked that he could calculate, or "reckon," before he could talk. However, as a college student, Gauss was torn between two loves: philology and mathematics. But inspired by some original mathematical achievements as a teenager and encouraged by the mathematician Wolfgang Bolyai, he found the choice not too difficult. At the age of twenty, Gauss settled on a career in mathematics. At twenty-two he completed a book on number theory, *Disquisitiones Arithmeticae.* Published in 1801, this text was recognized as a masterpiece and even today remains a classic in its field. Gauss's doctoral dissertation of 1799 also remains a memorable document. Using the theory of functions of a complex variable, he was the first to prove the so-called *fundamental theorem of algebra*—that is, every polynomial equation has at least one root.

Although Gauss was certainly recognized and respected as an outstanding mathematician during his lifetime, the full extent of his genius was not realized until the publication of his scientific diary in 1898, forty-four years after his death. Much to the chagrin of some nineteenth-century mathematicians, the diary revealed that Gauss had foreseen, sometimes by decades, many of their discoveries or, perhaps more accurately, rediscoveries. Oblivious to fame, he often pursued his mathematical researches like a child playing on a beach, simply for pleasure and self-satisfaction and not for the instruction that could be given to others through publication.

On any list of "greatest mathematicians who ever lived," Karl Friedrich Gauss must surely rank near or at the top. For his profound impact on so many branches of mathematics, Gauss is sometimes referred to as "the prince of mathematicians."

Wilhelm Jordan (1838–1922) Jordan, a German engineer, used this elimination method to solve linear systems in his 1888 text *Handbook of Geodosy.*

(iii) In consecutive nonzero rows, the first entry 1 in the lower row appears to the right of the 1 in the higher row.

(iv) Rows consisting of all zeros are at the bottom of the matrix.

Once the reduced row-echelon form has been attained, the solution of the system will be apparent by inspection. In terms of the equations in the original system, our goal is to simply make the coefficient of x_1 in the first equation* equal to 1 and then use multiples of that equation to eliminate x_1 from the remaining equations. The process is then repeated on the other variables.

To keep track of the row-operations on an augmented matrix we shall utilize the following notation:

$M_i(c)$ — multiply the ith row by c,

$A_{ij}(c)$ — multiply the ith row by c and add to the jth row,

$I_{ij}(c)$ — multiply the ith row by c and interchange with the jth row.

EXAMPLE 12 Solve

$$\begin{aligned} 2x + 6y + z &= 7 \\ x + 2y - z &= -1 \\ 5x + 7y - 4z &= 9. \end{aligned}$$

SOLUTION

$$\left(\begin{array}{ccc|c} 2 & 6 & 1 & 7 \\ 1 & 2 & -1 & -1 \\ 5 & 7 & -4 & 9 \end{array}\right) \xrightarrow{I_{12}(1)} \left(\begin{array}{ccc|c} 1 & 2 & -1 & -1 \\ 2 & 6 & 1 & 7 \\ 5 & 7 & -4 & 9 \end{array}\right) \xrightarrow[A_{13}(-5)]{A_{12}(-2)} \left(\begin{array}{ccc|c} 1 & 2 & -1 & -1 \\ 0 & 2 & 3 & 9 \\ 0 & -3 & 1 & 14 \end{array}\right)$$

$$\xrightarrow{M_2(\frac{1}{2})} \left(\begin{array}{ccc|c} 1 & 2 & -1 & -1 \\ 0 & 1 & \frac{3}{2} & \frac{9}{2} \\ 0 & -3 & 1 & 14 \end{array}\right) \xrightarrow[A_{23}(3)]{A_{21}(-2)} \left(\begin{array}{ccc|c} 1 & 0 & -4 & -10 \\ 0 & 1 & \frac{3}{2} & \frac{9}{2} \\ 0 & 0 & \frac{11}{2} & \frac{55}{2} \end{array}\right)$$

$$\xrightarrow{M_3(\frac{2}{11})} \left(\begin{array}{ccc|c} 1 & 0 & -4 & -10 \\ 0 & 1 & \frac{3}{2} & \frac{9}{2} \\ 0 & 0 & 1 & 5 \end{array}\right) \xrightarrow[A_{32}(-\frac{3}{2})]{A_{31}(4)} \left(\begin{array}{ccc|c} 1 & 0 & 0 & 10 \\ 0 & 1 & 0 & -3 \\ 0 & 0 & 1 & 5 \end{array}\right).$$

* We can always interchange equations so that the first equation contains the variable x_1.

The last matrix is in reduced row-echelon form and represents the system

$$\begin{aligned} x + 0y + 0z &= 10 \\ 0x + y + 0z &= -3 \\ 0x + 0y + z &= 5. \end{aligned}$$

Hence it is evident that the solution of the system is $x = 10$, $y = -3$, $z = 5$.

EXAMPLE 13

Solve

$$\begin{aligned} x + 3y - 2z &= -7 \\ 4x + y + 3z &= 5 \\ 2x - 5y + 7z &= 19. \end{aligned}$$

SOLUTION

$$\left(\begin{array}{ccc|c} 1 & 3 & -2 & -7 \\ 4 & 1 & 3 & 5 \\ 2 & -5 & 7 & 19 \end{array}\right) \xrightarrow{\substack{A_{12}(-4) \\ A_{13}(-2)}} \left(\begin{array}{ccc|c} 1 & 3 & -2 & -7 \\ 0 & -11 & 11 & 33 \\ 0 & -11 & 11 & 33 \end{array}\right)$$

$$\xrightarrow{\substack{M_2(-\frac{1}{11}) \\ M_3(-\frac{1}{11})}} \left(\begin{array}{ccc|c} 1 & 3 & -2 & -7 \\ 0 & 1 & -1 & -3 \\ 0 & 1 & -1 & -3 \end{array}\right) \xrightarrow{\substack{A_{21}(-3) \\ A_{23}(-1)}} \left(\begin{array}{ccc|c} 1 & 0 & 1 & 2 \\ 0 & 1 & -1 & -3 \\ 0 & 0 & 0 & 0 \end{array}\right)$$

In this case the last matrix in reduced row-echelon form implies that the original system of three equations in three unknowns is really equivalent to two equations in three unknowns. Since only z is common to both equations (the nonzero rows), we can assign its values arbitrarily. If we let $z = t$,, where t represents any real number, then we see that the system has infinitely many solutions: $x = 2 - t$, $y = -3 + t$, $z = t$. Geometrically the latter equations are the parametric equations for the line of intersection of the planes $x + 0y + z = 2$, $0x + y - z = -3$.

8.4.3 Eigenvalue Problem

Eigenvalues and Eigenvectors

We shall put the foregoing discussion of Gauss-Jordan elimination to immediate use to find the **eigenvectors** for an $n \times n$ matrix.

DEFINITION 8.13 Let $\mathbf{A}$ be an $n \times n$ matrix. A number λ is said to be an **eigenvalue** of $\mathbf{A}$ if there exists a *nonzero* solution vector $\mathbf{K}$ of the linear system

$$\mathbf{AK} = \lambda\mathbf{K}. \tag{6}$$

The solution vector $\mathbf{K}$ is said to be an **eigenvector** corresponding to the eigenvalue λ.

The word "eigenvalue" is a combination of German and English terms adapted from the German word *Eigenwert*, which, translated literally, is "proper value." Eigenvalues and eigenvectors are also called **characteristic values** and **characteristic vectors**, respectively.

Using properties of matrix algebra, we can write (6) in the alternative form

$$(\mathbf{A} - \lambda\mathbf{I})\mathbf{K} = \mathbf{0}, \tag{7}$$

where $\mathbf{I}$ is the multiplicative identity. If we let

$$\mathbf{K} = \begin{pmatrix} k_1 \\ k_2 \\ \vdots \\ k_n \end{pmatrix},$$

then (7) is the same as

$$\begin{aligned} (a_{11} - \lambda)k_1 + \quad a_{12}k_2 + \cdots + \quad a_{1n}k_n &= 0 \\ a_{21}k_1 + (a_{22} - \lambda)k_2 + \cdots + \quad a_{2n}k_n &= 0 \\ &\vdots \\ a_{n1}k_1 + \quad a_{n2}k_2 + \cdots + (a_{nn} - \lambda)k_n &= 0. \end{aligned} \tag{8}$$

Although an obvious solution of (8) is $k_1 = 0$, $k_2 = 0, \ldots, k_n = 0$, we are seeking only nontrivial solutions. Now it is known that a homogeneous system of n linear equations in n unknowns (that is, $b_i = 0$, $i = 1, 2, \ldots, n$ in (5)) has a nontrivial solution if and only if the determinant of the coefficient matrix is equal to zero. Thus to find a nonzero solution $\mathbf{K}$ for (7), we must have

$$\det(\mathbf{A} - \lambda\mathbf{I}) = 0. \tag{9}$$

Inspection of (8) shows that expansion of $\det(\mathbf{A} - \lambda\mathbf{I})$ by cofactors results in an nth degree polynomial in λ. The equation (9) is called the **characteristic**

equation of $\mathbf{A}$. Thus *the eigenvalues of* $\mathbf{A}$ *are the roots of the characteristic equation.* To find an eigenvector corresponding to an eigenvalue λ we simply solve the system of equations $(\mathbf{A} - \lambda\mathbf{I})\mathbf{K} = \mathbf{0}$ by applying Gauss-Jordan elimination to the augmented matrix $(\mathbf{A} - \lambda\mathbf{I}|\mathbf{0})$.

EXAMPLE 14

Find the eigenvalues and eigenvectors of

$$\mathbf{A} = \begin{pmatrix} 1 & 2 & 1 \\ 6 & -1 & 0 \\ -1 & -2 & -1 \end{pmatrix}.$$

SOLUTION To expand the determinant in the characteristic equation

$$\det(\mathbf{A} - \lambda\mathbf{I}) = \begin{vmatrix} 1-\lambda & 2 & 1 \\ 6 & -1-\lambda & 0 \\ -1 & -2 & -1-\lambda \end{vmatrix} = 0$$

we use the cofactors of the second row. It follows that

$$-\lambda^3 - \lambda^2 + 12\lambda = 0 \qquad \text{or} \qquad \lambda(\lambda + 4)(\lambda - 3) = 0.$$

Hence the eigenvalues are $\lambda_1 = 0$, $\lambda_2 = -4$, $\lambda_3 = 3$. To find the eigenvectors we must now reduce $(\mathbf{A} - \lambda\mathbf{I}|\mathbf{0})$ three times corresponding to the three distinct eigenvalues.

For $\lambda_1 = 0$ we have

$$(\mathbf{A} - 0\mathbf{I}|\mathbf{0}) = \left(\begin{array}{ccc|c} 1 & 2 & 1 & 0 \\ 6 & -1 & 0 & 0 \\ -1 & -2 & -1 & 0 \end{array}\right) \xRightarrow{\substack{A_{12}(-6) \\ A_{13}(1)}} \left(\begin{array}{ccc|c} 1 & 2 & 1 & 0 \\ 0 & -13 & -6 & 0 \\ 0 & 0 & 0 & 0 \end{array}\right)$$

$$\xRightarrow{M_2(-\frac{1}{13})} \left(\begin{array}{ccc|c} 1 & 2 & 1 & 0 \\ 0 & 1 & \frac{6}{13} & 0 \\ 0 & 0 & 0 & 0 \end{array}\right) \xRightarrow{A_{21}(-2)} \left(\begin{array}{ccc|c} 1 & 0 & \frac{1}{13} & 0 \\ 0 & 1 & \frac{6}{13} & 0 \\ 0 & 0 & 0 & 0 \end{array}\right).$$

Thus we see that $k_1 = (-1/13)k_3$ and $k_2 = (-6/13)k_3$. Choosing $k_3 = -13$ gives the eigenvector*

$$\mathbf{K}_1 = \begin{pmatrix} 1 \\ 6 \\ -13 \end{pmatrix}.$$

* Of course k_3 could be chosen as any nonzero number. In other words, a nonzero constant multiple of an eigenvector is also an eigenvector.

For $\lambda_2 = -4$,

$$(\mathbf{A} + 4\mathbf{I}\,|\,\mathbf{0}) = \left(\begin{array}{ccc|c} 5 & 2 & 1 & 0 \\ 6 & 3 & 0 & 0 \\ -1 & -2 & 3 & 0 \end{array}\right) \xrightarrow{I_{31}(-1)} \left(\begin{array}{ccc|c} 1 & 2 & -3 & 0 \\ 6 & 3 & 0 & 0 \\ 5 & 2 & 1 & 0 \end{array}\right)$$

$$\xrightarrow{\substack{A_{12}(-6) \\ A_{13}(-5)}} \left(\begin{array}{ccc|c} 1 & 2 & -3 & 0 \\ 0 & -9 & 18 & 0 \\ 0 & -8 & 16 & 0 \end{array}\right) \xrightarrow{\substack{M_2(-\frac{1}{9}) \\ M_3(-\frac{1}{8})}} \left(\begin{array}{ccc|c} 1 & 2 & -3 & 0 \\ 0 & 1 & -2 & 0 \\ 0 & 1 & -2 & 0 \end{array}\right)$$

$$\xrightarrow{\substack{A_{21}(-2) \\ A_{23}(-1)}} \left(\begin{array}{ccc|c} 1 & 0 & 1 & 0 \\ 0 & 1 & -2 & 0 \\ 0 & 0 & 0 & 0 \end{array}\right)$$

implies $k_1 = -k_3$ and $k_2 = 2k_3$. Choosing $k_3 = 1$ then yields the second eigenvector

$$\mathbf{K}_2 = \begin{pmatrix} -1 \\ 2 \\ 1 \end{pmatrix}.$$

Finally, for $\lambda_3 = 3$, Gauss-Jordan elimination gives

$$(\mathbf{A} - 3\mathbf{I}\,|\,\mathbf{0}) = \left(\begin{array}{ccc|c} -2 & 2 & 1 & 0 \\ 6 & -4 & 0 & 0 \\ -1 & -2 & -4 & 0 \end{array}\right) \Longrightarrow \left(\begin{array}{ccc|c} 1 & 0 & 1 & 0 \\ 0 & 1 & \frac{3}{2} & 0 \\ 0 & 0 & 0 & 0 \end{array}\right)$$

and so $k_1 = -k_3$ and $k_2 = (-3/2)k_3$. The choice of $k_3 = -2$ leads to the third eigenvector

$$\mathbf{K}_3 = \begin{pmatrix} 2 \\ 3 \\ -6 \end{pmatrix}.$$

When an $n \times n$ matrix $\mathbf{A}$ possesses n distinct eigenvalues $\lambda_1, \lambda_2, \ldots, \lambda_n$, it can be proved that a set of n linearly independent* eigenvectors $\mathbf{K}_1, \mathbf{K}_2, \ldots, \mathbf{K}_n$ can be found. However, when the characteristic equation has repeated roots it may not be possible to find n linearly independent eigenvectors for $\mathbf{A}$.

* Linear independence of column vectors is defined in exactly the same manner as for functions. See Definition 8.13.

EXAMPLE 15

Find the eigenvalues and eigenvectors of

$$\mathbf{A} = \begin{pmatrix} 3 & 4 \\ -1 & 7 \end{pmatrix}.$$

SOLUTION From the characteristic equation

$$\det(\mathbf{A} - \lambda\mathbf{I}) = \begin{vmatrix} 3-\lambda & 4 \\ -1 & 7-\lambda \end{vmatrix} = (\lambda - 5)^2 = 0,$$

we see $\lambda_1 = \lambda_2 = 5$ is an eigenvalue of multiplicity two. In the case of a 2×2 matrix there is no need to use Gauss-Jordan elimination. To find the eigenvector(s) corresponding to $\lambda_1 = 5$ we resort to the system $(\mathbf{A} - 5\mathbf{I}|\mathbf{0})$ in its equivalent form

$$\begin{aligned} -2k_1 + 4k_2 &= 0 \\ -k_1 + 2k_2 &= 0. \end{aligned}$$

It is apparent from this system that $k_1 = 2k_2$. Thus if we choose $k_2 = 1$, we find the single eigenvector

$$\mathbf{K}_1 = \begin{pmatrix} 2 \\ 1 \end{pmatrix}.$$

EXAMPLE 16

Find the eigenvalues and eigenvectors of

$$\mathbf{A} = \begin{pmatrix} 9 & 1 & 1 \\ 1 & 9 & 1 \\ 1 & 1 & 9 \end{pmatrix}.$$

SOLUTION The characteristic equation

$$\det(\mathbf{A} - \lambda\mathbf{I}) = \begin{vmatrix} 9-\lambda & 1 & 1 \\ 1 & 9-\lambda & 1 \\ 1 & 1 & 9-\lambda \end{vmatrix} = -(\lambda - 11)(\lambda - 8)^2 = 0$$

shows that $\lambda_1 = 11$ and that $\lambda_2 = \lambda_3 = 8$ is an eigenvalue of multiplicity two.

For $\lambda_1 = 11$ Gauss-Jordan elimination gives

$$(\mathbf{A} - 11\mathbf{I}|\mathbf{0}) = \left(\begin{array}{ccc|c} -2 & 1 & 1 & 0 \\ 1 & -2 & 1 & 0 \\ 1 & 1 & -2 & 0 \end{array}\right) \Rightarrow \left(\begin{array}{ccc|c} 1 & 0 & -1 & 0 \\ 0 & 1 & -1 & 0 \\ 0 & 0 & 0 & 0 \end{array}\right).$$

Hence $k_1 = k_3$ and $k_2 = k_3$. If $k_3 = 1$, then

$$\mathbf{K}_1 = \begin{pmatrix} 1 \\ 1 \\ 1 \end{pmatrix}.$$

Now for $\lambda_2 = 8$ we have

$$(\mathbf{A} - 8\mathbf{I}|\mathbf{0}) = \left(\begin{array}{ccc|c} 1 & 1 & 1 & 0 \\ 1 & 1 & 1 & 0 \\ 1 & 1 & 1 & 0 \end{array}\right) \Longrightarrow \left(\begin{array}{ccc|c} 1 & 1 & 1 & 0 \\ 0 & 0 & 0 & 0 \\ 0 & 0 & 0 & 0 \end{array}\right).$$

In the equation $k_1 + k_2 + k_3 = 0$, we are free to select two of the variables arbitrarily. Choosing, on the one hand, $k_2 = 1$, $k_3 = 0$ and, on the other, $k_2 = 0$, $k_3 = 1$, we obtain two linearly independent eigenvectors

$$\mathbf{K}_2 = \begin{pmatrix} -1 \\ 1 \\ 0 \end{pmatrix} \quad \text{and} \quad \mathbf{K}_3 = \begin{pmatrix} -1 \\ 0 \\ 1 \end{pmatrix}$$

corresponding to a single eigenvalue.

Exercises 8.4

Answers to odd-numbered problems begin on page A-23.

[8.4.1] **1.** If $\mathbf{A} = \begin{pmatrix} 4 & 5 \\ -6 & 9 \end{pmatrix}$ and $\mathbf{B} = \begin{pmatrix} -2 & 6 \\ 8 & -10 \end{pmatrix}$,

find **(a)** $\mathbf{A} + \mathbf{B}$, **(b)** $\mathbf{B} - \mathbf{A}$, **(c)** $2\mathbf{A} + 3\mathbf{B}$.

2. If $\mathbf{A} = \begin{pmatrix} -2 & 0 \\ 4 & 1 \\ 7 & 3 \end{pmatrix}$ and $\mathbf{B} = \begin{pmatrix} 3 & -1 \\ 0 & 2 \\ -4 & -2 \end{pmatrix}$,

find **(a)** $\mathbf{A} - \mathbf{B}$, **(b)** $\mathbf{B} - \mathbf{A}$, **(c)** $2(\mathbf{A} + \mathbf{B})$.

3. If $\mathbf{A} = \begin{pmatrix} 2 & -3 \\ -5 & 4 \end{pmatrix}$ and $\mathbf{B} = \begin{pmatrix} -1 & 6 \\ 3 & 2 \end{pmatrix}$,

find **(a)** $\mathbf{AB}$, **(b)** $\mathbf{BA}$, **(c)** $\mathbf{A}^2 = \mathbf{AA}$, **(d)** $\mathbf{B}^2 = \mathbf{BB}$.

4. If $\mathbf{A} = \begin{pmatrix} 1 & 4 \\ 5 & 10 \\ 8 & 12 \end{pmatrix}$ and $\mathbf{B} = \begin{pmatrix} -4 & 6 & -3 \\ 1 & -3 & 2 \end{pmatrix}$,

find **(a)** $\mathbf{AB}$, **(b)** $\mathbf{BA}$.

5. If $\mathbf{A} = \begin{pmatrix} 1 & -2 \\ -2 & 4 \end{pmatrix}$, $\mathbf{B} = \begin{pmatrix} 6 & 3 \\ 2 & 1 \end{pmatrix}$, and $\mathbf{C} = \begin{pmatrix} 0 & 2 \\ 3 & 4 \end{pmatrix}$,

find **(a)** $\mathbf{BC}$, **(b)** $\mathbf{A(BC)}$, **(c)** $\mathbf{C(BA)}$, **(d)** $\mathbf{A(B + C)}$.

6. If $\mathbf{A} = (5 \quad -6 \quad 7)$, $\mathbf{B} = \begin{pmatrix} 3 \\ 4 \\ -1 \end{pmatrix}$, and $\mathbf{C} = \begin{pmatrix} 1 & 2 & 4 \\ 0 & 1 & -1 \\ 3 & 2 & 1 \end{pmatrix}$,

find **(a)** $\mathbf{AB}$, **(b)** $\mathbf{BA}$, **(c)** $\mathbf{(BA)C}$, **(d)** $\mathbf{(AB)C}$.

7. If $\mathbf{A} = \begin{pmatrix} 4 \\ 8 \\ -10 \end{pmatrix}$ and $\mathbf{B} = (2 \quad 4 \quad 5)$,

find **(a)** $\mathbf{A}^T\mathbf{A}$, **(b)** $\mathbf{B}^T\mathbf{B}$, **(c)** $\mathbf{A} + \mathbf{B}^T$.

8. If $\mathbf{A} = \begin{pmatrix} 1 & 2 \\ 2 & 4 \end{pmatrix}$ and $\mathbf{B} = \begin{pmatrix} -2 & 3 \\ 5 & 7 \end{pmatrix}$,

find **(a)** $\mathbf{A} + \mathbf{B}^T$, **(b)** $2\mathbf{A}^T - \mathbf{B}^T$, **(c)** $\mathbf{A}^T(\mathbf{A} - \mathbf{B})$.

9. If $\mathbf{A} = \begin{pmatrix} 3 & 4 \\ 8 & 1 \end{pmatrix}$ and $\mathbf{B} = \begin{pmatrix} 5 & 10 \\ -2 & -5 \end{pmatrix}$,

find **(a)** $(\mathbf{AB})^T$, **(b)** $\mathbf{B}^T\mathbf{A}^T$.

10. If $\mathbf{A} = \begin{pmatrix} 5 & 9 \\ -4 & 6 \end{pmatrix}$ and $\mathbf{B} = \begin{pmatrix} -3 & 11 \\ -7 & 2 \end{pmatrix}$,

find **(a)** $\mathbf{A}^T + \mathbf{B}^T$, **(b)** $(\mathbf{A} + \mathbf{B})^T$.

In Problems 11–14 write the given sum as a single column matrix.

11. $4\begin{pmatrix} -1 \\ 2 \end{pmatrix} - 2\begin{pmatrix} 2 \\ 8 \end{pmatrix} + 3\begin{pmatrix} -2 \\ 3 \end{pmatrix}$

12. $3t\begin{pmatrix} 2 \\ t \\ -1 \end{pmatrix} + (t - 1)\begin{pmatrix} -1 \\ -t \\ 3 \end{pmatrix} - 2\begin{pmatrix} 3t \\ 4 \\ -5t \end{pmatrix}$

13. $\begin{pmatrix} 2 & -3 \\ 1 & 4 \end{pmatrix}\begin{pmatrix} -2 \\ 5 \end{pmatrix} - \begin{pmatrix} -1 & 6 \\ -2 & 3 \end{pmatrix}\begin{pmatrix} -7 \\ 2 \end{pmatrix}$

14. $\begin{pmatrix} 1 & -3 & 4 \\ 2 & 5 & -1 \\ 0 & -4 & -2 \end{pmatrix}\begin{pmatrix} t \\ 2t - 1 \\ -t \end{pmatrix} + \begin{pmatrix} -t \\ 1 \\ 4 \end{pmatrix} - \begin{pmatrix} 2 \\ 8 \\ -6 \end{pmatrix}$

In Problems 15–22 determine whether the given matrix is singular or nonsingular. If nonsingular, find $\mathbf{A}^{-1}$.

15. $\mathbf{A} = \begin{pmatrix} -3 & 6 \\ -2 & 4 \end{pmatrix}$

16. $\mathbf{A} = \begin{pmatrix} 2 & 5 \\ 1 & 4 \end{pmatrix}$

17. $\mathbf{A} = \begin{pmatrix} 4 & 8 \\ -3 & -5 \end{pmatrix}$

18. $\mathbf{A} = \begin{pmatrix} 7 & 10 \\ 2 & 2 \end{pmatrix}$

19. $\mathbf{A} = \begin{pmatrix} 2 & 1 & 0 \\ -1 & 2 & 1 \\ 1 & 2 & 1 \end{pmatrix}$

20. $\mathbf{A} = \begin{pmatrix} 3 & 2 & 1 \\ 4 & 1 & 0 \\ -2 & 5 & -1 \end{pmatrix}$

21. $\mathbf{A} = \begin{pmatrix} 2 & 1 & 1 \\ 1 & -2 & -3 \\ 3 & 2 & 4 \end{pmatrix}$

22. $\mathbf{A} = \begin{pmatrix} 4 & 1 & -1 \\ 6 & 2 & -3 \\ -2 & -1 & 2 \end{pmatrix}$

In Problems 23 and 24 show that the given matrix is nonsingular for every real value of t. Find $\mathbf{A}^{-1}(t)$.

23. $\mathbf{A}(t) = \begin{pmatrix} 2e^{-t} & e^{4t} \\ 4e^{-t} & 3e^{4t} \end{pmatrix}$

24. $\mathbf{A}(t) = \begin{pmatrix} 2e^t \sin t & -2e^t \cos t \\ e^t \cos t & e^t \sin t \end{pmatrix}$

In Problems 25–28 find $d\mathbf{X}/dt$.

25. $\mathbf{X} = \begin{pmatrix} 5e^{-t} \\ 2e^{-t} \\ -7e^{-t} \end{pmatrix}$

26. $\mathbf{X} = \begin{pmatrix} \frac{1}{2} \sin 2t - 4 \cos 2t \\ -3 \sin 2t + 5 \cos 2t \end{pmatrix}$

27. $\mathbf{X} = 2\begin{pmatrix} 1 \\ -1 \end{pmatrix} e^{2t} + 4\begin{pmatrix} 2 \\ 1 \end{pmatrix} e^{-3t}$

28. $\mathbf{X} = \begin{pmatrix} 5te^{2t} \\ t \sin 3t \end{pmatrix}$

29. Let
$$\mathbf{A}(t) = \begin{pmatrix} e^{4t} & \cos \pi t \\ 2t & 3t^2 - 1 \end{pmatrix}.$$

Find **(a)** $\dfrac{d\mathbf{A}}{dt}$, **(b)** $\displaystyle\int_0^2 \mathbf{A}(t)\, dt$, **(c)** $\displaystyle\int_0^t \mathbf{A}(s)\, ds$.

30. Let
$$\mathbf{A}(t) = \begin{pmatrix} \dfrac{1}{t^2+1} & 3t \\ t^2 & t \end{pmatrix} \quad \text{and} \quad \mathbf{B}(t) = \begin{pmatrix} 6t & 2 \\ \dfrac{1}{t} & 4t \end{pmatrix}.$$

Find **(a)** $\dfrac{d\mathbf{A}}{dt}$, **(b)** $\dfrac{d\mathbf{B}}{dt}$, **(c)** $\displaystyle\int_0^1 \mathbf{A}(t)\, dt$, **(d)** $\displaystyle\int_1^2 \mathbf{B}(t)\, dt$, **(e)** $\mathbf{A}(t)\mathbf{B}(t)$,

(f) $\dfrac{d}{dt}\mathbf{A}(t)\mathbf{B}(t)$, **(g)** $\displaystyle\int_1^t \mathbf{A}(s)\mathbf{B}(s)\, ds$.

[8.4.2] In Problems 31–38 solve the given system of equations by Gauss-Jordan elimination.

31.
$$\begin{aligned} x + y - 2z &= 14 \\ 2x - y + z &= 0 \\ 6x + 3y + 4z &= 1 \end{aligned}$$

32.
$$\begin{aligned} 5x - 2y + 4z &= 10 \\ x + y + z &= 9 \\ 4x - 3y + 3z &= 1 \end{aligned}$$

33.
$$\begin{aligned} y + z &= -5 \\ 5x + 4y - 16z &= -10 \\ x - y - 5z &= 7 \end{aligned}$$

34.
$$\begin{aligned} 3x + y + z &= 4 \\ 4x + 2y - z &= 7 \\ x + y - 3z &= 6 \end{aligned}$$

35.
$$\begin{aligned} 2x + y + z &= 4 \\ 10x - 2y + 2z &= -1 \\ 6x - 2y + 4z &= 8 \end{aligned}$$

36.
$$\begin{aligned} x + 2z &= 8 \\ x + 2y - 2z &= 4 \\ 2x + 5y - 6z &= 6 \end{aligned}$$

37.
$$\begin{aligned} x_1 + x_2 - x_3 - x_4 &= -1 \\ x_1 + x_2 + x_3 + x_4 &= 3 \\ x_1 - x_2 + x_3 - x_4 &= 3 \\ 4x_1 + x_2 - 2x_3 + x_4 &= 0 \end{aligned}$$

38.
$$\begin{aligned} 2x_1 + x_2 + x_3 &= 0 \\ x_1 + 3x_2 + x_3 &= 0 \\ 7x_1 + x_2 + 3x_3 &= 0 \end{aligned}$$

In Problems 39 and 40 use Gauss-Jordan elimination to demonstrate that the given system of equations has no solution.

39.
$$\begin{aligned} x + 2y + 4z &= 2 \\ 2x + 4y + 3z &= 1 \\ x + 2y - z &= 7 \end{aligned}$$

40.
$$\begin{aligned} x_1 + x_2 - x_3 + 3x_4 &= 1 \\ x_2 - x_3 - 4x_4 &= 0 \\ x_1 + 2x_2 - 2x_3 - x_4 &= 6 \\ 4x_1 + 7x_2 - 7x_3 &= 9 \end{aligned}$$

[8.4.3] In Problems 41–48 find the eigenvalues and eigenvectors of the given matrix.

41. $\begin{pmatrix} -1 & 2 \\ -7 & 8 \end{pmatrix}$

42. $\begin{pmatrix} 2 & 1 \\ 2 & 1 \end{pmatrix}$

43. $\begin{pmatrix} -8 & -1 \\ 16 & 0 \end{pmatrix}$

44. $\begin{pmatrix} 1 & 1 \\ \frac{1}{4} & 1 \end{pmatrix}$

45. $\begin{pmatrix} 5 & -1 & 0 \\ 0 & -5 & 9 \\ 5 & -1 & 0 \end{pmatrix}$

46. $\begin{pmatrix} 3 & 0 & 0 \\ 0 & 2 & 0 \\ 4 & 0 & 1 \end{pmatrix}$

47. $\begin{pmatrix} 0 & 4 & 0 \\ -1 & -4 & 0 \\ 0 & 0 & -2 \end{pmatrix}$

48. $\begin{pmatrix} 1 & 6 & 0 \\ 0 & 2 & 1 \\ 0 & 1 & 2 \end{pmatrix}$

In Problems 49 and 50 show that the given matrix has complex eigenvalues. Find the eigenvectors of the matrix.

49. $\begin{pmatrix} -1 & 2 \\ -5 & 1 \end{pmatrix}$

50. $\begin{pmatrix} 2 & -1 & 0 \\ 5 & 2 & 4 \\ 0 & 1 & 2 \end{pmatrix}$

Miscellaneous Problems

51. If $\mathbf{A}(t)$ is a 2×2 matrix of differentiable functions and $\mathbf{X}(t)$ is a 2×1 column matrix of differentiable functions, prove the product rule

$$\frac{d}{dt}[\mathbf{A}(t)\mathbf{X}(t)] = \mathbf{A}(t)\mathbf{X}'(t) + \mathbf{A}'(t)\mathbf{X}(t).$$

52. Derive formula (3). [*Hint:* Find a matrix $\mathbf{B} = \begin{pmatrix} b_{11} & b_{12} \\ b_{21} & b_{22} \end{pmatrix}$ for which $\mathbf{AB} = \mathbf{I}$. Solve for b_{11}, b_{12}, b_{21}, and b_{22}. Then show that $\mathbf{BA} = \mathbf{I}$.]

53. If $\mathbf{A}$ is nonsingular and $\mathbf{AB} = \mathbf{AC}$, show that $\mathbf{B} = \mathbf{C}$.

54. If $\mathbf{A}$ and $\mathbf{B}$ are nonsingular show that $(\mathbf{AB})^{-1} = \mathbf{B}^{-1}\mathbf{A}^{-1}$.

55. Let $\mathbf{A}$ and $\mathbf{B}$ be $n \times n$ matrices. In general, is $(\mathbf{A} + \mathbf{B})^2 = \mathbf{A}^2 + 2\mathbf{AB} + \mathbf{B}^2$?

8.5 Matrices and Systems of Linear First-Order Equations

8.5.1 Preliminary Theory

Matrix Form of a System

If $\mathbf{X}$, $\mathbf{A}(t)$, and $\mathbf{F}(t)$ denote the respective matrices

$$\mathbf{X} = \begin{pmatrix} x_1(t) \\ x_2(t) \\ \vdots \\ x_n(t) \end{pmatrix},$$

$$\mathbf{A}(t) = \begin{pmatrix} a_{11}(t) & a_{12}(t) & \cdots & a_{1n}(t) \\ a_{21}(t) & a_{22}(t) & \cdots & a_{2n}(t) \\ \vdots & & & \vdots \\ a_{n1}(t) & a_{n2}(t) & \cdots & a_{nn}(t) \end{pmatrix}, \qquad \mathbf{F}(t) = \begin{pmatrix} f_1(t) \\ f_2(t) \\ \vdots \\ f_n(t) \end{pmatrix},$$

then the system of linear first-order differential equations

$$\begin{aligned} \frac{dx_1}{dt} &= a_{11}(t)x_1 + a_{12}(t)x_2 + \cdots + a_{1n}(t)x_n + f_1(t) \\ \frac{dx_2}{dt} &= a_{21}(t)x_1 + a_{22}(t)x_2 + \cdots + a_{2n}(t)x_n + f_2(t) \\ &\vdots \\ \frac{dx_n}{dt} &= a_{n1}(t)x_1 + a_{n2}(t)x_2 + \cdots + a_{nn}(t)x_n + f_n(t) \end{aligned} \tag{1}$$

can be written as

$$\frac{d}{dt}\begin{pmatrix} x_1 \\ x_2 \\ \vdots \\ x_n \end{pmatrix} = \begin{pmatrix} a_{11}(t) & a_{12}(t) & \cdots & a_{1n}(t) \\ a_{21}(t) & a_{22}(t) & \cdots & a_{2n}(t) \\ \vdots & & & \vdots \\ a_{n1}(t) & a_{n2}(t) & \cdots & a_{nn}(t) \end{pmatrix}\begin{pmatrix} x_1 \\ x_2 \\ \vdots \\ x_n \end{pmatrix} + \begin{pmatrix} f_1(t) \\ f_2(t) \\ \vdots \\ f_n(t) \end{pmatrix}$$

or simply

$$\frac{d\mathbf{X}}{dt} = \mathbf{A}(t)\mathbf{X} + \mathbf{F}(t). \tag{2}$$

If the system is homogeneous, (2) becomes

$$\frac{d\mathbf{X}}{dt} = \mathbf{A}(t)\mathbf{X}. \tag{3}$$

Equations (2) and (3) are also written as $\mathbf{X}' = \mathbf{AX} + \mathbf{F}$ and $\mathbf{X}' = \mathbf{AX}$, respectively.

EXAMPLE 1

In matrix terms the nonhomogeneous system

$$\begin{aligned} \frac{dx}{dt} &= -2x + 5y + e^t - 2t \\ \frac{dy}{dt} &= 4x - 3y + 10t \end{aligned}$$

can be written as

$$\frac{d\mathbf{X}}{dt} = \begin{pmatrix} -2 & 5 \\ 4 & -3 \end{pmatrix}\mathbf{X} + \begin{pmatrix} e^t - 2t \\ 10t \end{pmatrix}$$

or

$$\mathbf{X}' = \begin{pmatrix} -2 & 5 \\ 4 & -3 \end{pmatrix}\mathbf{X} + \begin{pmatrix} 1 \\ 0 \end{pmatrix}e^t + \begin{pmatrix} -2 \\ 10 \end{pmatrix}t,$$

where $\mathbf{X} = \begin{pmatrix} x \\ y \end{pmatrix}$.

EXAMPLE 2

The matrix form of the homogeneous system

$$\begin{aligned} \frac{dx}{dt} &= 2x - 3y \\ \frac{dy}{dt} &= 6x + 5y \end{aligned} \qquad \text{is} \qquad \frac{d\mathbf{X}}{dt} = \begin{pmatrix} 2 & -3 \\ 6 & 5 \end{pmatrix}\mathbf{X},$$

where $\mathbf{X} = \begin{pmatrix} x \\ y \end{pmatrix}$.

DEFINITION 8.14 A **solution vector** on an interval I is any column matrix

$$\mathbf{X} = \begin{pmatrix} x_1(t) \\ x_2(t) \\ \vdots \\ x_n(t) \end{pmatrix}$$

whose entries are differentiable functions satisfying the system (2) on the interval.

EXAMPLE 3 Verify that

$$\mathbf{X}_1 = \begin{pmatrix} 1 \\ -1 \end{pmatrix} e^{-2t} = \begin{pmatrix} e^{-2t} \\ -e^{-2t} \end{pmatrix} \quad \text{and} \quad \mathbf{X}_2 = \begin{pmatrix} 3 \\ 5 \end{pmatrix} e^{6t} = \begin{pmatrix} 3e^{6t} \\ 5e^{6t} \end{pmatrix}$$

are solutions of

$$\mathbf{X}' = \begin{pmatrix} 1 & 3 \\ 5 & 3 \end{pmatrix} \mathbf{X} \tag{4}$$

on the interval $(-\infty, \infty)$.

SOLUTION We have

$$\mathbf{X}_1' = \begin{pmatrix} -2e^{-2t} \\ 2e^{-2t} \end{pmatrix}$$

and

$$\mathbf{AX}_1 = \begin{pmatrix} 1 & 3 \\ 5 & 3 \end{pmatrix} \begin{pmatrix} e^{-2t} \\ -e^{-2t} \end{pmatrix} = \begin{pmatrix} e^{-2t} - 3e^{-2t} \\ 5e^{-2t} - 3e^{-2t} \end{pmatrix} = \begin{pmatrix} -2e^{-2t} \\ 2e^{-2t} \end{pmatrix} = \mathbf{X}_1'.$$

Now

$$\mathbf{X}_2' = \begin{pmatrix} 18e^{6t} \\ 30e^{5t} \end{pmatrix}$$

and

$$\mathbf{AX}_2 = \begin{pmatrix} 1 & 3 \\ 5 & 3 \end{pmatrix} \begin{pmatrix} 3e^{6t} \\ 5e^{6t} \end{pmatrix} = \begin{pmatrix} 3e^{6t} + 15e^{6t} \\ 15e^{6t} + 15e^{6t} \end{pmatrix} = \begin{pmatrix} 18e^{6t} \\ 30e^{6t} \end{pmatrix} = \mathbf{X}_2'.$$

Much of the theory of systems of n linear first-order differential equations is similar to that of linear nth-order differential equations.

Initial-Value Problem

Let t_0 denote a point on an interval I and

$$\mathbf{X}(t_0) = \begin{pmatrix} x_1(t_0) \\ x_2(t_0) \\ \vdots \\ x_n(t_0) \end{pmatrix} \quad \text{and} \quad \mathbf{X}_0 = \begin{pmatrix} \gamma_1 \\ \gamma_2 \\ \vdots \\ \gamma_n \end{pmatrix},$$

where the γ_i, $i = 1, 2, \ldots, n$ are given constants. Then the problem

$$\begin{aligned} \textit{Solve:} \quad & \frac{d\mathbf{X}}{dt} = \mathbf{A}(t)\mathbf{X} + \mathbf{F}(t) \\ \textit{Subject to:} \quad & \mathbf{X}(t_0) = \mathbf{X}_0 \end{aligned} \tag{5}$$

is an **initial-value problem** on the interval.

THEOREM 8.3 Let the entries of the matrices $\mathbf{A}(t)$ and $\mathbf{F}(t)$ be functions continuous on a common interval I that contains the point t_0. Then there exists a unique solution of the initial-value problem (5) on the interval.

Homogeneous Systems

In the next several definitions and theorems, we are concerned only with homogeneous systems. Without stating it, we shall always assume that the a_{ij} and the f_i are continuous functions of t on some common interval I.

Superposition Principle

The following result is a **superposition principle** for solutions of linear systems.

THEOREM 8.4 Let $\mathbf{X}_1, \mathbf{X}_2, \ldots, \mathbf{X}_k$ be a set of solution vectors of the homogeneous system (3) on an interval I. Then the linear combination

$$\mathbf{X} = c_1\mathbf{X}_1 + c_2\mathbf{X}_2 + \cdots + c_k\mathbf{X}_k,$$

where the c_i, $i = 1, 2, \ldots, k$ are arbitrary constants, is also a solution on the interval.

It follows from Theorem 8.4 that a constant multiple of any solution vector of a homogeneous system of linear first-order differential equations is also a solution.

EXAMPLE 4

One solution of the system

$$\mathbf{X}' = \begin{pmatrix} 1 & 0 & 1 \\ 1 & 1 & 0 \\ -2 & 0 & -1 \end{pmatrix}\mathbf{X} \tag{6}$$

is

$$\mathbf{X}_1 = \begin{pmatrix} \cos t \\ -\frac{1}{2}\cos t + \frac{1}{2}\sin t \\ -\cos t - \sin t \end{pmatrix}.$$

For any constant c_1 the vector $\mathbf{X} = c_1\mathbf{X}_1$ is also a solution since

$$\frac{d\mathbf{X}}{dt} = \begin{pmatrix} -c_1 \sin t \\ \frac{1}{2}c_1 \sin t + \frac{1}{2}c_1 \cos t \\ c_1 \sin t - c_1 \cos t \end{pmatrix}$$

and
$$\mathbf{AX} = \begin{pmatrix} 1 & 0 & 1 \\ 1 & 1 & 0 \\ -2 & 0 & -1 \end{pmatrix} \begin{pmatrix} c_1 \cos t \\ -\frac{1}{2}c_1 \cos t + \frac{1}{2}c_1 \sin t \\ -c_1 \cos t - c_1 \sin t \end{pmatrix}$$

$$= \begin{pmatrix} -c_1 \sin t \\ \frac{1}{2}c_1 \cos t + \frac{1}{2}c_1 \sin t \\ -c_1 \cos t + c_1 \sin t \end{pmatrix}.$$

Inspection of the resulting matrices shows that $\mathbf{X}' = \mathbf{AX}$.

EXAMPLE 5

Consider the system (6) of Example 4. If

$$\mathbf{X}_2 = \begin{pmatrix} 0 \\ e^t \\ 0 \end{pmatrix}, \quad \text{then} \quad \mathbf{X}_2' = \begin{pmatrix} 0 \\ e^t \\ 0 \end{pmatrix}$$

and
$$\mathbf{AX}_2 = \begin{pmatrix} 1 & 0 & 1 \\ 1 & 1 & 0 \\ -2 & 0 & -1 \end{pmatrix} \begin{pmatrix} 0 \\ e^t \\ 0 \end{pmatrix}$$

$$= \begin{pmatrix} 0 \\ e^t \\ 0 \end{pmatrix} = \mathbf{X}_2'.$$

Thus we see that $\mathbf{X}_2$ is also a solution vector of (6). By the superposition principle the linear combination

$$\mathbf{X} = c_1\mathbf{X}_1 + c_2\mathbf{X}_2$$

$$= c_1 \begin{pmatrix} \cos t \\ -\frac{1}{2} \cos t + \frac{1}{2} \sin t \\ -\cos t - \sin t \end{pmatrix} + c_2 \begin{pmatrix} 0 \\ e^t \\ 0 \end{pmatrix}$$

is yet another solution of the system.

Linear Independence

We are primarily interested in linearly independent solutions of the homogeneous system (3).

DEFINITION 8.15 Let $\mathbf{X}_1, \mathbf{X}_2, \ldots, \mathbf{X}_k$ be a set of solution vectors of the homogeneous system (3) on an interval I. We say that the set is **linearly dependent** on the interval if there exist constants $c_1, c_2, \ldots, c_k$, not all zero, such that

$$c_1\mathbf{X}_1 + c_2\mathbf{X}_2 + \cdots + c_k\mathbf{X}_k = \mathbf{0}$$

for every t in the interval. If the set of vectors is not linearly dependent on the interval, it is said to be **linearly independent**.

The case when $k = 2$ should be clear; two solution vectors $\mathbf{X}_1$ and $\mathbf{X}_2$ are linearly dependent if one is a constant multiple of the other, and conversely. For $k > 2$ a set of solution vectors is linearly dependent if we can express at least one solution vector as a nontrivial linear combination of the remaining vectors.

EXAMPLE 6

It can be verified that

$$\mathbf{X}_1 = \begin{pmatrix} 3 \\ 1 \end{pmatrix} e^{t} \quad \text{and} \quad \mathbf{X}_2 = \begin{pmatrix} 1 \\ 1 \end{pmatrix} e^{-t}$$

are solution vectors of the system

$$\mathbf{X}' = \begin{pmatrix} 2 & -3 \\ 1 & -2 \end{pmatrix} \mathbf{X}. \tag{7}$$

Now $\mathbf{X}_1$ and $\mathbf{X}_2$ are linearly independent on the interval $(-\infty, \infty)$ since

$$c_1\mathbf{X}_1 + c_2\mathbf{X}_2 = \mathbf{0} \quad \text{or} \quad c_1\begin{pmatrix} 3 \\ 1 \end{pmatrix} e^{t} + c_2\begin{pmatrix} 1 \\ 1 \end{pmatrix} e^{-t} = \begin{pmatrix} 0 \\ 0 \end{pmatrix}$$

is equivalent to

$$\begin{aligned} 3c_1e^{t} + c_2e^{-t} &= 0 \\ c_1e^{t} + c_2e^{-t} &= 0. \end{aligned}$$

Solving this system for c_1 and c_2 immediately yields $c_1 = 0$ and $c_2 = 0$.

EXAMPLE 7

The vector $\mathbf{X}_3 = \begin{pmatrix} e^{t} + \cosh t \\ \cosh t \end{pmatrix}$ is also a solution of the system (7) given in Example 6. However, $\mathbf{X}_1$, $\mathbf{X}_2$, and $\mathbf{X}_3$ are linearly dependent since

$$\mathbf{X}_3 = \frac{1}{2}\mathbf{X}_1 + \frac{1}{2}\mathbf{X}_2.$$

Wronskian

As in our earlier consideration of the theory of a single ordinary differential equation we can introduce the concept of the **Wronskian** determinant as a test for linear independence. We state the following theorem without proof.

THEOREM 8.5 Let

$$\mathbf{X}_1 = \begin{pmatrix} x_{11} \\ x_{21} \\ \vdots \\ x_{n1} \end{pmatrix}, \mathbf{X}_2 = \begin{pmatrix} x_{12} \\ x_{22} \\ \vdots \\ x_{n2} \end{pmatrix}, \ldots, \mathbf{X}_n = \begin{pmatrix} x_{1n} \\ x_{2n} \\ \vdots \\ x_{nn} \end{pmatrix}$$

be n solution vectors of the homogeneous system (3) on an interval I. A necessary and sufficient condition that the set of solutions be linearly independent is that the Wronskian

$$W(\mathbf{X}_1, \mathbf{X}_2, \ldots, \mathbf{X}_n) = \begin{vmatrix} x_{11} & x_{12} & \cdots & x_{1n} \\ x_{21} & x_{22} & \cdots & x_{2n} \\ \vdots & & & \vdots \\ x_{n1} & x_{n2} & \cdots & x_{nn} \end{vmatrix} \neq 0 \qquad (8)$$

for every t in I.

In fact it can be shown that if $\mathbf{X}_1, \mathbf{X}_2, \ldots, \mathbf{X}_n$ are solution vectors of (3), then either

$$W(\mathbf{X}_1, \mathbf{X}_2, \ldots, \mathbf{X}_n) \neq 0$$

for every t in I or

$$W(\mathbf{X}_1, \mathbf{X}_2, \ldots, \mathbf{X}_n) = 0$$

for every t in the interval. Thus if we can show that $W \neq 0$ for some t_0 in I, then $W \neq 0$ for every t and hence the solutions are linearly independent on the interval.

Notice that, unlike our previous definition of the Wronskian, the determinant (8) does not involve differentiation.

EXAMPLE 8

In Example 3 we saw that

$$\mathbf{X}_1 = \begin{pmatrix} 1 \\ -1 \end{pmatrix} e^{-2t} \quad \text{and} \quad \mathbf{X}_2 = \begin{pmatrix} 3 \\ 5 \end{pmatrix} e^{-6t}$$

are solutions of the system (4). Clearly, $\mathbf{X}_1$ and $\mathbf{X}_2$ are linearly independent on $(-\infty, \infty)$ since neither vector is a constant multiple of the other. In addition, we have

$$W(\mathbf{X}_1, \mathbf{X}_2) = \begin{vmatrix} e^{-2t} & 3e^{6t} \\ -e^{-2t} & 5e^{6t} \end{vmatrix} = 8e^{4t} \neq 0$$

for all real values of t.

Fundamental Set of Solutions

DEFINITION 8.16 Any set $\mathbf{X}_1, \mathbf{X}_2, \ldots, \mathbf{X}_n$ of n linearly independent solution vectors of the homogeneous system (3) on an interval I is said to be a **fundamental set of solutions** on the interval.

THEOREM 8.6 There exists a fundamental set of solutions for the homogeneous system (3) on an interval I.

DEFINITION 8.17 Let $\mathbf{X}_1, \mathbf{X}_2, \ldots, \mathbf{X}_n$ be a fundamental set of solutions of the homogeneous system (3) on an interval I. The **general solution** of the system on the interval is defined to be

$$\mathbf{X} = c_1\mathbf{X}_1 + c_2\mathbf{X}_2 + \cdots + c_n\mathbf{X}_n,$$

where the c_i, $i = 1, 2, \ldots, n$ are arbitrary constants.

Although we shall not give the proof, it can be shown that, for appropriate choices of the constants $c_1, c_2, \ldots, c_n$, *any* solution of (3) on the interval I can be obtained from the general solution.

EXAMPLE 9

From Example 8 we know that

$$\mathbf{X}_1 = \begin{pmatrix} 1 \\ -1 \end{pmatrix} e^{-2t} \quad \text{and} \quad \mathbf{X}_2 = \begin{pmatrix} 3 \\ 5 \end{pmatrix} e^{6t}$$

are linearly independent solutions of (4) on $(-\infty, \infty)$. Hence $\mathbf{X}_1$ and $\mathbf{X}_2$ form a fundamental set of solutions on the interval. The general solution of the system on the interval is then

$$\mathbf{X} = c_1\mathbf{X}_1 + c_2\mathbf{X}_2 = c_1\begin{pmatrix} 1 \\ -1 \end{pmatrix} e^{-2t} + c_2\begin{pmatrix} 3 \\ 5 \end{pmatrix} e^{6t}. \tag{9}$$

EXAMPLE 10

The vectors

$$\mathbf{X}_1 = \begin{pmatrix} \cos t \\ -\frac{1}{2}\cos t + \frac{1}{2}\sin t \\ -\cos t - \sin t \end{pmatrix}, \quad \mathbf{X}_2 = \begin{pmatrix} 0 \\ 1 \\ 0 \end{pmatrix} e^t, \quad \text{and} \quad \mathbf{X}_3 = \begin{pmatrix} \sin t \\ -\frac{1}{2}\sin t - \frac{1}{2}\cos t \\ -\sin t + \cos t \end{pmatrix}$$

are solutions of the system (6)* in Example 4. Now

$$W(\mathbf{X}_1, \mathbf{X}_2, \mathbf{X}_3) = \begin{vmatrix} \cos t & 0 & \sin t \\ -\frac{1}{2}\cos t + \frac{1}{2}\sin t & e^t & -\frac{1}{2}\sin t - \frac{1}{2}\cos t \\ -\cos t - \sin t & 0 & -\sin t + \cos t \end{vmatrix}$$

$$= e^t \begin{vmatrix} \cos t & \sin t \\ -\cos t - \sin t & -\sin t + \cos t \end{vmatrix} = e^t \neq 0$$

for all real values of t. We conclude that $\mathbf{X}_1, \mathbf{X}_2$, and $\mathbf{X}_3$ form a fundamental set of solutions on $(-\infty, \infty)$. Thus the general solution of the system on the interval is

$$\mathbf{X} = c_1\mathbf{X}_1 + c_2\mathbf{X}_2 + c_3\mathbf{X}_3$$

$$= c_1 \begin{pmatrix} \cos t \\ -\frac{1}{2}\cos t + \frac{1}{2}\sin t \\ -\cos t - \sin t \end{pmatrix} + c_2 \begin{pmatrix} 0 \\ 1 \\ 0 \end{pmatrix} e^t + c_3 \begin{pmatrix} \sin t \\ -\frac{1}{2}\sin t - \frac{1}{2}\cos t \\ -\sin t + \cos t \end{pmatrix}.$$

Nonhomogeneous Systems

For nonhomogeneous systems a **particular solution** $\mathbf{X}_p$ on an interval I is any vector, free of arbitrary parameters, whose entries are functions satisfying the system (2).

EXAMPLE 11

Verify that the vector

$$\mathbf{X}_p = \begin{pmatrix} 3t - 4 \\ -5t + 6 \end{pmatrix}$$

is a particular solution of the nonhomogeneous system

$$\mathbf{X}' = \begin{pmatrix} 1 & 3 \\ 5 & 3 \end{pmatrix}\mathbf{X} + \begin{pmatrix} 12t - 11 \\ -3 \end{pmatrix} \tag{10}$$

on the interval $(-\infty, \infty)$.

---

* On pages 402 and 403 it was verified that $\mathbf{X}_1$ and $\mathbf{X}_2$ are solutions; it is left as an exercise to demonstrate that $\mathbf{X}_3$ is also a solution.

SOLUTION We have $\mathbf{X}'_p = \begin{pmatrix} 3 \\ -5 \end{pmatrix}$ and

$$\begin{aligned}
\begin{pmatrix} 1 & 3 \\ 5 & 3 \end{pmatrix}\mathbf{X}_p + \begin{pmatrix} 12t - 11 \\ -3 \end{pmatrix} &= \begin{pmatrix} 1 & 3 \\ 5 & 3 \end{pmatrix}\begin{pmatrix} 3t - 4 \\ -5t + 6 \end{pmatrix} + \begin{pmatrix} 12t - 11 \\ -3 \end{pmatrix} \\
&= \begin{pmatrix} (3t - 4) + 3(-5t + 6) \\ 5(3t - 4) + 3(-5t + 6) \end{pmatrix} + \begin{pmatrix} 12t - 11 \\ -3 \end{pmatrix} \\
&= \begin{pmatrix} -12t + 14 \\ -2 \end{pmatrix} + \begin{pmatrix} 12t - 11 \\ -3 \end{pmatrix} \\
&= \begin{pmatrix} 3 \\ -5 \end{pmatrix} = \mathbf{X}'_p.
\end{aligned}$$

THEOREM 8.7 Let $\mathbf{X}_1, \mathbf{X}_2, \ldots, \mathbf{X}_k$ be a set of solution vectors of the homogeneous system (3) on an interval I and let $\mathbf{X}_p$ be any solution vector of the nonhomogeneous system (2) on the same interval. Then

$$\mathbf{X} = c_1\mathbf{X}_1 + c_2\mathbf{X}_2 + \cdots + c_k\mathbf{X}_k + \mathbf{X}_p$$

is also a solution of the nonhomogeneous system on the interval for any constants $c_1, c_2, \ldots, c_k$.

DEFINITION 8.18 Let $\mathbf{X}_p$ be a given solution of the nonhomogeneous system (2) on an interval I, and let

$$\mathbf{X}_c = c_1\mathbf{X}_1 + c_2\mathbf{X}_2 + \cdots + c_n\mathbf{X}_n$$

denote the general solution on the same interval of the corresponding homogeneous system (3). The **general solution** of the nonhomogeneous system on the interval is defined to be

$$\mathbf{X} = \mathbf{X}_c + \mathbf{X}_p.$$

The general solution $\mathbf{X}_c$ of the homogeneous system (3) is called the **complementary function** of the nonhomogeneous system (2).

EXAMPLE 12

In Example 11 it was verified that a particular solution of the nonhomogeneous system (10) on $(-\infty, \infty)$ is

$$\mathbf{X}_p = \begin{pmatrix} 3t - 4 \\ -5t + 6 \end{pmatrix}.$$

The complementary function of (10) on the same interval, or general solution of

$$\mathbf{X}' = \begin{pmatrix} 1 & 3 \\ 5 & 3 \end{pmatrix}\mathbf{X},$$

was seen in Example 9 to be

$$\mathbf{X}_c = c_1\begin{pmatrix} 1 \\ -1 \end{pmatrix}e^{-2t} + c_2\begin{pmatrix} 3 \\ 5 \end{pmatrix}e^{6t}.$$

Hence by Definition 8.18,

$$\begin{aligned}\mathbf{X} &= \mathbf{X}_c + \mathbf{X}_p \\ &= c_1\begin{pmatrix} 1 \\ -1 \end{pmatrix}e^{-2t} + c_2\begin{pmatrix} 3 \\ 5 \end{pmatrix}e^{6t} + \begin{pmatrix} 3t - 4 \\ -5t + 6 \end{pmatrix}\end{aligned}$$

is the general solution of (10) on $(-\infty, \infty)$.

As one might expect, if $\mathbf{X}$ is *any* solution of the nonhomogeneous system (2) on an interval I, then it is always possible to find appropriate constants $c_1, c_2, \ldots, c_n$ so that $\mathbf{X}$ can be obtained from the general solution.

8.5.2 A Fundamental Matrix

If $\mathbf{X}_1, \mathbf{X}_2, \ldots, \mathbf{X}_n$ is a fundamental set of solutions of the homogeneous system (3) on an interval I, then its general solution on the interval is

$$\begin{aligned}\mathbf{X} &= c_1\mathbf{X}_1 + c_2\mathbf{X}_2 + \cdots + c_n\mathbf{X}_n \\ &= c_1\begin{pmatrix} x_{11} \\ x_{21} \\ \vdots \\ x_{n1} \end{pmatrix} + c_2\begin{pmatrix} x_{12} \\ x_{22} \\ \vdots \\ x_{n2} \end{pmatrix} + \cdots + c_n\begin{pmatrix} x_{1n} \\ x_{2n} \\ \vdots \\ x_{nn} \end{pmatrix} = \begin{pmatrix} c_1x_{11} + c_2x_{12} + \cdots + c_nx_{1n} \\ c_1x_{21} + c_2x_{22} + \cdots + c_nx_{2n} \\ \vdots \\ c_1x_{n1} + c_2x_{n2} + \cdots + c_nx_{nn} \end{pmatrix}.\end{aligned} \tag{11}$$

Observe that (11) can be written as the matrix product

$$\mathbf{X} = \begin{pmatrix} x_{11} & x_{12} & \cdots & x_{1n} \\ x_{21} & x_{22} & \cdots & x_{2n} \\ \vdots & & & \vdots \\ x_{n1} & x_{n2} & \cdots & x_{nn} \end{pmatrix}\begin{pmatrix} c_1 \\ c_2 \\ \vdots \\ c_n \end{pmatrix}. \tag{12}$$

We are led to the following definition.

DEFINITION 8.19 Let

$$\mathbf{X}_1 = \begin{pmatrix} x_{11} \\ x_{21} \\ \vdots \\ x_{n1} \end{pmatrix}, \mathbf{X}_2 = \begin{pmatrix} x_{12} \\ x_{22} \\ \vdots \\ x_{n2} \end{pmatrix}, \ldots, \mathbf{X}_n = \begin{pmatrix} x_{1n} \\ x_{2n} \\ \vdots \\ x_{nn} \end{pmatrix}$$

be a fundamental set of n solution vectors of the homogeneous system (3) on an interval I. The matrix

$$\mathbf{\Phi}(t) = \begin{pmatrix} x_{11} & x_{12} & \cdots & x_{1n} \\ x_{21} & x_{22} & \cdots & x_{2n} \\ \vdots & & & \vdots \\ x_{n1} & x_{n2} & \cdots & x_{nn} \end{pmatrix}$$

is said to be a **fundamental matrix** of the system on the interval.

EXAMPLE 13

The vectors

$$\mathbf{X}_1 = \begin{pmatrix} 1 \\ -1 \end{pmatrix} e^{-2t} = \begin{pmatrix} e^{-2t} \\ -e^{-2t} \end{pmatrix} \quad \text{and} \quad \mathbf{X}_2 = \begin{pmatrix} 3 \\ 5 \end{pmatrix} e^{6t} = \begin{pmatrix} 3e^{6t} \\ 5e^{6t} \end{pmatrix}$$

have been show to form a fundamental set of solutions of the system (4) on $(-\infty, \infty)$. A fundamental matrix of the system on the interval is then

$$\mathbf{\Phi}(t) = \begin{pmatrix} e^{-2t} & 3e^{6t} \\ -e^{-2t} & 5e^{6t} \end{pmatrix}. \tag{13}$$

The result given in (12) states that the general solution of any homogeneous system $\mathbf{X}' = \mathbf{A}(t)\mathbf{X}$ can always be written in terms of a fundamental matrix of the system: $\mathbf{X} = \mathbf{\Phi}(t)\mathbf{C}$, where $\mathbf{C}$ is an $n \times 1$ column vector of arbitrary constants.

EXAMPLE 14

The general solution given in (9) can be written

$$\mathbf{X} = \begin{pmatrix} e^{-2t} & 3e^{6t} \\ -e^{-2t} & 5e^{6t} \end{pmatrix} \begin{pmatrix} c_1 \\ c_2 \end{pmatrix}.$$

Furthermore, to say that $\mathbf{X} = \mathbf{\Phi}(t)\mathbf{C}$ is a solution of $\mathbf{X}' = \mathbf{A}(t)\mathbf{X}$ we mean

$$\mathbf{\Phi}'(t)\mathbf{C} = \mathbf{A}(t)\mathbf{\Phi}(t)\mathbf{C}$$

or

$$(\mathbf{\Phi}'(t) - \mathbf{A}(t)\mathbf{\Phi}(t))\mathbf{C} = \mathbf{0}.$$

Since the last equation is to hold for every t in the interval I and for every possible column matrix of constants $\mathbf{C}$, we must have

$$\mathbf{\Phi}'(t) - \mathbf{A}(t)\mathbf{\Phi}(t) = \mathbf{0}$$

or

$$\mathbf{\Phi}'(t) = \mathbf{A}(t)\mathbf{\Phi}(t). \tag{14}$$

This result will be useful in Section 8.8.

Fundamental Matrix Is Nonsingular

Comparison of Theorem 8.5 and Definition 8.19 shows that det $\mathbf{\Phi}(t)$ is the same as the Wronskian $W(\mathbf{X}_1, \mathbf{X}_2, \ldots, \mathbf{X}_n)$*. Hence the linear independence of the columns of $\mathbf{\Phi}(t)$ on an interval I guarantees that det $\mathbf{\Phi}(t) \neq 0$ for every t in the interval. That is, $\mathbf{\Phi}(t)$ is nonsingular on the interval.

THEOREM 8.8 Let $\mathbf{\Phi}(t)$ be a fundamental matrix of the homogeneous system (3) on an interval I. Then $\mathbf{\Phi}^{-1}(t)$ exists for every value of t in the interval.

EXAMPLE 15

For the fundamental matrix given in (13) we see that det $\mathbf{\Phi}(t) = 8e^{4t}$. It then follows from (3) of Section 8.4 that

$$\mathbf{\Phi}^{-1}(t) = \frac{1}{8e^{4t}}\begin{pmatrix} 5e^{6t} & -3e^{6t} \\ e^{-2t} & e^{-2t} \end{pmatrix}$$
$$= \begin{pmatrix} \frac{5}{8}e^{2t} & -\frac{3}{8}e^{2t} \\ \frac{1}{8}e^{-6t} & \frac{1}{8}e^{-6t} \end{pmatrix}.$$

Special Matrix

In some instances it is convenient to form another special $n \times n$ matrix, a matrix in which the column vectors $\mathbf{V}_i$ are solutions of $\mathbf{X}' = \mathbf{A}(t)\mathbf{X}$ that satisfy the conditions

$$\mathbf{V}_1(t_0) = \begin{pmatrix} 1 \\ 0 \\ \vdots \\ 0 \end{pmatrix}, \mathbf{V}_2(t_0) = \begin{pmatrix} 0 \\ 1 \\ \vdots \\ 0 \end{pmatrix}, \ldots, \mathbf{V}_n(t_0) = \begin{pmatrix} 0 \\ 0 \\ \vdots \\ 1 \end{pmatrix}. \tag{15}$$

* For this reason some texts will call $\mathbf{\Phi}(t)$ a *Wronski matrix*.

Here t_0 is an arbitrarily chosen point in the interval on which the general solution of the system is defined. We shall denote this special matrix by the symbol $\mathbf{\Psi}(t)$. Observe that $\mathbf{\Psi}(t)$ has the property

$$\mathbf{\Psi}(t_0) = \begin{pmatrix} 1 & 0 & 0 & \cdots & 0 \\ 0 & 1 & 0 & \cdots & 0 \\ \vdots & & & & \vdots \\ 0 & 0 & 0 & \cdots & 1 \end{pmatrix} = \mathbf{I}, \tag{16}$$

where $\mathbf{I}$ is the $n \times n$ multiplicative identity.

EXAMPLE 16 Find the matrix $\mathbf{\Psi}(t)$ satisfying $\mathbf{\Psi}(0) = \mathbf{I}$ for the system given in (4).

SOLUTION From (9) we know that the general solution of (4) is given by

$$\mathbf{X} = c_1\begin{pmatrix} 1 \\ -1 \end{pmatrix}e^{-2t} + c_2\begin{pmatrix} 3 \\ 5 \end{pmatrix}e^{6t}.$$

When $t = 0$ we first solve for constants c_1 and c_2 such that

$$c_1\begin{pmatrix} 1 \\ -1 \end{pmatrix} + c_2\begin{pmatrix} 3 \\ 5 \end{pmatrix} = \begin{pmatrix} 1 \\ 0 \end{pmatrix} \quad \text{or} \quad \begin{aligned} c_1 + 3c_2 &= 1 \\ -c_1 + 5c_2 &= 0. \end{aligned}$$

We find that $c_1 = 5/8$ and $c_2 = 1/8$. Hence we define the vector $\mathbf{V}_1$ to be the linear combination.

$$\mathbf{V}_1 = \frac{5}{8}\begin{pmatrix} 1 \\ -1 \end{pmatrix}e^{-2t} + \frac{1}{8}\begin{pmatrix} 3 \\ 5 \end{pmatrix}e^{6t}.$$

Again when $t = 0$ we wish to find another pair of constants c_1 and c_2 for which

$$c_1\begin{pmatrix} 1 \\ -1 \end{pmatrix} + c_2\begin{pmatrix} 3 \\ 5 \end{pmatrix} = \begin{pmatrix} 0 \\ 1 \end{pmatrix} \quad \text{or} \quad \begin{aligned} c_1 + 3c_2 &= 0 \\ -c_1 + 5c_2 &= 1. \end{aligned}$$

In this case we find $c_1 = -3/8$ and $c_2 = 1/8$. We then define

$$\mathbf{V}_2 = -\frac{3}{8}\begin{pmatrix} 1 \\ -1 \end{pmatrix}e^{-2t} + \frac{1}{8}\begin{pmatrix} 3 \\ 5 \end{pmatrix}e^{6t}.$$

Hence

$$\mathbf{\Psi}(t) = \begin{pmatrix} \frac{5}{8}e^{-2t} + \frac{3}{8}e^{6t} & -\frac{3}{8}e^{-2t} + \frac{3}{8}e^{6t} \\ -\frac{5}{8}e^{-2t} + \frac{5}{8}e^{6t} & \frac{3}{8}e^{-2t} + \frac{5}{8}e^{6t} \end{pmatrix}. \tag{17}$$

Observe that $\mathbf{\Psi}(0) = \begin{pmatrix} 1 & 0 \\ 0 & 1 \end{pmatrix} = \mathbf{I}$.

Note in the preceding example that since the columns of $\mathbf{\Psi}(t)$ are linear combinations of the solutions $\mathbf{\Psi}$ of $\mathbf{X}' = \mathbf{A}(t)\mathbf{X}$ we know from the superposition principle that each column is a solution of the system.

$\Psi(t)$ Is a Fundamental Matrix

From (16) it is seen that det $\mathbf{\Psi}(t_0) \neq 0$, and hence we conclude from Theorem 8.5 that the columns of $\mathbf{\Psi}(t)$ are linearly independent on the interval under consideration. Therefore $\mathbf{\Psi}(t)$ is a fundamental matrix. Also, it follows from Theorem 8.3 that $\mathbf{\Psi}(t)$ is the unique matrix, satisfying the condition $\mathbf{\Psi}(t_0) = \mathbf{I}$. Lastly, the fundamental matrices $\mathbf{\Phi}(t)$ and $\mathbf{\Psi}(t)$ are related by

$$\mathbf{\Psi}(t) = \mathbf{\Phi}(t)\mathbf{\Phi}^{-1}(t_0). \tag{18}$$

Equation (19) provides an alternative method for determining $\mathbf{\Phi}(t)$ (see Problem 37).

The answer to why anyone would want to form an obviously complicated looking fundamental matrix such as (17) will be answered in Sections 8.8 and 8.9.

Exercises 8.5

Answers to odd-numbered problems begin on page A–24.

[8.5.1] In Problems 1–6 write the given system in matrix form.

1. $\dfrac{dx}{dt} = 3x - 5y$

$\dfrac{dy}{dt} = 4x + 8y$

2. $\dfrac{dx}{dt} = 4x - 7y$

$\dfrac{dy}{dt} = 5x$

3. $\dfrac{dx}{dt} = -3x + 4y - 9z$

$\dfrac{dy}{dt} = 6x - y$

$\dfrac{dz}{dt} = 10x + 4y + 3z$

4. $\dfrac{dx}{dt} = x - y$

$\dfrac{dy}{dt} = x + 2z$

$\dfrac{dz}{dt} = -x + z$

5. $\dfrac{dx}{dt} = x - y + z + t - 1$

$\dfrac{dy}{dt} = 2x + y - z - 3t^2$

$\dfrac{dz}{dt} = x + y + z + t^2 - t + 2$

6. $\dfrac{dx}{dt} = -3x + 4y + e^{-t}\sin 2t$

$\dfrac{dy}{dt} = 5x + 9y + 4e^{-t}\cos 2t$

In Problems 7–10 write the given system without the use of matrices.

7. $\mathbf{X}' = \begin{pmatrix} 4 & 2 \\ -1 & 3 \end{pmatrix}\mathbf{X} + \begin{pmatrix} 1 \\ -1 \end{pmatrix}e^t$

8. $\mathbf{X}' = \begin{pmatrix} 7 & 5 & -9 \\ 4 & 1 & 1 \\ 0 & -2 & 3 \end{pmatrix}\mathbf{X} + \begin{pmatrix} 0 \\ 2 \\ 1 \end{pmatrix} e^{5t} - \begin{pmatrix} 8 \\ 0 \\ 3 \end{pmatrix} e^{-2t}$

9. $\dfrac{d}{dt}\begin{pmatrix} x \\ y \\ z \end{pmatrix} = \begin{pmatrix} 1 & -1 & 2 \\ 3 & -4 & 1 \\ -2 & 5 & 6 \end{pmatrix}\begin{pmatrix} x \\ y \\ z \end{pmatrix} + \begin{pmatrix} 1 \\ 2 \\ 2 \end{pmatrix} e^{-t} - \begin{pmatrix} 3 \\ -1 \\ 1 \end{pmatrix} t$

10. $\dfrac{d}{dt}\begin{pmatrix} x \\ y \end{pmatrix} = \begin{pmatrix} 3 & -7 \\ 1 & 1 \end{pmatrix}\begin{pmatrix} x \\ y \end{pmatrix} + \begin{pmatrix} 4 \\ 8 \end{pmatrix} \sin t + \begin{pmatrix} t-4 \\ 2t+1 \end{pmatrix} e^{4t}$

In Problems 11–16 verify that the vector $\mathbf{X}$ is a solution of the given system.

11. $\dfrac{dx}{dt} = 3x - 4y$

$\dfrac{dy}{dt} = 4x - 7y;\quad \mathbf{X} = \begin{pmatrix} 1 \\ 2 \end{pmatrix} e^{-5t}$

12. $\dfrac{dx}{dt} = -2x + 5y$

$\dfrac{dy}{dt} = -2x + 4y;\quad \mathbf{X} = \begin{pmatrix} 5 \cos t \\ 3 \cos t - \sin t \end{pmatrix} e^{t}$

13. $\mathbf{X}' = \begin{pmatrix} -1 & \frac{1}{4} \\ 1 & -1 \end{pmatrix}\mathbf{X};\quad \mathbf{X} = \begin{pmatrix} -1 \\ 2 \end{pmatrix} e^{-3t/2}$

14. $\mathbf{X}' = \begin{pmatrix} 2 & 1 \\ -1 & 0 \end{pmatrix}\mathbf{X};\quad \mathbf{X} = \begin{pmatrix} 1 \\ 3 \end{pmatrix} e^{t} + \begin{pmatrix} 4 \\ -4 \end{pmatrix} te^{t}$

15. $\dfrac{d\mathbf{X}}{dt} = \begin{pmatrix} 1 & 2 & 1 \\ 6 & -1 & 0 \\ -1 & -2 & -1 \end{pmatrix}\mathbf{X};\quad \mathbf{X} = \begin{pmatrix} 1 \\ 6 \\ -13 \end{pmatrix}$

16. $\mathbf{X}' = \begin{pmatrix} 1 & 0 & 1 \\ 1 & 1 & 0 \\ -2 & 0 & -1 \end{pmatrix}\mathbf{X};\quad \mathbf{X} = \begin{pmatrix} \sin t \\ -\frac{1}{2}\sin t - \frac{1}{2}\cos t \\ -\sin t + \cos t \end{pmatrix}$

In Problems 17–20 the given vectors are solutions of a system $\mathbf{X}' = \mathbf{A}\mathbf{X}$. Determine whether the vectors form a fundamental set on $(-\infty, \infty)$.

17. $\mathbf{X}_1 = \begin{pmatrix} 1 \\ 1 \end{pmatrix} e^{-2t},\quad \mathbf{X}_2 = \begin{pmatrix} 1 \\ -1 \end{pmatrix} e^{-6t}$

18. $\mathbf{X}_1 = \begin{pmatrix} 1 \\ -1 \end{pmatrix} e^{t},\quad \mathbf{X}_2 = \begin{pmatrix} 2 \\ 6 \end{pmatrix} e^{t} + \begin{pmatrix} 8 \\ -8 \end{pmatrix} te^{t}$

19. $\mathbf{X}_1 = \begin{pmatrix} 1 \\ -2 \\ 4 \end{pmatrix} + t\begin{pmatrix} 1 \\ 2 \\ 2 \end{pmatrix},\quad \mathbf{X}_2 = \begin{pmatrix} 1 \\ -2 \\ 4 \end{pmatrix},\quad \mathbf{X}_3 = \begin{pmatrix} 3 \\ -6 \\ 12 \end{pmatrix} + t\begin{pmatrix} 2 \\ 4 \\ 4 \end{pmatrix}$

20. $\mathbf{X}_1 = \begin{pmatrix} 1 \\ 6 \\ -13 \end{pmatrix}$, $\mathbf{X}_2 = \begin{pmatrix} 1 \\ -2 \\ -1 \end{pmatrix} e^{-4t}$, $\mathbf{X}_3 = \begin{pmatrix} 2 \\ 3 \\ -2 \end{pmatrix} e^{3t}$

In Problems 21–24 verify that the vector $\mathbf{X}_p$ is a particular solution of the given system.

21. $\dfrac{dx}{dt} = x + 4y + 2t - 7$

$\dfrac{dy}{dt} = 3x + 2y - 4t - 18$; $\mathbf{X}_p = \begin{pmatrix} 2 \\ -1 \end{pmatrix} t + \begin{pmatrix} 5 \\ 1 \end{pmatrix}$

22. $\mathbf{X}' = \begin{pmatrix} 2 & 1 \\ 1 & -1 \end{pmatrix} \mathbf{X} + \begin{pmatrix} -5 \\ 2 \end{pmatrix}$; $\mathbf{X}_p = \begin{pmatrix} 1 \\ 3 \end{pmatrix}$

23. $\mathbf{X}' = \begin{pmatrix} 2 & 1 \\ 3 & 4 \end{pmatrix} \mathbf{X} - \begin{pmatrix} 1 \\ 7 \end{pmatrix} e^t$; $\mathbf{X}_p = \begin{pmatrix} 1 \\ 1 \end{pmatrix} e^t + \begin{pmatrix} 1 \\ -1 \end{pmatrix} te^t$

24. $\mathbf{X}' = \begin{pmatrix} 1 & 2 & 3 \\ -4 & 2 & 0 \\ -6 & 1 & 0 \end{pmatrix} \mathbf{X} + \begin{pmatrix} -1 \\ 4 \\ 3 \end{pmatrix} \sin 3t$; $\mathbf{X}_p = \begin{pmatrix} \sin 3t \\ 0 \\ \cos 3t \end{pmatrix}$

25. Prove that the general solution of

$$\mathbf{X}' = \begin{pmatrix} 0 & 6 & 0 \\ 1 & 0 & 1 \\ 1 & 1 & 0 \end{pmatrix} \mathbf{X}$$

on the interval $(-\infty, \infty)$ is

$$\mathbf{X} = c_1 \begin{pmatrix} 6 \\ -1 \\ -5 \end{pmatrix} e^{-t} + c_2 \begin{pmatrix} -3 \\ 1 \\ 1 \end{pmatrix} e^{-2t} + c_3 \begin{pmatrix} 2 \\ 1 \\ 1 \end{pmatrix} e^{3t}.$$

26. Prove that the general solution of

$$\mathbf{X}' = \begin{pmatrix} -1 & -1 \\ -1 & 1 \end{pmatrix} \mathbf{X} + \begin{pmatrix} 1 \\ 1 \end{pmatrix} t^2 + \begin{pmatrix} 4 \\ -6 \end{pmatrix} t + \begin{pmatrix} -1 \\ 5 \end{pmatrix}$$

on the interval $(-\infty, \infty)$ is

$$\mathbf{X} = c_1 \begin{pmatrix} 1 \\ -1 - \sqrt{2} \end{pmatrix} e^{\sqrt{2}t} + c_2 \begin{pmatrix} 1 \\ -1 + \sqrt{2} \end{pmatrix} e^{-\sqrt{2}t} + \begin{pmatrix} 1 \\ 0 \end{pmatrix} t^2 + \begin{pmatrix} -2 \\ 4 \end{pmatrix} t + \begin{pmatrix} 1 \\ 0 \end{pmatrix}.$$

[8.5.2] In Problems 27–30 the indicated column vectors form a fundamental set of solutions for the given system on $(-\infty, \infty)$. Form a fundamental matrix $\mathbf{\Phi}(t)$ and compute $\mathbf{\Phi}^{-1}(t)$.

27. $\mathbf{X}' = \begin{pmatrix} 4 & 1 \\ 6 & 5 \end{pmatrix}\mathbf{X};\ \ \mathbf{X}_1 = \begin{pmatrix} 1 \\ -2 \end{pmatrix}e^{2t},\ \ \mathbf{X}_2 = \begin{pmatrix} 1 \\ 3 \end{pmatrix}e^{7t}$

28. $\mathbf{X}' = \begin{pmatrix} 2 & 3 \\ 3 & 2 \end{pmatrix}\mathbf{X};\ \ \mathbf{X}_1 = \begin{pmatrix} -1 \\ 1 \end{pmatrix}e^{-t},\ \ \mathbf{X}_2 = \begin{pmatrix} 1 \\ 1 \end{pmatrix}e^{5t}$

29. $\mathbf{X}' = \begin{pmatrix} 4 & 1 \\ -9 & -2 \end{pmatrix}\mathbf{X};\ \ \mathbf{X}_1 = \begin{pmatrix} -1 \\ 3 \end{pmatrix}e^{t},\ \ \mathbf{X}_2 = \begin{pmatrix} -1 \\ 3 \end{pmatrix}te^{t} + \begin{pmatrix} 0 \\ -1 \end{pmatrix}e^{t}$

30. $\mathbf{X}' = \begin{pmatrix} 3 & -2 \\ 5 & -3 \end{pmatrix}\mathbf{X};\ \ \mathbf{X}_1 = \begin{pmatrix} 2\cos t \\ 3\cos t + \sin t \end{pmatrix},\ \ \mathbf{X}_2 = \begin{pmatrix} -2\sin t \\ \cos t - 3\sin t \end{pmatrix}$

31. Find the fundamental matrix $\mathbf{\Psi}(t)$ satisfying $\mathbf{\Psi}(0) = \mathbf{I}$ for the system given in Problem 27.

32. Find the fundamental matrix $\mathbf{\Psi}(t)$ satisfying $\mathbf{\Psi}(0) = \mathbf{I}$ for the system given in Problem 28.

33. Find the fundamental matrix $\mathbf{\Psi}(t)$ satisfying $\mathbf{\Psi}(0) = \mathbf{I}$ for the system given in Problem 29.

34. Find the fundamental matrix $\mathbf{\Psi}(t)$ satisfying $\mathbf{\Psi}(\pi/2) = \mathbf{I}$ for the system given in Problem 30.

Miscellaneous Problems

35. If $\mathbf{X} = \mathbf{\Phi}(t)\mathbf{C}$ is the general solution of $\mathbf{X}' = \mathbf{AX}$, show that the solution of the initial-value problem $\mathbf{X}' = \mathbf{AX}$, $\mathbf{X}(t_0) = \mathbf{X}_0$ is $\mathbf{X} = \mathbf{\Phi}(t)\mathbf{\Phi}^{-1}(t_0)\mathbf{X}_0$.

36. Show that the solution of the initial-value problem given in Problem 35 is also given by $\mathbf{X} = \mathbf{\Psi}(t)\mathbf{X}_0$.

37. Show that $\mathbf{\Psi}(t) = \mathbf{\Phi}(t)\mathbf{\Phi}^{-1}(t_0)$. [*Hint:* Compare Problems 35 and 36.]

8.6 Homogeneous Linear Systems

8.6.1 Distinct Real Eigenvalues

For the remainder of this chapter we shall be concerned only with linear systems with real constant coefficients.

We saw in Example 9 of the preceding section that the general solution of the homogeneous system

$$\frac{dx}{dt} = x + 3y$$

$$\frac{dy}{dt} = 5x + 3y$$

is

$$\mathbf{X} = c_1\begin{pmatrix} 1 \\ -1 \end{pmatrix}e^{-2t} + c_2\begin{pmatrix} 3 \\ 5 \end{pmatrix}e^{6t}.$$

Since both solution vectors have the basic form

$$\mathbf{X}_i = \begin{pmatrix} k_1 \\ k_2 \end{pmatrix}e^{\lambda_i t}, \qquad i = 1, 2,$$

k_1 and k_2 constants, we are prompted to ask whether we can always find a solution of the form

$$\mathbf{X} = \begin{pmatrix} k_1 \\ k_2 \\ \vdots \\ k_n \end{pmatrix}e^{\lambda t} = \mathbf{K}e^{\lambda t} \tag{1}$$

for the general homogeneous linear first-order system

$$\mathbf{X}' = \mathbf{AX}, \tag{2}$$

where $\mathbf{A}$ is an $n \times n$ matrix of constants.

Eigenvalues and Eigenvectors

If (1) is to be a solution vector of (2), then $\mathbf{X}' = \mathbf{K}\lambda e^{\lambda t}$ so that the system becomes

$$\mathbf{K}\lambda e^{\lambda t} = \mathbf{AK}e^{\lambda t}.$$

After dividing out $e^{\lambda t}$ and rearranging, we obtain

$$\mathbf{AK} = \lambda\mathbf{K}$$

or

$$(\mathbf{A} - \lambda\mathbf{I})\mathbf{K} = \mathbf{0}. \tag{3}$$

Equation (3) is equivalent to the simultaneous algebraic equations (8) of Section 8.4. To find a nontrivial solution $\mathbf{X}$ of (2) we must find a nontrivial vector $\mathbf{K}$ satisfying (3). But in order that (3) have nontrivial solutions we must have

$$\det(\mathbf{A} - \lambda\mathbf{I}) = 0.$$

The latter equation is recognized as the characteristic equation of the matrix $\mathbf{A}$. In other words, $\mathbf{X} = \mathbf{K}e^{\lambda t}$ will be a solution of the system of differential equations (2) if and only if λ is an **eigenvalue** of $\mathbf{A}$ and $\mathbf{K}$ is an **eigenvector** corresponding to λ.

When the $n \times n$ matrix $\mathbf{A}$ possesses n distinct real eigenvalues $\lambda_1, \lambda_2, \ldots, \lambda_n$, then a set of n linearly independent eigenvectors $\mathbf{K}_1, \mathbf{K}_2, \ldots, \mathbf{K}_n$ can always be found and

$$\mathbf{X}_1 = \mathbf{K}_1 e^{\lambda_1 t}, \mathbf{X}_2 = \mathbf{K}_2 e^{\lambda_2 t}, \ldots, \mathbf{X}_n = \mathbf{K}_n e^{\lambda_n t}$$

is a fundamental set of solutions of (2) on $(-\infty, \infty)$.

THEOREM 8.9 Let $\lambda_1, \lambda_2, \ldots, \lambda_n$ be n distinct real eigenvalues of the coefficient matrix $\mathbf{A}$ of the homogeneous system (2) and let $\mathbf{K}_1, \mathbf{K}_2, \ldots, \mathbf{K}_n$ be the corresponding eigenvectors. Then the general solution of (2) on the interval $(-\infty, \infty)$ is given by

$$\mathbf{X} = c_1 \mathbf{K}_1 e^{\lambda_1 t} + c_2 \mathbf{K}_2 e^{\lambda_2 t} + \cdots + c_n \mathbf{K}_n e^{\lambda_n t}.$$

EXAMPLE 1

Solve

$$\begin{aligned} \frac{dx}{dt} &= 2x + 3y \\ \frac{dy}{dt} &= 2x + y. \end{aligned} \tag{4}$$

SOLUTION We first find the eigenvalues and eigenvectors of the matrix of coefficients.

The characteristic equation is

$$\det(\mathbf{A} - \lambda \mathbf{I}) = \begin{vmatrix} 2 - \lambda & 3 \\ 2 & 1 - \lambda \end{vmatrix} = \lambda^2 - 3\lambda - 4 = 0.$$

Since $\lambda^2 - 3\lambda - 4 = (\lambda + 1)(\lambda - 4)$, we see that the eigenvalues are $\lambda_1 = -1$ and $\lambda_2 = 4$.

Now for $\lambda_1 = -1$, (3) is equivalent to

$$\begin{aligned} 3k_1 + 3k_2 &= 0 \\ 2k_1 + 2k_2 &= 0. \end{aligned}$$

Thus $k_1 = -k_2$. By selecting $k_2 = -1$ the related eigenvector is then

$$\mathbf{K}_1 = \begin{pmatrix} 1 \\ -1 \end{pmatrix}.$$

For $\lambda_2 = 4$ we have

$$\begin{aligned} -2k_1 + 3k_2 &= 0 \\ 2k_1 - 3k_2 &= 0 \end{aligned}$$

so that $k_1 = 3k_2/2$, and therefore with $k_2 = 2$, the corresponding eigenvector is

$$\mathbf{K}_2 = \begin{pmatrix} 3 \\ 2 \end{pmatrix}.$$

Since the matrix of coefficients **A** is a 2×2 matrix and since we have found two linearly independent solutions of (4),

$$\mathbf{X}_1 = \begin{pmatrix} 1 \\ -1 \end{pmatrix} e^{-t} \quad \text{and} \quad \mathbf{X}_2 = \begin{pmatrix} 3 \\ 2 \end{pmatrix} e^{4t},$$

we conclude that the general solution of the system is

$$\begin{aligned} \mathbf{X} &= c_1 \mathbf{X}_1 + c_2 \mathbf{X}_2 \\ &= c_1 \begin{pmatrix} 1 \\ -1 \end{pmatrix} e^{-t} + c_2 \begin{pmatrix} 3 \\ 2 \end{pmatrix} e^{4t}. \end{aligned} \tag{5}$$

For the sake of review, the reader should keep firmly in mind that a solution of a system of first-order differential equations, when written in terms of matrices, is simply an alternative to the method that we employed in Section 8.1, namely, listing the individual functions and the relationships between the constants. By adding the vectors given in (5), we obtain

$$\begin{pmatrix} x(t) \\ y(t) \end{pmatrix} = \begin{pmatrix} c_1 e^{-t} + 3c_2 e^{4t} \\ -c_1 e^{-t} + 2c_2 e^{4t} \end{pmatrix}$$

and this in turn yields the more familiar statement

$$\begin{aligned} x(t) &= c_1 e^{-t} + 3c_2 e^{4t} \\ y(t) &= -c_1 e^{-t} + 2c_2 e^{4t} \end{aligned}$$

EXAMPLE 2 Solve

$$\begin{aligned} \frac{dx}{dt} &= -4x + y + z \\ \frac{dy}{dt} &= x + 5y - z \\ \frac{dz}{dt} &= y - 3z. \end{aligned} \tag{6}$$

SOLUTION Using the cofactors of the third row, we find

$$\det(\mathbf{A} - \lambda \mathbf{I}) = \begin{vmatrix} -4-\lambda & 1 & 1 \\ 1 & 5-\lambda & -1 \\ 0 & 1 & -3-\lambda \end{vmatrix} = -(\lambda + 3)(\lambda + 4)(\lambda - 5) = 0$$

and so the eigenvalues are $\lambda_1 = -3$, $\lambda_2 = -4$, $\lambda_3 = 5$.

Now for $\lambda_1 = -3$, Gauss-Jordan elimination gives

$$(\mathbf{A} + 3\mathbf{I}|\mathbf{0}) = \left(\begin{array}{ccc|c} -1 & 1 & 1 & 0 \\ 1 & 8 & -1 & 0 \\ 0 & 1 & 0 & 0 \end{array}\right) \Longrightarrow \left(\begin{array}{ccc|c} 1 & 0 & -1 & 0 \\ 0 & 1 & 0 & 0 \\ 0 & 0 & 0 & 0 \end{array}\right).$$

Therefore $k_1 = k_3$, $k_2 = 0$. The choice $k_3 = 1$ gives the eigenvector

$$\mathbf{K}_1 = \begin{pmatrix} 1 \\ 0 \\ 1 \end{pmatrix}. \tag{7}$$

Similarly, for $\lambda_2 = -4$,

$$(\mathbf{A} + 4\mathbf{I}|\mathbf{0}) = \left(\begin{array}{ccc|c} 0 & 1 & 1 & 0 \\ 1 & 9 & -1 & 0 \\ 0 & 1 & 1 & 0 \end{array}\right) \Longrightarrow \left(\begin{array}{ccc|c} 1 & 0 & -10 & 0 \\ 0 & 1 & 1 & 0 \\ 0 & 0 & 0 & 0 \end{array}\right)$$

implies $k_1 = 10k_3$, $k_2 = -k_3$. Choosing $k_3 = 1$ gives the second eigenvector

$$\mathbf{K}_2 = \begin{pmatrix} 10 \\ -1 \\ 1 \end{pmatrix}. \tag{8}$$

Finally, when $\lambda_3 = 5$, the augmented matrices

$$(\mathbf{A} - 5\mathbf{I}|\mathbf{0}) = \left(\begin{array}{ccc|c} -9 & 1 & 1 & 0 \\ 1 & 0 & -1 & 0 \\ 0 & 1 & -8 & 0 \end{array}\right) \Longrightarrow \left(\begin{array}{ccc|c} 1 & 0 & -1 & 0 \\ 0 & 1 & -8 & 0 \\ 0 & 0 & 0 & 0 \end{array}\right)$$

yield

$$\mathbf{K}_3 = \begin{pmatrix} 1 \\ 8 \\ 1 \end{pmatrix}. \tag{9}$$

Multiplying the vectors (7), (8), and (9) by e^{-3t}, e^{-4t}, and e^{5t}, respectively, gives three solutions of (6):

$$\mathbf{X}_1 = \begin{pmatrix} 1 \\ 0 \\ 1 \end{pmatrix} e^{-3t}, \quad \mathbf{X}_2 = \begin{pmatrix} 10 \\ -1 \\ 1 \end{pmatrix} e^{-4t}, \quad \mathbf{X}_3 = \begin{pmatrix} 1 \\ 8 \\ 1 \end{pmatrix} e^{5t}.$$

The general solution of the system is then

$$\mathbf{X} = c_1 \begin{pmatrix} 1 \\ 0 \\ 1 \end{pmatrix} e^{-3t} + c_2 \begin{pmatrix} 10 \\ -1 \\ 1 \end{pmatrix} e^{-4t} + c_3 \begin{pmatrix} 1 \\ 8 \\ 1 \end{pmatrix} e^{5t}.$$

8.6.2 Complex Eigenvalues

If $\lambda_1 = \alpha + i\beta$ and $\lambda_2 = \alpha - i\beta$, $i^2 = -1$

are complex eigenvalues of the coefficient matrix **A**, we can then certainly expect their corresponding eigenvectors to also have complex entries.*

For example, the characteristic equation of the system

$$\begin{aligned} \frac{dx}{dt} &= 6x - y \\ \frac{dy}{dt} &= 5x + 4y \end{aligned} \tag{10}$$

is $$\det(\mathbf{A} - \lambda\mathbf{I}) = \begin{vmatrix} 6 - \lambda & -1 \\ 5 & 4 - \lambda \end{vmatrix} = \lambda^2 - 10\lambda + 29 = 0.$$

From the quadratic formula we find

$$\lambda_1 = 5 + 2i, \qquad \lambda_2 = 5 - 2i.$$

Now for $\lambda_1 = 5 + 2i$ we must solve

$$\begin{aligned} (1 - 2i)k_1 - \quad k_2 &= 0 \\ 5k_1 - (1 + 2i)k_2 &= 0. \end{aligned}$$

Since $k_2 = (1 - 2i)k_1$† it follows, after choosing $k_1 = 1$, that one eigenvector is

$$\mathbf{K}_1 = \begin{pmatrix} 1 \\ 1 - 2i \end{pmatrix}.$$

Similarly, for $\lambda_2 = 5 - 2i$ we find the other eigenvector to be

$$\mathbf{K}_2 = \begin{pmatrix} 1 \\ 1 + 2i \end{pmatrix}.$$

Consequently two solutions of (17) are

$$\mathbf{X}_1 = \begin{pmatrix} 1 \\ 1 - 2i \end{pmatrix} e^{(5+2i)t} \quad \text{and} \quad \mathbf{X}_2 = \begin{pmatrix} 1 \\ 1 + 2i \end{pmatrix} e^{(5-2i)t}.$$

By the superposition principle another solution is

$$\mathbf{X} = c_1 \begin{pmatrix} 1 \\ 1 - 2i \end{pmatrix} e^{(5+2i)t} + c_2 \begin{pmatrix} 1 \\ 1 + 2i \end{pmatrix} e^{(5-2i)t}. \tag{11}$$

---

* When the characteristic equation has real coefficients, complex eigenvalues will always appear in conjugate pairs.

† Note that the second equation is simply $(1 + 2i)$ times the first.

Note that the entries in $\mathbf{K}_2$ corresponding to λ_2 are the conjugates of the entries in $\mathbf{K}_1$ corresponding to λ_1. The conjugate of λ_1 is, of course, λ_2. We write this as $\lambda_2 = \bar{\lambda}_1$ and $\mathbf{K}_2 = \bar{\mathbf{K}}_1$. We have illustrated the following general result.

THEOREM 8.10 Let $\mathbf{A}$ be the coefficient matrix having real entries of the homogeneous system (2), and let $\mathbf{K}$ be an eigenvector corresponding to the complex eigenvalue $\lambda_1 = \alpha + i\beta$, α and β real. Then

$$\mathbf{X}_1 = \mathbf{K}_1 e^{\lambda_1 t} \qquad \text{and} \qquad \mathbf{X}_2 = \bar{\mathbf{K}}_1 e^{\bar{\lambda}_1 t}$$

are solutions of (2).

It is desirable and relatively easy to rewrite a solution such as (11) in terms of real functions. Since

$$\begin{aligned} x &= c_1 e^{(5+2i)t} + c_2 e^{(5-2i)t} \\ y &= c_1(1-2i)e^{(5+2i)t} + c_2(1+2i)e^{(5-2i)t}, \end{aligned}$$

it follows from Euler's formula that

$$\begin{aligned} x &= e^{5t}[c_1 e^{2it} + c_2 e^{-2it}] \\ &= e^{5t}[(c_1 + c_2)\cos 2t + (c_1 i - c_2 i)\sin 2t] \\ y &= e^{5t}[(c_1(1-2i) + c_2(1+2i))\cos 2t + (c_1 i(1-2i) - c_2 i(1+2i))\sin 2t] \\ &= e^{5t}[(c_1 + c_2) - 2(c_1 i - c_2 i)]\cos 2t + e^{5t}[2(c_1 + c_2) + (c_1 i - c_2 i)]\sin 2t. \end{aligned}$$

If we replace $c_1 + c_2$ by C_1 and $c_1 i - c_2 i$ by C_2, then

$$\begin{aligned} x &= e^{5t}[C_1 \cos 2t + C_2 \sin 2t] \\ y &= e^{5t}[C_1 - 2C_2]\cos 2t + e^{5t}[2C_1 + C_2]\sin 2t, \end{aligned}$$

or, in terms of vectors,

$$\mathbf{X} = \begin{pmatrix} x \\ y \end{pmatrix} = C_1 \begin{pmatrix} \cos 2t \\ \cos 2t + 2\sin 2t \end{pmatrix} e^{5t} + C_2 \begin{pmatrix} \sin 2t \\ -2\cos 2t + \sin 2t \end{pmatrix} e^{5t}. \tag{12}$$

Here, of course, it can be verified that each vector in (12) is a solution of (10). In addition the solutions are linearly independent on the interval $(-\infty, \infty)$. We may further assume that C_1 and C_2 are completely arbitrary and real. Thus (12) is the general solution of (10).

The foregoing process can be generalized. Let $\mathbf{K}_1$ be an eigenvector of the matrix $\mathbf{A}$ corresponding to the complex eigenvalue $\lambda_1 = \alpha + i\beta$. Then $\mathbf{X}_1$ and $\mathbf{X}_2$ in Theorem 8.10 can be written as

$$\mathbf{K}_1 e^{\lambda_1 t} = \mathbf{K}_1 e^{\alpha t} e^{i\beta t} = \mathbf{K}_1 e^{\alpha t}(\cos \beta t + i \sin \beta t)$$
$$\bar{\mathbf{K}}_1 e^{\bar{\lambda}_1 t} = \bar{\mathbf{K}}_1 e^{\alpha t} e^{-i\beta t} = \bar{\mathbf{K}}_1 e^{\alpha t}(\cos \beta t - i \sin \beta t).$$

The foregoing equations then yield

$$\tfrac{1}{2}(\mathbf{K}_1 e^{\lambda_1 t} + \bar{\mathbf{K}}_1 e^{\bar{\lambda}_1 t}) = \tfrac{1}{2}(\mathbf{K}_1 + \bar{\mathbf{K}}_1)e^{\alpha t} \cos \beta t - \tfrac{i}{2}(-\mathbf{K}_1 + \bar{\mathbf{K}}_1)e^{\alpha t} \sin \beta t$$
$$\tfrac{i}{2}(-\mathbf{K}_1 e^{\lambda_1 t} + \bar{\mathbf{K}}_1 e^{\bar{\lambda}_1 t}) = \tfrac{i}{2}(-\mathbf{K}_1 + \bar{\mathbf{K}}_1)e^{\alpha t} \cos \beta t + \tfrac{1}{2}(\mathbf{K}_1 + \bar{\mathbf{K}}_1)e^{\alpha t} \sin \beta t.$$

For *any* complex number $z = a + ib$, we note that $\frac{1}{2}(z + \bar{z}) = a$ and $\frac{i}{2}(-z + \bar{z}) = b$ are *real* numbers. Therefore, the entries in the column vectors $\frac{1}{2}(\mathbf{K}_1 + \bar{\mathbf{K}}_1)$ and $\frac{i}{2}(-\mathbf{K}_1 + \bar{\mathbf{K}}_1)$ are real numbers. By defining

$$\mathbf{B}_1 = \tfrac{1}{2}[\mathbf{K}_1 + \bar{\mathbf{K}}_1] \qquad \text{and} \qquad \mathbf{B}_2 = \tfrac{i}{2}[-\mathbf{K}_1 + \bar{\mathbf{K}}_1], \tag{13}$$

we are led to the following theorem.

THEOREM 8.11 Let $\lambda_1 = \alpha + i\beta$ be a complex eigenvalue of the coefficient matrix $\mathbf{A}$ in the homogeneous system (2) and let $\mathbf{B}_1$ and $\mathbf{B}_2$ denote the column vectors defined in (13). Then

$$\begin{aligned} \mathbf{X}_1 &= (\mathbf{B}_1 \cos \beta t - \mathbf{B}_2 \sin \beta t)e^{\alpha t} \\ \mathbf{X}_2 &= (\mathbf{B}_2 \cos \beta t + \mathbf{B}_1 \sin \beta t)e^{\alpha t} \end{aligned} \tag{14}$$

are linearly independent solutions of (2) on $(-\infty, \infty)$.

The matrices $\mathbf{B}_1$ and $\mathbf{B}_2$ in (13) are often denoted by

$$\mathbf{B}_1 = \text{Re}(\mathbf{K}_1) \qquad \text{and} \qquad \mathbf{B}_2 = \text{Im}(\mathbf{K}_1) \tag{15}$$

since these vectors are, in turn, the *real* and *imaginary* parts of the eigenvector $\mathbf{K}_1$. For example, (12) follows from (14) with

$$\mathbf{K}_1 = \begin{pmatrix} 1 \\ 1 - 2i \end{pmatrix} = \begin{pmatrix} 1 \\ 1 \end{pmatrix} + i\begin{pmatrix} 0 \\ -2 \end{pmatrix}$$

$$\mathbf{B}_1 = \text{Re}(\mathbf{K}_1) = \begin{pmatrix} 1 \\ 1 \end{pmatrix} \qquad \text{and} \qquad \mathbf{B}_2 = \text{Im}(\mathbf{K}_1) = \begin{pmatrix} 0 \\ -2 \end{pmatrix}$$

EXAMPLE 3

Solve
$$\mathbf{X}' = \begin{pmatrix} 2 & 8 \\ -1 & -2 \end{pmatrix}\mathbf{X}.$$

SOLUTION First we obtain the eigenvalues from

$$\det(\mathbf{A} - \lambda\mathbf{I}) = \begin{vmatrix} 2-\lambda & 8 \\ -1 & -2-\lambda \end{vmatrix} = \lambda^2 + 4 = 0.$$

Thus the eigenvalues are $\lambda_1 = 2i$ and $\lambda_2 = \bar{\lambda}_1 = -2i$. For λ_1 we see that the system

$$\begin{aligned} (2-2i)k_1 + \qquad\quad 8k_2 &= 0 \\ -k_1 + (-2-2i)k_2 &= 0 \end{aligned}$$

gives $k_1 = -(2+2i)k_2$. By choosing $k_2 = -1$, we get

$$\mathbf{K}_1 = \begin{pmatrix} 2+2i \\ -1 \end{pmatrix} = \begin{pmatrix} 2 \\ -1 \end{pmatrix} + i\begin{pmatrix} 2 \\ 0 \end{pmatrix}.$$

Now from (15) we form

$$\mathbf{B}_1 = \operatorname{Re}(\mathbf{K}_1) = \begin{pmatrix} 2 \\ -1 \end{pmatrix} \quad \text{and} \quad \mathbf{B}_2 = \operatorname{Im}(\mathbf{K}_1) = \begin{pmatrix} 2 \\ 0 \end{pmatrix}$$

Since $\alpha = 0$, it follows from (14) that the general solution of the system is

$$\begin{aligned} \mathbf{X} &= c_1\left[\begin{pmatrix} 2 \\ -1 \end{pmatrix}\cos 2t - \begin{pmatrix} 2 \\ 0 \end{pmatrix}\sin 2t\right] + c_2\left[\begin{pmatrix} 2 \\ 0 \end{pmatrix}\cos 2t + \begin{pmatrix} 2 \\ -1 \end{pmatrix}\sin 2t\right] \\ &= c_1\begin{pmatrix} 2\cos 2t - 2\sin 2t \\ -\cos 2t \end{pmatrix} + c_2\begin{pmatrix} 2\cos 2t + 2\sin 2t \\ -\sin 2t \end{pmatrix}. \end{aligned}$$

EXAMPLE 4

Solve
$$\mathbf{X}' = \begin{pmatrix} 1 & 2 \\ -\frac{1}{2} & 1 \end{pmatrix}\mathbf{X}.$$

SOLUTION The solutions of the characteristic equation

$$\det(\mathbf{A} - \lambda\mathbf{I}) = \begin{vmatrix} 1-\lambda & 2 \\ -\frac{1}{2} & 1-\lambda \end{vmatrix} = \lambda^2 - 2\lambda + 2 = 0$$

are
$$\lambda_1 = 1 + i \quad \text{and} \quad \lambda_2 = \bar{\lambda}_1 = 1 - i.$$

Now an eigenvector associated with λ_1 is

$$\mathbf{K}_1 = \begin{pmatrix} 2 \\ i \end{pmatrix} = \begin{pmatrix} 2 \\ 0 \end{pmatrix} + i\begin{pmatrix} 0 \\ 1 \end{pmatrix}.$$

From (15) we find

$$\mathbf{B}_1 = \begin{pmatrix} 2 \\ 0 \end{pmatrix} \quad \text{and} \quad \mathbf{B}_2 = \begin{pmatrix} 0 \\ 1 \end{pmatrix}.$$

Thus (14) gives

$$\begin{aligned}\mathbf{X} &= c_1\left[\begin{pmatrix} 2 \\ 0 \end{pmatrix} \cos t - \begin{pmatrix} 0 \\ 1 \end{pmatrix} \sin t\right]e^t + c_2\left[\begin{pmatrix} 0 \\ 1 \end{pmatrix} \cos t + \begin{pmatrix} 2 \\ 0 \end{pmatrix} \sin t\right]e^t \\ &= c_1\begin{pmatrix} 2\cos t \\ -\sin t \end{pmatrix}e^t + c_2\begin{pmatrix} 2\sin t \\ \cos t \end{pmatrix}e^t.\end{aligned}$$

Alternative Method

When $\mathbf{A}$ is a 2×2 matrix having a complex eigenvalue $\lambda = \alpha + i\beta$, the general solution of the system can also be obtained from the assumption

$$\mathbf{X} = \begin{pmatrix} c_1 \\ c_2 \end{pmatrix} e^{\alpha t} \sin \beta t + \begin{pmatrix} c_3 \\ c_4 \end{pmatrix} e^{\alpha t} \cos \beta t$$

and then the substitution of $x(t)$ and $y(t)$ into one of the equations of the original system. This procedure is basically that of Section 8.1.

8.6.3 Repeated Eigenvalues

Up to this point we have not considered the case in which some of the n eigenvalues $\lambda_1, \lambda_2, \ldots, \lambda_n$ of an $n \times n$ matrix are repeated. For example, the characteristic equation of the coefficient matrix in

$$\mathbf{X}' = \begin{pmatrix} 3 & -18 \\ 2 & -9 \end{pmatrix}\mathbf{X} \tag{16}$$

is readily shown to be $(\lambda + 3)^2 = 0$, and therefore $\lambda_1 = \lambda_2 = -3$ is a root of *multiplicity two*. Now for this value we find the single eigenvector

$$\mathbf{K}_1 = \begin{pmatrix} 3 \\ 1 \end{pmatrix}$$

and so one solution of (16) is

$$\mathbf{X}_1 = \begin{pmatrix} 3 \\ 1 \end{pmatrix} e^{-3t}. \tag{17}$$

But since we are obviously interested in forming the general solution of the system, we need to pursue the question of finding a second solution.

In general, if m is a positive integer and $(\lambda - \lambda_1)^m$ is a factor of the characteristic equation, while $(\lambda - \lambda_1)^{m+1}$ is not a factor, then λ_1 is said to be an **eigenvalue of multiplicity m**. We distinguish two possibilities:

(i) For some $n \times n$ matrices $\mathbf{A}$ it may be possible to find m linearly independent eigenvectors $\mathbf{K}_1, \mathbf{K}_2, \ldots, \mathbf{K}_m$ corresponding to an eigenvalue λ_1 of multiplicity $m \le n$. In this case the general solution of the system contains the linear combination

$$c_1\mathbf{K}_1e^{\lambda_1 t} + c_2\mathbf{K}_2e^{\lambda_1 t} + \cdots + c_m\mathbf{K}_me^{\lambda_1 t}.$$

(ii) If there is only one eigenvector corresponding to the eigenvalue λ_1 of multiplicity m, then m linearly independent solutions of the form

$$\mathbf{X}_1 = \mathbf{K}_{11}e^{\lambda_1 t}$$

$$\mathbf{X}_2 = \mathbf{K}_{21}te^{\lambda_1 t} + \mathbf{K}_{22}e^{\lambda_1 t}$$

$$\vdots$$

$$\mathbf{X}_m = \mathbf{K}_{m1}\frac{t^{m-1}}{(m-1)!}e^{\lambda_1 t} + \mathbf{K}_{m2}\frac{t^{m-2}}{(m-2)!}e^{\lambda_1 t} + \cdots + \mathbf{K}_{mm}e^{\lambda_1 t},$$

where $\mathbf{K}_{ij}$ are column vectors, can always be found.

Eigenvalue of Multiplicity Two

We begin by considering eigenvalues of multiplicity two. In the first example we illustrate a matrix for which we can find two distinct eigenvectors corresponding to a double eigenvalue.

EXAMPLE 5

Solve
$$\mathbf{X}' = \begin{pmatrix} 1 & -2 & 2 \\ -2 & 1 & -2 \\ 2 & -2 & 1 \end{pmatrix}\mathbf{X}.$$

SOLUTION Expanding the determinant in the characteristic equation

$$\det(\mathbf{A} - \lambda\mathbf{I}) = \begin{vmatrix} 1-\lambda & -2 & 2 \\ -2 & 1-\lambda & -2 \\ 2 & -2 & 1-\lambda \end{vmatrix} = 0$$

yields $-(\lambda + 1)^2(\lambda - 5) = 0$. We see that $\lambda_1 = \lambda_2 = -1$ and $\lambda_3 = 5$.

For $\lambda_1 = -1$, Gauss-Jordan elimination gives immediately

$$(\mathbf{A} + \mathbf{I}|\mathbf{0}) = \left(\begin{array}{ccc|c} 2 & -2 & 2 & 0 \\ -2 & 2 & -2 & 0 \\ 2 & -2 & 2 & 0 \end{array}\right) \Longrightarrow \left(\begin{array}{ccc|c} 1 & -1 & 1 & 0 \\ 0 & 0 & 0 & 0 \\ 0 & 0 & 0 & 0 \end{array}\right).$$

From $k_1 - k_2 + k_3 = 0$ we can express, say, k_1 in terms of k_2 and k_3. By choosing $k_2 = 1$ and $k_3 = 0$ in $k_1 = k_2 - k_3$ we obtain $k_1 = 1$ and so one eigenvector is

$$\mathbf{K}_1 = \begin{pmatrix} 1 \\ 1 \\ 0 \end{pmatrix}.$$

But the choice $k_2 = 1$, $k_3 = 1$ implies $k_1 = 0$. Hence a second eigenvector is

$$\mathbf{K}_2 = \begin{pmatrix} 0 \\ 1 \\ 1 \end{pmatrix}.$$

Since neither eigenvector is a constant multiple of the other, we have found, corresponding to the same eigenvalue, two linearly independent solutions

$$\mathbf{X}_1 = \begin{pmatrix} 1 \\ 1 \\ 0 \end{pmatrix} e^{-t} \quad \text{and} \quad \mathbf{X}_2 = \begin{pmatrix} 0 \\ 1 \\ 1 \end{pmatrix} e^{-t}.$$

Lastly, for $\lambda_3 = 5$, the reduction

$$(\mathbf{A} - 5\mathbf{I}\,|\,\mathbf{0}) = \left(\begin{array}{ccc|c} -4 & -2 & 2 & 0 \\ -2 & -4 & -2 & 0 \\ 2 & -2 & -4 & 0 \end{array}\right) \Longrightarrow \left(\begin{array}{ccc|c} 1 & 0 & -1 & 0 \\ 0 & 1 & 1 & 0 \\ 0 & 0 & 0 & 0 \end{array}\right)$$

implies $k_1 = k_3$ and $k_2 = -k_3$. Picking $k_3 = 1$ gives $k_1 = 1$, $k_2 = -1$, and thus a third eigenvector is

$$\mathbf{K}_3 = \begin{pmatrix} 1 \\ -1 \\ 1 \end{pmatrix}.$$

We conclude that the general solution of the system is

$$\mathbf{X} = c_1 \begin{pmatrix} 1 \\ 1 \\ 0 \end{pmatrix} e^{-t} + c_2 \begin{pmatrix} 0 \\ 1 \\ 1 \end{pmatrix} e^{-t} + c_3 \begin{pmatrix} 1 \\ -1 \\ 1 \end{pmatrix} e^{5t}.$$

Second Solution

Now suppose λ_1 is an eigenvalue of multiplicity two and that there is only one eigenvector associated with this value. A second solution can be found of the form

$$\mathbf{X}_2 = \mathbf{K}te^{\lambda_1 t} + \mathbf{P}e^{\lambda_1 t}, \tag{18}$$

where $\mathbf{K} = \begin{pmatrix} k_1 \\ k_2 \\ \vdots \\ k_n \end{pmatrix}$ and $\mathbf{P} = \begin{pmatrix} p_1 \\ p_2 \\ \vdots \\ p_n \end{pmatrix}$.

To see this we substitute (18) into the system $\mathbf{X}' = \mathbf{AX}$ and simplify:

$$(\mathbf{AK} - \lambda_1\mathbf{K})te^{\lambda_1 t} + (\mathbf{AP} - \lambda_1\mathbf{P} - \mathbf{K})e^{\lambda_1 t} = \mathbf{0}.$$

Since this last equation is to hold for all values of t, we must have

$$(\mathbf{A} - \lambda_1\mathbf{I})\mathbf{K} = \mathbf{0} \tag{19}$$

and

$$(\mathbf{A} - \lambda_1\mathbf{I})\mathbf{P} = \mathbf{K}. \tag{20}$$

The first equation (19) simply states that $\mathbf{K}$ must be an eigenvector of $\mathbf{A}$ associated with λ_1. By solving (19), we find one solution $\mathbf{X}_1 = \mathbf{K}e^{\lambda_1 t}$. To find the second solution $\mathbf{X}_2$ we need only solve the additional system (20) for the vector $\mathbf{P}$.

EXAMPLE 6

Find the general solution of the system given in (16).

SOLUTION From (17) we know that $\lambda_1 = -3$ and that one solution is

$$\mathbf{X}_1 = \begin{pmatrix} 3 \\ 1 \end{pmatrix} e^{-3t}.$$

Identifying $\mathbf{K} = \begin{pmatrix} 3 \\ 1 \end{pmatrix}$ and $\mathbf{P} = \begin{pmatrix} p_1 \\ p_2 \end{pmatrix}$ it follows from (20) that we must now solve

$$(\mathbf{A} + 3\mathbf{I})\mathbf{P} = \mathbf{K} \quad \text{or} \quad \begin{pmatrix} 6 & -18 \\ 2 & -6 \end{pmatrix}\begin{pmatrix} p_1 \\ p_2 \end{pmatrix} = \begin{pmatrix} 3 \\ 1 \end{pmatrix}.$$

Multiplying out this last expression gives

$$\begin{aligned} 6p_1 - 18p_2 &= 3 \\ 2p_1 - 6p_2 &= 1. \end{aligned}$$

Since this system is obviously equivalent to one equation, we have an infinite number of choices for p_1 and p_2. For example, by choosing $p_1 = 1$, we find $p_2 = 1/6$. However, for simplicity, we shall choose $p_1 = 1/2$ so that $p_2 = 0$. Hence $\mathbf{P} = \begin{pmatrix} 1/2 \\ 0 \end{pmatrix}$. Thus from (18) we find

$$\mathbf{X}_2 = \begin{pmatrix} 3 \\ 1 \end{pmatrix} te^{-3t} + \begin{pmatrix} 1/2 \\ 0 \end{pmatrix} e^{-3t}.$$

The general solution of (16) is then

$$\mathbf{X} = c_1\begin{pmatrix} 3 \\ 1 \end{pmatrix} e^{-3t} + c_2\left[\begin{pmatrix} 3 \\ 1 \end{pmatrix} te^{-3t} + \begin{pmatrix} 1/2 \\ 0 \end{pmatrix} e^{-3t}\right].$$

Eigenvalues of Multiplicity Three

When a matrix **A** has only one eigenvector associated with an eigenvalue λ_1 of multiplicity three, we can find a second solution of form (18) and a third solution of the form

$$\mathbf{X}_3 = \mathbf{K}\frac{t^2}{2}e^{\lambda_1 t} + \mathbf{P}te^{\lambda_1 t} + \mathbf{Q}e^{\lambda_1 t}, \tag{21}$$

where $\mathbf{K} = \begin{pmatrix} k_1 \\ k_2 \\ \vdots \\ k_n \end{pmatrix}$, $\mathbf{P} = \begin{pmatrix} p_1 \\ p_2 \\ \vdots \\ p_n \end{pmatrix}$, and $\mathbf{Q} = \begin{pmatrix} q_1 \\ q_2 \\ \vdots \\ q_n \end{pmatrix}$.

By substituting (21) into the system $\mathbf{X}' = \mathbf{AX}$, we find the column vectors **K**, **P**, and **Q** must satisfy

$$(\mathbf{A} - \lambda_1\mathbf{I})\mathbf{K} = \mathbf{0}, \tag{22}$$

$$(\mathbf{A} - \lambda_1\mathbf{I})\mathbf{P} = \mathbf{K}, \tag{23}$$

and

$$(\mathbf{A} - \lambda_1\mathbf{I})\mathbf{Q} = \mathbf{P}. \tag{24}$$

Of course the solutions of (22) and (23) can be utilized in the formulation of the solutions $\mathbf{X}_1$ and $\mathbf{X}_2$.

EXAMPLE 7

Solve

$$\mathbf{X}' = \begin{pmatrix} 2 & 1 & 6 \\ 0 & 2 & 5 \\ 0 & 0 & 2 \end{pmatrix}\mathbf{X}.$$

SOLUTION The characteristic equation $(\lambda - 2)^3 = 0$ shows that $\lambda_1 = 2$ is an eigenvalue of multiplicity three. In succession we find that a solution of

$$(\mathbf{A} - 2\mathbf{I})\mathbf{K} = \mathbf{0} \quad \text{is} \quad \mathbf{K} = \begin{pmatrix} 1 \\ 0 \\ 0 \end{pmatrix};$$

a solution of

$$(\mathbf{A} - 2\mathbf{I})\mathbf{P} = \mathbf{K} \quad \text{is} \quad \mathbf{P} = \begin{pmatrix} 0 \\ 1 \\ 0 \end{pmatrix};$$

and finally a solution of

$$(\mathbf{A} - 2\mathbf{I})\mathbf{Q} = \mathbf{P} \quad \text{is} \quad \mathbf{Q} = \begin{pmatrix} 0 \\ -6/5 \\ 1/5 \end{pmatrix}.$$

We see from (18) and (21) that the general solution of the system is

$$\mathbf{X} = c_1\begin{pmatrix}1\\0\\0\end{pmatrix}e^{2t} + c_2\left[\begin{pmatrix}1\\0\\0\end{pmatrix}te^{2t} + \begin{pmatrix}0\\1\\0\end{pmatrix}e^{2t}\right] + c_3\left[\begin{pmatrix}1\\0\\0\end{pmatrix}\frac{t^2}{2}e^{2t} + \begin{pmatrix}0\\1\\0\end{pmatrix}te^{2t} + \begin{pmatrix}0\\-6/5\\1/5\end{pmatrix}e^{2t}\right].$$

Exercises 8.6

Answers to odd-numbered problems begin on page A–25.

[8.6.1] In Problems 1–12 find the general solution of the given system.

1. $\dfrac{dx}{dt} = x + 2y$

$\dfrac{dy}{dt} = 4x + 3y$

2. $\dfrac{dx}{dt} = 2y$

$\dfrac{dy}{dt} = 8x$

3. $\dfrac{dx}{dt} = -4x + 2y$

$\dfrac{dy}{dt} = -\dfrac{5}{2}x + 2y$

4. $\dfrac{dx}{dt} = \dfrac{1}{2}x + 9y$

$\dfrac{dy}{dt} = \dfrac{1}{2}x + 2y$

5. $\mathbf{X}' = \begin{pmatrix}10 & -5\\ 8 & -12\end{pmatrix}\mathbf{X}$

6. $\mathbf{X}' = \begin{pmatrix}-6 & 2\\ -3 & 1\end{pmatrix}\mathbf{X}$

7. $\dfrac{dx}{dt} = x + y - z$

$\dfrac{dy}{dt} = 2y$

$\dfrac{dz}{dt} = y - z$

8. $\dfrac{dx}{dt} = 2x - 7y$

$\dfrac{dy}{dt} = 5x + 10y + 4z$

$\dfrac{dz}{dt} = 5y + 2z$

9. $\mathbf{X}' = \begin{pmatrix}-1 & 1 & 0\\ 1 & 2 & 1\\ 0 & 3 & -1\end{pmatrix}\mathbf{X}$

10. $\mathbf{X}' = \begin{pmatrix}1 & 0 & 1\\ 0 & 1 & 0\\ 1 & 0 & 1\end{pmatrix}\mathbf{X}$

11. $\mathbf{X}' = \begin{pmatrix}-1 & -1 & 0\\ \frac{3}{4} & -\frac{3}{2} & 3\\ \frac{1}{8} & \frac{1}{4} & -\frac{1}{2}\end{pmatrix}\mathbf{X}$

12. $\mathbf{X}' = \begin{pmatrix}-1 & 4 & 2\\ 4 & -1 & -2\\ 0 & 0 & 6\end{pmatrix}\mathbf{X}$

In Problems 13 and 14 solve the given system subject to the indicated initial condition.

13. $\mathbf{X}' = \begin{pmatrix}\frac{1}{2} & 0\\ 1 & -\frac{1}{2}\end{pmatrix}\mathbf{X}, \quad \mathbf{X}(0) = \begin{pmatrix}3\\5\end{pmatrix}$

14. $\mathbf{X}' = \begin{pmatrix} 1 & 1 & 4 \\ 0 & 2 & 0 \\ 1 & 1 & 1 \end{pmatrix}\mathbf{X}, \quad \mathbf{X}(0) = \begin{pmatrix} 1 \\ 3 \\ 0 \end{pmatrix}$

[8.6.2] In Problems 15–26 find the general solution of the given system.

15. $\dfrac{dx}{dt} = 6x - y$
$\dfrac{dy}{dt} = 5x + 2y$

16. $\dfrac{dx}{dt} = x + y$
$\dfrac{dy}{dt} = -2x - y$

17. $\dfrac{dx}{dt} = 5x + y$
$\dfrac{dy}{dt} = -2x + 3y$

18. $\dfrac{dx}{dt} = 4x + 5y$
$\dfrac{dy}{dt} = -2x + 6y$

19. $\mathbf{X}' = \begin{pmatrix} 4 & -5 \\ 5 & -4 \end{pmatrix}\mathbf{X}$

20. $\mathbf{X}' = \begin{pmatrix} 1 & -8 \\ 1 & -3 \end{pmatrix}\mathbf{X}$

21. $\dfrac{dx}{dt} = z$
$\dfrac{dy}{dt} = -z$
$\dfrac{dz}{dt} = y$

22. $\dfrac{dx}{dt} = 2x + y + 2z$
$\dfrac{dy}{dt} = 3x + 6z$
$\dfrac{dz}{dt} = -4x - 3z$

23. $\mathbf{X}' = \begin{pmatrix} 1 & -1 & 2 \\ -1 & 1 & 0 \\ -1 & 0 & 1 \end{pmatrix}\mathbf{X}$

24. $\mathbf{X}' = \begin{pmatrix} 4 & 0 & 1 \\ 0 & 6 & 0 \\ -4 & 0 & 4 \end{pmatrix}\mathbf{X}$

25. $\mathbf{X}' = \begin{pmatrix} 2 & 5 & 1 \\ -5 & -6 & 4 \\ 0 & 0 & 2 \end{pmatrix}\mathbf{X}$

26. $\mathbf{X}' = \begin{pmatrix} 2 & 4 & 4 \\ -1 & -2 & 0 \\ -1 & 0 & -2 \end{pmatrix}\mathbf{X}$

In Problem 27 and 28 solve the given system subject to the indicated initial condition.

27. $\mathbf{X}' = \begin{pmatrix} 1 & -12 & -14 \\ 1 & 2 & -3 \\ 1 & 1 & -2 \end{pmatrix}\mathbf{X}, \quad \mathbf{X}(0) = \begin{pmatrix} 4 \\ 6 \\ -7 \end{pmatrix}$

28. $\mathbf{X}' = \begin{pmatrix} 6 & -1 \\ 5 & 4 \end{pmatrix}\mathbf{X}, \quad \mathbf{X}(0) = \begin{pmatrix} -2 \\ 8 \end{pmatrix}$

[8.6.3] In Problems 29–38 find the general solution of the given system.

29. $\dfrac{dx}{dt} = 3x - y$
$\dfrac{dy}{dt} = 9x - 3y$

30. $\dfrac{dx}{dt} = -6x + 5y$
$\dfrac{dy}{dt} = -5x + 4y$

31. $\dfrac{dx}{dt} = -x + 3y$
$\dfrac{dy}{dt} = -3x + 5y$

32. $\dfrac{dx}{dt} = 12x - 9y$
$\dfrac{dy}{dt} = 4x$

33. $\dfrac{dx}{dt} = 3x - y - z$
$\dfrac{dy}{dt} = x + y - z$
$\dfrac{dz}{dt} = x - y + z$

34. $\dfrac{dx}{dt} = 3x + 2y + 4z$
$\dfrac{dy}{dt} = 2x + 2z$
$\dfrac{dz}{dt} = 4x + 2y + 3z$

35. $\mathbf{X}' = \begin{pmatrix} 5 & -4 & 0 \\ 1 & 0 & 2 \\ 0 & 2 & 5 \end{pmatrix}\mathbf{X}$

36. $\mathbf{X}' = \begin{pmatrix} 1 & 0 & 0 \\ 0 & 3 & 1 \\ 0 & -1 & 1 \end{pmatrix}\mathbf{X}$

37. $\mathbf{X}' = \begin{pmatrix} 1 & 0 & 0 \\ 2 & 2 & -1 \\ 0 & 1 & 0 \end{pmatrix}\mathbf{X}$

38. $\mathbf{X}' = \begin{pmatrix} 4 & 1 & 0 \\ 0 & 4 & 1 \\ 0 & 0 & 4 \end{pmatrix}\mathbf{X}$

In Problems 39 and 40 solve the given system subject to the indicated initial condition.

39. $\mathbf{X}' = \begin{pmatrix} 2 & 4 \\ -1 & 6 \end{pmatrix}\mathbf{X}, \quad \mathbf{X}(0) = \begin{pmatrix} -1 \\ 6 \end{pmatrix}$

40. $\mathbf{X}' = \begin{pmatrix} 0 & 0 & 1 \\ 0 & 1 & 0 \\ 1 & 0 & 0 \end{pmatrix}\mathbf{X}, \quad \mathbf{X}(0) = \begin{pmatrix} 1 \\ 2 \\ 5 \end{pmatrix}$

Miscellaneous Problems

If $\mathbf{\Phi}(t)$ is a fundamental matrix of the system, the initial-value problem $\mathbf{X}' = \mathbf{AX}$, $\mathbf{X}(t_0) = \mathbf{X}_0$ has the solution $\mathbf{X} = \mathbf{\Phi}(t)\mathbf{\Phi}^{-1}(t_0)\mathbf{X}_0$ (see Problem 35 of Section 8.5). In Problems 41 and 42 use this result to solve the given system subject to the indicated initial condition.

41. $\mathbf{X}' = \begin{pmatrix} 4 & 3 \\ 3 & -4 \end{pmatrix}\mathbf{X}, \quad \mathbf{X}(0) = \begin{pmatrix} 1 \\ 1 \end{pmatrix}$

42. $\mathbf{X}' = \begin{pmatrix} -2/25 & 1/50 \\ 2/25 & -2/25 \end{pmatrix}\mathbf{X}, \quad \mathbf{X}(0) = \begin{pmatrix} 25 \\ 0 \end{pmatrix}$

In Problems 43 and 44 find a solution of the given system of the form $\mathbf{X} = t^{\lambda}\mathbf{K}$, $t > 0$, where $\mathbf{K}$ is a column vector of constants.

43. $t\mathbf{X}' = \begin{pmatrix} 1 & 3 \\ -1 & 5 \end{pmatrix}\mathbf{X}$ **44.** $t\mathbf{X}' = \begin{pmatrix} 2 & -2 \\ 2 & 7 \end{pmatrix}\mathbf{X}$

[O] 8.7 Undetermined Coefficients

The methods of **undetermined coefficients** and **variation of parameters** can both be adapted to the solution of a nonhomogeneous linear system $\mathbf{X}' = \mathbf{AX} + \mathbf{F}(t)$. Of these two methods, variation of parameters is the more powerful technique. However, there are a few instances when the method of undetermined coefficients gives a quick means of finding a particular solution $\mathbf{X}_p$.

EXAMPLE 1

Solve the system $\mathbf{X}' = \begin{pmatrix} -1 & 2 \\ -1 & 1 \end{pmatrix}\mathbf{X} + \begin{pmatrix} -8 \\ 3 \end{pmatrix}$ on $(-\infty, \infty)$.

SOLUTION We first solve the homogeneous system

$$\mathbf{X}' = \begin{pmatrix} -1 & 2 \\ -1 & 1 \end{pmatrix}\mathbf{X}.$$

The characteristic equation

$$\det(\mathbf{A} - \lambda\mathbf{I}) = \begin{vmatrix} -1-\lambda & 2 \\ -1 & 1-\lambda \end{vmatrix} = \lambda^2 + 1 = 0$$

yields the complex eigenvalues $\lambda_1 = i$ and $\lambda_2 = \bar{\lambda}_1 = -i$. By the procedures of the last section, we find

$$\mathbf{X}_c = c_1\begin{pmatrix} \cos t + \sin t \\ \cos t \end{pmatrix} + c_2\begin{pmatrix} \cos t - \sin t \\ -\sin t \end{pmatrix}.$$

Now since $\mathbf{F}(t)$ is a constant vector, we shall assume a constant particular solution vector $\mathbf{X}_p = \begin{pmatrix} a_1 \\ b_1 \end{pmatrix}$. Substituting this latter assumption into the original system leads to

$$0 = -a_1 + 2b_1 - 8$$
$$0 = -a_1 + b_1 + 3.$$

Solving this system of algebraic equations gives $a_1 = 14$ and $b_1 = 11$, and so $\mathbf{X}_p = \begin{pmatrix} 14 \\ 11 \end{pmatrix}$. The general solution of the system is

$$\mathbf{X} = c_1\begin{pmatrix} \cos t + \sin t \\ \cos t \end{pmatrix} + c_2\begin{pmatrix} \cos t - \sin t \\ -\sin t \end{pmatrix} + \begin{pmatrix} 14 \\ 11 \end{pmatrix}.$$

EXAMPLE 2

Solve the system

$$\frac{dx}{dt} = 6x + y + 6t \qquad \text{on } (-\infty, \infty).$$
$$\frac{dy}{dt} = 4x + 3y - 10t + 4$$

SOLUTION We first solve the homogeneous system

$$\frac{dx}{dt} = 6x + y$$
$$\frac{dy}{dt} = 4x + 3y$$

by the method of Section 8.6. The eigenvalues are determined from

$$\det(\mathbf{A} - \lambda\mathbf{I}) = \begin{vmatrix} 6-\lambda & 1 \\ 4 & 3-\lambda \end{vmatrix} = \lambda^2 - 9\lambda + 14 = 0.$$

Since $\lambda^2 - 9\lambda + 14 = (\lambda - 2)(\lambda - 7)$, we have $\lambda_1 = 2$ and $\lambda_2 = 7$. It is then easily verified that the respective eigenvectors of the coefficient matrix are

$$\begin{pmatrix} 1 \\ -4 \end{pmatrix} \quad \text{and} \quad \begin{pmatrix} 1 \\ 1 \end{pmatrix}.$$

Consequently the complementary function is

$$\mathbf{X}_c = c_1\begin{pmatrix} 1 \\ -4 \end{pmatrix}e^{2t} + c_2\begin{pmatrix} 1 \\ 1 \end{pmatrix}e^{7t}.$$

Because $\mathbf{F}(t)$ can be written as

$$\mathbf{F}(t) = \begin{pmatrix} 6 \\ -10 \end{pmatrix}t + \begin{pmatrix} 0 \\ 4 \end{pmatrix},$$

we shall try to find a particular solution of the system possessing the *same* form:

$$\mathbf{X}_p = \begin{pmatrix} a_2 \\ b_2 \end{pmatrix}t + \begin{pmatrix} a_1 \\ b_1 \end{pmatrix}.$$

In matrix terms we must have

$$\mathbf{X}_p' = \begin{pmatrix} 6 & 1 \\ 4 & 3 \end{pmatrix}\mathbf{X}_p + \begin{pmatrix} 6 \\ -10 \end{pmatrix}t + \begin{pmatrix} 0 \\ 4 \end{pmatrix}$$

or
$$\begin{pmatrix} a_2 \\ b_2 \end{pmatrix} = \begin{pmatrix} 6 & 1 \\ 4 & 3 \end{pmatrix}\left[\begin{pmatrix} a_2 \\ b_2 \end{pmatrix}t + \begin{pmatrix} a_1 \\ b_1 \end{pmatrix}\right] + \begin{pmatrix} 6 \\ -10 \end{pmatrix}t + \begin{pmatrix} 0 \\ 4 \end{pmatrix}$$

$$\begin{pmatrix} 0 \\ 0 \end{pmatrix} = \begin{pmatrix} (6a_2 + b_2 + 6)t + 6a_1 + b_1 - a_2 \\ (4a_2 + 3b_2 - 10)t + 4a_1 + 3b_1 - b_2 + 4 \end{pmatrix}.$$

From this last identity we conclude that

$$\begin{aligned} 6a_2 + b_2 + 6 &= 0 \\ 4a_2 + 3b_2 - 10 &= 0 \end{aligned} \quad \text{and} \quad \begin{aligned} 6a_1 + b_1 - a_2 &= 0 \\ 4a_1 + 3b_1 - b_2 + 4 &= 0. \end{aligned}$$

Solving the first two equations simultaneously yields $a_2 = -2$ and $b_2 = 6$. Substituting these values into the last two equations and solving for a_1 and b_1 gives $a_1 = -4/7$, $b_1 = 10/7$. It follows, therefore, that a particular solution vector is

$$\mathbf{X}_p = \begin{pmatrix} -2 \\ 6 \end{pmatrix}t + \begin{pmatrix} -4/7 \\ 10/7 \end{pmatrix},$$

and so the general solution of the system on $(-\infty, \infty)$ is

$$\begin{aligned} \mathbf{X} &= \mathbf{X}_c + \mathbf{X}_p \\ &= c_1\begin{pmatrix} 1 \\ -4 \end{pmatrix}e^{2t} + c_2\begin{pmatrix} 1 \\ 1 \end{pmatrix}e^{7t} + \begin{pmatrix} -2 \\ 6 \end{pmatrix}t + \begin{pmatrix} -4/7 \\ 10/7 \end{pmatrix}. \end{aligned}$$

EXAMPLE 3

Determine the form of the particular solution vector $\mathbf{X}_p$ for

$$\frac{dx}{dt} = 5x + 3y - 2e^{-t} + 1$$

$$\frac{dy}{dt} = -x + y + e^{-t} - 5t + 7.$$

SOLUTION Proceeding in the usual manner we find

$$\mathbf{X}_c = c_1\begin{pmatrix} 1 \\ -1 \end{pmatrix}e^{2t} + c_2\begin{pmatrix} 3 \\ -1 \end{pmatrix}e^{4t}.$$

Now since
$$\mathbf{F}(t) = \begin{pmatrix} -2 \\ 1 \end{pmatrix} e^{-t} + \begin{pmatrix} 0 \\ -5 \end{pmatrix} t + \begin{pmatrix} 1 \\ 7 \end{pmatrix},$$

we assume a particular solution of the form

$$\mathbf{X}_p = \begin{pmatrix} a_3 \\ b_3 \end{pmatrix} e^{-t} + \begin{pmatrix} a_2 \\ b_2 \end{pmatrix} t + \begin{pmatrix} a_1 \\ b_1 \end{pmatrix}.$$

Remark: The method of undetermined coefficients is not as simple as the last three examples would seem to indicate. As in Section 4.4 the method can be applied only when the entries in the matrix $\mathbf{F}(t)$ are constants, polynomials, exponential functions, sines and cosines, or finite sums and products of these functions. There are further difficulties. The assumption for $\mathbf{X}_p$ is actually predicated on a prior knowledge of the complementary function $\mathbf{X}_c$. For example, if $\mathbf{F}(t)$ is a constant vector and $\lambda = 0$ is an eigenvalue, then $\mathbf{X}_c$ contains a constant vector. In this case $\mathbf{X}_p$ is *not* a constant vector as in Example 1 but rather

$$\mathbf{X}_p = \begin{pmatrix} a_2 \\ b_2 \end{pmatrix} t + \begin{pmatrix} a_1 \\ b_1 \end{pmatrix}.$$

See Problem 11.

Similiarly, in Example 3, if we replace e^{-t} in $\mathbf{F}(t)$ by e^{2t} ($\lambda = 2$ is an eigenvalue), then the correct form of the particular solution is

$$\mathbf{X}_p = \begin{pmatrix} a_4 \\ b_4 \end{pmatrix} te^{2t} + \begin{pmatrix} a_3 \\ b_3 \end{pmatrix} e^{2t} + \begin{pmatrix} a_2 \\ b_2 \end{pmatrix} t + \begin{pmatrix} a_1 \\ b_1 \end{pmatrix}.$$

Rather than pursue these difficulties we turn our attention now to the method of variation of parameters.

Exercises 8.7

Answers to odd-numbered problems begin on page A–26.

In Problems 1–8 use the method of undetermined coefficients to solve the given system on $(-\infty, \infty)$.

1. $$\frac{dx}{dt} = 2x + 3y - 7$$
$$\frac{dy}{dt} = -x - 2y + 5$$

2. $$\frac{dx}{dt} = 5x + 9y + 2$$
$$\frac{dy}{dt} = -x + 11y + 6$$

3. $\dfrac{dx}{dt} = x + 3y - 2t^2$

$\dfrac{dy}{dt} = 3x + y + t + 5$

4. $\dfrac{dx}{dt} = x - 4y + 4t + 9e^{6t}$

$\dfrac{dy}{dt} = 4x + y - t + e^{6t}$

5. $\mathbf{X}' = \begin{pmatrix} 4 & \frac{1}{3} \\ 9 & 6 \end{pmatrix}\mathbf{X} + \begin{pmatrix} -3 \\ 10 \end{pmatrix}e^{4t}$

6. $\mathbf{X}' = \begin{pmatrix} -1 & 5 \\ -1 & 1 \end{pmatrix}\mathbf{X} + \begin{pmatrix} \sin t \\ -2\cos t \end{pmatrix}$

7. $\mathbf{X}' = \begin{pmatrix} 1 & 1 & 1 \\ 0 & 2 & 3 \\ 0 & 0 & 5 \end{pmatrix}\mathbf{X} + \begin{pmatrix} 1 \\ -1 \\ 2 \end{pmatrix}e^{4t}$

8. $\mathbf{X}' = \begin{pmatrix} 0 & 0 & 5 \\ 0 & 5 & 0 \\ 5 & 0 & 0 \end{pmatrix}\mathbf{X} + \begin{pmatrix} 5 \\ -10 \\ 40 \end{pmatrix}$

9. Solve $\mathbf{X}' = \begin{pmatrix} -1 & -2 \\ 3 & 4 \end{pmatrix}\mathbf{X} + \begin{pmatrix} 3 \\ 3 \end{pmatrix}$ subject to $\mathbf{X}(0) = \begin{pmatrix} -4 \\ 5 \end{pmatrix}$.

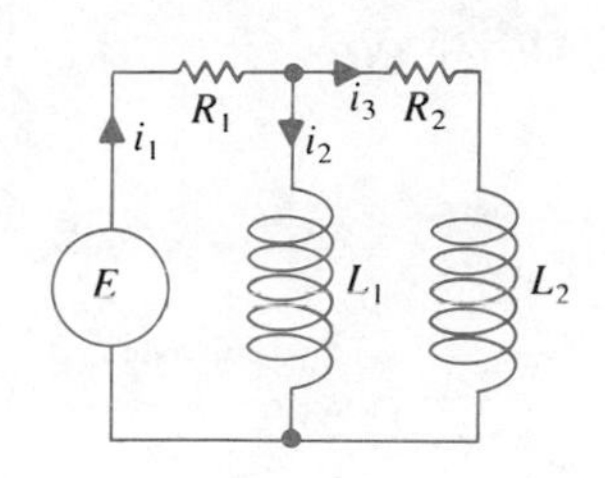

Figure 8.11

10. **(a)** Show that the system of differential equations for the currents $i_2(t)$ and $i_3(t)$ in the electrical network shown in Figure 8.11 is

$$\frac{d}{dt}\begin{pmatrix} i_2 \\ i_3 \end{pmatrix} = \begin{pmatrix} -R_1/L_1 & -R_1/L_1 \\ -R_1/L_2 & -(R_1 + R_2)/L_2 \end{pmatrix}\begin{pmatrix} i_2 \\ i_3 \end{pmatrix} + \begin{pmatrix} E/L_1 \\ E/L_2 \end{pmatrix}.$$

(b) Solve the system in part (a) if $R_1 = 2$ ohms, $R_2 = 3$ ohms, $L_1 = 1$ henry, $L_2 = 1$ henry, $E = 60$ volts, $i_2(0) = 0$, and $i_3(0) = 0$.

(c) Determine the current $i_1(t)$.

Miscellaneous Problems

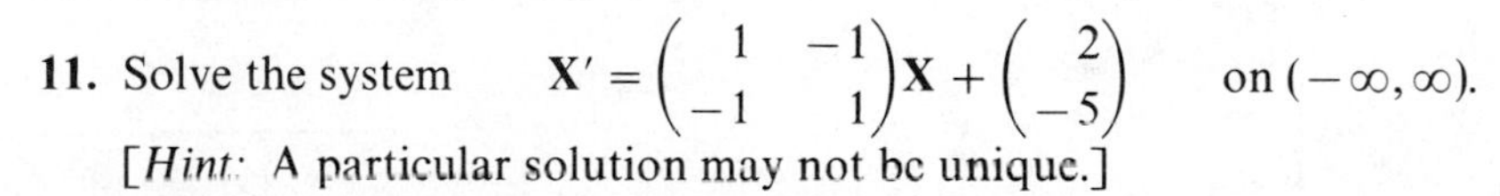

11. Solve the system $\mathbf{X}' = \begin{pmatrix} 1 & -1 \\ -1 & 1 \end{pmatrix}\mathbf{X} + \begin{pmatrix} 2 \\ -5 \end{pmatrix}$ on $(-\infty, \infty)$.

[*Hint:* A particular solution may not be unique.]

8.8 Variation of Parameters

In Section 8.5 we saw that the general solution of a homogeneous system $\mathbf{X}' = \mathbf{AX}$ can be written as the product

$$\mathbf{X} = \mathbf{\Phi}(t)\mathbf{C},$$

where $\mathbf{\Phi}(t)$ is a fundamental matrix of the system and $\mathbf{C}$ is an $n \times 1$ column vector of constants. As in the procedure of Section 4.5, we ask whether it is possible to replace $\mathbf{C}$ by a column matrix of functions

$$\mathbf{U}(t) = \begin{pmatrix} u_1(t) \\ u_2(t) \\ \vdots \\ u_n(t) \end{pmatrix}$$

so that
$$\mathbf{X}_p = \mathbf{\Phi}(t)\mathbf{U}(t) \tag{1}$$

is a particular solution of the nonhomogeneous system

$$\mathbf{X}' = \mathbf{AX} + \mathbf{F}(t). \tag{2}$$

By the product rule* the derivative of (1) is

$$\mathbf{X}_p' = \mathbf{\Phi}(t)\mathbf{U}'(t) + \mathbf{\Phi}'(t)\mathbf{U}(t). \tag{3}$$

Substituting (3) and (1) into (2) gives

$$\mathbf{\Phi}(t)\mathbf{U}'(t) + \mathbf{\Phi}'(t)\mathbf{U}(t) = \mathbf{A\Phi}(t)\mathbf{U}(t) + \mathbf{F}(t). \tag{4}$$

Now recall from (14) of Section 8.5 that $\mathbf{\Phi}'(t) = \mathbf{A\Phi}(t)$. Thus (4) becomes

$$\mathbf{\Phi}(t)\mathbf{U}'(t) + \mathbf{A\Phi}(t)\mathbf{U}(t) = \mathbf{A\Phi}(t)\mathbf{U}(t) + \mathbf{F}(t).$$

or
$$\mathbf{\Phi}(t)\mathbf{U}'(t) = \mathbf{F}(t). \tag{5}$$

Multiplying both sides of equation (5) by $\mathbf{\Phi}^{-1}(t)$ gives

$$\mathbf{U}'(t) = \mathbf{\Phi}^{-1}(t)\mathbf{F}(t)$$

or
$$\mathbf{U}(t) = \int \mathbf{\Phi}^{-1}(t)\mathbf{F}(t)\,dt.$$

Hence by assumption (1) we conclude that a particular solution of (2) is given by

$$\mathbf{X}_p = \mathbf{\Phi}(t)\int \mathbf{\Phi}^{-1}(t)\mathbf{F}(t)\,dt. \tag{6}$$

To calculate the indefinite integral of the column matrix $\mathbf{\Phi}^{-1}(t)\mathbf{F}(t)$ in (6) we integrate each entry. Thus the general solution of the system (2) is $\mathbf{X} = \mathbf{X}_c + \mathbf{X}_p$ or

$$\mathbf{X} = \mathbf{\Phi}(t)\mathbf{C} + \mathbf{\Phi}(t)\int \mathbf{\Phi}^{-1}(t)\mathbf{F}(t)\,dt. \tag{7}$$

---

* See Problem 51 of Section 8.4. Note that the order of the products is very important. Since $\mathbf{U}(t)$ is a column matrix, the products $\mathbf{U}'(t)\mathbf{\Phi}(t)$ and $\mathbf{U}(t)\mathbf{\Phi}'(t)$ are not defined.

EXAMPLE 1 Find the general solution of the nonhomogeneous system

$$\mathbf{X}' = \begin{pmatrix} -3 & 1 \\ 2 & -4 \end{pmatrix}\mathbf{X} + \begin{pmatrix} 3t \\ e^{-t} \end{pmatrix} \tag{8}$$

on the interval $(-\infty, \infty)$.

SOLUTION We first solve the homogeneous system

$$\mathbf{X}' = \begin{pmatrix} -3 & 1 \\ 2 & -4 \end{pmatrix}\mathbf{X}. \tag{9}$$

The characteristic equation of the coefficient matrix is

$$\det(\mathbf{A} - \lambda\mathbf{I}) = \begin{vmatrix} -3-\lambda & 1 \\ 2 & -4-\lambda \end{vmatrix} = (\lambda + 2)(\lambda + 5) = 0$$

and so the eigenvalues are $\lambda_1 = -2$ and $\lambda_2 = -5$. By the usual method we find the eigenvectors corresponding to λ_1 and λ_2 are, respectively,

$$\begin{pmatrix} 1 \\ 1 \end{pmatrix} \quad \text{and} \quad \begin{pmatrix} 1 \\ -2 \end{pmatrix}.$$

The solution vectors of the system (9) are then

$$\mathbf{X}_1 = \begin{pmatrix} 1 \\ 1 \end{pmatrix} e^{-2t} \quad \text{and} \quad \mathbf{X}_2 = \begin{pmatrix} 1 \\ -2 \end{pmatrix} e^{-5t}.$$

Next we form

$$\mathbf{\Phi}(t) = \begin{pmatrix} e^{-2t} & e^{-5t} \\ e^{-2t} & -2e^{-5t} \end{pmatrix} \quad \text{and} \quad \mathbf{\Phi}^{-1}(t) = \begin{pmatrix} \frac{2}{3}e^{2t} & \frac{1}{3}e^{2t} \\ \frac{1}{3}e^{5t} & -\frac{1}{3}e^{5t} \end{pmatrix}.$$

From (6) we then obtain

$$\begin{aligned}
\mathbf{X}_p &= \mathbf{\Phi}(t) \int \mathbf{\Phi}^{-1}(t)\mathbf{F}(t)\,dt \\
&= \begin{pmatrix} e^{-2t} & e^{-5t} \\ e^{-2t} & -2e^{-5t} \end{pmatrix} \int \begin{pmatrix} \frac{2}{3}e^{2t} & \frac{1}{3}e^{2t} \\ \frac{1}{3}e^{5t} & -\frac{1}{3}e^{5t} \end{pmatrix}\begin{pmatrix} 3t \\ e^{-t} \end{pmatrix} dt \\
&= \begin{pmatrix} e^{-2t} & e^{-5t} \\ e^{-2t} & -2e^{-5t} \end{pmatrix} \int \begin{pmatrix} 2te^{2t} + \frac{1}{3}e^{t} \\ te^{5t} - \frac{1}{3}e^{4t} \end{pmatrix} dt \\
&= \begin{pmatrix} e^{-2t} & e^{-5t} \\ e^{-2t} & -2e^{-5t} \end{pmatrix}\begin{pmatrix} te^{2t} - \frac{1}{2}e^{2t} + \frac{1}{3}e^{t} \\ \frac{1}{5}te^{5t} - \frac{1}{25}e^{5t} - \frac{1}{12}e^{4t} \end{pmatrix} \\
&= \begin{pmatrix} \frac{6}{5}t - \frac{27}{50} + \frac{1}{4}e^{-t} \\ \frac{3}{5}t - \frac{21}{50} + \frac{1}{2}e^{-t} \end{pmatrix}.
\end{aligned}$$

Hence from (7) the general solution of (8) on the interval is

$$\mathbf{X} = \begin{pmatrix} e^{-2t} & e^{-5t} \\ e^{-2t} & -2e^{-5t} \end{pmatrix}\begin{pmatrix} c_1 \\ c_2 \end{pmatrix} + \begin{pmatrix} \frac{6}{5}t - \frac{27}{50} + \frac{1}{4}e^{-t} \\ \frac{3}{5}t - \frac{21}{50} + \frac{1}{2}e^{-t} \end{pmatrix}$$

$$= c_1\begin{pmatrix} 1 \\ 1 \end{pmatrix}e^{-2t} + c_2\begin{pmatrix} 1 \\ -2 \end{pmatrix}e^{-5t} + \begin{pmatrix} \frac{6}{5} \\ \frac{3}{5} \end{pmatrix}t - \begin{pmatrix} \frac{27}{50} \\ \frac{21}{50} \end{pmatrix} + \begin{pmatrix} \frac{1}{4} \\ \frac{1}{2} \end{pmatrix}e^{-t}.$$

The general solution of (2) on an interval can be written in the alternative manner

$$\mathbf{X} = \mathbf{\Phi}(t)\mathbf{C} + \mathbf{\Phi}(t)\int_{t_0}^{t} \mathbf{\Phi}^{-1}(s)\mathbf{F}(s)\,ds, \tag{10}$$

where t and t_0 are points in the interval. This last form is useful in solving (2) subject to an initial condition $\mathbf{X}(t_0) = \mathbf{X}_0$. Substituting $t = t_0$ in (10) yields

$$\mathbf{X}_0 = \mathbf{\Phi}(t_0)\mathbf{C},$$

from which we see immediately that

$$\mathbf{C} = \mathbf{\Phi}^{-1}(t_0)\mathbf{X}_0.$$

We conclude that the solution of the initial-value problem is given by

$$\mathbf{X} = \mathbf{\Phi}(t)\mathbf{\Phi}^{-1}(t_0)\mathbf{X}_0 + \mathbf{\Phi}(t)\int_{t_0}^{t} \mathbf{\Phi}^{-1}(s)\mathbf{F}(s)\,ds. \tag{11}$$

Recall from Section 8.5 that an alternative way of forming a fundamental matrix is to choose its column vectors $\mathbf{V}_i$ in such a manner that

$$\mathbf{V}_1(t_0) = \begin{pmatrix} 1 \\ 0 \\ \vdots \\ 0 \end{pmatrix}, \quad \mathbf{V}_2(t_0) = \begin{pmatrix} 0 \\ 1 \\ \vdots \\ 0 \end{pmatrix}, \quad \ldots, \quad \mathbf{V}_n(t_0) = \begin{pmatrix} 0 \\ 0 \\ \vdots \\ 1 \end{pmatrix}. \tag{12}$$

This fundamental matrix is denoted by $\mathbf{\Psi}(t)$. As a consequence of (12) we know that $\mathbf{\Psi}(t)$ has the property

$$\mathbf{\Psi}(t_0) = \mathbf{I}. \tag{13}$$

But since $\mathbf{\Psi}(t)$ is nonsingular for all values of t in an interval, (13) implies

$$\mathbf{\Psi}^{-1}(t_0) = \mathbf{I}. \tag{14}$$

Thus when $\mathbf{\Psi}(t)$ is used rather than $\mathbf{\Phi}(t)$, it follows from (14) that (11) can be written

$$\mathbf{X} = \mathbf{\Psi}(t)\mathbf{X}_0 + \mathbf{\Psi}(t)\int_{t_0}^{t} \mathbf{\Psi}^{-1}(s)\mathbf{F}(s)\,ds. \tag{15}$$

Exercises 8.8

Answers to odd-numbered problems begin on page A–26.

In Problems 1–20 use variation of parameters to solve the given system.

1. $\dfrac{dx}{dt} = 3x - 3y + 4$
   $\dfrac{dy}{dt} = 2x - 2y - 1$

2. $\dfrac{dx}{dt} = 2x - y$
   $\dfrac{dy}{dt} = 3x - 2y + 4t$

3. $\mathbf{X}' = \begin{pmatrix} 3 & -5 \\ \frac{3}{4} & -1 \end{pmatrix}\mathbf{X} + \begin{pmatrix} 1 \\ -1 \end{pmatrix}e^{t/2}$

4. $\mathbf{X}' = \begin{pmatrix} 2 & -1 \\ 4 & 2 \end{pmatrix}\mathbf{X} + \begin{pmatrix} \sin 2t \\ 2\cos 2t \end{pmatrix}e^{2t}$

5. $\mathbf{X}' = \begin{pmatrix} 0 & 2 \\ -1 & 3 \end{pmatrix}\mathbf{X} + \begin{pmatrix} 1 \\ -1 \end{pmatrix}e^{t}$

6. $\mathbf{X}' = \begin{pmatrix} 0 & 2 \\ -1 & 3 \end{pmatrix}\mathbf{X} + \begin{pmatrix} 2 \\ e^{-3t} \end{pmatrix}$

7. $\mathbf{X}' = \begin{pmatrix} 1 & 8 \\ 1 & -1 \end{pmatrix}\mathbf{X} + \begin{pmatrix} 12 \\ 12 \end{pmatrix}t$

8. $\mathbf{X}' = \begin{pmatrix} 1 & 8 \\ 1 & -1 \end{pmatrix}\mathbf{X} + \begin{pmatrix} e^{-t} \\ te^{t} \end{pmatrix}$

9. $\mathbf{X}' = \begin{pmatrix} 3 & 2 \\ -2 & -1 \end{pmatrix}\mathbf{X} + \begin{pmatrix} 2e^{-t} \\ e^{-t} \end{pmatrix}$

10. $\mathbf{X}' = \begin{pmatrix} 3 & 2 \\ -2 & -1 \end{pmatrix}\mathbf{X} + \begin{pmatrix} 1 \\ 1 \end{pmatrix}$

11. $\mathbf{X}' = \begin{pmatrix} 0 & -1 \\ 1 & 0 \end{pmatrix}\mathbf{X} + \begin{pmatrix} \sec t \\ 0 \end{pmatrix}$

12. $\mathbf{X}' = \begin{pmatrix} 1 & -1 \\ 1 & 1 \end{pmatrix}\mathbf{X} + \begin{pmatrix} 3 \\ 3 \end{pmatrix}e^{t}$

13. $\mathbf{X}' = \begin{pmatrix} 1 & -1 \\ 1 & 1 \end{pmatrix}\mathbf{X} + \begin{pmatrix} \cos t \\ \sin t \end{pmatrix}e^{t}$

14. $\mathbf{X}' = \begin{pmatrix} 2 & -2 \\ 8 & -6 \end{pmatrix}\mathbf{X} + \begin{pmatrix} 1 \\ 3 \end{pmatrix}\dfrac{e^{-2t}}{t}$

15. $\mathbf{X}' = \begin{pmatrix} 0 & 1 \\ -1 & 0 \end{pmatrix}\mathbf{X} + \begin{pmatrix} 0 \\ \sec t \tan t \end{pmatrix}$

16. $\mathbf{X}' = \begin{pmatrix} 0 & 1 \\ -1 & 0 \end{pmatrix}\mathbf{X} + \begin{pmatrix} 1 \\ \cot t \end{pmatrix}$

17. $\mathbf{X}' = \begin{pmatrix} 1 & 2 \\ -\frac{1}{2} & 1 \end{pmatrix}\mathbf{X} + \begin{pmatrix} \csc t \\ \sec t \end{pmatrix}e^{t}$

18. $\mathbf{X}' = \begin{pmatrix} 1 & -2 \\ 1 & -1 \end{pmatrix}\mathbf{X} + \begin{pmatrix} \tan t \\ 1 \end{pmatrix}$

19. $\mathbf{X}' = \begin{pmatrix} 1 & 1 & 0 \\ 1 & 1 & 0 \\ 0 & 0 & 3 \end{pmatrix}\mathbf{X} + \begin{pmatrix} e^{t} \\ e^{2t} \\ te^{3t} \end{pmatrix}$

20. $\mathbf{X}' = \begin{pmatrix} 3 & -1 & -1 \\ 1 & 1 & -1 \\ 1 & -1 & 1 \end{pmatrix}\mathbf{X} + \begin{pmatrix} 0 \\ t \\ 2e^{t} \end{pmatrix}$

In Problems 21 and 22 use (11) to solve the given system subject to the indicated initial condition.

21. $\mathbf{X}' = \begin{pmatrix} 3 & -1 \\ -1 & 3 \end{pmatrix}\mathbf{X} + \begin{pmatrix} 4e^{2t} \\ 4e^{4t} \end{pmatrix}, \quad \mathbf{X}(0) = \begin{pmatrix} 1 \\ 1 \end{pmatrix}$

22. $\mathbf{X}' = \begin{pmatrix} 1 & -1 \\ 1 & -1 \end{pmatrix}\mathbf{X} + \begin{pmatrix} 1/t \\ 1/t \end{pmatrix}, \quad \mathbf{X}(1) = \begin{pmatrix} 2 \\ -1 \end{pmatrix}$

In Problems 23 and 24 use (15) to solve the given system subject to the indicated initial condition. Use the results of Problems 31 and 34 of Section 8.5.

23. $\mathbf{X}' = \begin{pmatrix} 4 & 1 \\ 6 & 5 \end{pmatrix}\mathbf{X} + \begin{pmatrix} 50e^{7t} \\ 0 \end{pmatrix}, \quad \mathbf{X}(0) = \begin{pmatrix} 5 \\ -5 \end{pmatrix}$

24. $\mathbf{X}' = \begin{pmatrix} 3 & -2 \\ 5 & -3 \end{pmatrix}\mathbf{X} + \begin{pmatrix} 2 \\ 3 \end{pmatrix}, \quad \mathbf{X}(\pi/2) = \begin{pmatrix} 0 \\ 0 \end{pmatrix}$

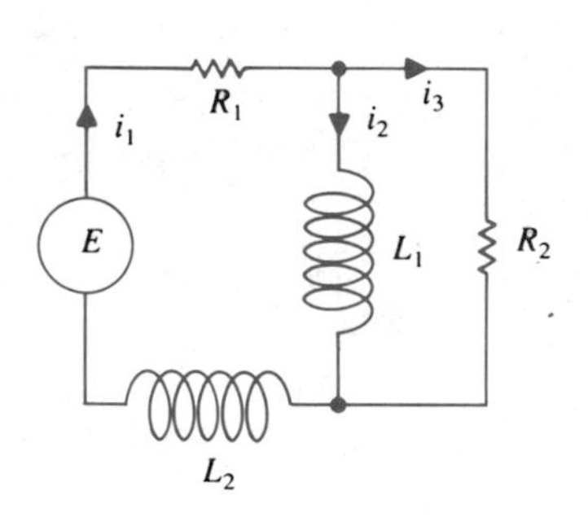

Figure 8.12

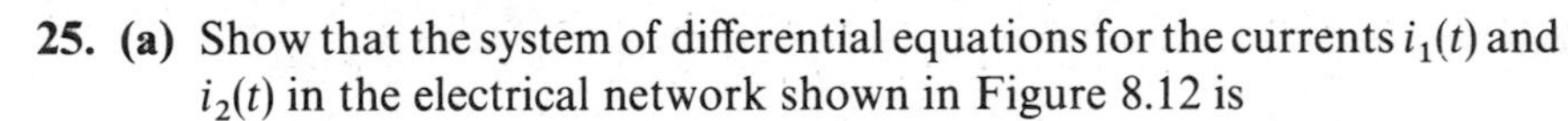

25. (a) Show that the system of differential equations for the currents $i_1(t)$ and $i_2(t)$ in the electrical network shown in Figure 8.12 is

$$\frac{d}{dt}\begin{pmatrix} i_1 \\ i_2 \end{pmatrix} = \begin{pmatrix} -(R_1 + R_2)/L_2 & R_2/L_2 \\ R_2/L_1 & -R_2/L_1 \end{pmatrix}\begin{pmatrix} i_1 \\ i_2 \end{pmatrix} + \begin{pmatrix} E/L_2 \\ 0 \end{pmatrix}.$$

(b) Solve the system in part (a) if $R_1 = 8$ ohms, $R_2 = 3$ ohms, $L_1 = 1$ henry, $L_2 = 1$ henry, $E(t) = 100 \sin t$ volts, $i_1(0) = 0$, and $i_2(0) = 0$.

[O] 8.9 Matrix Exponential

Matrices can be utilized in an entirely different manner to solve a homogeneous system of linear first-order differential equations.

Recall that the simple linear first-order differential equation

$$x' = ax,$$

where a is a constant, has the general solution

$$x = ce^{at}.$$

It seems natural then to ask whether we can define a matrix exponential $e^{t\mathbf{A}}$ so that the homogeneous system

$$\mathbf{X}' = \mathbf{AX},$$

where $\mathbf{A}$ is an $n \times n$ matrix of constants, has a solution

$$\mathbf{X} = e^{t\mathbf{A}}\mathbf{C}. \tag{1}$$

Since **C** is to be an $n \times 1$ column vector of arbitrary constants, we want $e^{t\mathbf{A}}$ to be an $n \times n$ matrix. While the complete development of the meaning of the **matrix exponential** would necessitate a more thorough investigation of matrix algebra, one means of computing $e^{t\mathbf{A}}$ is given in the following definition.

DEFINITION 8.20 For any $n \times n$ matrix **A**,

$$e^{t\mathbf{A}} = \sum_{n=0}^{\infty} \frac{(t\mathbf{A})^n}{n!}$$

$$= \mathbf{I} + t\mathbf{A} + \frac{t^2}{2!}\mathbf{A}^2 + \frac{t^3}{3!}\mathbf{A}^3 + \cdots. \tag{2}$$

It can be shown that the series given in (2) converges to an $n \times n$ matrix for every value of t. Also, $\mathbf{A}^2 = \mathbf{AA}$, $\mathbf{A}^3 = \mathbf{A}(\mathbf{A}^2)$, and so on.

Now the general solution of the single differential equation

$$x' = ax + f(t),$$

where a is a constant, can be expressed as

$$x = x_c + x_p = ce^{at} + e^{at}\int_{t_0}^{t} e^{-as}f(s)\,ds.$$

For systems of linear first-order differential equations, it can be shown that the general solution of

$$\mathbf{X}' = \mathbf{AX} + \mathbf{F}(t),$$

where **A** is an $n \times n$ matrix of constants, is

$$\mathbf{X} = \mathbf{X}_c + \mathbf{X}_p = e^{t\mathbf{A}}\mathbf{C} + e^{t\mathbf{A}}\int_{t_0}^{t} e^{-s\mathbf{A}}\mathbf{F}(s)\,ds. \tag{3}$$

The matrix exponential $e^{t\mathbf{A}}$ is always nonsingular and $e^{-s\mathbf{A}} = (e^{s\mathbf{A}})^{-1}$. In practice, $e^{-s\mathbf{A}}$ can be obtained from $e^{t\mathbf{A}}$ by replacing t by $-s$.

Additional Properties

From (2) it is seen that

$$e^{\mathbf{0}} = \mathbf{I}. \tag{4}$$

Also, formal termwise differentiation of (2) shows that

$$\frac{d}{dt}e^{t\mathbf{A}} = \mathbf{A}e^{t\mathbf{A}}. \tag{5}$$

If we denote the matrix exponential by $\mathbf{\Psi}(t)$, then (5) and (4) are equivalent

to $$\mathbf{\Psi}'(t) = \mathbf{A}\mathbf{\Psi}(t) \tag{6}$$

and $$\mathbf{\Psi}(0) = \mathbf{I}, \tag{7}$$

respectively. The notation here is chosen deliberately. Comparing (6) with (14) of Section 8.5 reveals that $e^{t\mathbf{A}}$ is a fundamental matrix of the system $\mathbf{X}' = \mathbf{AX}$. It is precisely this formulation of the fundamental matrix that was discussed on page 394 of Section 8.5.

By multiplying the series defining $e^{t\mathbf{A}}$ and $e^{-s\mathbf{A}}$, we can prove that

$$e^{t\mathbf{A}}e^{-s\mathbf{A}} = e^{(t-s)\mathbf{A}} \qquad \text{or equivalently} \qquad \mathbf{\Psi}(t)\mathbf{\Psi}^{-1}(s) = \mathbf{\Psi}(t-s).^*$$

This last result enables us to relate (3) to (10) of the preceding section

$$\mathbf{X} = e^{t\mathbf{A}}\mathbf{C} + \int_{t_0}^{t} e^{t\mathbf{A}}e^{-s\mathbf{A}}\mathbf{F}(s)\,ds$$

$$= e^{t\mathbf{A}}\mathbf{C} + \int_{t_0}^{t} e^{(t-s)\mathbf{A}}\mathbf{F}(s)\,ds \tag{8}$$

$$\mathbf{X} = \mathbf{\Psi}(t)\mathbf{C} + \int_{t_0}^{t} \mathbf{\Psi}(t-s)\mathbf{F}(s)\,ds. \tag{9}$$

Equation (9) possesses a form simpler than (10) of Section 8.8. In other words, there is no need to compute $\mathbf{\Psi}^{-1}$; we need only replace t by $t - s$ in $\mathbf{\Psi}(t)$.

Exercises 8.9

Answers to odd-numbered problems begin on page A–27.

In Problems 1 and 2 use (2) to compute $e^{t\mathbf{A}}$ and $e^{-t\mathbf{A}}$.

1. $\mathbf{A} = \begin{pmatrix} 0 & 1 \\ 1 & 0 \end{pmatrix}$ **2.** $\mathbf{A} = \begin{pmatrix} 1 & 0 \\ 0 & 2 \end{pmatrix}$

In Problems 3 and 4 use (1) to find the general solution of each system.

3. $\mathbf{X}' = \begin{pmatrix} 0 & 1 \\ 1 & 0 \end{pmatrix}\mathbf{X}$ **4.** $\mathbf{X}' = \begin{pmatrix} 1 & 0 \\ 0 & 2 \end{pmatrix}\mathbf{X}$

In Problems 5–8 use (3) to find the general solution of each system.

5. $\mathbf{X}' = \begin{pmatrix} 0 & 1 \\ 1 & 0 \end{pmatrix}\mathbf{X} + \begin{pmatrix} 1 \\ 1 \end{pmatrix}$ **6.** $\mathbf{X}' = \begin{pmatrix} 0 & 1 \\ 1 & 0 \end{pmatrix}\mathbf{X} + \begin{pmatrix} \cosh t \\ \sinh t \end{pmatrix}$

---

* Although $e^{t\mathbf{A}}e^{-s\mathbf{A}} = e^{(t-s)\mathbf{A}}$, it is interesting to note that $e^{\mathbf{A}}e^{\mathbf{B}}$ is, in general, not the same as $e^{\mathbf{A}+\mathbf{B}}$ for $n \times n$ matrices $\mathbf{A}$ and $\mathbf{B}$.

7. $\mathbf{X}' = \begin{pmatrix} 1 & 0 \\ 0 & 2 \end{pmatrix}\mathbf{X} + \begin{pmatrix} t \\ e^{4t} \end{pmatrix}$

8. $\mathbf{X}' = \begin{pmatrix} 1 & 0 \\ 0 & 2 \end{pmatrix}\mathbf{X} + \begin{pmatrix} 3 \\ -1 \end{pmatrix}$

Miscellaneous Problems

Let **P** denote a matrix whose columns are eigenvectors $\mathbf{K}_1, \mathbf{K}_2, \ldots, \mathbf{K}_n$ corresponding to *distinct* eigenvalues $\lambda_1, \lambda_2, \ldots, \lambda_n$ of an $n \times n$ matrix **A**. Then it can be shown that $\mathbf{A} = \mathbf{PDP}^{-1}$, where **D** is defined by

$$\mathbf{D} = \begin{pmatrix} \lambda_1 & 0 & \ldots & 0 \\ 0 & \lambda_2 & \ldots & 0 \\ \vdots & & & \vdots \\ 0 & 0 & \ldots & \lambda_n \end{pmatrix}. \tag{10}$$

In Problems 9 and 10 verify the above result for the given matrix.

9. $\mathbf{A} = \begin{pmatrix} 2 & 1 \\ -3 & 6 \end{pmatrix}$

10. $\mathbf{A} = \begin{pmatrix} 2 & 1 \\ 1 & 2 \end{pmatrix}$

11. Suppose $\mathbf{A} = \mathbf{PDP}^{-1}$, where **D** is defined in (10). Use (2) to show that $e^{t\mathbf{A}} = \mathbf{P}e^{t\mathbf{D}}\mathbf{P}^{-1}$.

12. Use (2) to show that

$$e^{t\mathbf{D}} = \begin{pmatrix} e^{\lambda_1 t} & 0 & \cdots & 0 \\ 0 & e^{\lambda_2 t} & \cdots & 0 \\ \vdots & & & \vdots \\ 0 & 0 & \cdots & e^{\lambda_n t} \end{pmatrix},$$

where **D** is defined in (10).

In Problems 13 and 14 use the results of Problems 9–12 to solve the given system.

13. $\mathbf{X}' = \begin{pmatrix} 2 & 1 \\ -3 & 6 \end{pmatrix}\mathbf{X}$

14. $\mathbf{X}' = \begin{pmatrix} 2 & 1 \\ 1 & 2 \end{pmatrix}\mathbf{X}$

CHAPTER 8 SUMMARY

Throughout this chapter we have considered **systems of linear** differential equations.

For linear systems probably the most basic technique of solution consists of rewriting the entire system in **operator notation** and then using **systematic elimination** to obtain single differential equations in one dependent variable that can be solved by the usual procedures. Also, **determinants** can usually be utilized to accomplish the same result. Once all dependent variables have been determined, it is necessary to use the system itself to find various relationships between the parameters.

When initial conditions are specified, the **Laplace transform** can be used to reduce the system to simultaneous algebraic equations in the transformed functions.

A first-order system in **normal form** in *two* dependent variables is any system

$$\begin{aligned}\frac{dx}{dt} &= a_{11}(t)x + a_{12}(t)y + f_1(t)\\ \frac{dy}{dt} &= a_{21}(t)x + a_{22}(t)y + f_2(t),\end{aligned} \tag{1}$$

where the coefficients $a_{ij}(t)$, $f_1(t)$, and $f_2(t)$ are continuous on some common interval I. When $f_1(t) = 0$, $f_2(t) = 0$, the system is said to be **homogeneous**; otherwise it is said to be **nonhomogeneous**. Any linear second-order differential equation can be expressed in this form.

Using matrices, the system (1) can be written compactly as

$$\frac{d\mathbf{X}}{dt} = \mathbf{A}(t)\mathbf{X} + \mathbf{F}(t), \tag{2}$$

where

$$\mathbf{X} = \begin{pmatrix} x \\ y \end{pmatrix}, \qquad \mathbf{A}(t) = \begin{pmatrix} a_{11}(t) & a_{12}(t) \\ a_{21}(t) & a_{22}(t) \end{pmatrix}, \qquad \text{and} \qquad \mathbf{F}(t) = \begin{pmatrix} f_1(t) \\ f_2(t) \end{pmatrix}.$$

The **general solution of the homogeneous system** in two dependent variables

$$\frac{d\mathbf{X}}{dt} = \mathbf{A}(t)\mathbf{X} \tag{3}$$

is defined to be the linear combination

$$\mathbf{X} = c_1\mathbf{X}_1 + c_2\mathbf{X}_2, \tag{4}$$

where $\mathbf{X}_1$ and $\mathbf{X}_2$ form a **fundamental set of solutions** of (3) on I. The **general solution of the nonhomogeneous system** (2) is defined to be

$$\mathbf{X} = \mathbf{X}_c + \mathbf{X}_p,$$

where $\mathbf{X}_c$ is defined by (4) and $\mathbf{X}_p$ is *any* solution vector of (2).

To solve a homogeneous system (3) we determine the **eigenvalues** of the coefficient matrix **A** and then find the corresponding **eigenvectors**.

To solve a nonhomogeneous system we first solve the associated homogeneous system. A particular solution vector $\mathbf{X}_p$ of the nonhomogeneous system is found by either **undetermined coefficients** or **variation of parameters**.

A **fundamental matrix** of a homogeneous system (3) in two dependent variables is defined to be

$$\mathbf{\Phi}(t) = \begin{pmatrix} x_1 & x_2 \\ y_1 & y_2 \end{pmatrix}. \tag{5}$$

The columns in (5) are obtained from two linearly independent solution vectors $\mathbf{X}_1$ and $\mathbf{X}_2$ of (3). In terms of matrices, the method of variation of parameters leads to a particular solution given by:

$$\mathbf{X}_p = \mathbf{\Phi}(t) \int \mathbf{\Phi}^{-1}(t)\mathbf{F}(t)\,dt.$$

The general solution of (2) on an interval is

$$\mathbf{X} = \mathbf{\Phi}(t)\mathbf{C} + \mathbf{\Phi}(t) \int \mathbf{\Phi}^{-1}(t)\mathbf{F}(t)\,dt,$$

where $\mathbf{C}$ is a column matrix containing two arbitrary constants. The matrix $\mathbf{\Phi}^{-1}(t)$ is called the **multiplicative inverse** of $\mathbf{\Phi}(t)$; the multiplicative inverse satisfies $\mathbf{\Phi}(t)\mathbf{\Phi}^{-1}(t) = \mathbf{\Phi}^{-1}(t)\mathbf{\Phi}(t) = \mathbf{I}$, where $\mathbf{I}$ is the 2×2 **multiplicative identity**.

CHAPTER 8 REVIEW EXERCISES

Answers to odd-numbered problems begin on page A–27.

Answer Problems 1–12 without referring back to the text. Fill in the blank or answer true/false.

1. Every second-order linear differential equation can be expressed as a system of two linear first-order differential equations. ______
2. If $\mathbf{A} = \begin{pmatrix} 1 \\ 2 \end{pmatrix}$ and $\mathbf{B} = (3 \quad 4)$, then $\mathbf{AB} =$ ______ and $\mathbf{BA} =$ ______.
3. If $\mathbf{A} = \begin{pmatrix} 1 & 2 \\ 3 & 4 \end{pmatrix}$, then $\mathbf{A}^{-1} =$ ______.
4. If $\mathbf{A}$ is a nonsingular matrix for which $\mathbf{AB} = \mathbf{AC}$, then $\mathbf{B} = \mathbf{C}$. ______
5. If $\mathbf{X}_1$ is a solution of $\mathbf{X}' = \mathbf{AX}$ and $\mathbf{X}_2$ is a solution of $\mathbf{X}' = \mathbf{AX} + \mathbf{F}$, then $\mathbf{X} = \mathbf{X}_1 + \mathbf{X}_2$ is a solution of $\mathbf{X}' = \mathbf{AX} + \mathbf{F}$. ______
6. A fundamental matrix $\mathbf{\Phi}$ of a system $\mathbf{X}' = \mathbf{AX}$ is always nonsingular. ______
7. Let $\mathbf{A}$ be an $n \times n$ matrix. The eigenvalues of $\mathbf{A}$ are the nonzero solutions of $\det(\mathbf{A} - \lambda\mathbf{I}) = 0$. ______
8. A nonzero constant multiple of an eigenvector is an eigenvector corresponding to the same eigenvalue. ______
9. An $n \times 1$ column vector $\mathbf{K}$ with all zero entries is never an eigenvector of an $n \times n$ matrix $\mathbf{A}$. ______
10. Let $\mathbf{A}$ be an $n \times n$ matrix with real entries. If λ is a complex eigenvalue, then $\bar{\lambda}$ is also an eigenvalue of $\mathbf{A}$. ______

11. An $n \times n$ matrix $\mathbf{A}$ always possesses n linearly independent eigenvectors. ______

12. The augmented matrix

$$\begin{pmatrix} 1 & 1 & 1 & | & 2 \\ 0 & 1 & 0 & | & 3 \\ 0 & 0 & 0 & | & 0 \end{pmatrix}$$

is in reduced row-echelon form. ______

In Problems 13–16 use systematic elimination or determinants to solve the given system.

13. $x' + y' = 2x + 2y + 1$

$x' + 2y' = y + 3$

14. $\dfrac{dx}{dt} = 2x + y + t - 2$

$\dfrac{dy}{dt} = 3x + 4y - 4t$

15. $(D - 2)x - y = -e^t$

$-3x + (D - 4)y = -7e^t$

16. $(D + 2)x + (D + 1)y = \sin 2t$

$5x + (D + 3)y = \cos 2t$

In Problems 17 and 18 use the Laplace transform to solve each system.

17. $x' + y = t$

$4x + y' = 0$

$x(0) = 1, \quad y(0) = 2$

18. $x'' + y'' = e^{2t}$

$2x' + y'' = -e^{2t}$

$x(0) = 0, \quad y(0) = 0$

$x'(0) = 0, \quad y'(0) = 0$

19. **(a)** Write as one column matrix $\mathbf{X}$:

$$\begin{pmatrix} 3 & 1 & 1 \\ -1 & 2 & -1 \\ 0 & -2 & 4 \end{pmatrix} \begin{pmatrix} t \\ t^2 \\ t^3 \end{pmatrix} - \begin{pmatrix} 2 \\ -2 \\ -1 \end{pmatrix} + 2t \begin{pmatrix} 1 \\ 0 \\ 4 \end{pmatrix} + 2t^2 \begin{pmatrix} 1 \\ -1 \\ 7 \end{pmatrix}.$$

(b) Find $d\mathbf{X}/dt$.

20. Write the differential equation

$$3y^{(4)} - 5y'' + 9y = 6e^t - 2t$$

as a system of first-order equations in linear normal form.

21. Write the system

$$(2D^2 + D)y - D^2x = \ln t$$

$$D^2y + (D + 1)x = 5t - 2$$

as a system of first-order equations in linear normal form.

22. Verify that the general solution of the system

$$\frac{dx}{dt} = y$$

$$\frac{dy}{dt} = -x + 2y - 2\cos t$$

on the interval $(-\infty, \infty)$ is

$$\mathbf{X} = c_1\begin{pmatrix}1\\1\end{pmatrix}e^t + c_2\left\{\begin{pmatrix}1\\1\end{pmatrix}te^t + \begin{pmatrix}0\\1\end{pmatrix}e^t\right\} + \begin{pmatrix}\sin t\\ \cos t\end{pmatrix}.$$

In Problems 23–28 use the concept of eigenvalues and eigenvectors to solve each system.

23. $\dfrac{dx}{dt} = 2x + y$, $\quad \dfrac{dy}{dt} = -x$

24. $\dfrac{dx}{dt} = -4x + 2y$, $\quad \dfrac{dy}{dt} = 2x - 4y$

25. $\mathbf{X}' = \begin{pmatrix}1 & 2\\ -2 & 1\end{pmatrix}\mathbf{X}$

26. $\mathbf{X}' = \begin{pmatrix}-2 & 5\\ -2 & 4\end{pmatrix}\mathbf{X}$

27. $\mathbf{X}' = \begin{pmatrix}1 & 1 & 1\\ 1 & 1 & 1\\ 1 & 1 & 1\end{pmatrix}\mathbf{X}$

28. $\mathbf{X}' = \begin{pmatrix}1 & -1 & 1\\ 0 & 1 & 3\\ 4 & 3 & 1\end{pmatrix}\mathbf{X}$

In Problems 29–33 use either undetermined coefficients or variation of parameters to solve the given system.

29. $\mathbf{X}' = \begin{pmatrix}2 & 8\\ 0 & 4\end{pmatrix}\mathbf{X} + \begin{pmatrix}2\\ 16t\end{pmatrix}$

30. $\dfrac{dx}{dt} = x + 2y$, $\quad \dfrac{dy}{dt} = -\dfrac{1}{2}x + y + e^t\tan t$

31. $\mathbf{X}' = \begin{pmatrix}-1 & 1\\ -2 & 1\end{pmatrix}\mathbf{X} + \begin{pmatrix}1\\ \cot t\end{pmatrix}$

32. $\mathbf{X}' = \begin{pmatrix}3 & 1\\ -1 & 1\end{pmatrix}\mathbf{X} + \begin{pmatrix}-2\\ 1\end{pmatrix}e^{2t}$

CHAPTER 9

Numerical Methods

IMPORTANT TERMS

lineal elements
Isocline
slope field
Euler's method
step size
absolute error
relative error
percentage relative error
improved Euler's method
Three-term Taylor method
First and second-order Runge-Kutta methods
Fourth-order Runge-Kutta method
predictor
corrector
Milne's method
multistep methods
single-step methods
truncation error
round-off error
accumulated round-off error
propagation error

A differential equation does not have to possess a solution; and even if a solution exists, we need not always be able to find explicit or implicit solutions of the equation. In many instance, particularly in the study of nonlinear equations, we have to be content with an *approximation* to a solution.

In this chapter we shall study five numerical methods for approximating solutions of differential equations.

9.1 Direction Fields

Introduction

If a solution of a differential equation exists it represents a locus of points (points connected by a smooth curve) in the Cartesian plane. Beginning in Section 9.2 we shall consider numerical procedures that utilize the differential equation to obtain a sequence of distinct points whose coordinates approximate the coordinates of the points on the actual solution curve (see Figure 9.1).

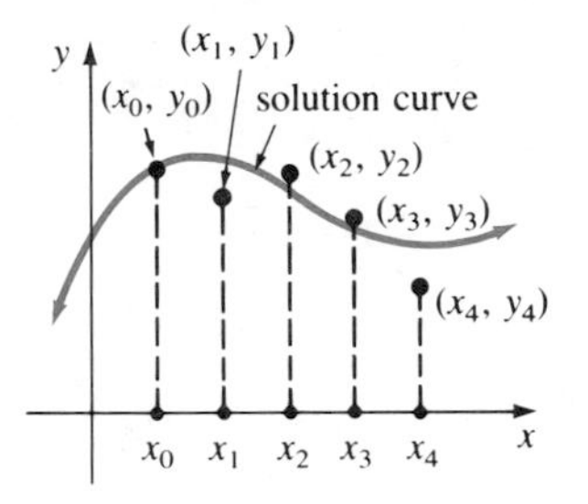

Figure 9.1

Throughout this chapter we shall confine our attention to first-order differential equations $dy/dx = f(x, y)$. As we saw in the preceding chapter, higher-order equations can always be reduced to a system of first-order equations. In Section 9.5 we shall see that the numerical methods developed for first-order equations are easily adapted to systems of first-order equations.

We begin with the study of direction fields. Although it is not a numerical method, the concept of a direction field enables us to obtain a rough sketch of a solution of a first-order differential equation without solving it.

Lineal Elements

Suppose for the moment that we do not know the general solution of the simple equation $y' = y$. Specifically, the differential equation implies that slopes of tangent lines to a solution curve are given by the function $f(x, y) = y$. When $f(x, y)$ is held constant, that is, when

$$y = c, \tag{1}$$

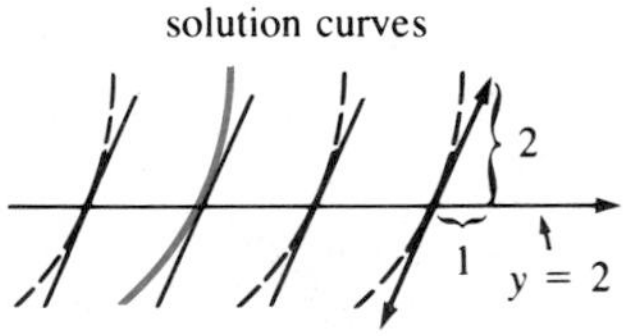

where c is any constant, we are in effect stating that the slope of the tangents to the solution curves is the same constant value along a horizontal line. For example, for $y = 2$ let us draw a sequence of short line segments, or **lineal elements**, each having slope 2 and its midpoint on the line. As shown in Figure 9.2, the solution curves pass through this horizontal line at every point tangent to the lineal elements.

Figure 9.2

Isoclines and Direction Fields

Equation (1) represents a one-parameter family of horizontal lines. In general, any member of the family $f(x, y) = c$ is called an **isocline**, which literally means a curve along which the inclination (of the tangents) is the same. As the parameter c is varied, we obtain a collection of isoclines on which the lineal elements are judiciously constructed. The totality of these lineal elements is called a **direction field**, **slope field**, or **lineal element field** of the differential equation $y' = f(x, y)$. As we see in Figure 9.3(a), the direction field suggests the "flow pattern" for the family of solution curves of the differential equation $y' = y$. In particular, if we want the one solution passing through the point $(0, 1)$, then, as indicated in Figure 9.3(b), we construct a curve through this point and passing through the isoclines with the appropriate slopes.

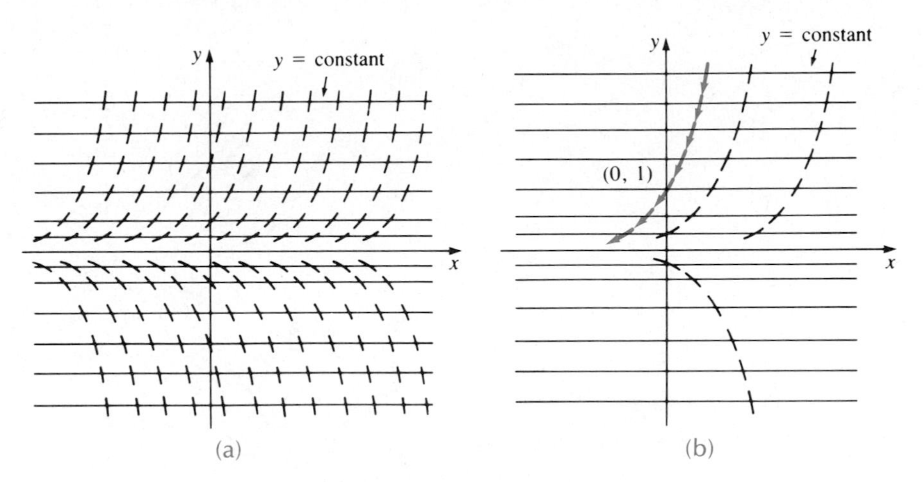

Figure 9.3

EXAMPLE 1

Determine the isoclines for the differential equation

$$\frac{dy}{dx} = 4x^2 + 9y^2.$$

SOLUTION For $c > 0$ the isoclines are the curves

$$4x^2 + 9y^2 = c.$$

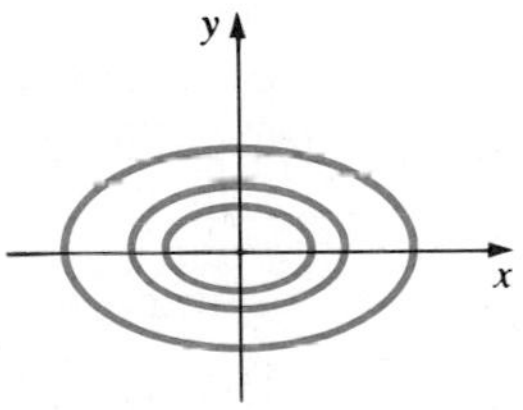

As Figure 9.4 shows, the curves are a concentric family of ellipses with major axis along the x-axis.

Figure 9.4

EXAMPLE 2

Sketch the direction field and indicate several possible members of the family of solution curves for

$$\frac{dy}{dx} = \frac{x}{y}.$$

SOLUTION Before sketching the direction field corresponding to the isoclines $x/y = c$ or $y = x/c$, we note that the differential equation gives the following information:

(a) If a solution curve crosses the x-axis ($y = 0$), it does so tangent to a vertical lineal element at every point except possibly (0, 0).

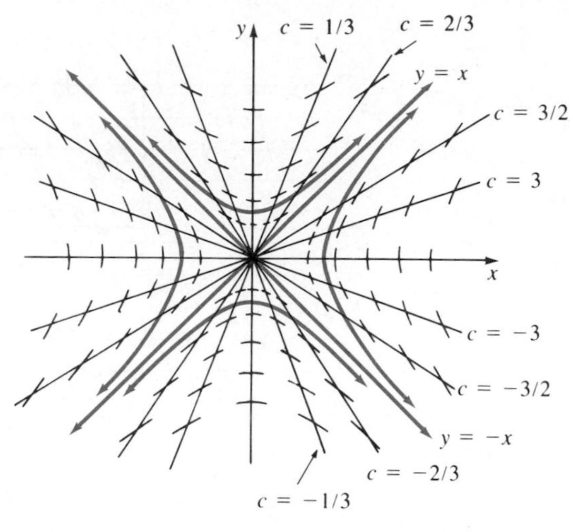

Figure 9.5

(b) If a solution curve crosses the y-axis ($x = 0$), it does so tangent to a horizontal lineal element at every point except possibly (0, 0).

(c) The lineal elements corresponding to the isoclines $c = 1$ and $c = -1$ are collinear with the lines $y = x$ and $y = -x$, respectively. Indeed, it is easily verified that these isoclines are both particular solutions of the given differential equation. However, it should be noted that *in general* isoclines are themselves not solutions to a differential equation.*

Figure 9.5 shows the direction field and several possible solution curves. Remember, on any particular isocline all the lineal elements are parallel. Also, the lineal elements may be drawn in such a manner as to suggest the flow of a particular curve.† In other words imagine the isoclines so close together that if the lineal elements were connected, we would have a polygonal curve suggestive of the shape of a smooth curve.

EXAMPLE 3

In Section 2.1 we indicated that the differential equation

$$\frac{dy}{dx} = x^2 + y^2$$

* When the isoclines are straight lines it is easy to determine which, if any, of these isoclines are also particular solutions of the differential equation (see Problems 21–26).

† Alternatively, the lineal elements can be drawn uniformly spaced on the isocline.

cannot be solved in terms of elementary functions. Use a direction field to locate an approximate solution satisfying $y(0) = 1$.

SOLUTION The isoclines are concentric circles defined by

$$x^2 + y^2 = c, \qquad c > 0.$$

By choosing $c = 1/4$, $c = 1$, $c = 9/4$, and $c = 4$, we obtain the circles with radii $1/2$, 1, $3/2$, and 2 shown in Figure 9.6(a). The lineal elements superimposed on each circle have slope corresponding to the particular value of c. It seems plausible from inspection of Figure 9.6(a) that a solution curve of the given initial-value problem might have the shape given in Figure 9.6(b). Unfortunately, we are not able to obtain any formula that describes this curve.

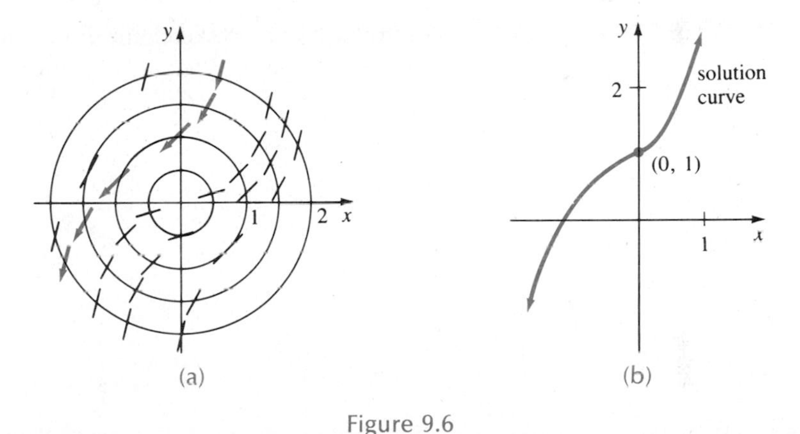

Figure 9.6

The concept of the direction field is used primarily to establish the existence of, and possibly to locate an approximate solution curve for, a first-order differential equation that cannot be solved by the usual standard techniques. However, the preceding discussion is of little value in determining specific values of a solution $y(x)$ at given points. For example, if we want to know the approximate value of $y(0.5)$ for the solution of

$$\frac{dy}{dx} = x^2 + y^2$$

$$y(0) = 1,$$

then Figure 9.6(b) can do nothing more for us than indicate that $y(0.5)$ may be in the same "ball park" as $y = 2$.

Exercises 9.1

Answers to odd-numbered problems begin on page A–27.

In Problems 1–10 identify the isoclines for the given differential equation.

1. $\dfrac{dy}{dx} = x + 4$

2. $\dfrac{dy}{dx} = 2x + y$

3. $\dfrac{dy}{dx} = x^2 - y^2$

4. $\dfrac{dy}{dx} = y - x^2$

5. $y' = \sqrt{x^2 + y^2 + 2y + 1}$

6. $y' = (x^2 + y^2)^{-1}$

7. $\dfrac{dy}{dx} = y(x + y)$

8. $\dfrac{dy}{dx} = y + e^x$

9. $\dfrac{dy}{dx} = \dfrac{y - 1}{x - 2}$

10. $\dfrac{dy}{dx} = \dfrac{x - y}{x + y}$

In Problems 11–18 sketch the direction field for the given differential equation and indicate several possible solution curves.

11. $y' = x$

12. $y' = x + y$

13. $y\dfrac{dy}{dx} = -x$

14. $\dfrac{dy}{dx} = \dfrac{1}{y}$

15. $\dfrac{dy}{dx} = xy$

16. $\dfrac{dy}{dx} = 1 - xy$

17. $y' = y - \cos\dfrac{\pi}{2}x$

18. $y' = 1 - \dfrac{y}{x}$

Miscellaneous Problems

19. Formally show that the isoclines for the differential equation

$$\frac{dy}{dx} = \frac{\alpha x + \beta y}{\gamma x + \delta y}$$

are straight lines through the origin.

20. Show that $y = cx$ is a solution of the differential equation in Problem 19 if and only if $(\beta - \gamma)^2 + 4\alpha\delta \geq 0$.

In Problems 21–26 find those isoclines that are also solutions of the given differential equation (see Problems 19 and 20).

EXAMPLE 4

The isoclines of the differential equation

$$y' = 2x + y \tag{2}$$

are the straight lines

$$2x + y = c. \tag{3}$$

A line in this latter family will be a solution of the differential equation whenever its slope is the same as c. In other words both the original equation and the line will satisfy $y' = c$. Since the slope of (3) is -2 if we choose $c = -2$, then $2x + y = -2$ is a solution of (2).

21. $y' = 3x + 2y$

22. $y' = 6x - 2y$

23. $y' = \dfrac{2x}{y}$

24. $y' = \dfrac{2y}{x + y}$

25. $\dfrac{dy}{dx} = \dfrac{4x + 3y}{y}$

26. $\dfrac{dy}{dx} = \dfrac{5x + 10y}{-4x + 3y}$

9.2 The Euler Methods

9.2.1 Euler's Method

One of the simplest techniques for approximating solutions of differential equations is known as **Euler's method**, or the method of **tangent lines**. Suppose we wish to approximate the solution of the initial-value problem

$$y' = f(x, y)$$
$$y(x_0) = y_0.$$

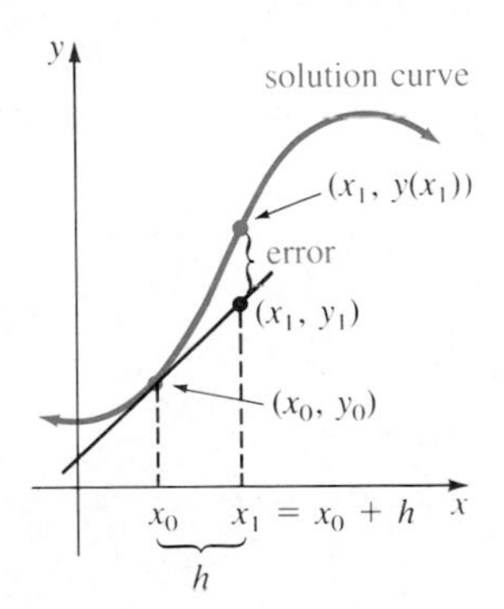

Figure 9.7

If h is a positive increment on the x-axis, then as Figure 9.7 shows, we can find a point $(x_1, y_1) = (x_0 + h, y_1)$ on the line tangent to the unknown solution curve at (x_0, y_0).

By the point-slope form of the equation of a line, we have

$$\frac{y_1 - y_0}{(x_0 + h) - x_0} = y'_0 \qquad \text{or} \qquad y_1 = y_0 + hy'_0,$$

where $y'_0 = f(x_0, y_0)$. If we label $x_0 + h$ by x_1, then the point (x_1, y_1) on the tangent line is an approximation to the point $(x_1, y(x_1))$ on the solution curve—that is, $y_1 \approx y(x_1)$. Of course the accuracy of the approximation depends heavily on the size of the increment h. Usually we must choose this **step size** to be "reasonably small."

Assuming a uniform (constant) value of h, we can obtain a succession of points $(x_1, y_1), (x_2, y_2), \ldots, (x_n, y_n)$, which we hope are close to the points

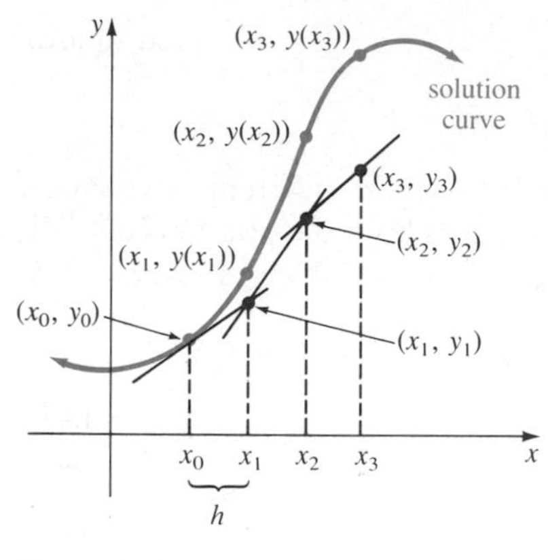

Figure 9.8

$(x_1, y(x_1)), (x_2, y(x_2)), \ldots, (x_n, y(x_n))$ (see Figure 9.8). Now using (x_1, y_1), we can obtain the value of y_2, which is the ordinate of a point on a new "tangent" line. We have

$$\frac{y_2 - y_1}{h} = y_1' \qquad \text{or} \qquad \begin{aligned} y_2 &= y_1 + hy_1' \\ &= y_1 + hf(x_1, y_1). \end{aligned}$$

In general it follows that

$$\begin{aligned} y_{n+1} &= y_n + hy_n' \\ &= y_n + hf(x_n, y_n), \end{aligned} \tag{1}$$

where $x_n = x_0 + nh$.

As an example, suppose we try the iteration scheme (1) on a differential equation for which we know the explicit solution; in this way we can compare the estimated values y_n and the true values $y(x_n)$.

EXAMPLE 1

Consider the initial-value problem

$$\begin{aligned} y' &= 2xy \\ y(1) &= 1. \end{aligned}$$

Use the Euler method to obtain an approximation to $y(1.5)$ using first $h = 0.1$ and then $h = 0.05$.

SOLUTION We first identify $f(x, y) = 2xy$ so that (1) becomes

$$y_{n+1} = y_n + h(2x_n y_n).$$

Then for $h = 0.1$ we find

$$\begin{aligned} y_1 &= y_0 + (0.1)(2x_0 y_0) \\ &= 1 + (0.1)[2(1)(1)] = 1.2, \end{aligned}$$

which is an estimate to the value of $y(1.1)$. However, if we use $h = 0.05$, it takes *two* iterations to reach $x = 1.1$. We have

$$\begin{aligned} y_1 &= 1 + (0.05)[2(1)(1)] = 1.1 \\ y_2 &= 1.1 + (0.05)[2(1.05)(1.1)] = 1.2155. \end{aligned}$$

Here we note that $y_1 \approx y(1.05)$ and $y_2 \approx y(1.1)$. The remainder of the calculations are summarized in Tables 9.1 and 9.2. Each entry is rounded to four decimal places.

Table 9.1 Euler's Method with $h = 0.1$

x_n	y_n	True Value	Abs. Error	% Rel. Error
1.00	1.0000	1.0000	0.0000	0.00
1.10	1.2000	1.2337	0.0337	2.73
1.20	1.4640	1.5527	0.0887	5.71
1.30	1.8154	1.9937	0.1784	8.95
1.40	2.2874	2.6117	0.3244	12.42
1.50	2.9278	3.4904	0.5625	16.12

Table 9.2 Euler's Method with $h = 0.05$

x_n	y_n	True Value	Abs. Error	% Rel. Error
1.00	1.0000	1.0000	0.0000	0.00
1.05	1.1000	1.1079	0.0079	0.72
1.10	1.2155	1.2337	0.0182	1.47
1.15	1.3492	1.3806	0.0314	2.27
1.20	1.5044	1.5527	0.0483	3.11
1.25	1.6849	1.7551	0.0702	4.00
1.30	1.8955	1.9937	0.0982	4.93
1.35	2.1419	2.2762	0.1343	5.90
1.40	2.4311	2.6117	0.1806	6.92
1.45	2.7714	3.0117	0.2403	7.98
1.50	3.1733	3.4904	0.3171	9.08

In Example 1 the true values were calculated from the known solution $y = e^{x^2-1}$. Also, the **absolute error** is defined to be

$$|\text{true value}—\text{approximation}|.$$

The **relative error** and the **percentage relative error** are, in turn,

$$\frac{|\text{true value}-\text{approximation}|}{\text{true value}}$$

and
$$\frac{|\text{true value} - \text{approximation}|}{\text{true value}} \times 100 = \frac{\text{absolute error}}{\text{true value}} \times 100.$$

It should be apparent that in the case of the step size $h = 0.1$ a 16%, relative error in the calculation of the approximation to $y(1.5)$ is totally unacceptable. At the expense of doubling the number of calculations, a slight improvement in accuracy is obtained by halving the step size to $h = 0.05$.

Of course in many instances we may not know the solution of a particular differential equation, or for that matter whether a solution of an initial-value problem actually exists. The following nonlinear equation does possess a solution in closed form, but we leave it as an exercise for the reader to find it (see Problem 1).

EXAMPLE 2

Use the Euler method to obtain the approximate value of $y(0.5)$ for the solution of

$$y' = (x + y - 1)^2$$
$$y(0) = 2.$$

SOLUTION For $n = 0$ and $h = 0.1$, we have

$$y_1 = y_0 + (0.1)(x_0 + y_0 - 1)^2$$
$$= 2 + (0.1)(1)^2 = 2.1.$$

The remaining calculations are summarized in Tables 9.3 and 9.4 for $h = 0.1$ and $h = 0.05$, respectively.

Table 9.3 Euler's Method with $h = 0.1$

x_n	y_n
0.00	2.0000
0.10	2.1000
0.20	2.2440
0.30	2.4525
0.40	2.7596
0.50	3.2261

Table 9.4 Euler's Method with $h = 0.05$

x_n	y_n
0.00	2.0000
0.05	2.0500
0.10	2.1105
0.15	2.1838
0.20	2.2727
0.25	2.3812
0.30	2.5142
0.35	2.6788
0.40	2.8845
0.45	3.1455
0.50	3.4823

We may want greater accuracy than that displayed, say, in Table 9.2, and so we could try a step size even smaller than $h = 0.05$. However, rather than resorting to this extra labor, it probably would be more advantageous to employ an alternative numerical procedure. The Euler formula by itself, though attractive in its simplicity, is seldom used in serious calculations.

9.2.2 Improved Euler's Method

The formula

$$y_{n+1} = y_n + h\frac{f(x_n, y_n) + f(x_{n+1}, y^*_{n+1})}{2},$$
$$\text{where} \quad y^*_{n+1} = y_n + hf(x_n, y_n). \tag{2}$$

is known as the **improved Euler formula,** or **Heun's formula**. The values $f(x_n, y_n)$ and $f(x_{n+1}, y^*_{n+1})$ are approximations to the slope of the curve at

$(x_n, y(x_n))$ and $(x_{n+1}, y(x_{n+1}))$ and, consequently, the quotient

$$\frac{f(x_n, y_n) + f(x_{n+1}, y_{n+1}^*)}{2}$$

can be interpreted as an average slope on the interval between x_n and x_{n+1}.

The equations in (2) can be readily visualized. In Figure 9.9 we show the case in which $n = 0$. Note that

$$f(x_0, y_0) \quad \text{and} \quad f(x_1, y_1^*)$$

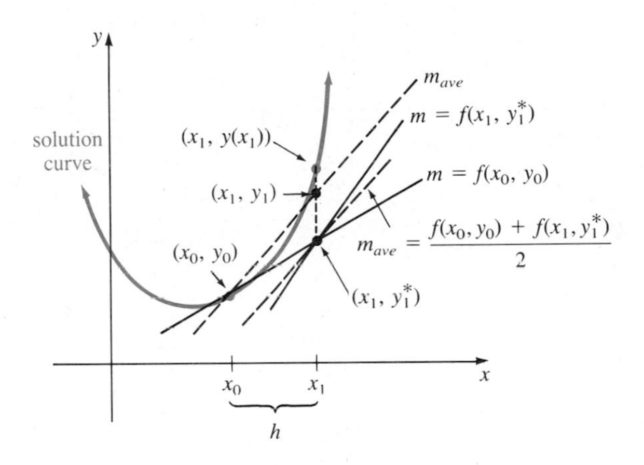

Figure 9.9

are slopes of the indicated straight lines passing through the points (x_0, y_0) and (x_1, y_1^*), respectively. By taking an average of these slopes, we obtain the slope of the dashed skew lines. Rather than advancing along the line with slope $m = f(x_0, y_0)$ to the point with ordinate y_1^* obtained by the usual Euler method, we advance instead along the line through (x_0, y_0) with slope m_{ave} until we reach x_1. It seems plausible from inspection of the figure that y_1 is an improvement over y_1^*.

We might also say that the value of

$$y_1^* = y_0 + hf(x_0, y_0)$$

predicts a value of $y(x_1)$, whereas

$$y_1 = y_0 + h\frac{f(x_0, y_0) + f(x_1, y_1^*)}{2}$$

corrects this estimate.

EXAMPLE 3

Use the improved Euler formula to obtain the approximate value of $y(1.5)$ for the solution of the initial-value problem in Example 1. Compare the results for $h = 0.1$ and $h = 0.05$.

SOLUTION For $n = 0$ and $h = 0.1$, we first compute

$$y_1^* = y_0 + (0.1)(2x_0y_0) = 1.2.$$

Then from (2)

$$y_1 = y_0 + (0.1)\frac{2x_0y_0 + 2x_1y_1^*}{2}$$

$$= 1 + (0.1)\frac{2(1)(1) + 2(1.1)(1.2)}{2} = 1.232.$$

The comparative values of the calculations for $h = 0.1$ and $h = 0.5$ are given in Tables 9.5 and 9.6, respectively.

Table 9.5 Improved Euler's Method with $h = 0.1$

x_n	y_n	True Value	Abs. Error	% Rel. Error
1.00	1.0000	1.0000	0.0000	0.00
1.10	1.2320	1.2337	0.0017	0.14
1.20	1.5479	1.5527	0.0048	0.31
1.30	1.9832	1.9937	0.0106	0.53
1.40	2.5908	2.6117	0.0209	0.80
1.50	3.4509	3.4904	0.0394	1.13

Table 9.6 Improved Euler's Method with $h = 0.05$

x_n	y_n	True Value	Abs. Error	% Rel. Error
1.00	1.0000	1.0000	0.0000	0.00
1.05	1.1077	1.1079	0.0002	0.02
1.10	1.2332	1.2337	0.0004	0.04
1.15	1.3798	1.3806	0.0008	0.06
1.20	1.5514	1.5527	0.0013	0.08
1.25	1.7531	1.7551	0.0020	0.11
1.30	1.9909	1.9937	0.0029	0.14
1.35	2.2721	2.2762	0.0041	0.18
1.40	2.6060	2.6117	0.0057	0.22
1.45	3.0038	3.0117	0.0079	0.26
1.50	3.4795	3.4904	0.0108	0.31

A brief word of caution is in order here. We cannot compute all the values of y_n^* first and then substitute these values in the first formula of (2). In other words we cannot use the data in Table 9.1 to help construct the values in Table 9.5. Why not?

EXAMPLE 4

Use the improved Euler formula to obtain the approximate value of $y(0.5)$ for the solution of the initial-value problem in Example 2.

SOLUTION For $n = 0$ and $h = 0.1$, we have

$$y_1^* = y_0 + (0.1)(x_0 + y_0 - 1)^2 = 2.1$$

and so

$$y_1 = y_0 + (0.1)\frac{(x_0 + y_0 - 1)^2 + (x_1 + y_1^* - 1)^2}{2}$$

$$= 2 + (0.1)\frac{1 + 1.44}{2} = 2.122.$$

The remaining calculations are summarized in Tables 9.7 and 9.8 for $h = 0.1$ and $h = 0.05$, respectively.

Table 9.7 Improved Euler's Method with $h = 0.1$

x_n	y_n
0.00	2.0000
0.10	2.1220
0.20	2.3049
0.30	2.5858
0.40	3.0378
0.50	3.8254

Table 9.8 Improved Euler's Method with $h = 0.05$

x_n	y_n
0.00	2.0000
0.05	2.0553
0.10	2.1228
0.15	2.2056
0.20	2.3075
0.25	2.4342
0.30	2.5931
0.35	2.7953
0.40	3.0574
0.45	3.4057
0.50	3.8840

BASIC Programs

The following BASIC programs were used to obtain the approximations in this section. These programs can be utilized, if desired, in solving the problems in Exercises 9.2.

```
100 REM EULER'S METHOD TO SOLVE Y' = FNF(X, Y)
110 REM DEFINE FNF(X, Y) HERE
120 REM GET INPUTS
130 PRINT
140 INPUT "STEP SIZE = ",H
150 INPUT "NUMBER OF STEPS = ",N
160 INPUT "XO = ",X
170 INPUT "YO = ",Y
180 PRINT
190 REM SET UP TABLE
200 PRINT "X", "Y"
210 PRINT
220 PRINT X, Y
230 REM COMPUTE X AND Y VALUES
240 FOR I = 1 TO N
250 Y = Y + H * FNF(X, Y)
260 X = X + H
270 PRINT X, Y
280 NEXT I
290 END
```

```
100 REM IMPROVED EULER'S METHOD TO SOLVE Y' = FNF(X, Y)
110 REM DEFINE FNF(X, Y) HERE
120 REM GET INPUTS
```

```
130 PRINT
140 INPUT "STEP SIZE = ",H
150 INPUT "NUMBER OF STEPS = ",N
160 INPUT "XO = ",X
170 INPUT "YO = ",Y
180 PRINT
190 REM SET UP TABLE
200 PRINT "X", "Y"
210 PRINT
220 PRINT X, Y
230 REM COMPUTE X AND Y VALUES
240 FOR I = 1 TO N
250 FVAL = FNF(X, Y)
260 Y = Y + H * (FVAL + FNF(X + H, Y + H * FVAL))/2
270 X = X + H
280 PRINT X, Y
290 NEXT I
300 END
```

Exercises 9.2

Answers to odd-numbered problems begin on page A–28.

1. Solve the initial-value problem

$$y' = (x + y - 1)^2, \qquad y(0) = 2$$

in terms of elementary functions.

2. Let $y(x)$ be the solution of the initial-value problem given in Problem 1. Rounded to four decimal places, compute the exact values of $y(0.1)$, $y(0.2)$, $y(0.3)$, $y(0.4)$, and $y(0.5)$. Compare these values with the entries in Tables 9.3, 9.4, 9.7, and 9.8.

Given the initial-value problems in Problems 3–12, use the Euler formula to obtain a four-decimal approximation to the indicated value. First use **(a)** $h = 0.1$ and then **(b)** $h = 0.05$.

3. $y' = 2x - 3y + 1, \quad y(1) = 5; \quad y(1.5)$

4. $y' = 4x - 2y, \quad y(0) = 2; \quad y(0.5)$

5. $y' = 1 + y^2, \quad y(0) = 0; \quad y(0.5)$

6. $y' = x^2 + y^2, \quad y(0) = 1; \quad y(0.5)$

7. $y' = e^{-y}, \quad y(0) = 0; \quad y(0.5)$

8. $y' = x + y^2, \quad y(0) = 0; \quad y(0.5)$

9. $y' = (x - y)^2, \quad y(0) = 0.5; \quad y(0.5)$

10. $y' = xy + \sqrt{y}, \quad y(0) = 1; \quad y(0.5)$

11. $y' = xy^2 - \dfrac{y}{x}$, $y(1) = 1$; $y(1.5)$

12. $y' = y - y^2$, $y(0) = 0.5$; $y(0.5)$

13. As parts **(a)**–**(e)** of this problem, repeat the calculations of Problems 3, 5, 7, 9, and 11 using the improved Euler formula.

14. As parts **(a)**–**(e)** of this problem, repeat the calculations of Problems 4, 6, 8, 10, and 12 using the improved Euler formula.

15. Although it may not be obvious from the differential equation, its solution could "behave badly" near a point x at which we wish to approximate $y(x)$. Numerical procedures may then give widely differing results near this point. Let $y(x)$ be the solution of the initial-value problem

$$y' = x^2 + y^3, \qquad y(1) = 1.$$

Using the step size $h = 0.1$, compare the results obtained from the Euler formula with the results from the improved Euler formula in the approximation of $y(1.4)$.

EXAMPLE 5

The improved Euler formula can be used to obtain a sequence of approximations to $y(x_n)$ at a *fixed value* of x_n. If the basic Euler formula is denoted by

$$y_{n+1,1} = y_n + hf(x_n, y_n), \tag{3}$$

then we can define

$$y_{n+1,k+1} = y_n + h\frac{f(x_n, y_n) + f(x_{n+1}, y_{n+1,k})}{2} \tag{4}$$

for $n \geq 0$, $k \geq 1$. For $n = 0$ and $k = 1, 2, 3, \ldots$, equations (3) and (4) yield the sequence of values

$$y_{1,1}, y_{1,2}, y_{1,3}, y_{1,4}, \ldots,$$

which are all approximations to $y(x)$ at $x = x_1$. For example, $y_{1,2}$ corresponds to the *original* improved Euler formula given in (2) and

$$y_{1,3} = y_0 + h\frac{f(x_0, y_0) + f(x_1, y_{1,2})}{2}.$$

It might be conjectured that, since an average of two slopes (formula (2)) yields an improved approximation to $y(x)$ at a point, an average including an average (formula (4)) *may* even give better results (see Problem 16).

16. Consider the initial-value problem

$$y' = 2xy, \qquad y(1) = 1.$$

Convince yourself that "more" is not necessarily better by computing the values $y_{1,1}, y_{1,2}, y_{1,3}, y_{1,4}, y_{1,5}$ with $h = 0.1$. By using the exact value of $y(1.1)$ (see Table 9.1), compute the percentage relative error at each step of the calculation.

Miscellaneous Problems

17. Derive the basic Euler formula by integrating both sides of the equation $y' = f(x, y)$ on the integral, $x_n \leq x \leq x_{n+1}$. Approximate the integral of the right side by replacing the function $f(x, y)$ by its value at the left end point of the interval of integration.

18. By following the procedure outlined in Problem 17, derive the improved Euler formula. [*Hint:* Replace the integrand of the right side by the average of its values at the endpoints of the interval of integration.]

9.3 The Three-Term Taylor Method

The numerical method considered in this section, the **three-term Taylor method**, is more of theoretical interest than of practical importance since the results obtained using the following formula (5) will not differ substantially from those obtained using the improved Euler method.

In the study of numerical solutions of differential equations, many computational algorithms can be derived from a Taylor series expansion. Recall from calculus that the form of this expansion about a point $x = a$ is

$$y(x) = y(a) + y'(a)\frac{(x-a)}{1!} + y''(a)\frac{(x-a)^2}{2!} + \cdots. \tag{1}$$

It is understood that the function $y(x)$ possesses derivatives of all orders and that the series (1) converges in some interval defined by $|x - a| < R$. Notice, in particular, that if we set $a = x_n$ and $x = x_n + h$, then (1) becomes

$$y(x_n + h) = y(x_n) + y'(x_n)h + y''(x_n)\frac{h^2}{2} + \cdots. \tag{2}$$

Euler's Method Revisited

Furthermore, let us now assume that the function $y(x)$ is a solution of the first-order differential equation

$$y' = f(x, y).$$

If we then truncate the series (2) after, say, two terms, we obtain the approximation

$$y(x_n + h) \approx y(x_n) + y'(x_n)h$$

or

$$y(x_n + h) \approx y(x_n) + f(x_n, y(x_n))h. \tag{3}$$

Observe that we can obtain the Euler formula

$$y_{n+1} = y_n + hf(x_n, y_n) \tag{4}$$

of the preceding section by replacing $y(x_n + h)$ and $y(x_n)$ in (3) by their approximations y_{n+1} and y_n, respectively. The approximation symbol $\approx$ is replaced by an equality since we are defining the left side of (4) by the numbers obtained from the right-hand member.

Taylor's Method

By retaining three terms in the series (2), we can write

$$y(x_n + h) \approx y(x_n) + y'(x_n)h + y''(x_n)\frac{h^2}{2}.$$

After using the replacements noted in the preceding material, it follows that

$$y_{n+1} = y_n + y'_n h + y''_n \frac{h^2}{2}. \tag{5}$$

The second derivative y'' can be obtained by differentiating $y' = f(x, y)$.

At this point let us reexamine the two initial-value problems of the preceding section.

EXAMPLE 1

Use the three-term Taylor formula to obtain the approximate value of $y(1.5)$ for the solution of

$$y' = 2xy$$

$$y(1) = 1.$$

Compare the results for $h = 0.1$ and $h = 0.05$.

SOLUTION Since $y' = 2xy$, it follows by the product rule that $y'' = 2xy' + 2y$. Thus, for example, when $h = 0.1$, $n = 0$, we can first calculate

$$\begin{aligned} y'_0 &= 2x_0 y_0 \\ &= 2(1)(1) = 2, \end{aligned}$$

and then

$$\begin{aligned} y''_0 &= 2x_0 y'_0 + 2y_0 \\ &= 2(1)(2) + 2(1) = 6. \end{aligned}$$

Hence (5) becomes

$$y_1 = y_0 + y_0'(0.1) + y_0''\frac{(0.1)^2}{2}$$

$$= 1 + 2(0.1) + 6(0.005) = 1.23.$$

The results of the iteration, along with the comparative exact values, are summarized in Tables 9.9 and 9.10.

Table 9.9 Three-Term Taylor Method with $h = 0.1$

x_n	y_n	True Value	Abs. Error	% Rel. Error
1.00	1.0000	1.0000	0.0000	0.00
1.10	1.2300	1.2337	0.0037	0.30
1.20	1.5427	1.5527	0.0100	0.65
1.30	1.9728	1.9937	0.0210	1.05
1.40	2.5721	2.6117	0.0396	1.52
1.50	3.4188	3.4904	0.0715	2.05

Table 9.10 Three-Term Taylor Method with $h = 0.05$

x_n	y_n	True Value	Abs. Error	% Rel. Error
1.00	1.0000	1.0000	0.0000	0.00
1.05	1.1075	1.1079	0.0004	0.04
1.10	1.2327	1.2337	0.0010	0.08
1.15	1.3788	1.3806	0.0018	0.13
1.20	1.5499	1.5527	0.0028	0.18
1.25	1.7509	1.7551	0.0041	0.23
1.30	1.9879	1.9937	0.0059	0.29
1.35	2.2681	2.2762	0.0081	0.36
1.40	2.6006	2.6117	0.0111	0.43
1.45	2.9967	3.0117	0.0150	0.50
1.50	3.4702	3.4904	0.0202	0.58

EXAMPLE 2

Use the three-term Taylor formula to obtain the approximate value of $y(0.5)$ for the solution of

$$y' = (x + y - 1)^2, \qquad y(0) = 2.$$

SOLUTION In this case we compute y'' by the power rule. We have

$$y'' = 2(x + y - 1)(1 + y').$$

The results are summarized in Table 9.11 and 9.12 for $h = 0.1$ and $h = 0.05$, respectively.

Table 9.11 Three-term Taylor's Method with $h = 0.1$

x_n	y_n
0.00	2.0000
0.10	2.1200
0.20	2.2992
0.30	2.5726
0.40	3.0077
0.50	3.7511

Table 9.12 Three-term Taylor's Method with $h = 0.05$

x_n	y_n
0.00	2.0000
0.05	2.0550
0.10	2.1222
0.15	2.2045
0.20	2.3058
0.25	2.4315
0.30	2.5890
0.35	2.7889
0.40	3.0475
0.45	3.3898
0.50	3.8574

A comparison of the last two examples with the corresponding results obtained from the improved Euler method shows no startling dissimilarities. In fact when $f(x, y)$ is linear in both variables x and y, the Taylor method gives the *same* values of y_n as the improved Euler method for a given value of h. See Problems 12 and 13.

BASIC Program

A BASIC program for the three-term Taylor method follows.

```
100 REM THREE-TERM TAYLOR'S METHOD TO SOLVE Y' = FNF(X, Y)
110 REM THE DERIVATIVE OF FNF(X, Y) IS DENOTED BY FNDF(X, Y)
120 REM DEFINE FNF(X, Y) HERE
130 REM DEFINE FNDF(X, Y) HERE
140 REM GET INPUTS
150 PRINT
160 INPUT "STEP SIZE = ",H
170 INPUT "NUMBER OF STEPS = ",N
180 INPUT "XO = ",X
190 INPUT "YO = ",Y
200 PRINT
210 REM SET UP TABLE
220 PRINT "X", "Y"
```

```
230 PRINT
240 PRINT X, Y
250 REM COMPUTE X AND Y VALUES
260 FOR I = 1 TO N
270 FVAL = FNF(X, Y)
280 Y = Y + H * FNF(X, Y) + H * H * FNDF(X, Y)/2
290 X = X + H
300 PRINT X, Y
310 NEXT I
320 END
```

Exercises 9.3

Answers to odd-numbered problems begin on page A–30.

Given the initial-value problems in Problems 1–10, use the three-term Taylor formula to obtain a four-decimal approximation to the indicated value. First use **(a)** $h = 0.1$ and then **(b)** $h = 0.05$.

1. $y' = 2x - 3y + 1, \quad y(1) = 5; \quad y(1.5)$

2. $y' = 4x - 2y, \quad y(0) = 2; \quad y(0.5)$

3. $y' = 1 + y^2, \quad y(0) = 0; \quad y(0.5)$

4. $y' = x^2 + y^2, \quad y(0) = 1; \quad y(0.5)$

5. $y' = e^{-y}, \quad y(0) = 0; \quad y(0.5)$

6. $y' = x + y^2, \quad y(0) = 0; \quad y(0.5)$

7. $y' = (x - y)^2, \quad y(0) = 0.5; \quad y(0.5)$

8. $y' = xy + \sqrt{y}, \quad y(0) = 1; \quad y(0.5)$

9. $y' = xy^2 - \dfrac{y}{x}, \quad y(1) = 1; \quad y(1.5)$

10. $y' = y - y^2, \quad y(0) = 0.5; \quad y(0.5)$

11. Let $y(x)$ be the solution of the initial-value problem

$$y' = x^2 + y^3, \qquad y(1) = 1.$$

Use $h = 0.1$ and the three-term Taylor formula to obtain an approximation to $y(1.4)$. Compare your answer with the results obtained in Problem 15 of Section 9.2.

Miscellaneous Problems

12. Consider the differential equation $y' = f(x, y)$, where f is linear in x and y. In this case prove that the improved Euler formula is the same as the three-term Taylor formula. [*Hint:* Recall from calculus that a Taylor series for a function g of two variables is

$$g(a + h, b + k) = g(a, b) + g_x(a, b)h + g_y(a, b)k + \frac{1}{2}(h^2 g_{xx} + 2hk g_{xy} + k^2 g_{yy})\Big|_{(a,b)}$$
$$+ \text{ terms involving higher-order derivatives.}$$

Apply this result to $f(x_n + h, y_n + hf(x_n, y_n))$ in the improved Euler formula. Also use the fact that $y''(x) = \frac{d}{dx}y'(x) = f_x + f_y y'$.]

13. Compare the approximate values of $y(1.5)$ for

$$y' = x + y - 1, \qquad y(1) = 5,$$

using the three-term Taylor method and the improved Euler method with $h = 0.1$. Solve the initial-value problem and compute the true values $y(x_n)$, $n = 0, 1, \ldots, 5$.

9.4 The Runge-Kutta Method

Probably one of the most popular as well as accurate numerical procedures used in obtaining approximate solutions to differential equations is the **fourth-order Runge-Kutta method.*** As the name suggests there are Runge-Kutta's methods of different orders.

For the moment let us consider a **second-order** procedure. This consists of finding constants a, b, α, and β such that the formula

$$y_{n+1} = y_n + ak_1 + bk_2, \tag{1}$$

where

$$k_1 = hf(x_n, y_n)$$
$$k_2 = hf(x_n + \alpha h, y_n + \beta k_1), \tag{2}$$

---

* **Carl D. T. Runge** (1856–1927) A German professor of applied mathematics at the University of Göttingen, Runge devised this numerical method around 1895. (M. W. Kutta expanded on this work in 1901.) In mathematics Runge also did notable work in the field of Diophantine equations. In physics Runge is remembered for his work on the Zeeman effect (the spectral lines in the emission spectrum of an element are affected by the presence of a magnetic field).

Martin W. Kutta (1867–1944) Also a German applied mathematician, Kutta made significant contributions to the field of aerodynamics.

agrees with a Taylor series expansion to as many terms as possible. The obvious purpose is to achieve the accuracy of the Taylor method without the necessity of having to compute higher-order derivatives. Now it can be shown that whenever the constants satisfy

$$a + b = 1, \qquad b\alpha = \frac{1}{2}, \qquad b\beta = \frac{1}{2},$$

then (1) agrees with a Taylor expansion out to the h^2, or third term. It should be of interest to observe that when $a = 1/2$, $b = 1/2$, $\alpha = 1$, $\beta = 1$, then (1) reduces to the improved Euler method. Thus we can conclude that the three-term Taylor formula is essentially equivalent to the improved Euler formula. Also, the basic Euler method is a **first-order** Runge-Kutta procedure.

Notice too that the sum $ak_1 + bk_2$, $a + b = 1$ in equation (1) is simply a *weighted average* of k_1 and k_2. The numbers k_1 and k_2 are multiples of approximations to the slope at two different points.

Fourth-Order Runge-Kutta Formula

The **fourth-order** Runge-Kutta method consists of determining appropriate constants so that a formula such as

$$y_{n+1} = y_n + ak_1 + bk_2 + ck_3 + dk_4$$

agrees with a Taylor expansion out to h^4 or the fifth term. As in (2) the k_i are constant multiples of $f(x, y)$ evaluated at select points. The derivation of the actual method is tedious, to say the least, so we state the results:

$$\begin{aligned} y_{n+1} &= y_n + \tfrac{1}{6}(k_1 + 2k_2 + 2k_3 + k_4), \\ k_1 &= hf(x_n, y_n) \\ k_2 &= hf(x_n + \tfrac{1}{2}h, y_n + \tfrac{1}{2}k_1) \\ k_3 &= hf(x_n + \tfrac{1}{2}h, y_n + \tfrac{1}{2}k_2) \\ k_4 &= hf(x_n + h, y_n + k_3). \end{aligned} \tag{3}$$

The reader is advised to look carefully at the formulas in (3); note that k_2 depends on k_1, k_3 depends on k_2, and so on. Also, k_2 and k_3 are approximations to the slope at the midpoint of the interval between x_n and $x_{n+1} = x_n + h$.

EXAMPLE 1

Use the Runge-Kutta method with $h = 0.1$ to obtain an approximation to $y(1.5)$ for the solution of

$$y' = 2xy$$

$$y(1) = 1.$$

SOLUTION For the sake of illustration let us compute the case when $n = 0$. From (3) we find

$$\begin{aligned} k_1 &= (0.1)f(x_0, y_0) \\ &= (0.1)(2x_0y_0) = 0.2, \\ k_2 &= (0.1)f\left(x_0 + \frac{1}{2}(0.1), y_0 + \frac{1}{2}(0.2)\right) \\ &= (0.1)2\left(x_0 + \frac{1}{2}(0.1)\right)\left(y_0 + \frac{1}{2}(0.2)\right) = 0.231, \\ k_3 &= (0.1)f\left(x_0 + \frac{1}{2}(0.1), y_0 + \frac{1}{2}(0.231)\right) \\ &= (0.1)2\left(x_0 + \frac{1}{2}(0.1)\right)\left(y_0 + \frac{1}{2}(0.231)\right) = 0.234255, \\ k_4 &= (0.1)f(x_0 + 0.1, y_0 + 0.234255) \\ &= (0.1)2(x_0 + 0.1)(y_0 + 0.234255) = 0.2715361, \end{aligned}$$

and, therefore,

$$\begin{aligned} y_1 &= y_0 + \frac{1}{6}(k_1 + 2k_2 + 2k_3 + k_4) \\ &= 1 + \frac{1}{6}(0.2 + 2(0.231) + 2(0.234255) + 0.2715361) = 1.23367435. \end{aligned}$$

When we round to the usual four decimal places, we obtain

$$y_1 = 1.2337.$$

The accompanying table should convince the student why the Runge-Kutta method is so popular. Of course there is no need to use any smaller step size.

Table 9.13 Runge-Kutta's Method with $h = 0.1$

x_n	y_n	True Value	Abs. Error	% Rel. Error
1.00	1.0000	1.0000	0.0000	0.00
1.10	1.2337	1.2337	0.0000	0.00
1.20	1.5527	1.5527	0.0000	0.00
1.30	1.9937	1.9937	0.0000	0.00
1.40	2.6116	2.6117	0.0001	0.00
1.50	3.4902	3.4904	0.0001	0.00

The reader might be interested in inspecting Tables 9.14 and 9.15 at this point. These tables compare the results obtaining from the various formulas that we have examined applied to the two specific problems

$$y' = 2xy, \qquad y(1) = 1,$$
$$y' = (x + y - 1)^2, \qquad y(0) = 2,$$

that we have considered throughout the last three sections.

Table 9.14 $y' = 2xy, \quad y(1) = 1$

Comparison of Numerical Methods with $h = 0.1$

x_n	Euler	Improved Euler	Three-Term Taylor	Runge-Kutta	True Value
1.00	1.0000	1.0000	1.0000	1.0000	1.0000
1.10	1.2000	1.2320	1.2300	1.2337	1.2337
1.20	1.4640	1.5479	1.5427	1.5527	1.5527
1.30	1.8154	1.9832	1.9728	1.9937	1.9937
1.40	2.2874	2.5908	2.5721	2.6116	2.6117
1.50	2.9278	3.4509	3.4188	3.4902	3.4904

Comparison of Numerical Methods with $h = 0.05$

x_n	Euler	Improved Euler	Three-Term Taylor	Runge-Kutta	True Value
1.00	1.0000	1.0000	1.0000	1.0000	1.0000
1.05	1.1000	1.1077	1.1075	1.1079	1.1079
1.10	1.2155	1.2332	1.2327	1.2337	1.2337
1.15	1.3492	1.3798	1.3788	1.3806	1.3806
1.20	1.5044	1.5514	1.5499	1.5527	1.5527
1.25	1.6849	1.7531	1.7509	1.7551	1.7551
1.30	1.8955	1.9909	1.9879	1.9937	1.9937
1.35	2.1419	2.2721	2.2681	2.2762	2.2762
1.40	2.4311	2.6060	2.6006	2.6117	2.6117
1.45	2.7714	3.0038	2.9967	3.0117	3.0117
1.50	3.1733	3.4795	3.4702	3.4903	3.4904

Table 9.15 $y' = (x + y - 1)^2, \quad y(0) = 2$

Comparison of Numerical Methods with $h = 0.1$

x_n	Euler	Improved Euler	Three-Term Taylor	Runge-Kutta	True Value
0.00	2.0000	2.0000	2.0000	2.0000	2.0000
0.10	2.1000	2.1220	2.1200	2.1230	2.1230
0.20	2.2440	2.3049	2.2992	2.3085	2.3085
0.30	2.4525	2.5858	2.5726	2.5958	2.5958
0.40	2.7596	3.0378	3.0077	3.0649	3.0650
0.50	3.2261	3.8254	3.7511	3.9078	3.9082

Table 9.15 *(continued)*

Comparison of Numerical Methods with $h = 0.05$					
x_n	Euler	Improved Euler	Three-Term Taylor	Runge-Kutta	True Value
0.00	2.0000	2.0000	2.0000	2.0000	2.0000
0.05	2.0500	2.0553	2.0550	2.0554	2.0554
0.10	2.1105	2.1228	2.1222	2.1230	2.1230
0.15	2.1838	2.2056	2.2045	2.2061	2.2061
0.20	2.2727	2.3075	2.3058	2.3085	2.3085
0.25	2.3812	2.4342	2.4315	2.4358	2.4358
0.30	2.5142	2.5931	2.5890	2.5958	2.5958
0.35	2.6788	2.7953	2.7889	2.7998	2.7997
0.40	2.8845	3.0574	3.0475	3.0650	3.0650
0.45	3.1455	3.4057	3.3898	3.4189	3.4189
0.50	3.4823	3.8840	3.8574	3.9082	3.9082

BASIC Program

A listing of a BASIC program for the Runge-Kutta method is given here.

```
100 REM RUNGE-KUTTA METHOD TO SOLVE Y' = FNF(X, Y)
110 REM DEFINE FNF(X, Y) HERE
120 REM GET INPUTS
130 PRINT
140 INPUT "STEP SIZE =",H
150 INPUT "NUMBER OF STEPS =",N
160 INPUT "XO =",X
170 INPUT "YO =",Y
180 PRINT
190 REM SET UP TABLE
200 PRINT "X", "Y"
210 PRINT
220 PRINT X, Y
230 REM COMPUTE X AND Y VALUES
240 FOR I = 1 TO N
250 K1 = H * FNF(X, Y)
260 K2 = H * FNF(X + H/2, Y + K1/2)
270 K3 = H * FNF(X + H/2, Y + K2/2)
280 K4 = H * FNF(X + H, Y + K3)
290 Y = Y + (K1 + 2 * K2 + 2 * K3 + K4)/6
300 X = X + H
310 PRINT X, Y
320 NEXT I
330 END
```

Exercises 9.4

Answers to odd-numbered problems begin on page A–31.

Given the initial-value problems in Problems 1–10, use the Runge-Kutta method with $h = 0.1$ to obtain a four-decimal approximation to the indicated value.

1. $y' = 2x - 3y + 1, \quad y(1) = 5; \quad y(1.5)$

2. $y' = 4x - 2y, \quad y(0) = 2; \quad y(0.5)$

3. $y' = 1 + y^2, \quad y(0) = 0; \quad y(0.5)$

4. $y' = x^2 + y^2, \quad y(0) = 1; \quad y(0.5)$

5. $y' = e^{-y}, \quad y(0) = 0; \quad y(0.5)$

6. $y' = x + y^2, \quad y(0) = 0; \quad y(0.5)$

7. $y' = (x - y)^2, \quad y(0) = 0.5; \quad y(0.5)$

8. $y' = xy + \sqrt{y}, \quad y(0) = 1; \quad y(0.5)$

9. $y' = xy^2 - y/x, \quad y(1) = 1; \quad y(1.5)$

10. $y' = y - y^2, \quad y(0) = 0.5; \quad y(0.5)$

11. If air resistance is proportional to the square of the instantaneous velocity, the velocity v of a mass m dropped from a height h is determined from

$$m\frac{dv}{dt} = mg - kv^2, \qquad k > 0$$

(see Problem 8, Chapter 3 Review Exercises). If $v(0) = 0$, $k = 0.125$, $m = 5$ slugs, and $g = 32$ ft/sec^2, use the Runge-Kutta method to find an approximation to the velocity of the falling mass at $t = 5$ sec. Use $h = 1$.

12. Solve the initial-value problem in Problem 11 by one of the methods of Chapter 2. Find the true value $v(5)$.

13. A mathematical model for the area A (in cm^2) that a colony of bacteria occupies (*B. dendroides*) is given by*

$$\frac{dA}{dt} = A(2.128 - 0.0432A).$$

If $A(0) = 0.24$ cm^2, use the Runge-Kutta method to complete the following table. Use $h = 0.5$.

---

* See V. A. Kostitzin, *Mathematical Biology*, London: Harrap, 1939.

t (days)	1	2	3	4	5
A (observed)	2.78	13.53	36.30	47.50	49.40
A (approximated)					

14. Solve the initial-value problem in Problem 13. Compute the values $A(1)$, $A(2)$, $A(3)$, $A(4)$, and $A(5)$. [*Hint:* See Section 3.3.]

15. Let $y(x)$ be the solution of the initial-value problem

$$y' = x^2 + y^3, \qquad y(1) = 1.$$

Determine whether the Runge-Kutta formula can be used to obtain an approximation for $y(1.4)$. Use $h = 0.1$

Miscellaneous Problems

16. Consider the differential equation $y' = f(x)$. In this case show that the fourth-order Runge-Kutta method reduces to Simpson's rule for the integral of $f(x)$ on the interval $x_n \leq x \leq x_{n+1}$.

[O] 9.5 Milne's Method, Second-Order Equations, Errors

There are many additional formulas that can be applied to obtain approximations to solutions of differential equations. Although it is not our intention to survey the vast field of numerical methods, one additional formula deserves mention. The **Milne method**, like the improved Euler formula, is a predictor-corrector method. By first using the **predictor**

$$y^*_{n+1} = y_{n-3} + \frac{4h}{3}(2y'_n - y'_{n-1} + 2y'_{n-2}), \tag{1}$$

where $n \geq 3$ and

$$y'_n = f(x_n, y_n)$$
$$y'_{n-1} = f(x_{n-1}, y_{n-1})$$
$$y'_{n-2} = f(x_{n-2}, y_{n-2}),$$

we are then able to substitute the value of y^*_{n+1} into the **corrector**

$$y_{n+1} = y_{n-1} + \frac{h}{3}(y'_{n+1} + 4y'_n + y'_{n-1}), \tag{2}$$

where

$$y'_{n+1} = f(x_{n+1}, y^*_{n+1}).$$

Notice that formula (1) requires that we must know y_0, y_1, y_2, and y_3 in order to obtain y_4. Usually these last three values are computed by an accurate method such as the Runge-Kutta formula.

Since the Milne predictor-corrector formulas demand that we know more than just y_n to compute y_{n+1}, the procedure is called a **multistep**, or **continuing**, method. The Euler formulas, the three-term Taylor, and the Runge-Kutta formulas are examples of **single-step**, or **starting**, methods.

Higher-Order Equations

The numerical procedures that we have discussed in this chapter were applied only to the first-order equation $dy/dx = f(x, y)$ subject to an initial condition $y(x_0) = y_0$. To approximate a solution to, say, a second-order equation

$$\frac{d^2y}{dx^2} = f(x, y, y'), \tag{3}$$

we first reduce the equation to a system of first-order equations. If we let $y' = u$, equation (3) becomes

$$\begin{aligned} y' &= u \\ u' &= f(x, y, u). \end{aligned} \tag{4}$$

We now apply a particular method to *each* equation in the resulting system. For example, the basic Euler formulas would be

$$\begin{aligned} y_{n+1} &= y_n + hu_n \\ u_{n+1} &= u_n + hf(x_n, y_n, u_n). \end{aligned} \tag{5}$$

EXAMPLE 1

Use the Euler method to obtain the approximate value of $y(0.2)$, where $y(x)$ is the solution of the initial-value problem

$$y'' + xy' + y = 0$$

$$y(0) = 1, \qquad y'(0) = 2.$$

SOLUTION In terms of the substitution $y' = u$, the equation is equivalent to the system

$$\begin{aligned} y' &= u \\ u' &= -xu - y. \end{aligned}$$

Thus from (5) we obtain

$$\begin{aligned} y_{n+1} &= y_n + hu_n \\ u_{n+1} &= u_n + h[-x_n u_n - y_n]. \end{aligned}$$

Using the step size $h = 0.1$, we find

$$\begin{aligned} y_1 &= y_0 + (0.1)u_0 \\ &= 1 + (0.1)2 = 1.2, \end{aligned}$$

$$\begin{aligned} u_1 &= u_0 + (0.1)[-x_0u_0 - y_0] \\ &= 2 + (0.1)[-(0)(2) - 1] = 1.9, \\ y_2 &= y_1 + (0.1)u_1 \\ &= 1.2 + (0.1)(1.9) = 1.39, \\ u_2 &= u_1 + (0.1)[-x_1u_1 - y_1] \\ &= 1.9 + (0.1)[-(0.1)(1.9) - 1.2] = 1.761. \end{aligned}$$

In other words $y(0.2) \approx 1.39$ and $y'(0.2) \approx 1.761$.

Errors

In a serious and detailed study of numerical solutions of differential equations, we would have to pay close attention to the various sources of errors. For some kinds of computation, accumulation of errors might reduce the accuracy of an approximation to the point of being useless.

By using only three terms of a Taylor series to approximate the value of a function, the method itself naturally will be a source of error. As we have seen, the Euler formula is essentially two terms of a Taylor series expansion; by advancing along a tangent line, we do not necessarily get to a point on or even near the solution curve. The errors inherent to these methods are known as **truncation errors.***

Any calculator or computer can compute only to, at most, a finite number of decimal places. Suppose for the sake of illustration that we have a calculator that can display six digits while carrying eight digits internally. If we multiply two numbers, each having six decimals, then the product actually contains twelve decimal places. But the number that we see is rounded to six decimal places, while the machine has stored a number rounded to eight decimal places. In one calculation such as this, the **round-off error** may not be significant, but a problem could arise if many calculations are performed with rounded numbers. The effects of round-off can be minimized on a computer if it has a double-precision capabilities.

When iterating a formula such as

$$y_{n+1} = y_n + hf(x_n, y_n),$$

we obtain a sequence of values

$$y_1, y_2, y_3, \ldots.$$

The value of y_1 is, of course, in error, and unfortunately y_2 depends on y_1. Thus y_2 must also be in error. In turn y_3 inherits an error from y_2. The error resulting from the inheritance of errors in preceding calculations is known as

* This kind of error is also known as **formula error** or **discretization error**.

propagation error. To make matters worse, formulas can be **unstable**. This means that errors occurring in the early stage of calculation are not only propagated but are also *compounded* at each step of the iteration. The error may grow so fast that the subsequent approximations are completely overwhelmed. Under certain circumstances the corrector formula in Milne's method is unstable.

Exercises 9.5

Answers to odd-numbered problems begin on page A–32.

1. Use Milne's predictor-corrector method to approximate the value of $y(0.4)$, where $y(x)$ is a solution of

$$y' = x + y - 1, \qquad y(0) = 1.$$

Obtain the values of y_1, y_2, and y_3 from the Runge-Kutta formula using $h = 0.1$.

2. (a) Generalize the improved Euler formula to the system (4).

(b) Use the result of part (a) and $h = 0.1$ to approximate $y(0.2)$, where $y(x)$ is the solution of the initial-value problem in Example 1.

In Problems 3 and 4 use the Euler method (5) and $h = 0.1$ to obtain an approximation to the indicated value.

3. $y'' - x^2 y = 0$
$y(2) = 2, \quad y'(2) = 4;$
$y(2.2)$

4. $y'' - 2y' + xy = 0$
$y(0) = -1, \quad y'(0) = 5;$
$y(0.3)$

5. (a) Generalize the Euler method to systems of the form

$$x' = f(x, y, t)$$
$$y' = g(x, y, t)$$

(b) Use the results of part (a) and $h = 0.1$ to approximate the values of $x(0.2)$, $y(0.2)$, where $x(t)$ and $y(t)$ are solutions of

$$x' = x + y$$
$$y' = x - y,$$
$$x(0) = 1, \quad y(0) = 2.$$

6. Consider the recurrence formula

$$y_{n+1} = k(1 - y_n),$$

where $n = 0, 1, 2, \ldots$, and k is a constant. Suppose that the initial value y_0 has an absolute error $\varepsilon = y_0 - y$, where y is the true value. Show that the formula is unstable for increasing n when $|k| > 1$ and stable when $|k| < 1$.

CHAPTER 9 SUMMARY

A solution of a differential equation may exist and yet we may not be able to determine it in terms of the familiar elementary functions. A way of convincing oneself that a first-order equation $y' = f(x, y)$ possesses a solution passing through a specific point (x_0, y_0) is to sketch the **direction field** associated with the equation. The equation $f(x, y) = c$ determines the **isoclines**, or curves of constant inclination. This means that every solution curve passing through a particular isocline does so with the same slope. The direction field is the totality of short line segments throughout two-dimensional space that have midpoints on the isoclines and that possess slope equal to the value of the parameter c. A carefully plotted sequence of these **lineal elements** can suggest the shape of a solution curve passing through the given point (x_0, y_0).

At best, a direction field can give only the crudest form of an approximation to a numerical value of the solution $y(x)$ of the initial-value problem when x is close to x_0.

To obtain the approximate values of $y(x)$, wc used **Euler's formula**:

$$y_{n+1} + y_n + hf(x_n, y_n);$$

the **improved Euler formula**:

$$y_{n+1} = y_n + h\frac{f(x_n, y_n) + f(x_{n+1}, y^*_{n+1})}{2},$$

where

$$y^*_{n+1} = y_n + hf(x_n, y_n);$$

the **three-term Taylor formula**:

$$y_{n+1} = y_n + y'_n h + y''_n \frac{h^2}{2};$$

the **Runge-Kutta formula**:

$$y_{n+1} = y_n + \frac{1}{6}(k_1 + 2k_2 + 2k_3 + k_4),$$

where

$$k_1 = hf(x_n, y_n)$$

$$k_2 = hf\left(x_n + \frac{1}{2}h, y_n + \frac{1}{2}k_1\right)$$

$$k_3 = hf\left(x_n + \frac{1}{2}h, y_n + \frac{1}{2}k_2\right)$$

$$k_4 = hf(x_n + h, y_n + k_3);$$

and the **Milne formulas**:

$$y^*_{n+1} = y_{n-3} + \frac{4h}{3}(2y'_n - y'_{n-1} + 2y'_{n-2})$$

$$y_{n+1} = y_{n-1} + \frac{h}{3}(y'_{n+1} + 4y'_n + y'_{n-1}),$$

where $$y'_{n+1} = f(x_{n+1}, y^*_{n+1}).$$

In each of the preceding formulas, the number h is the length of a uniform step. In other words

$$x_1 = x_0 + h,\ x_2 = x_1 + h = x_0 + 2h, \ldots, x_n = x_0 + nh.$$

Euler's method consists of approximately the solution curve by a sequence of straight lines. The improved Euler and Runge-Kutta methods use the idea of averaging slopes.

The first four methods are known as **single-step**, or **starting**, **methods**, while Milne's method is an example of a **multi-step**, or **continuing**, **method**. To use the latter method, we must first compute y_1, y_2, and y_3 by some starting method such as the improved Euler or the Runge-Kutta formulas. Euler's method is not generally used if we desire accuracy to several decimal places. The improved Euler and Milne formulas are also particular examples of a class of approximating formulas known as **predictor-corrector** formulas. For example, when using the improved Euler formula, the value y^*_{n+1} obtained from the basic Euler formula is the predicated value, which is then corrected through the new formula.

To obtain numerical approximations to higher-order differential equations, we can reduce the differential equation to a system of first-order equations. We then apply a particular numerical technique to each equation of the system.

CHAPTER 9 REVIEW EXERCISES

Answers to odd-numbered problems begin on page A–32.

In Problems 1–2 sketch the direction field for the given differential equation. Indicate several possible solution curves.

1. $ydx - xdy = 0$ **2.** $y' = 2x - y$

In Problems 3–6 construct a table comparing the indicated values of $y(x)$ using the Euler, improved Euler, three-term Taylor, and Runge-Kutta methods. Compute to four rounded decimal places. Use $h = 0.1$ and $h = 0.05$.

3. $y' = 2\ln xy, \quad y(1) = 2;$

$y(1.1), y(1.2), y(1.3), y(1.4), y(1.5)$

4. $y' = \sin x^2 + \cos y^2, \quad y(0) = 0;$

$y(0.1), y(0.2), y(0.3), y(0.4), y(0.5)$

5. $y' = \sqrt{x + y}, \quad y(0.5) = 0.5$

$y(0.6), y(0.7), y(0.8), y(0.9), y(1.0)$

6. $y' = xy + y^2, \quad y(1) = 0$

$y'(1.1), y(1.2), y(1.3), y(1.4), y(1.5)$

7. Use the Euler method to obtain the approximate value of $y(0.2)$, where $y(x)$ is the solution of the initial-value problem

$$y'' - (2x + 1)y = 0$$

$$y(0) = 3, \qquad y'(0) = 1.$$

First use one step with $h = 0.2$ and then repeat the calculations using $h = 0.1$.

8. Use Milne's predictor-corrector method to approximate the value of $y(0.4)$, where $y(x)$ is the solution of

$$y' = 4x - 2y$$

$$y(0) = 2.$$

Use the Runge-Kutta formula and $h = 0.1$ to obtain the values of y_1, y_2, and y_3.

CHAPTER 10

Partial Differential Equations

IMPORTANT TERMS

orthogonal functions
orthogonal set of functions
square norm
norm
orthonormal set of functions
normalized
orthogonality with respect to a weight function
generalized Fourier series
Fourier series
Fourier coefficients
periodic extension
even function
odd function
cosine series
sine series
half-range expansions
linear partial differential equation
homogeneous equation
nonhomogeneous equation
method of separation of variables
superposition principle
wave equation
heat equation
Laplace's equation
boundary conditions
initial conditions
boundary-value problem
eigenvalues
eigenfunctions
insulated boundary

Throughout the preceding chapters, our attention has been focused on finding general solutions of ordinary differential equations. Also, we were primarily concerned with the theory and application of linear equations of order $n \leq 2$. In this chapter we shall limit our consideration to a special kind of linear partial differential. However, we shall make no attempt to find or even pursue the concept of a general solution of such an equation. The emphasis will be on a specific procedure used in solving certain problems in the mathematical physics of temperature distributions and vibrations. The mathematical models for these problems are relatively simple second-order partial differential equations. By the so-called method of separation of variables particular solutions can be found for these differential equations by solving associated ordinary differential equations.

10.1 Orthogonal Functions

Recall from calculus that in 3-space the **dot**, or **inner**, **product** of two vectors $\mathbf{u} = a_1\mathbf{i} + b_1\mathbf{j} + c_1\mathbf{k}$ and $\mathbf{v} = a_2\mathbf{i} + b_2\mathbf{j} + c_2\mathbf{k}$ is given by $\mathbf{u} \cdot \mathbf{v} = a_1a_2 + b_1b_2 + c_1c_2$. The concept of an inner product has been generalized to spaces where the vectors are functions. An **integral inner product** of two functions f_1 and f_2 defined on an interval $[a, b]$ is given by

$$(f_1, f_2) = \int_a^b f_1(x)f_2(x)\,dx,$$

provided the integral exists.

DEFINITION 10.1 Two real-valued functions f_1 and f_2 are said to be **orthogonal** on an interval $[a, b]$ if*

$$(f_1, f_2) = \int_a^b f_1(x)f_2(x)\,dx = 0. \tag{1}$$

EXAMPLE 1

$f_1(x) = x^2$ and $f_2(x) = x^3$ are orthogonal on $[-1, 1]$ since

$$\begin{aligned}\int_{-1}^{1} f_1(x)f_2(x)\,dx &= \int_{-1}^{1} x^2 \cdot x^3\,dx \\ &= \frac{1}{6}x^6\Big|_{-1}^{1} \\ &= \frac{1}{6}[1 - (-1)^6] \\ &= 0.\end{aligned}$$

Unlike vector analysis where the word "orthogonal" is a synonym for "perpendicular," in this present context the term "orthogonal" and condition (1) have no geometric significance.

---

* The interval could also be $(-\infty, \infty)$, $[0, \infty)$, and so on.

DEFINITION 10.2 A set of real-valued functions

$$\phi_0(x), \phi_1(x), \phi_2(x), \ldots,$$

is said to be **orthogonal** on an interval $[a, b]$ if

$$(\phi_m, \phi_n) = \int_a^b \phi_m(x)\phi_n(x)\,dx = 0, \qquad m \neq n. \tag{2}$$

The number

$$\|\phi_n(x)\|^2 = \int_a^b \phi_n^2(x)\,dx \tag{3}$$

is called the **square norm** and

$$\|\phi_n(x)\| = \sqrt{\int_a^b \phi_n^2(x)\,dx}$$

is the **norm** of the function $\phi_n(x)$. If $\{\phi_n(x)\}$ is an orthogonal set on $[a, b]$ with the property that $\|\phi_n(x)\| = 1$ for $n = 0, 1, 2, \ldots$, then $\{\phi_n(x)\}$ is said to be an **orthonormal set** on the interval.

EXAMPLE 2

Show that the set $1, \cos x, \cos 2x, \ldots,$ is orthogonal on the interval $[-\pi, \pi]$.

SOLUTION If we make the identification $\phi_0(x) = 1$ and $\phi_n(x) = \cos nx$, we must then show $\int_{-\pi}^{\pi} \phi_0(x)\phi_n(x)\,dx = 0$, $n \neq 0$ and $\int_{-\pi}^{\pi} \phi_m(x)\phi_n(x)\,dx = 0$, $m \neq n$. We have in the first case

$$\begin{aligned}(\phi_0, \phi_n) = \int_{-\pi}^{\pi} \phi_0(x)\phi_n(x)\,dx &= \int_{-\pi}^{\pi} \cos nx\,dx\\ &= \frac{1}{n}\sin nx\Big|_{-\pi}^{\pi} = \frac{1}{n}\Big[\sin n\pi - \sin(-n\pi)\Big]\\ &= 0, \qquad n \neq 0,\end{aligned}$$

and in the second,

$$\begin{aligned}(\phi_m, \phi_n) = \int_{-\pi}^{\pi} \phi_m(x)\phi_n(x)\,dx &= \int_{-\pi}^{\pi} \cos mx \cos nx\,dx\\ &= \frac{1}{2}\int_{-\pi}^{\pi}\Big[\cos(m+n)x + \cos(m-n)x\Big]\,dx\\ &= \frac{1}{2}\left[\frac{\sin(m+n)x}{m+n} + \frac{\sin(m-n)x}{m-n}\right]_{-\pi}^{\pi}\\ &= 0, \qquad m \neq n.\end{aligned}$$

EXAMPLE 3

Find the norms of each function in the orthogonal set given in Example 2.

SOLUTION For $\phi_0(x) = 1$, we have from (2)

$$\|\phi_0(x)\|^2 = \int_{-\pi}^{\pi} dx = 2\pi$$

so that $\|\phi_0(x)\| = \sqrt{2\pi}$. For $\phi_n(x) = \cos nx$, $n > 0$, it follows that

$$\|\phi_n(x)\|^2 = \int_{-\pi}^{\pi} \cos^2 nx\, dx = \frac{1}{2}\int_{-\pi}^{\pi} [1 + \cos 2nx]\, dx = \pi.$$

Thus for $n > 0$, $\|\phi_n(x)\| = \sqrt{\pi}$.

Any orthogonal set of nonzero functions $\{\phi_n(x)\}$, $n = 0, 1, 2, \ldots$, can be **normalized**, that is, made into an orthonormal set, by dividing each function by its norm.

EXAMPLE 4

It follows from Examples 2 and 3 that the set

$$\frac{1}{\sqrt{2\pi}}, \frac{\cos x}{\sqrt{\pi}}, \frac{\cos 2x}{\sqrt{\pi}}, \ldots,$$

is orthonormal on $[-\pi, \pi]$.

DEFINITION 10.3 A set of functions $\{\phi_n(x)\}$, $n = 0, 1, 2, \ldots$, is said to be **orthogonal with respect to a weight function** $w(x)$ on an interval $[a, b]$ if

$$\int_a^b w(x)\phi_m(x)\phi_n(x)\, dx = 0, \qquad m \neq n.$$

EXAMPLE 5

The set $1, \cos x, \cos 2x, \ldots,$ is orthogonal with respect to the constant weight function $w(x) = 1$ on the interval $[-\pi, \pi]$.

Generalized Fourier Series

Suppose $\{\phi_n(x)\}$ is an infinite orthogonal set of functions on an interval $[a, b]$. We ask: If $y = f(x)$ is a function defined on the open interval (a, b), is it possible to determine a set of coefficients c_n, $n = 0, 1, 2, \ldots$, for which

$$f(x) = c_0\phi_0(x) + c_1\phi_1(x) + \cdots + c_n\phi_n(x) + \cdots ? \tag{4}$$

Multiplying (4) by $\phi_m(x)$ and integrating over the interval give

$$\int_a^b f(x)\phi_m(x)\,dx = c_0 \int_a^b \phi_0(x)\phi_m(x)\,dx + c_1 \int_a^b \phi_1(x)\phi_m(x)\,dx + \cdots + c_n \int_a^b \phi_n(x)\phi_m(x)\,dx + \cdots.$$

By orthogonality, each term on the right-hand side of the last equation is zero except when $m = n$. In this case we have

$$\int_a^b f(x)\phi_n(x)\,dx = c_n \int_a^b \phi_n^2(x)\,dx.$$

It follows that the required coefficients are

$$c_n = \frac{\int_a^b f(x)\phi_n(x)\,dx}{\int_a^b \phi_n^2(x)\,dx}, \qquad n = 0, 1, 2, \ldots.$$

In other words, if

$$f(x) = \sum_{n=0}^{\infty} c_n\phi_n(x), \tag{5}$$

then

$$c_n = \frac{\int_a^b f(x)\phi_n(x)\,dx}{\|\phi_n(x)\|^2}. \tag{6}$$

If $\{\phi_n(x)\}$ is orthogonal with respect to a weight function $w(x)$ on $[a, b]$ then multiplication of (4) by $w(x)\phi_m(x)$ and integration yield

$$c_n = \frac{\int_a^b f(x)w(x)\phi_n(x)\,dx}{\|\phi_n(x)\|^2}, \tag{7}$$

where

$$\|\phi_n(x)\|^2 = \int_a^b w(x)\phi_n^2(x)\,dx. \tag{8}$$

The series (5) with coefficients given by either (6) or (7) is called a **generalized Fourier series**.

We note that the procedure outlined for determining the c_n was *formal*; that is, basic questions on whether a series expansion such as (4) is actually possible were ignored.

Throughout the remainder of this chapter we shall assume that each orthogonal set is **complete**. This means that the only function orthogonal to each member of the set is the zero function.

Exercises 10.1

Answers to odd-numbered problems begin on page A-33.

In Problems 1–6 show that the given functions are orthogonal on the indicated interval.

1. $f_1(x) = x, \quad f_2(x) = x^2, \quad [-2, 2]$

2. $f_1(x) = x^3$, $f_2(x) = x^2 + 1$, $[-1, 1]$

3. $f_1(x) = e^x$, $f_2(x) = xe^{-x} - e^{-x}$, $[0, 2]$

4. $f_1(x) = \cos x$, $f_2(x) = \sin^2 x$, $[0, \pi]$

5. $f_1(x) = x$, $f_2(x) = \cos 2x$, $[-\pi/2, \pi/2]$

6. $f_1(x) = e^x$, $f_2(x) = \sin x$, $[\pi/4, 5\pi/4]$

In Problems 7–12 show that the given set of functions is orthogonal on the indicated interval. Find the norm of each function in the set.

7. $\sin x, \sin 3x, \sin 5x, \ldots,$ $[0, \pi/2]$

8. $\cos x, \cos 3x, \cos 5x, \ldots,$ $[0, \pi/2]$

9. $\{\sin nx\}$, $n = 1, 2, 3, \ldots,$ $[0, \pi]$

10. $\left\{\sin \frac{n\pi}{p} x\right\}$, $n = 1, 2, 3, \ldots,$ $[0, p]$

11. $\left\{1, \cos \frac{n\pi}{p} x\right\}$, $n = 1, 2, 3, \ldots,$ $[0, p]$

12. $\left\{1, \cos \frac{n\pi}{p} x, \sin \frac{m\pi}{p} x\right\}$, $n = 1, 2, 3, \ldots,$
$m = 1, 2, 3, \ldots,$ $[-p, p]$

In Problems 13 and 14 verify by direct integration that the functions are orthogonal with respect to the indicated weight function on the given interval.

13. $H_0(x) = 1$, $H_1(x) = 2x$, $H_2(x) = 4x^2 - 2$;
$w(x) = e^{-x^2}$, $(-\infty, \infty)$

14. $L_0(x) = 1$, $L_1(x) = -x + 1$, $L_2(x) = \frac{1}{2}x^2 - 2x + 1$;
$w(x) = e^{-x}$, $[0, \infty)$

Miscellaneous Problems

15. Let $\{\phi_n(x)\}$ be an orthogonal set of functions on $[a, b]$ such that $\phi_0(x) = 1$. Show that $\int_a^b \phi_n(x)\,dx = 0$ for $n = 1, 2, \ldots$.

16. Let $\{\phi_n(x)\}$ be an orthogonal set of functions on $[a, b]$ such that $\phi_0(x) = 1$ and $\phi_1(x) = x$. Show that $\int_a^b (\alpha x + \beta)\phi_n(x)\,dx = 0$ for $n = 2, 3, \ldots,$ and any constants α and β.

17. Let $\{\phi_n(x)\}$ be an orthogonal set of functions on $[a, b]$. Show that $\|\phi_m(x) + \phi_n(x)\|^2 = \|\phi_m(x)\|^2 + \|\phi_n(x)\|^2$, $m \neq n$.

18. From Problem 1 we know that $f_1(x) = x$ and $f_2(x) = x^2$ are orthogonal on $[-2, 2]$. Find constants c_1 and c_2 such that $f_3(x) = x + c_1 x^2 + c_2 x^3$ is orthogonal to both f_1 and f_2 on the same interval.

19. Show that the set $\{\sin \lambda_n x\}$, $n = 1, 2, 3, \ldots,$ is orthogonal on $[0, 1]$ when the λ_n are the positive roots of $\tan \lambda = -\lambda$. Find the value of $\|\sin \lambda_n x\|^2$.

10.2 Fourier Series

The set of functions

$$1, \cos\frac{\pi}{p}x, \cos\frac{2\pi}{p}x, \ldots, \sin\frac{\pi}{p}x, \sin\frac{2\pi}{p}x, \sin\frac{3\pi}{p}x, \ldots, \tag{1}$$

is orthogonal on the interval $[-p, p]$ (see Problem 12, Section 10.1). Suppose f is a function defined on the open interval $(-p, p)$ that can be expanded in the trigonometric series

$$f(x) = \frac{a_0}{2} + \sum_{n=1}^{\infty}\left(a_n \cos\frac{n\pi}{p}x + b_n \sin\frac{n\pi}{p}x\right). \tag{2}$$

Then the coefficients $a_0, a_1, a_2, \ldots, b_1, b_2, \ldots,$ can be determined as follows.*

Integrating both sides of (2) from $-p$ to p gives

$$\int_{-p}^{p} f(x)\,dx = \frac{a_0}{2}\int_{-p}^{p} dx + \sum_{n=1}^{\infty}\left(a_n \int_{-p}^{p} \cos\frac{n\pi}{p}x\,dx + b_n \int_{-p}^{p} \sin\frac{n\pi}{p}x\,dx\right). \tag{3}$$

Since each function $\cos(n\pi x/p)$, $\sin(n\pi x/p)$, $n > 1$, is orthogonal to 1 on the interval, the right side of (3) reduces to a single term and, consequently,

$$\int_{-p}^{p} f(x)\,dx = \frac{a_0}{2}\int_{-p}^{p} dx = \frac{a_0}{2}x\Big|_{-p}^{p} = pa_0.$$

Solving for a_0 yields

$$a_0 = \frac{1}{p}\int_{-p}^{p} f(x)\,dx. \tag{4}$$

Now multiply (2) by $\cos(m\pi x/p)$ and integrate:

$$\int_{-p}^{p} f(x)\cos\frac{m\pi}{p}x\,dx = \frac{a_0}{2}\int_{-p}^{p}\cos\frac{m\pi}{p}x\,dx + \sum_{n=1}^{\infty}\left(a_n\int_{-p}^{p}\cos\frac{m\pi}{p}x\cos\frac{n\pi}{p}x\,dx + b_n\int_{-p}^{p}\cos\frac{m\pi}{p}x\sin\frac{n\pi}{p}x\,dx\right). \tag{5}$$

---

* We have chosen to write the coefficient of 1 in the series (2) as $a_0/2$ rather than a_0. This is for convenience only; the formula for a_n will then reduce to a_0 when $n = 0$.

Now
$$\int_{-p}^{p} \cos\frac{m\pi}{p}x\,dx = 0, \qquad m > 0,$$
$$\int_{-p}^{p} \cos\frac{m\pi}{p}x\cos\frac{n\pi}{p}x\,dx \begin{cases} =0, & m \neq n \\ =p, & m = n, \end{cases}$$

and
$$\int_{-p}^{p} \cos\frac{m\pi}{p}x\sin\frac{n\pi}{p}x\,dx = 0$$

so (5) reduces to
$$\int_{-p}^{p} f(x)\cos\frac{n\pi}{p}x\,dx = a_n p.$$

Therefore
$$a_n = \frac{1}{p}\int_{-p}^{p} f(x)\cos\frac{n\pi}{p}x\,dx. \tag{6}$$

Finally, if we multiply (2) by $\sin(m\pi x/p)$, integrate, and make use of the results

$$\int_{-p}^{p} \sin\frac{m\pi}{p}x\,dx = 0, \qquad m > 0$$
$$\int_{-p}^{p} \sin\frac{m\pi}{p}x\cos\frac{n\pi}{p}x\,dx = 0$$
$$\int_{-p}^{p} \sin\frac{m\pi}{p}x\sin\frac{n\pi}{p}x\,dx \begin{cases} =0, & m \neq n \\ =p, & m = n, \end{cases}$$

we find that
$$b_n = \frac{1}{p}\int_{-p}^{p} f(x)\sin\frac{n\pi}{p}x\,dx. \tag{7}$$

The trigonometric series (2) with coefficients a_0, a_n, and b_n defined by (4), (6), and (7), respectively, is said to be the **Fourier series** of the function f.* The coefficients obtained from (4), (6), and (7) are referred to as **Fourier coefficients** of f.

---

* **Jean-Baptiste Joseph Fourier** (1766–1830) A French mathematical physicist, Fourier used such trigonometric series in his investigations into the theory of heat, and they appear throughout his 1822 treatise *Théorie analytique de la chaleur*. However, Fourier did not "invent" Fourier series. The development of the theory of expanding functions in trigonometric series was due principally to Daniel Bernoulli and Leonhard Euler. The integral formulas that define the coefficients a_0, a_n, and b_n were discovered by Euler in 1777. Today, Fourier series, the Fourier integral, and the Fourier transform comprise a branch of mathematical analysis that is invaluable in the study of wave phenomena.

A friend and confident of Napoleon, Fourier served in the emperor's retinue during the latter's 1798 campaign to "civilize" Egypt. Fourier is also remembered for his patronage of the young Jean François Champollion, who was the first to decipher Egyptian hieroglyphics through his work on the Rosetta stone.

As in the discussion of generalized Fourier series in the preceding section, the underlying assumption that f can be represented by such series (2) and the subsequent determination of the coefficients corresponding to this assumption was strictly formal. We assumed that f was integrable on the interval and that (2), as well as the series obtained by multiplying (2) by $\cos(m\pi x/p)$, converged in such a manner as to permit term-by-term integration. Until (2) is shown to be convergent for a given function f, the equality sign is not to be taken in a strict or literal sense.* We summarize the results.

The **Fourier series** of a function f defined on the interval $(-p, p)$ is given by

$$f(x) = \frac{a_0}{2} + \sum_{n=1}^{\infty} \left(a_n \cos \frac{n\pi}{p} x + b_n \sin \frac{n\pi}{p} x \right), \tag{8}$$

where
$$a_0 = \frac{1}{p} \int_{-p}^{p} f(x)\, dx \tag{9}$$

$$a_n = \frac{1}{p} \int_{-p}^{p} f(x) \cos \frac{n\pi}{p} x\, dx \tag{10}$$

$$b_n = \frac{1}{p} \int_{-p}^{p} f(x) \sin \frac{n\pi}{p} x\, dx. \tag{11}$$

EXAMPLE 1

Expand
$$f(x) = \begin{cases} 0, & -\pi < x < 0 \\ \pi - x, & 0 < x < \pi \end{cases} \tag{12}$$

in a Fourier series.

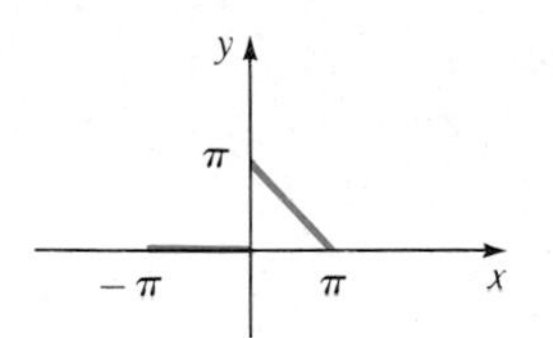

Figure 10.1

SOLUTION The graph of f is given in Figure 10.1. With $p = \pi$, we have from (9) and (10) that

$$a_0 = \frac{1}{\pi} \int_{-\pi}^{\pi} f(x)\, dx = \frac{1}{\pi} \left[\int_{-\pi}^{0} 0\, dx + \int_{0}^{\pi} (\pi - x)\, dx \right]$$
$$= \frac{1}{\pi} \left[\pi x - \frac{x^2}{2} \right]_0^{\pi} = \frac{\pi}{2},$$

* Some texts use the symbol $\sim$ in place of $=$. In view of the fact that most functions in applications are of a type sufficient to guarantee convergence of the series we shall use the equality symbol.

$$a_n = \frac{1}{\pi}\int_{-\pi}^{\pi} f(x)\cos nx\,dx = \frac{1}{\pi}\left[\int_{-\pi}^{0} 0\,dx + \int_{0}^{\pi}(\pi - x)\cos nx\,dx\right]$$
$$= \frac{1}{\pi}\left[(\pi - x)\frac{\sin nx}{n}\Big|_0^{\pi} + \frac{1}{n}\int_0^{\pi}\sin nx\,dx\right]$$
$$= -\frac{1}{n\pi}\frac{\cos nx}{n}\Big|_0^{\pi}$$
$$= \frac{-\cos n\pi + 1}{n^2\pi} = \frac{1 - (-1)^n}{n^2\pi}.$$

In like manner we find from (11) that

$$b_n = \frac{1}{\pi}\int_0^{\pi}(\pi - x)\sin nx\,dx = \frac{1}{n}.$$

Therefore
$$f(x) = \frac{\pi}{4} + \sum_{n=1}^{\infty}\left\{\frac{1 - (-1)^n}{n^2\pi}\cos nx + \frac{1}{n}\sin nx\right\}. \tag{13}$$

Note that a_n defined by (10) reduces to a_0 given by (9) when we set $n = 0$. But as the last example shows, this may not be the case *after* the integral for a_n is evaluated.

Convergence of a Fourier Series

The following theorem gives sufficient conditions for convergence of a Fourier series of $f(x)$.

THEOREM 10.1 Let f and f' be piecewise continuous on the interval $(-p, p)$; that is, let f and f' be continuous except at a finite number of points in the interval and have only finite discontinuities at these points. Then the Fourier series of f on the interval converges to $f(x)$ at a point of continuity. At a point of discontinuity, the Fourier series will converge to the average

$$\frac{f(x+) + f(x-)}{2},$$

where $f(x+)$ and $f(x-)$ denote the limit of f at x from the right and from the left, respectively.*

---

* This means that for x a point in the interval and $h > 0$,

$$f(x+) = \lim_{h\to 0} f(x + h), \qquad f(x-) = \lim_{h\to 0} f(x - h).$$

EXAMPLE 2

The function (12) given in the preceding example satisfies the conditions of Theorem 10.1. Thus for every x in the interval $(-\pi, \pi)$, except at $x = 0$, the series (13) will converge to $f(x)$. At $x = 0$ the function is discontinuous and so the series (13) will converge to

$$\frac{f(0+) + f(0-)}{2} = \frac{\pi + 0}{2}$$
$$= \frac{\pi}{2}.$$

Periodic Extension

Observe that the functions in the basic set (1) have a common period $2p$. Hence the right side of (2) is periodic. We conclude that a Fourier series not only represents the function on the interval $(-p, p)$ but also gives the **periodic extension** of f outside this interval. We can now apply Theorem 10.1 to the periodic extension of f, or we may assume from the outset that the given function is periodic with period $2p$ (that is, $f(x + 2p) = f(x)$). When f is piecewise continuous and the right- and left-hand derivatives exist at $x = -p$ and $x = p$, respectively, then the series (8) will converge to the average $[f(p-) + f(-p+)]/2$ at these endpoints and to this value extended periodically to $\pm 3p$, $\pm 5p$, $\pm 7p$, and so on.

EXAMPLE 3

The Fourier series (13) converges to the periodic extension of (12) onto the entire x-axis. The solid dots given in Figure 10.2 represent the value

$$\frac{f(0+) + f(0-)}{2} = \frac{\pi}{2}$$

at $0, \pm 2\pi, \pm 4\pi, \ldots$. At $\pm\pi, \pm 3\pi, \pm 5\pi, \ldots$, the series will converge to the value

$$\frac{f(\pi-) + f(-\pi+)}{2} = 0.$$

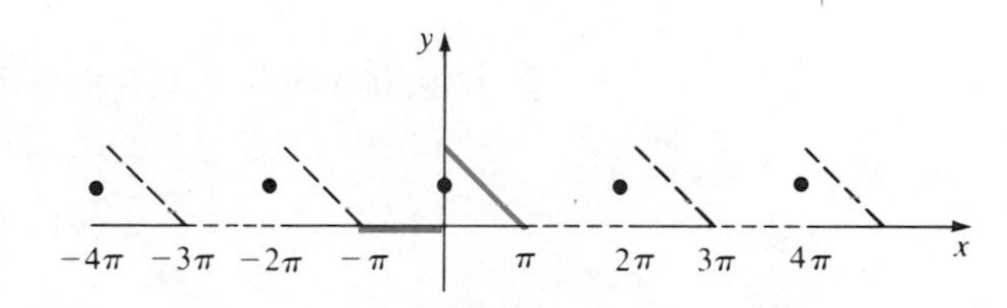

Figure 10.2

Exercises 10.2

Answers to odd-numbered problems begin on page A–33.

In Problems 1–16 find the Fourier series of f on the given interval.

1. $f(x) = \begin{cases} 0, & -\pi < x < 0 \\ 1, & 0 \le x < \pi \end{cases}$

2. $f(x) = \begin{cases} -1, & -\pi < x < 0 \\ 2, & 0 \le x < \pi \end{cases}$

3. $f(x) = \begin{cases} 1, & -1 < x < 0 \\ x, & 0 \le x < 1 \end{cases}$

4. $f(x) = \begin{cases} 0, & -1 < x < 0 \\ x, & 0 \le x < 1 \end{cases}$

5. $f(x) = \begin{cases} 0, & -\pi < x < 0 \\ x^2, & 0 \le x < \pi \end{cases}$

6. $f(x) = \begin{cases} \pi^2 & -\pi < x < 0 \\ \pi^2 - x^2, & 0 \le x < \pi \end{cases}$

7. $f(x) = x + \pi, \quad -\pi < x < \pi$

8. $f(x) = 3 - 2x, \quad -\pi < x < \pi$

9. $f(x) = \begin{cases} 0, & -\pi < x < 0 \\ \sin x, & 0 \le x < \pi \end{cases}$

10. $f(x) = \begin{cases} 0, & -\pi/2 < x < 0 \\ \cos x, \quad 0 & 0 \le x < \pi/2 \end{cases}$

11. $f(x) = \begin{cases} 0, & -2 < x < -1 \\ -2, & -1 \le x < 0 \\ 1, & 0 \le x < 1 \\ 0, & 1 \le x < 2 \end{cases}$

12. $f(x) = \begin{cases} 0, & -2 < x < 0 \\ x, & 0 \le x < 1 \\ 1, & 1 \le x < 2 \end{cases}$

13. $f(x) = \begin{cases} 1, & -5 < x < 0 \\ 1 + x, & 0 \le x < 5 \end{cases}$

14. $f(x) = \begin{cases} 2 + x, & -2 < x < 0 \\ 2, & 0 \le x < 2 \end{cases}$

15. $f(x) = e^x, \quad -\pi < x < \pi$

16. $f(x) = \begin{cases} 0, & -\pi < x < 0 \\ e^x - 1, & 0 \le x < \pi \end{cases}$

17. Use the result of Problem 5 to show

$$\frac{\pi^2}{6} = 1 + \frac{1}{2^2} + \frac{1}{3^2} + \frac{1}{4^2} + \cdots \quad \text{and} \quad \frac{\pi^2}{12} = 1 - \frac{1}{2^2} + \frac{1}{3^2} - \frac{1}{4^2} + \cdots.$$

18. Use Problem 17 to find a series giving the numerical value of $\dfrac{\pi^2}{8}$.

19. Use the result of Problem 7 to show

$$\frac{\pi}{4} = 1 - \frac{1}{3} + \frac{1}{5} - \frac{1}{7} + \cdots.$$

20. Use the result of Problem 9 to show

$$\frac{\pi}{4} = \frac{1}{2} + \frac{1}{1 \cdot 3} - \frac{1}{3 \cdot 5} + \frac{1}{5 \cdot 7} - \frac{1}{7 \cdot 9} + \cdots.$$

10.3 Fourier Cosine and Sine Series

Even and Odd Functions

The reader may recall that a function f is said to be **even** if

$$f(-x) = f(x),$$

whereas, if

$$f(-x) = -f(x),$$

then f is said to be an **odd** function.

EXAMPLE 1

(a) $f(x) = x^2$ is even since

$$f(-x) = (-x)^2 = x^2$$
$$= f(x) \qquad \text{(see Figure 10.3).}$$

(b) $f(x) = x^3$ is odd since

$$f(-x) = (-x)^3 = -x^3$$
$$= -f(x) \qquad \text{(see Figure 10.4).}$$

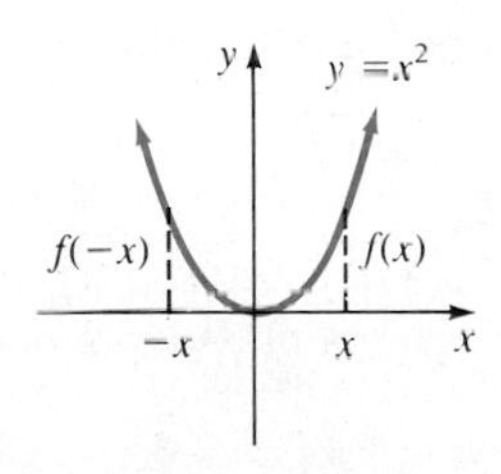

Figure 10.3

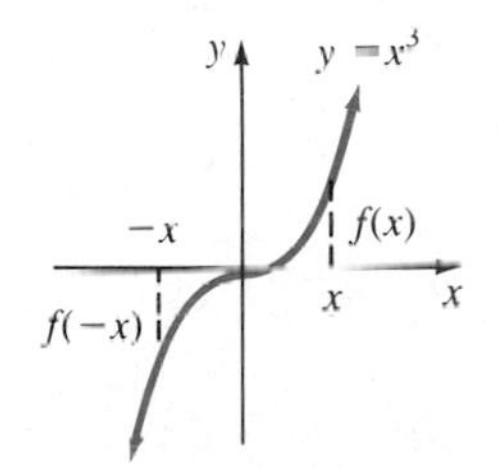

Figure 10.4

As illustrated in Figures 10.3 and 10.4, the graph of an even function is symmetric with respect to the y-axis and the graph of an odd function is symmetric with respect to the origin.

EXAMPLE 2

Since $\cos(-x) = \cos x$ and $\sin(-x) = -\sin x$, the cosine and sine are even and odd functions, respectively.

Properties of Even and Odd Functions

The proofs of the following properties are left as exercises.

(i) The product of two even functions is even.

(ii) The product of two odd functions is even.

(iii) The product of an even function and an odd function is odd.

(iv) If f is even, then $\int_{-a}^{a} f(x)\,dx = 2\int_0^a f(x)\,dx$.

(v) If f is odd, then $\int_{-a}^{a} f(x)\,dx = 0$.

Cosine and Sine Series

If f is an even function on $(-p, p)$, then in view of the foregoing properties, the coefficients (9), (10), and (11) of Section 10.2 become

$$a_0 = \frac{1}{p}\int_{-p}^{p} f(x)\,dx = \frac{2}{p}\int_0^p f(x)\,dx$$

$$a_n = \frac{1}{p}\int_{-p}^{p} \underbrace{f(x)\cos\frac{n\pi}{p}x}_{\text{even}}\,dx = \frac{2}{p}\int_0^p f(x)\cos\frac{n\pi}{p}x\,dx$$

$$b_n = \frac{1}{p}\int_{-p}^{p} \underbrace{f(x)\sin\frac{n\pi}{p}x}_{\text{odd}}\,dx = 0.$$

Similarly, when f is odd on the interval $(-p, p)$,

$$a_n = 0,\ n = 0, 1, 2, \ldots, \qquad b_n = \frac{2}{p}\int_0^p f(x)\sin\frac{n\pi}{p}x\,dx.$$

We summarize the results:

The Fourier series of an even function on the interval $(-p, p)$ is the **cosine series**

$$f(x) = \frac{a_0}{2} + \sum_{n=1}^{\infty} a_n \cos\frac{n\pi}{p}x, \tag{1}$$

where
$$a_0 = \frac{2}{p}\int_0^p f(x)\,dx \tag{2}$$

$$a_n = \frac{2}{p}\int_0^p f(x)\cos\frac{n\pi}{p}x\,dx. \tag{3}$$

The Fourier series of an odd function on the interval $(-p, p)$ is the **sine series**

$$f(x) = \sum_{n=1}^{\infty} b_n \sin\frac{n\pi}{p}x, \tag{4}$$

where
$$b_n = \frac{2}{p}\int_0^p f(x)\sin\frac{n\pi}{p}x\,dx. \tag{5}$$

EXAMPLE 3

Expand $f(x) = x, \quad -2 < x < 2,$ in a Fourier series.

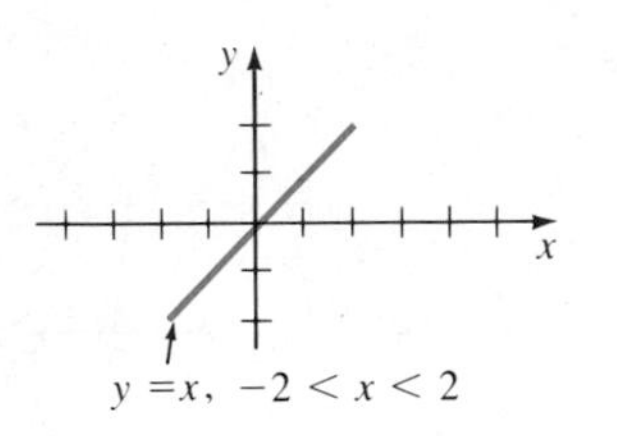

Figure 10.5

SOLUTION We expand f in a sine series since inspection of Figure 10.5 shows that the function is odd on the interval $(-2, 2)$.

With the identifications $2p = 4$, $p = 2$, $2/p = 1$, we can write (5) as

$$b_n = \int_0^2 x \sin \frac{n\pi}{2} x \, dx.$$

Integration by parts then yields

$$b_n = \frac{4(-1)^{n+1}}{n\pi}$$

Therefore

$$f(x) = \frac{4}{\pi} \sum_{n=1}^{\infty} \frac{(-1)^{n+1}}{n} \sin \frac{n\pi}{2} x. \tag{6}$$

EXAMPLE 4

The function in Example 3 satisfies the conditions of Theorem 10.1. Hence the series (6) converges to the function on $(-2, 2)$ and the periodic extension (of period 4) given in Figure 10.6.

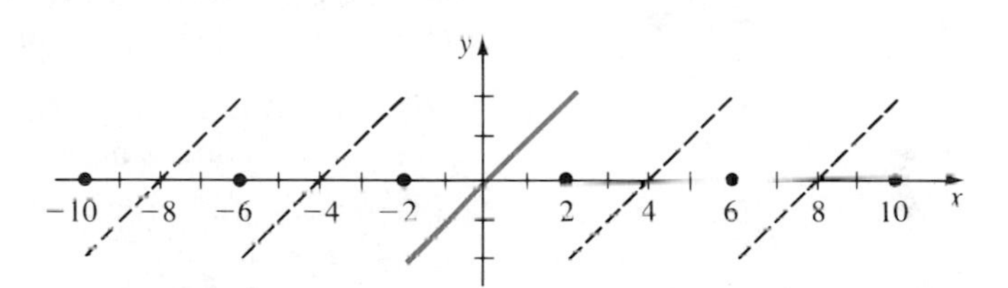

Figure 10.6

EXAMPLE 5

The function

$$f(x) = \begin{cases} -1, & -\pi < x < 0 \\ 1, & 0 \le x < \pi \end{cases}$$

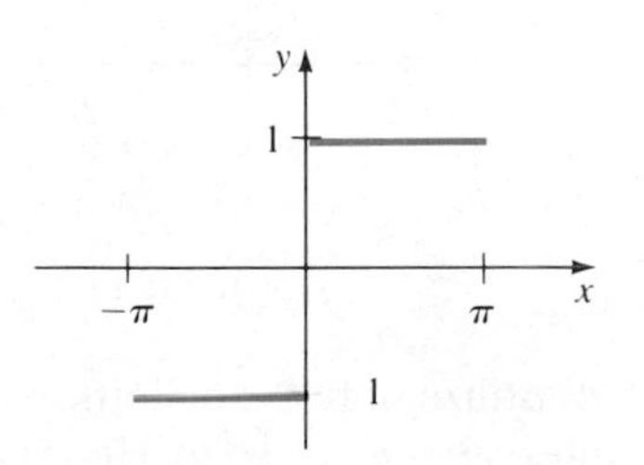

Figure 10.7

shown in Figure 10.7 is odd on the interval $(-\pi, \pi)$. With $p = \pi$ we have from (5),

$$b_n = \frac{2}{\pi} \int_0^{\pi} (1) \sin nx \, dx = \frac{2}{\pi} \frac{1 - (-1)^n}{n}$$

and so

$$f(x) = \frac{2}{\pi} \sum_{n=1}^{\infty} \frac{1 - (-1)^n}{n} \sin nx. \tag{7}$$

Sequence of Partial Sums

It is interesting to see how the sequence of partial sums of a Fourier series approximates a function. In Figure 10.8 the graph of the function f in Example 5 is compared with the graphs of the first three partial sums of (7).

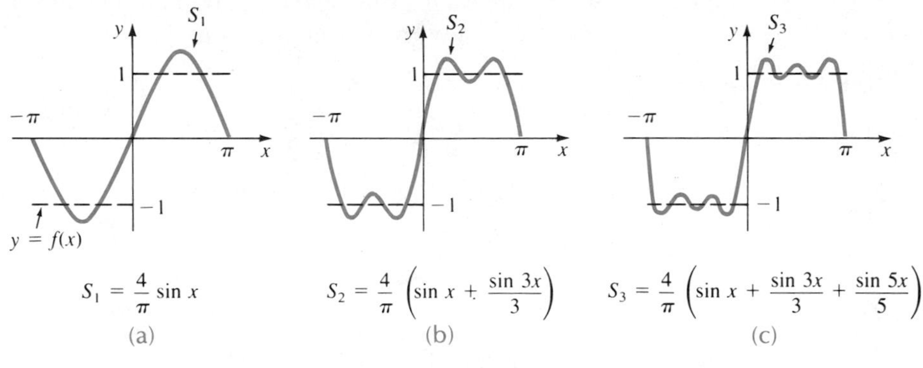

Figure 10.8

Half-range Expansions

Throughout the preceding discussion it was understood that a function f was defined on an interval with the origin as midpoint, that is, $-p < x < p$. However, in many instances we are interested in representing a function that is defined only for $0 < x < L$ by a trigonometric series. This can be done in many different ways by supplying an arbitrary *definition* of the function to the interval $-L < x < 0$. For brevity we consider the three most important cases.

If $y = f(x)$ is defined on the interval $0 < x < L$,

(i) reflect the graph of the function about the y-axis onto $-L < x < 0$. The function is now even on $-L < x < L$ (see Figure 10.9);

(ii) reflect the graph of the function through the origin onto $-L < x < 0$. The function is now odd on $-L < x < L$ (see Figure 10.10);

(iii) define f on $-L < x < 0$ by $f(x) = f(x + L)$ (see Figure 10.11).

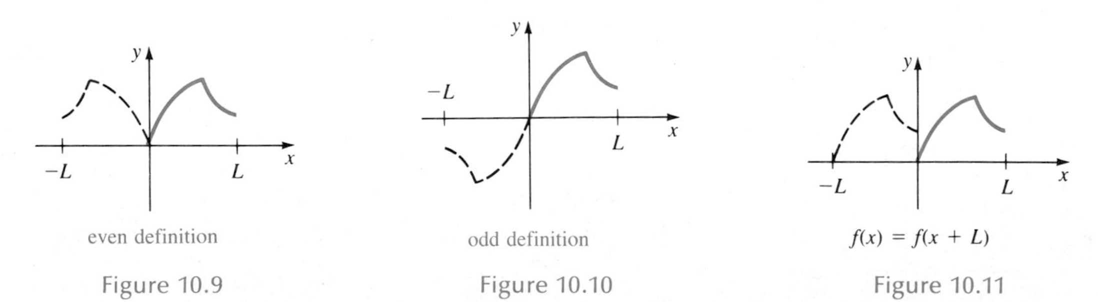

Figure 10.9 Figure 10.10 Figure 10.11

Note that the coefficients of the series (1) and (4) utilize only the definition of the function on $0 < x < p$ (that is, half of the interval $-p < x < p$). Hence in practice there is no actual need to make the reflections described in (i) and (ii); if f is defined on $0 < x < L$, we simply identify the half-period as the length of the interval $p = L$. The coefficient formulas (2), (3), and (5) and the corresponding series will yield either an even or odd periodic extension of period $2L$ of the original function. The cosine and sine series obtained in this manner are known as **half-range expansions**. Lastly, in case (iii), we are defining the functional values on the interval $-L < x < 0$ to be the same as the values on $0 < x < L$. As in the previous two cases, there is no real need to do this. It can be shown that the set of functions in (1) of Section 10.2 is orthogonal on $a \le x \le a + 2p$ for any real number a. By choosing $a = -p$, we obtain the limits of integration in (9), (10), and (11) of that section. But for $a = 0$ the limits of integration are from $x = 0$ to $x = 2p$. Thus if f is defined over the interval $0 < x < L$, we identify $2p = L$ or $p = L/2$. The resulting Fourier series will give the periodic extension of f with period L. In this manner the values to which the series converges on $-L < x < 0$ will be the same as on $0 < x < L$.

EXAMPLE 6

Expand $$f(x) = x^2, \qquad 0 < x < L$$

(a) in a cosine series, **(b)** in a sine series, **(c)** in a Fourier series.

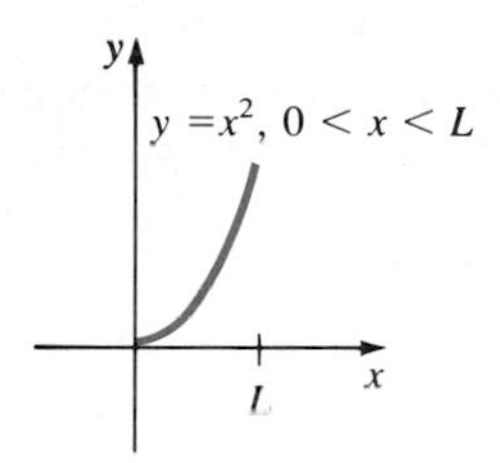

Figure 10.12

SOLUTION The graph of the function is given in Figure 10.12.

(a) We have $$a_0 = \frac{2}{L}\int_0^L x^2\,dx = \frac{2}{3}L^2,$$

and, integrating by parts,

$$\begin{aligned} a_n &= \frac{2}{L}\int_0^L x^2 \cos\frac{n\pi}{L}x\,dx \\ &= \frac{2}{L}\left[\frac{Lx^2 \sin\frac{n\pi}{L}x}{n\pi}\bigg|_0^L - \frac{2L}{n\pi}\int_0^L x\sin\frac{n\pi}{L}x\,dx\right] \\ &= -\frac{4}{n\pi}\left[-\frac{Lx\cos\frac{n\pi}{L}x}{n\pi}\bigg|_0^L + \frac{L}{n\pi}\int_0^L \cos\frac{n\pi}{L}x\,dx\right] \\ &= \frac{4L^2(-1)^n}{n^2\pi^2}. \end{aligned}$$

Thus $$f(x) = \frac{L^2}{3} + \frac{4L^2}{\pi^2}\sum_{n=1}^{\infty}\frac{(-1)^n}{n^2}\cos\frac{n\pi}{L}x. \tag{8}$$

(b) In this case $$b_n = \frac{2}{L}\int_0^L x^2 \sin\frac{n\pi}{L}x\,dx.$$

After integrating by parts we find

$$b_n = \frac{2L^2(-1)^{n+1}}{n\pi} + \frac{4L^2}{n^3\pi^3}[(-1)^n - 1].$$

Thus $$f(x) = \frac{2L^2}{\pi} \sum_{n=1}^{\infty} \left\{ \frac{(-1)^{n+1}}{n} + \frac{2}{n^3\pi^2} [(-1)^n - 1] \right\} \sin \frac{n\pi}{L} x. \tag{9}$$

(c) With $p = L/2$, $1/p = 2/L$, and $n\pi/p = 2n\pi/L$, we have

$$a_0 = \frac{2}{L} \int_0^L x^2 \, dx = \frac{2}{3} L^2,$$

$$a_n = \frac{2}{L} \int_0^L x^2 \cos \frac{2n\pi}{L} x \, dx = \frac{L^2}{n^2\pi^2},$$

$$b_n = \frac{2}{L} \int_0^L x^2 \sin \frac{2n\pi}{L} x \, dx = -\frac{L^2}{n\pi}.$$

Therefore

$$f(x) = \frac{L^2}{3} + \frac{L^2}{\pi} \sum_{n=1}^{\infty} \left\{ \frac{1}{n^2\pi} \cos \frac{2n\pi}{L} x - \frac{1}{n} \sin \frac{2n\pi}{L} x \right\}. \tag{10}$$

The series (8), (9), and (10) converge to the $2L$-periodic even extension of f, to the $2L$-periodic odd extension of f, and to the L-periodic extension of f, respectively. The graphs of these periodic extensions are shown in Figure 10.13.

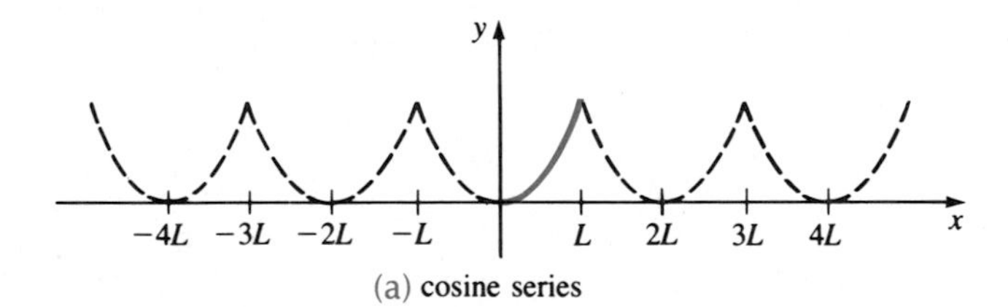

(a) cosine series

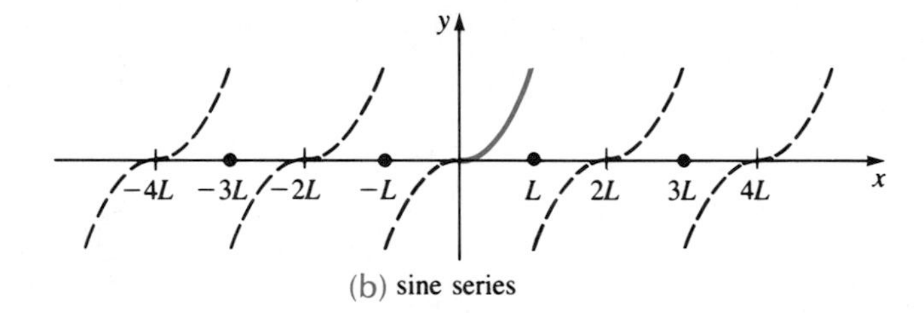

(b) sine series

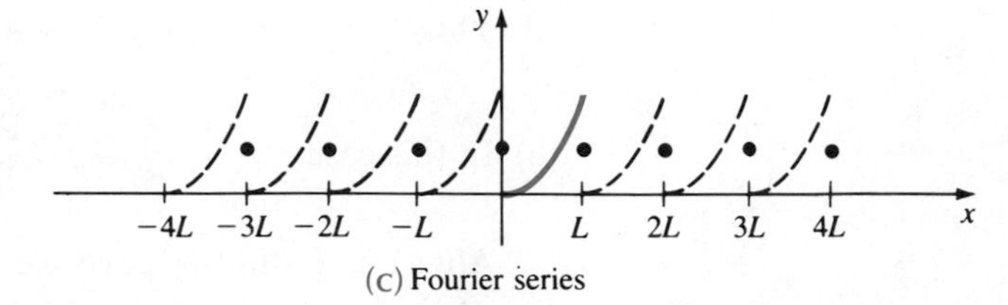

(c) Fourier series

Figure 10.13

Exercises 10.3

Answers to odd-numbered problems begin on page A–34.

In Problems 1–10 determine whether the function is even, odd, or neither.

1. $f(x) = \sin 3x$

2. $f(x) = x \cos x$

3. $f(x) = x^2 + x$

4. $f(x) = x^3 - 4x$

5. $f(x) = e^{|x|}$

6. $f(x) = |x^5|$

7. $f(x) = \begin{cases} x^2, & -1 < x \le 0 \\ -x^2, & 0 < x < 1 \end{cases}$

$f(x) = \begin{cases} x + 5, & -2 < x < 0 \\ -x + 5, & 0 \le x < 2 \end{cases}$

9. $f(x) = x^3, 0 \le x \le 2$

10. $f(x) = 2|x| - 1$

In Problems 11–24 expand the given function in an appropriate cosine or sine series.

11. $f(x) = \begin{cases} -1, & -\pi < x < 0 \\ 1 & 0 \le x < \pi \end{cases}$

12. $f(x) = \begin{cases} 1, & -2 < x < -1 \\ 0, & -1 < x < 1 \\ 1, & 1 < x < 2 \end{cases}$

13. $f(x) = |x|, \quad -\pi < x < \pi$

14. $f(x) = x, \quad -\pi < x < \pi$

15. $f(x) = x^2, \quad -1 < x < 1$

16. $f(x) = x|x|, \quad -1 < x < 1$

17. $f(x) = \pi^2 - x^2, \quad -\pi < x < \pi$

18. $f(x) = x^3, \quad -\pi < x < \pi$

19. $f(x) = \begin{cases} x - 1, & -\pi < x < 0 \\ x + 1, & 0 \le x < \pi \end{cases}$

20. $f(x) = \begin{cases} x + 1, & -1 < x < 0 \\ x - 1, & 0 \le x < 1 \end{cases}$

21. $f(x) = \begin{cases} 1, & -2 < x < -1 \\ -x, & -1 \le x < 0 \\ x, & 0 \le x < 1 \\ 1, & 1 \le x < 2 \end{cases}$

22. $f(x) = \begin{cases} -\pi, & -2\pi < x < \pi \\ x, & -\pi \le x < \pi \\ \pi, & \pi \le x < 2\pi \end{cases}$

23. $f(x) = |\sin x|, \quad -\pi < x < \pi$

24. $f(x) = \cos x, \quad -\frac{\pi}{2} < x < \frac{\pi}{2}$

In Problems 25–34 find the half-range cosine and sine expansions of the given function.

25. $f(x) = \begin{cases} 1, & 0 < x < \frac{1}{2} \\ 0, & \frac{1}{2} \le x < 1 \end{cases}$

26. $f(x) = \begin{cases} 0, & 0 < x < \frac{1}{2} \\ 1, & \frac{1}{2} \le x < 1 \end{cases}$

27. $f(x) = \cos x, \quad 0 < x < \frac{\pi}{2}$

28. $f(x) = \sin x, \quad 0 < x < \pi$

29. $f(x) = \begin{cases} x, & 0 < x < \frac{\pi}{2} \\ \pi - x, & \frac{\pi}{2} \le x < \pi \end{cases}$

30. $f(x) = \begin{cases} 0, & 0 < x < \pi \\ x - \pi, & \pi \le x < 2\pi \end{cases}$

31. $f(x) = \begin{cases} x, & 0 < x < 1 \\ 1, & 1 \le x < 2 \end{cases}$

32. $f(x) = \begin{cases} 1, & 0 < x < 1 \\ 2 - x, & 1 \le x < 2 \end{cases}$

33. $f(x) = x^2 + x, \quad 0 < x < 1$

34. $f(x) = x(2 - x), \quad 0 < x < 2$

In Problems 35–38 expand the given function in a Fourier series.

35. $f(x) = x^2, \quad 0 < x < 2\pi$

36. $f(x) = x, \quad 0 < x < \pi$

37. $f(x) = x + 1, \quad 0 < x < 1$

38. $f(x) = 2 - x, \quad 0 < x < 2$

Miscellaneous Problems

39. Prove Property (i).

40. Prove Property (ii).

41. Prove Property (iii).

42. Prove Property (iv).

43. Prove Property (v).

44. Prove that any function f can be written as a sum of an even and an odd function. [*Hint:* Use the identity

$$f(x) = \frac{f(x) + f(-x)}{2} + \frac{f(x) - f(-x)}{2}.]$$

45. Find the Fourier series of

$$f(x) = \begin{cases} 0, & -\pi < x < 0 \\ x, & 0 \le x < \pi \end{cases}$$

using the identify $f(x) = (|x| + x)/2$, $-\pi < x < \pi$ and the results of Problems 13 and 14. Observe that $|x|/2$ and $x/2$ are even and odd, respectively, on the interval (see Problem 44).

10.4 Partial Differential Equations

Linear Equations

In this brief introduction to partial differential equations, we shall be interested in **linear** equations in two variables:

$$A\frac{\partial^2 u}{\partial x^2} + B\frac{\partial^2 u}{\partial x \partial y} + C\frac{\partial^2 u}{\partial y^2} + D\frac{\partial u}{\partial x} + E\frac{\partial u}{\partial y} + Fu = G,$$

where $A, B, C, \ldots, G$ are functions of x and y. When $G(x, y) = 0$, the equation is said to be **homogeneous**; otherwise it is **nonhomogeneous**.

Solution by Integration

Recall from calculus, integration of a partial derivative results in an arbitrary function rather than a constant. For example, the solution of $\partial u/\partial x = 0$ is $u = f(y)$, where f is a differentiable function.

EXAMPLE 1

The second-order equation

$$\frac{\partial^2 u}{\partial y^2} = 0 \tag{1}$$

can be solved by integrating twice with respect to y

$$\frac{\partial u}{\partial y} = f(x),$$

$$u = yf(x) + g(x), \tag{2}$$

where f and g are arbitrary differentiable functions.

EXAMPLE 2

Solve
$$\frac{\partial^2 u}{\partial x \partial y} + \frac{\partial u}{\partial y} = 1. \tag{3}$$

SOLUTION If we let $v = \partial u/\partial y$, the equation becomes

$$\frac{\partial v}{\partial x} + v = 1.$$

By treating this latter equation as we would an ordinary linear first-order equation, we see that an integrating factor is e^x. Therefore

$$\frac{\partial}{\partial x}(e^x v) = e^x$$

yields
$$v = 1 + F(y)e^{-x},$$

where F is arbitrary. Using the original substitution and integrating with respect to y then give

$$u = y + f(y)e^{-x} + g(x), \tag{4}$$

where we have written $f(y) = \int F(y)\,dy$.

EXAMPLE 3

Solve
$$\frac{\partial^2 u}{\partial x^2} - y^2 u = e^x. \tag{5}$$

SOLUTION We solve the equation as we would a nonhomogeneous linear second-order ordinary differential equation by first solving

$$\frac{\partial^2 u}{\partial x^2} - y^2 u = 0.$$

Treating y as a constant, it follows from Section 4.3 that

$$u_c = f(y)e^{xy} + g(y)e^{-xy}.$$

To find a particular solution we use undetermined coefficients and assume

$$u_p = A(y)e^x.$$

Substituting this latter function into the given equation yields

$$Ae^x - y^2 Ae^x = e^x$$

and so $A(y) = 1/(1 - y^2)$. Hence a solution of the equation is

$$u = f(y)e^{xy} + g(y)e^{-xy} + \frac{e^x}{1 - y^2}. \tag{6}$$

Since (1), (3), and (5) are second-order and (2), (4), and (6) involve two arbitrary functions, each could be called a *general solution.* But we shall leave unanswered the question whether each solution yields every function satisfying the equation. Furthermore, it is not our intention in this section to focus upon those procedures leading to a general solution. Most partial differential equations cannot be solved as readily as the preceding three examples. However, in many applications involving linear partial differential equations, it is sufficient to obtain particular solutions.

Separation of Variables

For a linear partial differential equation, it is sometimes possible to find particular solutions in the form of a product

$$u(x\ y) = X(x)Y(y) \tag{7}$$

The use of the product (7), called the **method of separation of variables**, may enable us to reduce a partial differential equation to several *ordinary* differential equations. To this end we note

$$\frac{\partial u}{\partial x} = X'Y, \qquad \frac{\partial u}{\partial y} = XY'$$

and

$$\frac{\partial^2 u}{\partial x^2} = X''Y, \qquad \frac{\partial^2 u}{\partial y^2} = XY'',$$

where the primes denote ordinary differentiation.

EXAMPLE 4 Find product solutions of the equation

$$\frac{\partial^2 u}{\partial x^2} = 4\frac{\partial u}{\partial y}. \tag{8}$$

SOLUTION If $u = X(x)Y(y)$, then (8) becomes

$$X''Y = 4XY'.$$

After dividing both sides by $4XY$, we have separated the variables:

$$\frac{X''}{4X} = \frac{Y'}{Y}.$$

Since the left-hand side of the last equation is independent of y and is identically equal to the right-hand side, which is independent of x, we conclude that both sides must be a constant (see Problem 37). In practice it is convenient to write this real constant as either λ^2 or $-\lambda^2$. We distinguish the following cases.

CASE I Using $\lambda^2 > 0$, the equalities

$$\frac{X''}{4X} = \frac{Y'}{Y} = \lambda^2$$

lead to $\quad X'' - 4\lambda^2 X = 0 \quad$ and $\quad Y' - \lambda^2 Y = 0.$

These latter equations have the solutions

$$X = c_1 \cosh 2\lambda x + c_2 \sinh 2\lambda x \quad \text{and} \quad Y - c_3 e^{\lambda^2 y},$$

respectively.* Thus a particular solution of (8) is

$$\begin{aligned} u &= XY \\ &= (c_1 \cosh 2\lambda x + c_2 \sinh 2\lambda x)(c_3 e^{\lambda^2 y}) \\ &= A_1 e^{\lambda^2 y} \cosh 2\lambda x + B_1 e^{\lambda^2 y} \sinh 2\lambda x, \end{aligned} \tag{9}$$

where $A_1 = c_1 c_3$ and $B_1 = c_2 c_3$.

CASE II Using $-\lambda^2 < 0$, the equalities

$$\frac{X''}{4X} = \frac{Y'}{Y} = -\lambda^2$$

give $\quad X'' + 4\lambda^2 X = 0 \quad$ and $\quad Y' + \lambda^2 Y = 0.$

---

* Recall that X can be written in the alternative form $X = c_1 e^{-2\lambda x} + c_2 e^{2\lambda x}$

Since the solutions of these equations are

$$X = c_4 \cos 2\lambda x + c_5 \sin 2\lambda x \qquad \text{and} \qquad Y = c_6 e^{-\lambda^2 y},$$

respectively, another solution of (8) is

$$u = A_2 e^{-\lambda^2 y} \cos 2\lambda x + B_2 e^{-\lambda^2 y} \sin 2\lambda x, \tag{10}$$

where $A_2 = c_4 c_6$ and $B_2 = c_5 c_6$.

CASE III If $\lambda^2 = 0$, it follows that

$$X'' = 0 \qquad \text{and} \qquad Y' = 0.$$

In this case $\quad X = c_7 x + c_8 \quad$ and $\quad Y = c_9$

so that

$$u = A_3 x + B_3, \tag{11}$$

where $A_3 = c_7 c_9$ and $B_3 = c_8 c_9$.

It is left as an exercise to verify that (9), (10), and (11) satisfy the given equation.

EXAMPLE 5 Find a product solution of

$$k \frac{\partial^2 u}{\partial x^2} = \frac{\partial u}{\partial t}, \qquad k > 0, \tag{12}$$

satisfying the conditions $u(0, t) = 0$, $u(L, t) = 0$.

SOLUTION If $u = XT$, we can write the given equation as

$$\frac{X''}{X} = \frac{T'}{kT} = -\lambda^2, \tag{13}$$

which leads to

$$X'' + \lambda^2 X = 0 \tag{14}$$
$$T' + k\lambda^2 T = 0$$

and

$$X = c_1 \cos \lambda x + c_2 \sin \lambda x \tag{15}$$
$$T = c_3 e^{-k\lambda^2 t},$$

respectively. Now since

$$u(0, t) = X(0)T(t) = 0$$
$$u(L, t) = X(L)T(t) = 0,$$

we must have $X(0) = 0$ and $X(L) = 0$. These are boundary conditions for the ordinary differential equation (14). Applying the first of these conditions in (15) immediately gives $c_1 = 0$. Therefore

$$X = c_2 \sin \lambda x.$$

The second boundary condition now implies

$$X(L) = c_2 \sin \lambda L = 0.$$

If $c_2 = 0$, then $X = 0$ so that $u = 0$. To obtain a nontrivial solution u, we must have $c_2 \neq 0$, and so the last equation is satisfied when

$$\sin \lambda L = 0.$$

This implies that $\lambda L = n\pi$ or $\lambda = n\pi/L$, $n = 1, 2, 3, \ldots$.

Thus
$$u = (c_2 \sin \lambda x)(c_3 e^{-k\lambda^2 t})$$
$$= A_n e^{-k(n^2\pi^2/L^2)t} \sin \frac{n\pi}{L} x$$

satisfies the given equation and both side conditions. The coefficient $c_2 c_3$ is rewritten as A_n to emphasize the fact that a different solution is obtained for each n.* The reader should verify that using $\lambda^2 \geq 0$ in (13) does not lead to a solution of (12) that satisfies $u(0, t) = 0$ and $u(L, t) = 0$.

Superposition Principle

The following theorem is analogous to Theorem 4.3 and is known as the **superposition principle**.

THEOREM 10.2 If $u_1, u_2, \ldots, u_k$ are solutions of a homogeneous linear partial differential equation, then the linear combination

$$u = c_1 u_1 + c_2 u_2 + \cdots + c_k u_k,$$

where the c_i, $i = 1, 2, \ldots, k$ are constants, is also a solution.

In the next section we shall make the assumption that whenever we have an infinite set

$$u_1, u_2, u_3, \ldots,$$

* Note that when $n = 0$, $\sin 0 = 0$ so that $u = 0$. Also, if n is a negative integer, say, $n = -k$, $k = 1, 2, \ldots$, we can use the trigonometric identity $\sin(-\theta) = -\sin\theta$ to rewrite $\sin(-k\pi x/L)$ as $-\sin k\pi x/L$. The factor of -1 can be absorbed in the arbitrary constant A_n. Thus, in this case, we need only consider the solutions obtained for the positive integers.

of solutions of a homogeneous linear equation, we can construct yet another solution u by forming the infinite series

$$u = \sum_{k=1}^{\infty} u_k.$$

EXAMPLE 6

In view of the superposition principle, the function defined by the series

$$u = \sum_{n=1}^{\infty} A_n e^{-k(n^2\pi^2/L^2)t} \sin \frac{n\pi}{L} x \tag{16}$$

must also, although formally, satisfy equation (12) in Example 5. Notice, too, that (16) satisfies the conditions $u(0, t) = 0$ and $u(L, t) = 0$.

Exercises 10.4

Answers to odd-numbered problems begin on page A–35.

In Problems 1–12 solve the given partial differential equation.

1. $\dfrac{\partial u}{\partial x} + y = 0$

2. $\dfrac{\partial u}{\partial y} = 2xy$

3. $\dfrac{\partial u}{\partial y} + 2u = e^y$

4. $x\dfrac{\partial u}{\partial x} + u = 4ye^{2x}$

5. $\dfrac{\partial^2 u}{\partial x^2} = 8xy^2 + 1$

6. $\dfrac{\partial^2 u}{\partial y^2} + \sin(xy) = 0$

7. $\dfrac{\partial^2 u}{\partial x \partial y} = 1$

8. $\dfrac{\partial^2 u}{\partial x \partial y} = 2x + 4y$

9. $\dfrac{\partial^2 u}{\partial x \partial y} - \dfrac{\partial u}{\partial y} = 6xe^x$

10. $\dfrac{\partial^2 u}{\partial x \partial y} + 2y\dfrac{\partial u}{\partial x} = 4xy$

11. $\dfrac{\partial^2 u}{\partial y^2} - x^2 u = xe^{4y}$

12. $\dfrac{\partial^2 u}{\partial x^2} + y^2 u = y \sin 2x$

In Problems 13 and 14 solve the given equation subject to the indicated conditions.

13. $\dfrac{\partial^2 u}{\partial x^2} = 6x; \quad u(0, y) = y, \quad u(1, y) = y^2 + 1$

14. $y\dfrac{\partial^2 u}{\partial y^2} + \dfrac{\partial u}{\partial y} = 0; \quad u(x, 1) = x^2, \quad u(x, e) = 1$

In Problems 15–30 determine whether the method of separation of variables is applicable to the given equation. If so, find the product solutions.

15. $\dfrac{\partial u}{\partial x} = \dfrac{\partial u}{\partial y}$

16. $\dfrac{\partial u}{\partial x} + 3\dfrac{\partial u}{\partial y} = 0$

17. $\dfrac{\partial u}{\partial x} + \dfrac{\partial u}{\partial y} = u$

18. $\dfrac{\partial u}{\partial x} = \dfrac{\partial u}{\partial y} + u$

19. $x\dfrac{\partial u}{\partial x} = y\dfrac{\partial u}{\partial y}$

20. $y\dfrac{\partial u}{\partial x} + x\dfrac{\partial u}{\partial y} = 0$

21. $\dfrac{\partial^2 u}{\partial x^2} + \dfrac{\partial^2 u}{\partial x \partial y} + \dfrac{\partial^2 u}{\partial y^2} = 0$

22. $y\dfrac{\partial^2 u}{\partial x \partial y} + u = 0$

23. $k\dfrac{\partial^2 u}{\partial x^2} - u = \dfrac{\partial u}{\partial t}, \quad k > 0$

24. $k\dfrac{\partial^2 u}{\partial x^2} = \dfrac{\partial u}{\partial t}, \quad k > 0$

25. $a^2\dfrac{\partial^2 u}{\partial x^2} = \dfrac{\partial^2 u}{\partial t^2}$

26. $a^2\dfrac{\partial^2 u}{\partial x^2} = \dfrac{\partial^2 u}{\partial t^2} + 2k\dfrac{\partial u}{\partial t}, \quad k > 0$

27. $\dfrac{\partial^2 u}{\partial x^2} + \dfrac{\partial^2 u}{\partial y^2} = 0$

28. $x^2\dfrac{\partial^2 u}{\partial x^2} + \dfrac{\partial^2 u}{\partial y^2} = 0$

29. $\dfrac{\partial^2 u}{\partial x^2} + \dfrac{\partial^2 u}{\partial y^2} = u$

30. $a^2\dfrac{\partial^2 u}{\partial x^2} - g = \dfrac{\partial^2 u}{\partial t^2}, \quad g$ a constant

In Problems 31–34 find product solutions satisfying the given equations and the indicated conditions.

31. $k\dfrac{\partial^2 u}{\partial x^2} = \dfrac{\partial u}{\partial t}, \quad k > 0; \quad \left.\dfrac{\partial u}{\partial x}\right|_{x=0} = 0, \quad \left.\dfrac{\partial u}{\partial x}\right|_{x=5} = 0$

32. $a^2\dfrac{\partial^2 u}{\partial x^2} = \dfrac{\partial^2 u}{\partial t^2}; \quad u(0, t) = 0, u(2, t) = 0, \quad \left.\dfrac{\partial u}{\partial t}\right|_{t=0} = 0$

33. $\dfrac{\partial^2 u}{\partial x^2} + \dfrac{\partial^2 u}{\partial y^2} = 0; \quad u(0, y) = 0, \quad u(x, 0) = 0, \quad u(x, 1) = 0$

34. $\dfrac{\partial^2 u}{\partial x^2} + \dfrac{\partial^2 u}{\partial y^2} = 0; \quad \left.\dfrac{\partial u}{\partial x}\right|_{x=0} = 0, \quad \left.\dfrac{\partial u}{\partial x}\right|_{x=\pi} = 0, \quad u(x, 0) = 0$

Miscellaneous Problems

35. Show that the equation

$$\frac{\partial u}{\partial t} = k\left(\frac{\partial^2 u}{\partial r^2} + \frac{1}{r}\frac{\partial u}{\partial r}\right), \qquad k > 0$$

possesses the product solution

$$u = e^{-k\lambda^2 t}\left(A_1 J_0(\lambda r) + B_2 Y_0(\lambda r)\right).$$

36. Find a product solution of

$$\frac{\partial^2 u}{\partial t^2} = a^2\left(\frac{\partial^2 u}{\partial r^2} + \frac{1}{r}\frac{\partial u}{\partial r}\right).$$

37. Prove that each side of the separated form of equation (8) is constant.

10.5 Boundary-Value Problems

Special Equations

The following linear partial differential equations

$$k\frac{\partial^2 u}{\partial x^2} = \frac{\partial u}{\partial t}, \qquad k > 0, \tag{1}$$

$$a^2\frac{\partial^2 u}{\partial x^2} = \frac{\partial^2 u}{\partial t^2}, \tag{2}$$

$$\frac{\partial^2 u}{\partial x^2} + \frac{\partial^2 u}{\partial y^2} = 0 \tag{3}$$

play an important role in many areas of physics and engineering. Equations (1) and (2) are known as the **one-dimensional heat equation** and the **one-dimensional wave equation**, respectively. *One-dimensional* refers to the fact that x denotes a spatial dimension, whereas t usually represents time. Equation (3) is called **Laplace's equation.**

We conclude this chapter by utilizing the method of separation of variables to solve several applied problems, each of which is described by one of the above equations along with certain side conditions. These side conditions consist of:

(i) Boundary conditions: u or $\partial u/\partial x$ specified at $x =$ constant; u or $\partial u/\partial y$ specified at $y =$ constant, and

(ii) Initial conditions: u at $t = 0$ for equation (1), or, u and $\partial u/\partial t$ at $t = 0$ for equation (2).

The collective mathematical description of such a problem is known as a **boundary-value problem.**

Equation (1) occurs in the theory of heat flow (that is, heat transferred by conduction) in a rod or a thin wire. The function $u(x, t)$ is temperature in the rod. Problems in mechanical vibrations often lead to the wave equation (2). For our purposes the solution $u(x, t)$ of (2) will represent the small displacements of an idealized vibrating string. Lastly, the solution $u(x, y)$ of

Laplace's equation (3) can be interpreted as the steady-state (that is, time independent) temperature distribution in a thin flat plate. For a derivation of the heat and wave equations, the reader is referred to the alternate edition of this text.*

Although we shall confine our attention to solving the problems described in the preceding discussion, we note that the analysis of a wide variety of diverse phenomena yield equations (1), (2), or (3), or their generalizations involving a greater number of spatial variables. For example, (1) is sometimes called the **diffusion equation** since the diffusion of dissolved substances in solution is analogous to the flow of heat in a solid. The function $u(x, t)$ satisfying the partial differential equation in this case represents the concentration of the liquid. Similarly, equation (1) arises in the study of the flow of electricity in a long cable or a transmission line. In this setting, (1) is known as a **telegraph equation**. It can be shown that under certain assumptions the current and the voltage in the line are functions satisfying two equations identical in form with equation (1). The wave equation (2) also appears in the theory of high frequency transmission lines, fluid mechanics, acoustics, and elasticity. Laplace's equation (3) is encountered in engineering problems, in static displacements of membranes, and most often in problems dealing with potentials such as electrostatic, gravitational, and velocity potentials in fluid mechanics.

10.5.1 Heat Equation

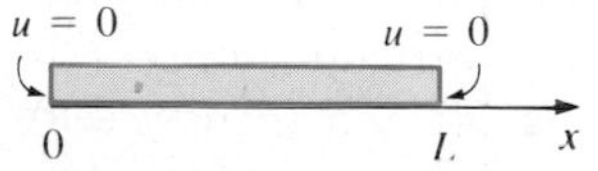

$u = 0$ $u = 0$

0 L x

Figure 10.14

Consider a thin rod of length L with an initial temperature of $f(x)$ throughout and whose ends are held at a constant temperature of zero degrees for all time (see Figure 10.14). If

- the flow of heat takes place only in the x-direction,
- no heat escapes from the lateral surface of the rod,
- no heat is being generated in the rod,
- the rod is homogeneous, that is, its density per length is constant,
- its specific heat and thermal conductivity are constant,

then the temperature $u(x, t)$ in the rod is given by the solution of the boundary-value problem

$$k\frac{\partial^2 u}{\partial x^2} = \frac{\partial u}{\partial t}, \quad k > 0, \quad 0 < x < L, \quad t > 0, \tag{4}$$

$$u(0, t) = 0, \quad u(L, t) = 0, \quad t > 0, \tag{5}$$

$$u(x, 0) = f(x), \quad 0 < x < L. \tag{6}$$

---

* See Dennis G. Zill, *Differential Equations with Boundary-Value Problems*, second edition (Boston: PWS-KENT Publishing Company, 1989).

The constant k is proportional to the thermal conductivity and is called the **thermal diffusivity**.

Solution

Using the product $u = XT$ and $-\lambda^2$ as a separation constant leads to

$$\frac{X''}{X} = \frac{T'}{kT} = -\lambda^2$$

and

$$X'' + \lambda^2 X = 0, \qquad X(0) = 0, \qquad X(L) = 0, \tag{7}$$

$$T' + k\lambda^2 T = 0. \tag{8}$$

We have already obtained the solution of equation (4) subject to the boundary conditions (5) by solving (7) (see Example 5, Section 10.4). It was seen that (7) possessed a nontrivial solution only if the value of the parameter λ took on the values

$$\lambda = \frac{n\pi}{L}, \qquad n = 1, 2, 3, \ldots. \tag{9}$$

The corresponding solutions of (7) were then

$$X = c_1 \sin \frac{n\pi}{L} x, \qquad n = 1, 2, 3, \ldots. \tag{10}$$

Eigenvalues and Eigenfunctions

The values (9) for which (7) possess a nontrivial solution are known as **characteristic values**, or more commonly as **eigenvalues**. The solutions (10) are called **characteristic functions**, or **eigenfunctions**. We note that for a choice of λ, other than those given in (9), the only solution of (7) is the zero function $X = 0$. In turn, this would imply that a function satisfying (4) and (5) is $u = 0$. However, $u = 0$ is *not* a solution of the original boundary-value problem when $f(x) \neq 0$. We naturally assume this last condition.

Since the solution of (8) is

$$T = c_3 e^{-k\lambda^2 t} = c_3 e^{-k(n^2\pi^2/L^2)t},$$

the products

$$u_n = XT = A_n e^{-k(n^2\pi^2/L^2)t} \sin \frac{n\pi}{L} x \tag{11}$$

satisfy the partial differential equation (4) and the boundary conditions (5) for each value of the positive integer n. For convenience we have replaced the constant $c_1 c_3$ by A_n. In order that the functions given in (11) satisfy the initial condition (6), we would have to choose the constant coefficients A_n in such a manner that

$$u_n(x, 0) = f(x) = A_n \sin \frac{n\pi}{L} x. \tag{12}$$

In general, we would not expect condition (12) to be satisfied for an arbitrary, but reasonable, choice of f. Therefore, we are forced to admit that (11) is *not a solution of the given problem.* However, by the superposition principle, the function

$$u(x, t) = \sum_{n=1}^{\infty} u_n = \sum_{n=1}^{\infty} A_n e^{-k(n^2\pi^2/L^2)t} \sin \frac{n\pi}{L} x \tag{13}$$

also satisfies (4) and (5). Substituting $t = 0$ in (13) implies

$$u(x, 0) = f(x) = \sum_{n=1}^{\infty} A_n \sin \frac{n\pi}{L} x.$$

This last expression is recognized as the half-range expression of f in a sine series. Thus, if we make the identification $A_n = b_n$, $n = 1, 2, 3, \ldots$, it follows from (5) of Section 10.3 that

$$A_n = \frac{2}{L} \int_0^L f(x) \sin \frac{n\pi}{L} x \, dx.$$

We conclude that the solution of the boundary-value problem described in (4), (5), and (6) is given by the infinite series

$$u(x, t) = \frac{2}{L} \sum_{n=1}^{\infty} \left(\int_0^L f(x) \sin \frac{n\pi}{L} x \, dx \right) e^{-k(n^2\pi^2/L^2)t} \sin \frac{n\pi}{L} x.$$

Insulated Boundaries

In the problem just described, the ends or boundaries of the rod could be **insulated**. At an insulated boundary the normal derivative of the temperature is zero. This fact follows from an empirical law that states the flux of heat across a surface (the time rate of flow of heat per unit area) is proportional to the value of the directional derivative of the temperature normal (perpendicular) to the surface.

EXAMPLE 1

Give the boundary-value problem for the temperature u in a horizontal rod of length L if its ends are insulated and if its initial temperature throughout is given by $f(x)$, $0 < x < L$.

SOLUTION The ends of the rod are surfaces perpendicular to the x-axis and so $\partial u/\partial x = 0$ at both boundaries. Hence the temperature in the rod is given by the solution of

$$k \frac{\partial^2 u}{\partial x^2} = \frac{\partial u}{\partial t}, \qquad k > 0, \quad 0 < x < L, \quad t > 0.$$

$$\left.\frac{\partial u}{\partial x}\right|_{x=0} = 0, \qquad \left.\frac{\partial u}{\partial x}\right|_{x=L} = 0, \quad t > 0,$$

$$u(x, 0) = f(x), \qquad 0 < x < L.$$

The solution of the problem given in Example 1 is left as an exercise (see Problem 3).

10.5.2 Wave Equation

In the next example we consider the transverse vibrations of a string stretched between two points, say, $x = 0$ and $x = L$. As shown in Figure 10.15, the motion takes place in the xy-plane in such a manner that each point of the string moves in a direction perpendicular to the x-axis. If $u(x, t)$ denotes the displacement of the string measured from the x-axis for $t > 0$, then u satisfies equation (2) under the following assumptions:

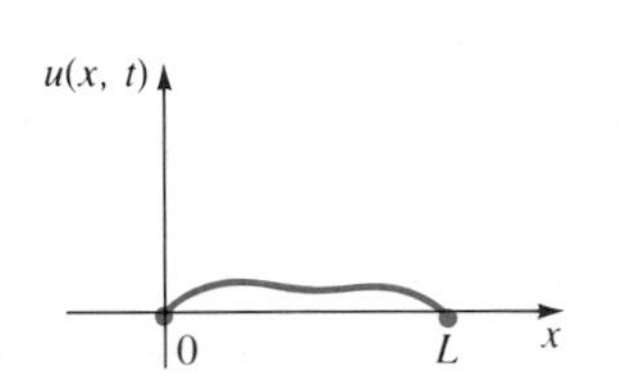

Figure 10.15

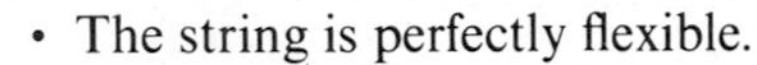

- The string is perfectly flexible.
- The string is homogeneous, that is, its mass per unit length is constant.
- The displacements u are small compared to the length of the string.
- The tension of the string is constant.
- The tension is large compared with the force of gravity.
- No other forces act on the string.

Thus a typical boundary-value problem is

$$a^2 \frac{\partial^2 u}{\partial x^2} = \frac{\partial^2 u}{\partial t^2}, \qquad 0 < x < L, \quad t > 0, \tag{14}$$

$$u(0, t) = 0, \qquad u(L, t) = 0, \quad t \geq 0, \tag{15}$$

$$u(x, 0) = f(x), \qquad \left.\frac{\partial u}{\partial t}\right|_{t=0} = g(x), \quad 0 < x < L. \tag{16}$$

The boundary conditions (15) simply state that the string is secured at the endpoints for all time. At $t = 0$, the functions f and g given in (16) specify the initial configuration and the initial velocity of each point of the string, respectively. It is implicit in this context that f is continuous and $f(0) = 0$, $f(L) = 0$.

Solution

Separating variables in (14) gives

$$\frac{X''}{X} = \frac{T''}{a^2 T} = -\lambda^2$$

so that

$$X'' + \lambda^2 X = 0$$

$$T'' + \lambda^2 a^2 T = 0$$

and therefore
$$X = c_1 \cos \lambda x + c_2 \sin \lambda x$$
$$T = c_3 \cos \lambda at + c_4 \sin \lambda at.^*$$

As before, the boundary conditions (15) translate into $X(0) = 0$ and $X(L) = 0$. In turn we find

$$c_1 = 0 \qquad \text{and} \qquad c_2 \sin \lambda L = 0.$$

This last equation yields the eigenvalues $\lambda = n\pi/L$, $n = 1, 2, 3, \ldots$. The corresponding eigenfunctions are

$$X = c_2 \sin \frac{n\pi}{L} x, \quad n = 1, 2, 3, \ldots.$$

Thus solutions of equation (14) satisfying the boundary conditions (15) are

$$u_n = \left(A_n \cos \frac{n\pi a}{L} t + B_n \sin \frac{n\pi a}{L} t \right) \sin \frac{n\pi}{L} x,$$

and
$$u(x, t) = \sum_{n=1}^{\infty} \left(A_n \cos \frac{n\pi a}{L} t + B_n \sin \frac{n\pi a}{L} t \right) \sin \frac{n\pi}{L} x. \tag{17}$$

Setting $t = 0$ in (17) gives

$$u(x, 0) = f(x) = \sum_{n=1}^{\infty} A_n \sin \frac{n\pi}{L} x,$$

which is a half-range expansion for f in a sine series. As in the discussion of the heat equation, we can write $A_n = b_n$,

$$A_n = \frac{2}{L} \int_0^L f(x) \sin \frac{n\pi}{L} x \, dx. \tag{18}$$

To determine B_n we differentiate (17) with respect to t and then set $t = 0$:

$$\frac{\partial u}{\partial t} = \sum_{n=1}^{\infty} \left(-A_n \frac{n\pi a}{L} \sin \frac{n\pi a}{L} t + B_n \frac{n\pi a}{L} \cos \frac{n\pi a}{L} t \right) \sin \frac{n\pi}{L} x$$

$$\left. \frac{\partial u}{\partial t} \right|_{t=0} = g(x) = \sum_{n=1}^{\infty} \left(B_n \frac{n\pi a}{L} \right) \sin \frac{n\pi}{L} x.$$

In order for this last series to be the half-range since expansion of g on the interval, the *total* coefficient $B_n n\pi a/L$ must be given by the form of (5) in

---

* You should convince yourself that a different choice of separation constant does not lead to a solution of the problem.

Section 10.3—that is,

$$B_n \frac{n\pi a}{L} = \frac{2}{L}\int_0^L g(x) \sin \frac{n\pi}{L} x \, dx,$$

from which we obtain

$$B_n = \frac{2}{n\pi a}\int_0^L g(x) \sin \frac{n\pi}{L} x \, dx. \tag{19}$$

The solution of the problem consists of the series (17), with A_n and B_n defined by (18) and (19), respectively.

We note that when the string is released from *rest*, then $g(x) = 0$ for every x in $0 \le x \le L$ and, consequently, $B_n = 0$.

10.5.3 Laplace's Equation

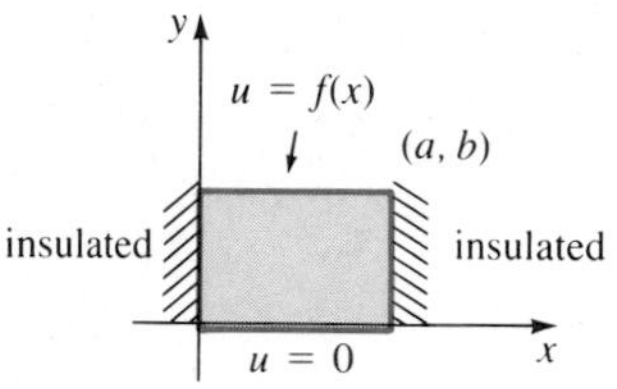

y
u = f(x)
(a, b)
insulated
insulated
u = 0
x

Figure 10.16

Suppose we wish to find the steady-state temperature $u(x, y)$ in a rectangular plate with boundary conditions indicated in Figure 10.16. When no heat escapes from the lateral faces of the plate, the problem is

$$\frac{\partial^2 u}{\partial x^2} + \frac{\partial^2 u}{\partial y^2} = 0, \qquad 0 < x < a, \quad 0 < y < b$$

$$\left.\frac{\partial u}{\partial x}\right|_{x=0} = 0, \qquad \left.\frac{\partial u}{\partial x}\right|_{x=a} = 0, \quad 0 < y < b$$

$$u(x, 0) = 0, \qquad u(x, b) = f(x), \quad 0 < x < a.$$

Solution

Separation of variables leads to

$$\frac{X''}{X} = -\frac{Y''}{Y} = -\lambda^2,$$

$$X'' + \lambda^2 X = 0 \tag{20}$$

$$Y'' - \lambda^2 Y = 0, \tag{21}$$

$$X = c_1 \cos \lambda x + c_2 \sin \lambda x \tag{22}$$

$$Y = c_3 \cosh \lambda y + c_4 \sinh \lambda y,^* \tag{23}$$

where the first three boundary conditions translate into $X'(0) = 0$, $X'(a) = 0$, and $Y(0) = 0$. Differentiating X and setting $x = 0$ imply $c_2 = 0$ and, therefore, $X = c_1 \cos \lambda x$. Differentiating again and then setting $x = a$ give

---

* There are problems in which a solution in terms of real exponential functions is more useful (see Problems 21 and 22).

$-c_1\lambda \sin \lambda a = 0$. This last condition is satisfied when $\lambda = 0$ or when $\lambda a = n\pi$, or $\lambda = n\pi/a$, $n = 1, 2, \ldots$. Observe that $\lambda = 0$ implies (20) is $X'' = 0$. The general solution of this equation is given by the linear function $X = c_1 + c_2 x$ and *not* by (22). In this case the boundary conditions $X'(0) = 0$, $X'(a) = 0$ demand that $X = c_1$. Unlike in the previous two examples, we are forced to conclude $\lambda = 0$ is an eigenvalue in this example. Corresponding $\lambda = 0$ with $n = 0$, the eigenfunctions are

$$X = c_1, \quad n = 0 \qquad \text{and} \qquad X = c_1 \cos \frac{n\pi}{a} x, \quad n = 1, 2, \ldots.$$

Finally, the condition that $Y(0) = 0$ dictates that $c_3 = 0$ in (23) when $\lambda > 0$. However, when $\lambda = 0$, equation (21) becomes $Y'' = 0$ and thus the solution is given by $Y = c_3 + c_4 y$ rather than by (23). But $Y(0) = 0$ implies again that $c_3 = 0$ and so $Y = c_4 y$. Thus product solutions of the equation satisfying the first three boundary conditions are

$$A_0 y, \quad n = 0 \qquad \text{and} \qquad A_n \sinh \frac{n\pi}{a} y \cos \frac{n\pi}{a} x, \quad n = 1, 2, \ldots.$$

The superposition principle yields another solution

$$u(x, y) = A_0 y + \sum_{n=1}^{\infty} A_n \sinh \frac{n\pi}{a} y \cos \frac{n\pi}{a} x. \tag{24}$$

Substituting $y = b$ in (24) gives

$$u(x, b) = f(x) = A_0 b + \sum_{n=1}^{\infty} \left(A_n \sinh \frac{n\pi}{a} b \right) \cos \frac{n\pi}{a} x,$$

which, in this case, is a half-range expansion of f in a cosine series. If we make the identifications $A_0 b = a_0/2$ and $A_n \sinh(n\pi b/a) = a_n$, $n = 1, 2, 3, \ldots$, it follows from (2) and (3) of Section 10.3 that

$$2A_0 b = \frac{2}{a} \int_0^a f(x)\, dx$$

$$A_0 = \frac{1}{ab} \int_0^a f(x)\, dx, \tag{25}$$

and

$$A_n \sinh \frac{n\pi}{a} b = \frac{2}{a} \int_0^a f(x) \cos \frac{n\pi}{a} x\, dx$$

$$A_n = \frac{2}{a \sinh \frac{n\pi}{a} b} \int_0^a f(x) \cos \frac{n\pi}{a} x\, dx. \tag{26}$$

The formal solution of this problem consists of the series given in (24), where A_0 and A_n are defined by (25) and (26), respectively.

Exercises 10.5

Answers to odd-numbered problems begin on page A–36.

[10.5.1] In Problems 1 and 2 solve the heat equation (1) subject to the given conditions. Assume a rod of length L.

1. $u(0, t) = 0, \quad u(L, t) = 0$

$$u(x, 0) = \begin{cases} 1, & 0 < x < \dfrac{L}{2} \\ 0, & \dfrac{L}{2} < x < L \end{cases}$$

2. $u(0, t) = 0, \quad u(L, t) = 0$

$u(x, 0) = x(L - x)$

3. Find the temperature $u(x, t)$ in a rod of length L if the initial temperature is $f(x)$ throughout and if the ends $x = 0$ and $x = L$ are insulated.

4. Solve Problem 3 if $L = 2$ and $f(x) = \begin{cases} x, & 0 < x < 1 \\ 0, & 1 < x < 2 \end{cases}$.

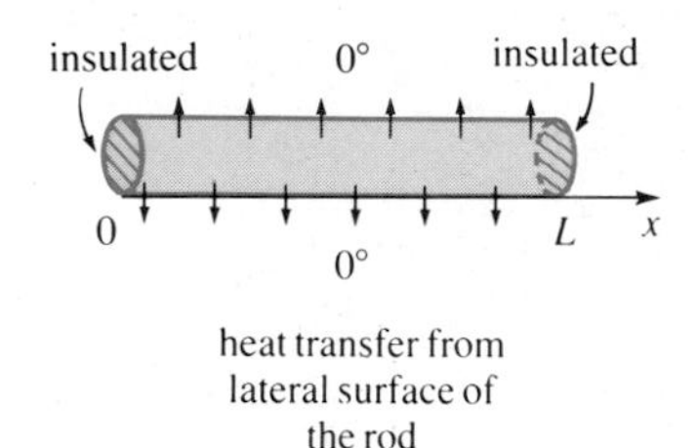

Figure 10.17

5. Suppose heat is lost from the lateral surface of a thin rod of length L into a surrounding medium at temperature zero. If the linear law of heat transfer applies, then the heat equation takes on the form $k\dfrac{\partial^2 u}{\partial x^2} - hu = \dfrac{\partial u}{\partial t}$, $0 < x < L$, $t > 0$, h a constant. Find the temperature $u(x, t)$ if the initial temperature is $f(x)$ throughout and the ends $x = 0$ and $x = L$ are insulated (see Figure 10.17).

6. Solve Problem 5 if the ends $x = 0$ and $x = L$ are held at temperature zero..

[10.5.2] In Problems 7–13 solve the wave equation (2) subject to the given conditions.

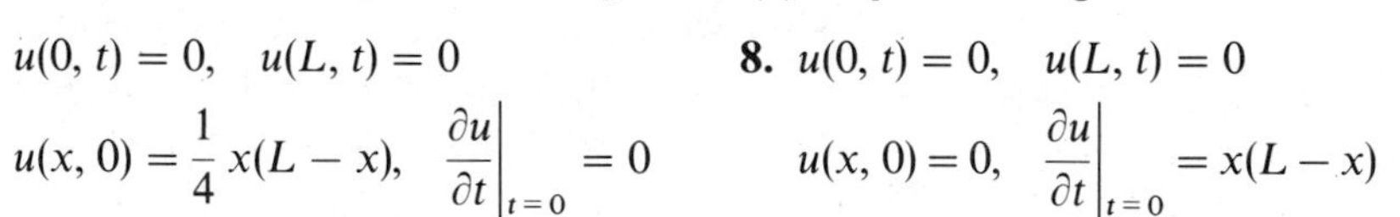

7. $u(0, t) = 0, \quad u(L, t) = 0$

$u(x, 0) = \dfrac{1}{4}x(L - x), \quad \left.\dfrac{\partial u}{\partial t}\right|_{t=0} = 0$

8. $u(0, t) = 0, \quad u(L, t) = 0$

$u(x, 0) = 0, \quad \left.\dfrac{\partial u}{\partial t}\right|_{t=0} = x(L - x)$

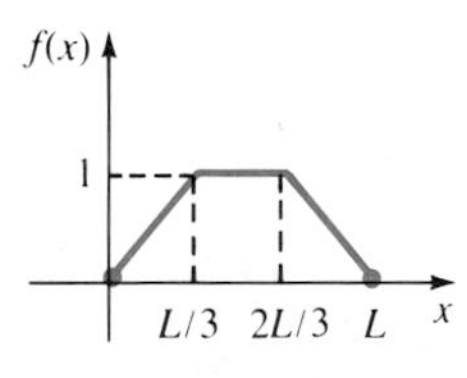

Figure 10.18

9. $u(0, t) = 0, \quad u(L, t) = 0$

$u(x, 0)$, as specified in Figure 10.18.

$\left.\dfrac{\partial u}{\partial t}\right|_{t=0} = 0$

10. $u(0, t) = 0, \quad u(\pi, t) = 0$

$u(x, 0) = \dfrac{1}{6}x(\pi^2 - x^2)$

$\left.\dfrac{\partial u}{\partial t}\right|_{t=0} = 0$

11. $u(0, t) = 0, \quad u(\pi, t) = 0$

$u(x, 0) = 0$

$\left.\dfrac{\partial u}{\partial t}\right|_{t=0} = \sin x$

12. $u(0, t) = 0, \quad u(1, t) = 0$

$u(x, 0) = 0.01 \sin 3\pi x$

$\left.\dfrac{\partial u}{\partial t}\right|_{t=0} = 0$

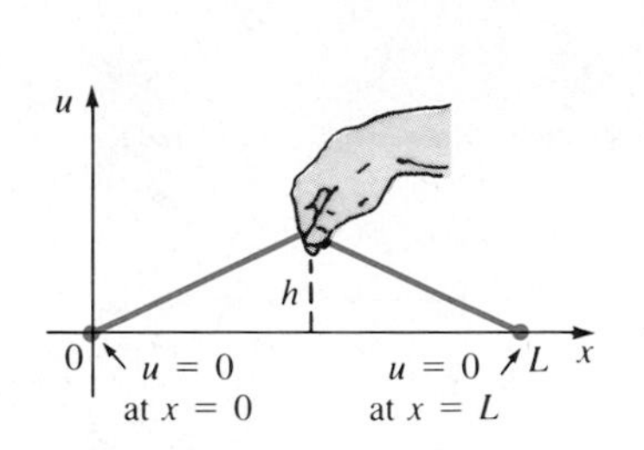

Figure 10.19

13. $u(0, t) = 0, \quad u(L, t) = 0$

$$u(x, 0) = \begin{cases} \dfrac{2hx}{L}, & 0 < x < \dfrac{L}{2} \\ 2h\left(1 - \dfrac{x}{L}\right), & \dfrac{L}{2} \le x < L, \end{cases}$$

$$\left.\frac{\partial u}{\partial t}\right|_{t=0} = 0$$

The constant h is positive but small compared to L. This is referred to as the "plucked string" problem (see Figure 10.19).

14. A string is stretched and secured on the x-axis at $x = 0$ and $x = \pi$ for $t > 0$. If the transverse vibrations take place in a medium imparting a resistance proportional to the instantaneous velocity, then the wave equation takes on the form

$$\frac{\partial^2 u}{\partial x^2} = \frac{\partial^2 u}{\partial t^2} + 2\beta\frac{\partial u}{\partial t}, \quad 0 < \beta < 1, \quad t > 0.$$

Find the displacement $u(x, t)$ if the string starts from rest from the initial displacement $f(x)$.

[10.5.3] In Problems 15–20 find the steady-state temperature for a rectangular plate with boundary conditions as given.

15. $u(0, y) = 0, \quad u(a, y) = 0$
$u(x, 0) = 0, \quad u(x, b) = f(x)$

16. $u(0, y) = 0, \quad u(a, y) = 0$
$\left.\dfrac{\partial u}{\partial y}\right|_{y=0} = 0, \quad u(x, b) = f(x)$

17. $u(0, y) = 0, \quad u(a, y) = 0$
$u(x, 0) = f(x), \quad u(x, b) = 0$

18. $\left.\dfrac{\partial u}{\partial x}\right|_{x=0} = 0, \quad \left.\dfrac{\partial u}{\partial x}\right|_{x=a} = 0$
$u(x, 0) = x, \quad u(x, b) = 0.$

19. $u(0, y) = 0, \quad u(1, y) = 1 - y$
$\left.\dfrac{\partial u}{\partial y}\right|_{y=0} = 0, \quad \left.\dfrac{\partial u}{\partial y}\right|_{y=1} = 0$

20. $u(0, y) = g(y), \quad \left.\dfrac{\partial u}{\partial x}\right|_{x=1} = 0$
$\left.\dfrac{\partial u}{\partial y}\right|_{y=0} = 0, \quad \left.\dfrac{\partial u}{\partial y}\right|_{y=\pi} = 0$

In Problems 21 and 22 find the steady-state temperature in the given semi-infinite plate extending in the positive y-direction. In each case assume $u(x, y)$ is bounded as $y \to \infty$.

21.

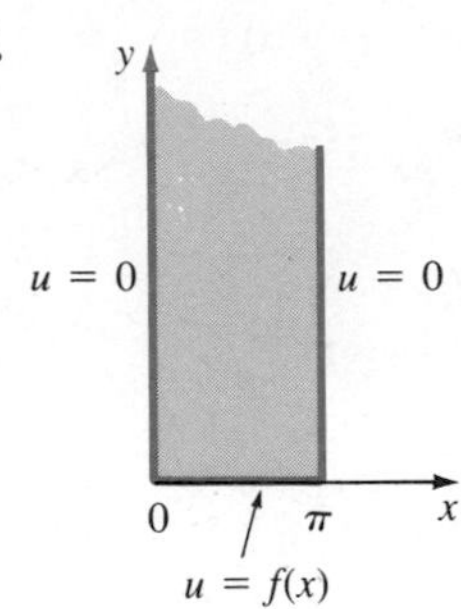

Figure 10.20

22.

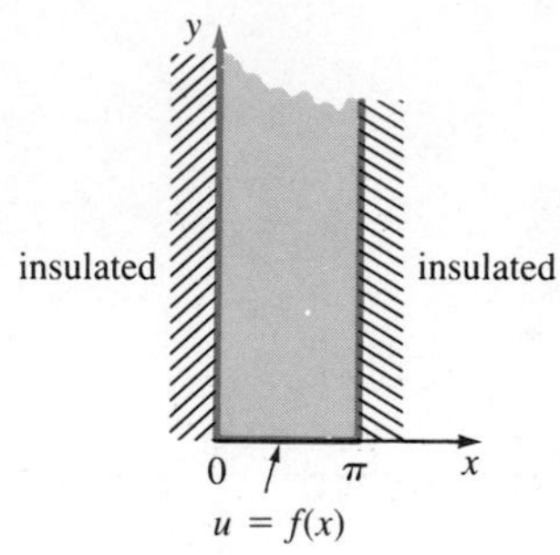

Figure 10.21

Miscellaneous Problems

23. Solve the boundary-value problem consisting of the heat equation (1) and the conditions

$$u(0, t) = 100 \qquad u(L, t) = 100$$
$$u(x, 0) = f(x), \qquad 0 < x < L$$

by means of the substitution $v = u - 100$.

CHAPTER 10 SUMMARY

A set of functions $\{\phi_n(x)\}$, $n = 0, 1, 3, \ldots$, is said to be **orthogonal** on an interval $[a, b]$ if

$$\int_a^b \phi_m(x)\phi_n(x)\,dx \begin{cases} =0, & m \neq n, \\ \neq 0, & m = n. \end{cases}$$

A function f defined on the interval $(-p, p)$ can formally be expanded in terms of the trigonometric functions in the orthogonal set

$$1, \cos\frac{\pi}{p}x, \cos\frac{2\pi}{p}x, \ldots, \sin\frac{\pi}{p}x, \sin\frac{2\pi}{p}x, \ldots.$$

We say
$$f(x) = \frac{a_0}{2} + \sum_{n=1}^{\infty}\left(a_n\cos\frac{n\pi}{p}x + b_n\sin\frac{n\pi}{p}x\right),$$

where
$$a_0 = \frac{1}{p}\int_{-p}^{p} f(x)\,dx,$$

$$a_n = \frac{1}{p}\int_{-p}^{p} f(x)\cos\frac{n\pi}{p}x\,dx, \qquad b_n = \frac{1}{p}\int_{-p}^{p} f(x)\sin\frac{n\pi}{p}x\,dx,$$

is the **Fourier series** corresponding to f. The coefficients are obtained using the concept of orthogonality. If f is an even function on the interval, then $b_n = 0$,

$n = 1, 2, 3, \ldots$, and

$$a_0 = \frac{2}{p}\int_0^p f(x)\,dx, \qquad a_n = \frac{2}{p}\int_0^p f(x)\cos\frac{n\pi}{p}x\,dx. \tag{1}$$

Similarly, if f is odd on the interval, then $a_n = 0$, $n = 0, 1, 2, 3, \ldots$,

and
$$b_n = \frac{2}{p}\int_0^p f(x)\sin\frac{n\pi}{p}x\,dx. \tag{2}$$

When f is defined on an interval $(0, L)$, it can be neither even nor odd. Nonetheless we can expand f in a cosine or a sine series. By defining $p = L$, we obtain a cosine series by using the coefficients (1) and a sine series by using (2). Such series are known as **half-range expansions.**

A particular solution of a linear partial differential equation in two variables can possibly be found by assuming a solution in the form of a product $u = XY$, where X is a function of x only and Y is a function of y only. If applicable, this **method of separation of variables** leads to two ordinary differential equations.

A **boundary-value problem** consists of finding a function that satisfies a partial differential equation as well as side conditions consisting of perhaps both boundary conditions and initial conditions. We applied the method of separation of variables to obtain solutions of certain boundary-value problems involving the heat equation, the wave equation, and Laplace's equation. The procedure consisted of five basic steps.

(i) Separate the variables.

(ii) Solve the separated ordinary differential equations and find the eigenvalues and eigenfunctions of the problem.

(iii) Form the products u_n.

(iv) Use the superposition principle to form an infinite series of the functions u_n.

(v) After using a boundary, or the initial condition(s), obtain the coefficients in the series by making an appropriate identification with a half-range sine or cosine expansion.

CHAPTER 10 REVIEW EXERCISES

Answers to odd-numbered problems begin on page A–36.

In Problems 1–6 fill in the blank or answer true/false.

1. The functions $f(x) = x^2 - 1$ and $f(x) = x^5$ are orthogonal on $[-\pi, \pi]$. ______

2. The product of an odd function with an odd function is ______.

3. To expand $f(x) = |x| + 1$, $-\pi < x < \pi$, in an appropriate Fourier series we would use a ______ series.

4. Since $f(x) = x^2$, $0 < x < 2$, is not an even function, it cannot be expanded in a Fourier cosine series. ______

5. The Fourier series of $f(x) = \begin{cases} 3, & -\pi < x < 0 \\ 0, & 0 < x < \pi \end{cases}$ will converge to ______ at $x = 0$.

6. Separation of variables will always yield particular solutions of a homogeneous linear partial differential equation. ______

7. Show that the set

$$\sin \frac{\pi}{2L} x, \quad \sin \frac{3\pi}{2L} x, \quad \sin \frac{5\pi}{2L} x, \ldots,$$

is orthogonal on the interval $[0, L]$.

8. Find the norm of each function in Problem 7. Construct an orthonormal set.

9. Expand $f(x) = |x| - x$, $-1 < x < 1$, in a Fourier series.

10. Expand $f(x) = 2x^2 - 1$, $-1 < x < 1$, in a Fourier series.

11. Expand $f(x) = e^{-x}$, $0 < x < 1$, in a cosine series.

12. Expand the function given in Problem 11 in a sine series.

13. Solve: $\dfrac{\partial u}{\partial x} = xye^x + 1$

14. Solve: $\dfrac{\partial^2 u}{\partial x \partial y} = 8y^3 \cos 6x$

15. Solve: $\dfrac{\partial u}{\partial x} + yu = 1$

16. Solve: $\dfrac{\partial^2 u}{\partial x^2} - y^2 \dfrac{\partial u}{\partial x} = 0$

17. Use separation of variables to find product solutions of

$$\frac{\partial^2 u}{\partial x \partial y} = u.$$

18. Use separation of variables to find product solutions of

$$\frac{\partial^2 u}{\partial x^2} + \frac{\partial^2 u}{\partial y^2} + 2\frac{\partial u}{\partial x} + 2\frac{\partial u}{\partial y} = 0.$$

Is it possible to choose a constant of separation so that both X and Y are oscillatory functions?

19. Find the steady-state temperature $u(x, y)$ in the square plate shown in Figure 10.22. The boundary conditions are as indicated.

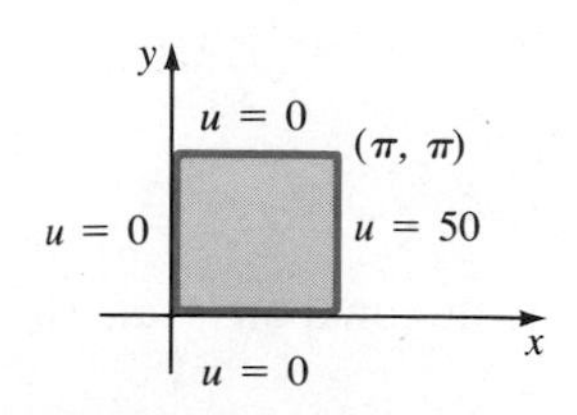

Figure 10.22

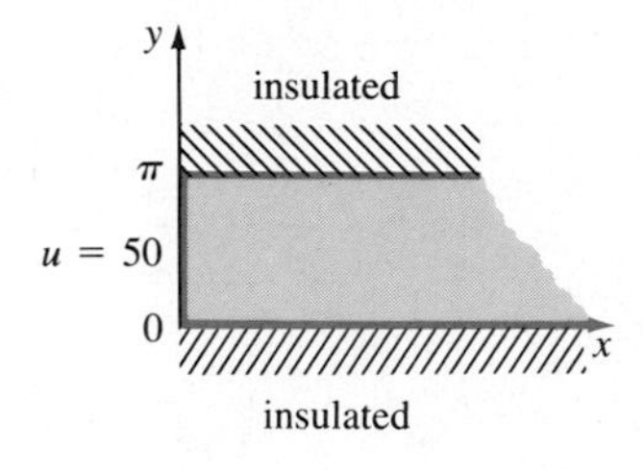

Figure 10.23

20. Find the steady-state temperature $u(x, y)$ in the semi-infinite plate of width π shown in Figure 10.23. The boundary conditions are as indicated.

21. Solve Problem 20 if the boundaries $y = 0$ and $y = \pi$ are held at a constant zero degrees for all time.

22. **(a)** Use the substitutions to $\xi = x + at$ and $\eta = x - at$ and the chain rule to show that the wave equation

$$a^2 \frac{\partial^2 u}{\partial x^2} = \frac{\partial^2 u}{\partial t^2} \qquad \text{becomes} \qquad \frac{\partial^2 u}{\partial \xi \partial \eta} = 0.$$

(b) By integrating show that the last equation has the solution $u = f(\xi) + g(\eta)$, where f and g are arbitrary twice differentiable functions. Hence a solution of the wave equation is given by $u(x, t) = f(x + at) + g(x - at)$. This is known as **d'Alembert's solution*** of the wave equation. The solution can be interpreted as a superposition of two traveling waves: $f(x + at)$ is a wave moving to the left and $g(x - at)$ is a wave moving to the right. Each wave moves with a velocity a.

(c) Verify that $u(x, t) = f(x + at) + g(x - at)$ satisfies the wave equation.

---

* **Jean le Rond D'Alembert** (1717–1783) Educated in law and medicine, the Frenchman D'Alembert devoted his life to the study of physics and mathematics. D'Alembert was also a collaborator with the famous philosopher Denis Diderot on the latter's *Encyclopédie.*

Appendices

IMPORTANT TERMS

Gamma function
generalized factorial function
minor
cofactor
Cramer's rule
complex number
real part
imaginary part
conjugate
modulus
polar form
argument

I Uniqueness Proof

Consider the initial-value problem

$$a_2 y'' + a_1 y' + a_0 y = g(x) \tag{1}$$

$$y(0) = y_0, \qquad y'(0) = y'_0, \tag{2}$$

where a_2, a_1, and a_0 are positive constants and $g(x)$ is continuous for all x. To show that this problem has a unique solution, we first prove that $y = 0$ is the only solution of the problem

$$a_2 y'' + a_1 y' + a_0 y = 0 \tag{3}$$

$$y(0) = 0, \qquad y'(0) = 0. \tag{4}$$

Multiplying (3) by y' and integrating over the interval $0 \le x \le t$ give

$$a_2 \int_0^t y''y'\,dx + a_1 \int_0^t (y')^2\,dx + a_0 \int_0^t yy'\,dx = \int_0^t 0\,dx$$

$$\frac{a_2}{2}[y'(t)^2 - y'(0)^2] + a_1 \int_0^t (y')^2\,dx + \frac{a_0}{2}[y(t)^2 - y(0)^2] = 0.$$

In view of the imposed initial conditions (4), we find

$$\frac{a_2}{2}(y')^2 + a_1 \int_0^t (y')^2\,dx + \frac{a_0}{2} y^2 = 0$$

for every t. It follows that the sum of three nonnegative quantities can be zero only when $y = 0$.

To return to the original problem let us now suppose that y_1 and y_2 are two *different* solutions of equation (1) and that both solutions satisfy the initial conditions (2). If we define the function $u = y_1 - y_2$, then

$$u(0) = y_1(0) - y_2(0) = y_0 - y_0 = 0$$

$$u'(0) = y'_1(0) - y'_2(0) = y'_0 - y'_0 = 0$$

and

$$\begin{aligned} a_2 u'' + a_1 u' + a_0 u &= a_2[y''_1 - y''_2] + a_1[y'_1 - y'_2] + a_0[y_1 - y_2] \\ &= (a_2 y''_1 + a_1 y'_1 + a_0 y_1) - (a_2 y''_2 + a_1 y'_2 + a_0 y_2) \\ &= g(x) - g(x) = 0. \end{aligned}$$

Hence u satisfies the initial-value problem

$$a_2u'' + a_1u' + a_0u = 0$$
$$u(0) = 0, \qquad u'(0) = 0.$$

Since the only solution of this last problem is $u = 0$, we have $y_1 - y_2 = 0$ or $y_1 = y_2$.

II Gamma Function

Euler's integral definition of the **Gamma function*** is

$$\Gamma(x) = \int_0^\infty t^{x-1}e^{-t}\,dt. \tag{1}$$

Convergence of the integral requires that $x - 1 > -1$ or $x > 0$. The recurrence relation

$$\Gamma(x + 1) = x\Gamma(x), \tag{2}$$

which we have seen in Section 6.4, can be obtained from (1) by employing integration by parts. Now when $x = 1$

$$\Gamma(1) = \int_0^\infty e^{-t}\,dt = 1$$

and thus (2) gives

$$\Gamma(2) = 1\Gamma(1) = 1$$
$$\Gamma(3) = 2\Gamma(2) = 2 \cdot 1$$
$$\Gamma(4) = 3\Gamma(3) = 3 \cdot 2 \cdot 1,$$

and so on. In this manner it is seen that when n is a positive integer

$$\Gamma(n + 1) = n!.$$

* This function was first defined by Leonhard Euler in his text *Institutiones calculi integralis* published in 1768.

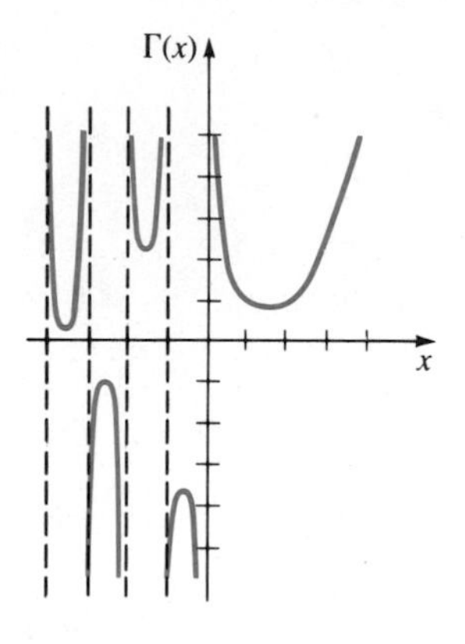

Figure A.1

For this reason the Gamma function is often called the **generalized factorial function**.

Although the integral form (1) does not converge for $x < 0$, it can be shown by means of alternative definitions that the Gamma function is defined for all real and complex numbers *except* $x = -n$, $n = 0, 1, 2, \ldots$. As a consequence (2) is actually valid for $x \neq -n$. Considered as a function of a real variable x, the graph of $\Gamma(x)$ is as given in Figure A.1. Observe that the nonpositive integers correspond to vertical asymptotes of the graph.

In Problems 27–33 of Exercises 6.4 we utilized the fact that $\Gamma(\frac{1}{2}) = \sqrt{\pi}$. This result can be derived from (1) by setting $x = 1/2$:

$$\Gamma(\tfrac{1}{2}) = \int_0^\infty t^{-1/2} e^{-t}\, dt. \tag{3}$$

By letting $t = u^2$, (3) can be written as

$$\Gamma(\tfrac{1}{2}) = 2\int_0^\infty e^{-u^2}\, du.$$

But

$$\int_0^\infty e^{-u^2}\, du = \int_0^\infty e^{-v^2}\, dv$$

and so

$$[\Gamma(\tfrac{1}{2})]^2 = \left(2\int_0^\infty e^{-u^2}\, du\right)\left(2\int_0^\infty e^{-v^2}\, dv\right)$$
$$= 4\int_0^\infty \int_0^\infty e^{-(u^2+v^2)}\, du\, dv.$$

Switching to polar coordinates $u = r\cos\theta$, $v = r\sin\theta$ enables us to evaluate the double integral,

$$4\int_0^\infty \int_0^\infty e^{-(u^2+v^2)}\, du\, dv = 4\int_0^{\pi/2} \int_0^\infty e^{-r^2} r\, dr\, d\theta = \pi.$$

Hence

$$[\Gamma(\tfrac{1}{2})]^2 = \pi \quad \text{or} \quad \Gamma(\tfrac{1}{2}) = \sqrt{\pi}.$$

EXAMPLE 1

Evaluate $\Gamma(-1/2)$.

SOLUTION In view of (2) it follows that with $x = -1/2$

$$\Gamma\left(\frac{1}{2}\right) = -\frac{1}{2}\Gamma\left(-\frac{1}{2}\right).$$

Therefore

$$\Gamma\left(-\frac{1}{2}\right) = -2\Gamma\left(\frac{1}{2}\right)$$
$$= -2\sqrt{\pi}.$$

Appendix II Exercises

Answers to odd-numbered problems begin on page A–37.

1. Evaluate
   (a) $\Gamma(5)$ (b) $\Gamma(7)$ (c) $\Gamma(-3/2)$ (d) $\Gamma(-5/2)$
2. Use (1) and the fact that $\Gamma(6/5) = 0.92$ to evaluate $\int_0^\infty x^5 e^{-x^5}\,dx$. [*Hint:* Let $t = x^5$.]
3. Use (1) and the fact that $\Gamma(5/3) = 0.89$ to evaluate $\int_0^\infty x^4 e^{-x^3}\,dx$.
4. Evaluate $\int_0^1 x^3\left(\ln \frac{1}{x}\right)^3 dx$. [*Hint:* Let $t = -\ln x$.]
5. Use the fact that $\Gamma(x) > \int_0^1 t^{x-1}e^{-t}\,dt$ to show that $\Gamma(x)$ is unbounded as $x \to 0^+$.
6. Use (1) to derive (2) for $x > 0$.

III Table of Laplace Transforms

$f(t)$	$\mathscr{L}\{f(t)\} = F(s)$
1. 1	$\frac{1}{s}$
2. t^n, $n = 1, 2, 3, \ldots$	$\frac{n!}{s^{n+1}}$
3. $t^{-1/2}$	$\sqrt{\frac{\pi}{s}}$
4. e^{at}	$\frac{1}{s-a}$
5. $\sin kt$	$\frac{k}{s^2+k^2}$
6. $\cos kt$	$\frac{s}{s^2+k^2}$
7. $\sinh kt$	$\frac{k}{s^2-k^2}$
8. $\cosh kt$	$\frac{s}{s^2-k^2}$

$f(t)$	$\mathscr{L}\{f(t)\} = F(s)$
9. $e^{at}f(t)$	$F(s-a)$
10. $f(t-a)\mathscr{U}(t-a),\ a>0$	$e^{-as}F(s)$
11. $t^n f(t),\quad n = 1, 2, 3, \ldots$	$(-1)^n \dfrac{d^n}{ds^n} F(s)$
12. $f^{(n)}(t),\quad n = 1, 2, 3, \ldots$	$s^n F(s) - s^{n-1}f(0) - \cdots - f^{(n-1)}(0)$
13. $\int_0^t f(\tau)g(t-\tau)\,d\tau$	$F(s)G(s)$
14. $\delta(t-t_0)$	e^{-st_0}
15. $t^n e^{at},\quad n = 1, 2, 3, \ldots$	$\dfrac{n!}{(s-a)^{n+1}}$
16. $e^{at}\sin kt$	$\dfrac{k}{(s-a)^2+k^2}$
17. $e^{at}\cos kt$	$\dfrac{s-a}{(s-a)^2+k^2}$
18. $t\sin kt$	$\dfrac{2ks}{(s^2+k^2)^2}$
19. $t\cos kt$	$\dfrac{s^2-k^2}{(s^2+k^2)^2}$
20. $\sin kt - kt\cos kt$	$\dfrac{2k^3}{(s^2+k^2)^2}$
21. $\sin kt + kt\cos kt$	$\dfrac{2ks^2}{(s^2+k^2)^2}$
22. $\sinh kt - \sin kt$	$\dfrac{2k^3}{s^4-k^4}$
23. $\cosh kt - \cos kt$	$\dfrac{2k^2 s}{s^4-k^4}$
24. $1-\cos kt$	$\dfrac{k^2}{s(s^2+k^2)}$
25. $kt - \sin kt$	$\dfrac{k^3}{s^2(s^2+k^2)}$
26. $\dfrac{a\sin bt - b\sin at}{ab(a^2-b^2)}$	$\dfrac{1}{(s^2+a^2)(s^2+b^2)}$
27. $\dfrac{\cos bt - \cos at}{a^2-b^2}$	$\dfrac{s}{(s^2+a^2)(s^2+b^2)}$

IV Review of Determinants and Cramer's Rule

The determinant of a 2×2 matrix $\mathbf{A}$ is defined by

$$\begin{vmatrix} a_{11} & a_{12} \\ a_{21} & a_{22} \end{vmatrix} = a_{11}a_{22} - a_{12}a_{21}.$$

Minors and Cofactors

For an $n \times n$ matrix $\mathbf{A}$, let a_{ij} be the entry in the ith row and jth column. The **minor** M_{ij} associated with a_{ij} is the determinant of the $(n-1) \times (n-1)$ matrix obtained by deleting the ith row and the jth column of the matrix. The **cofactor** C_{ij} associated with a_{ij} is a *signed minor*, specifically

$$C_{ij} = (-1)^{i+j}M_{ij}.$$

EXAMPLE 1

The cofactors of the entries in the 3×3 matrix

$$\begin{pmatrix} 2 & 4 & 7 \\ 1 & 2 & 3 \\ 1 & 5 & 3 \end{pmatrix} \tag{1}$$

are

$$C_{11} = (-1)^{1+1}M_{11} = \begin{vmatrix} 2 & 3 \\ 5 & 3 \end{vmatrix} = -9 \qquad C_{12} = (-1)^{1+2}M_{12} = (-1)\begin{vmatrix} 1 & 3 \\ 1 & 3 \end{vmatrix} = 0 \qquad C_{13} = (-1)^{1+3}M_{13} = \begin{vmatrix} 1 & 2 \\ 1 & 5 \end{vmatrix} = 3$$

$$C_{21} = (-1)^{2+1}M_{21} = (-1)\begin{vmatrix} 4 & 7 \\ 5 & 3 \end{vmatrix} = 23 \qquad C_{22} = (-1)^{2+2}M_{22} = \begin{vmatrix} 2 & 7 \\ 1 & 3 \end{vmatrix} = -1 \qquad C_{23} = (-1)^{2+3}M_{23} = (-1)\begin{vmatrix} 2 & 4 \\ 1 & 5 \end{vmatrix} = -6$$

$$C_{31} = (-1)^{3+1}M_{31} = \begin{vmatrix} 4 & 7 \\ 2 & 3 \end{vmatrix} = -2 \qquad C_{32} = (-1)^{3+2}M_{32} = (-1)\begin{vmatrix} 2 & 7 \\ 1 & 3 \end{vmatrix} = 1 \qquad C_{33} = (-1)^{3+3}M_{33} = \begin{vmatrix} 2 & 4 \\ 1 & 2 \end{vmatrix} = 0$$

Expansion by Cofactors

It can be proved that a determinant can be **expanded** in terms of cofactors:

> Multiply the entries a_{ij} in any row (or column) by their corresponding cofactors C_{ij} and add the n products.

Thus a 3×3 determinant* can be expanded into three 2×2 determinants, a 4×4 determinant can be expanded into four 3×3 determinants, and so on.

EXAMPLE 2

Evaluate the determinant of the matrix in (1).

SOLUTION Expanding by the first row gives

$$\begin{vmatrix} 2 & 4 & 7 \\ 1 & 2 & 3 \\ 1 & 5 & 3 \end{vmatrix} = 2\begin{vmatrix} 2 & 3 \\ 5 & 3 \end{vmatrix} + 4(-1)\begin{vmatrix} 1 & 3 \\ 1 & 3 \end{vmatrix} + 7\begin{vmatrix} 1 & 2 \\ 1 & 5 \end{vmatrix} = 3.$$

Alternatively, we can expand the determinant by, say, the second column:

$$\begin{vmatrix} 2 & 4 & 7 \\ 1 & 2 & 3 \\ 1 & 5 & 3 \end{vmatrix} = 4(-1)\begin{vmatrix} 1 & 3 \\ 1 & 3 \end{vmatrix} + 2\begin{vmatrix} 2 & 7 \\ 1 & 3 \end{vmatrix} + 5(-1)\begin{vmatrix} 2 & 7 \\ 1 & 3 \end{vmatrix} = 3.$$

We note that if a determinant has a row (or a column) containing many zero entries, then wisdom dictates that we expand the determinant by that row (or column).

Cramer's Rule

Determinants are sometimes useful in solving algebraic systems of n linear equations in n unknowns:

$$\begin{aligned} a_{11}x_1 + a_{12}x_2 + \cdots + a_{1n}x_n &= b_1 \\ a_{21}x_1 + a_{22}x_2 + \cdots + a_{2n}x_n &= b_2 \\ \vdots \qquad\qquad & \quad \vdots \\ a_{n1}x_1 + a_{n2}x_2 + \cdots + a_{nn}x_n &= b_n. \end{aligned} \tag{2}$$

Let $\mathbf{A}$ be the matrix of coefficients of (2) and let

$$\det \mathbf{A} = \begin{vmatrix} a_{11} & a_{12} & \cdots & a_{1n} \\ a_{21} & a_{22} & \cdots & a_{2n} \\ \vdots & & & \vdots \\ a_{n1} & a_{n2} & \cdots & a_{nn} \end{vmatrix}.$$

---

* Even though a determinant of a matrix of numbers is a number, it is sometimes convenient to refer to it as if it were an array.

If

$$\det \mathbf{A}_k = \begin{vmatrix} a_{11} & a_{12} & \cdots & a_{1,k-1} & b_1 & a_{1,k+1} & \cdots & a_{1n} \\ a_{21} & a_{22} & \cdots & a_{2,k-1} & b_2 & a_{2,k+1} & \cdots & a_{2n} \\ \vdots & & & \vdots & \vdots & \vdots & & \vdots \\ a_{n1} & a_{n2} & \cdots & a_{n,k-1} & b_n & a_{n,k+1} & \cdots & a_{nn} \end{vmatrix}$$

(kth column ↓ at the b column)

is the same as det **A** except that its kth column has been replaced by the column

$$\begin{matrix} b_1 \\ b_2 \\ \vdots \\ b_n, \end{matrix}$$

then (2) has the unique solution

$$x_1 = \frac{\det \mathbf{A}_1}{\det \mathbf{A}}, \; x_2 = \frac{\det \mathbf{A}_2}{\det \mathbf{A}}, \ldots, x_n = \frac{\det \mathbf{A}_n}{\det \mathbf{A}}. \tag{3}$$

whenever $\det \mathbf{A} \neq 0$. This method of solving (2) by determinants is known as **Cramer's rule.***

EXAMPLE 3 Solve the system

$$\begin{aligned} 3x + 2y + z &= 7 \\ x - y + 3z &= 3 \\ 5x + 4y - 2z &= 1 \end{aligned}$$

by Cramer's rule.

SOLUTION The solution requires calculation of four determinants:

$$\det \mathbf{A} = \begin{vmatrix} 3 & 2 & 1 \\ 1 & -1 & 3 \\ 5 & 4 & -2 \end{vmatrix} = 13, \quad \det \mathbf{A}_1 = \begin{vmatrix} 7 & 2 & 1 \\ 3 & -1 & 3 \\ 1 & 4 & -2 \end{vmatrix} = -39$$

* This rule was named after **Gabriel Cramer** (1704–1752), a Swiss mathematician who was the first to publish this result in 1750.

$$\det \mathbf{A}_2 = \begin{vmatrix} 3 & 7 & 1 \\ 1 & 3 & 3 \\ 5 & 1 & -2 \end{vmatrix} = 78, \quad \det \mathbf{A}_3 = \begin{vmatrix} 3 & 2 & 7 \\ 1 & -1 & 3 \\ 5 & 4 & 1 \end{vmatrix} = 52.$$

Hence (3) gives

$$x = \frac{\det \mathbf{A}_1}{\det \mathbf{A}} = -3, \quad y = \frac{\det \mathbf{A}_2}{\det \mathbf{A}} = 6, \quad z = \frac{\det \mathbf{A}_3}{\det \mathbf{A}} = 4.$$

Homogeneous Systems

If $b_i = 0$, $i = 1, 2, \ldots, n$, then the system of equations (2) is said to be **homogeneous**. If at least one of the b_i is not zero, the system is **nonhomogeneous**. Now if $\det \mathbf{A} \neq 0$, (3) implies that the only solution of a homogeneous system is $x_1 = 0, x_2 = 0, \ldots, x_n = 0$. If $\det \mathbf{A} = 0$, then a homogeneous system of n linear equations in n unknowns has infinitely many solutions. These solutions can be found by solving the system through elimination. If $\det \mathbf{A} = 0$, then a nonhomogeneous system may either have infinitely many solutions or no solution at all.

Appendix IV Exercises

Answers to odd-numbered problems begin on page A–37.

In Problems 1–8 evaluate the given determinant.

1. $\begin{vmatrix} 2 & 4 & 6 \\ -1 & 5 & 1 \\ 0 & 2 & -3 \end{vmatrix}$

2. $\begin{vmatrix} 1 & 4 & 2 \\ -2 & 6 & 3 \\ 9 & 8 & 4 \end{vmatrix}$

3. $\begin{vmatrix} 2 & 0 & 5 \\ 0 & 7 & 9 \\ -6 & 1 & 4 \end{vmatrix}$

4. $\begin{vmatrix} 79 & 81 & 40 \\ 22 & 16 & 59 \\ 0 & 0 & 0 \end{vmatrix}$

5. $\begin{vmatrix} 1 & 2 & 3 & 4 \\ 1 & 1 & 0 & 0 \\ 8 & 7 & 0 & 0 \\ 9 & 5 & 3 & 0 \end{vmatrix}$

6. $\begin{vmatrix} 1 & 0 & 9 & 0 & 3 \\ 2 & 1 & 7 & 0 & 0 \\ 0 & 0 & 2 & 0 & 0 \\ -1 & 1 & 5 & 2 & 2 \\ 2 & 2 & 8 & 1 & 1 \end{vmatrix}$

7. $\begin{vmatrix} e^t & e^{3t} & e^{-t} \\ e^t & 3e^{3t} & -e^{-t} \\ e^t & 9e^{3t} & e^{-t} \end{vmatrix}$

8. $\begin{vmatrix} e^{2t} & \sin t & \cos t \\ 2e^{2t} & \cos t & -\sin t \\ 4e^{2t} & -\sin t & -\cos t \end{vmatrix}$

In Problems 9–12 use Cramer's rule to solve the given system of equations.

9. $2x + y = 1$
$3x + 2y = -2$

10. $5x + 4y = -1$
$10x - 6y = 5$

11. $\begin{aligned} x + 2y + z &= 8 \\ 2x - 2y + 2z &= 7 \\ x - 4y + 3z &= 1 \end{aligned}$

12. $\begin{aligned} 4x + 3y + 2z &= 8 \\ -x \quad\quad + 2z &= 12 \\ 3x + 2y + z &= 3 \end{aligned}$

13. Given the system

$$\begin{aligned} x - y + 2z &= 0 \\ 2x + y - z &= 0 \\ 4x - y + 3x &= 0. \end{aligned}$$

(a) If **A** denotes the matrix of coefficients, show that det $\mathbf{A} = 0$.

(b) Show that the system has infinitely many solutions.

(c) Explain the geometric significance of the system.

14. Given the system

$$\begin{aligned} a_1x + b_1y &= c_1 \\ a_2x + b_2y &= c_2. \end{aligned}$$

Let **A** denote the matrix of coefficients.

(a) Explain the geometric significance of det $\mathbf{A} \neq 0$.

(b) Explain the geometric significance of det $\mathbf{A} = 0$.

V Complex Numbers and Euler's Formula

A **complex number** is any expression of the form

$$z = a + bi, \qquad \text{where} \qquad i^2 = -1.$$

The real numbers a and b are called the **real** and **imaginary parts** of z, respectively. In practice the symbol i is written $i = \sqrt{-1}$. The number $\bar{z} = a - bi$ is called the **conjugate** of z.

EXAMPLE 1

From the properties of radicals,

$$\sqrt{-25} = \sqrt{25}\sqrt{-1} = 5i.$$

EXAMPLE 2

The conjugates of the complex numbers $z_1 = 4 + 5i$ and $z_2 = 3 - 2i$ are, in turn, $\bar{z}_1 = 4 - 5i$ and $\bar{z}_2 = 3 + 2i$.

Sum, Difference, and Product

The **sum**, **difference**, and **product** of two complex numbers $z_1 = a_1 + b_1 i$ and $z_2 = a_2 + b_2 i$ are defined as follows:

(i) $z_1 + z_2 = (a_1 + a_2) + (b_1 + b_2)i$
(ii) $z_1 - z_2 = (a_1 - a_2) + (b_1 - b_2)i$
(iii) $z_1 z_2 = (a_1 a_2 - b_1 b_2) + (a_1 b_2 + b_1 a_2)i.$

In other words, to add or subtract two complex numbers we simply add or subtract the corresponding real and imaginary parts. To multiply two complex numbers we use the distributive law and the fact that $i^2 = -1$.

EXAMPLE 3

If $z_1 = 4 + 5i$ and $z_2 = 3 - 2i$ then

$$\begin{aligned}
z_1 + z_2 &= (4 + 3) + (5 + (-2))i = 7 + 3i \\
z_1 - z_2 &= (4 - 3) + (5 - (-2))i = 1 + 7i \\
z_1 z_2 &= (4 + 5i)(3 - 2i) \\
&= (4 + 5i)3 + (4 + 5i)(-2i) \\
&= 12 + 15i - 8i - 10i^2 \\
&= (12 + 10) + (15 - 8)i = 22 + 7i.
\end{aligned}$$

The product of a complex number $z = a + bi$ and its conjugate $\bar{z} = a - bi$ is the real number

$$z\bar{z} = a^2 + b^2 \tag{1}$$

Quotient

The quotient of two complex numbers z_1 and z_2 is found by multiplying numerator and denominator of z_1/z_2 by the conjugate of the denominator z_2 and using (1). The next example illustrates the procedure.

EXAMPLE 4

$$\begin{aligned}
\frac{z_1}{z_2} &= \frac{4 + 5i}{3 - 2i} \\
&= \frac{4 + 5i}{3 - 2i}\,\frac{3 + 2i}{3 + 2i} \\
&= \frac{12 + 15i + 8i + 10i^2}{9 + 4} = \frac{2}{13} + \frac{23}{13}i
\end{aligned}$$

Geometric Interpretation

By taking a and b to be the x and y coordinates of a point in the plane, a complex number $z = a + bi$ can be interpreted as a **vector** from the origin terminating at (a, b) (see Figure A.2). The length of the vector is called the **modulus**

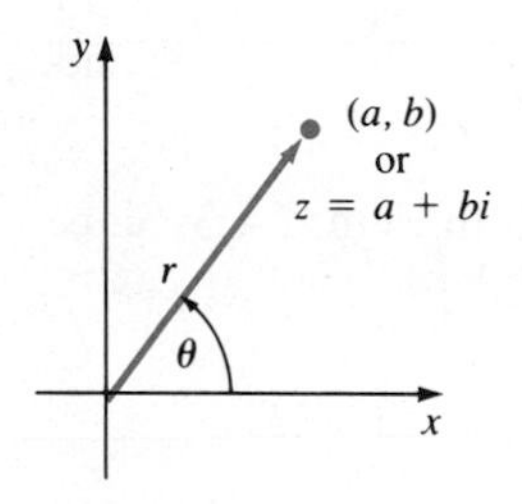

Figure A.2

of z and is written as r or $|z|$. From the Pythagorean theorem it follows that

$$r = |z| = \sqrt{a^2 + b^2}.$$

If θ is an angle the vector makes with the positive x-axis, then from Figure A.2 we see

$$a = r\cos\theta \qquad \text{and} \qquad b = r\sin\theta.$$

Thus

$$z = r\cos\theta + ir\sin\theta \qquad \text{or} \qquad z = r(\cos\theta + i\sin\theta). \tag{2}$$

This latter form is called the **polar form** of the complex number z. The angle θ is called an **argument** of z.

Euler's Formula

The power series $e^X = \sum_{n=0}^{\infty} \frac{X^n}{n!}$ is known to converge for all real and complex numbers. If we let $X = i\theta$, θ a real number, then

$$e^{i\theta} = \sum_{n=0}^{\infty} \frac{(i\theta)^n}{n!} = 1 + i\theta + \frac{i^2\theta^2}{2!} + \frac{i^3\theta^3}{3!} + \frac{i^4\theta^4}{4!} + \frac{i^5\theta^5}{5!} + \frac{i^6\theta^6}{6!} + \frac{i^7\theta^7}{7!} + \cdots. \tag{3}$$

Now $i^2 = -1, i^3 = -i, i^4 = 1, i^5 = i$, and so on. Thus (3) can be separated into real and imaginary parts

$$e^{i\theta} = \left(1 - \frac{\theta^2}{2!} + \frac{\theta^4}{4!} - \frac{\theta^6}{6!} + \cdots\right) + i\left(\theta - \frac{\theta^3}{3!} + \frac{\theta^5}{5!} - \frac{\theta^7}{7!} + \cdots\right). \tag{4}$$

But from calculus we recall that

$$\cos\theta = \sum_{n=0}^{\infty} \frac{(-1)^n}{(2n)!}\theta^{2n} \qquad \text{and} \qquad \sin\theta = \sum_{n=0}^{\infty} \frac{(-1)^n}{(2n+1)!}\theta^{2n+1},$$

where each series converges for every real number θ. Hence (4) can be written as

$$e^{i\theta} = \cos\theta + i\sin\theta. \tag{5}$$

This last result is known as **Euler's formula**. Note that in view of (5), the polar form (2) of a complex number can be expressed in the compact form

$$z = re^{i\theta}. \tag{6}$$

EXAMPLE 5

Find the polar form (6) of $z = 1 - i$.

SOLUTION The graph of the complex number is given in Figure A.3. Since $a = 1$ and $b = -1$, the modulus of z is

$$r = \sqrt{1^2 + (-1)^2} = \sqrt{2}.$$

As seen in the Figure A.3, $\tan\theta = -1$ and so we can take an argument of z to be $\theta = -\pi/4$. Therefore, the polar form of the number is

$$z = \sqrt{2}e^{-i\pi/4}.$$

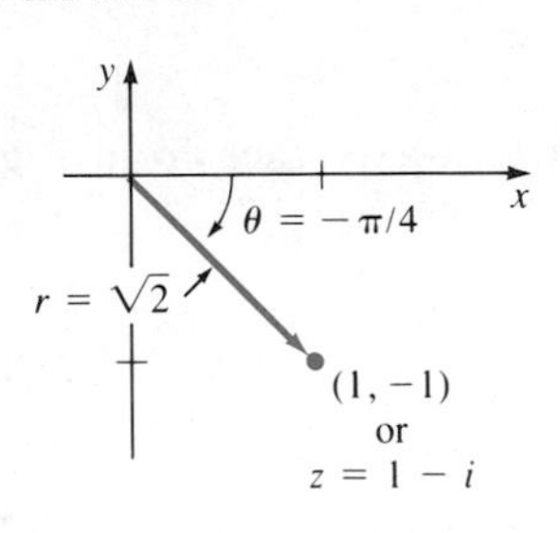

Figure A.3

Appendix V Exercises

Answers to odd-numbered problems begin on page A–37.

In Problems 1–10 let $z_1 = 2 - i$ and $z_2 = 5 + 3i$. Perform the indicated operation.

1. $z_1 + \bar{z}_2$ **2.** $4z_1 + z_2$ **3.** $2z_1 - 3z_2$ **4.** $z_1 z_2$ **5.** $(z_1)^2$

6. $\bar{z}_1(i + z_2)$ **7.** z_1/z_2 **8.** z_2/z_1 **9.** $1/z_2$ **10.** z_1/i

In Problems 11–20 write the given complex number in the polar form (6).

11. $z = i$

12. $z = -4i$

13. $z = i^2$

14. $z = 6i^5$

15. $z = 2 + 2i$

16. $z = -\sqrt{5} - \sqrt{5}i$

17. $z = 6 + 6\sqrt{3}i$

18. $z = -10\sqrt{3} + 10i$

19. $z = i(1 - \sqrt{3}i)$

20. $z = -7 + 7i$

In Problems 21 and 22 express the given complex number in polar form in the form $z = a + bi$.

21. $z = 8e^{-i\pi}$

22. $z = 2e^{i7\pi/4}$

23. Prove **DeMoivre's theorem:*** For any positive integer n,

$$[r(\cos\theta + i\sin\theta)]^n = r^n[\cos n\theta + i\sin n\theta].$$

24. Use DeMoivre's theorem of Problem 23 to evaluate $(1 + i)^{10}$.

25. Use Euler's formula to show that

$$\cos\theta = \frac{e^{i\theta} + e^{-i\theta}}{2} \quad \text{and} \quad \sin\theta = \frac{e^{i\theta} - e^{-i\theta}}{2i}.$$

---

* This theorem was named after the French mathematician **Abraham DeMoivre** (1667–1754).

Answers to Odd-Numbered Problems

Exercises 1.1, Page 10

1. linear, second-order **3.** nonlinear, first-order

5. linear, fourth-order **7.** nonlinear, second-order

9. linear, third-order

11. $2y' + y = 2(-\frac{1}{2})e^{-x/2} + e^{-x/2} = 0$

13. $\dfrac{dy}{dx} - 2y - e^{3x} = (3e^{3x} + 20e^{2x})$

$-2(e^{3x} + 10e^{2x}) - e^{3x} = 0$

15. $y' - 25 - y^2 = 25\sec^2 5x - 25(1 + \tan^2 5x)$
$= 25\sec^2 5x - 25\sec^2 5x = 0$

17. $y' + y - \sin x = \frac{1}{2}\cos x + \frac{1}{2}\sin x$
$-10e^{-x} + \frac{1}{2}\sin x - \frac{1}{2}\cos x + 10e^{-x} - \sin x = 0$

19. $yx^2 = -1$ implies $d(yx^2) = 0$
or $2yx\,dx + x^2\,dy = 0$

21. $y - 2xy' - y(y')^2 = y - 2x\dfrac{c_1}{2y} - y\dfrac{c_1^2}{4y^2}$

$= \dfrac{y^2 - (c_1 x + (c_1^2/4))}{y}$

$= \dfrac{y^2 - y^2}{y} = 0$

23. $y' - \dfrac{1}{x}y - 1 = 1 + \ln x - \ln x - 1 = 0$

25. $\dfrac{d}{dt}\ln\dfrac{2 - X}{1 - X} = 1$

$\left[\dfrac{-1}{2 - X} + \dfrac{1}{1 - X}\right]\dfrac{dX}{dt} = 1$

simplifies to $\dfrac{dX}{dt} = (2 - X)(1 - X)$

27. The differential of $c_1 = xe^{y/x}/(x + y)^2$ is
$\{(x + y)^2[xe^{y/x}(x\,dy - y\,dx/x^2) + e^{y/x}\,dx]$
$- xe^{y/x}2(x + y)(dx + dy)\}/(x + y)^4 = 0.$
Multiplying by $-x^2(x + y)^3e^{-y/x}$ and simplifying yield $(x^2 + y^2)dx + (x^2 - xy)dy = 0$.

29. $y'' - 6y' + 13y = 5e^{3x}\cos 2x - 12e^{3x}\sin 2x$
$+ 12e^{3x}\sin 2x - 18e^{3x}\cos 2x + 13e^{3x}\cos 2x = 0$

31. $y'' = \cosh x + \sinh x = y$

33. $y'' + (y')^2 = -\dfrac{1}{(x + c_1)^2} + \dfrac{1}{(x + c_1)^2} = 0$

35. $x\dfrac{d^2y}{dx^2} + 2\dfrac{dy}{dx} = x(2c_2x^{-3}) + 2(-c_2x^{-2}) = 0$

37. $x^2y'' - 3xy' + 4y$
$= x^2(5 + 2\ln x) - 3x(3x + 2x\ln x)$
$+ 4(x^2 + x^2\ln x) = 9x^2 - 9x^2$
$+ 6x^2\ln x - 6x^2\ln x = 0$

39. $y''' - 3y'' + 3y' - y$
$= x^2e^x + 6xe^x + 6e^x - 3x^2e^x - 12xe^x$
$- 6e^x + 3x^2e^x + 6xe^x - x^2e^x = 0$

41. For $x < 0$
$xy' - 2y = x(-2x) - 2(-x^2) = 0;$
for $x \geq 0$
$xy' - 2y = x(2x) - 2(x^2) = 0.$

43. $\left(\dfrac{dy}{dx}\right)^3 + 2x\dfrac{dy}{dx} - 2y - 1$

$= t^3 - 3t^3 + 2t^3 + 1 - 1 = 0$

45. $y - xy' - (y')^2 + \ln y'$
$= -t^2 - \ln t + 1 + 2t^2 - 1 - t^2 + \ln t = 0$

47. $y - xy' - (y')^2 = cx + c^2 - x(c) - c^2 = 0;$
$k = -1/4$

49. $y = -1$ **51.** $m = 2$ and $m = 3$

53. $m = \dfrac{1 \pm \sqrt{5}}{2}$

55. For $y = x^2$
$x^2y'' - 4xy' + 6y = x^2(2) - 4x(2x) + 6x^2$
$= 8x^2 - 8x^2 = 0;$
for $y = x^3$
$x^2y'' - 4xy' + 6y = x^2(6x) - 4x(3x^2) + 6x^3$
$= 12x^3 - 12x^3 = 0;$
yes; yes.

57. **(a)** $y = 0$; **(b)** no real solution;
(c) $y = 1$ or $y = -1$

Exercises 1.2, Page 27

1. $xy' = y - 2$ **3.** $y' + y = 0$ **5.** $2(x + 1)y' = y$

7. $(x^2 - y)y' = xy$ **9.** $y'' - y' = 0$

11. $y'' + \omega^2 y = 0$ **13.** $y'' - k^2y = 0$

15. $y'' - 8y' + 16y = 0$ **17.** $xy'' + y' = 0$

19. $y''' - 6y'' + 11y' - 6y = 0$

21. $(1 + \cos\theta)\dfrac{dr}{d\theta} + r\sin\theta = 0$ **23.** $xy' - y = 0$

25. $2xyy' = y^2 - x^2$ **27.** $2xy' = y$

29. $yy'' + (y')^2 = 0$ **31.** $L\dfrac{di}{dt} + Ri = E(t)$

33. $\dfrac{dv}{dt} + \dfrac{k}{m}v^2 = g$ **35.** $\dfrac{dy}{dx} = -\dfrac{y}{\sqrt{s^2 - y^2}}$

37.
$$mx'' = -k\cos\theta = -k\cdot\frac{1}{v}\frac{dx}{dt} = -|c|\frac{dx}{dt},$$
$$my'' = -mg - k\sin\theta = -mg - k\cdot\frac{1}{v}\frac{dy}{dt} = -mg - |c|\frac{dy}{dt}$$

39. $\dfrac{dx}{dt} = r - kx, \quad k > 0$

41. Using $\tan\phi = \dfrac{x}{y}$, $\tan\left(\dfrac{\pi}{2} - \theta\right) = \dfrac{dy}{dx}$,

$\tan\theta = \dfrac{dx}{dy}$, and $\tan\phi = \tan 2\theta$

$$= \frac{2\tan\theta}{1 - \tan^2\theta},$$

we obtain $x\left(\dfrac{dx}{dy}\right)^2 + 2y\dfrac{dx}{dy} = x$.

43. By combining Newton's second law of motion with his law of gravitation, we obtain

$$m\frac{d^2y}{dt^2} = -k_1\frac{mM}{y^2},$$

where M is the mass of the earth and k_1 is a constant of proportionality. Dividing by m gives

$$\frac{d^2y}{dt^2} = -\frac{k}{y^2},$$

where $k = k_1 M$. The constant k is gR^2, where R is the radius of the earth. This follows from the fact that on the surface of the earth $y = R$ so that

$$k_1\frac{mM}{R^2} = mg$$

$$k_1 M = gR^2 \quad \text{or} \quad k = gR^2.$$

If $t = 0$ is the time at which burnout occurs, then

$$y(0) = R + y_B,$$

where y_B is the distance from the earth's surface to the rocket at the time of burnout, and

$$y'(0) = V_B$$

is the corresponding velocity at that time.

Chapter 1 Review, Page 32

1. ordinary, first-order, nonlinear

3. partial, second-order **9.** $y = x^2$ **11.** $y = \dfrac{x^2}{2}$

13. $y = 0$; $y = e^x$ **15.** $x < 0$ or $x > 1$

19. $(x - 2)y' = y - 1$

21. Water lost in time Δt = change in volume of water

$$\tfrac{1}{4}\sqrt{2gh}\,\Delta t = -\pi r^2\,\Delta h,$$

where Δh is the change in height for a small change in time Δt and r is the radius of A_1. Dividing by Δt gives

$$\frac{\Delta h}{\Delta t} = -\frac{\sqrt{2gh}}{4\pi r^2}.$$

Now by similar triangles it follows that $r = 2h/5$ at any time. Thus as $t \to 0$

$$\frac{dh}{dt} = -\frac{25\sqrt{2g}}{16\pi}h^{-3/2}.$$

Exercises 2.1, Page 39

1. half planes defined by either $y > 0$ or $y < 0$

3. half planes defined by either $x > 0$ or $x < 0$

5. the regions defined by either $y > 2$, $y < -2$, or $-2 < y < 2$

7. any region not containing $(0, 0)$

9. the entire xy-plane **11.** $y = 0$, $y = x^3$

13. There is some interval around $x = 0$ on which the unique solution is $y = 0$.

15. $xy' - y = x(c) - cx = 0$ for every c; $y = 0$, $y = x$. No, the given function is nondifferentiable at $x = 0$.

17. yes 19. no

Exercises 2.2, Page 47

1. $y = -\frac{1}{5}\cos 5x + c$ 3. $y = \frac{1}{3}e^{-3x} + c$

5. $y = x + 5\ln|x+1| + c$ 7. $y = cx^4$

9. $y^{-2} = 2x^{-1} + c$ 11. $-3 + 3x\ln|x| = xy^3 + cx$

13. $-3e^{-2y} = 2e^{3x} + c$ 15. $2 + y^2 = c(4 + x^2)$

17. $y^2 = x - \ln|x+1| + c$

19. $\frac{x^3}{3}\ln x - \frac{1}{9}x^3 = \frac{y^2}{2} + 2y + \ln|y| + c$

21. $S = ce^{kr}$ 23. $\frac{P}{1-P} = ce^t$ or $P = \frac{ce^t}{1+ce^t}$

25. $4\cos y = 2x + \sin 2x + c$

27. $-2\cos x + e^y + ye^{-y} + e^{-y} = c$

29. $(e^x + 1)^{-2} + 2(e^y + 1)^{-1} = c$

31. $(y+1)^{-1} + \ln|y+1| = \frac{1}{2}\ln\left|\frac{x+1}{x-1}\right| + c$

33. $y - 5\ln|y+3| = x - 5\ln|x+4| + c$
or $\left(\frac{y+3}{x+4}\right)^5 = c_1 e^{y-x}$

35. $-\cot y = \cos x + c$ 37. $y = \sin\left(\frac{x^2}{2} + c\right)$

39. $-y^{-1} = \tan^{-1}(e^x) + c$

41. $(1 + \cos x)(1 + e^y) = 4$

43. $\sqrt{y^2+1} = 2x^2 + \sqrt{2}$

45. $x = \tan(4y - 3\pi/4)$ 47. $xy = e^{-(1+1/x)}$

49. **(a)** $y = 3\frac{1-e^{6x}}{1+e^{6x}}$, **(b)** $y = 3$,
(c) $y = 3\frac{2-e^{6x-2}}{2+e^{6x-2}}$

51. $y = 1$ 53. $y = -x - 1 + \tan(x + c)$

55. $2y - 2x + \sin 2(x + y) = c$

57. $4(y - 2x + 3) = (x + c)^2$

Exercises 2.3, Page 55

1. homogeneous of degree 3

3. homogeneous of degree 2

5. not homogeneous

7. homogeneous of degree 0

9. homogeneous of degree -2

11. $x\ln|x| + y = cx$

13. $(x - y)\ln|x - y| = y + c(x - y)$

15. $x + y\ln|x| = cy$

17. $\ln(x^2 + y^2) + 2\tan^{-1}(y/x) = c$

19. $4x = y(\ln|y| - c)^2$ 21. $y^9 = c(x^3 + y^3)^2$

23. $(y/x)^2 = 2\ln|x| + c$ 25. $e^{2x/y} = 8\ln|y| + c$

27. $x\cos(y/x) = c$ 29. $y + x = cx^2e^{y/x}$

31. $y^3 + 3x^3\ln|x| = 8x^3$ 33. $y^2 = 4x(x + y)^2$

35. $\ln|x| = e^{y/x} - 1$

37. $4x\ln\left|\frac{y}{x}\right| + x\ln x + y - x = 0$

39. $3x^{3/2}\ln x + 3x^{1/2}y + 2y^{3/2} = 5x^{3/2}$

41. $(x + y)\ln|y| + x = 0$

43. $\ln|y| = -2(1 - x/y)^{1/2} + \sqrt{2}$

45. $(y+1)^2 + 2(y+1)(x-2) - (x-2)^2 = c$

47. By homogeneity the equation can be written as

$$M(x/y, 1)\,dx + N(x/y, 1)\,dy = 0.$$

With $v = x/y$, it follows that

$$M(v, 1)(v\,dy + y\,dv) + N(v, 1)\,dy = 0$$
$$[vM(v, 1) + N(v, 1)]\,dy + yM(v, 1)\,dv = 0$$

or

$$\frac{dy}{y} + \frac{M(v, 1)\,dv}{vM(v, 1) + N(v, 1)} = 0.$$

49. $$\frac{dy}{dx} = -\frac{M(x, y)}{N(x, y)} = -\frac{y^nM(x/y, 1)}{y^nN(x/y, 1)} = -\frac{M(x/y, 1)}{N(x/y, 1)} = G(x/y)$$

Exercises 2.4, Page 63

1. $x^2 - x + \frac{3}{2}y^2 + 7y = c$ 3. $\frac{5}{2}x^2 + 4xy - 2y^4 = c$

5. $x^2y^2 - 3x + 4y = c$

7. not exact, but is homogeneous

9. $xy^3 + y^2\cos x - \frac{1}{2}x^2 = c$ 11. not exact

13. $xy - 2xe^x + 2e^x - 2x^3 = c$

15. $x + y + xy - 3\ln|xy| = c$

17. $x^3y^3 - \tan^{-1}3x = c$
19. $-\ln|\cos x| + \cos x \sin y = c$
21. $y - 2x^2y - y^2 - x^4 = c$
23. $x^4y - 5x^3 - xy + y^3 = c$
25. $\frac{1}{3}x^3 + x^2y + xy^2 - y = \frac{4}{3}$
27. $4xy + x^2 - 5x + 3y^2 - y = 8$
29. $y^2 \sin x - x^3y - x^2 + y \ln y - y = 0$
31. $k = 10$ 33. $k = 1$
35. $M(x, y) = ye^{xy} + y^2 - (y/x^2) + h(x)$
37. $M(x, y) = 6xy^3$
$N(x, y) = 4y^3 + 9x^2y^2$
$\partial M/\partial y = 18xy^2 = \partial N/\partial x$
solution is $3x^2y^3 + y^4 = c$
39. $M(x, y) = -x^2y^2 \sin x + 2xy^2 \cos x$
$N(x, y) = 2x^2y \cos x$
$\partial M/\partial y = -2x^2y \sin x + 4xy \cos x = \partial N/\partial x$
solution is $x^2y^2 \cos x = c$
41. $M(x, y) = 2xy^2 + 3x^2$
$N(x, y) = 2x^2y$
$\partial M/\partial y = 4xy = \partial N/\partial x$
solution is $x^2y^2 + x^3 = c$
43. A separable first-order differential equation can be written $h(y)\,dy - g(x)\,dx = 0$. Identifying $M(x, y) = -g(x)$ and $N(x, y) = h(y)$, it follows that $\partial M/\partial y = 0 = \partial N/\partial x$.

Exercises 2.5, Page 74

1. $y = ce^{5x}, \quad -\infty < x < \infty$
3. $y = \frac{1}{3} + ce^{-4x}, \quad -\infty < x < \infty$
5. $y = \frac{1}{4}e^{3x} + ce^{-x}, \quad -\infty < x < \infty$
7. $y = \frac{1}{3} + ce^{-x^3}, \quad -\infty < x < \infty$
9. $y = x^{-1} \ln x + cx^{-1}, \quad 0 < x < \infty$
11. $x = -\frac{4}{5}y^2 + cy^{-1/2}, \quad 0 < y < \infty$
13. $y = -\cos x + \dfrac{\sin x}{x} + \dfrac{c}{x}, \quad 0 < x < \infty$
15. $y = \dfrac{c}{e^x + 1}, \quad -\infty < x < \infty$
17. $y = \sin x + c \cos x, \quad -\pi/2 < x < \pi/2$
19. $y = \frac{1}{7}x^3 - \frac{1}{5}x + cx^{-4}, \quad 0 < x < \infty$
21. $y = \dfrac{1}{2x^2}e^x + \dfrac{c}{x^2}e^{-x}, \quad 0 < x < \infty$
23. $y = \sec x + c \csc x, \quad 0 < x < \pi/2$
25. $x = \dfrac{1}{2}e^y - \dfrac{1}{2y}e^y + \dfrac{1}{4y^2}e^y + \dfrac{c}{y^2}e^{-y}, \quad 0 < y < \infty$
27. $y = e^{-3x} + \dfrac{c}{x}e^{-3x}, \quad 0 < x < \infty$
29. $x = 2y^6 + cy^4, \quad 0 < y < \infty$
31. $y = e^{-x} \ln(e^x + e^{-x}) + ce^{-x}, \quad -\infty < x < \infty$
33. $x = \dfrac{1}{y} + \dfrac{c}{y}e^{-y^2}, \quad 0 < y < \infty$
35. $(\sec\theta + \tan\theta)r = \theta - \cos\theta + c, \quad -\pi/2 < \theta < \pi/2$
37. $y = \frac{5}{3}(x + 2)^{-1} + c(x + 2)^{-4}, \quad -2 < x < \infty$
39. $y = 10 + ce^{-\sinh x}, \quad -\infty < x < \infty$
41. $y = 4 - 2e^{-5x}, \quad -\infty < x < \infty$
43. $i(t) = E/R + (i_0 - E/R)e^{-Rt/L}, \quad -\infty < t < \infty$
45. $y = \sin x \cos x - \cos x, \quad -\pi/2 < x < \pi/2$
47. $T(t) = 50 + 150e^{kt}, \quad -\infty < t < \infty$
49. $(x + 1)y = x \ln x - x + 21, \quad 0 < x < \infty$
51. $y = \dfrac{2x}{x - 2}, \quad 2 < x < \infty$
53. $x = \frac{1}{2}y + 8/y, \quad 0 < y < \infty$
55. $y = \begin{cases} \frac{1}{2}(1 - e^{-2x}), & 0 \le x \le 3, \\ \frac{1}{2}(e^6 - 1)e^{-2x}, & x > 3 \end{cases}$
57. $y = \begin{cases} \frac{1}{2} + \frac{3}{2}e^{-x^2}, & 0 \le x < 1 \\ (\frac{1}{2}e + \frac{3}{2})e^{-x^2}, & x \ge 1 \end{cases}$

Exercises 2.6, Page 80

1. $y^3 = 1 + cx^{-3}$ 3. $y^{-3} = x + \frac{1}{3} + ce^{3x}$
5. $e^{x/y} = cx$ 7. $y^{-3} = -\frac{9}{5}x^{-1} + \frac{49}{5}x^{-6}$
9. $x^{-1} = 2 - y^2 - e^{-y^2/2}$, the equation is Bernoulli in the variable x.
11. $y = 2 + \dfrac{1}{ce^{-3x} - 1/3}$
13. $y = \dfrac{2}{x} + \dfrac{1}{cx^{-3} - x/4}$
15. $y = -e^x + \dfrac{1}{ce^{-x} - 1}$
17. $y = -2 + \dfrac{1}{ce^{-x} - 1}$
19. $y = cx + 1 - \ln c; \quad y = 2 + \ln x$

21. $y = cx - c^3$; $27y^2 = 4x^3$

23. $y = cx - e^c$; $y = x \ln x - x$

25. If $y = y_1 + u$, then $y' = y_1' + u'$ and so $dy/dx = P(x) + Q(x)y + R(x)y^2$ becomes
$y_1' + u' = P + Q(y_1 + u) + R(y_1 + u)^2$
$y_1' + u' = P + Qy_1 + Ry_1^2 + Qu + 2y_1Ru + Ru^2$.
Since y_1 is a solution of the Ricatti equation, we obtain

$$u' - (Q + 2y_1R)u = Ru^2.$$

27. If $y = \dfrac{w'}{w}$, then $y' = \dfrac{ww'' - (w')^2}{w^2}$ and so

$y' + y^2 - Q(x)y - P(x) = 0$ becomes

$$\frac{ww'' - (w')^2}{w^2} + \frac{(w')^2}{w^2} - Q\frac{w'}{w} - P = 0.$$

The last equation simplifies to

$$w'' - Qw' - Pw = 0.$$

29. If $y = cx + f(c)$, then $y' = c$ and $f(c) = f(y')$. Therefore $y - xy' - f(y') = y - cx - f(c) = 0$.

Exercises 2.7, Page 84

1. $x^2e^{2y} = 2x \ln x - 2x + c$
3. $e^{-x} = y \ln|y| + cy$
5. $-e^{-y/x^4} = x^2 + c$
7. $x^2 + y^2 = x - 1 + ce^{-x}$
9. $\ln(\tan y) = x + cx^{-1}$
11. $x^3y^3 = 2x^3 - 9\ln|x| + c$
13. $e^y = -e^{-x}\cos x + ce^{-x}$
15. $y^2 \ln x = ye^y - e^y + c$
17. $y = \ln|\cos(c_1 - x)| + c_2$
19. $y = -\dfrac{1}{c_1}(1 - c_1^2x^2)^{1/2} + c_2$
21. The given equation is a Clairaut equation in $u = y'$. The solution is $y = c_1x^2/2 + x + c_1^3x + c_2$.
23. $y = c_1 + c_2x^2$
25. $\frac{1}{3}y^3 - c_1y = x + c_2$
27. $y = -\sqrt{1 - x^2}$

Exercises 2.8, Page 87

1. $y_1(x) = 1 - x$

$y_2(x) = 1 - \dfrac{x}{1!} + \dfrac{x^2}{2!}$

$y_3(x) = 1 - \dfrac{x}{1!} + \dfrac{x^2}{2!} - \dfrac{x^3}{3!}$

$y_4(x) = 1 - \dfrac{x}{1!} + \dfrac{x^2}{2!} - \dfrac{x^3}{3!} + \dfrac{x^4}{4!}$;

$y_n(x) \to e^{-x}$ as $n \to \infty$

3. $y_1(x) = 1 + x^2$

$y_2(x) = 1 + \dfrac{x^2}{1!} + \dfrac{x^4}{2!}$

$y_3(x) = 1 + \dfrac{x^2}{1!} + \dfrac{x^4}{2!} + \dfrac{x^6}{3!}$

$y_4(x) = 1 + \dfrac{x^2}{1!} + \dfrac{x^4}{2!} + \dfrac{x^6}{3!} + \dfrac{x^8}{4!}$

$y_n(x) \to e^{x^2}$ as $n \to \infty$

5. $y_1(x) = y_2(x) = y_3(x) = y_4(x) = 0$; $y_n(x) \to 0$ as $n \to \infty$

7. **(a)** $y_1(x) = x$
$y_2(x) = x + \frac{1}{3}x^3$
$y_3(x) = x + \frac{1}{3}x^3 + \frac{2}{15}x^5 + \frac{1}{63}x^7$
(b) $y = \tan x$
(c) The Maclaurin series expansion of $\tan x$ is $x + \frac{1}{3}x^3 + \frac{2}{15}x^5 + \frac{17}{315}x^7 + \cdots$, $|x| < \pi/2$.

Chapter 2 Review, Page 89

1. the regions defined by $x^2 + y^2 > 25$ and $x^2 + y^2 < 25$
3. false
5. **(a)** linear in x
(b) homogeneous, exact, linear in y
(c) Clairaut
(d) Bernoulli in x
(e) separable
(f) separable, Ricatti
(g) linear in x
(h) homogeneous
(i) Bernoulli
(j) homogeneous, exact, Bernoulli
(k) separable, homogeneous, exact, linear in x and in y

(l) exact, linear in y
(m) homogeneous
(n) separable
(o) Clairaut
(p) Ricatti

7. $2y^2 \ln y - y^2 = 4xe^x - 4e^x - 1$

9. $2y^2 + x^2 = 9x^6$ **11.** $e^{xy} - 4y^3 = 5$

13. $y = \frac{1}{4} - 320(x^2 + 4)^{-4}$

15. $y = \dfrac{1}{x^4 - x^4 \ln|x|}$ **17.** $x^2 - \sin \dfrac{1}{y^2} = c$

19. $y_1(x) = 1 + x + \frac{1}{3}x^3$
$y_2(x) = 1 + x + x^2 + \frac{2}{3}x^3 + \frac{1}{6}x^4 + \frac{2}{15}x^5 + \frac{1}{63}x^7$

Exercises 3.1, Page 96

1. $x^2 + y^2 = c_2^2$ **3.** $2y^2 + x^2 = c_2$

5. $2 \ln|y| = x^2 + y^2 + c_2$ **7.** $y^2 = 2x + c_2$

9. $2x^2 + 3y^2 = c_2$ **11.** $x^3 + y^3 = c_2$

13. $y^2 \ln|y| + x^2 = c_2 y^2$ **15.** $y^2 - x^2 = c_2 x$

17. $2y^2 = 2 \ln|x| + x^2 + c_2$

19. $y = \frac{1}{4} - \frac{1}{6}x^2 + c_2 x^{-4}$ **21.** $2y^3 = 3x^2 + c_2$

23. $2 \ln(\cosh y) + x^2 = c_2$ **25.** $y^{5/3} = x^{5/3} + c_2$

27. $y = 2 - x + 3e^{-x}$ **29.** $r = c_2 \sin\theta$

31. $r^2 = c_2 \cos 2\theta$ **33.** $r = c_2 \csc\theta$

35. Let β be the angle of inclination, measured from the positive x-axis, of the tangent line to a member of the given family, and ϕ the angle of inclination of the tangent to a trajectory. At the point where the curves intersect, the angle between the tangents is α. From the following figures we conclude that there exist two

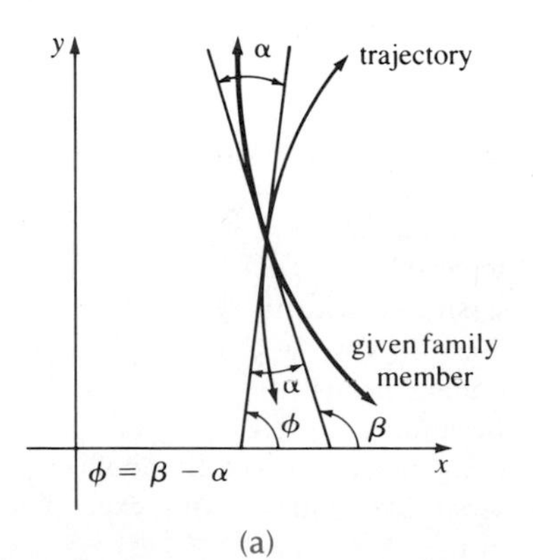

(a)

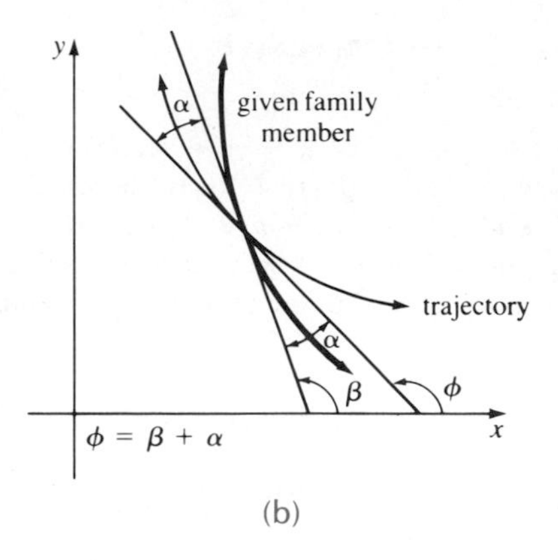

(b)

possible cases and that $\phi = \beta \pm \alpha$. Thus the slope of the tangent line to a trajectory is

$$\frac{dy}{dx} = \tan\phi = \tan(\beta \pm \alpha)$$

$$= \frac{\tan\beta \pm \tan\alpha}{1 \mp \tan\beta \tan\alpha}$$

$$= \frac{f(x, y) \pm \tan\alpha}{1 \mp f(x, y)\tan\alpha}$$

37. $\mp \dfrac{2}{\sqrt{3}} \tan^{-1}(y/x) + \ln c_2(x^2 + y^2) = 0$

39. Since the given equation is quadratic in c_1, it follows from the quadratic formula that

$$c_1 = -x \pm \sqrt{x^2 + y^2}.$$

Differentiating this last expression and solving for dy/dx give

$$\frac{dy}{dx} = \frac{-x + \sqrt{x^2 + y^2}}{y}$$

and

$$\frac{dy}{dx} = \frac{-x - \sqrt{x^2 + y^2}}{y}.$$

These two equations correspond to choosing $c_1 > 0$ and $c_1 < 0$ in the given family, respectively. Forming the product of these derivatives yields

$$\left(\frac{dy}{dx}\right)_{(1)} \cdot \left(\frac{dy}{dx}\right)_{(2)} = \frac{x^2 - x^2 - y^2}{y^2} = -1.$$

This shows that the family is self-orthogonal.

41. The differential equation of the orthogonal family is $(x-y)\,dx + (x+y)\,dy = 0$. The verification follows by substituting $x = c_2e^{-t}\cos t$ and $y = c_2e^{-t}\sin t$ into the equation.

Exercises 3.2, Page 107

1. 7.9 years; 10 years 3. 760 5. 11 hours

7. 136.5 hours

9. $I(15) = 0.00098I_0$ or $I(15)$ is approximately 0.1% of I_0.

11. 15,600 years

13. $T(1) = 36.67$ degrees; approximately 3.06 minutes

15. $i(t) = \frac{3}{5} - \frac{3}{5}e^{-500t}$; $i \to \frac{3}{5}$ as $t \to \infty$.

17. $q(t) = \frac{1}{100} - \frac{1}{100}e^{-50t}$; $i(t) = \frac{1}{2}e^{-50t}$

19. $i(t) = \begin{cases} 60 - 60e^{-t/10}, & 0 \le t \le 20, \\ 60(e^2 - 1)e^{-t/10}, & t > 20 \end{cases}$

21. $A(t) = 200 - 170e^{-t/50}$

23. $A(t) = 1000 - 1000e^{-t/100}$ 25. 64.38 lb

27. $v(t) = \dfrac{mg}{k} + \left(v_0 - \dfrac{mg}{k}\right)e^{-kt/m}$;

$v \to \dfrac{mg}{k}$ as $t \to \infty$;

$$s(t) = \frac{mg}{k}t - \frac{m}{k}\left(v_0 - \frac{mg}{k}\right)e^{-kt/m} + \frac{m}{k}\left(v_0 - \frac{mg}{k}\right) + s_0$$

29. $E(t) = E_0e^{-(t-t_1)/RC}$

31. **(a)** $P(t) = P_0e^{(k_1-k_2)t}$
   (b) $k_1 > k_2$, births surpass deaths so population increases.
   $k_1 = k_2$, a constant population since number of births equals the number of deaths.
   $k_1 < k_2$, deaths surpass births so population decreases.

33. From $r^2\,d\theta = \dfrac{L}{M}\,dt$ we get

$$A = \frac{1}{2}\int_{\theta_1}^{\theta_2} r^2\,d\theta = \frac{1}{2}\frac{L}{M}\int_a^b dt$$

$$= \frac{1}{2}\frac{L}{M}(b-a).$$

Exercises 3.3, Page 120

1. 1834; 2000 3. 1,000,000; 52.9 months

5. **(a)** Separating variables gives

$$\frac{dP}{P(a - b\ln P)} = dt,$$

so that

$$\begin{aligned} -(1/b)\ln|a - b\ln P| &= t + c_1 \\ a - b\ln P &= c_2e^{-bt} && (e^{-bc_1} = c_2) \\ \ln P &= (a/b) - ce^{-bt} && (c_2/b = c) \\ P(t) &= e^{a/b}\cdot e^{-ce^{-bt}} \end{aligned}$$

(b) If $P(0) = P_0$, then

$$P_0 = e^{a/b}e^{-c} = e^{a/b-c}$$

and so

$$\ln P_0 = (a/b) - c$$
$$c = (a/b) - \ln P_0.$$

7. 29.3 grams; $X \to 60$ as $t \to \infty$; 0 grams of A and 30 grams of B.

9. For $\alpha \neq \beta$ the differential equation separates as

$$\frac{1}{\alpha-\beta}\left[-\frac{1}{\alpha - X} + \frac{1}{\beta - X}\right]dx = k\,dt.$$

It follows immediately that

$$\frac{1}{\alpha-\beta}[\ln|\alpha - X| - \ln|\beta - X|] = kt + c$$

or

$$\frac{1}{\alpha-\beta}\ln\left|\frac{\alpha - X}{\beta - X}\right| = kt + c.$$

For $\alpha = \beta$ the equation can be written as

$$(\alpha - X)^{-2}\,dX = k\,dt.$$

It follows that $(\alpha - X)^{-1} = kt + c$ or

$$X = \alpha - \frac{1}{kt + c}.$$

11. $v^2 = (2gR^2/y) + v_0^2 - 2gR$.
We note that as y increases, v decreases. In particular, if $v_0^2 - 2gR < 0$, then there must be some value of y

for $v = 0$; the rocket stops and returns to earth under the influence of gravity. However, if $v_0^2 - 2gR \geq 0$, then $v > 0$ for all values of y. Hence we should have $v_0 \geq \sqrt{2gR}$. Using the values $R = 4000$ miles, $g = 32$ ft/sec^2, 1 ft $= 1/5280$ mi, 1 sec $= 1/3600$ hr, it follows that $v_0 \geq 25{,}067$ mi/hr.

13. Using the condition $y'(1) = 0$, we find

$$\frac{dy}{dx} = \frac{1}{2}\left[x^{v_1/v_2} - x^{-v_1/v_2}\right].$$

Now if $v_1 = v_2$, $y = \frac{1}{4}x^2 - \frac{1}{2}\ln x - \frac{1}{4}$; if $v_1 \neq v_2$, then

$$y = \frac{1}{2}\left[\frac{x^{1+(v_1/v_2)}}{1 + \dfrac{v_1}{v_2}} - \frac{x^{1-(v_1/v_2)}}{1 - \dfrac{v_1}{v_2}}\right] + \frac{v_1 v_2}{v_2^2 - v_1^2}.$$

15. $2h^{1/2} = -\frac{1}{25}t + 2\sqrt{20}$; $t = 50\sqrt{20}$ sec

17. To evaluate the indefinite integral of the left side of

$$\frac{\sqrt{100 - y^2}}{y}\,dy = -dx$$

we use the substitution $y = 10\cos\theta$. It follows that

$$x = 10\ln\left(\frac{10 + \sqrt{100 - y^2}}{y}\right) - \sqrt{100 - y^2}.$$

19. Under the substitution $w = x^2$, the differential equation becomes

$$w = y\frac{dw}{dy} + \frac{1}{4}\left(\frac{dw}{dy}\right)^2,$$

which is Clairaut's equation. The solution is

$$x^2 = cy + \frac{c^2}{4}.$$

If $2c_1 = c$, then we recognize

$$x^2 = 2c_1 y + c_1^2$$

as describing a family of parabolas.

21. $-\gamma \ln y + \delta y = \alpha \ln x - \beta x + c$

23. **(a)** The equation $2\dfrac{d^2\theta}{dt^2}\dfrac{d\theta}{dt} + 2\dfrac{g}{l}\sin\theta\dfrac{d\theta}{dt} = 0$ is the same as

$$\frac{d}{dt}\left(\frac{d\theta}{dt}\right)^2 + 2\frac{g}{l}\sin\theta\frac{d\theta}{dt} = 0.$$

Integrating this last equation with respect to t and using the initial conditions give the result.

(b) From (a)

$$dt = \sqrt{\frac{l}{2g}}\frac{d\theta}{\sqrt{\cos\theta - \cos\theta_0}}.$$

Integrating this last equation gives the time for the pendulum to move from $\theta = \theta_0$ to $\theta = 0$,

$$t = \sqrt{\frac{l}{2g}}\int_0^{\theta_0}\frac{d\theta}{\sqrt{\cos\theta - \cos\theta_0}}.$$

The period is the total time T to go from $\theta = \theta_0$ to $\theta = -\theta_0$ and back again to $\theta = \theta_0$. This is

$$T = 4\sqrt{\frac{l}{2g}}\int_0^{\theta_0}\frac{d\theta}{\sqrt{\cos\theta - \cos\theta_0}} = 2\sqrt{\frac{2l}{g}}\int_0^{\theta_0}\frac{d\theta}{\sqrt{\cos\theta - \cos\theta_0}}.$$

Chapter 3 Review, Page 126

1. $y^3 + 3/x = c_2$ **3.** $2(y-2)^2 + (x-1)^2 = c_2^2$

5. $P(45) = 8.99$ billion

7. $x(t) = \dfrac{\alpha c_1 e^{\alpha k_1 t}}{1 + c_1 e^{\alpha k_1 t}}$, $y(t) = c_2(1 + c_1 e^{\alpha k_1 t})^{k_2/k_1}$

Exercises 4.1, Page 148

1. $y = \frac{1}{2}e^x - \frac{1}{2}e^{-x}$ **3.** $y = \frac{3}{5}e^{4x} + \frac{2}{5}e^{-x}$

5. $y = 3x - 4x\ln x$ **7.** $y = 0$, $y = x^2$

9. **(a)** $y = e^x\cos x - e^x\sin x$
(b) no solution
(c) $y = e^x\cos x + e^{-\pi/2}e^x\sin x$
(d) $y = c_2 e^x\sin x$, where c_2 is arbitrary

11. $(-\infty, 2)$ **13.** $\lambda = n$, $n = 1, 2, 3, \ldots$

15. dependent 17. dependent 19. dependent

21. independent

23. $W(x^{1/2}, x^2) = \frac{3}{2}x^{3/2} \neq 0$ on $(0, \infty)$

25. $W(\sin x, \csc x) = -2 \cot x$.
$W = 0$ only at $x = \pi/2$ in the interval.

27. $W(e^x, e^{-x}, e^{4x}) = -30e^{4x} \neq 0$ on $(-\infty, \infty)$

29. no

31. (a) $y'' - 2y^3 = \dfrac{2}{x^3} - 2\left(\dfrac{1}{x}\right)^3 = 0$

(b) $y'' - 2y^3 = \dfrac{2c}{x^3} - 2\dfrac{c^3}{x^3} = \dfrac{2}{x^3}c(1 - c^2) \neq 0$

for $c \neq 0, \pm 1$

33. The functions satisfy the differential equation and are linearly independent on the interval since $W(e^{-3x}, e^{4x}) = 7e^x \neq 0$; $y = c_1e^{-3x} + c_2e^{4x}$.

35. The functions satisfy the differential equation and are linearly independent on the interval since $W(e^x \cos 2x, e^x \sin 2x) = 2e^{2x} \neq 0$; $y = c_1e^x \cos 2x + c_2e^x \sin 2x$.

37. The functions satisfy the differential equation and are linearly independent on the interval since $W(x^3, x^4) = x^6 \neq 0$; $y = c_1x^3 + c_2x^4$.

39. The functions satisfy the differential equation and are linearly independent on the interval since $W(x, x^{-2}, x^{-2} \ln x) = 9x^{-6} \neq 0$; $y = c_1x + c_2x^{-2} + c_3x^{-2} \ln x$.

41. e^{2x} and e^{5x} form a fundamental set of solutions of the homogeneous equation; $6e^x$ is a particular solution of the nonhomogeneous equation.

43. e^{2x} and xe^{2x} form a fundamental set of solutions of the homogeneous equation; $x^2e^{2x} + x - 2$ is a particular solution of the nonhomogeneous equation.

45. (a) The accompanying graphs show that y_1 and y_2 are not multiples of one another.

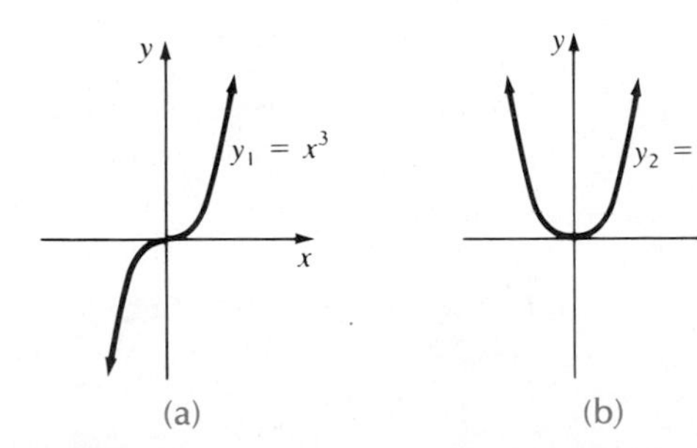

Also,

$$x^2y_1'' - 4xy_1' + 6y_1 = x^2(6x) - 4x(3x^2) + 6x^3 = 12x^3 - 12x^3 = 0.$$

For $x \geq 0$ the demonstration that y_2 is a solution of the equation is exactly as given above for y_1. For $x < 0$, $y_2 = -x^3$ and so

$$x^2y_2'' - 4xy_2' + 6y_2 = x^2(-6x) - 4x(-3x^2) + 6(-x^3) = -12x^3 + 12x^3 = 0.$$

(b) For $x \geq 0$

$$W(y_1, y_2) = \begin{vmatrix} x^3 & x^3 \\ 3x^2 & 3x^2 \end{vmatrix} = 3x^5 - 3x^5 = 0.$$

For $x < 0$

$$W(y_1, y_2) = \begin{vmatrix} x^3 & -x^3 \\ 3x^2 & -3x^2 \end{vmatrix} = -3x^5 + 3x^5 = 0.$$

Thus $W(y_1, y_2) = 0$ for every real value of x.

(c) No, $a_2(x) = x^2$ is zero at $x = 0$.

(d) Since $Y_1 = y_1$, we need only show

$$x^2Y_2'' - 4xY_2' + 6Y_2 = x^2(2) - 4x(2x) + 6x^2 = 8x^2 - 8x^2 = 0$$

and $W(x^3, x^2) = -x^4$. Thus Y_1 and Y_2 are linearly independent solutions on the interval.

(e) $Y_1 = x^3$, $Y_2 = x^2$, or $y_2 = |x|^3$

(f) Neither; we form a general solution on an interval for which $a_2(x) \neq 0$ for every x in the interval. The linear combination

$$y = c_1Y_1 + c_2Y_2$$

would be a general solution of the equation on, say, the interval $(0, \infty)$.

47. (a) Since y_1 and y_2 are solutions of the given differential equation we have

$$a_2(x)y_1'' + a_1(x)y_1' + a_0(x)y_1 = 0$$

and $$a_2(x)y_2'' + a_1(x)y_2' + a_0(x)y_2 = 0.$$

Now multiply the first equation by y_2 and the second by y_1 and subtract the first from the second:

$$a_2(x)[y_1y_2'' - y_2y_1''] + a_1(x)[y_1y_2' - y_2y_1'] = 0.$$

Now it is easily verified that

$$\frac{dW}{dx} = \frac{d}{dx}(y_1y_2' - y_2y_1') = y_1y_2'' - y_2y_1'',$$

and so it follows that

$$a_2(x)\frac{dW}{dx} + a_1(x)W = 0.$$

(b) Since this last equation is a linear first-order differential equation the integrating factor is

$$e^{\int (a_1(x)/a_2(x))\,dx}.$$

Therefore, from

$$\frac{d}{dx}[e^{\int (a_1(x)/a_2(x))\,dx}W] = 0$$

we obtain $W = ce^{-\int (a_1(x)/a_2(x))\,dx}$.

(c) Substituting $x = x_0$ in the given result, we find $c = W(x_0)$.

(d) Since an exponential function is never zero when $W(x_0) \neq 0$, it follows from part (c) that $W \neq 0$. On the other hand, if $W(x_0) = 0$, we have immediately that $W = 0$.

49. From part (c) of Problem 47 we have

$$W(y_1, y_2) = W(y_1(x_0), y_2(x_0))e^{-\int_{x_0}^{x} dt/t}$$

$$= \begin{vmatrix} k_1 & k_3 \\ k_2 & k_4 \end{vmatrix} e^{-\ln(x/x_0)}$$

$$= (k_1k_4 - k_3k_2)\left(\frac{x_0}{x}\right).$$

51. Let $y = y_1 + y_2$. Then

$$\begin{aligned} a_2y'' + a_1y' + a_0y &\\ &= a_2[y_1'' + y_2''] + a_1[y_1' + y_2'] + a_0[y_1 + y_2] \\ &= [a_2y_1'' + a_1y_1' + a_0y_1] + [a_2y_1'' + a_1y_1' + a_0y_1] \\ &= E_1 + E_2. \end{aligned}$$

Exercises 4.2, Page 157

1. $y_2 = e^{-5x}$ **3.** $y_2 = xe^{2x}$ **5.** $y_2 = \sin 4x$
7. $y_2 = \sinh x$ **9.** $y_2 = xe^{2x/3}$ **11.** $y_2 = x^4 \ln|x|$
13. $y_2 = 1$ **15.** $y_2 = x^2 + x + 2$
17. $y_2 = x\cos(\ln x)$ **19.** $y_2 = x$ **21.** $y_2 = x \ln x$
23. $y_2 = x^3$ **25.** $y_2 = x^2$ **27.** $y_2 = 3x + 2$
29. $y_2 = \frac{1}{2}[\tan x \sec x + \ln|\sec x + \tan x|]$
31. $y_2 = e^{2x}$, $y_p = -\frac{1}{2}$ **33.** $y_2 = e^{2x}$, $y_p = \frac{5}{2}e^{3x}$

Exercises 4.3, Page 165

1. $y = c_1 + c_2e^{-x/4}$ **3.** $y = c_1e^{-6x} + c_2e^{6x}$
5. $y = c_1 \cos 3x + c_2 \sin 3x$ **7.** $y = c_1e^{3x} + c_2e^{-2x}$
9. $y = c_1e^{-4x} + c_2xe^{-4x}$
11. $y = c_1e^{(-3+\sqrt{29})x/2} + c_2e^{(-3-\sqrt{29})x/2}$
13. $y = c_1e^{2x/3} + c_2e^{-x/4}$
15. $y = e^{2x}(c_1 \cos x + c_2 \sin x)$
17. $y = e^{-x/3}\left(c_1 \cos \frac{\sqrt{2}}{3}x + c_2 \sin \frac{\sqrt{2}}{3}x\right)$
19. $y = c_1 + c_2e^{-x} + c_3e^{5x}$
21. $y = c_1e^{x} + e^{-x/2}\left(c_2 \cos \frac{\sqrt{3}}{2}x + c_3 \sin \frac{\sqrt{3}}{2}x\right)$
23. $y = c_1e^{-x} + c_2e^{3x} + c_3xe^{3x}$
25. $y = c_1e^{x} + e^{-x}(c_2 \cos x + c_3 \sin x)$
27. $y = c_1e^{-x} + c_2xe^{-x} + c_3x^2e^{-x}$
29. $y = c_1 + c_2x + e^{-x/2}\left(c_3 \cos \frac{\sqrt{3}}{2}x + c_4 \sin \frac{\sqrt{3}}{2}x\right)$
31. $y = c_1 \cos \frac{\sqrt{3}}{2}x + c_2 \sin \frac{\sqrt{3}}{2}x + c_3x \cos \frac{\sqrt{3}}{2}x + c_4x \sin \frac{\sqrt{3}}{2}x$
33. $y = c_1 + c_2e^{-2x} + c_3e^{2x} + c_4 \cos 2x + c_5 \sin 2x$
35. $y = c_1e^{x} + c_2xe^{x} + c_3e^{-x} + c_4xe^{-x} + c_5e^{-5x}$
37. $y = 2\cos 4x - \frac{1}{2}\sin 4x$
39. $y = -\frac{3}{4}e^{-5x} + \frac{3}{4}e^{-x}$
41. $y = -e^{x/2}\cos(x/2) + e^{x/2}\sin(x/2)$ **43.** $y = 0$
45. $y = e^{2(x-1)} - e^{x-1}$
47. $y = \frac{5}{36} - \frac{5}{36}e^{-6x} + \frac{1}{6}xe^{-6x}$

49. $y = -\frac{1}{6}e^{2x} + \frac{1}{6}e^{-x}\cos\sqrt{3}x - \frac{\sqrt{3}}{6}e^{-x}\sin\sqrt{3}x$

51. $y = 2 - 2e^x + 2xe^x - \frac{1}{2}x^2e^x$ 53. $y = e^{5x} - xe^{5x}$

55. $y = -2\cos x$

57. $\frac{d^3y}{dx^3} + 6\frac{d^2y}{dx^2} - 15\frac{dy}{dx} - 100y = 0$

59. $y = c_1e^x + e^{4x}(c_2\cos x + c_3\sin x)$

61. $y'' - 3y' - 18y = 0$ 63. $y''' - 17y'' = 0$

65. $y = e^{-\sqrt{2}x/2}\left(c_1\cos\frac{\sqrt{2}}{2}x + c_2\sin\frac{\sqrt{2}}{2}x\right) + e^{\sqrt{2}x/2}\left(c_3\cos\frac{\sqrt{2}}{2}x + c_4\sin\frac{\sqrt{2}}{2}x\right)$

Exercises 4.4, Page 178

1. $(3D-2)(3D+2)$ 3. $(D-6)(D+2)$

5. $D(D+5)^2$ 7. $(D-1)(D-2)(D+5)$

9. $D(D+2)(D^2-2D+4)$ 11. D^4

13. $D(D-2)$ 15. D^2+4 17. $D^3(D^2+16)$

19. $(D+1)(D-1)^3$ 21. $D(D^2-2D+5)$

23. $y = c_1e^{-3x} + c_2e^{3x} - 6$

25. $y = c_1 + c_2e^{-x} + 3x$

27. $y = c_1e^{2x} + c_2xe^{-2x} + \frac{1}{2}x + 1$

29. $y = c_1 + c_2x + c_3e^{-x} + \frac{2}{3}x^4 - \frac{8}{3}x^3 + 8x^2$

31. $y = c_1e^{-3x} + c_2e^{4x} + \frac{1}{7}xe^{4x}$

33. $y = c_1e^{-x} + c_2e^{3x} - e^x + 3$

35. $y = c_1\cos 5x + c_2\sin 5x + \frac{1}{4}\sin x$

37. $y = c_1e^{-3x} + c_2xe^{-3x} - \frac{1}{49}xe^{4x} + \frac{2}{343}e^{4x}$

39. $y = c_1e^{-x} + c_2e^x + \frac{1}{6}x^3e^x - \frac{1}{4}x^2e^x + \frac{1}{4}xe^x - 5$

41. $y = e^x(c_1\cos 2x + c_2\sin 2x) + \frac{1}{3}e^x\sin x$

43. $y = c_1\cos 5x + c_2\sin 5x - 2x\cos 5x$

45. $y = e^{-x/2}\left(c_1\cos\frac{\sqrt{3}}{2}x + c_2\sin\frac{\sqrt{3}}{2}x\right) + \sin x + 2\cos x - x\cos x$

47. $y = c_1 + c_2x + c_3e^{-8x} + \frac{11}{256}x^2 + \frac{7}{32}x^3 - \frac{1}{16}x^4$

49. $y = c_1e^x + c_2xe^x + c_3x^2e^x + \frac{1}{6}x^3e^x + x - 13$

51. $y = c_1 + c_2x + c_3e^x + c_4xe^x + \frac{1}{2}x^2e^x + \frac{1}{2}x^2$

53. $y = c_1e^{x/2} + c_2e^{-x/2} + c_3\cos\frac{x}{2} + c_4\sin\frac{x}{2} + \frac{1}{8}xe^{x/2}$

55. $y = \frac{5}{8}e^{-8x} + \frac{5}{8}e^{8x} - \frac{1}{4}$

57. $y = -\frac{41}{125} + \frac{41}{125}e^{5x} - \frac{1}{10}x^2 + \frac{9}{25}x$

59. $y = -\pi\cos x - \frac{11}{3}\sin x - \frac{8}{3}\cos 2x + 2x\cos x$

61. $y = 2e^{2x}\cos 2x - \frac{3}{64}e^{2x}\sin 2x + \frac{1}{8}x^3 + \frac{3}{16}x^2 + \frac{3}{32}x$

63. $y_p = A\,xe^x + B\,e^x\cos 2x + C\,e^x\sin 2x + E\,xe^x\cos 2x + F\,xe^x\sin 2x$

65. The operators do not commute.

Exercises 4.5, Page 186

1. $y = c_1\cos x + c_2\sin x + x\sin x + \cos x\ln|\cos x|$; $(-\pi/2, \pi/2)$

3. $y = c_1\cos x + c_2\sin x + \frac{1}{2}\sin x - \frac{1}{2}x\cos x = c_1\cos x + c_3\sin x - \frac{1}{2}x\cos x$; $(-\infty, \infty)$

5. $y = c_1\cos x + c_2\sin x + \frac{1}{2} - \frac{1}{6}\cos 2x$; $(-\infty, \infty)$

7. $y = c_1e^x + c_2e^{-x} + \frac{1}{4}xe^x - \frac{1}{4}xe^{-x} = c_1e^x + c_2e^{-x} + \frac{1}{2}x\sinh x$; $(-\infty, \infty)$

9. $y = c_1e^{2x} + c_2e^{-2x} + \frac{1}{4}\left(e^{2x}\ln|x| - e^{-2x}\int_{x_0}^{x}\frac{e^{4t}}{t}\,dt\right)$, $x_0 > 0$; $(0, \infty)$

11. $y = c_1e^{-x} + c_2e^{-2x} + (e^{-x} + e^{-2x})\times\ln(1 + e^x)$; $(-\infty, \infty)$

13. $y = c_1e^{-2x} + c_2e^{-x} - e^{-2x}\sin e^x$; $(-\infty, \infty)$

15. $y = c_1e^x + c_2xe^x - \frac{1}{2}e^x\ln(1 + x^2) + xe^x\tan^{-1}x$; $(-\infty, \infty)$

17. $y = c_1e^{-x} + c_2xe^{-x} + \frac{1}{2}x^2e^{-x}\ln x - \frac{3}{4}x^2e^{-x}$; $(0, \infty)$

19. $y = c_1e^x\cos 3x + c_2e^x\sin x - \frac{1}{27}e^x\cos 3x\ln|\sec 3x + \tan 3x|$; $(-\pi/6, \pi/6)$

21. $y = c_1 + c_2\cos x + c_3\sin x - \ln|\cos x| - \sin x\ln|\sec x + \tan x|$; $(-\pi/2, \pi/2)$

23. $y = c_1e^x + c_2e^{2x} + c_3e^{-x} + \frac{1}{8}e^{3x}$; $(-\infty, \infty)$

25. $y = \frac{1}{4}e^{-x/2} + \frac{3}{4}e^{x/2} + \frac{1}{8}x^2e^{x/2} - \frac{1}{4}xe^{x/2}$

27. $y = \frac{4}{9}e^{-4x} + \frac{25}{36}e^{2x} - \frac{1}{4}e^{-2x} + \frac{1}{9}e^{-x}$

29. $y = c_1x + c_2x\ln x + \frac{2}{3}x(\ln x)^3$

31. $y = c_1x^{-1/2}\cos x + c_2x^{-1/2}\sin x + x^{-1/2}$

33. $y = c_1 + c_2e^x + c_3e^{-x} + \frac{1}{4}x^2e^x - \frac{3}{4}xe^x$

Chapter 4 Review, Page 189

1. $y = 0$.
3. False, the functions $f_1(x) = 0$ and $f_2(x) = e^x$ are linearly dependent on $(-\infty, \infty)$ but f_2 is not a constant multiple of f_1.
5. $(-\infty, 0)$; $(0, \infty)$ 7. false 9. $y_p = A + Bxe^x$
11. $y_2 = \sin 2x$ 13. $y = c_1 e^{(1+\sqrt{3})x} + c_2 e^{(1-\sqrt{3})x}$
15. $y = c_1 + c_2 e^{-5x} + c_3 x e^{-5x}$
17. $y = c_1 e^{-x/3} + e^{-3x/2}\left(c_2 \cos \frac{\sqrt{7}}{2} x + c_3 \sin \frac{\sqrt{7}}{2} x\right)$
19. $y = e^{3x/2}\left(c_1 \cos \frac{\sqrt{11}}{2} x + c_2 \sin \frac{\sqrt{11}}{2} x\right) + \frac{4}{5}x^3 + \frac{36}{25}x^2 + \frac{46}{125}x - \frac{222}{625}$
21. $y = c_1 + c_2 e^{2x} + c_3 e^{3x} + \frac{1}{5}\sin x - \frac{1}{5}\cos x + \frac{4}{3}x$
23. $y = e^x(c_1 \cos x + c_2 \sin x) - e^x \cos x \ln|\sec x + \tan x|$
25. $y = \frac{1}{2}\cos x + \frac{1}{2}\sin x + \frac{1}{2}\sec x$

Exercises 5.1, Page 199

1. A weight of 4 lb ($\frac{1}{8}$ slug), attached to a spring, is released from a point 3 units above the equilibrium position with an initial upward velocity of 2 ft/sec. The spring constant is 3 lb/ft.
3. $x(t) = 2\sqrt{2}\sin\left(5t - \frac{\pi}{4}\right)$
5. $x(t) = \sqrt{5}\sin(\sqrt{2}t + 3.6052)$
7. $x(t) = \frac{\sqrt{101}}{10}\sin(10t + 1.4711)$ 9. 8 lb
11. $\sqrt{2}\pi/8$ 13. $x(t) = -\frac{1}{4}\cos 4\sqrt{6}t$
15. (a) $x(\pi/12) = -1/4$; $x(\pi/8) = -1/2$; $x(\pi/6) = -1/4$; $x(\pi/4) = 1/2$; $x(9\pi/32) = \sqrt{2}/4$
    (b) 4 ft/sec; downward
    (c) $t = (2n + 1)\pi/16$, $n = 0, 1, 2, \ldots$
17. (a) the 20 kg mass
    (b) the 20 kg mass; the 50 kg mass
    (c) $t = n\pi$, $n = 0, 1, 2, \ldots$; at the equilibrium position; the 50-kg mass is moving upward whereas the 20-kg mass is moving upward when n is even and downward when n is odd.
19. $x(t) = \frac{1}{2}\cos 2t + \frac{3}{4}\sin 2t = \frac{\sqrt{13}}{4}\sin(2t + 0.5880)$
21. (a) $x(t) = -\frac{2}{3}\cos 10t + \frac{1}{2}\sin 10t = \frac{5}{6}\sin(10t - 0.927)$
    (b) 5/6 ft; $\pi/5$
    (c) 15 cycles
    (d) 0.721 sec
    (e) $(2n + 1)\pi/20 + 0.0927$, $n = 0, 1, 2, \ldots$
    (f) $x(3) = -0.597$ ft
    (g) $x'(3) = -5.814$ ft/sec
    (h) $x''(3) = 59.702$ ft/sec^2
    (i) $\pm 8\frac{1}{3}$ ft/sec
    (j) $0.1451 + n\pi/5$; $0.3545 + n\pi/5$, $n = 0, 1, 2, \ldots$
    (k) $0.3545 + n\pi/5$, $n = 0, 1, 2, \ldots$
23. 120 lb/ft; $x(t) = \frac{\sqrt{3}}{12}\sin 8\sqrt{3}t$
25. Using $x(t) = c_1 \cos \omega t + c_2 \sin \omega t$, $x(0) = x_0$ and $x'(0) = v_0$, we find $c_1 = x_0$ and $c_2 = v_0/\omega$. The result follows from $A = \sqrt{c_1^2 + c_2^2}$.
27. $x(t) = 2\sqrt{2}\cos\left(5t + \frac{5\pi}{4}\right)$
29. When $\omega t + \phi = (2m + 1)\pi/2$, $|x''| = A\omega^2$. But $T = 2\pi/\omega$ implies $\omega = 2\pi/T$ and $\omega^2 = 4\pi^2/T^2$. Therefore, the magnitude of the acceleration is $|x''| = 4\pi^2 A/T^2$.

Exercises 5.2, Page 211

1. A 2-lb weight is attached to a spring whose constant is 1 lb/ft. The system is damped with a resisting force numerically equal to 2 times the instantaneous velocity. The weight starts from the equilibrium position with an upward velocity of 1.5 ft/sec.
3. $\frac{1}{4}$ sec; $\frac{1}{2}$ sec, $x(\frac{1}{2}) = e^{-2}$, that is, the weight is approximately 0.14 ft below the equilibrium position.
5. (a) $x(t) = \frac{4}{3}e^{-2t} - \frac{1}{3}e^{-8t}$
    (b) $x(t) = -\frac{2}{3}e^{-2t} + \frac{5}{3}e^{-8t}$
7. (a) $x(t) = e^{-2t}[-\cos 4t - \frac{1}{2}\sin 4t]$
    (b) $x(t) = \frac{\sqrt{5}}{2}e^{-2t}\sin(4t + 4.249)$
    (c) $t = 1.294$ sec
9. (a) $\beta > 5/2$
    (b) $\beta = 5/2$
    (c) $0 < \beta < 5/2$

11. $x(t) = \frac{2}{7}e^{-7t}\sin 7t$ 13. $v_0 > 2$ ft/sec

15. Suppose $\gamma = \sqrt{\omega^2 - \lambda^2}$ then the derivative of $x(t) = Ae^{-\lambda t}\sin(\gamma t + \phi)$ is

$$x'(t) = Ae^{-\lambda t}[\gamma\cos(\gamma t + \phi) - \lambda\sin(\gamma t + \phi)].$$

So $x'(t) = 0$ implies

$$\tan(\gamma t + \phi) = \gamma/\lambda$$

from which it follows that

$$t = \frac{1}{\gamma}\left[\tan^{-1}\frac{\gamma}{\lambda} + k\pi - \phi\right].$$

The difference between the t values between two successive maxima (or minima) is then

$$t_{k+2} - t_k = (k+2)(\pi/\gamma) - k(\pi/\gamma) = 2\pi/\gamma.$$

17. $$t^*_{k+1} - t^*_k = \frac{(2k+3)\pi/2 - \phi}{\sqrt{\omega^2-\lambda^2}} - \frac{(2k+1)\pi/2 - \phi}{\sqrt{\omega^2-\lambda^2}} = \frac{\pi}{\sqrt{\omega^2-\lambda^2}}$$

19. Let the quasi period $2\pi/\sqrt{\omega^2 - \lambda^2}$ be denoted by T_q. From equation (15) we find

$$x_n/x_{n+2} = x(t)/x(t + T_q) = \frac{e^{-\lambda t}\sin(\sqrt{\omega^2-\lambda^2}\,t + \phi)}{e^{-\lambda(t+T_q)}\sin(\sqrt{\omega^2-\lambda^2}(t+T_q) + \phi)} = e^{\lambda T_q}$$

since

$$\sin(\sqrt{\omega^2-\lambda^2}\,t + \phi) = \sin(\sqrt{\omega^2-\lambda^2}(t + T_q) + \phi).$$

Therefore

$$\ln(x_n/x_{n+2}) = \lambda T_q = 2\pi\lambda/\sqrt{\omega^2 - \lambda^2}.$$

Exercises 5.3, Page 221

1. $$x(t) = e^{-t/2}\left(-\frac{4}{3}\cos\frac{\sqrt{47}}{2}t - \frac{64}{3\sqrt{47}}\sin\frac{\sqrt{47}}{2}t\right) + \frac{10}{3}(\cos 3t + \sin 3t)$$

3. $x(t) = \frac{1}{4}e^{-4t} + te^{-4t} - \frac{1}{4}\cos 4t$

5. $x(t) = -\frac{1}{2}\cos 4t + \frac{9}{4}\sin 4t + \frac{1}{2}e^{-2t}\cos 4t - 2e^{-2t}\sin 4t$

7. $m\dfrac{d^2x}{dt^2} = -k(x - h) - \beta\dfrac{dx}{dt}$ or

$$\frac{d^2x}{dt^2} + 2\lambda\frac{dx}{dt} + \omega^2 x = \omega^2 h(t), \quad \text{where}$$

$$2\lambda = \beta/m \quad \text{and} \quad \omega^2 = k/m.$$

9. **(a)** $x(t) = \frac{2}{3}\sin 4t - \frac{1}{3}\sin 8t$
(b) $t = n\pi/4,\ \ n = 0, 1, 2, \ldots$
(c) $t = \pi/6 + n\pi/2,\ \ n = 0, 1, 2, \ldots$ and $t = \pi/3 + n\pi/2,\ \ n = 0, 1, 2, \ldots$
(d) $\sqrt{3}/2$ cm, $\ -\sqrt{3}/2$ cm
(e)

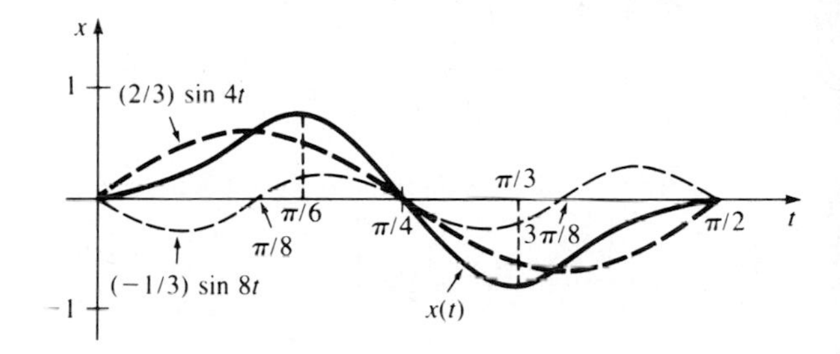

11. **(a)** $g'(\gamma) = 0$ implies $\gamma(\gamma^2 - \omega^2 + 2\lambda^2) = 0$ so that either $\gamma = 0$ or $\gamma = \sqrt{\omega^2 - 2\lambda^2}$. The first derivative test can be used to verify that $g(\gamma)$ is a maximum at the latter value.
(b) $g(\sqrt{\omega^2 - 2\lambda^2}) = F_0/2\lambda\sqrt{\omega^2 - \lambda^2}$

13. $$x_p = -5\cos 2t + 5\sin 2t = 5\sqrt{2}\sin\left(2t - \frac{\pi}{4}\right)$$

15. **(a)** $$x(t) = x_c + x_p = c_1\cos\omega t + c_2\sin\omega t + \frac{F_0}{\omega^2 - \gamma^2}\cos\gamma t,$$
where the initial conditions imply that $c_1 = -F_0/(\omega^2 - \gamma^2)$ and $c_2 = 0$.
(b) By L'Hôpital's rule the given limit is the same as

$$\lim_{\gamma\to\omega}\frac{F_0(-t\sin\gamma t)}{-2\gamma} = \frac{F_0}{2\omega}t\sin\omega t.$$

17. $x(t) = -\cos 2t - \frac{1}{8}\sin 2t + \frac{3}{4}t\sin 2t + \frac{5}{4}t\cos 2t$

19. **(a)** Recall that

$$\cos(u - v) = \cos u \cos v + \sin u \sin v.$$
$$\cos(u + v) = \cos u \cos v - \sin u \sin v.$$

Subtracting gives

$$\sin u \sin v = \tfrac{1}{2}[\cos(u - v) - \cos(u + v)].$$

Setting $u = \frac{1}{2}(\gamma - \omega)t$ and $v = \frac{1}{2}(\gamma + \omega)t$ then gives

$$\sin \tfrac{1}{2}(\gamma - \omega)t \sin \tfrac{1}{2}(\gamma + \omega)t = \tfrac{1}{2}[\cos \omega t - \cos \gamma t],$$

from which the result follows.

(b) For small, ε, $\gamma \approx \omega$ so $\gamma + \omega \approx 2\gamma$ and therefore

$$\frac{-2F_0}{(\omega + \gamma)(\omega - \gamma)} \sin \frac{1}{2}(\gamma - \omega)t \sin \frac{1}{2}(\gamma + \omega)t$$
$$\approx \frac{F_0}{2\gamma\varepsilon} \sin \varepsilon t \sin \frac{1}{2}(2\gamma)t.$$

(c) By L'Hôpital's rule the given limit is the same as

$$\lim_{\varepsilon \to 0} \frac{F_0 t \cos \varepsilon t \sin \gamma t}{2\gamma} = \frac{F_0}{2\gamma} t \sin \gamma t = \frac{F_0}{2\omega} t \sin \omega t.$$

Exercises 5.4, Page 230

1. $q(t) = -\dfrac{15}{4} \cos 4t + \dfrac{15}{4}$; $i(t) = 15 \sin 4t$

3. underdamped **5.** 4.568 coulombs; 0.0669 sec

7. $q(t) = 10 - 10e^{-3t}(\cos 3t + \sin 3t)$; $i(t) = 60e^{-3t} \sin 3t$; 10.432 coulombs

9. $i_p = \dfrac{100}{13} \cos t - \dfrac{150}{13} \sin t$

13. $q(t) = -\frac{1}{2}e^{-10t}(\cos 10t + \sin 10t) + \frac{3}{2}$; $\frac{3}{2}$ coulombs

15. Show that $dZ/dC = 0$ when $C = 1/L\gamma^2$. At this value, Z is a minimum and, correspondingly, the amplitude E_0/Z is a maximum.

17. $q(t) = \left(q_0 - \dfrac{E_0 C}{1 - \gamma^2 LC}\right) \cos \dfrac{t}{\sqrt{LC}} + \sqrt{LC}\, i_0 \sin \dfrac{t}{\sqrt{LC}} + \dfrac{E_0 C}{1 - \gamma^2 LC} \cos \gamma t$;

$$i(t) = i_0 \cos \frac{t}{\sqrt{LC}} - \frac{1}{\sqrt{LC}}\left(q_0 - \frac{E_0 C}{1 - \gamma^2 LC}\right) \sin \frac{t}{\sqrt{LC}} - \frac{E_0 C\gamma}{1 - \gamma^2 LC} \sin \gamma t$$

19. $\theta(t) = \dfrac{1}{2} \cos 4t + \dfrac{\sqrt{3}}{2} \sin 4t$; 1; $\pi/2$; $2/\pi$

Chapter 5 Review, Page 233

1. 8 ft **3.** 5/4 m

5. False; there could be an impressed force driving the system.

7. overdamped **9.** 9/2 lb/ft

11. $x(t) = -\frac{2}{3}e^{-2t} + \frac{1}{3}e^{-4t}$ **13.** $0 < m \le 2$

15. $\gamma = 8\sqrt{3}/3$

17. $x(t) = e^{-4t} \times \left(\dfrac{26}{17} \cos 2\sqrt{2}\,t + \dfrac{28\sqrt{2}}{17} \sin 2\sqrt{2}\,t\right) + \dfrac{8}{17} e^{-t}$

19. $q(t) = -\dfrac{1}{150} \sin 100t + \dfrac{1}{75} \sin 50t$;

$i(t) = -\dfrac{2}{3} \cos 100t + \dfrac{2}{3} \cos 50t$;

$t = n\pi/50$, $n = 0, 1, 2, \ldots$

Exercises 6.1, Page 244

1. $y = c_1 x^{-1} + c_2 x^2$ **3.** $y = c_1 + c_2 \ln x$

5. $y = c_1 \cos(2 \ln x) + c_2 \sin(2 \ln x)$

7. $y = c_1 x^{(2 - \sqrt{6})} + c_2 x^{(2 + \sqrt{6})}$

9. $y_1 = c_1 \cos(\frac{1}{5} \ln x) + c_2 \sin(\frac{1}{5} \ln x)$

11. $y = c_1 x^{-2} + c_2 x^{-2} \ln x$

13. $y = x[c_1 \cos(\ln x) + c_2 \sin(\ln x)]$

15. $y = x^{-1/2}\left[c_1 \cos\left(\dfrac{\sqrt{3}}{6} \ln x\right) + c_2 \sin\left(\dfrac{\sqrt{3}}{6} \ln x\right)\right]$

17. $y = c_1 x^3 + c_2 \cos(\sqrt{2} \ln x) + c_3 \sin(\sqrt{2} \ln x)$

19. $y = c_1 x^{-1} + c_2 x^2 + c_3 x^4$

21. $y = c_1 + c_2 x + c_3 x^2 + c_4 x^{-3}$ **23.** $y = 2 - 2x^{-2}$

25. $y = \cos(\ln x) + 2\sin(\ln x)$

27. $y = 2(-x)^{1/2} - 5(-x)^{1/2}\ln(-x)$

29. $y = c_1 + c_2 \ln x + \dfrac{x^2}{4}$

31. $y = c_1x^{-1/2} + c_2x^{-1} + \frac{1}{15}x^2 - \frac{1}{6}x$

33. $y = c_1x + c_2x\ln x + x(\ln x)^2$

35. $y = c_1x^{-1} + c_2x^{-8} + \frac{1}{30}x^2$

37. $y = x^2[c_1\cos(3\ln x) + c_2\sin(3\ln x)] + \frac{4}{13} + \frac{3}{10}x$

39. $y = c_1x^2 + c_2x^{-10} - \frac{1}{7}x^{-3}$

41. $y = c_1(x-1)^{-1} + c_2(x-1)^4$

43. $y = c_1\cos(\ln(x+2)) + c_2\sin(\ln(x+2))$

Exercises 6.2, Page 256

1. $y = ce^{-x};\quad y = c_0 \sum_{n=0}^{\infty} \dfrac{(-1)^n}{n!}x^n$

3. $y = ce^{x^3/3};\quad y = c_0 \sum_{n=0}^{\infty} \dfrac{1}{n!}\left(\dfrac{x^3}{3}\right)^n$

5. $y = c/(1-x);\quad y = c_0 \sum_{n=0}^{\infty} x^n$

7. $y = C_1\cos x + C_2\sin x;$

$$y = c_0 \sum_{n=0}^{\infty} \frac{(-1)^n}{(2n)!}x^{2n} + c_1 \sum_{n=0}^{\infty} \frac{(-1)^n}{(2n+1)!}x^{2n+1}$$

9. $y = C_1 + C_2e^x;$

$$y = c_0 + c_1 \sum_{n=1}^{\infty} \frac{x^n}{n!} = c_0 - c_1 + c_1 \sum_{n=0}^{\infty} \frac{x^n}{n!}$$
$$= c_0 - c_1 + c_1e^x$$

11. $$y_1(x) = c_0\left[1 + \frac{1}{3\cdot 2}x^3 + \frac{1}{6\cdot 5\cdot 3\cdot 2}x^6 + \frac{1}{9\cdot 8\cdot 6\cdot 5\cdot 3\cdot 2}x^9 + \cdots\right]$$
$$y_2(x) = c_1\left[x + \frac{1}{4\cdot 3}x^4 + \frac{1}{7\cdot 6\cdot 4\cdot 3}x^7 + \frac{1}{10\cdot 9\cdot 7\cdot 6\cdot 4\cdot 3}x^{10} + \cdots\right]$$

13. $$y_1(x) = c_0\left[1 - \frac{1}{2}x^2 - \frac{3}{4!}x^4 - \frac{21}{6!}x^6 - \cdots\right]$$
$$y_2(x) = c_1\left[x + \frac{1}{3!}x^3 + \frac{5}{5!}x^5 + \frac{45}{7!}x^7 + \cdots\right]$$

15. $$y_1(x) = c_0\left[1 - \frac{1}{3!}x^3 + \frac{4^2}{6!}x^6 - \frac{7^2\cdot 4^2}{9!}x^9 + \cdots\right]$$
$$y_2(x) = c_1\left[x - \frac{2^2}{4!}x^4 + \frac{5^2\cdot 2^2}{7!}x^7 - \frac{8^2\cdot 5^2\cdot 2^2}{10!}x^{10} + \cdots\right]$$

17. $y_1(x) = c_0;\quad y_2(x) = c_1 \sum_{n=1}^{\infty} \dfrac{1}{n}x^n$

19. $y_1(x) = c_0 \sum_{n=0}^{\infty} x^{2n};\quad y_2(x) = c_1 \sum_{n=0}^{\infty} x^{2n+1}$

21. $$y_1(x) = c_0\left[1 + \frac{1}{4}x^2 - \frac{7}{4\cdot 4!}x^4 + \frac{23\cdot 7}{8\cdot 6!}x^6 - \cdots\right]$$
$$y_2(x) = c_1\left[x - \frac{1}{6}x^3 + \frac{14}{2\cdot 5!}x^5 - \frac{34\cdot 14}{4\cdot 7!}x^7 - \cdots\right]$$

23. $y_1(x) = c_0[1 + \frac{1}{2}x^2 + \frac{1}{6}x^3 + \frac{1}{6}x^4 + \cdots]$
$y_2(x) = c_1[x + \frac{1}{2}x^2 + \frac{1}{2}x^3 + \frac{1}{4}x^4 + \cdots]$

25. $y(x) = -2[1 + \frac{1}{2!}x^2 + \frac{1}{3!}x^3 + \frac{1}{4!}x^4 + \cdots] + 6x$
$= 8x - 2e^x$

27. $y(x) = 3 - 12x^2 + 4x^4$

29. $y_1(x) = c_0[1 - \frac{1}{6}x^3 + \frac{1}{120}x^5 + \cdots]$
$y_2(x) = c_1[x - \frac{1}{12}x^4 + \frac{1}{180}x^6 + \cdots]$

31. $y_1(x) = c_0[1 - \frac{1}{2}x^2 + \frac{1}{6}x^3 - \frac{1}{40}x^5 + \cdots]$
$y_2(x) = c_1[x - \frac{1}{6}x^3 + \frac{1}{12}x^4 - \frac{1}{60}x^5 + \cdots]$

33. $$y_1(x) = c_0\left[1 + \frac{1}{3!}x^3 + \frac{4}{6!}x^6 + \frac{7\cdot 4}{9!}x^9 + \cdots\right] + c_1\left[x + \frac{2}{4!}x^4 + \frac{5\cdot 2}{7!}x^7 + \frac{8\cdot 5\cdot 2}{10!}x^{10} + \cdots\right]$$
$$+ \frac{1}{2!}x^2 + \frac{3}{5!}x^5 + \frac{6\cdot 3}{8!}x^8 + \frac{9\cdot 6\cdot 3}{11!}x^{11} + \cdots$$

Exercises 6.3, Page 275

1. $x = 0$, irregular singular point

3. $x = -3$, regular singular point; $x = 3$, irregular singular point

5. $x = 0, 2i, -2i$, regular singular points

7. $x = -3, 2$, regular singular points

9. $x = 0$, irregular singular point; $x = -5, 5, 2$, regular singular points

11. $r_1 = \frac{3}{2}, r_2 = 0;$

$$y(x) = C_1 x^{3/2}\left[1 - \frac{2}{5}x + \frac{2^2}{7\cdot 5\cdot 2}x^2 - \frac{2^3}{9\cdot 7\cdot 5\cdot 3!}x^3 + \cdots\right] + C_2\left[1 + 2x - 2x^2 + \frac{2^3}{3\cdot 3!}x^3 - \cdots\right]$$

13. $r_1 = \frac{7}{8}, r_2 = 0;$

$$y(x) = C_1 x^{7/8}\left[1 - \frac{2}{15}x + \frac{2^2}{23\cdot 15\cdot 2}x^2 - \frac{2^3}{31\cdot 23\cdot 15\cdot 3!}x^3 + \cdots\right] + C_2\left[1 - 2x + \frac{2^2}{9\cdot 2}x^2 - \frac{2^3}{17\cdot 9\cdot 3!}x^3 + \cdots\right]$$

15. $r_1 = \frac{1}{3}, r_2 = 0;$

$$y(x) = C_1 x^{1/3}\left[1 + \frac{1}{3}x + \frac{1}{3^2\cdot 2}x^2 + \frac{1}{3^3\cdot 3!}x^3 + \cdots\right] + C_2\left[1 + \frac{1}{2}x + \frac{1}{5\cdot 2}x^2 + \frac{1}{8\cdot 5\cdot 2}x^3 + \cdots\right]$$

17. $r_1 = \frac{5}{2}, r_2 = 0;$

$$y(x) = C_1 x^{5/2}\left[1 + \frac{2\cdot 2}{7}x + \frac{2^2\cdot 3}{9\cdot 7}x^2 + \frac{2^3\cdot 4}{11\cdot 9\cdot 7}x^3 + \cdots\right] + C_2\left[1 + \frac{1}{3}x - \frac{1}{6}x^2 - \frac{1}{6}x^3 - \cdots\right]$$

19. $r_1 = \frac{2}{3}, r_2 = \frac{1}{3};$

$$y(x) = C_1 x^{2/3}[1 - \tfrac{1}{2}x + \tfrac{5}{28}x^2 - \tfrac{1}{21}x^3 + \cdots] + C_2 x^{1/3}[1 - \tfrac{1}{2}x + \tfrac{1}{5}x^2 - \tfrac{7}{120}x^3 + \cdots]$$

21. $r_1 = 1, r_2 = -\frac{1}{2};$

$$y(x) = C_1 x\left[1 + \frac{1}{5}x + \frac{1}{5\cdot 7}x^2 + \frac{1}{5\cdot 7\cdot 9}x^3 + \cdots\right] + C_2 x^{-1/2}\left[1 + \frac{1}{2}x + \frac{1}{2\cdot 4}x^2 + \frac{1}{2\cdot 4\cdot 6}x^3 + \cdots\right]$$

23. $r_1 = 0, r_2 = -1;$

$$y(x) = C_1 x^{-1}\sum_{n=0}^{\infty}\frac{1}{(2n)!}x^{2n} + C_2 x^{-1}\sum_{n=0}^{\infty}\frac{1}{(2n+1)!}x^{2n+1} = \frac{1}{x}[C_1\cosh x + C_2\sinh x]$$

25. $r_1 = 4, r_2 = 0;$

$$y(x) = C_1\left[1 + \frac{2}{3}x + \frac{1}{3}x^2\right] + C_2\sum_{n=0}^{\infty}(n+1)x^{n+4}$$

27. $r_1 = r_2 = 0;$

$$y(x) = C_1 y_1(x) + C_2\left[y_1(x)\ln x + y_1(x) \times \left(-x + \frac{1}{4}x^2 - \frac{1}{3\cdot 3!}x^3 + \frac{1}{4\cdot 4!}x^4 - \cdots\right)\right],$$

where $y_1(x) = \sum_{n=0}^{\infty}\frac{1}{n!}x^n = e^x$

29. $r_1 = r_2 = 0;$
$y(x) = C_1 y_1(x) + C_2[y_1(x)\ln x + y_1(x) \times (2x + \frac{5}{4}x^2 + \frac{23}{27}x^3 + \cdots)],$

where $y_1(x) = \sum_{n=0}^{\infty}\frac{(-1)^n}{(n!)^{2n}}x^n$

31. $r_1 = r_2 = 1;$
$$y(x) = C_1 xe^{-x} + C_2 xe^{-x} \times \left[\ln x + x + \frac{1}{4}x^2 + \frac{1}{3\cdot 3!}x^3 + \cdots\right]$$

33. $r_1 = 2, r_2 = 0;$

$$y(x) = C_1 x^2 + C_2\left[\frac{1}{2}x^2\ln x - \frac{1}{2} + x - \frac{1}{3!}x^3 + \cdots\right]$$

35. The method of Frobenius yields only the trivial solution $y(x) = 0$.

37. There is a regular singular point at ∞.

39. There is a regular singular point at ∞.

41. The assumption $y = \sum_{n=0}^{\infty} c_n x^{n+r}$ leads to $c_n[(n+r)(n+r+2) - 8] = 0$ for $n \geq 0$. For $n = 0$ and $c_0 \neq 0$ we have $r^2 + 2r - 8 = 0$ and so $r_1 = 2$, $r_2 = -4$. For these values we are forced to take $c_n = 0$ for $n > 0$. Hence a solution exists of the form $y = c_0 x^r$. It follows that the general solution on $0 < x < \infty$ is $y = C_1 x^2 + C_2 x^{-4}$.

43. $r(r-1) + \frac{5}{3}r - \frac{1}{3} = 0;\ r_1 = \frac{1}{3}, r_2 = -1$

Exercises 6.4, Page 285

1. $y = c_1 J_{1/3}(x) + c_2 J_{-1/3}(x)$

3. $y = c_1 J_{5/2}(x) + c_2 J_{-5/2}(x)$

5. $y = c_1 J_0(x) + c_2 Y_0(x)$

7. $y = c_1 J_2(3x) + c_2 Y_2(3x)$

9. After using the change of variables, the differential equation becomes

$$x^2 v'' + xv' + (\lambda^2 x^2 - \tfrac{1}{4})v = 0.$$

Since the solution of the last equation is

$$v = c_1 J_{1/2}(\lambda x) + c_2 J_{-1/2}(\lambda x),$$

we find

$$y = c_1 x^{-1/2} J_{1/2}(\lambda x) + c_2 x^{-1/2} J_{-1/2}(\lambda x).$$

11. After substituting into the differential equation, we find

$$\begin{aligned} xy'' + (1+2n)y' + xy &= x^{-n-1}[x^2 J_n'' + xJ_n' + (x^2 - n^2)J_n] \\ &= x^{-n-1} \cdot 0 = 0. \end{aligned}$$

13. From Problem 10 with $n = \frac{1}{2}$ we find $y = x^{1/2} J_{1/2}(x)$; from Problem 11 with $n = -\frac{1}{2}$ we find $y = x^{1/2} J_{-1/2}(x)$.

15. From Problem 10 with $n = -1$ we find $y = x^{-1} J_{-1}(x)$; from Problem 11 with $n = 1$ we find $y = x^{-1} J_1(x)$ but since $J_{-1}(x) = -J_1(x)$, no new solution results.

17. From Problem 12 with $\lambda = 1$ and $\nu = \pm 3/2$ we find $y = \sqrt{x}\, J_{3/2}(x)$ and $y = \sqrt{x}\, J_{-3/2}(x)$.

19. Using the hint, we can write

$$\begin{aligned} xJ_\nu'(x) &= -\nu \sum_{n=0}^{\infty} \frac{(-1)^n}{n!\Gamma(1+\nu+n)} \left(\frac{x}{2}\right)^{2n+\nu} \\ &\quad + 2 \sum_{n=0}^{\infty} \frac{(-1)^n (n+\nu)}{n!(n+\nu)\Gamma(n+\nu)} \left(\frac{x}{2}\right)^{2n+\nu} \\ &= -\nu \sum_{n=0}^{\infty} \frac{(-1)^n}{n!\Gamma(1+\nu+n)} \left(\frac{x}{2}\right)^{2n+\nu} \\ &\quad + x \sum_{n=0}^{\infty} \frac{(-1)^n}{n!\Gamma(n+\nu)} \left(\frac{x}{2}\right)^{2n+\nu-1} \\ &= -\nu J_\nu(x) + xJ_{\nu-1}(x). \end{aligned}$$

21. Subtracting the equations

$$\begin{aligned} xJ_\nu'(x) &= \nu J_\nu(x) - xJ_{\nu+1}(x) \\ xJ_\nu'(x) &= -\nu J_\nu(x) + xJ_{\nu-1}(x) \end{aligned}$$

gives $\quad 2\nu J_\nu(x) = xJ_{\nu+1}(x) + xJ_{\nu-1}(x).$

23. The result from the given example

$$xJ_\nu'(x) - \nu J_\nu(x) = -xJ_{\nu+1}(x)$$

is a linear first-order differential equation in $J_\nu(x)$. Dividing by x, we find the integrating factor is $x^{-\nu}$. It follows from this theory that after multiplying both sides of the equation by the integrating factor, we must have

$$\frac{d}{dx}[x^{-\nu} J_\nu(x)] = -x^{-\nu} J_{\nu+1}(x).$$

25. From Problem 22, $\dfrac{d}{dr}[rJ_1(r)] = rJ_0(r)$. Therefore

$$\begin{aligned} \int_0^x rJ_0(r)\,dr &= \int_0^x \frac{d}{dr}[rJ_1(r)]\,dr = rJ_1(r)\Big|_0^x \\ &= xJ_1(x). \end{aligned}$$

27. $J_{-1/2}(x) = \sqrt{\dfrac{2}{\pi x}} \cos x$

29. $J_{-3/2}(x) = \sqrt{\dfrac{2}{\pi x}} \left[-\sin x - \dfrac{\cos x}{x} \right]$

31. $J_{-5/2}(x) = \sqrt{\dfrac{2}{\pi x}} \left[\dfrac{3}{x} \sin x + \left(\dfrac{3}{x^2} - 1\right) \cos x \right]$

33. $J_{-7/2}(x) = \sqrt{\dfrac{2}{\pi x}} \left[\left(1 - \dfrac{15}{x^2}\right) \sin x + \left(\dfrac{6}{x} - \dfrac{15}{x^3}\right) \cos x \right]$

35. $y = c_1 I_\nu(x) + c_2 I_{-\nu}(x)$, $\nu \neq$ integer

37. Since $1/\Gamma(1 - m + n) = 0$ when $n \leq m - 1$, m a positive integer,

$$\begin{aligned} J_{-m}(x) &= \sum_{n=0}^{\infty} \frac{(-1)^n}{n!\Gamma(1-m+n)} \left(\frac{x}{2}\right)^{2n-m} \\ &= \sum_{n=m}^{\infty} \frac{(-1)^n}{n!\Gamma(1-m+n)} \left(\frac{x}{2}\right)^{2n-m} \end{aligned}$$

$$J_{-m}(x) = \sum_{k=0}^{\infty} \frac{(-1)^{k+m}}{(k+m)!\Gamma(1+k)} \left(\frac{x}{2}\right)^{2k+m} \quad (n = k+m)$$

$$= (-1)^m \sum_{k=0}^{\infty} \frac{(-1)^k}{\Gamma(1+k+m)k!} \left(\frac{x}{2}\right)^{2k+m}$$

$$= (-1)^m J_m(x).$$

39. **(a)** $P_6(x) = \frac{1}{16}(231x^6 - 315x^4 + 105x^2 - 5)$
$P_7(x) = \frac{1}{16}[429x^7 - 693x^5 + 315x^3 - 35x]$
(b) $y = P_6(x)$ satisfies $(1 - x^2)y'' - 2xy' + 42y = 0$.
$y = P_7(x)$ satisfies $(1 - x^2)y'' - 2xy' + 56y = 0$.

41. If $x = \cos\theta$, then $\dfrac{dy}{d\theta} = \dfrac{dy}{dx}\dfrac{dx}{d\theta} = -\sin\theta\dfrac{dy}{dx}$ and $\dfrac{d^2y}{d\theta^2} = \sin^2\theta\dfrac{d^2y}{dx^2} - \cos\theta\dfrac{dy}{dx}$. Now the original equation can be written as

$$\frac{d^2y}{d\theta^2} + \frac{\cos\theta}{\sin\theta}\frac{dy}{d\theta} + n(n+1)y = 0,$$

and so

$$\sin^2\theta\frac{d^2y}{dx^2} - 2\cos\theta\frac{dy}{dx} + n(n+1)y = 0.$$

Since $x = \cos\theta$ and $\sin^2\theta = 1 - \cos^2\theta = 1 - x^2$, we obtain

$$(1 - x^2)\frac{d^2y}{dx^2} - 2x\frac{dy}{dx} + n(n+1)y = 0.$$

43. By the binomial theorem we have formally

$$(1 - 2xt + t^2)^{-1/2} = 1 + \frac{1}{2}(2xt - t^2) + \frac{1\cdot 3}{2^2 2!} \times (2xt - t^2)^2 + \cdots.$$

Grouping by powers of t, we then find

$$(1 - 2xt + t^2)^{-1/2} = 1\cdot t^0 + x\cdot t + \tfrac{1}{2}(3x^2 - 1)t^2 + \cdots$$
$$= P_0(x)t^0 + P_1(x)t + P_2(x)t^2 + \cdots.$$

45. For $k = 1$, $P_2(x) = \frac{1}{2}[3xP_1(x) - P_0(x)] = \frac{1}{2}(3x^2 - 1)$.
For $k = 2$, $P_3(x) = \frac{1}{3}[5xP_2(x) - 2P_1(x)] = \frac{1}{3}[5x\cdot\frac{1}{2}(3x^2 - 1) - 2x] = \frac{1}{2}(5x^3 - 3x)$.
For $k = 3$, $P_4(x) = \frac{1}{4}[7xP_3(x) - 3P_2(x)] = \frac{1}{4}[7x\cdot\frac{1}{2}(5x^3 - 3x) - \frac{3}{2}(3x^2 - 1)] = \frac{1}{8}(35x^4 - 30x^2 + 3)$.

47. For $n = 0, 1, 2, 3$, the value of the integral is $2, \frac{2}{3}, \frac{2}{5}$, and $\frac{2}{7}$, respectively. In general,

$$\int_{-1}^{1} P_n^2(x)\,dx = \frac{2}{2n+1}, \quad n = 0, 1, 2, \ldots.$$

49. y_2 is obtained from (4) of Section 4.2.

Chapter 6 Review, Page 292

1. $y = c_1x^{-1/3} + c_2x^{1/2}$

3. $y(x) = c_1x^2 + c_2x^3 + x^4 - x^2\ln x$

5. The singular points are $x = 0$, $x = -1 + \sqrt{3}i$, $x = -1 - \sqrt{3}i$. All other finite values of x, real or complex, are ordinary points.

7. RSP $x = 0$; ISP $x = 5$

9. RSP $x = -3, x = 3$; ISP $x = 0$ **11.** $|x| < \infty$

13. $$y_1(x) = c_0\left[1 - \frac{1}{3\cdot 2}x^3 + \frac{1}{6\cdot 5\cdot 3\cdot 2}x^6 - \frac{1}{9\cdot 8\cdot 6\cdot 5\cdot 3\cdot 2}x^9 + \cdots\right]$$
$$y_2(x) = c_1\left[x - \frac{1}{4\cdot 3}x^4 + \frac{1}{7\cdot 6\cdot 4\cdot 3}x^7 - \frac{1}{10\cdot 9\cdot 7\cdot 6\cdot 4\cdot 3}x^{10} + \cdots\right]$$

15. $y_1(x) = c_0[1 + \frac{3}{2}x^2 + \frac{1}{2}x^3 + \frac{5}{8}x^4 + \cdots]$
$y_2(x) = c_1[x + \frac{1}{2}x^3 + \frac{1}{4}x^4 + \cdots]$

17. $r_1 = 1, r_2 = -\frac{1}{2}$;
$$y(x) = C_1x\left[1 + \frac{1}{5}x + \frac{1}{7\cdot 5\cdot 2}x^2 + \frac{1}{9\cdot 7\cdot 5\cdot 3\cdot 2}x^3 + \cdots\right] + C_2x^{-1/2}\left[1 - x - \frac{1}{2}x^2 - \frac{1}{3^2\cdot 2}x^3 - \cdots\right]$$

19. $r_1 = 3, r_2 = 0$;
$$y_1(x) = C_3\left[x^3 + \frac{5}{4}x^4 + \frac{11}{8}x^5 + \cdots\right]$$
$$y(x) = C_1y_1(x) + C_2\left[-\frac{1}{36}y_1(x)\ln x + y_1(x) \times \left(-\frac{1}{3}\frac{1}{x^3} + \frac{1}{4}\frac{1}{x^2} + \frac{1}{16}\frac{1}{x} + \cdots\right)\right]$$

21. $r_1 = r_2 = 0$; $y(x) = C_1 e^x + C_2 e^x \ln x$

23. $y(x) = c_0\left[1 - \frac{1}{2^2}x^2 + \frac{1}{2^4(1 \cdot 2)^2}x^4 - \frac{1}{2^6(1 \cdot 2 \cdot 3)^2}x^6 + \cdots\right]$

Exercises 7.1, Page 304

1. $\frac{2}{s}e^{-s} - \frac{1}{s}$ 3. $\frac{1}{s^2} - \frac{1}{s^2}e^{-s}$ 5. $\frac{1 + e^{-s\pi}}{s^2 + 1}$

7. $\frac{e^7}{s - 1}$ 9. $\frac{1}{(s - 4)^2}$ 11. $\frac{1}{s^2 + 2s + 2}$

13. $\frac{s^2 - 1}{(s^2 + 1)^2}$ 15. $\frac{48}{s^5}$ 17. $\frac{4}{s^2} - \frac{10}{s}$

19. $\frac{2}{s^3} + \frac{6}{s^2} - \frac{3}{s}$ 21. $\frac{6}{s^4} + \frac{6}{s^3} + \frac{3}{s^2} + \frac{1}{s}$

23. $\frac{1}{s} + \frac{1}{s - 4}$ 25. $\frac{1}{s} + \frac{2}{s - 2} + \frac{1}{s - 4}$

27. $\frac{8}{s^3} - \frac{15}{s^2 + 9}$

29. Use $\sinh kt = \frac{e^{kt} - e^{-kt}}{2}$ to show that

$$\mathscr{L}\{\sinh kt\} = \frac{k}{s^2 - k^2}$$

31. $\frac{1}{2(s - 2)} - \frac{1}{2s}$ 33. $\frac{2}{s^2 + 16}$

35. $\frac{1}{2}\left(\frac{s}{s^2 + 9} + \frac{s}{s^2 + 1}\right)$ 37. $\frac{1}{2}\left(\frac{3}{s^2 + 9} - \frac{1}{s^2 + 1}\right)$

39. The result follows by letting $u = st$ in

$$\mathscr{L}\{t^\alpha\} = \int_0^\infty t^\alpha e^{-st}\,dt.$$

41. $\frac{\frac{1}{2}\Gamma(\frac{1}{2})}{s^{3/2}} = \frac{\sqrt{\pi}}{2s^{3/2}}$

43. On $0 \le t \le 1$, $e^{-st} \ge e^{-s}(s > 0)$. Therefore,

$$\int_0^1 e^{-st}\frac{1}{t^2}\,dt \ge e^{-s}\int_0^1 \frac{1}{t^2}\,dt.$$

The latter integral diverges.

Exercises 7.2, Page 311

1. $\frac{1}{2}t^2$ 3. $1 + 3t + \frac{3}{2}t^2 + \frac{1}{6}t^3$ 5. $t - 1 + e^{2t}$

7. $\frac{1}{4}e^{-t/4}$ 9. $\frac{5}{3}\sin 3t$ 11. $\cos\frac{t}{2}$ 13. $\frac{1}{4}\sinh 4t$

15. $2\cos 3t - 2\sin 3t$ 17. $\frac{1}{3} - \frac{1}{3}e^{-3t}$

19. $\frac{3}{4}e^{-3t} + \frac{1}{4}e^t$

21. $-\frac{1}{3}e^{-t} + \frac{8}{15}e^{2t} - \frac{1}{5}e^{-3t}$ 23. $\frac{1}{4}t - \frac{1}{8}\sin 2t$

25. $\frac{1}{4}e^{-2t} + \frac{1}{4}\cos 2t + \frac{1}{4}\sin 2t$

27. $\frac{1}{3}\sin t - \frac{1}{3}\sin 2t$ 29. $1/s$

Exercises 7.3, Page 326

1. $\frac{1}{(s - 10)^2}$ 3. $\frac{6}{(s + 2)^4}$

5. $\frac{3}{(s - 1)^2 + 9}$ 7. $\frac{3}{(s - 5)^2 - 9}$

9. $\frac{1}{(s - 2)^2} + \frac{2}{(s - 3)^2} + \frac{1}{(s - 4)^2}$

11. $\frac{1}{2}\left[\frac{1}{s + 1} - \frac{s + 1}{(s + 1)^2 + 4}\right]$ 13. $\frac{1}{2}t^2e^{-2t}$

15. $e^{3t}\sin t$ 17. $e^{-2t}\cos t - 2e^{-2t}\sin t$

19. $e^{-t} - te^{-t}$ 21. $5 - t - 5e^{-t} - 4te^{-t} - \frac{3}{2}t^2e^{-t}$

23. $\frac{e^{-s}}{s^2}$ 25. $\frac{e^{-2s}}{s^2} + 2\frac{e^{-2s}}{s}$ 27. $\frac{s}{s^2 + 4}e^{-\pi s}$

29. $\frac{6e^{-s}}{(s - 1)^4}$ 31. $\frac{1}{2}(t - 2)^2\mathscr{U}(t - 2)$

33. $-\sin t\,\mathscr{U}(t - \pi)$ 35. $\mathscr{U}(t - 1) - e^{-(t-1)}\mathscr{U}(t - 1)$

37. $\frac{s^2 - 4}{(s^2 + 4)^2}$ 39. $\frac{6s^2 + 2}{(s^2 - 1)^3}$

41. $\frac{12s - 24}{[(s - 2)^2 + 36]^2}$ 43. $\frac{1}{2}t\sin t$

45. $f(t) = 2 - 4\mathscr{U}(t - 3)$;

$$\mathscr{L}\{f(t)\} = \frac{2}{s} - \frac{4}{s}e^{-3s}$$

47. $f(t) = t^2\mathscr{U}(t - 1)$
$= (t - 1)^2\mathscr{U}(t - 1) + 2(t - 1)\mathscr{U}(t - 1) + \mathscr{U}(t - 1)$;

$$\mathscr{L}\{f(t)\} = 2\frac{e^{-s}}{s^3} + 2\frac{e^{-s}}{s^2} + \frac{e^{-s}}{s}$$

49. $f(t) = t - t\mathcal{U}(t-2)$
$= t - (t-2)\mathcal{U}(t-2) - 2\mathcal{U}(t-2);$

$$\mathcal{L}\{f(t)\} = \frac{1}{s^2} - \frac{e^{-2s}}{s^2} - 2\frac{e^{-2s}}{s}$$

51. $f(t) = \mathcal{U}(t-a) - \mathcal{U}(t-b);$

$$\mathcal{L}\{f(t)\} = \frac{e^{-as}}{s} - \frac{e^{-bs}}{s}$$

53.

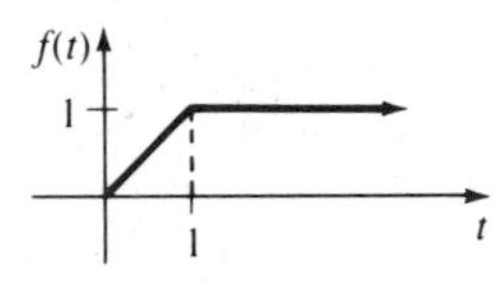

55. $\dfrac{e^{-t} - e^{3t}}{t}$ **57.** $\dfrac{\sin 2t}{t}$

59. Since $f'(t) = e^t$, $f(0) = 1$, it follows from Theorem 7.8 that $\mathcal{L}\{e^t\} = s\mathcal{L}\{e^t\} - 1$. Solving gives $\mathcal{L}\{e^t\} = 1/(s-1)$.

61. $\dfrac{s+1}{s[(s+1)^2+1]}$ **63.** $\dfrac{1}{s^2(s-1)}$ **65.** $\dfrac{3s^2+1}{s^2(s^2+1)^2}$

67. $\dfrac{6}{s^5}$ **69.** $\dfrac{48}{s^8}$ **71.** $\dfrac{s-1}{(s+1)[(s-1)^2+1]}$

73. $\displaystyle\int_0^t f(\tau)e^{-5(t-\tau)}\,d\tau$ **75.** $1 - e^{-t}$

77. $-\frac{1}{3}e^{-t} + \frac{1}{3}e^{2t}$ **79.** $\frac{1}{4}t\sin 2t$

81. $\dfrac{(1-e^{-as})^2}{s(1-e^{-2as})} = \dfrac{1-e^{-as}}{s(1+e^{-as})}$ **83.** $\dfrac{a}{s}\left(\dfrac{1}{bs} - \dfrac{1}{e^{bs}-1}\right)$

85. $\dfrac{\coth(\pi s/2)}{s^2+1}$ **87.** $\dfrac{1}{s^2+1}$

89. The result follows from letting $u = t - \tau$ in the first integral.

91. The result follows from $\cosh at = (e^{at} + e^{-at})/2$ and the first translation theorem.

93. The result follows from letting $u = at$, $a > 0$.

95. $\ln\dfrac{s+1}{s-1}$

Exercises 7.4, Page 339

1. $y = -1 + e^t$ **3.** $y = te^{-4t} + 2e^{-4t}$

5. $y = \frac{4}{3}e^{-t} - \frac{1}{3}e^{-4t}$ **7.** $y = \frac{1}{9}t + \frac{2}{27} - \frac{2}{27}e^{3t} + \frac{10}{9}te^{3t}$

9. $y = \frac{1}{20}t^5e^{2t}$ **11.** $y = \cos t - \frac{1}{2}\sin t - \frac{1}{2}t\cos t$

13. $y = \frac{1}{2} - \frac{1}{2}e^t\cos t + \frac{1}{2}e^t\sin t$

15. $y = -\frac{8}{9}e^{-t/2} + \frac{1}{9}e^{-2t} + \frac{5}{18}e^t + \frac{1}{2}e^{-t}$ **17.** $y = \cos t$

19. $y = [5 - 5e^{-(t-1)}]\mathcal{U}(t-1)$

21. $y = -\frac{1}{4} + \frac{1}{2}t + \frac{1}{4}e^{-2t} - \frac{1}{4}\mathcal{U}(t-1) - \frac{1}{2}(t-1)\mathcal{U}(t-1) + \frac{1}{4}e^{-2(t-1)}\mathcal{U}(t-1)$

23. $y = \cos 2t - \frac{1}{6}\sin 2(t-2\pi)\mathcal{U}(t-2\pi) + \frac{1}{3}\sin(t-2\pi)\mathcal{U}(t-2\pi)$

25. $y = \sin t + [1 - \cos(t-\pi)]\mathcal{U}(t-\pi) - [1 - \cos(t-2\pi)\mathcal{U}(t-2\pi)$

27. $y = (e+1)te^{-t} + (e-1)e^{-t}$

29. $f(t) = \sin t$

31. $f(t) = -\frac{1}{8}e^{-t} + \frac{1}{8}e^t + \frac{3}{4}te^t + \frac{1}{4}t^2e^t$

33. $f(t) = e^{-t}$

35. $f(t) = \frac{3}{8}e^{2t} + \frac{1}{8}e^{-2t} + \frac{1}{2}\cos 2t + \frac{1}{4}\sin 2t$

37. $y = \sin t - \frac{1}{2}t\sin t$

39. $i(t) = 20{,}000[te^{-100t} - (t-1)e^{-100(t-1)}\mathcal{U}(t-1)]$

41. $q(t) = \dfrac{E_0C}{1-kRC}(e^{-kt} - e^{-t/RC});$

$$q(t) = \frac{E_0}{R}te^{-t/RC}$$

43. $$i(t) = \frac{1}{101}e^{-10t} - \frac{1}{101}\cos t + \frac{10}{101}\sin t - \frac{10}{101}e^{-10(t-3\pi/2)}\mathcal{U}\left(t - \frac{3\pi}{2}\right) + \frac{10}{101}\cos\left(t - \frac{3\pi}{2}\right)\mathcal{U}\left(t - \frac{3\pi}{2}\right) + \frac{1}{101}\sin\left(t - \frac{3\pi}{2}\right)\mathcal{U}\left(t - \frac{3\pi}{2}\right)$$

45. $$i(t) = \frac{t}{R} + \frac{L}{R^2}(e^{-Rt/L} - 1) + \frac{1}{R}\sum_{n=1}^{\infty}(e^{-R(t-n)/L} - 1)\mathcal{U}(t-n);$$

For $0 \le t < 2$,

$$i(t) = \begin{cases} \dfrac{t}{R} + \dfrac{L}{R^2}(e^{-Rt/L} - 1), & 0 \le t < 1 \\ \dfrac{t}{R} + \dfrac{L}{R^2}(e^{-Rt/L} - 1) + \dfrac{1}{R}(e^{-R(t-1)/L} - 1), & 1 \le t < 2 \end{cases}$$

47. $q(t) = \frac{3}{5}e^{-10t} + 6te^{-10t} - \frac{3}{5}\cos 10t$;
$i(t) = -60te^{-10t} + 6\sin 10t$;
steady-state current is $6 \sin 10t$.

49. $$q(t) = \frac{E_0}{L\left(k^2 + \frac{1}{LC}\right)}[e^{-kt} - \cos(t/\sqrt{LC})] + \frac{kE_0\sqrt{C/L}}{k^2 + \frac{1}{LC}}\sin(t/\sqrt{LC})$$

51. $x(t) = -\frac{3}{2}e^{-7t/2}\cos\frac{\sqrt{15}}{2}t - \frac{7\sqrt{15}}{10}e^{-7t/2}\sin\frac{\sqrt{15}}{2}t$

53. $y(x) = \frac{w_0}{EI}\left(\frac{L^2}{4}x^2 - \frac{L}{6}x^3 + \frac{1}{24}x^4\right)$

55. $$y(x) = \frac{w_0L^2}{16EI}x^2 - \frac{w_0L}{12EI}x^3 + \frac{w_0}{24EI}x^4 - \frac{w_0}{24EI}\left(x - \frac{L}{2}\right)^4\mathscr{U}\left(x - \frac{L}{2}\right)$$

57. $y = \frac{1}{3}t^3 + \frac{1}{2}ct^2$

Exercises 7.5, Page 348

1. $y = e^{3(t-2)}\mathscr{U}(t-2)$

3. $y = \sin t + \sin t\,\mathscr{U}(t - 2\pi)$

5. $y = -\cos t\,\mathscr{U}\left(t - \frac{\pi}{2}\right) + \cos t\,\mathscr{U}\left(t - \frac{3\pi}{2}\right)$

7. $y = \frac{1}{2} - \frac{1}{2}e^{-2t} + [\frac{1}{2} - \frac{1}{2}e^{-2(t-1)}]\mathscr{U}(t-1)$

9. $y = e^{-2(t-2\pi)}\sin t\,\mathscr{U}(t - 2\pi)$

11. $y = e^{-2t}\cos 3t + \frac{2}{3}e^{-2t}\sin 3t$
$+ \frac{1}{3}e^{-2(t-\pi)}\sin 3(t-\pi)\mathscr{U}(t-\pi)$
$+ \frac{1}{3}e^{-2(t-3\pi)}\sin 3(t-3\pi)\mathscr{U}(t-3\pi)$

13. $$y = \begin{cases} \frac{P_0}{EI}\left(\frac{L}{4}x^2 - \frac{1}{6}x^3\right), & 0 \le x < L/2, \\ \frac{P_0L^2}{4EI}\left(\frac{1}{2}x - \frac{L}{12}\right), & L/2 \le x \le L \end{cases}$$

15. $$\mathscr{L}\{\delta_a(t - t_0)\} = \int_{t_0-a}^{t_0+a}\frac{1}{2a}e^{-st}\,dt = \frac{1}{2a}\left(-\frac{e^{-st}}{s}\right)\Bigg|_{t_0-a}^{t_0+a} = \frac{-1}{2sa}[e^{-s(t_0+a)} - e^{-s(t_0-a)}] = \frac{e^{-st_0}}{2sa}(e^{sa} - e^{-sa})$$

17. $y = e^{-t}\cos t + e^{-(t-3\pi)}\sin t\,\mathscr{U}(t - 3\pi)$

19. $i(t) = \frac{1}{L}e^{-Rt/L}$; no

Chapter 7 Review, Page 349

1. $\frac{1}{s^2} - \frac{2}{s^2}e^{-s}$ **3.** false **5.** true **7.** $\frac{1}{s+7}$

9. $\frac{2}{s^2+4}$ **11.** $\frac{4s}{(s^2+4)^2}$ **13.** $\frac{1}{6}t^5$ **15.** $\frac{1}{2}t^2e^{5t}$

17. $e^{5t}\cos 2t + \frac{5}{2}e^{5t}\sin 2t$ **19.** -5

21. $e^{-ks}F(s-a)$

23. $f(t) = t - (t-1)\mathscr{U}(t-1) - \mathscr{U}(t-4)$;
$$\mathscr{L}\{f(t)\} = \frac{1}{s^2} - \frac{1}{s^2}e^{-s} - \frac{1}{s}e^{-4s};$$
$$\mathscr{L}\{e^tf(t)\} = \frac{1}{(s-1)^2} - \frac{1}{(s-1)^2}e^{-(s-1)} - \frac{1}{s-1}e^{-4(s-1)}$$

25. $f(t) = 2 + (t-2)\mathscr{U}(t-2)$;
$$\mathscr{L}\{f(t)\} = \frac{2}{s} + \frac{1}{s^2}e^{-2s};$$
$$\mathscr{L}\{e^tf(t)\} = \frac{2}{s-1} + \frac{1}{(s-1)^2}e^{-2(s-1)}$$

27. $y = 5te^t + \frac{1}{2}t^2e^t$

29. $y = -\frac{2}{125} - \frac{2}{25}t - \frac{1}{5}t^2 + \frac{127}{125}e^{5t}$
$-[-\frac{37}{125} - \frac{12}{25}(t-1) - \frac{1}{5}(t-1)^2$
$+ \frac{37}{125}e^{5(t-1)}]\mathscr{U}(t-1)$

31. $y = 1 + t + \frac{1}{2}t^2$

33. $y = -9 + 2t + 9e^{-t/5}$

Exercises 8.1, Page 361

1. $x = c_1e^t + c_2te^t$
$y = (c_1 - c_2)e^t + c_2te^t$

3. $x = c_1\cos t + c_2\sin t + t + 1$
$y = c_1\sin t - c_2\cos t + t - 1$

5. $x = \frac{1}{2}c_1\sin t + \frac{1}{2}c_2\cos t - 2c_3\sin\sqrt{6}t$
$-2c_4\cos\sqrt{6}t$
$y = c_1\sin t + c_2\cos t + c_3\sin\sqrt{6}t + c_4\cos\sqrt{6}t$

7. $x = c_1e^{2t} + c_2e^{-2t} + c_3 \sin 2t + c_4 \cos 2t + \frac{1}{5}e^t$
$y = c_1e^{2t} + c_2e^{-2t} - c_3 \sin 2t - c_4 \cos 2t - \frac{1}{5}e^t$

9. $x = c_1 - c_2 \cos t + c_3 \sin t + \frac{17}{15}e^{3t}$
$y = c_1 + c_2 \sin t + c_3 \cos t - \frac{4}{15}e^{3t}$

11. $x = c_1e^t + c_2e^{-t/2} \cos \frac{\sqrt{3}}{2}t + c_3e^{-t/2} \sin \frac{\sqrt{3}}{2}t$

$$y = \left(-\frac{3}{2}c_2 - \frac{\sqrt{3}}{2}c_3\right)e^{-t/2} \cos \frac{\sqrt{3}}{2}t + \left(\frac{\sqrt{3}}{2}c_2 - \frac{3}{2}c_3\right)e^{-t/2} \sin \frac{\sqrt{3}}{2}t$$

13. $x = c_1e^{4t} + \frac{4}{3}e^t$
$y = -\frac{3}{4}c_1e^{4t} + c_2 + 5e^t$

15. $x = c_1 + c_2t + c_3e^t + c_4e^{-t} - \frac{1}{2}t^2$
$y = (c_1 - c_2 + 2) + (c_2 + 1)t + c_4e^{-t} - \frac{1}{2}t^2$

17. $x = c_1e^t + c_2e^{-t/2} \sin \frac{\sqrt{3}}{2}t + c_3e^{-t/2} \cos \frac{\sqrt{3}}{2}t$

$$y = c_1e^t + \left(-\frac{1}{2}c_2 - \frac{\sqrt{3}}{2}c_3\right)e^{-t/2} \sin \frac{\sqrt{3}}{2}t + \left(\frac{\sqrt{3}}{2}c_2 - \frac{1}{2}c_3\right)e^{-t/2} \cos \frac{\sqrt{3}}{2}t$$

$$z = c_1e^t + \left(-\frac{1}{2}c_2 + \frac{\sqrt{3}}{2}c_3\right)e^{-t/2} \sin \frac{\sqrt{3}}{2}t + \left(-\frac{\sqrt{3}}{2}c_2 - \frac{1}{2}c_3\right)e^{-t/2} \cos \frac{\sqrt{3}}{2}t$$

19. $x = -6c_1e^{-t} - 3c_2e^{-2t} + 2c_3e^{3t}$
$y = c_1e^{-t} + c_2e^{-2t} + c_3e^{3t}$
$z = 5c_1e^{-t} + c_2e^{-2t} + c_3e^{3t}$

21. $x = -c_1e^{-t} + c_2 + \frac{1}{3}t^3 - 2t^2 + 5t$
$y = c_1e^{-t} + 2t^2 - 5t + 5$

23. $x = e^{-3t+3} - te^{-3t+3}$
$y = -e^{-3t+3} + 2te^{-3t+3}$

25. $Dx - Dy = 0$
$(D - 1)x - y = 0$

Exercises 8.2, Page 369

1. $x = -\frac{1}{3}e^{-2t} + \frac{1}{3}e^t$
$y = \frac{1}{3}e^{-2t} + \frac{2}{3}e^t$

3. $x = -\cos 3t - \frac{5}{3} \sin 3t$
$y = 2 \cos 3t - \frac{7}{3} \sin 3t$

5. $x = -2e^{3t} + \frac{5}{2}e^{2t} - \frac{1}{2}$
$y = \frac{8}{3}e^{3t} - \frac{5}{2}e^{2t} - \frac{1}{6}$

7. $x = -\frac{1}{2}t - \frac{3}{4}\sqrt{2} \sin \sqrt{2}t$
$y = -\frac{1}{2}t + \frac{3}{4}\sqrt{2} \sin \sqrt{2}t$

9. $x = 8 + \frac{2}{3!}t^3 + \frac{1}{4!}t^4$

$y = -\frac{2}{3!}t^3 + \frac{1}{4!}t^4$

11. $x = \frac{1}{2}t^2 + t + 1 - e^{-t}$
$y = -\frac{1}{3} + \frac{1}{3}e^{-t} + \frac{1}{3}te^{-t}$

13. $x_1 = \frac{1}{5} \sin t + \frac{2\sqrt{6}}{15} \sin \sqrt{6}t + \frac{2}{5} \cos t - \frac{2}{5} \cos \sqrt{6}t$

$x_2 = \frac{2}{5} \sin t - \frac{\sqrt{6}}{15} \sin \sqrt{6}t + \frac{4}{5} \cos t + \frac{1}{5} \cos \sqrt{6}t$

15. $i_2 = \frac{100}{9} - \frac{100}{9}e^{-900t}$
$i_3 = \frac{80}{9} - \frac{80}{9}e^{-900t}$
$i_1 = 20 - 20e^{-900t}$

17. $i_2 = -\frac{20}{13}e^{-2t} + \frac{375}{1469}e^{-15t} + \frac{145}{113} \cos t + \frac{85}{113} \sin t$
$i_3 = \frac{30}{13}e^{-2t} + \frac{250}{1469}e^{-15t} - \frac{280}{113} \cos t + \frac{810}{113} \sin t$

19. $i_1 = \frac{6}{5} - \frac{6}{5}e^{-100t} \cos 100t$
$i_2 = \frac{6}{5} - \frac{6}{5}e^{-100t} \cos 100t - \frac{6}{5}e^{-100t} \sin 100t$

21. $q = 50e^{-t} \sin (t - 1)\mathscr{U}(t - 1)$

Exercises 8.3, Page 376

1. $x_1' = x_2$
$x_2' = -4x_1 + 3x_2 + \sin 3t$

3. $x_1' = x_2$
$x_2' = x_3$
$x_3' = 10x_1 - 6x_2 + 3x_3 + t^2 + 1$

5. $x_1' = x_2$
$x_2' = x_3$
$x_3' = x_4$
$x_4' = -x_1 - 4x_2 + 2x_3 + t$

7. $x_1' = x_2$

$x_2' = \frac{t}{t + 1}x_1$

9. $x' = -2x + y + 5t$
$y' = 2x + y - 2t$

11. $Dx = t^2 + 5t - 2$
$Dy = -x + 5t - 2$

13. The system is degenerate.

15. $Dx = u$
$Dy = v$
$Du = w$
$Dv = 10t^2 - 4u + 3v$
$Dw = 4x + 4v - 3w$

17. $x_1 = \frac{25}{2}e^{-t/25} + \frac{25}{2}e^{-3t/25}$
$x_2 = \frac{25}{4}e^{-t/25} - \frac{25}{4}e^{-3t/25}$

19. $x_1' = \frac{1}{50}x_2 - \frac{3}{50}x_1$
$x_2' = \frac{3}{50}x_1 - \frac{7}{100}x_2 + \frac{1}{100}x_3$
$x_3' = \frac{1}{20}x_2 - \frac{1}{20}x_3$

Exercises 8.4, Page 395

1. **(a)** $\begin{pmatrix} 2 & 11 \\ 2 & -1 \end{pmatrix}$ **(b)** $\begin{pmatrix} -6 & 1 \\ 14 & -19 \end{pmatrix}$ **(c)** $\begin{pmatrix} 2 & 28 \\ 12 & -12 \end{pmatrix}$

3. **(a)** $\begin{pmatrix} -11 & 6 \\ 17 & -22 \end{pmatrix}$ **(b)** $\begin{pmatrix} -32 & 27 \\ -4 & -1 \end{pmatrix}$

(c) $\begin{pmatrix} 19 & -18 \\ -30 & 31 \end{pmatrix}$ **(d)** $\begin{pmatrix} 19 & 6 \\ 3 & 22 \end{pmatrix}$

5. **(a)** $\begin{pmatrix} 9 & 24 \\ 3 & 8 \end{pmatrix}$ **(b)** $\begin{pmatrix} 3 & 8 \\ -6 & -16 \end{pmatrix}$ **(c)** $\begin{pmatrix} 0 & 0 \\ 0 & 0 \end{pmatrix}$

(d) $\begin{pmatrix} -4 & -5 \\ 8 & 10 \end{pmatrix}$

7. **(a)** 180 **(b)** $\begin{pmatrix} 4 & 8 & 10 \\ 8 & 16 & 20 \\ 10 & 20 & 25 \end{pmatrix}$ **(c)** $\begin{pmatrix} 6 \\ 12 \\ -5 \end{pmatrix}$

9. **(a)** $\begin{pmatrix} 7 & 38 \\ 10 & 75 \end{pmatrix}$ **(b)** $\begin{pmatrix} 7 & 38 \\ 10 & 75 \end{pmatrix}$

11. $\begin{pmatrix} -14 \\ 1 \end{pmatrix}$ **13.** $\begin{pmatrix} -38 \\ -2 \end{pmatrix}$ **15.** singular

17. nonsingular; $\mathbf{A}^{-1} = \frac{1}{4}\begin{pmatrix} -5 & -8 \\ 3 & 4 \end{pmatrix}$

19. nonsingular; $\mathbf{A}^{-1} = \frac{1}{2}\begin{pmatrix} 0 & -1 & 1 \\ 2 & 2 & -2 \\ -4 & -3 & 5 \end{pmatrix}$

21. nonsingular; $\mathbf{A}^{-1} = -\frac{1}{9}\begin{pmatrix} -2 & -2 & -1 \\ -13 & 5 & 7 \\ 8 & -1 & -5 \end{pmatrix}$

23. $\det \mathbf{A}(t) = 2e^{3t} \neq 0$ for every value of t;

$$\mathbf{A}^{-1}(t) = \frac{1}{2e^{3t}}\begin{pmatrix} 3e^{4t} & -e^{4t} \\ -4e^{-t} & 2e^{-t} \end{pmatrix}$$

25. $\dfrac{d\mathbf{X}}{dt} = \begin{pmatrix} -5e^{-t} \\ -2e^{-t} \\ 7e^{-t} \end{pmatrix}$

27. $\dfrac{d\mathbf{X}}{dt} = 4\begin{pmatrix} 1 \\ -1 \end{pmatrix}e^{2t} - 12\begin{pmatrix} 2 \\ 1 \end{pmatrix}e^{-3t}$

29. **(a)** $\begin{pmatrix} 4e^{4t} & -\pi \sin \pi t \\ 2 & 6t \end{pmatrix}$ **(b)** $\begin{pmatrix} \frac{1}{4}e^{8} - \frac{1}{4} & 0 \\ 4 & 6 \end{pmatrix}$

(c) $\begin{pmatrix} \frac{1}{4}e^{4t} - \frac{1}{4} & (1/\pi) \sin \pi t \\ t^2 & t^3 - t \end{pmatrix}$

31. $x = 3, y = 1, z = -5$

33. $x = 2 + 4t, y = -5 - t, z = t$

35. $x = -\frac{1}{2}, y = \frac{3}{2}, z = \frac{7}{2}$

37. $x_1 = 1, x_2 = 0, x_3 = 2, x_4 = 0$

41. $\lambda_1 = 6, \lambda_2 = 1,\quad \mathbf{K}_1 = \begin{pmatrix} 2 \\ 7 \end{pmatrix},\quad \mathbf{K}_2 = \begin{pmatrix} 1 \\ 1 \end{pmatrix}$

43. $\lambda_1 = \lambda_2 = -4,\quad \mathbf{K}_1 = \begin{pmatrix} 1 \\ -4 \end{pmatrix}$

45. $\lambda_1 = 0, \lambda_2 = 4, \lambda_3 = -4,$

$\mathbf{K}_1 = \begin{pmatrix} 9 \\ 45 \\ 25 \end{pmatrix},\quad \mathbf{K}_2 = \begin{pmatrix} 1 \\ 1 \\ 1 \end{pmatrix},\quad \mathbf{K}_3 = \begin{pmatrix} 1 \\ 9 \\ 1 \end{pmatrix}$

47. $\lambda_1 = \lambda_2 = \lambda_3 = -2,$

$\mathbf{K}_1 = \begin{pmatrix} 2 \\ -1 \\ 0 \end{pmatrix},\quad \mathbf{K}_2 = \begin{pmatrix} 0 \\ 0 \\ 1 \end{pmatrix}$

49. $\lambda_1 = 3i, \lambda_2 = -3i,$

$\mathbf{K}_1 = \begin{pmatrix} 1 - 3i \\ 5 \end{pmatrix},\quad \mathbf{K}_2 = \begin{pmatrix} 1 + 3i \\ 5 \end{pmatrix}$

51. $\dfrac{d}{dt}\begin{pmatrix} a_{11}(t) & a_{12}(t) \\ a_{21}(t) & a_{22}(t) \end{pmatrix}\begin{pmatrix} x_1(t) \\ x_2(t) \end{pmatrix}$

$= \dfrac{d}{dt}\begin{pmatrix} a_{11}(t)x_1(t) + a_{12}(t)x_2(t) \\ a_{21}(t)x_1(t) + a_{22}(t)x_2(t) \end{pmatrix}$

$= \begin{pmatrix} a_{11}(t)x_1'(t) + a_{11}'(t)x_1(t) + a_{12}(t)x_2'(t) + a_{12}'(t)x_2(t) \\ a_{21}(t)x_1'(t) + a_{21}'(t)x_1(t) + a_{22}(t)x_2'(t) + a_{22}'(t)x_2(t) \end{pmatrix}$

$= \begin{pmatrix} a_{11}(t)x_1'(t) + a_{12}(t)x_2'(t) + a_{11}'(t)x_1(t) + a_{12}'(t)x_2(t) \\ a_{21}(t)x_1'(t) + a_{22}(t)x_2'(t) + a_{21}'(t)x_1(t) + a_{22}'(t)x_2(t) \end{pmatrix}$

$= \begin{pmatrix} a_{11}(t) & a_{12}(t) \\ a_{21}(t) & a_{22}(t) \end{pmatrix}\begin{pmatrix} x_1'(t) \\ x_2'(t) \end{pmatrix} + \begin{pmatrix} a_{11}'(t) & a_{12}'(t) \\ a_{21}'(t) & a_{22}'(t) \end{pmatrix}\begin{pmatrix} x_1(t) \\ x_2(t) \end{pmatrix}$

$= \mathbf{A}(t)\mathbf{X}'(t) + \mathbf{A}'(t)\mathbf{X}(t)$

53. Since $\mathbf{A}^{-1}$ exists, $\mathbf{AB} = \mathbf{AC}$ implies $\mathbf{A}^{-1}(\mathbf{AB}) = \mathbf{A}^{-1}(\mathbf{AC})$, $(\mathbf{A}^{-1}\mathbf{A})\mathbf{B} = (\mathbf{A}^{-1}\mathbf{A})\mathbf{C}$, $\mathbf{IB} = \mathbf{IC}$ or $\mathbf{B} = \mathbf{C}$.

55. No, since in general $\mathbf{AB} \neq \mathbf{BA}$.

Exercises 8.5, Page 413

1. $\mathbf{X}' = \begin{pmatrix} 3 & -5 \\ 4 & 8 \end{pmatrix}\mathbf{X}$, where $\mathbf{X} = \begin{pmatrix} x \\ y \end{pmatrix}$

3. $\mathbf{X}' = \begin{pmatrix} -3 & 4 & -9 \\ 6 & -1 & 0 \\ 10 & 4 & 3 \end{pmatrix}\mathbf{X}$, where $\mathbf{X} = \begin{pmatrix} x \\ y \\ z \end{pmatrix}$

5. $\mathbf{X}' = \begin{pmatrix} 1 & -1 & 1 \\ 2 & 1 & -1 \\ 1 & 1 & 1 \end{pmatrix}\mathbf{X} + \begin{pmatrix} 0 \\ -3t^2 \\ t^2 \end{pmatrix} + \begin{pmatrix} t \\ 0 \\ -t \end{pmatrix}$

$+ \begin{pmatrix} -1 \\ 0 \\ 2 \end{pmatrix}$, where $\mathbf{X} = \begin{pmatrix} x \\ y \\ z \end{pmatrix}$

7. $\dfrac{dx}{dt} = 4x + 2y + e^t$

$\dfrac{dy}{dt} = -x + 3y - e^t$

9. $\dfrac{dx}{dt} = x - y + 2z + e^{-t} - 3t$

$\dfrac{dy}{dt} = 3x - 4y + z + 2e^{-t} + t$

$\dfrac{dz}{dt} = -2x + 5y + 6z + 2e^{-t} - t$

11. $\mathbf{X}' = \begin{pmatrix} -5e^{-5t} \\ -10e^{-5t} \end{pmatrix}$

$\begin{pmatrix} 3 & -4 \\ 4 & -7 \end{pmatrix}\mathbf{X} = \begin{pmatrix} 3-8 \\ 4-14 \end{pmatrix}e^{-5t} = \begin{pmatrix} -5 \\ -10 \end{pmatrix}e^{-5t} = \mathbf{X}'$

13. $\mathbf{X}' = \begin{pmatrix} \frac{3}{2} \\ 3 \end{pmatrix}$

$\begin{pmatrix} -1 & \frac{1}{4} \\ 1 & -1 \end{pmatrix}\mathbf{X} = \begin{pmatrix} 1 + \frac{1}{2} \\ 1 + 2 \end{pmatrix}e^{-3t/2} = \begin{pmatrix} \frac{3}{2} \\ 3 \end{pmatrix}e^{-3t/2} = \mathbf{X}'$

15. $\dfrac{d\mathbf{X}}{dt} = \begin{pmatrix} 0 \\ 0 \\ 0 \end{pmatrix}$

$\begin{pmatrix} 1 & 2 & 1 \\ 6 & -1 & 0 \\ -1 & -2 & -1 \end{pmatrix}\mathbf{X} = \begin{pmatrix} 1 + 12 - 13 \\ 6 - 6 \\ -1 - 12 + 13 \end{pmatrix} = \begin{pmatrix} 0 \\ 0 \\ 0 \end{pmatrix} = \dfrac{d\mathbf{X}}{dt}$

17. Yes; $W(\mathbf{X}_1, \mathbf{X}_2) = -2e^{-8t} \neq 0$ implies $\mathbf{X}_1$ and $\mathbf{X}_2$ are linearly independent on $(-\infty, \infty)$.

19. No; $W(\mathbf{X}_1, \mathbf{X}_2, \mathbf{X}_3) =$

$\begin{vmatrix} 1 + t & 1 & 3 + 2t \\ -2 + 2t & -2 & -6 + 4t \\ 4 + 2t & 4 & 12 + 4t \end{vmatrix} = 0$ for every t.

The solution vectors are linearly dependent on $(-\infty, \infty)$. Note that $\mathbf{X}_3 = 2\mathbf{X}_1 + \mathbf{X}_2$.

21. $\dfrac{d\mathbf{X}_p}{dt} = \begin{pmatrix} 2 \\ -1 \end{pmatrix}$

$\begin{pmatrix} 1 & 4 \\ 3 & 2 \end{pmatrix}\mathbf{X}_p + \begin{pmatrix} 2 \\ -4 \end{pmatrix}t - \begin{pmatrix} 7 \\ 18 \end{pmatrix}$

$= \begin{pmatrix} (2-4)t + 9 + 2t - 7 \\ (6-2)t + 17 - 4t - 18 \end{pmatrix}$

$= \begin{pmatrix} 2 \\ -1 \end{pmatrix} = \dfrac{d\mathbf{X}_p}{dt}$

23. $\mathbf{X}_p' = \begin{pmatrix} 2e^t + te^t \\ -te^t \end{pmatrix}$

$\begin{pmatrix} 2 & 1 \\ 3 & 4 \end{pmatrix}\mathbf{X}_p - \begin{pmatrix} 1 \\ 7 \end{pmatrix}e^t = \begin{pmatrix} 3e^t + te^t - e^t \\ 7e^t - te^t - 7e^t \end{pmatrix}$

$= \begin{pmatrix} 2e^t + te^t \\ -te^t \end{pmatrix} = \mathbf{X}_p'$

25. Let $\mathbf{X}_1 = \begin{pmatrix} 6 \\ -1 \\ -5 \end{pmatrix}e^{-t}$, $\mathbf{X}_2 = \begin{pmatrix} -3 \\ 1 \\ 1 \end{pmatrix}e^{-2t}$,

$\mathbf{X}_3 = \begin{pmatrix} 2 \\ 1 \\ 1 \end{pmatrix}e^{3t}$, and $\mathbf{A} = \begin{pmatrix} 0 & 6 & 0 \\ 1 & 0 & 1 \\ 1 & 1 & 0 \end{pmatrix}$. Then

$\mathbf{AX}_1 = \begin{pmatrix} -6 \\ 1 \\ 5 \end{pmatrix}e^{-t} = \mathbf{X}_1'$,

$\mathbf{AX}_2 = \begin{pmatrix} 6 \\ -2 \\ -2 \end{pmatrix}e^{-2t} = \mathbf{X}_2'$,

$\mathbf{AX}_3 = \begin{pmatrix} 6 \\ 3 \\ 3 \end{pmatrix}e^{3t} = \mathbf{X}_3'$ and

$W(\mathbf{X}_1, \mathbf{X}_2, \mathbf{X}_3) = \begin{vmatrix} 6e^{-t} & -3e^{-2t} & 2e^{3t} \\ -e^{-t} & e^{-2t} & e^{3t} \\ -5e^{-t} & e^{-2t} & e^{3t} \end{vmatrix}$

$= 20 \neq 0.$

Therefore, $\mathbf{X}_1$, $\mathbf{X}_2$, $\mathbf{X}_3$ form a fundamental set of solutions of $\mathbf{X}' = \mathbf{AX}$ on $(-\infty, \infty)$. By definition

$$\mathbf{X} = c_1\mathbf{X}_1 + c_2\mathbf{X}_2 + c_3\mathbf{X}_3$$

is the general solution.

27. $\mathbf{\Phi}(t) = \begin{pmatrix} e^{2t} & e^{7t} \\ -2e^{2t} & 3e^{7t} \end{pmatrix}$,

$$\mathbf{\Phi}^{-1}(t) = \frac{1}{5e^{9t}}\begin{pmatrix} 3e^{7t} & -e^{7t} \\ 2e^{2t} & e^{2t} \end{pmatrix}$$

29. $\mathbf{\Phi}(t) = \begin{pmatrix} -e^{t} & -te^{t} \\ 3e^{t} & 3te^{t} - e^{t} \end{pmatrix}$,

$$\mathbf{\Phi}^{-1}(t) = \frac{1}{e^{2t}}\begin{pmatrix} 3te^{t} - e^{t} & te^{t} \\ -3e^{t} & -e^{t} \end{pmatrix}$$

31. $\mathbf{\Psi}(t) = \begin{pmatrix} \frac{3}{5}e^{2t} + \frac{2}{5}e^{7t} & -\frac{1}{5}e^{2t} + \frac{1}{5}e^{7t} \\ -\frac{6}{5}e^{2t} + \frac{6}{5}e^{7t} & \frac{2}{5}e^{2t} + \frac{3}{5}e^{7t} \end{pmatrix}$

33. $\mathbf{\Psi}(t) = \begin{pmatrix} 3te^{t} + e^{t} & te^{t} \\ -9te^{t} & -3te^{t} + e^{t} \end{pmatrix}$

35. $\mathbf{X}(t_0) = \mathbf{\Phi}(t_0)\mathbf{C}$ implies $\mathbf{C} = \mathbf{\Phi}^{-1}(t_0)\mathbf{X}(t_0)$. Substituting in $\mathbf{X} = \mathbf{\Phi}(t)\mathbf{C}$ gives $\mathbf{X} = \mathbf{\Phi}(t)\mathbf{\Phi}^{-1}(t_0)\mathbf{X}_0$.

37. Comparing $\mathbf{X} = \mathbf{\Phi}(t)\mathbf{\Phi}^{-1}(t_0)\mathbf{X}_0$ and $\mathbf{X} = \mathbf{\Psi}(t)\mathbf{X}_0$ implies $[\mathbf{\Psi}(t) - \mathbf{\Phi}(t)\mathbf{\Phi}^{-1}(t_0)]\mathbf{X}_0 = \mathbf{0}$. Since this last equation is to hold for any $\mathbf{X}_0$, we conclude $\mathbf{\Psi}(t) = \mathbf{\Phi}(t)\mathbf{\Phi}^{-1}(t_0)$.

Exercises 8.6, Page 430

1. $\mathbf{X} = c_1\begin{pmatrix} 1 \\ 2 \end{pmatrix}e^{5t} + c_2\begin{pmatrix} 1 \\ -1 \end{pmatrix}e^{-t}$

3. $\mathbf{X} = c_1\begin{pmatrix} 2 \\ 1 \end{pmatrix}e^{-3t} + c_2\begin{pmatrix} 2 \\ 5 \end{pmatrix}e^{t}$

5. $\mathbf{X} = c_1\begin{pmatrix} 5 \\ 2 \end{pmatrix}e^{8t} + c_2\begin{pmatrix} 1 \\ 4 \end{pmatrix}e^{-10t}$

7. $\mathbf{X} = c_1\begin{pmatrix} 1 \\ 0 \\ 0 \end{pmatrix}e^{t} + c_2\begin{pmatrix} 2 \\ 3 \\ 1 \end{pmatrix}e^{2t} + c_3\begin{pmatrix} 1 \\ 0 \\ 2 \end{pmatrix}e^{-t}$

9. $\mathbf{X} = c_1\begin{pmatrix} -1 \\ 0 \\ 1 \end{pmatrix}e^{-t} + c_2\begin{pmatrix} 1 \\ 4 \\ 3 \end{pmatrix}e^{3t} + c_3\begin{pmatrix} 1 \\ -1 \\ 3 \end{pmatrix}e^{-2t}$

11. $\mathbf{X} = c_1\begin{pmatrix} 4 \\ 0 \\ -1 \end{pmatrix}e^{-t} + c_2\begin{pmatrix} -12 \\ 6 \\ 5 \end{pmatrix}e^{-t/2} + c_3\begin{pmatrix} 4 \\ 2 \\ -1 \end{pmatrix}e^{-3t/2}$

13. $\mathbf{X} = 3\begin{pmatrix} 1 \\ 1 \end{pmatrix}e^{t/2} + 2\begin{pmatrix} 0 \\ 1 \end{pmatrix}e^{-t/2}$

15. $\mathbf{X} = c_1\begin{pmatrix} \cos t \\ 2\cos t + \sin t \end{pmatrix}e^{4t} + c_2\begin{pmatrix} \sin t \\ 2\sin t - \cos t \end{pmatrix}e^{4t}$

17. $\mathbf{X} = c_1\begin{pmatrix} \cos t \\ -\cos t - \sin t \end{pmatrix}e^{4t} + c_2\begin{pmatrix} \sin t \\ -\sin t + \cos t \end{pmatrix}e^{4t}$

19. $\mathbf{X} = c_1\begin{pmatrix} 5\cos 3t \\ 4\cos 3t + 3\sin 3t \end{pmatrix}$

$$+ c_2\begin{pmatrix} 5\sin 3t \\ 4\sin 3t - 3\cos 3t \end{pmatrix}$$

21. $\mathbf{X} = c_1\begin{pmatrix} 1 \\ 0 \\ 0 \end{pmatrix} + c_2\begin{pmatrix} -\cos t \\ \cos t \\ \sin t \end{pmatrix} + c_3\begin{pmatrix} \sin t \\ -\sin t \\ \cos t \end{pmatrix}$

23. $\mathbf{X} = c_1\begin{pmatrix} 0 \\ 2 \\ 1 \end{pmatrix}e^{t} + c_2\begin{pmatrix} \sin t \\ \cos t \\ \cos t \end{pmatrix}e^{t} + c_3\begin{pmatrix} \cos t \\ -\sin t \\ -\sin t \end{pmatrix}e^{t}$

25. $\mathbf{X} = \begin{pmatrix} 28 \\ -5 \\ 25 \end{pmatrix}e^{2t} + c_2\begin{pmatrix} 5\cos 3t \\ -4\cos 3t - 3\sin 3t \\ 0 \end{pmatrix}e^{-2t}$

$$+ c_3\begin{pmatrix} 5\sin 3t \\ -4\sin 3t + 3\cos 3t \\ 0 \end{pmatrix}e^{-2t}$$

27. $\mathbf{X} = -\begin{pmatrix} 25 \\ -7 \\ 6 \end{pmatrix}e^{t} - \begin{pmatrix} \cos 5t - 5\sin 5t \\ \cos 5t \\ \cos 5t \end{pmatrix}$

$$+ 6\begin{pmatrix} 5\cos 5t + \sin 5t \\ \sin 5t \\ \sin 5t \end{pmatrix}$$

29. $\mathbf{X} = c_1\begin{pmatrix} 1 \\ 3 \end{pmatrix} + c_2\left\{\begin{pmatrix} 1 \\ 3 \end{pmatrix}t + \begin{pmatrix} \frac{1}{4} \\ -\frac{1}{4} \end{pmatrix}\right\}$

31. $\mathbf{X} = c_1\begin{pmatrix} 1 \\ 1 \end{pmatrix}e^{2t} + c_2\left\{\begin{pmatrix} 1 \\ 1 \end{pmatrix}te^{2t} + \begin{pmatrix} -\frac{1}{3} \\ 0 \end{pmatrix}e^{2t}\right\}$

33. $\mathbf{X} = c_1\begin{pmatrix} 1 \\ 1 \\ 1 \end{pmatrix}e^{t} + c_2\begin{pmatrix} 1 \\ 1 \\ 0 \end{pmatrix}e^{2t} + c_3\begin{pmatrix} 1 \\ 0 \\ 1 \end{pmatrix}e^{2t}$

35. $\mathbf{X} = c_1\begin{pmatrix}-4\\-5\\2\end{pmatrix} + c_2\begin{pmatrix}2\\0\\-1\end{pmatrix}e^{5t} + c_3\left\{\begin{pmatrix}2\\0\\-1\end{pmatrix}te^{5t} + \begin{pmatrix}-\frac{1}{2}\\-\frac{1}{2}\\-1\end{pmatrix}e^{5t}\right\}$

37. $\mathbf{X} = c_1\begin{pmatrix}0\\1\\1\end{pmatrix}e^t + c_2\left\{\begin{pmatrix}0\\1\\1\end{pmatrix}te^t + \begin{pmatrix}0\\1\\0\end{pmatrix}e^t\right\} + c_3\left\{\begin{pmatrix}0\\1\\1\end{pmatrix}\frac{t^2}{2}e^t + \begin{pmatrix}0\\1\\0\end{pmatrix}te^t + \begin{pmatrix}\frac{1}{2}\\0\\0\end{pmatrix}e^t\right\}$

39. $\mathbf{X} = -7\begin{pmatrix}2\\1\end{pmatrix}e^{4t} + 13\begin{pmatrix}2t+1\\t+1\end{pmatrix}e^{4t}$

41. $\mathbf{X} = \begin{pmatrix}\frac{6}{5}e^{5t} - \frac{1}{5}e^{-5t}\\ \frac{2}{5}e^{5t} + \frac{3}{5}e^{-5t}\end{pmatrix}$

43. $\mathbf{X} = c_1t^2\begin{pmatrix}3\\1\end{pmatrix} + c_2t^4\begin{pmatrix}1\\1\end{pmatrix}$

Exercises 8.7, Page 436

1. $\mathbf{X} = c_1\begin{pmatrix}-1\\1\end{pmatrix}e^{-t} + c_2\begin{pmatrix}-3\\1\end{pmatrix}e^t + \begin{pmatrix}-1\\3\end{pmatrix}$

3. $\mathbf{X} = c_1\begin{pmatrix}1\\-1\end{pmatrix}e^{-2t} + c_2\begin{pmatrix}1\\1\end{pmatrix}e^{4t} + \begin{pmatrix}-\frac{1}{4}\\\frac{3}{4}\end{pmatrix}t^2 + \begin{pmatrix}\frac{1}{4}\\-\frac{1}{4}\end{pmatrix}t + \begin{pmatrix}-2\\\frac{3}{4}\end{pmatrix}$

5. $\mathbf{X} = c_1\begin{pmatrix}1\\-3\end{pmatrix}e^{3t} + c_2\begin{pmatrix}1\\9\end{pmatrix}e^{7t} + \begin{pmatrix}\frac{55}{36}\\-\frac{19}{4}\end{pmatrix}e^t$

7. $\mathbf{X} = c_1\begin{pmatrix}1\\0\\0\end{pmatrix}e^t + c_2\begin{pmatrix}1\\1\\0\end{pmatrix}e^{2t} + c_3\begin{pmatrix}1\\2\\2\end{pmatrix}e^{5t} - \begin{pmatrix}\frac{3}{2}\\\frac{7}{2}\\2\end{pmatrix}e^{4t}$

9. $\mathbf{X} = 13\begin{pmatrix}1\\-1\end{pmatrix}e^t + 2\begin{pmatrix}-4\\6\end{pmatrix}e^{2t} + \begin{pmatrix}-9\\6\end{pmatrix}$

11. $\mathbf{X} = c_1\begin{pmatrix}1\\1\end{pmatrix} + c_2\begin{pmatrix}1\\-1\end{pmatrix}e^{2t} + \begin{pmatrix}-\frac{3}{2}\\-\frac{3}{2}\end{pmatrix}t + \begin{pmatrix}-\frac{5}{2}\\1\end{pmatrix}$

Exercises 8.8, Page 441

1. $\mathbf{X} = c_1\begin{pmatrix}1\\1\end{pmatrix} + c_2\begin{pmatrix}3\\2\end{pmatrix}e^t - \begin{pmatrix}11\\11\end{pmatrix}t - \begin{pmatrix}15\\10\end{pmatrix}$

3. $\mathbf{X} = c_1\begin{pmatrix}2\\1\end{pmatrix}e^{t/2} + c_2\begin{pmatrix}10\\3\end{pmatrix}e^{3t/2} - \begin{pmatrix}\frac{13}{2}\\\frac{13}{4}\end{pmatrix}te^{t/2} - \begin{pmatrix}\frac{15}{2}\\\frac{9}{4}\end{pmatrix}e^{t/2}$

5. $\mathbf{X} = c_1\begin{pmatrix}2\\1\end{pmatrix}e^t + c_2\begin{pmatrix}1\\1\end{pmatrix}e^{2t} + \begin{pmatrix}3\\3\end{pmatrix}e^t + \begin{pmatrix}4\\2\end{pmatrix}te^t$

7. $\mathbf{X} = c_1\begin{pmatrix}4\\1\end{pmatrix}e^{3t} + c_2\begin{pmatrix}-2\\1\end{pmatrix}e^{-3t} + \begin{pmatrix}-12\\0\end{pmatrix}t - \begin{pmatrix}\frac{4}{3}\\\frac{4}{3}\end{pmatrix}$

9. $\mathbf{X} = c_1\begin{pmatrix}1\\-1\end{pmatrix}e^t + c_2\begin{pmatrix}-t\\\frac{1}{2}-t\end{pmatrix}e^t + \begin{pmatrix}\frac{1}{2}\\-2\end{pmatrix}e^{-t}$

11. $\mathbf{X} = c_1\begin{pmatrix}\cos t\\\sin t\end{pmatrix} + c_2\begin{pmatrix}\sin t\\-\cos t\end{pmatrix} + \begin{pmatrix}\cos t\\\sin t\end{pmatrix}t + \begin{pmatrix}-\sin t\\\cos t\end{pmatrix}\ln|\cos t|$

13. $\mathbf{X} = c_1\begin{pmatrix}\cos t\\\sin t\end{pmatrix}e^t + c_2\begin{pmatrix}\sin t\\-\cos t\end{pmatrix}e^t + \begin{pmatrix}\cos t\\\sin t\end{pmatrix}te^t$

15. $\mathbf{X} = c_1\begin{pmatrix}\cos t\\-\sin t\end{pmatrix} + c_2\begin{pmatrix}\sin t\\\cos t\end{pmatrix} + \begin{pmatrix}\cos t\\-\sin t\end{pmatrix}t + \begin{pmatrix}-\sin t\\\sin t\tan t\end{pmatrix} - \begin{pmatrix}\sin t\\\cos t\end{pmatrix}\ln|\cos t|$

17. $\mathbf{X} = c_1\begin{pmatrix}2\sin t\\\cos t\end{pmatrix}e^t + c_2\begin{pmatrix}2\cos t\\-\sin t\end{pmatrix}e^t + \begin{pmatrix}3\sin t\\\frac{3}{2}\cos t\end{pmatrix}te^t + \begin{pmatrix}\cos t\\-\frac{1}{2}\sin t\end{pmatrix}e^t\ln|\sin t| + \begin{pmatrix}2\cos t\\-\sin t\end{pmatrix}e^t\ln|\cos t|$

19. $\mathbf{X} = c_1\begin{pmatrix}1\\-1\\0\end{pmatrix} + c_2\begin{pmatrix}1\\1\\0\end{pmatrix}e^{2t} + c_3\begin{pmatrix}0\\0\\1\end{pmatrix}e^{3t} + \begin{pmatrix}-\frac{1}{4}e^{2t} + \frac{1}{2}te^{2t}\\-e^t + \frac{1}{4}e^{2t} + \frac{1}{2}te^{2t}\\\frac{1}{2}t^2e^{3t}\end{pmatrix}$

21. $\mathbf{X} = \begin{pmatrix}2\\2\end{pmatrix}te^{2t} + \begin{pmatrix}-1\\1\end{pmatrix}e^{2t} + \begin{pmatrix}-2\\2\end{pmatrix}te^{4t} + \begin{pmatrix}2\\0\end{pmatrix}e^{4t}$

23. $\mathbf{X} = \begin{pmatrix}-2\\4\end{pmatrix}e^{2t} + \begin{pmatrix}7\\-9\end{pmatrix}e^{7t} + \begin{pmatrix}20\\60\end{pmatrix}te^{7t}$

25. $\begin{pmatrix}i_1\\i_2\end{pmatrix} = 2\begin{pmatrix}1\\3\end{pmatrix}e^{-2t} + \frac{6}{29}\begin{pmatrix}3\\-1\end{pmatrix}e^{-12t} + \begin{pmatrix}\frac{332}{29}\\\frac{276}{29}\end{pmatrix}\sin t - \begin{pmatrix}\frac{76}{29}\\\frac{168}{29}\end{pmatrix}\cos t$

Exercises 8.9, Page 444

1. $\begin{pmatrix} \cosh t & \sinh t \\ \sinh t & \cosh t \end{pmatrix}$

3. $\mathbf{X} = \begin{pmatrix} \cosh t & \sinh t \\ \sinh t & \cosh t \end{pmatrix}\begin{pmatrix} c_1 \\ c_2 \end{pmatrix} = c_1\begin{pmatrix} \cosh t \\ \sinh t \end{pmatrix} + c_2\begin{pmatrix} \sinh t \\ \cosh t \end{pmatrix}$

5. $\mathbf{X} = c_1\begin{pmatrix} \cosh t \\ \sinh t \end{pmatrix} + c_2\begin{pmatrix} \sinh t \\ \cosh t \end{pmatrix} - \begin{pmatrix} 1 \\ 1 \end{pmatrix}$

7. $\mathbf{X} = c_1\begin{pmatrix} 1 \\ 0 \end{pmatrix}e^t + c_2\begin{pmatrix} 0 \\ 1 \end{pmatrix}e^{2t} + \begin{pmatrix} -t-1 \\ \frac{1}{2}e^{4t} \end{pmatrix}$

9. $\mathbf{P} = \begin{pmatrix} 1 & 1 \\ 1 & 3 \end{pmatrix}$, $\mathbf{P}^{-1} = \begin{pmatrix} \frac{3}{2} & -\frac{1}{2} \\ -\frac{1}{2} & \frac{1}{2} \end{pmatrix}$,

$\mathbf{D} = \begin{pmatrix} 3 & 0 \\ 0 & 5 \end{pmatrix}$

$\mathbf{PDP}^{-1} = \begin{pmatrix} 2 & 1 \\ -3 & 6 \end{pmatrix}$

11. $e^{t\mathbf{A}} = e^{\mathbf{PDP}^{-1}}$

$= \mathbf{I} + t\mathbf{PDP}^{-1} + \frac{t^2}{2!}(\mathbf{PDP}^{-1})^2 + \cdots$

$= \mathbf{PP}^{-1} + t\mathbf{PDP}^{-1} + \frac{t^2}{2!}\mathbf{PD}^2\mathbf{P}^{-1} + \cdots$

$= \mathbf{P}\left[\mathbf{I} + t\mathbf{D} + \frac{t^2}{2!}\mathbf{D}^2 + \cdots\right]\mathbf{P}^{-1}$

$= \mathbf{P}e^{t\mathbf{D}}\mathbf{P}^{-1}$

13. $\mathbf{X} = \begin{pmatrix} \frac{3}{2}e^{3t} - \frac{1}{2}e^{5t} & -\frac{1}{2}e^{3t} + \frac{1}{2}e^{5t} \\ \frac{3}{2}e^{3t} - \frac{3}{2}e^{5t} & -\frac{1}{2}e^{3t} + \frac{3}{2}e^{5t} \end{pmatrix}\begin{pmatrix} c_1 \\ c_2 \end{pmatrix}$

Chapter 8 Review, Page 447

1. true **3.** $\begin{pmatrix} -2 & 1 \\ \frac{3}{2} & -\frac{1}{2} \end{pmatrix}$ **5.** true **7.** false

9. true **11.** false

13. $x = -c_1e^t - \frac{3}{2}c_2e^{2t} + \frac{5}{2}$
$y = c_1e^t + c_2e^{2t} - 3$

15. $x = c_1e^t + c_2e^{5t} + te^t$
$y = -c_1e^t + 3c_2e^{5t} - te^t + 2e^t$

17. $x = -\frac{1}{4} + \frac{9}{8}e^{-2t} + \frac{1}{8}e^{2t}$
$y = t + \frac{9}{4}e^{-2t} - \frac{1}{4}e^{2t}$

19. (a) $\begin{pmatrix} t^3 + 3t^2 + 5t - 2 \\ -t^3 - t + 2 \\ 4t^3 + 12t^2 + 8t + 1 \end{pmatrix}$ (b) $\begin{pmatrix} 3t^2 + 6t + 5 \\ -3t^2 - 1 \\ 12t^2 + 24t + 8 \end{pmatrix}$

21. $Dx = u$
$Dy = v$
$Du = -2u + v - 2x - \ln t + 10t - 4$
$Dv = -u - x + 5t - 2$

23. $\mathbf{X} = c_1\begin{pmatrix} 1 \\ -1 \end{pmatrix}e^t + c_2\left\{\begin{pmatrix} 1 \\ -1 \end{pmatrix}te^t + \begin{pmatrix} 0 \\ 1 \end{pmatrix}e^t\right\}$

25. $\mathbf{X} = c_1\begin{pmatrix} \cos 2t \\ -\sin 2t \end{pmatrix}e^t + c_2\begin{pmatrix} \sin 2t \\ \cos 2t \end{pmatrix}e^t$

27. $\mathbf{X} = c_1\begin{pmatrix} -1 \\ 1 \\ 0 \end{pmatrix} + c_2\begin{pmatrix} -1 \\ 0 \\ 1 \end{pmatrix} + c_3\begin{pmatrix} 1 \\ 1 \\ 1 \end{pmatrix}e^{3t}$

29. $\mathbf{X} = c_1\begin{pmatrix} 1 \\ 0 \end{pmatrix}e^{2t} + c_2\begin{pmatrix} 4 \\ 1 \end{pmatrix}e^{4t} + \begin{pmatrix} 16 \\ -4 \end{pmatrix}t + \begin{pmatrix} 11 \\ -1 \end{pmatrix}$

31. $\mathbf{X} = c_1\begin{pmatrix} \cos t \\ \cos t - \sin t \end{pmatrix} + c_2\begin{pmatrix} \sin t \\ \sin t + \cos t \end{pmatrix} - \begin{pmatrix} 1 \\ 1 \end{pmatrix} + \begin{pmatrix} \sin t \\ \sin t + \cos t \end{pmatrix}\ln|\csc t - \cot t|$

Exercises 9.1, Page 456

1. a family of vertical lines $x = c - 4$

3. a family of hyperbolas $x^2 - y^2 = c$

5. a family of circles $x^2 + (y + 1)^2 = c^2$ with center at $(0, -1)$

7. a family of hyperbolas $xy + y^2 = c$

9. a family of straight lines $y = c(x - 2) + 1$ passing through $(2, 1)$

11.

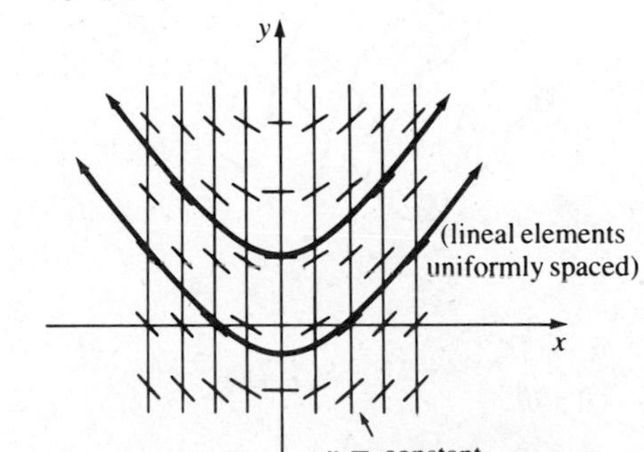

13.

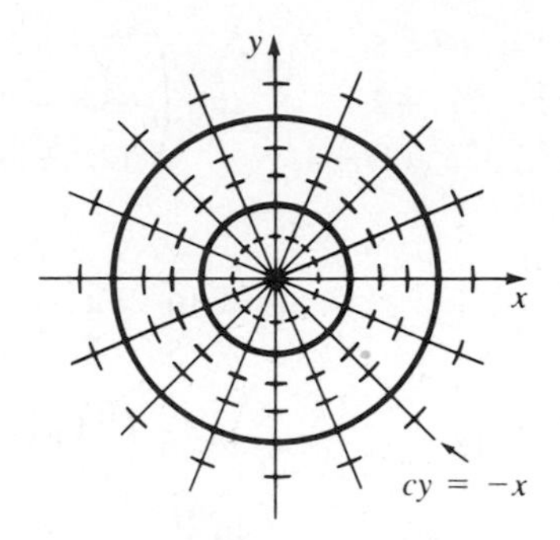

15.

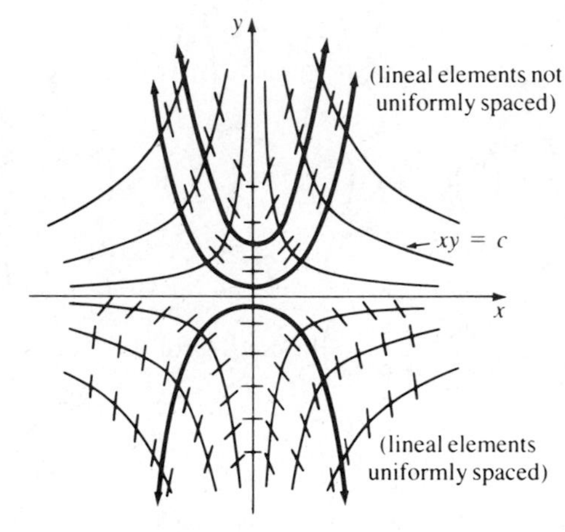

17.

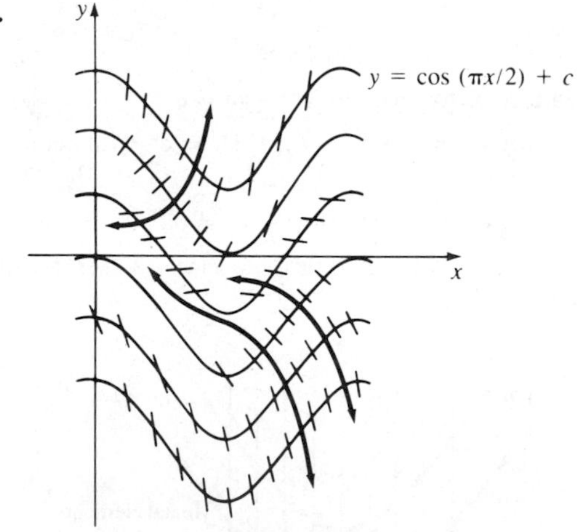

19. $y = \dfrac{\alpha - c\gamma}{c\delta - \beta}\,x$ **21.** $3x + 2y = -\frac{3}{2}$

23. $y = \pm\sqrt{2}x$ **25.** $y = 4x;\ y = -x$

Exercises 9.2, Page 464

1. $y = 1 - x + \tan(x + \pi/4)$

3. (a)

x_n	y_n
1.00	5.0000
1.10	3.8000
1.20	2.9800
1.30	2.4260
1.40	2.0582
1.50	1.8207

(b)

x_n	y_n
1.00	5.0000
1.05	4.4000
1.10	3.8950
1.15	3.4707
1.20	3.1151
1.25	2.8179
1.30	2.5702
1.35	2.3647
1.40	2.1950
1.45	2.0557
1.50	1.9424

5. (a)

x_n	y_n
0.00	0.0000
0.10	0.1000
0.20	0.2010
0.30	0.3050
0.40	0.4143
0.50	0.5315

(b)

x_n	y_n
0.00	0.0000
0.05	0.0500
0.10	0.1001
0.15	0.1506
0.20	0.2018
0.25	0.2538
0.30	0.3070
0.35	0.3617
0.40	0.4183
0.45	0.4770
0.50	0.5384

7. (a)

x_n	y_n
0.00	0.0000
0.10	0.1000
0.20	0.1905
0.30	0.2731
0.40	0.3492
0.50	0.4198

(b)

x_n	y_n
0.00	0.0000
0.05	0.0500
0.10	0.0976
0.15	0.1429
0.20	0.1863
0.25	0.2278
0.30	0.2676
0.35	0.3058
0.40	0.3427
0.45	0.3782
0.50	0.4124

9. (a)

x_n	y_n
0.00	0.5000
0.10	0.5250
0.20	0.5431
0.30	0.5548
0.40	0.5613
0.50	0.5639

(b)

x_n	y_n
0.00	0.5000
0.05	0.5125
0.10	0.5232
0.15	0.5322
0.20	0.5395
0.25	0.5452
0.30	0.5496
0.35	0.5527
0.40	0.5547
0.45	0.5559
0.50	0.5565

11. (a)

x_n	y_n
1.00	1.0000
1.10	1.0000
1.20	1.0191
1.30	1.0588
1.40	1.1231
1.50	1.2194

(b)

x_n	y_n
1.00	1.0000
1.05	1.0000
1.10	1.0049
1.15	1.0147
1.20	1.0298
1.25	1.0506
1.30	1.0775
1.35	1.1115
1.40	1.1538
1.45	1.2057
1.50	1.2696

13. (a) $h = 0.1$

x_n	y_n
1.00	5.0000
1.10	3.9900
1.20	3.2545
1.30	2.7236
1.40	2.3451
1.50	2.0801

$h = 0.05$

x_n	y_n
1.00	5.0000
1.05	4.4475
1.10	3.9763
1.15	3.5751
1.20	3.2342
1.25	2.9452
1.30	2.7009
1.35	2.4952
1.40	2.3226
1.45	2.1786
1.50	2.0592

(b) $h = 0.1$

x_n	y_n
0.00	0.0000
0.10	0.1005
0.20	0.2030
0.30	0.3098
0.40	0.4234
0.50	0.5470

$h = 0.05$

x_n	y_n
0.00	0.0000
0.05	0.0501
0.10	0.1004
0.15	0.1512
0.20	0.2028
0.25	0.2554
0.30	0.3095
0.35	0.3652
0.40	0.4230
0.45	0.4832
0.50	0.5465

(c) $h = 0.1$

x_n	y_n
0.00	0.0000
0.10	0.0952
0.20	0.1822
0.30	0.2622
0.40	0.3363
0.50	0.4053

$h = 0.05$

x_n	y_n
0.00	0.0000
0.05	0.0488
0.10	0.0953
0.15	0.1397
0.20	0.1823
0.25	0.2231
0.30	0.2623
0.35	0.3001
0.40	0.3364
0.45	0.3715
0.50	0.4054

(d) $h = 0.1$

x_n	y_n
0.00	0.5000
0.10	0.5215
0.20	0.5362
0.30	0.5449
0.40	0.5490
0.50	0.5503

$h = 0.05$

x_n	y_n
0.00	0.5000
0.05	0.5116
0.10	0.5214
0.15	0.5294
0.20	0.5359
0.25	0.5408
0.30	0.5444
0.35	0.5469
0.40	0.5484
0.45	0.5492
0.50	0.5495

(e) $h = 0.1$

x_n	y_n
1.00	1.0000
1.10	1.0095
1.20	1.0404
1.30	1.0967
1.40	1.1866
1.50	1.3260

$h = 0.05$

x_n	y_n
1.00	1.0000
1.05	1.0024
1.10	1.0100
1.15	1.0228
1.20	1.0414
1.25	1.0663
1.30	1.0984
1.35	1.1389
1.40	1.1895
1.45	1.2526
1.50	1.3315

15.

x_n	Euler	Improved Euler
1.0	1.0000	1.0000
1.1	1.2000	1.2469
1.2	1.4938	1.6668
1.3	1.9711	2.6427
1.4	2.9060	8.7989

17. Using $y' = f(x, y)$ gives

$$\int_{x_n}^{x_{n+1}} y'\,dx = \int_{x_n}^{x_{n+1}} f(x, y)\,dx$$

$$\begin{aligned} y(x_{n+1}) - y(x_n) &\approx f(x_n, y_n)(x_{n+1} - x_n) \\ &= hf(x_n, y_n) \\ y(x_{n+1}) &\approx y(x_n) + hf(x_n, y_n). \end{aligned}$$

We write this as

$$y_{n+1} = y_n + hf(x_n, y_n).$$

Exercises 9.3, Page 470

1. (a)

x_n	y_n	x_n	y_n
1.00	5.0000	1.30	2.7236
1.10	3.9900	1.40	2.3451
1.20	3.2545	1.50	2.0801

(b)

x_n	y_n	x_n	y_n
1.00	5.0000	1.30	2.7009
1.05	4.4475	1.35	2.4952
1.10	3.9763	1.40	2.3226
1.15	3.5751	1.45	2.1786
1.20	3.2342	1.50	2.0592
1.25	2.9452		

3. (a)

x_n	y_n
0.00	0.0000
0.10	0.1000
0.20	0.2020
0.30	0.3082
0.40	0.4211
0.50	0.5438

(b)

x_n	y_n
0.00	0.0000
0.05	0.0500
0.10	0.1003
0.15	0.1510
0.20	0.2025
0.25	0.2551
0.30	0.3090
0.35	0.3647
0.40	0.4223
0.45	0.4825
0.50	0.5456

5. (a)

x_n	y_n
0.00	0.0000
0.10	0.0950
0.20	0.1818
0.30	0.2617
0.40	0.3357
0.50	0.4046

(b)

x_n	y_n
0.00	0.0000
0.05	0.0488
0.10	0.0952
0.15	0.1397
0.20	0.1822
0.25	0.2230
0.30	0.2622
0.35	0.2999
0.40	0.3363
0.45	0.3714
0.50	0.4053

7. (a)

x_n	y_n	x_n	y_n
0.00	0.5000	0.30	0.5438
0.10	0.5213	0.40	0.5475
0.20	0.5355	0.50	0.5482

(b)

x_n	y_n	x_n	y_n
0.00	0.5000	0.30	0.5441
0.05	0.5116	0.35	0.5466
0.10	0.5213	0.40	0.5480
0.15	0.5293	0.45	0.5487
0.20	0.5357	0.50	0.5490
0.25	0.5406		

9. (a)

x_n	y_n
1.00	1.0000
1.10	1.0100
1.20	1.0410
1.30	1.0969
1.40	1.1857
1.50	1.3226

(b)

x_n	y_n
1.00	1.0000
1.05	1.0025
1.10	1.0101
1.15	1.0229
1.20	1.0415
1.25	1.0663
1.30	1.0983
1.35	1.1387
1.40	1.1891
1.45	1.2518
1.50	1.3301

11.

x_n	Euler	Improved Euler	Three-Term Taylor
1.0	1.0000	1.0000	1.0000
1.1	1.2000	1.2469	1.2400
1.2	1.4938	1.6668	1.6345
1.3	1.9711	2.6427	2.4600
1.4	2.9060	8.7988	5.6353

13.

x_n	Improved Euler	Three-Term Taylor	True Value
1.00	5.0000	5.0000	5.0000
1.10	5.5300	5.5300	5.5310
1.20	6.1262	6.1262	6.1284
1.30	6.7954	6.7954	6.7992
1.40	7.5454	7.5454	7.5510
1.50	8.3847	8.3847	8.3923

Exercises 9.4, Page 476

1.

x_n	y_n
1.00	5.0000
1.10	3.9724
1.20	3.2284
1.30	2.6945
1.40	2.3163
1.50	2.0533

3.

x_n	y_n
0.00	0.0000
0.10	0.1003
0.20	0.2027
0.30	0.3093
0.40	0.4228
0.50	0.5463

5.

x_n	y_n
0.00	0.0000
0.10	0.0953
0.20	0.1823
0.30	0.2624
0.40	0.3365
0.50	0.4055

7.

x_n	y_n
0.00	0.5000
0.10	0.5213
0.20	0.5358
0.30	0.5443
0.40	0.5482
0.50	0.5493

9.

x_n	y_n
1.00	1.0000
1.10	1.0101
1.20	1.0417
1.30	1.0989
1.40	1.1905
1.50	1.3333

11. $v(5) \approx 35.7678$

13.

1.93	12.50	36.46	47.23	49.00

15.

x_n	y_n
1.00	1.0000
1.10	1.2511
1.20	1.6934
1.30	2.9425
1.40	903.0283

Exercises 9.5, Page 480

1.

x_n	y_n
0.0	1.0000
0.1	1.0052
0.2	1.0214
0.3	1.0499
0.4	1.0918

3. $y(2.2) \approx 2.88$

5. $x_{n+1} = x_n + hf(x_n, y_n, t_n)$
$y_{n+1} = y_n + hg(x_n, y_n, t_n)$;
$x(0.2) \approx 1.62, \quad y(0.2) \approx 1.84$

Chapter 9 Review, Page 482

1. All isoclines $y = cx$ are solutions of the differential equation.

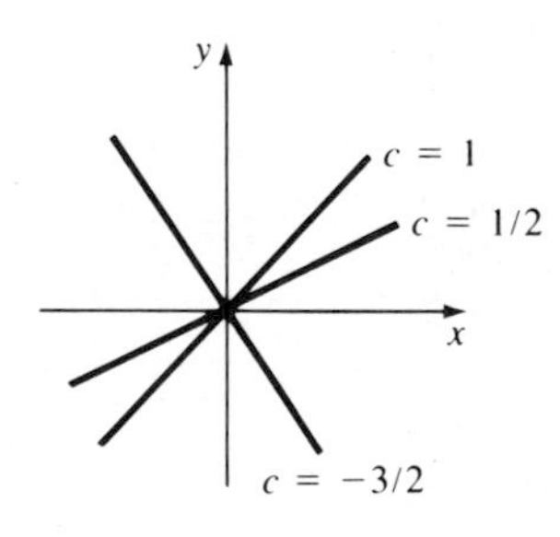

3. Comparison of Numerical Methods with $h = 0.1$

x_n	Euler	Improved Euler	3-Term Taylor	Runge-Kutta
1.00	2.0000	2.0000	2.0000	2.0000
1.10	2.1386	2.1549	2.1556	2.1556
1.20	2.3097	2.3439	2.3453	2.3454
1.30	2.5136	2.5672	2.5694	2.5695
1.40	2.7504	2.8246	2.8277	2.8278
1.50	3.0201	3.1157	3.1198	3.1197

Comparison of Numerical Methods with $h = 0.05$

x_n	Euler	Improved Euler	3-Term Taylor	Runge-Kutta
1.00	2.0000	2.0000	2.0000	2.0000
1.05	2.0693	2.0735	2.0735	2.0736
1.10	2.1469	2.1554	2.1556	2.1556
1.15	2.2329	2.2459	2.2462	2.2462
1.20	2.3272	2.3450	2.3454	2.3454
1.25	2.4299	2.4527	2.4532	2.4532
1.30	2.5410	2.5689	2.5695	2.5695
1.35	2.6604	2.6937	2.6944	2.6944
1.40	2.7883	2.8269	2.8278	2.8278
1.45	2.9245	2.9686	2.9696	2.9696
1.50	3.0690	3.1187	3.1198	3.1197

5. Comparison of Numerical Methods with $h = 0.1$

x_n	Euler	Improved Euler	3-Term Taylor	Runge-Kutta
0.50	0.5000	0.5000	0.5000	0.5000
0.60	0.6000	0.6048	0.6050	0.6049
0.70	0.7095	0.7191	0.7195	0.7194
0.80	0.8283	0.8427	0.8433	0.8431
0.90	0.9559	0.9752	0.9759	0.9757
1.00	1.0921	1.1163	1.1172	1.1169

Comparison of Numerical Methods with $h = 0.05$

x_n	Euler	Improved Euler	3-Term Taylor	Runge-Kutta
0.50	0.5000	0.5000	0.5000	0.5000
0.55	0.5500	0.5512	0.5512	0.5512
0.60	0.6024	0.6049	0.6049	0.6049
0.65	0.6573	0.6609	0.6610	0.6610
0.70	0.7144	0.7193	0.7194	0.7194
0.75	0.7739	0.7800	0.7802	0.7801
0.80	0.8356	0.8430	0.8431	0.8431
0.85	0.8996	0.9082	0.9083	0.9083
0.90	0.9657	0.9755	0.9757	0.9757
0.95	1.0340	1.0451	1.0453	1.0452
1.00	1.1044	1.1168	1.1170	1.1169

7. $h = 0.2$: $y(0.2) \approx 3.2$
$h = 0.1$: $y(0.2) \approx 3.23$

Exercises 10.1, Page 489

1. $$\int_{-2}^{2} x \cdot x^2\, dx = \frac{x^4}{4}\Big|_{-2}^{2} = 4 - 4 = 0$$

3. $$\int_{0}^{2} e^x(xe^{-x} - e^{-x})\, dx = \int_{0}^{2} (x-1)\, dx = \frac{x^2}{2} - x\Big|_{0}^{2} = 0$$

5. $$\int_{-\pi/2}^{\pi/2} x\cos 2x\, dx = \left[\frac{1}{2}x\sin 2x + \frac{1}{4}\cos 2x\right]_{-\pi/2}^{\pi/2} = \left[\frac{\pi}{4}\sin\pi + \frac{1}{4}\cos\pi\right] - \left[-\frac{\pi}{4}\sin(-\pi) + \frac{1}{4}\cos(-\pi)\right] = -\frac{1}{4} + \frac{1}{4} = 0$$

7. $$\int_{0}^{\pi/2} \sin(2m+1)x\sin(2n+1)x\, dx = \frac{1}{2}\int_{0}^{\pi/2}[\cos 2(m-n)x - \cos 2(m+n+1)x]\, dx = \frac{1}{4}\left[\frac{\sin 2(m-n)x}{m-n} - \frac{\sin 2(m+n+1)x}{m+n+1}\right]_{0}^{\pi/2}$$
$= 0,\quad m \neq n;\quad \sqrt{\pi}/2$

9. $$\int_{0}^{\pi} \sin mx\sin nx\, dx = \frac{1}{2}\int_{0}^{\pi}[\cos(m-n)x - \cos(m+n)x]\, dx = \frac{1}{2}\left[\frac{\sin(m-n)x}{m-n} - \frac{\sin(m+n)x}{m+n}\right]_{0}^{\pi}$$
$= 0,\quad m \neq n;\quad \sqrt{\pi/2}$

11. $$\int_{0}^{p} \cos\frac{n\pi}{p}x\, dx = \frac{p}{n\pi}\sin\frac{n\pi}{p}x\Big|_{0}^{p} = 0,\quad n \neq 0;$$
$$\int_{0}^{p} \cos\frac{m\pi}{p}x\cos\frac{n\pi}{p}x\, dx = \frac{1}{2}\int_{0}^{p}\left[\cos\frac{(m-n)\pi}{p}x + \cos\frac{(m+n)\pi}{p}x\right]dx = \frac{p}{2\pi}\left[\frac{\sin\frac{(m-n)\pi}{p}x}{m-n} + \frac{\sin\frac{(m+n)\pi}{p}x}{m+n}\right]_{0}^{p}$$
$= 0,\quad m \neq n;\quad \|1\| = \sqrt{p},\quad \left\|\cos\frac{n\pi}{p}x\right\| = \sqrt{p/2}$

13. For example,
$$\int_{-\infty}^{\infty} e^{-x^2}H_0(x)H_1(x)\, dx = \int_{-\infty}^{\infty} e^{-x^2}(2x)\, dx = -\int_{-\infty}^{0} e^{-x^2}(-2x\, dx) - \int_{0}^{\infty} e^{-x^2}(-2x\, dx) = -e^{-x^2}\Big|_{-\infty}^{0} - e^{-x^2}\Big|_{0}^{\infty} = -1 - (-1) = 0.$$
The results
$$\int_{-\infty}^{\infty} e^{-x^2}H_0(x)H_2(x)\, dx = 0$$
and
$$\int_{-\infty}^{\infty} e^{-x^2}H_1(x)H_2(x)\, dx = 0$$
follow from integration by parts.

15. $$\int_{a}^{b} \phi_n(x)\, dx = \int_{a}^{b} \phi_0(x)\phi_n(x)\, dx = 0 \quad \text{for} \quad n = 1, 2, 3, \ldots$$

17. $$\|\phi_m(x) + \phi_n(x)\|^2 = \int_{a}^{b}[\phi_m(x) + \phi_n(x)]^2\, dx = \int_{a}^{b}\phi_m^2(x)\, dx + 2\underbrace{\int_{a}^{b}\phi_m(x)\phi_n(x)\, dx}_{\text{zero by orthogonality}} + \int_{a}^{b}\phi_n^2(x)\, dx = \int_{a}^{b}\phi_m^2(x)\, dx + \int_{a}^{b}\phi_n^2(x)\, dx = \|\phi_m(x)\|^2 + \|\phi_n(x)\|^2$$

19. Proceed as in Problem 7 and use the fact that $\sin\lambda = -\lambda\cos\lambda$;
$$\|\sin\lambda_n x\|^2 = \frac{1}{2}(1 + \cos^2\lambda_n)$$

Exercises 10.2, Page 496

1. $$f(x) = \frac{1}{2} + \frac{1}{\pi}\sum_{n=1}^{\infty}\frac{1-(-1)^n}{n}\sin nx$$

3. $f(x) = \frac{3}{4} + \sum_{n=1}^{\infty} \left\{ \frac{(-1)^n - 1}{n^2\pi^2} \cos n\pi x - \frac{1}{n\pi} \sin n\pi x \right\}$

5. $f(x) = \frac{\pi^2}{6} + \sum_{n=1}^{\infty} \left\{ \frac{2(-1)^n}{n^2} \cos nx + \left(\frac{(-1)^{n+1}\pi}{n} + \frac{2}{\pi n^3} [(-1)^n - 1] \right) \sin nx \right\}$

7. $f(x) = \pi + 2 \sum_{n=1}^{\infty} \frac{(-1)^{n+1}}{n} \sin n\pi$

9. $f(x) = \frac{1}{\pi} + \frac{1}{2} \sin x + \frac{1}{\pi} \sum_{n=2}^{\infty} \frac{(-1)^n + 1}{1 - n^2} \cos nx$

11. $f(x) = -\frac{1}{4} + \frac{1}{\pi} \sum_{n=1}^{\infty} \left\{ -\frac{1}{n} \sin \frac{n\pi}{2} \cos \frac{n\pi}{2} x + \frac{3}{n} \left(1 - \cos \frac{n\pi}{2} \right) \sin \frac{n\pi}{2} x \right\}$

13. $f(x) = \frac{9}{4} + 5 \sum_{n=1}^{\infty} \left\{ \frac{(-1)^n - 1}{n^2\pi^2} \cos \frac{n\pi}{5} x + \frac{(-1)^{n+1}}{n\pi} \sin \frac{n\pi}{5} x \right\}$

15. $f(x) = \frac{2 \sinh \pi}{\pi} \left[\frac{1}{2} + \sum_{n=1}^{\infty} \frac{(-1)^n}{1 + n^2} (\cos nx - n \sin nx) \right]$

17. At the endpoint $x = \pi$, the series will converge to

$$\frac{f(\pi-) + f(-\pi+)}{2} = \frac{\pi^2}{2}.$$

Substituting $x = \pi$ into the series gives

$$\frac{\pi^2}{6} + 2 \sum_{n=1}^{\infty} \frac{1}{n^2}.$$

Equating the two results then yields

$$\frac{\pi^2}{6} = \sum_{n=1}^{\infty} \frac{1}{n^2} = 1 + \frac{1}{2^2} + \frac{1}{3^2} + \frac{1}{4^2} + \cdots.$$

Now at $x = 0$ the series converges to

$$f(0) = 0 = \frac{\pi^2}{6} + \sum_{n=1}^{\infty} \frac{2(-1)^n}{n^2}.$$

This implies

$$\frac{\pi^2}{12} = \sum_{n=1}^{\infty} \frac{(-1)^{n+1}}{n^2} = 1 - \frac{1}{2^2} + \frac{1}{3^2} - \frac{1}{4^2} + \cdots.$$

19. Set $x = \pi/2$.

Exercises 10.3, Page 503

1. odd **3.** neither even nor odd **5.** even
7. odd **9.** neither even nor odd

11. $f(x) = \frac{2}{\pi} \sum_{n=1}^{\infty} \frac{1 - (-1)^n}{n} \sin nx$

13. $f(x) = \frac{\pi}{2} + \frac{2}{\pi} \sum_{n=1}^{\infty} \frac{(-1)^n - 1}{n^2} \cos nx$

15. $f(x) = \frac{1}{3} + \frac{4}{\pi^2} \sum_{n=1}^{\infty} \frac{(-1)^n}{n^2} \cos n\pi x$

17. $f(x) = \frac{2\pi^2}{3} + 4 \sum_{n=1}^{\infty} \frac{(-1)^{n+1}}{n^2} \cos nx$

19. $f(x) = \frac{2}{\pi} \sum_{n=1}^{\infty} \frac{1 - (-1)^n(1 + \pi)}{n} \sin nx$

21. $f(x) = \frac{3}{4} + \frac{4}{\pi^2} \sum_{n=1}^{\infty} \frac{\cos \frac{n\pi}{2} - 1}{n^2} \cos \frac{n\pi}{2} x$

23. $f(x) = \frac{2}{\pi} + \frac{2}{\pi} \sum_{n=2}^{\infty} \frac{1 + (-1)^n}{1 - n^2} \cos nx$

25. $f(x) = \frac{1}{2} + \frac{2}{\pi} \sum_{n=1}^{\infty} \frac{\sin \frac{n\pi}{2}}{n} \cos n\pi x$

$f(x) = \frac{2}{\pi} \sum_{n=1}^{\infty} \frac{1 - \cos \frac{n\pi}{2}}{n} \sin n\pi x$

27. $f(x) = \frac{2}{\pi} + \frac{4}{\pi} \sum_{n=1}^{\infty} \frac{(-1)^n}{1 - 4n^2} \cos 2nx$

$f(x) = \frac{8}{\pi} \sum_{n=1}^{\infty} \frac{n}{4n^2 - 1} \sin 2nx$

29. $f(x) = \frac{\pi}{4} + \frac{2}{\pi} \sum_{n=1}^{\infty} \frac{2 \cos \frac{n\pi}{2} - (-1)^n - 1}{n^2} \cos nx$

$f(x) = \frac{4}{\pi} \sum_{n=1}^{\infty} \frac{\sin \frac{n\pi}{2}}{n^2} \sin nx$

31. $f(x) = \frac{3}{4} + \frac{4}{\pi^2} \sum_{n=1}^{\infty} \frac{\cos \frac{n\pi}{2} - 1}{n^2} \cos \frac{n\pi}{2} x$

$f(x) = \sum_{n=1}^{\infty} \left\{ \frac{4}{n^2\pi^2} \sin \frac{n\pi}{2} - \frac{2}{n\pi} (-1)^n \right\} \sin \frac{n\pi}{2} x$

33. $f(x) = \frac{5}{6} + \frac{2}{\pi^2} \sum_{n=1}^{\infty} \frac{3(-1)^n - 1}{n^2} \cos n\pi x$

$f(x) = 4 \sum_{n=1}^{\infty} \left\{ \frac{(-1)^{n+1}}{n\pi} + \frac{(-1)^n - 1}{n^3\pi^3} \right\} \sin n\pi x$

35. $f(x) = \frac{4\pi^2}{3} + 4 \sum_{n=1}^{\infty} \left\{ \frac{1}{n^2} \cos nx - \frac{\pi}{n} \sin nx \right\}$

37. $f(x) = \frac{3}{2} - \frac{1}{\pi} \sum_{n=1}^{\infty} \frac{1}{n} \sin 2n\pi x$

39. Let f and g be even functions. Define $F(x) = f(x)g(x)$. Then

$$F(-x) = f(-x)g(-x) = f(x)g(x) = F(x).$$

41. Let f be an even function and g be an odd function. Define $F(x) = f(x)g(x)$. Then

$$\begin{aligned} F(-x) &= f(-x)g(-x) \\ &= f(x)[-g(x)] = -f(x)g(x) \\ &= -F(x). \end{aligned}$$

43. Let f be an odd function. Then

$$\int_{-a}^{a} f(x)\,dx = \int_{-a}^{0} f(x)\,dx + \int_{0}^{a} f(x)\,dx = I_1 + I_2.$$

In I_1 let $-x = t$ and $-dx = dt$ so that

$$\begin{aligned} I_1 &= \int_{a}^{0} f(-t)(-dt) = \int_{0}^{a} f(-t)\,dt \\ &= -\int_{0}^{a} f(t)\,dt = -I_2. \end{aligned}$$

Therefore $I_1 + I_2 = 0$.

45. Adding the results of Problems 13 and 14 and dividing by 2 give:

$$\frac{\pi}{4} + \sum_{n=1}^{\infty} \left\{ \frac{(-1)^n - 1}{\pi n^2} \cos nx + \frac{(-1)^{n+1}}{n} \sin nx \right\}$$

Exercises 10.4, Page 510

1. $u = -xy + f(y)$ 3. $u = \frac{1}{3}e^y + f(x)e^{-2y}$

5. $u = \frac{4}{3}x^3y^2 + \frac{1}{2}x^2 + xf(y) + g(y)$

7. $u = xy + f(x) + g(y)$

9. $u = 3x^2ye^x + f(y)e^x + g(x)$

11. $u = f(x)e^{xy} + g(x)e^{-xy} + \frac{xe^{4y}}{16 - x^2}$

13. $u = x^3 + x(y^2 - y) + y$

15. The possible cases can be summarized in one form $u = c_1 e^{c_2(x+y)}$, where c_1 and c_2 are constants.

17. $u = c_1 e^{y + c_2(x+y)}$ 19. $u = c_1(xy)^{c_2}$

21. Not separable

23. $u = e^{-t}[A_1 e^{k\lambda^2 t} \cosh \lambda x + B_1 e^{k\lambda^2 t} \sinh \lambda x]$
$u = e^{-t}[A_2 e^{-k\lambda^2 t} \cos \lambda x + B_2 e^{-k\lambda^2 t} \sin \lambda x]$
$u = (c_7 x + c_8)c_9 e^{-t}$

25. $u = (c_1 \cosh \lambda x + c_2 \sinh \lambda x)(c_3 \cosh \lambda at + c_4 \sinh \lambda at)$
$u = (c_5 \cos \lambda x + c_6 \sin \lambda x)(c_7 \cos \lambda at + c_8 \sin \lambda at)$
$u = (c_9 x + c_{10})(c_{11} t + c_{12})$

27. $u = (c_1 \cosh \lambda x + c_2 \sinh \lambda x)(c_3 \cos \lambda y + c_4 \sin \lambda y)$
$u = (c_5 \cos \lambda x + c_6 \sin \lambda x)(c_7 \cosh \lambda y + c_8 \sinh \lambda y)$
$u = (c_9 x + c_{10})(c_{11} y + c_{12})$

29. For $\lambda^2 > 0$ there are three possibilities:
$u = (c_1 \cosh \lambda x + c_2 \sinh \lambda x)(c_3 \cosh \sqrt{1 - \lambda^2}\, y + c_4 \sinh \sqrt{1 - \lambda^2}\, y)$, $\lambda^2 < 1$,
$u = (c_1 \cosh \lambda x + c_2 \sinh \lambda x)(c_3 \cos \sqrt{\lambda^2 - 1}\, y + c_4 \sin \sqrt{\lambda^2 - 1}\, y)$, $\lambda^2 > 1$,
$u = (c_1 \cosh x + c_2 \sinh x)(c_3 y + c_4)$, $\lambda^2 = 1$.
The results for the case $-\lambda^2 < 0$ are similar. For $\lambda^2 = 0$ we have
$u = (c_1 x + c_2)(c_3 \cosh y + c_4 \sinh y)$

31. $u = A_n e^{-k(n^2\pi^2/25)t} \cos \frac{n\pi}{5} x$, $n = 0, 1, 2, \ldots$

33. $u = A_n \sinh n\pi x \sin n\pi y$, $n = 1, 2, 3, \ldots$

35. Using $-\lambda^2$ as a separation constant, we obtain

$$\begin{aligned} T' + k\lambda^2 T &= 0 \\ rR'' + R' + \lambda^2 rR &= 0. \end{aligned}$$

This last equation can be written as

$$r^2R'' + rR' + \lambda^2 r^2 R = 0,$$

which we recognize as Bessel's equation with $\nu = 0$. The solutions of the respective equations are as indicated in the problem.

37. Differentiate each side with respect to x:

$$\frac{d}{dx}\left[\frac{X''}{4X}\right] = \frac{d}{dx}\left[\frac{Y'}{Y}\right].$$

Since the right-hand side is zero, we have

$$\frac{d}{dx}\left[\frac{X''}{4X}\right] = 0.$$

This implies $X''/4X$ is a constant. Similarly, if we differentiate both sides with respect to y, we can show that Y'/Y is a constant.

Exercises 10.5, Page 520

1. $u(x, t) = \dfrac{2}{\pi}\sum_{n=1}^{\infty} \dfrac{-\cos\frac{n\pi}{2} + 1}{n} e^{-k(n^2\pi^2/L^2)t} \sin\dfrac{n\pi}{L}x$

3. $u(x, t) = \dfrac{1}{L}\int_0^L f(x)\,dx + \dfrac{2}{L}\sum_{n=1}^{\infty}\left(\int_0^L f(x)\cos\dfrac{n\pi}{L}x\,dx\right) \times e^{-k(n^2\pi^2/L^2)t}\cos\dfrac{n\pi}{L}x$

5. $u(x, t) = e^{-ht}\left[\dfrac{1}{L}\int_0^L f(x)\,dx + \dfrac{2}{L}\sum_{n=1}^{\infty}\left(\int_0^L f(x)\cos\dfrac{n\pi}{L}x\,dx\right)e^{-k(n^2\pi^2/L^2)t}\cos\dfrac{n\pi}{L}x\right]$

7. $u(x, t) = \dfrac{L^2}{\pi^3}\sum_{n=1}^{\infty}\dfrac{1-(-1)^n}{n^3}\cos\dfrac{n\pi a}{L}t\,\sin\dfrac{n\pi}{L}x$

9. $u(x, t) = \dfrac{6\sqrt{3}}{\pi^2}\left(\cos\dfrac{\pi a}{L}t\,\sin\dfrac{\pi}{L}x - \dfrac{1}{5^2}\cos\dfrac{5\pi a}{L}t\,\sin\dfrac{5\pi}{L}x + \dfrac{1}{7^2}\cos\dfrac{7\pi a}{L}t\,\sin\dfrac{7\pi}{L}x - \cdots\right)$

11. $u(x, t) = \dfrac{1}{a}\sin at\,\sin x$

13. $u(x, t) = \dfrac{8h}{\pi^2}\sum_{n=1}^{\infty}\dfrac{\sin\frac{n\pi}{2}}{n^2}\cos\dfrac{n\pi a}{L}t\,\sin\dfrac{n\pi}{L}x$

15. $u(x, y) = \dfrac{2}{a}\sum_{n=1}^{\infty}\left(\dfrac{1}{\sinh\frac{n\pi}{a}b}\int_0^a f(x)\sin\dfrac{n\pi}{a}x\,dx\right) \times \sinh\dfrac{n\pi}{a}y\,\sin\dfrac{n\pi}{a}x$

17. $u(x, y) = \dfrac{2}{a}\sum_{n=1}^{\infty}\left(\dfrac{1}{\sinh\frac{n\pi}{a}b}\int_0^a f(x)\sin\dfrac{n\pi}{a}x\,dx\right) \times \sinh\dfrac{n\pi}{a}(b-y)\,\sin\dfrac{n\pi}{a}x$

19. $u(x, y) = \dfrac{1}{2}x + \dfrac{2}{\pi^2}\sum_{n=1}^{\infty}\dfrac{1-(-1)^n}{n^2\sinh n\pi} \times \sinh n\pi x\,\cos n\pi y$

21. $u(x, y) = \dfrac{2}{\pi}\sum_{n=1}^{\infty}\left(\int_0^{\pi} f(x)\sin nx\,dx\right)e^{-ny}\sin nx$

23. $u(x, t) = 100 + \dfrac{2}{L}\sum_{n=1}^{\infty}\left(\int_0^L (f(x)-100)\sin\dfrac{n\pi}{L}x\,dx\right) \times e^{-k(n^2\pi^2/L^2)t}\sin\dfrac{n\pi}{L}x$

Chapter 10 Review, Page 523

1. true **3.** cosine **5.** $\frac{3}{2}$

7. $\displaystyle\int_0^L \sin\frac{(2m+1)\pi}{2L}x\,\sin\frac{(2n+1)\pi}{2L}x\,dx$

$$= \frac{1}{2}\int_0^L\left[\cos\frac{(m-n)\pi}{L}x - \cos\frac{(m+n+1)\pi}{L}x\right]dx$$

$$= \frac{L}{2\pi}\left[\frac{\sin\frac{(m-n)\pi}{L}x}{m-n} - \frac{\sin\frac{(m+n+1)\pi}{L}x}{m+n+1}\right]_0^L$$

$= 0,\quad m \neq n.$

9. $f(x) = \dfrac{1}{2} + \dfrac{2}{\pi}\sum_{n=1}^{\infty}\left\{\dfrac{1}{n^2\pi}[(-1)^n - 1]\cos n\pi x + \dfrac{2}{n}(-1)^n\sin n\pi x\right\}$

11. $f(x) = 1 - e^{-1} + 2\sum_{n=1}^{\infty}\dfrac{1-(-1)^n e^{-1}}{1+n^2\pi^2}\cos n\pi x$

13. $u = xye^x - ye^x + x + f(y)$

15. $u = \frac{1}{y} + f(y)e^{-xy}$ 17. $u = c_1 e^{(c_2 x + y/c_2)}$

19. $u(x, y) = \frac{100}{\pi} \sum_{n=1}^{\infty} \frac{1 - (-1)^n}{n \sinh n\pi} \sinh nx \sin ny$

21. $u(x, y) = \frac{100}{\pi} \sum_{n=1}^{\infty} \frac{1 - (-1)^n}{n} e^{-nx} \sin ny$

Appendix II Exercises, Page 531

1. (a) 24 (b) 720 (c) $4\sqrt{\pi}/3$ (d) $-8\sqrt{\pi}/15$

3. 0.297

5. $\Gamma(x) > \int_0^1 t^{x-1}e^{-t}\,dt > e^{-1}\int_0^1 t^{x-1}\,dt = \frac{1}{xe}$ for $x > 0$. As $x \to 0^+$, $1/x \to +\infty$.

Appendix IV Exercises, Page 536

1. -58 3. 248 5. 12 7. $16e^{3t}$

9. $x = 4$, $y = -7$ 11. $x = 4$, $y = 3/2$, $z = 1$

13. Let $z = t$, t any real number. The solution is $x = -t/3$, $y = 5t/3$, $z = t$; the system represents the line of intersection of two planes.

Appendix V Exercises, Page 540

1. $7 - 4i$ 3. $-11 - 11i$ 5. $3 - 4i$

7. $\frac{7}{34} - \frac{11}{34}i$ 9. $\frac{5}{34} - \frac{3}{34}i$ 11. $z = e^{i\pi/2}$

13. $z = e^{i\pi}$ 15. $z = 2\sqrt{2}e^{i\pi/4}$ 17. $z = 12e^{i\pi/3}$

19. $z = 2e^{i\pi/6}$ 21. $z = -8$

Index